Animal Sciences

LIST OF BOOKS

Additional McGraw-Hill titles of interest:

Erikson/Akridge/Barnard/Downey: *Agribusiness Management,* 3e

Ferris: *Agricultural Commodity Market Analysis,* 1e

Jones: *Introduction to Plant Systematics,* 2e

Kay/Edwards: *Farm Management,* 4e

Schaffner/Schroder/Earle: *Food Marketing Management: An International Perspective,* 1e

Seitz/Nelson/Halcrow: *Economics of Resources, Agriculture, and Food,* 2e

Thein/Graveel: *Laboratory Manual for Soil and Environmental Science,* 2e

Fourth Edition

ANIMAL SCIENCES

The Biology, Care, and Production of Domestic Animals

John R. Campbell
President Emeritus
Oklahoma State University

Formerly Dean, College of Agriculture
University of Illinois

Formerly Professor of Dairy Science
University of Missouri

M. Douglas Kenealy
Professor of Animal Science
Iowa State University

Karen L. Campbell
Professor of Veterinary Clinical Medicine
Section Head, Dermatology and Specialty Medicine
University of Illinois

Successor to
The Science of Animals That Serve Humanity
John R. Campbell and John F. Lasley (1913–1994)

Boston   Burr Ridge, IL   Dubuque, IA   Madison, WI   New York   San Francisco   St. Louis
Bangkok   Bogotá   Caracas   Kuala Lumpur   Lisbon   London   Madrid   Mexico City
Milan   Montreal   New Delhi   Santiago   Seoul   Singapore   Sydney   Taipei   Toronto

McGraw-Hill Higher Education

A Division of The **McGraw-Hill** Companies

ANIMAL SCIENCES: THE BIOLOGY, CARE, AND PRODUCTION OF DOMESTIC
ANIMALS, FOURTH EDITION

Published by McGraw-Hill, a business unit of The McGraw-Hill Companies, Inc., 1221
Avenue of the Americas, New York, NY 10020. Copyright © 2003, 1985 by The McGraw-Hill
Companies, Inc. All rights reserved. Formerly published under the title of *The Science of
Animals That Serve Mankind,* copyright © 1975, 1969 by McGraw-Hill, Inc. All rights
reserved. No part of this publication may be reproduced or distributed in any form or by any
means, or stored in a database or retrieval system, without the prior written consent of The
McGraw-Hill Companies, Inc., including, but not limited to, in any network or other electronic
storage or transmission, or broadcast for distance learning.

Some ancillaries, including electronic and print components, may not be available to
customers outside the United States.

This book is printed on acid-free paper.

2 3 4 5 6 7 8 9 0 CCW/CCW 0 9 8 7 6 5 4 3 2

ISBN 0–07–366175–9

Publishers: *Edward E. Bartell/Margaret J. Kemp*
Marketing manager: *Heather K. Wagner*
Senior project manager: *Kay J. Brimeyer*
Senior production supervisor: *Laura Fuller*
Designer: *K. Wayne Harms*
Cover image: *Courtesy of Colorado State University, Fort Collins. ©George Seidel, Ph.D.,
used by permission*
Media technology producer: *Judi David*
Compositor: *Precision Graphics*
Typeface: *10/12 Times Roman*
Printer: *Courier Westford*

Library of Congress Control Number: 2002100216

This book is respectfully dedicated

To our students from whom we gained the inspiration to prepare the text,
and
To all who possess a passion for and commitment to sharing learning experiences with those who enjoy the presence and seek to better understand the biology and care of companion animals, horses, and animals that provide invaluable food, service, and pleasure to billions of people around the world.

CONTENTS

CHAPTER 21

The Nutritional Contributions of Minerals to Humans and Animals   356

CHAPTER 22

Animal Disease and the Health of Humans   370

CHAPTER 23

Selected Insects and Parasites of Significance to Humans and Animals   398

CHAPTER 24

Ethology and Animal Behavior   416

APPENDICES

PREFACE

This 24-chapter book is aimed to serve as a text for college students and others desiring a comprehensive introduction to the biology, care, and production of domestic animals and freshwater fish raised to provide food, as well as companionship and recreation for billions of humans around the globe. We hope it communicates our enthusiasm for this exciting field of science.

The fourth edition of *Animal Sciences: The Biology, Care, and Production of Domestic Animals* (formerly titled *The Science of Animals That Serve Humanity*) includes three new chapters: Companion Animals, Aquaculture, and State of Being of Domestic Animals. All chapters have been thoroughly revised and updated by disciplinary experts to include recent scientific advances in the animal sciences. Important new materials include the life cycles of food-producing animals, retail cuts of meat, food safety, recent exciting applications of technology in reproductive physiology, expanded discussion of animal excreta, new discussions of *Escherichia coli* O157:H7, Bovine Spongiform Encephalopathy (BSE) and the new variant Creutzfeldt–Jakob Disease, as well as Cat Scratch and Lyme diseases.

The initial chapter presents materials related to the history and important economic aspects of animal agriculture in the United States and the world. The second chapter highlights breeds and life cycles of livestock and poultry. Presented in subsequent chapters is information pertaining to the nutritional contributions of animal products to humans; food preservation and safety; and the principles of animal genetics, anatomy, and physiology. Included are the principles of digestion, growth, senescence, lactation, egg laying, reproduction, ecology, and stress responses of animals. Other materials related to animal disease and public health, parasites of importance to domestic animals, insects and their biological control, and ethology and animal behavior are included. Because of the increasing interest in horses major revisions have been made in that chapter as well.

A survey conducted by the publisher of our teaching colleagues at other schools revealed that different teachers assign higher priority to certain chapters than to others. Moreover, most instructors have their own preferences as to the sequence of subject matter presented. Therefore, each chapter (and even each major section in most chapters) stands on its own. This enables teachers to arrange topics for study in keeping with their preferred course outlines. And, because the book includes more than can be covered in most introductory courses, much of the material will be useful in upper-level courses taught in the various disciplines of animal science.

More than 350 illustrations aid in comprehending scientific concepts as well as adding to reading pleasure. Included are four color pages of breeds of horses, cats, and dogs plus four color pages of the retail cuts of meats (beef, lamb, pork, and veal). To stimulate thought each chapter begins with a historical or philosophical quotation.

> Quotations preserve for humanity not only the beauty of literature, but also the wisdom of philosophy, the counsel of experience, and the inspiration of achievement.
>
> **Lewis Copeland (1888–1949)**
> **American editor and publisher**

To better understand materials presented, a glossary has been included. Glossary words/terms are set in **boldface** type the first time they appear in a particular chapter. Study questions are provided at the end of each chapter so students can test their knowledge and understanding of the materials presented.

Authors of a book of this magnitude rely on the talents and professional expertise of many people. Indeed, scientists who contributed to research, teaching, and public service over the years—those who generated the data and information recorded here—are too numerous to name, impossible to repay. But we are especially indebted to the more than 40 animal science and veterinary medicine faculty colleagues at Berea College; the Universities of California, Florida, Illinois, Kentucky, and Missouri; Colorado State University, Iowa State University, Kentucky State University, Oklahoma State University, and Texas A&M University who reviewed chapter manuscripts and contributed immensely to this new edition. They are acknowledged in the respective chapters. Special appreciation is expressed to the authors named at the beginning of the three new chapters. Additionally, our abundant gratitude is expressed to Benita Bale for her efficient clerical assistance and editorial suggestions. We acknowledge, as well, contributions of the late John F. Lasley, Professor of Animal Science, University of Missouri, to previous editions.

Finally, we thank members of the McGraw-Hill editorial and production team: Ed Bartell, Kassi Radomski, Kay Brimeyer, Heather Wagner, Judi David, Laura Fuller, K. Wayne Harms, and John Leland for their contributions to clarity, consistency, design, and other important aspects of this book.

John R. Campbell
M. Douglas Kenealy
Karen L. Campbell

ANIMAL AGRICULTURE

A land poor in livestock is never rich, and a land rich in livestock is never poor.

Arab philosopher

1.1 HISTORY AND DEVELOPMENT OF ANIMAL AGRICULTURE

The success of human beings on earth is attributable largely to the animals that have fed, clothed, and carried them and cultivated their fields. Archaeological studies found both nomadic and place-based herding and flocking activities among early Middle Eastern, African, and Asian farmers. Interestingly, Eastern Hemisphere explorers did not leave their home surroundings for the New World without animal companions by their side. Apparently, the first travelers to the North American continent, later known as Native Americans, brought dogs across what would become the Alaskan frontier and eventually populated the North and South American continents. By the thirteenth century, and possibly as early as the eleventh century, Vikings from northern Europe brought small **ruminants**[1] to colony sites in Greenland and Newfoundland on the eastern coast of North America. In the fifteenth and sixteenth centuries, when Columbus and his fellow mariners came to America, they brought cattle, horses, sheep, and swine. They, as well as later colonists arriving in North America in the seventeenth century, found Native Americans using wild poultry for food and domesticated dogs for both food and labor. Within two centuries horses brought originally to the Americas by the European explorers had transformed tribes of the western plains through service in travel and battle.

Development of the field of animal science began with the **domestication** of animals in the Neolithic (New Stone Age) period, when the world population is estimated to have been a meager 5 million. This era also marked the first step toward civilization of the most primitive tribes of humans. It was the beginning of humanity's transformation from the savage to the civilized way of life—from nomads, or wanderers, to, eventually, urban dwellers. The herding of animals became indicative of the superiority of one tribe over another. Historically, the great **livestock** countries of the world have supported the most advanced civilizations and have been the most progressive and powerful.

Throughout most of their existence, humans were nomadic, their numbers small, their technologies rudimentary. Agricultural research is quite new when viewed in terms of the perhaps millions of years of human existence. The domestication of animals that serve humanity is only about 10,000 to 14,000 years old. With each succeeding step in their advancing civilizations, people have become more dependent on animals and their products.

Animal agriculture utilizes biological processes to produce animal products useful to humans. Animal science has traditionally embraced all disciplines in the biological and physical sciences that influence animal life, and more recently has begun to include areas of the liberal arts influenced by animals. For example, teachers and researchers have, for many years, discussed both the nutrition of animals and the impact of animal products on human nutrition. But, only recently have animal scientists begun to teach students about the influence animals and their production have had on art, literature, history, and philosophy.

Supporting successful animal production are years of experience and scientific research. From this research, Texas Longhorns were replaced by meat-type **steers;** Arkansas **Razorbacks** were replaced by improved meat-type hogs; black, brown, and spotted sheep were replaced by improved **mutton-** and wool-type sheep; inefficient **poultry** were replaced by fast-gaining birds with high **feed conversion**—birds that now convert 2 pounds (lb) of feed

---

[1]Words set in boldface type are defined in the Glossary at the back of the book.

into 1 lb of meat; and **cows** originally selected for meat, milk, and **draft** in many countries have been replaced by high-producing dairy cows.

Just as they do today, animals served humanity in early times in many ways other than as food. They provided leather and wool for clothing, bones for tools, and **dung** for fertilizer and fuel. They were a means of transportation and were used for entertainment and in religious offerings. Today portions of various animals are used in the manufacture of certain pharmaceuticals, fuel, fertilizer, oil, gelatin, glue, and other industrial products. Catgut (commonly made from sheep intestines) is used for violin and tennis racket strings and for sutures. Pigskin, the hide of swine, is synonymous with football in the United States, and has been used in human skin grafting as well. Animals serve humanity as subjects in experiments to advance medical and scientific research. Painters enjoy brushing farm scenes that include animal life.

> He who looks on his cattle merely as meat and milk has lost the art of living. There is beauty in the scene of feeding cattle equaling that of music or theater. Both associate man with the meaningful things of life.
>
> **A. P. Schultze**

Since the early days of their domestication, animals have also served as companions to humans. The hunting dog slept at the foot of its master around the campfire, further serving as watchdog. More recently humans have recognized the comforting nature of companionship with animals as exemplified by rehabilitation programs utilizing companion animals in nursing homes. In short, animals contribute greatly to both the mental and physical health and well-being of humans (Chapter 4).

> The power bestowed on the horse, the dog, the ox, the sheep, the cat, and many species of domestic fowls, of supporting almost every climate, was given expressly to enable them to follow man throughout all parts of the globe in order that we might obtain their services, and they our protection.
>
> **Sir Charles Lyell (1797–1875)**
> **English author**

1.2 DOMESTICATION OF ANIMALS

Domesticated cattle have long been a hallmark of civilization. Where beef and dairy animals were raised, humans enjoyed improved health and prosperity. Cultivation of plants and domestication of animals began at approximately the same place, but at different times. The place was the hills of southwestern Asia in the Zagros, Lebanese, and Palestinian mountains.[2] The first agriculturalists and stock raisers were people of the Mediterranean.

[2]Anthropologists at the University of Massachusetts found evidence that animal husbandry may have begun some 15,000 years ago in east Africa. The findings include bones and teeth of cattle at three separate sites in the Kenya highlands, about 25 miles from Nairobi. Using modern radiocarbon dating techniques, the Massachusetts scientists were able to identify the animals and to determine approximately when they lived. Archaeological examinations indicated that the cattle must have been domesticated, because there probably were never wild cattle in the area. Moreover, other studies have shown that tsetse flies, the primary cause of cattle deaths in Africa (Chapter 23), would have wiped out any wild animals that were roaming the area freely. The assumption that civilization began in Africa long before the Iron Age is supported by discoveries of 18,000-year-old domesticated grain crops in Africa.

	When, B.P.,*			
Species	**Years**	**Where**	**Why**	**How**
Dog	12,000	Old and New Worlds	Pet, companion	Wolf or jackal
Goat	8500–9000	Old World	Food, milk, and clothing	Wild goat
Pig	8000–9000	Old World	Food and sport	European wild boar
Sheep	6000–7000	Old World	Food, milk, and clothing	European mouflon and Asiatic urial
Cattle	6000–6500	Old World	Religious reasons	Aurochs
Cat	6000	Egypt	Pet, companion	African bush cat
Chickens	5000–5500	India, Sumatra, and Java	Cockfights, shows, food, and religion	Jungle fowl
Horse	4000–5000	Old World	Transportation	Wild horse
Ducks	4000?	Probably China	Food and feathers	Wild duck
Geese	3000?	Greece and Italy	Food and feathers	Wild goose
Turkeys	1000?	Mexico or North America	Food and feathers	Wild Turkey

TABLE 1.1 The Domestication of Animals

*B.P. means *before present*.

The shift from food gathering to food cultivation began about 10,000 to 16,000 years ago. Mortars and pestles used for grinding grain have been found that verify this belief. Domestication of animals came somewhat later. The first animals to be domesticated may have been the dog (some 12,000 years ago) and goat (probably 8500 to 9000 years ago). The probable time of domestication for selected animals, together with other information, is given in Table 1.1.

Some plants and a few animals were domesticated in the New World (the Americas), but most were domesticated in the Old World. Animals domesticated in the New World include the llama, alpaca, vicuna, Andean guinea pig, and turkey. Dogs and bees and bee products were common to both the New and Old Worlds.

The distribution of livestock and **cropping** systems in the world is limited by the potential of geographic areas for types of plant production. According to the Food and Agriculture Organization of the United Nations (**FAO,** see Section 1.4.1) approximately 10 percent of the world's land mass is tilled, approximately 25 percent is suitable for **forage** crops and grazing, and 65 percent is either too dry, wet, or mountainous for major cropping or animal production.

Several important crops of today were first cultivated in the New World. Among these are the white and sweet potatoes, chili, sunflower, peanut, common bean and other varieties of bean, pumpkin, gourd, squash, tomato, pineapple, tobacco, and Indian corn, or maize. In 1965 in Athens, Ohio, an ear of maize was discovered that yielded radiocarbon dated 280 B.C. Tobacco, Indian corn, and the white potato probably were the New World's greatest contributions to crops of the world. The impact of introducing the white potato into Ireland was sensational. In the latter part of the seventeenth century, Ireland had a population of nearly 2 million living in hunger. Then the white potato was introduced from the New World. This crop rapidly became popular because the soil and **climate** were ideal for its growth and production. The white potato produced more food per acre in that country than had ever been produced before. By 1835 the population of Ireland had increased from 2 to 8 million persons, largely because of the increased food supply. Then came a potato crop failure, resulting in a great famine. It is said that 2 million people died of starvation; another 2 million migrated to other countries. Since that time the population of Ireland has been nearly stabilized at approximately 4 million, and the white potato is still an important food crop.

Reports of famines are as old as the twelfth chapter of Genesis when Abraham went down to Egypt "and there was a famine in the land." In 1125 A.D. a famine reduced by one-half the population of Germany. Hungary experienced a serious famine in 1505. England records a terrible famine in 1586, and in 1870 to 1872 Persia lost one-fourth of its population to hunger.

Some 10 million Chinese died of starvation in 1877 to 1878. Famines in India claimed 3 million lives in 1769 to 1770, 1.5 million in 1865 to 1866, and 0.5 million in 1877. In 1891 to 1892 a Russian famine brought severe hardship to an estimated 27 million people.

One of the great historical events in Europe during the twentieth century was the Russian Revolution of 1917. Included on its banner was the inscription "Bread and Peace." These two words are related and have always been important to the welfare and perpetuation of the human race.

> When the price of rice goes higher than a common man can pay, Heaven ordains a new ruler.
>
> **Old Chinese proverb**

Indian corn has become one of the greatest crops in the history of the world, especially in the Corn Belt area of the United States. More than 9.8 billion bushels of corn are produced annually in that region, and it is a chief source of energy for growing and **finishing** millions of livestock. It is also an important source of food for humans in much of the world. Through the development of a high-protein variety having a better balance of amino acids, corn promises to become an even more important food crop for humans. Moreover, the discovery of high-lysine corn has spurred the quest for a similar gene in wheat, rice, and grain sorghum. Perhaps genetic engineering of food crops will result in other research findings of considerable consequence for people.

The development of cities began with the cultivation of crops and domestication of animals. The first small cities appeared about 5500 years ago. Growth of cities was rather slow over the next few centuries. Before 1850 no society could be described as predominantly urbanized. By 1900 Great Britain was the only nation that was highly urbanized. Today, all industrial nations are highly urbanized and continue that trend. The rapid movement from farms to cities has left the task of food production in the hands of fewer and fewer people and has resulted in an important problem of proper food distribution, as well as socioeconomic challenges for small rural communities. This problem points out the immense importance to the urbanized economy of a viable infrastructure for agriculture, including an excellent transportation system.

No commercially important new **species** of animal or crop has been domesticated in recent years. However, some crops and animals that originated in one country have been introduced into others, where they became popular and productive. Important crops introduced into the United States during the twentieth century include Korean lespedeza and soybeans. Examples of livestock that have been introduced into the United States are Brahman, Charolais, Limousin, and Simmental cattle and Landrace swine (Chapter 2). The introduction of several large-framed breeds of European cattle (often termed "exotics") created significant change in the American beef industry in the 1970s.

The major efforts of both plant and animal breeders in the 1900s were directed toward more efficient production of a better-quality product through close attention to breeding, feeding, and management methods. Specific breeding methods used will be discussed in Chapter 9.

1.3 HISTORY OF AGRICULTURAL EDUCATION AND RESEARCH

The signing of the Land-Grant College Act on July 2, 1862, by President Lincoln began a new era in **agriculture.**[3] Sponsoring the bill was Justin Smith Morrill, a representative from Vermont. The Morrill Act proposed that a portion of federally owned land be sold and the proceeds used to establish at least one college in each state, the main goal of which would be to teach branches of learning related to agricultural and mechanical arts. This was to be done without the exclusion of other scientific and classical studies. The United States now has 105 land-grant colleges and universities.

The Morrill Act of 1862, in its attempt to democratize higher education, did not exclude African Americans. However, southern customs, traditions, and laws requiring racial segregation prevented those newly emancipated citizens from becoming full participants in the new educational venture. Twenty-eight years after passage of the 1862 legislation authorizing the establishment of the land-grant college and university system, Justin Smith Morrill—by then serving in the U.S. Senate—introduced the Morrill College Aid Act (passed August 30, 1890), which included the provision that became known as the "separate-but-equal" policy. Tuskegee University, headed by Booker T. Washington, and 17 other so-called 1890 land-grant institutions were funded through the 1890 Morrill Act.

---

[3]John R. Campbell, 1998, *Reclaiming a Lost Heritage . . . Land-Grant and Other Higher Education Initiatives for the Twenty-First Century,* Michigan State University Press, East Lansing, MI 48823-5202.

In October 1994 Congress passed legislation conferring land-grant status on 29 Native American tribal colleges located in 12 states. The legislation also provided monies that go to the Cooperative Extension Service of the 1862 land-grant institutions in states that have tribal colleges. The 1862 institutions use the funds to cooperate with tribal colleges in setting up joint agricultural extension programs focused on the needs of Native Americans.

The name *land-grant* comes from the granting of land by the federal government for the establishment of these colleges. Names and addresses of the agricultural colleges and experimental stations in the United States and its territories are given in Appendix D.

Teaching agriculture in the early years was difficult. No textbooks on agricultural subjects were available, and there were no bulletins or circulars that instructors could assign for student reading or use for lecture material. Little or no formal research had been done. Therefore, principles of agricultural production based on research could not be taught. That situation was completely different from the present one in which there usually is a choice of more than one textbook for each course, and articles on recent research are available in agricultural journals regularly (often monthly) to individuals and libraries in print form or electronically on the Internet.

There was a clear need for research in agriculture long before passage of legislation authorizing development of the land-grant system. George Washington, in his 1796 message to Congress, requested a board of agriculture with one of its purposes to be the encouragement of experimentation. However, it was not until passage of the Hatch Act in 1887 that land-grant colleges of agriculture had federal funding for the establishment of agricultural experiment stations in all states and territories of the United States. From these experiment stations and the United States Department of Agriculture (**USDA**) has come a wealth of information on agricultural subjects. This information has helped in producing animals and crops more efficiently, thereby helping make the United States the world's richest agricultural nation.

It was recognized that information gained from agricultural research must be made available to people on farms and ranches as well as to students. To facilitate the dissemination of research information to user groups, the Smith-Lever Act was passed by Congress in 1914. The legislation provided for cooperative financing of the present-day county (parish) agent system under the direction of land-grant colleges. This system has as its main objective the conveying of new research information to people on farms, ranches, and communities where they can put it into practice.

The Smith-Hughes Act (passed by Congress in 1917) made federal funds available, if matched by state funds, for the study of vocational agriculture and vocational home economics and for education in the trades and industry. Land-grant colleges are often designated as the institutions to train teachers in these subjects.

The threefold objective of agricultural education, then, is (1) gaining knowledge through research, (2) teaching established principles to students of agriculture in high schools and colleges, and (3) disseminating new information directly to the agricultural producer, to be applied with minimal delay. In twenty-first century language, these three missions have been codified in the words *discovery, learning,* and *engagement.* And, we should note that each of the three important missions/services improves the overall contributions and efficiency of the other two. What has been accomplished in efficient agricultural production in the past has been amazing. Indeed, the land-grant college and university system has changed human destiny in remarkably commendable ways. What will be accomplished in the future could be even more phenomenal. And it will need to be if the increasing world population is to be fed properly.

1.4 ANIMAL AGRICULTURE AND THE WORLD ECONOMY

> Let us never forget that the cultivation of the earth is the most important labor of man. When tillage begins, other arts follow. The farmers, therefore, are the founders of human civilization.
> **Daniel Webster (1782–1852)**
> **American statesman and orator**

Agriculture is the world's oldest and largest primary industry. It plays a vital role in the economic life of virtually all nations regardless of their state of development. It employs more than one-half of the world's population. In developing countries, more than two-thirds of the people live on farms and ranches. The basic necessities of life—food, clothing, and shelter—are supplied by people on the land.

An imminent and great challenge to human ingenuity is the problem of the uneven distribution of various populations relative to agricultural resources and national income. As is shown in Figure 1.1, more than 50 percent of the world's people live in Asia, and yet that portion of the earth yields only 28 percent of the world's agricultural production and has only about 12 percent of the world's income, whereas North America, with 6 percent of the world's population, produces 22 percent and has about 40 percent of the world's income. These extremes make it difficult to foresee the fulfillment of the aim of the Universal Declaration of Human Rights adopted by the General Assembly of the United Nations: "A common standard of achievement for all peoples of all nations."

Concurrently, major developments in Asian agriculture have had both positive and negative impacts. During the mid-twentieth century farmers in the People's Republic of China struggled to feed a rapidly increasing population. Following an agricultural revolution in the 1970s, China made significant strides in the production and processing of food. By century end China not only was reasonably food self-sufficient, but it also exported various food commodities. But according to Dr. Dennis Keeney, past Director of the Leopold Center for Sustainable Agriculture, China's agricultural policy came at an enormous cost to the land, with approximately one-third of its topsoil being lost due to water and wind erosion between 1970 and 2000.

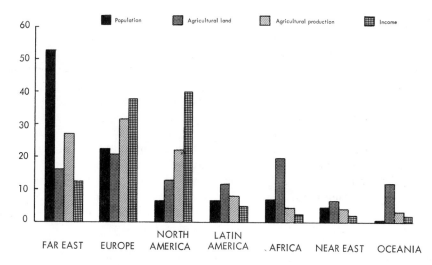

Figure 1.1 Regional distribution of the world's population, agricultural land, income, and agricultural production.
(Agriculture in the World Economy, FAO, Rome.)

1.4.1 Sources of Multidisciplinary Information on Agricultural Issues and Food Policy

Food and Agriculture Organization (FAO)

The FAO of the United Nations was formally created in 1945. It collects, analyzes, interprets, and disseminates information relating to worldwide trends in population, nutrition, food, and agriculture. It promotes and recommends national and international action with respect to scientific research, the improvement of education, administration, and the spread of public knowledge of nutritional and agricultural science; the conservation of natural resources and the adoption of improved methods of agricultural production; the improvement of processing, marketing, and distributing food and agricultural products; the adoption of policies for the provision of adequate national and international agricultural credit; and the adoption of international policies with respect to agricultural commodity arrangements. Additionally, FAO furnishes technical assistance to governments when requested. It consists of some 180 member nations and is headquartered near the Colosseum in Rome. Information pertaining to its activities and publications may be obtained by writing FAO, Viale delle Terme di Caracalla, 00100 Rome, Italy, or by accessing the website http://www.fao.org/. FAO publications may be obtained in the United States from UNIPUB, 4611/F Assembly Drive, Lanham, MD 20706-4391 (Internet: http://www.bernan.com), or by accessing the website http://www.fao.org/CATALOG/GIPHOME.HTM.

Council for Agricultural Science and Technology (CAST)

The Council for Agricultural Science and Technology was formed in 1972. The mission of CAST is to identify food and fiber, environmental and other agricultural issues, and to interpret related scientific research information for legislators, regulators, and the media for use in making public policy decisions. CAST has published numerous interdisciplinary task force reports written by diverse teams of scholarly scientists with expertise in disciplines related to the subjects reported.

CAST membership is comprised of individuals as well as commercial organizations and 38 scientific societies, including the American Dairy Science Association, American Society of Animal Science, American Meat Science Association, American Society for Nutritional Sciences, American Veterinary Medical Association, Entomological Society of America, Institute of Food Technologists, and Poultry Science Association. CAST may be contacted for publications or general information at Council for Agricultural Science and Technology, 4420 West Lincoln Way, Ames, IA 50014-3447, or by accessing its website http://www.cast-science.org.

1.4.2 World Population Trends

It took all the years from humanity's first appearance on earth until Christ's time to reach an estimated world population of one-quarter billion. The population doubled to one-half billion by 1600 and doubled again to 1 billion by 1830 (Figure 1.2). The world's population now exceeds 6 billion people and is growing at a rate of about 1.3 percent annually. A population doubles in about 60 years at this rate (Table 1.2). It is estimated that the world population will approximate 7.7 billion by the year 2020.

The world population doubled during the 100-year period from 1830 to 1930. It increased another billion during the 30-year period from 1930 to 1960 (Figure 1.2). It took only ~15 years to add the fourth billion in 1975. The fifth billion was added by 1989 and the sixth billion by 2000. Each year the world population increases by some 86 million people, enough to populate a new nation larger than Argentina, Canada, France, Great Britain, or Mexico.

The current annual growth rate of the U.S. population is about 0.7 percent. Largely because of legalized abortion, Japan has the lowest population growth rate in Asia, 0.3 percent, and other countries having planned parenthood and/or legalized abortions (e.g., Belgium, Hungary, Sweden, and Great Britain) have rates of 0.5 percent or less.

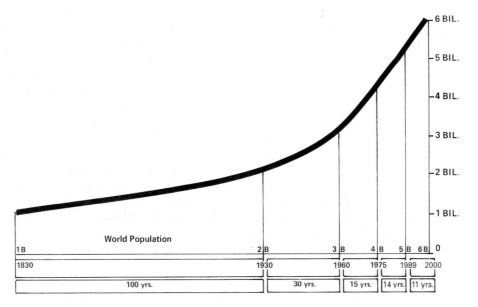

Figure 1.2 World population increases, in billions, since 1830. After an estimated 2 million years of human life, world population reached 1 billion in 1830. Since 1830 each successive billion has been added in fewer and fewer years.
(FAO data.)

TABLE 1.2	The Relation Between Annual Increase and Time Required to Double a Population
Annual Increase, %	**Doubling Time, Years**
0.5	139
0.8	87
1.0	70
2.0	35
3.0	2.3
4.0	1.7

Populations of the world's developing countries are growing faster than those of industrialized nations (Figure 1.3). Unfortunately, areas with the *least food available* and the most underdeveloped transportation infrastructure have the greatest population increases. The seriousness of "the stork outrunning the plow" in many developing countries is made even worse by those countries' lack of funds to import foods. Moreover, estimates of world population increases by the FAO project that the situation will become even more serious in the years ahead. Developed countries have both opportunity and obligation to share their agricultural and industrial technology and expertise in assisting less fortunate countries.

> I know of no pursuit in which more real and important service can be rendered any country than by improving its agriculture.
> **George Washington (1732–1799)**
> **American general**
> **First President of the United States of America (1789–1797)**

We cannot live in a world divided between, on one hand, two-thirds who do not eat properly and, knowing the causes of their hunger, revolt, and, on the other, one-third who eat well—

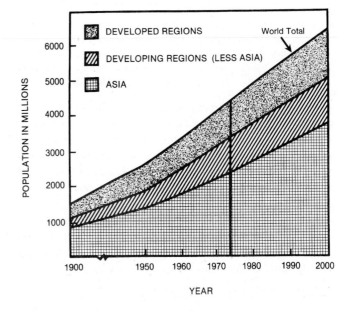

Figure 1.3 World population 1900 to 2000. Note that the largest population increases continue to be in Asia and other developing regions of the world.
(FAO, Rome.)

sometimes too much—but who can sleep no longer for fear of revolt on the part of the two-thirds who do not have enough to eat.
> **David Rockefeller**

Effective measures for the control of malaria, yellow fever, smallpox, cholera, and other infectious diseases brought sharp reductions in the death rate and the concomitant increase in global life expectancy in many countries during the twentieth century. This resulted in a substantial global population increase (Figure 1.2 and Table 1.3). Thailand is a classic exam-

TABLE 1.3	Changes in Global Population and Life Expectancy During the Twentieth Century

Year	Average Global Life Expectancy, Years	Global Population, Billions
1900	30	1.5
2000	65	6.0

Source: FAO, Rome.

ple of how declining death rates increase population growth. The death rate decreased from 30 per thousand after World War II to below 20 per thousand in the 1950s and is now about 10 per thousand. In 1937 life expectancy in Thailand was approximately 35 years, but currently exceeds 65 years.

Approximately 40 percent of the world's population is under 15 years of age. This means there are more young people who soon will be reproducing—adding to the world's population—than there are old people who soon will be dying—subtracting from it. This fact points to the global need for bringing birth and death rates closer together. A high birthrate can lead to a high death rate due to hunger. Malthus warned of this in 1798:

> I wish to make two postula. First, that food is necessary for the existence of man. Secondly, that passion between the sexes is necessary, and will remain nearly in its present state. I therefore conclude that the power of population is indefinitely greater than the power in the earth to produce subsistence for man.

Some believe the ever-expanding growth of the human population is the most significant event in history. Most agree that next to the pursuit of peace, the greatest challenge to humanity is the race between food production and population increases. The world food:hunger equation is affected most assuredly by the fact that people multiply—land does not.

> Mankind's future is at stake in a formidable race between population growth and famine.
>
> **Arnold J. Toynbee (1889–1975)**
> **English historian**

1.4.3 World Animal Production Trends

World food production is presently inadequate to ensure a balanced diet for all people of all lands. Providing food to meet caloric needs is not enough. Adequate protein is also required for normal maintenance of body tissues and functions and additionally for growth, maturation, pregnancy, lactation, and recovery from disease. Supplies of protein are particularly scarce and costly[4] for the populations of most developing nations.[5]

---

[4]Poverty is perhaps the world's greatest cause of hunger. In 38 nations, the average annual per capita income is $100 or less. People in these countries have little money to purchase food or fertilizer. The underdeveloped countries have approximately 55 percent of the world's land but only 10 percent of the commercial fertilizer needed to increase food production.

[5]According to CAST scientists, "Global demand for meat . . . is predicted to be 63 percent greater in 2020 than in 1993, with 88 percent of the increase in developing countries and nearly 50 percent of that in China." *Animal Agriculture and Global Food Supply,* Task Force Report No. 135, July 1999.

Malnutrition is the world's number one health problem. It adversely affects mental and physical development, productivity, and the span of working years, all of which significantly influence the economic potential of people. However, malnutrition does not arouse the sense of urgency that accompanies an outbreak of a contagious disease such as smallpox.

Estimates compiled by the Economic and Social Council of the United Nations indicate that more than 500 million children and perhaps an equal number of adults throughout the world are malnourished. An estimated 20 million people starve to death annually. Although there are varying degrees of starvation, it is accepted that starvation results in adults when daily caloric intake is consistently below 1600 calories. Child starvation is demonstrated when individuals are below 60 percent of standard body weight for their age. Any death that would not have occurred had the individual been properly nourished is considered to be due to starvation regardless of the ultimate cause of death. About one-fourth of the children in the developing countries die before 5 years of age, mostly from nutrition-related causes. More than 100,000 children go blind annually for lack of vitamin A. These children's lack of protein, calories, vitamins, and minerals is reflected in retarded physical growth and development, and for many of them mental development, learning, and behavior will also be impaired. (An estimated 80 percent of brain cell growth occurs during the first 2 years of life.) Protein and caloric deficiencies also affect the health and economic productivity of adult populations.

If the nutritional status of the world's hungry masses is to improve, food production and distribution must increase at an unprecedented rate. What are the prospects of this happening? Can animal agriculture be expected to contribute further to the health and well-being of humanity? Foods of animal origin provide high-quality protein, vitamins, minerals, and other dietary essentials (Chapter 3). Meat, milk, eggs, and wool (including the feed that goes into their production) represent about two-fifths of the value of the world's agricultural output. Meat and milk constitute about 85 percent of the major livestock products (Figure 1.4). In the following sections the worldwide base for the production of animal protein will be considered.

Livestock and Poultry

The livestock and poultry base for the world's food production consists of about 14.1 billion chickens, 1.5 billion cattle and buffalo, 1.1 billion sheep, 913 million pigs, 829 million ducks, and 710 million goats. Horses and mules number approximately 80 million. The largest number of cattle is in the Far East where they serve mainly as draft animals and suppliers of milk.

Recent studies indicate that present cattle and sheep numbers could be doubled with intensified management of tropical and subtropical pastures, coupled with increased yields of grain and forage crops. Moreover, swine and poultry numbers are increasing and could be increased even more if feed were made available. The question arises: Will land, seed, fertilizer, water (including water supplied by the weather), and the ability to cultivate be in large enough supply?

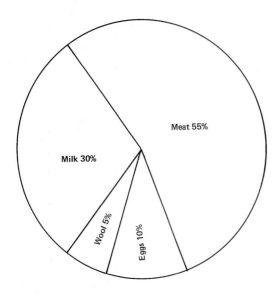

Figure 1.4 Worldwide output (percent of dollar value) of major livestock products. Farmers' total estimated income from these products was $207 billion in 2000.
(Compiled from various FAO publications.)

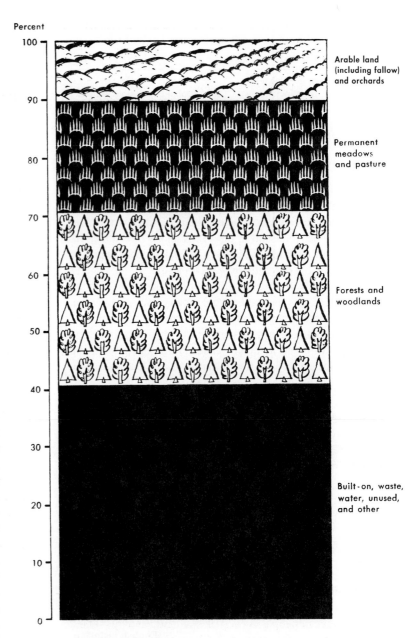

Figure 1.5 Classification of the world's land resources.
(Agriculture in the World Economy, FAO, Rome.)

Not all of the earth's surface is suitable for agriculture. If all of the present food-producing areas and all areas of potential crop-growing soil were combined into a single band encircling the earth, it would occupy only about 5 percent of the total surface. As is shown in Figure 1.5, the world's land presently is classified as approximately 10 percent arable, 20 percent pasture and meadow, 30 percent forest, and 40 percent nonproductive (desert, rock, etc.). It is estimated that one-third of the forest land has potential for agricultural production. Additionally, the yield per acre may be increased in much of the present productive arable land; however, cropland per person is diminishing. The per capita grain-producing area in North America, for example, declined from 1.7 to 1.0 acres between 1940 and 2000. In the United States about 40 percent of the grain crop is exported.

Virtually all nutrients consumed by humans come from two sources: plants (food grains, fruits, vegetables, and nuts) and animal products (meat, milk, eggs, and aquatic foods). Plant sources provide about 84 percent of all calories and about two-thirds of all protein consumed by humans worldwide. Animal products and fish provide about one-sixth of the food energy and one-third of the protein consumed by people globally.

If water can be made available for expanded irrigation, large acreages will become available worldwide for crop production. Most human food comes directly or indirectly from grain, and there is over 3 times as much arid land within 300 miles of the sea as the world now employs in grain production. The desert is human beings' greatest land bank: it offers more than 8 million square miles of space for human occupation and use. This potentially wondrous rich bank may someday turn green when scientists develop economical means of tapping desalinized seawater for irrigation. The majority of this potentially arable land is in Africa, South America, and Asia—continents in serious need of additional food.

Fish

The 20 leading fish-catching countries in the world are given in Figure 1.6. Our marine resources could yield a significantly increased supply of high-quality protein. World fish production and consumption have increased substantially during each of the past three decades. Marine biologists believe an annual global fish yield of 160 million metric tons is attainable. (The current annual fish harvest is about 82 million metric tons.)

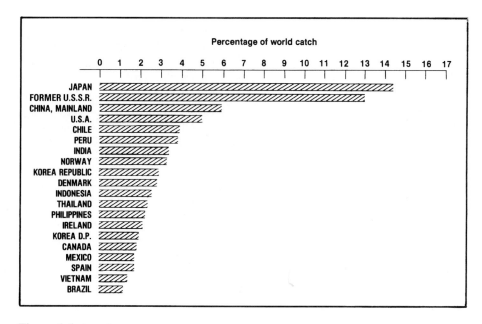

Figure 1.6 Contributions of 20 countries to the world catch of fish, crustaceans, and mollusks. These 20 countries supply an estimated 77 percent of the total world catch. *(FAO, Yearbook of Fishery Statistics, Rome.)*

Fish at present provide about 15 percent of the world's animal protein.

Slightly over half of the world's fish catch is used directly as food for people; the balance is fed to cattle, swine, poultry, and pet animals. Fish constitutes an average of only 1 percent of people's total diet throughout the world. Japan however depends on the sea for nearly half its protein for human consumption. There is a fish flour made from waste products of the sea that, once its fat has been removed, can be added to wheat flour to provide a balanced protein diet needed in many parts of the world, especially in the tropics.

Development of **aquaculture** offers another means of supplying humans with fish protein. Fish were domesticated in ponds by wealthy people as early as 3000 B.C., and by 400 B.C. raising fish was commonplace in China and Persia. Recently scientists of Israel have been producing as much as 9000 kilogram (kg) of fish annually per hectare of lake or pond.

Aquaculture

Fish is the only major class of foods that is still gathered, in large part, from the wild. But this is changing. American farmers are now producing fish under managed conditions, just as they have produced meat, milk, and eggs. This is called aquaculture (Chapter 6).

The growing and harvesting of channel catfish is expanding rapidly in the United States. Moreover, the channel catfish has an even greater potential for commercial production because it can be grown in all parts of the United States, and it reproduces and grows well under managed conditions. A well-managed catfish farm can yield well over 100,000 lb of fish per person-year of labor annually. Yields of 5000 lb/acre are commonly harvested from earthen ponds that average about 10

acres in size. It takes about 18 months for a channel catfish to grow to its market size of 1 lb.

Fish are highly efficient in converting feed into edible human proteins. They gain about a pound of weight for every 1.5 lb of feed, approximating the feed conversion of chickens and exceeding that of beef or swine. Moreover, through research on genetics, disease, and nutrition, the efficiency of feed utilization of fish is expected to increase significantly in the years ahead.

1.4.4 Availability of Animal Protein

> A hungry people listens not to reason nor are its demands turned aside by prayers.
>
> **Lucius Annaeus Seneca (4 B.C.–A.D. 65)**
> **Roman statesman, dramatist, philosopher**

The foremost reason for maintaining our animal populations is to provide a nutritious and desirable form of food for human consumption. It has been well established that, nutritionally, animal proteins are superior to vegetable proteins for humans. This superiority results largely from the better balance of amino acids in animal products. However, when animals are kept for the purpose of producing meat, milk, and eggs, food cannot be produced as efficiently as when grains are consumed directly by people (Figure 1.7). Daily per capita grain usage of developing and developed countries is approximately 1 and 5 lb, respectively. In developed countries, most grain is consumed indirectly via animal products. The dairy cow is the most efficient among farm animals in converting feed into both protein and energy; poultry and swine follow (Table 1.4). Beef cattle and sheep are the least efficient of the farm animals in this respect. It should be noted, however, that

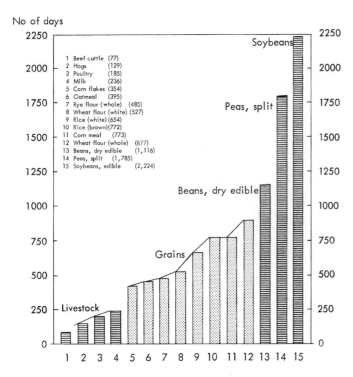

No of days

1 Beef cattle (77)
2 Hogs (129)
3 Poultry (185)
4 Milk (236)
5 Corn flakes (354)
6 Oatmeal (395)
7 Rye flour (whole) (485)
8 Wheat flour (white) (527)
9 Rice (white) (654)
10 Rice (brown)(772)
11 Corn meal (773)
12 Wheat flour (whole) (877)
13 Beans, dry edible (1,116)
14 Peas, split (1,785)
15 Soybeans, edible (2,224)

Figure 1.7 Number of days of protein requirement produced by 1 acre via selected foods.
(After L. H. Bean, Protein Advisory Group News Bull., 6, WHO/FAE/UNICEF.)

TABLE 1.4	Estimate of Relative Percentages of Feed Nutrients Converted into Edible Products by Selected Farm Animals		
Animal Product	Energy Conversion, %	Protein Conversion, %	Gross Edible Product Output as Percent of Feed Intake
Milk	20	30	90
Chicken (broilers)	10	25	45
Eggs	15	20	33
Pork	15	20	30
Turkey	10	20	29
Beef	8	15	10
Lamb	6	10	7

Source: Adapted from R. E. Hodgson, "Place of Animals in World Agriculture," *J. Dairy Sci.* 54(1971):442–447.

much of the feed utilized by animals, particularly the ruminant, cannot be utilized directly by people. The ruminant, in fact, is actually a creator of food nutrients in that it can synthesize **essential amino acids** and B-complex vitamins. *Ruminant is the name given to a* **herbivorous** *animal that chews its* **cud** *and has split hooves.* The ox, sheep, cow, llama, deer,

goat, antelope, and giraffe are ruminants (Chapter 19). The central role of ruminants in feeding a hungry world is outlined schematically in Figure 1.8.

The people of North America, Oceania, and Europe are blessed with an abundance of animal protein, whereas the balance of the earth's people obtain the larger portion of their dietary protein from other sources (Figure 1.9). Because a diet containing more livestock products requires more land and thus feeds fewer persons per acre, it may be anticipated that, with projected population increases ahead, animal proteins will be a premium. However, supplementing predominantly grain diets with animal products will improve the nutrient balance of populations consuming these diets. Moreover, most people prefer animal foods to plant foods, and as

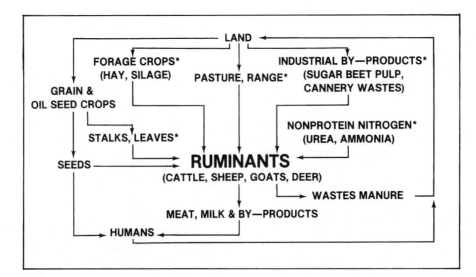

Figure 1.8 The central role of ruminant animals in human nutrition is shown. Items marked with an asterisk are utilized by ruminants but not used directly as food by humans.
(From Ruminants as Food Producers—Now and for the Future, Council for Agricultural Science & Technology, Special Pub. No.4, 1975.)

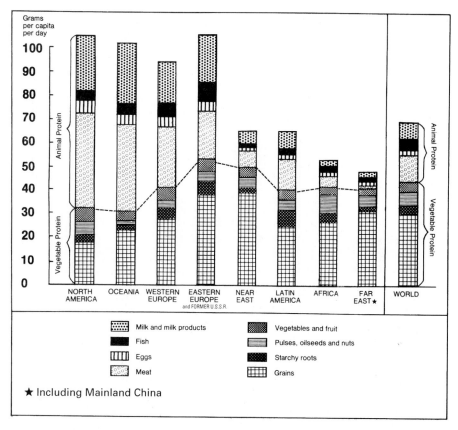

Figure 1.9 Regional per capita protein supplied from major food groups. *(FAO, Rome.)*

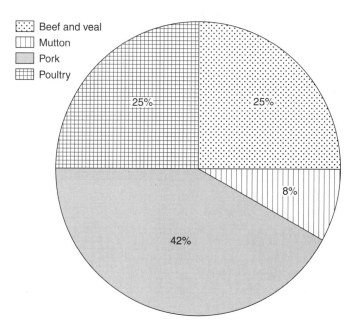

Figure 1.10 World meat production in 2000, percent by major species. *(From World Agricultural Output and World Animal Review.)*

their economic status improves, their consumption of this type of food will increase. This, coupled with the accelerating rate of population increase and the awakening of a world conscience in human nutrition, offers an enormous potential to the producers of animal protein worldwide. Present world meat production (percent by major species) is depicted in Figure 1.10.

1.4.5 Animals as Competitors with Humans for Food

The degree to which animals compete with humans for food is of special interest in a hungry world. Ruminants (particularly cattle, sheep, and goats) can consume and utilize large quantities of roughage, pasturage, and forage. This fact is of special significance in the harvesting of such feeds from rough, hilly lands in which crop production is impractical. Moreover, in semiarid areas, the only practical means of harvest is by livestock.

Although animals are not especially efficient converters of plant nitrogen into protein, an advisory committee of the United Nations stated, "Livestock will nevertheless be an increasingly important source of protein in the developing countries." This statement suggests that efforts to increase the efficiency of producing milk and meat from cattle, meat from sheep and swine, and eggs and meat from poultry should be expanded throughout

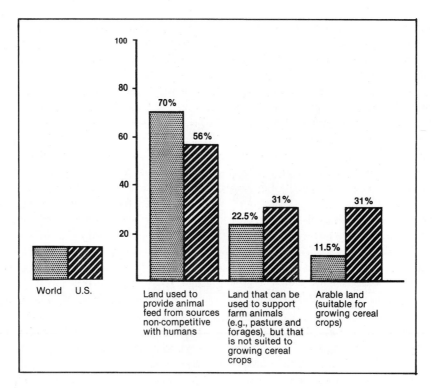

Figure 1.11 Land use in the United States and the world. Note that only 11.5 percent of the world's land is suited to the raising of cereal crops (bar graphs at right), whereas 22.5 percent of the world's land can be used for pasturing animals but is not suited for crops (center bar graphs).

(Data are from various FAO and USDA publications.)

the developing world. This provides further opportunities for developed countries to share their knowledge related to genetic selection for efficient feed converters, increased productivity by use of certain feed additives and balanced rations, and greater use of protective measures against livestock diseases and parasites.

Contrary to the commonly held view, the ruminant does not necessarily compete directly with humans for food. Instead, because the herbivore can utilize cellulose material that people cannot use, the ruminant should be regarded as an animal without which considerable quantities of high-quality food would be denied to humans. The most obvious function for these herbivores is their use of natural grazing lands that are not suited to cultivation (Figure 1.11).

It is possible to realize good production from cattle and sheep fed solely forage rations. Moreover, it has been demonstrated that dairy cows and goats can maintain themselves and produce milk at creditable levels when fed **rations** consisting entirely of forages, pasture hay, and **silage.** An example of "grassland dairy farming" is the dairy production system of New Zealand.

Beef cattle and sheep can be raised from birth to market (or to an age suitable for finishing) with little feed other than the mother's milk and forages. It has been estimated that in a beef cow-calf operation, a cow and her offspring can be maintained, with the calves raised and fed to a slaughter weight of 454 kg (1000 lb), on rations consisting largely of forages supplemented with a molasses-urea mixture or, in the finishing period, with

corn and cob meal and urea. The yield of meat protein from this operation approximately equals the yield when protein of oilseeds is fed to cattle. This feed regimen would free more plant protein for humans and illustrates how the production of meat need not compete with people for plant protein.

Most of the world's best lands are already farmed; future agricultural developments are more likely to result from intensification of the management of marginal lands. Much such land is best suited to forage production, which suggests a globally expanded animal agriculture in the years ahead.

1.4.6 Energy and Efficiency in Animal Production

The price consumers pay for food is influenced heavily by the cost of energy. More than 8 billion gallons of petroleum fuels and oils are used annually to raise, harvest, process, store and handle, transport, refrigerate, retail, and cook food consumed in the United States. Late twentieth and early twenty-first century threats to the U.S. energy supply have focused considerable attention on methods and levels of food production because petroleum is vital for motive power in modern agriculture; natural gas is needed to produce fertilizers and other chemicals; and energy is needed to irrigate and to produce farm machinery, vehicles, and other items used in modern food production. Future developments in animal agriculture will be greatly impacted by both the availability and the cost of fossil fuels. (Depending on the cost of crop production, ethanol made from corn and/or other renewable crops and natural resources, as well as oils from soybeans and/or other crops, may be practical to derive energy needed in production agriculture from agriculture per se.) High prices of petroleum products will be reflected in higher costs of crop production, which in turn will likely result in higher-priced animal feeds, especially grain crops. Increasing worldwide demands for U.S. grains also can be expected to support increasing grain prices. Therefore, to produce meat and milk at marketable prices, it is probable that more and more ruminants will be fed proportionally greater quantities of forages in future years, a practice already common throughout most of the world.

Consumption of energy—a useful index of both resource consumption and impact on the environment—is growing worldwide at an annual rate of about 5 percent. This increase in energy consumption has caused some to advocate a return to the sole use of animal power and organic fertilizers to produce people's food supply. It is true that animals currently provide an estimated three-fourths of the draft power for the world's agriculture. Additionally, many countries use minimal commercial fertilizer. However, an estimated one-fourth of the total human food supply is attributable to the use of chemical fertilizer.

United States corn yields increased approximately 200 percent during the second half of the twentieth century. Increased use of fertilizer accounted for an estimated one-half of the increase. It is possible, of course, to use green manures to reduce the high energy demand of chemical fertilizer. (For example, planting sweet clover in the fall and plowing it under 1 year later adds about 150 lb of nitrogen per acre to the soil.) This practice, however, removes land from the cropping system for a full year. Studies at the University of Illinois indicate that dried sewage sludge can be used to add nitrogen and other important plant nutrients to soil in support of crop production.

The displacement of 25 million horses by tractors in the United States during the first half of the twentieth century released some 70 million acres once used to grow oats, corn, hay, and other feedstuffs needed to feed the nation's horses. This land now can be used to feed an equal number of beef and dairy cattle that provide meat and milk for humans. Moreover, the increased labor efficiency that accompanied mechanization of farming (1 hour, h, of farm labor now produces more than 16 times as much food as it did in 1920) released millions of people for other forms of employment, thereby adding billions to the annual U.S. gross domestic product as well as providing goods and services important to our way of life.

If the United States returned to using animal power in agriculture, as is done in much of the world (Figure 1.12), it would need approximately 60 million draft animals, more than 20 times the number presently available. It would require about two decades to acquire this number, and it would take about 180 million acres (73 million hectares) of prime farmland to raise enough crops to feed them. This is about the amount of land now in grain crops in the Corn Belt states of Iowa, Illinois, Indiana, Minnesota, Missouri, Nebraska, and Ohio plus Wisconsin, Michigan, and California. To revert to labor-intensive agriculture would require an estimated 30 million farm workers, more than one-fourth of the working population of the United States.

Of course, efficiency cannot be measured solely in terms of agricultural yields. Important, too, is the amount of energy required for the production of a given kind and quantity of food. Additional research is needed to determine the energy inputs required to produce various foods at varying levels of productivity.

Figure 1.12 Animals provide a major source of power in the cultivation, harvesting, and marketing of food crops, as well as in providing transportation, in much of the world.
(Photograph by John R. Campbell in Latin America.)

1.4.7 Utilizing Animal Wastes

Current annual U.S. livestock manure production is estimated at 1.8 billion tons (10 to 12 times that of humans), of which more than 50 percent is produced in feedlots and confinement situations. Cattle and horses excrete about 15 lb of solids daily, pigs 1.7 lb, sheep 2.5 lb, chickens 0.12 lb, and turkeys 0.35 lb. Expressed another way, meat-producing animals excrete about 10 lb (including moisture) for each pound of weight gained, dairy cows excrete more than 1 lb of feces for each pound of milk produced, and laying hens excrete approximately 5 lb of feces per pound of eggs produced. Other animal wastes are associated with the processing of meat animals and with the manufacture of dairy products.

Animal wastes actually are resources out of place. Research and economies will help put them into a place and/or form in which they will provide useful functions, some as fertilizers, others as feedstuffs.

The meat-slaughtering and meat-packing industry, for example, has converted animal wastes into useful products. Typical by-products and their uses include edible fats—tallow and grease; meat scraps and blood—tankage and other animal feeds; bone—bone meal; intestines—sausage casings and surgical thread; glands—pharmaceutical products; and feathers—feather meal for animal feed.

The processing of milk and milk products requires large quantities of water. Soluble milk solids precipitate and accumulate in pipes and tanks. These milk solids can be reclaimed and used as animal feed, as food supplements, and as a growth medium for microorganisms used in producing certain pharmaceuticals.

The liquid waste whey is a by-product of cheese manufacture. Liquid whey from cheddar cheese production contains over half the nutrients from milk, and the acid whey from cottage cheese manufacture contains approximately 70 percent of the nutrients of the nonfat milk used. Heretofore, nearly half the whey has been discharged to waste treatment plants and to streams or other bodies of water. This loss of milk nutrients increases the cost of cheese production and the national cost of water pollution control. Complete utilization of whey would eliminate that source of pollution and concurrently increase the human food supply. Expanded uses of whey in the future include blending whey powder with basic food materials to produce new and/or less expensive foods such as processed cheese foods, fruit sherbets, custards, and bakery goods. Studies at the University of Missouri include recycling whey with various papers and/or nonprotein nitrogen through ruminants in the production of meat and milk.

Whey can be dehydrated by roller drying and spray drying. Whey powder contains about 11 to 13 percent protein, 70 to 75 percent lactose, and 7 to 8 percent minerals. Research under way is aimed at fractionating whey solids into pure protein and lactose.

Farm animals, particularly ruminants, will serve people in the future when their products that have previously been considered useless waste materials are recycled. Only 21.8 percent of the total animal waste nitrogen is identified as amino acids; the balance comes from nonprotein nitrogen compounds (31.6 percent from urea). Therefore, recycling animal wastes can be expected to be most effective in ruminants, animals that can utilize nonprotein nitrogen.

Researchers at the University of Arizona and Texas A&M University cooperated with scientists at the General Electric Company in using thermophilic bacteria to convert feedlot cattle manure into a high-protein livestock feed supplement. Gas containing 60 to 80 percent methane can be produced during the anaerobic digestion of animal wastes. Approximately 8 cubic feet (ft^3) of gas can be produced per pound of volatile solids added to the digester when cattle, swine, and poultry wastes are digested. The process holds great promise as a potential source of energy. Several studies have shown that dried animal wastes can be recycled. The utilization of dried chicken manure as part of the feed for chickens and ruminants has been demonstrated. Recent research shows considerable promise for recycling cattle manure ensiled with chopped corn and other forages.

Studies at the University of Arkansas showed that the amino acids of hydrolyzed feather meal are more than 95 percent utilized by the chick. Other studies indicate that amino acids of properly hydrolyzed hog hair are available to the chick.

The nutritional value of poultry litter in the diets of cattle and sheep has been studied extensively at the University of Arkansas, Virginia Polytechnic Institute and State University, and other agricultural research stations. On a dry matter basis, broiler litter averages about 31 percent crude protein (23 percent digestible protein) and is a rich source of calcium and phosphorus. Feeding poultry waste to finish cattle has not affected carcass grade or meat flavor, nor has it affected the flavor or composition of milk when fed to lactating cows. One possible problem caused by feeding animal waste is copper toxicity, observed in sheep fed broiler litter containing high levels of copper. (Three heavy metals—arsenic, copper, and selenium—are commonly added in very low amounts to livestock and poultry rations.) Additional research is needed in this area.

Notwithstanding recent developments, the most effective agricultural use of animal wastes continues to be disposal on land in a crop production cycle. However, dried animal wastes have considerable potential as soil conditioners and fertilizers for the home gardener, florist, and nursery operator. Animal manure is high in salts and may contain certain pesticides, drugs, and/or toxic metabolites. Consideration of these aspects must be made in future studies. Effective animal-waste management can be achieved only when food production is maintained with minimal threat to environmental quality at economical prices.

1.5 ANIMAL AGRICULTURE AND THE U.S. ECONOMY

On the front page of the December 20, 1920, issue of *The New York Times* there appeared this prediction:

The United States will have a population of 197,000,000 people, the maximum which its continental territory can sustain in about the year 2100; Professor Raymond Pearl of the Johns Hopkins School of Hygiene and Public Health estimated in a Lowell Institute lecture last night. To support such a population, he said 260 trillion calories of food a year would be needed, and judging from the production of the last seven years, when the maximum population was reached, it would be necessary to import about half the calories necessary for sustenance.

The U.S. population passed the 197 million mark in 1966, exceeded 280 million in 2001. And, as a result of American agriculture's productivity advancing more in the past 50 years than in all previous years of its history, there is no problem in feeding that number. In 1800, 94 percent of the people in the United States lived and worked on farms, only 6 percent being urban dwellers; whereas in 2001 the reverse was more nearly true. In fact, the farm population now represents only about 2 percent and the urban about 98 percent. Japan is experiencing a similar, although somewhat delayed, trend. In 1900 some 85 percent of Japan's people lived on farms and only 15 percent were urban dwellers. But by 2000, just one century later, the reverse was true. It is interesting that China feeds nearly one-fourth of the world's population with only about 8 percent of the world's arable land. In terms of acreage, China is about the size of the United States, but it feeds, domestically, about 4.6 times as many people. Approximately 80 percent of China's 1.3 billion people live on farms and in communes.

Meat, milk, eggs, wool, mohair, leather, and other fibers have long been staples of the American economy. The magnitude of providing Americans with these products and substances is made evident by awareness of the following facts:

Land Area

Animals are the largest users of our nation's landmass. Some 64 percent of the total land area of the United States is devoted to the production of animal feeds (36 percent for grazing, 28 percent for the production of **hay** and other forage crops and grain).

Comparative Cash Income

The sale of livestock and their products accounts for about 51 percent of the total cash income of U.S. farmers. (Of the 50 states, 32 derive half or more of their cash farm receipts from the sale of livestock and their products, which amounted to approximately $96 billion in 2000.)

Livestock and Poultry Inventory and Production

There are more than 180 million farm mammals and over 1.9 billion chickens and turkeys on U.S. farms and ranches. In 2000 the United States produced 25.4 billion lb of beef, 18.6 billion lb of pork, 30.8 billion lb of broilers, 5.3 billion lb of turkeys, 7.1 billion dozen eggs, and 168 billion lb of milk. Each dollar of sales of animals and their products generates multiple dollars of business activity in farm supply, food businesses, transportation, and related industries.

Creator of Employment

Approximately one of every six jobs in private employment is related to agriculture. Millions of persons are employed in producing, processing, packaging, transporting, marketing, and distributing meat, milk, eggs, and other animal products and by-products.

Farm Purchases

Altogether, U.S. farmers and ranchers spent over $111 billion in the production of food and fiber in 2000. They purchased more than $30 billion of manufactured inputs (fertilizer, lime, pesticides, petroleum fuel and oil, and pharmaceuticals). Another $46.7 billion went for feed and seed; $14.2 billion was paid in interest; $14 billion went for farm vehicles, machinery, and equipment; $8.2 billion for fuels and lubricants; $7.6 billion for taxes; and $3.1 billion for electricity. Total U.S. agricultural production expenses in 2000 were $193.6 billion. Think of the large number of jobs dependent directly on farm purchases. Noteworthy is the fact that there is a close association between prosperity on farms and ranches and prosperity in cities.

One and a third centuries ago the United States was an undeveloped country with great natural resources. Its agriculture was primitive, its population small. With approximately 90 percent of the population gaining its livelihood from farming and with minimal use of technology, much early agriculture exploited the land resources. Farmers moved from old to new soil as fertility decreased. As land frontiers expanded, the realization developed that agriculture greatly affects economic development. Indeed, a great industrial system depends in large part on an efficient agriculture releasing humanpower to the labor pool and concurrently providing a dependable source of food and fiber.

At the beginning of the twentieth century 7 of every 10 American workers were producing goods and 3 in 10 workers were providing services for others. Today the figures have reversed with approximately 7.5 of 10 American workers providing services and 2.5 in 10 producing goods.

Food Production

Nearly one-half the total food supply of people is contributed by mammalian, **avian,** and aquatic life. Farmers and ranchers of the United States represent less than 0.1 percent of the world's population and yet annually produce approximately 22 percent of the meat (about 80 billion lb), 32 percent of the fluid milk (about 168 billion lb), and 29 percent of the eggs (about 7 billion dozen) of the world. The United States is the world's largest poultry producer, generating roughly 41 percent of the world output.

Recreation

Although the horse no longer provides our power and transportation, its popularity for recreational activities is at an all-time high. There are an estimated 6.9 million horses in the United States, on which their owners annually spend an estimated $8.5 billion for feed and **tack.** Also, more people attend horse races annually than see baseball games (both minor and major leagues) or automobile races, the number two and number three spectator sports, respectively. Dog racing attracts more spectators than professional football, basketball, or hockey. Although dog racing is legalized in only 10 states, it attracts some 21 million spectators annually in the United States.

The productivity of agriculture is a major source of strength in the progressive American economy. In no other nation do consumers have so varied and nutritious a diet or buy their food for so small a fraction of their disposable income (about 11 percent).

Companion Animals

A 1996 national survey estimated that 58.2 million U.S. households (nearly two-thirds) owned a companion animal. Animals considered companion animals in the study included dogs, cats, birds, horses, hamsters, guinea pigs, gerbils, rabbits, and fish. Approximately one-third of the U.S. households owned some 53 million dogs, and 27 million U.S. households owned an estimated 59 million cats. The feeding and care of companion animals contribute to multibillion-dollar pet-related industries. The production of dog food is a $10.1 billion, 7 million ton industry in the United States with substantial additional tonnage manufactured in Japan and European countries using ingredients purchased from the United States. Cat owners in the United States annually spend about $2 billion on cat food and $300 million on cat litter. Vet-

erinary expenditures were approximately $8 billion for dogs and $4 billion for cats in 1998. Additional billions are spent on pet-care products. Expenditures related to the feeding and care of 6.9 million horses, 56 million fish, 5 million rabbits, 2 million hamsters, 764,000 gerbils, and millions of other household pets are indeed substantial. That companion animals contribute greatly to the mental health and well-being of millions of children and adults alike is of great importance (Chapter 4).

1.5.1 Economy of Animal Products

United States consumers spend about 33 percent of their food budget for meat, poultry, and fish; 14 percent for milk, cheese, and ice cream; and 4 percent for eggs. Total expenditures for animal products represent about 51 percent. However, consumers receive substantially more than 51 percent of their food nutrients from animal products. Moreover, as noted in Figure 1.13, farmers receive a greater proportion of the consumer retail food dollar through animal products than through other foods. North Americans are blessed with an average of 1 quart (qt) of milk and 1 egg per capita daily, in contrast to the undernourished people of Asia who have only 1 qt of milk every 14 days and 1 egg every 10 days per capita (Chapter 3).

Although expenditures for food in the United States total more than $548 billion, Americans earn, in wages, a week's food supply by early afternoon on Monday. A century ago peo-

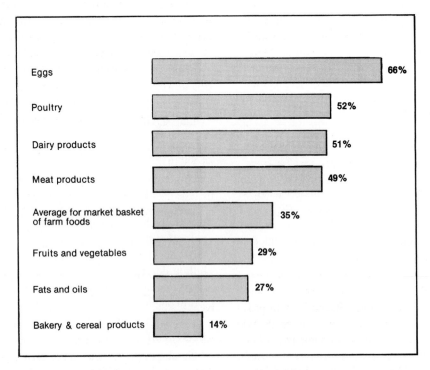

Figure 1.13 Farm share of retail food prices, based on the payment to farmers for the farm products equivalent to foods in the market basket and the retail price. Fruits and vegetables include both fresh and processed foods.
(USDA.)

ple of the United States paid approximately 60 percent of their disposable income for food. Today they pay only about 11 percent, less than that spent for food in any other nation of the world. It should be appreciated that about 30 percent of the money spent in supermarkets goes for nonfood items.

1.5.2 The United States as Food Supplier to the World

The United States is the world's largest exporter of agricultural products. Its agricultural exports go to more than 120 countries. Major regional and country foreign markets in 2000 were Asia ($22.1 billion); Latin America and Caribbean ($10.6 billion); western Europe ($6.7 billion); European Union ($6.4 billion); Africa ($2.3 billion); the former U.S.S.R., now the Commonwealth of Independent States and Russia ($1.6 billion); and eastern Europe ($1.0 billion). Japan imports about half its total food requirements and currently is the largest market ($9.4 billion in 2000) for U.S. agricultural exports. Other large individual country importers of U.S. agricultural exports in 2000 were Canada ($7.5 billion), Mexico ($6.4 billion), China ($2.7 billion), Republic of Korea ($2.6 billion), and Taiwan ($2.0 billion). Farm exports in 2000 were valued at $51 billion (approximately 40 percent of U.S. cropland production and about 11 percent of animal agriculture production is exported). United States farm exports are expected to increase in future years because of favorable prices, tight world supply of grain, improving world economic conditions, and expansion of trade with several countries.

The United States exports significant numbers of breeding animals (especially beef and dairy cattle, goats, and swine). These were valued at $608 million in 2000 and made important contributions to the genetic improvement of the livestock industries of other countries. Additionally, the United States supplies more than one-third of all protein meal consumed abroad. American agricultural abundance and technologies are powerful forces for world peace. Our food and farm products are helping relieve hunger and promote economic growth in many developing countries.

Expansion of U.S. agricultural exports benefits other sectors of the economy. For example, it is estimated that for every $100 increase in production of feed grains, wheat, rice, and oilseeds for export, an additional $110 output occurs in other sectors of the economy. These include transportation, storage, handling, and marketing.

In the more affluent countries, such as the United States and Canada, animal products could become more expensive to produce as world demands for cereals and protein supplements increase. Additional research is needed to discover ways and means of producing meat and milk more efficiently with proportionately greater amounts of forages and cereal by-products. Moreover, increased public concern with environmental pollution as well as animal state-of-being aspects of intensive animal agriculture can be expected (Chapter 7). There will be increasing competition from modified or engineered foods, such as textured vegetable protein analogues, which simulate animal food products in appearance, taste, and nutritive value.

1.6 SUMMARY

Food ranks first among the needs of the human race, and is humanity's most important renewable resource. Human anxieties about food are as old as the human race. So are people's interest in animals, which have contributed to human welfare since prehistoric times. Domestication of animals was an important part of agricultural growth and development. Today, the primary importance of domestic animals for people is as a source of food and other products, but animals are also important for companionship and other purposes (Chapter 4).

Of all the ills afflicting the human race, none seem more solvable—but concurrently more intractable—than hunger. The keenest competition for food is not between people of different regions or economic groups, but rather between people and the pests and diseases that attack food crops and plague animal agricultural production. Worldwide, insects and diseases claim an estimated one-fourth of all crops before they are harvested, and another estimated 15 to 20 percent are lost to insects and other pests following harvest. Similarly, diseases and internal and external parasites sharply reduce the potential productivity of animal agriculture (Chapter 23).

Food production is the nation's, indeed the world's, largest business. Agriculture is an indispensable base for the U.S. and world economies. Producing, processing, and distributing food employs more people than the automobile, steel, transportation, and utilities industries combined. More than two-thirds of the people of developing countries are directly involved in food production.

American agriculture has advanced more in the past 50 years than in all previous years of American history. However, evidence presented in this chapter illustrates clearly that agriculture is on a balance (Figure 1.14). From the beginning of agriculture, it has been a constant struggle to maintain this balance and thereby provide people with satisfaction of their wants and needs. Development of new and improved breeds of livestock and varieties of plants, coupled with continuing research, has increased the likelihood of humans achieving this goal, although there is an ever-present threat that the precarious balance will tip.

The balance may be disturbed by a number of forces. These include the ever-increasing population, reduction of the amount of acreage available per person, and the threat of economic distress, represented by low purchasing power. Disease, poor selection of breeds, inefficient management of resources, nutritional imbalances, predators, weather, and the threat or actuality of war inflict their various individual and combined effects on the comfort and nutritional well-being of people on a global basis.

The world is short on food *and* time. Today's hungry countries must compress the progress of decades into years if they are to adequately feed their increasing populations. There are three basic benchmarks to which the rate of increase in food production can be usefully related: (1) the rate of increase needed to keep pace with population growth, (2) the rate of increase needed to attain target rates of economic growth while

Agriculture on a balance

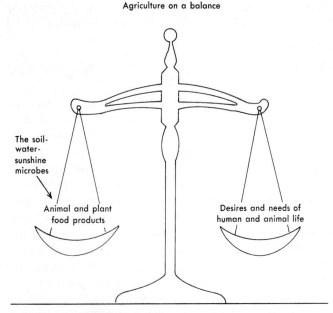

The soil-water-sunshine microbes

Animal and plant food products

Desires and needs of human and animal life

Things that may disturb (shift) the balance that agriculture can achieve in nature

1. Disease
2. Wide birth:death ratio
3. Wide population:acreage ratio
4. Economic distress (low purchasing power)
5. Poor selection of animal and/or plant breeding stock
6. Inefficient management and use of resources
7. Nutritional imbalances
8. Predators
9. Weather
10. War

Figure 1.14 Diagram depicting the important role of agriculture in the delicate balance of nature.

maintaining stable prices, and (3) the rate of increase needed to eliminate the serious malnutrition common to many developing countries.

There is a deep and growing concern throughout the world over the outcome of the food:population race. During the next 11 years the world must prepare to feed an additional billion people. Significantly, approximately four-fifths of the billion persons will be added in the already food-deficient developing countries. This growing imbalance between food and people threatens the economic and political stability of developing countries.

The present generation is the first to possess the capability to essentially eliminate hunger. It will earn the gratitude of future generations by perceiving and acting on this possibility. Efforts to meet world food deficits should not overlook the need for the development of a balanced agricultural economy in each country. A viable animal agriculture is of vital importance in this respect. Animal production is important to maximal effective use of available natural resources as human needs continue to be served through animal agriculture.

Conditions favoring animal production include (1) the requirements of an expanding human population; (2) the merits and/or special qualities of animal products; (3) the need for ani-

mals as a source of power; (4) the need for animals for mental health and personal satisfaction; (5) the role of animals in maintaining soil fertility and water quality and conservation; (6) the flexibility of animals as transformers of feed into food and other useful products; and (7) the economic, social, and institutional forces that favor greater utilization of animal products and the practices of animal husbandry.

The key to sustained animal agriculture is the proper use of natural resources: air, water, land, and energy. Farm animals are kept for food production under a broad diversity of systems. The extremes vary from ruminants grazing rangeland to highly integrated confinement systems for poultry and swine.

In this chapter present trends in human and animal populations were surveyed. An attempt was made to present an overview of animal agriculture. It was noted that because forages and roughages can be grown on land where tillage is impossible or impractical, ruminants are not necessarily competitive with humans for agronomic crops. Instead, by utilizing the grasses of such rough land, ruminants provide an added source of nutritious food for humans. It was noted that malnutrition is both a consequence and a cause of underdevelopment. Improved nutrition among children of developing countries is important to their growth and subsequent contributions to their respective national economy.

> Changes of diet are more important than changes of dynasty or even of religion.
>
> **George Orwell (1903–1950)**
> **English author**

Inadequate nutrition is one of humanity's oldest problems. Today the greatest problem facing people worldwide is not nuclear warfare, pollution, taxes, or inflation; instead, it is the problem of not enough to eat. Current political and military problems will fade as the importance of world food supplies comes into sharper focus.

> Give a hungry child a cup of milk and he will be nourished for a day, but give his family a heifer and show them how to care for it and they will drink milk the rest of their lives.
>
> **M. E. Boyer**

The flow of life is a continuum, delicate in its balances, intricate in its nuances, demanding in its observances. For animal agriculture, indeed for all agriculture, to survive, it must be in harmony with its natural surroundings. It is up to people to control their agriculture and thereby feed themselves and their kind, who will inherit the earth. People must somehow combat and defeat their foes and achieve an agricultural balance in which their wants and needs are met by an ample supply of nutritious animal and plant foods. Human needs are best served through a viable, dynamic animal agriculture.

The balance of the book is dedicated to an understanding of the animals that serve humanity—their contributions to people and society—and the scientific disciplines that describe them and the production practices/systems under which they are managed and cared for. To the authors it is an exciting, enlightening, interesting story. Please enjoy!

STUDY QUESTIONS

1. Did you read the Preface to this book? If not, please do so.

2. Where were food-producing animals first domesticated? When?

3. Which farm animals were domesticated first?

4. Of what significance were the Land-Grant College Acts of 1862 and 1890?

5. What act of Congress provided for the establishment of state agricultural experiment stations? When?

6. What was the Smith-Lever Act of 1914? The Smith-Hughes Act of 1917?

7. What is the threefold objective of agricultural education?

8. What proportion of the world population is engaged in agriculture? Is this true in the United States? Why? Of what significance is this to our way of life?

9. Is there an even distribution (and ratio) of people, land, agricultural production, and income throughout the world? Is the position of the United States favorable?

10. Which major areas of the world are increasing the fastest in population? Is their food supply adequate? Why should *we* be concerned about *their* problems? Explain.

11. Which animal products are produced in the greatest amounts throughout the world?

12. Can world livestock numbers be significantly increased? What are some possible constraints?

13. How does the United States rank in the world production of animal products?

14. Is the ocean fish supply presently being depleted by humans? What is aquaculture? Discuss its potential as a source of human food.

15. Can more people be fed if (a) they consume cereal grains (corn, oats, wheat) themselves or (b) the grain is first fed to animals and then people consume their products? Which is preferred? Which farm animal is most efficient in converting feed into food protein and energy?

16. What is a ruminant? (See also the Glossary.)

17. Do the people in the United States obtain most of their dietary protein from animal or plant sources? What about people in other regions of the world?

18. Which countries are generally the most progressive: (a) those consuming largely animal protein or (b) those consuming largely plant protein? Which source of protein do you prefer? Why?

19. Is the U.S. population increasing faster or more slowly than was predicted in 1920? What about food production?

20. What proportion of the total land area of the United States is used to produce animal feeds?

21. How much of the U.S. farmer dollar (income) is derived from the production and sale of livestock/poultry and their products?

22. Of each 100 jobs in the United States, approximately how many are related to agriculture?

23. United States farmers represent what proportion of the world's population? Are they producing their share of animal products? What proportion of U.S. crop production and animal production is exported annually?

24. What major spectator sports in the U.S. are closely related to agriculture?

25. What proportion of the U.S. consumer dollar is spent for animal products? Are animal products a good purchase as far as dietary nutrients are concerned?

26. Which country in the world has the least expensive food in relation to wages earned? Why?

27. What country exports the largest amount of agricultural products?

28. Define agriculture. (See also the Glossary.) Discuss agriculture's impact on the U.S. economy.

29. Are animals necessarily in direct competition with humans for food? Discuss.

30. Cite examples of how animals may serve humans by recycling waste products.

31. What is the FAO? CAST? What are some of their functions and activities?

2

BREEDS AND LIFE CYCLES OF LIVESTOCK AND POULTRY

The sculptor works with lifeless clay that responds to his every touch, reflecting his degree of genius. The livestock breeder works with flesh and blood. His art is affected by the laws of heredity. The drag of Nature is his constant deterrent. Ideas, perceptions, facts, and counsel are his guides. His rewards for developing breeds and creating superior seedstock are financial returns, public recognition, and personal satisfaction.

F. S. Idtse

2.1 INTRODUCTION

This chapter includes illustrations and discussions of selected types and breeds of food-producing animals and reviews development of livestock and poultry species (Part I). Additionally, it highlights life cycles and discusses guidelines for food animal production in the United States (Part II). Selected species and management practices of companion animals are discussed in Chapter 4, horses are discussed in Chapter 5, aquaculture in Chapter 6.

Part I: Breeds of Livestock and Poultry

2.2 DEVELOPMENT OF BREEDS

A **breed** is a group of animals possessing certain characteristics common to the individuals within the group that distinguish them from other groups of animals within the same species. These characteristics are the trademarks of the breed and are transmitted from one *generation* to another. Hundreds of breeds of livestock, poultry, and companion animals have been developed throughout the world and new breeds are being developed on a continuing basis.

Older breeds originated many decades, if not centuries, ago. Examples of old breeds include the Arabian horse and the Chinese Meishan hog. Breeds originated, as a general rule, within a geographic region from animals that often had few observable common characteristics. Eventually breeders selected and mated animals for specific objectives such as coat or feather color, presence or absence of horns, and spe-

cific body conformation. Most breeds of livestock were originally developed for the purpose of improving and standardizing animals so that breed examples assumed a particular form or performed a certain function. Breeding and selecting for specific characteristics dates back to the early history of domestication of animals, but it was not until after the 1800s that most registry associations were formed and breed associations were organized (Table 2.1).

Careful selection by producers resulted eventually in a group of animals having enough common characteristics to be identified as a breed. Such animals are referred to as **purebreds,** meaning they arose from mating animals within the breed. Individuals within that breed possess certain identifiable characteristics such as coat or feather color. For example, Hereford cattle have a red body color and white face. Their offspring have the same coloring. The red coat color, however, may vary from a light yellowish red to a dark red. The white face color may also be associated with a large amount of white on the face, head, shoulders, and neck or with a small amount of white on those body parts. The expression of these colors is determined by genes, and because breed members are not completely **homozygous,** variability in expression commonly occurs. Similarly, variation in most specific traits within breeds occurs regularly.

The black coat color of Angus cattle is another illustration of the lack of genetic purity within a breed. Most Angus cattle are black, but of all Angus calves born in the United States to black parents approximately 1 of 200 is red. To produce this proportion of red calves at birth about 13 percent of the black parents must be carriers of the red gene, or **heterozygous** (*Bb*). The genetics of this phenomenon will be discussed in Chapter 8.

TABLE 2.1	Selected Breeds of Livestock in North America

Breed	Color (Swine, Cattle) / Use (Sheep, Horses)	Country or State of Origin	Date of Origin of Herd Book or Association	Breed	Color (Swine, Cattle) / Use (Sheep, Horses)	Country or State of Origin	Date of Origin of Herd Book or Association
Swine							
American Landrace	White (W)	Denmark	1950	Minnesota No. 2	B and W	Minnesota	1948
Berkshire	Black (B) and W	England	1884	Montana No. 1	B	Montana	1948
Chester White	W	Pennsylvania	1894	Palouse	W	Washington (state)	1956
Duroc	Red (R)	New York and New Jersey	1872	Poland China	B and W	Ohio	1860
Hampshire	B and W (belt)	England or United States	1893	Spotted Poland China	B and W (spotted)	Indiana	1912
Hereford	R and W	United States	1934	Tamworth	R	England	1897
Minnesota No. 1	R	Minnesota	1946	Yorkshire	W	England	1893
Beef Cattle							
Aberdeen Angus	B, R	Scotland	1862	Hereford	R, white face	England	1846
American Brahman	Steel gray	India	1924	Polled Hereford	R, white face	United States and Canada	1900
Beefmaster	Many colors	United States	1949	Limousin	Wheat to rust	France	1886
Brangus	B	United States	1949	Santa Gertrudis	R	Texas	1951
Charbray	Dun	United States	1949	Shorthorn	B, R, W, roan,	England	1835
Charolais	W	France	1864	Polled Shorthorn	R, W, roan	United States	1889
Galloway	B	Scotland	1878	Simmental	B, R, white face, spotted	Switzerland	1969
Dairy Cattle							
Ayrshire	R and W or mahogany and W	Scotland	1875	Jersey	Light R or fawn	Jersey Isle	1868
Brown Swiss	Gray brown	Switzerland	1880	Milking Shorthorn	R, W, roan	England	1912
Guernsey	Fawn and W	Guernsey Isle	1877	Red Poll	R	England	1874
Holstein–Friesian	B and W, R and W	Netherlands	1885				
Sheep							
Cheviot	Mutton	England and Scotland	1891	Oxford	Mutton	England	1888
Corriedale	Mutton and wool	New Zealand	1911	Rambouillet	Wool	France	1889
Dorset	Mutton	England	1891	Shropshire	Mutton	England	1883
Hampshire	Mutton	England	1889	Southdown	Mutton	England	1882
Merino	Wool	Spain	1879	Suffolk	Mutton	England	1892
Horses							
American Quarter Horse	Saddle	Southwestern United States	1940	Palomino	Saddle	United States	1941
American Saddle Horse	Saddle	America	1891	Percheron	Draft	France	1876*
American Trotter or Standardbred	Trotter	America	1871	Pinto	Saddle	United States	1956
Appaloosa	Saddle	United States	1938	Shetland pony	Driving and saddle	Shetland Isle	1888*
Arabian	Saddle	Arabia	1908*	Shire	Draft	England	1878
Belgian	Draft	Belgium	1887*	Suffolk	Draft	England	1880
Clydesdale	Draft	Scotland	1877	Tennessee Walking Horse	Saddle	Tennessee	1935
Hackney	Carriage	England (Norfolk)	1891*	Thoroughbred	Racing	England	1791
Morgan	Saddle	America (Vermont)	1909				

*When herd book was established in the United States.

2.2.1 Breed Differences

Breeds within the same species differ in many traits including coat color, conformation, milk production, egg production, and rate and efficiency of gain. *Breed differences are genetic differences.* The statement is often made that there are greater differences among individuals within a breed than among breeds. This is to be expected because there are greater differences among individuals within a breed than there are among breed averages. The average of all individuals within a breed (breed average) varies less, of course, than the total population of all individuals of all breeds. The reduced variation is a direct function of the number of individuals included in a given breed. For most traits individuals of equal merit exist in all breeds. Therefore, in selecting for most traits it is more important to select the best individuals within a breed than it is to choose the best breed.

The differences among breeds within a species are largely due to the kind of *genes* each breed possesses and to the way these genes express themselves alone or in different combinations. Breeds may differ genetically for several reasons. Two breeds may differ because one breed may be *homozygous* for each of several pairs of genes (*AABBccddEE*), and the other breed may be homozygous for the opposite gene of a pair or of several pairs (*aabbCCDDee*). Thus one breed may have a gene the other lacks. Complete homozygosity within a breed in which the genes of a pair are always the same (e.g., *aa*) probably seldom occurs. Instead, what is more likely is that for a pair of contrasting genes (*allelomorphs*), such as *A* and *a,* one breed possesses a high percentage of *A* (e.g., 85 percent) and a small percentage of *a* (e.g., 15 percent); whereas the second breed possesses a low percentage of *A* (e.g., 15 percent) and a high percentage of *a* (e.g., 85 percent). Hence the main difference between the two breeds may be that they differ in the frequency with which a gene or genes is found in those two breeds. Principles of genetics are discussed in Chapter 8.

Breeds of animals within the same species are different genetically as are individuals within a given breed. Breeds may differ from one another because they have started from a different group of ancestors, because humans have selected them for different purposes, or because they have drifted apart genetically as a result of being kept separate for many generations.

Efficiency of production within a breed may be increased by identifying and mating the best individuals. For other traits improvement comes more quickly when two or more families or strains are formed within a breed and then crossed to take advantage of **hybrid vigor (heterosis)** in the progeny. Because different breeds within a species are unlike in many of the genes they carry, crossing them results in more pairs of unlike genes (heterozygous) in their crossbred offspring. This is conducive to the expression of heterosis. Hybrid vigor is discussed further in Chapter 9.

2.2.2 New Breeds

New breeds resulted from selection for a specific trait that was discovered in an existing breed such as absence of horns in cattle (e.g., Polled Hereford), a recessive gene color (e.g., Red and White Holsteins), or even abnormal structural traits (e.g., the short-legged Munchkin breed of cats). Two important new breeds developed in the United States a half century ago are Santa Gertrudis cattle and the Minnesota Number 1 breed of swine.

New breeds may also be developed by crossing established breeds to capitalize on strengths of each parent breed. Selection is then concentrated on the desired traits for several generations after the crossing and a new breed may be developed. For example, the King Ranch developed the Santa Gertrudis breed of cattle in Texas by crossing the Shorthorn and Brahman breeds. Shorthorns possessed desirable beef qualities but lacked a high degree of resistance to heat and insects. Brahmans were not as desirable in beef qualities but possessed a high degree of tolerance to heat and insects. By crossing the two breeds and selecting within the crossed offspring for desirable beef qualities and heat and insect tolerance, a new breed was developed that combined the desirable characteristics of both parent breeds. The same principle has been used to develop new breeds in other species.

The term *synthetic breed* refers to new breeds developed by crossing two or more older breeds in much the same way the Santa Gertrudis breed was developed. Researchers from the United States Department of Agriculture (USDA) and from commercial companies have developed synthetic breeds of poultry, sheep, and swine. The object is to develop superior new breeds that perform well in specific performance traits and then cross the new breeds in a systematic manner to obtain heterosis in the crossbred offspring. The breeds to be crossed are usually unrelated genetically. The further apart the parents are in relationship the more hybrid vigor can be expected in the crossbred offspring.

The term *exotic breed* among beef cattle groups is used to characterize purebred cattle that have been shipped from their country of origin to another country in which they are not native. The Simmental, Limousin, and Chianina breeds are examples of exotic breeds of cattle shipped to North America during the second half of the twentieth century.

2.2.3 The Pedigree

A *pedigree* is a record of an individual's ancestry. Usually a pedigree includes only the names and registration numbers of the ancestors of a selected animal. However, pedigrees may include selected production (e.g., weight gains and carcass data) or performance information for the ancestors and their progeny and close relatives. A four-generation pedigree of a Hereford female is shown in Figure 2.1. The name and registration number of the **sire** (father) are placed at the top of the pedigree, those of the **dam** (mother) are placed at the bottom. This pedigree shows the names and numbers both of the *paternal* and *maternal* grandparents, great-grandparents, and great-great-grandparents. Because all ancestors were registered as purebreds, the offspring is a purebred.

2.2.4 Pedigree or Record Associations

Each pure breed has a pedigree or record association. This association is an organization of breeders who cooperate to improve the breed, preserve its purity, and promote the interests of animal producers. Generally each breed association has an execu-

Figure 2.1 Pedigree of a purebred beef cow showing the format and the names and registration numbers of ancestors.

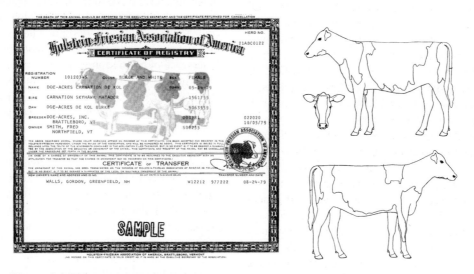

Figure 2.2 Sample registration certificate for a purebred animal (left). As a means of further identifying registered animals, certain breeds require either a photograph or a sketch of the animal (right).

(Courtesy of Holstein–Friesian Association of America.)

tive secretary and a board of directors who are responsible for the association's business. The business includes recording pedigrees and for a fee issuing certificates of ownership and registry (Figure 2.2) for each registered animal. Funds from these fees support the association's work. Breed associations adopt a standard of perfection for the breed, stating specifically the desirable and distinguishing breed characteristics. They also have rules that set forth requirements for admission of an ani-

mal to the registry. The association generally publishes a herd/registry book that gives the name, registration number, breeder, owner, *sire, dam,* date of birth, and other information pertaining to each animal registered. Breed associations engage in various activities to promote the breed such as offering prizes at fairs and advertising in agricultural publications. Many breeds are also promoted by a special magazine, journal, or periodical dedicated solely to activities of the specific breed.

2.2.5 Form and Function of Breeds

Results of selection and development of breeds of animals for a particular form or function (called **type**) are evident among many breeds of farm animals. Certain body conformations are, in a general way, best suited for specific types of production. As examples, draft purposes call for strong-boned heavy-muscled animals whereas meat purposes stress muscle over skeletal structure. Selection for milk or egg production would not necessarily be enhanced by either muscular or skeletal development. As breeders selected animals for particular purposes very definite "types" developed aligned with the intended use of the breed of livestock or poultry.

Cattle breeds in the United States have diverged dramatically into meat-type (beef) or milking-type (dairy) breeds (Figure 2.3). Meanwhile, in much of the world breeds of cattle remain as **dual-purpose** (meat and milk) or **triple-purpose** (meat, milk, and draft). Although most breeds of swine in the United States are solely meat-type, a more lard-type hog exists in countries such as Ukraine and China. Horses have been developed for many purposes including draft, racing, and stock (or cattle) work (Chapter 5). Breeds of sheep in the United States are generally categorized as either wool or mutton (meat) type, although milking breeds are common in Asia. In the poultry industry breeds are now of little commercial consequence. Nearly all commercially produced chickens are either of egg-laying (layer) or meat (**broiler**) type. Other breeds have been developed for fighting (game birds) and for ornamental purposes. Companion animal breeds also fall into specific type categories and are discussed in Chapter 4.

Figure 2.3 A Hereford female of excellent beef type (top) is compared with a Holstein cow of excellent dairy character (bottom). Note the difference in body conformation.

(Courtesy of American Hereford Association and Holstein-Friesian Association of America.)

2.3 BREEDS OF LIVESTOCK AND POULTRY

Numerous breeds of livestock and poultry are present in North America and more are being developed or imported from other countries. Some breeds are present in larger numbers than others because they have been here longer and/or have become popular because of their particular breed characteristics and production traits.

2.3.1 Breeds of Swine

New breeds of swine developed in the United States include the Minnesota Number 1, Minnesota Number 2, Montana Number 1, and the Palouse. These new breeds were developed from crosses of certain older breeds with the Landrace. The Landrace breed was first introduced into the United States several decades ago, but in order to introduce it an agreement was reached that the breed would not be sold and reproduced as a pure one because European swine breeders wanted to protect their purebred herds from competition with United States swine breeders. Some of the first imported Landrace lived in the USDA research facilities at Beltsville, Maryland. Another group was at Ames, Iowa, in the Iowa Experiment Station herd. Because it was not possible to breed and distribute purebred Landrace they were crossed with some of the older, established breeds such as the Chester White, Black Poland, and Tamworth, and new breeds were developed from these crosses that carried a considerable percentage of Landrace blood. It was agreed that these new breeds could be reproduced and distributed within the United States free of restrictions.

Many commercial swine producers are currently using a **three-breed cross** of Hampshires, Yorkshires, and Durocs. Other older breeds are sometimes used for crossing but not as often as the three breeds mentioned above. Selection for faster, more efficient weight gains and for more lean and less fat in the various breeds of swine during the past several years has been highly successful. This has been reflected in thinner backfat as shown in Figure 2.4. Photographs of selected breeds of swine in the United States are shown in Figure 2.5.

2.3.2 Breeds of Beef Cattle

New breeds of beef cattle have been developed in the United States, mostly from Brahman crosses. These include the Brangus, Beefmaster, Charbray, and Santa Gertrudis breeds. Other new breeds are currently being developed that do not contain Brahman blood. Breeds referred to as *exotic breeds* include the recently imported Blonde d'Aquitane, Gelbvieh, Maine-Anjou, Marchigiana, Normande, Pinzgauer, Romagnola, Simmental, and Tarantaise. Established beef breeds that have enjoyed considerable popularity in the United States over the past 60 years include the Angus, Hereford, and Shorthorn. Photographs of selected beef breeds of the United States are shown in Figure 2.6 (see page 26).

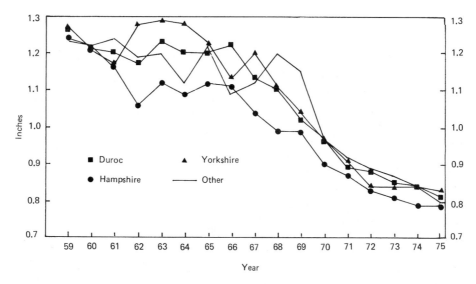

Figure 2.4 Trend in thickness of backfat, in inches at 220 lb body weight, among four major breeds of swine at the Missouri Boar Evaluation Station, 1959 to 1975.
(From Mo. Agr. Exp. Sta. Res. Bull. 1021.)

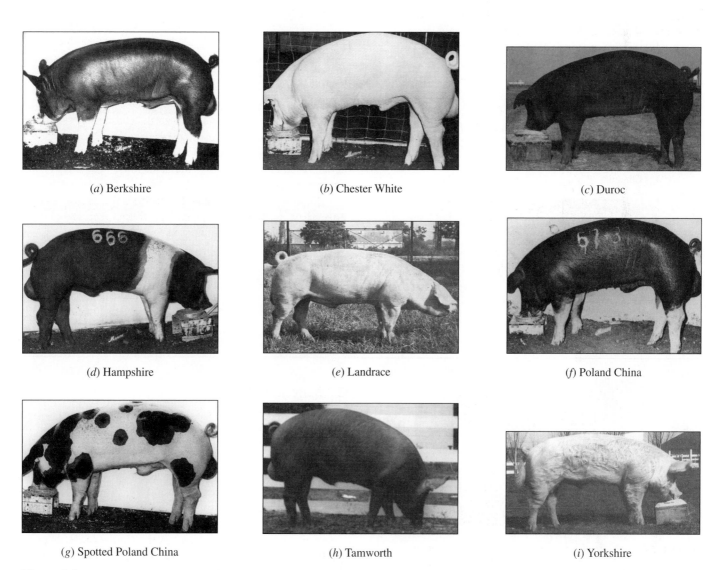

(*a*) Berkshire   (*b*) Chester White   (*c*) Duroc

(*d*) Hampshire   (*e*) Landrace   (*f*) Poland China

(*g*) Spotted Poland China   (*h*) Tamworth   (*i*) Yorkshire

Figure 2.5 Selected breeds of swine in the United States.
(Courtesy of the respective breed associations.)

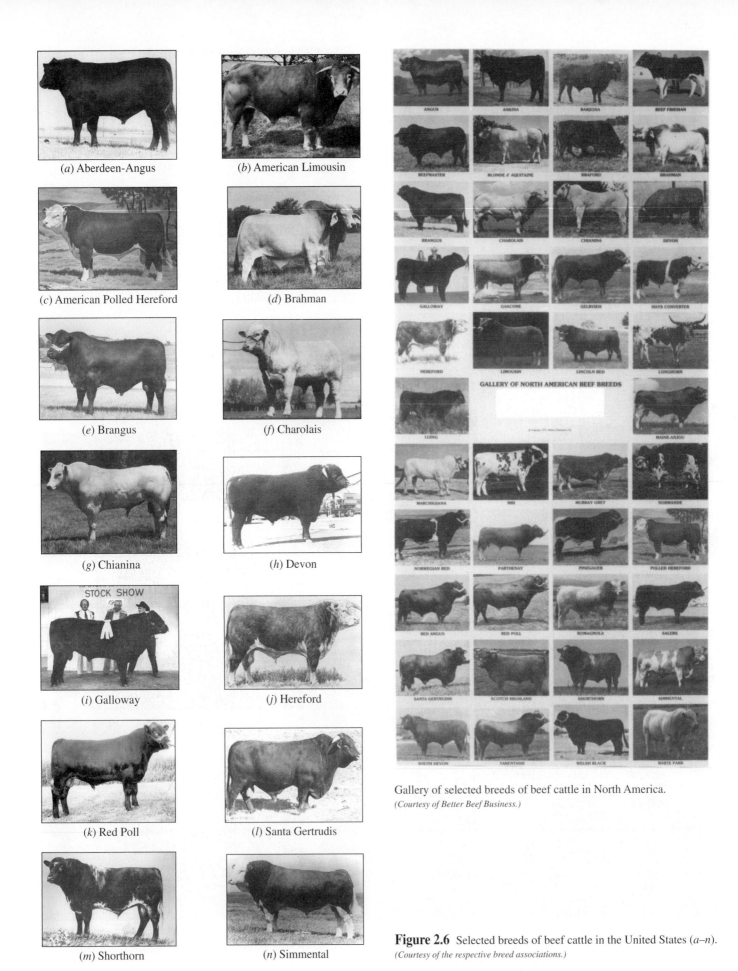

(a) Aberdeen-Angus

(b) American Limousin

(c) American Polled Hereford

(d) Brahman

(e) Brangus

(f) Charolais

(g) Chianina

(h) Devon

(i) Galloway

(j) Hereford

(k) Red Poll

(l) Santa Gertrudis

(m) Shorthorn

(n) Simmental

Gallery of selected breeds of beef cattle in North America.
(Courtesy of Better Beef Business.)

Figure 2.6 Selected breeds of beef cattle in the United States (*a–n*).
(Courtesy of the respective breed associations.)

2.3.3 Breeds of Dairy Cattle

The dairy cattle breeds of North America have been established on this continent for many decades. The Holstein–Friesian breed is present in the largest numbers. Other breeds of dairy cattle in the United States and Canada are the Ayrshire, Brown Swiss, Guernsey, Jersey, and Milking Shorthorn. Photographs of these six major dairy breeds are presented in Figure 2.7.

2.3.4 Breeds of Sheep

The sheep population in the United States has gradually decreased in recent years so that the number of registered sheep now approximates 100,000 head. Breeds showing the greatest decrease in numbers include the Shropshire, Southdown, Hampshire, Corriedale, Merino, and Cheviot. The Suffolk breed has shown the greatest increase in recent years; it numbered approximately 45,000 in 2000, approximately twice the population of a half century ago. Photographs of selected breeds of sheep in the United States are shown in Figure 2.8.

2.3.5 Breeds of Horses

Horses were first brought to America by Columbus on his second voyage in 1493. Spanish colonists and explorers later introduced additional animals. Horses were first used in the United States for either draft or riding purposes; few were

(a) Ayrshire

(b) Brown Swiss

(c) Guernsey

(d) Holstein-Friesian

(e) Jersey

(f) Milking Shorthorn

Figure 2.7 Major breeds of dairy cattle of North America.
(Courtesy of the respective dairy breed associations.)

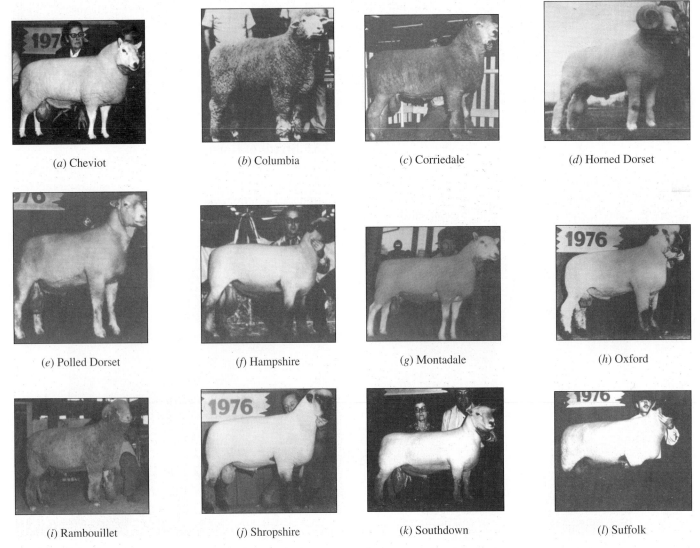

(a) Cheviot   *(b)* Columbia   *(c)* Corriedale   *(d)* Horned Dorset

(e) Polled Dorset   *(f)* Hampshire   *(g)* Montadale   *(h)* Oxford

(i) Rambouillet   *(j)* Shropshire   *(k)* Southdown   *(l)* Suffolk

Figure 2.8 Selected breeds of sheep in the United States.
(Courtesy of Sheep Breeder and Sheepman, Columbia, MO.)

from pure breeds. With the advent of tractors in the late 1930s and early 1940s the horse population declined markedly. It has, however, increased in recent years. Light horses for recreational purposes have become popular in the United States. Photographs of selected U.S. breeds are shown in Figure 2.9 and in Color Plate D in Chapter 4. Chapter 5 is devoted entirely to horses and how they serve humanity.

2.3.6 Breeds of Chickens and Turkeys

There are about 200 breeds and varieties of chickens and turkeys listed in the *American Standard of Perfection,* a publication of the American Poultry Association that lists and describes the recognized breeds and varieties of fowls. Selected breeds and varieties of fowl are listed in Table 2.2. Of these only the Single Comb White Leghorn, White Plymouth Rock, New Hampshire, and White Cornish (Figure 2.10 *a* to *d* on page 30) have been used in recent years to develop the strains of birds used by the poultry industry to produce table eggs and broilers (fryers). Similarly, the Bronze and White Holland breeds have served as the parent stock for developing the strains used to produce broad-breasted, white-feathered market turkeys (Figure 2.11 *a* to *c* on page 30). Examples of typical egg-type lines used to produce white- and brown-shelled eggs are given in Figure 2.12 *a* and *b* (see page 30). Typical female and male breeders used to produce commercial broiler chicks are shown in Figure 2.13 *a* and *b* (see page 31).

Because of the high reproductive rate of chickens and turkeys relatively small numbers of birds are needed to produce hatching eggs. Therefore, breeders can be highly selective in deciding which individuals will be mated to maintain the breeds and lines and to replenish breeding stocks. Also favoring rapid genetic progress in the breeding of poultry is the relatively short generation interval (see Chapter 9).

(a) American Quarter Horse

(b) American Saddle Horse

(c) American Albino Horse

(d) American Paint Horse

(e) American Paso Fino

(f) American Trotter or Standardbred

(g) Appaloosa

(h) Arabian

(i) Tennessee Walking Horse

(j) Thoroughbred stallion

Figure 2.9 Selected breeds of horses in the United States.
(Courtesy of the respective breed associations.)

TABLE 2.2	Selected Breeds and Varieties of Fowl			
American fowl (29 varieties)	**Asiatic fowl (16 varieties)**	**Belgian, Dutch, and German fowl (20 varieties)**	**English fowl (18 varieties)**	**Mediterranean fowl (16 varieties)**
New Hampshire	Dark Brahma	Silver Campine	White Cornish	Ancona
Barred Plymouth Rock	Partridge Cochin	Silver Penciled Hamburg	Silver Gray Dorking	Black Leghorn
Silver Laced Wyandotte	Light Brahma	Golden Penciled Hamburg	Buff Orpington	Andalusian
Buff Plymouth Rock	White Cochin	Silver Spangled Hamburg	White Dorking	Brown Leghorn
White Wyandotte	Buff Cochin		White Orpington	Black Minorca
Partridge Plymouth Rock	Black Langshan		Black Orpington	White Leghorn
Buff Wyandotte	Black Cochin			White Minorca
White Plymouth Rock	White Langshan			Buff Leghorn
Columbian Wyandotte				
Rhode Island Red				
Silver Penciled Wyandotte				
Dominique				

(a)

(b)

(c)

(d)

Figure 2.10 Most lines of chickens presently used for producing market eggs and poultry meat have been developed from breeds and varieties of (*a*) the White Leghorn, (*b*) the White Plymouth Rock, (*c*) the New Hampshire, and (*d*) the White Cornish.

(Courtesy of Watt Publishing Company.)

(a)   (b)

(c)

Figure 2.11 (*a*) Modern white-feathered broad-breasted turkeys have been developed from highly selected lines originating from the (*b*) Bronze and (*c*) White Holland breeds.

(Part a, Courtesy of Nicholas Turkey Breeding Farms, Inc.; b and c, Watt Publishing Company.)

(a)   (b)

Figure 2.12 The DeKalb XL Link is typical of hybrid and strain-cross egg-type chickens used to produce white table eggs (*a*), and the DeKalb Sex-Sal-Link G is a typical egg-type line used to produce brown eggs (*b*).

(Courtesy of DeKalb AgResearch, Inc.)

(a)   (b)

Figure 2.13 Typical female (*a*) and male (*b*) breeders used to produce commercial broiler chicks.
(Courtesy of Hubbard Poultry Farms.)

Part II: Life Cycles of Livestock and Poultry

2.4 BEEF LIFE CYCLE

From prehistoric days Egypt concentrated on the breeding of two animals, the greyhound and a polled, a hornless cattle . . . . The tomb of Huy, who was in charge of Tutankhamen's royal herds, portrays his men branding King Tut's cattle, while the tomb of Auta, of the Fifth Dynasty, before 2625 B.C., shows a bull branded 113 on the left rump.

Mari Sandoz
The Cattlemen (1958)

2.4.1 Historical Perspective of the Beef Industry

The American beef industry evolved from two fronts. Cattle were not indigenous to what we know as "the Americas." European explorers brought them in the fifteenth and sixteenth centuries. Cattle production on the east coast began with colonization in the early seventeenth century. The first efforts at propagation failed because of the inability of the colonists to store sufficient food to survive the cold winters in North America. Livestock were consumed as food leaving few for building herds. In 1611 colonists finally passed a winter with enough surviving cattle to begin beef production. Cattle brought to the northeastern colonies were primarily of British heritage, or *Bos taurus.* The first imports were used for dual or triple purposes, meaning they served as sources of milk, meat, and draft. The first colonists did not believe the horse could fulfill all three needs and was, therefore, considered less useful. As living conditions improved the horse replaced the ox as beast of burden for transporting farm produce and eventually for tilling the soil. Devon cattle were among the first breeds imported to the colonies. In the 1700s Shorthorn were imported followed in the 1800s by Hereford and Angus.

Spanish explorers brought European continental cattle and other livestock to Central America as early as the second voyage of Columbus in 1493. Coronado drove large herds of European longhorn cattle to supply meat and leather for his fellow explorers of western America early in the sixteenth century. In the ensuing centuries large herds of wild escaped cattle, as well as ranch or mission-managed Spanish cattle, increased in Mexico and what became the southwestern United States. Early ranchers in the southwest were able to "round up" the wild, and sometimes not so wild, longhorns that provided the basis for developing a ranching system. Thus began the western American beef industry.

Eventually these two beginnings of beef production systems came together as Americans laid claim to all of what is now the United States and moved westward. Great cattle drives served to spread stock from the Texas area to the northwestern grazing states. Subsequent drives of **finished** animals to railroad marketing points in the midwestern United States such as Abilene, Kansas, and Kansas City, Missouri, are ingrained in the mystique of beef production as well as the nation's history. The purebred industry grew strong and dominated beef production into the mid-1900s. Crossbreeding became more popular early in the twentieth century.

Two other breeding trends had significant impact during mid-twentieth century. First, in the 1930s producers became interested in using *Bos indicus,* or **zebu,** cattle in crossbreeding to improve performance in the hotter, more humid southern states. These zebu, or "hump-backed" cattle, were originally from countries such as India where cattle evolved with looser, lighter-colored skin, with a tolerance to heat and significant resistance to insects through skin secretions. Eventually the strengths of these cattle were infused into breeding programs with existing breeds of beef cattle to create new breeds as described earlier in Section 2.2.2.

Second, in the 1960s producers implemented a significant change in cattle conformation with imported European breeds. The smaller-framed American cattle were rapidly exchanged through breeding programs for larger-framed leaner, European types by importing semen and cattle from various breeds. Charolais began the revolution but soon the Chianina, Simmental, Limousin, and other breeds from throughout western Europe became popular. American producers called these breeds *exotics* because of their foreign heritage and dramatic difference in body conformation. As we begin the twenty-first century these exotic cattle and the changes they brought about continue to have a significant impact on the North American beef industry.

The U.S. beef industry has had several phases of development with dominance in beef production moving from one area of the country to another. By the end of the twentieth century cattle numbers had become concentrated in two groups of states: **range** states with large areas of grazing lands, and states with heavy feed grain production. Table 2.3 presents 2001 beef cattle numbers in the United States. Although the range for average number of livestock per operation is somewhat narrow when all farms and ranches are counted (e.g., Texas averaged

TABLE 2.3	Number of Beef Cattle and Operations by State, 2001		
Rank	**State**	**Millions of Cattle**	**Number of Operations**
1	Texas	13.7	133,000
2	Kansas	6.7	29,000
3	Nebraska	6.6	23,000
4	California	5.1	14,200
5	Oklahoma	5.0	50,000
6	Missouri	4.2	58,000
7	South Dakota	4.0	17,000
8	Iowa	3.6	27,000
9	Wisconsin	3.4	12,000
10	Colorado	3.2	11,400
Total	**All states**	**97.4**	**830,880**

100; Iowa averaged 89), in 1995 Texas had 1500 operations with 1000 or more cattle while Iowa had only 400. The USDA reported 1995 U.S. gross income from meat species to be $44.6 billion. Beef cattle accounted for 76 percent of that income ($34.0 billion).

Cattle utilized in U.S. beef production resulted primarily from crossbreeding including *Bos taurus* and *Bos indicus* parentage. When measured by registry numbers predominant breeds currently used are Angus, Charolais, Hereford, and Polled Hereford. Approximately 10 other breeds have significant impact on beef production. More than 50 breeds are recognized in the United States.

2.4.2 The Season for New Life

Traditionally beef calves are born in the spring to take advantage of favorable weather and renewed forage availability for the cow's milk production. The desire is to have a beef calf crop that is consistent in age, size, and conformation for feeder calf sales or to enter the **feedlot.** Where climate permits calving begins in early March with April and May being the major birthing months.

2.4.3 Birth to Weaning

Management is critical to newborn calf survival. **Colostrum,** the first lactation product of the dam, is an extremely critical link to calf survival. It is also important for other newborn mammals. Colostrum has a concentration of protective antibodies, protein, energy, and minerals. To be protected by colostral antibodies calves should receive colostrum soon after birth. See Chapter 15 for additional information concerning the composition and value of colostrum. Under normal pasture calving systems it is nearly impossible to determine colostrum consumption by calves. Therefore, good herdspersons monitor milk intakes of newborns closely during the calving season. Some administer colostrum to calves within the first 6 h after birth using an esophageal feeder.

Shortly after birth calves should have their navels dipped in an antiseptic solution to prevent infection. Calves born in more confined (**drylot**) facilities may be exposed to a greater number of environmental microorganisms, but even those born on pasture should have their navels dipped.

Individual calf identification is critical to tracking the life and production of an animal. Identification may be divided into two categories: identification for proof of ownership, and identification for individual animal management and record keeping. Proof of ownership for cattle is done primarily by hot **brand.** Cold brand and tattoo techniques are also used. Identification for management purposes should be by a method easily read or recognized in the ordinary working environment. Numbered ear tags are the most common devices used in beef cattle, but cold branding and neck chain tags are also used. The newest technology has allowed individual animal identification by computer chip implanted under the skin of the calf. The chip may be read by a handheld receiver, which also may accept keyed input about the animal from the herd manager. These devices may be downloaded to a personal computer to create useful management data and reports.

For livestock improvement, record keeping is an important adjunct to sound husbandry practices. Information to be considered for calves includes date of birth, birth weight, parentage, and ease of birth. Record keeping per se is of little importance but *using the information* to identify top-performing individuals for breeding programs, to follow improvements of the herd, and to help achieve performance goals represents the payoff.

Most newborn beef calves weigh 70 to 80 lb although some exceed 100 lb. From birth to weaning most calves are nourished by their dam's milk. **Creep feeding,** a process of providing a balanced grain addition directly to the calf diet, may be used to increase calf weight gains and/or to assist when pasture conditions are poor or cows are in thin body condition and challenged to secrete an abundance of milk. Producers should project the economic costs and returns before implementing creep feeding.

In **commercial** cattle production in the United States male calves are castrated at an early age. In Europe and many other areas male calves are fed to slaughter as intact bulls (**bullocks**). Intact males gain faster and leaner than their castrated counterparts. Marketing and grading systems in the United States are unfavorable for intact males, especially as they reach sexual maturity, thus minimizing the number sold as bullocks. Therefore, the current standard for the U.S. beef industry is to market **steers.** Castration should be done as early as the producer's management system permits. If calves are handled to record weight, **implant** growth stimulants, or vaccinate, these events provide logical times to castrate. Otherwise it is recommended that male calves be castrated at less than 1 month of age to minimize stress. See Chapter 13 for information regarding castration procedures for livestock.

Horned stock to enter the feedlot should be dehorned at an early age, preferably at the time of castration or in conjunction with other early management procedures. At an early age, when the horns are essentially in the "button" stage, they may be removed by hot iron, caustic paste, or a simple surgical procedure. At later stages the horns must be removed by saw or large cutting tool, which subjects animals to significantly more stress and potential infection.

Calfhood vaccination programs should begin with the recommendation of the producer's veterinarian. If calves will be sold they should be vaccinated at least 1 month prior to sale. Standard weaning age is about 7 months but this may be affected by calving season and the management program used for cows. Under traditional spring calving programs, weaning date is designed to remove calves from cows as the fall season begins. The calf then enters a **feeder** or development program while its dam often begins grazing late-season grasses or gleaning grain **stover** following harvest of grains. Calf weight at weaning is impacted by genetics and the previous management system utilized. General goals for weaning weights are in the range of 450 to 500 lb.

2.4.4 Replacement Animals

Calves for **replacement** purposes receive initial scrutiny for selection at weaning. Criteria should be based on **performance records** and **type** (Section 2.2.5). Calves' records are compared on an adjusted 205-day performance approximately equaling the common 7-month weaning age. Heifers meeting minimum standards to continue toward reproductive replacement use are maintained on a lean growth diet, which carefully matches energy intake to calf body type and weight. The diet should include good quality forages and have minimal dependence on grains. Females have greater reproductive and lactation performance if they are grown and maintained in reasonably lean body condition. Gains of 1.25 to 1.75 lb per day are recommended. Gain in this range is well below the maximum achieved for slaughter cattle fed high-grain diets.

Following weaning bull calves intended for breeding purposes usually proceed to a feedlot "trial" period. This trial may last up to 160 days and be designed to test the bull's genetic capacity for body weight gain. The resultant record is called yearling performance because it combines the pre- and post-weaning record (i.e., 205-day weaning record plus 160-day feedlot test = 365-day, or yearling, record). Heifers or bulls failing to meet performance standards for replacement commonly enter a feedlot program.

Young bulls meeting all minimum standards may be used in breeding programs at 15 to 18 months of age. Heifers reach puberty at approximately 12 months. Those entering the breeding herd should be bred at about 15 months to calve at 24 months. Size at breeding is more important than age; heifers should have achieved 60 to 65 percent of mature body weight before being inseminated. Bred heifers should continue being fed good-quality forage diets and not be overfed. Ideal calving weight for first-calf heifers is about 75 percent of mature weight. Heifers should be monitored carefully at calving because they are more likely to have calving difficulty (**dystocia**) than mature cows. Dystocia is related both to heifer size at calving and to sire genetics for calf size.

2.4.5 Feedlot Animals

Feeder animals commonly go directly to a feedlot after weaning, although some may enter an alternate **stocker** program. Stocker cattle consume lower energy forages supporting a low rate of gain, often from 1 to 2 lb per day. Weight gain will be primarily skeleton and lean muscle. Later, when the stocker animal moves to the feedlot, it will be lean but have a large digestive and body frame capacity. Subsequent feedlot gains can approach and even exceed 4 lb daily. In numerous areas of the world slaughter cattle are fed solely on range, or forage diets, never receiving grain. Cattle fed in this manner commonly exceed 3 years of age when slaughtered. The choice of a stocker-type program is based upon feed availability and economics of using late-season forages or row-crop residue feeds.

Cattle entering the feedlot directly after weaning may undergo considerable stress because of weaning, nutritional change, and other environmental factors. To reduce stress cattle should be **preconditioned.** Preconditioning guidelines include (1) weaning cattle 30 days or more before shipment, (2) providing a feedlot preparation ration that contains concentrates during the period from weaning until shipment, and (3) following specified vaccination programs. Some producers may use the term **backgrounding,** which generally refers to more than the usual 30-day minimum on feedlot-type diets before shipping.

Cattle entering the feedlot may be moved only a few hundred yards, as on some midwestern farms, or they may travel hundreds of miles from a distant ranch. To minimize nutritional stress after arrival at the feedlot cattle often receive a high-forage diet to provide bulk for the **rumen** and an ample supply of clean water. After an appropriate adjustment period cattle will begin consuming balanced grain mixes and move quickly to high-**concentrate** diets. This move is often in phases, thus facilitating change from high-forage entry diets to finishing diets of up to 90 percent concentrates.

Cattle are marketed by **finish** and projected **grade** rather than age. High-quality genetics, nutrition, and management may combine to produce market-ready slaughter beef as early as 13 to 15 months of age. In 1996, 36.6 million feedlot cattle were slaughtered in the United States at an average slaughter weight of 1169 lb.

2.4.6 Completing the Cycle

Replacement bulls and heifers continue to mature until approximately 5 to 6 years of age. During this time care must be taken to provide for adequate growth, reproductive performance, and lactation although the daily nutritional needs for growth in these later years do not greatly exceed those for body maintenance. Cattle can be fed adequately with good-quality forages supplemented with a proper balance of minerals and vitamins. If cows lose condition during winter from cold stress or poor-quality forages, supplementation of higher-energy feeds may be recommended. Cows can often be maintained in good health and body condition up to and exceeding 10 years. Please note, as described in Chapter 9, that younger cattle waiting to enter the herd should have greater genetic worth and that extension of the **generation interval** with older cattle may slow genetic progress.

Ideally a young female should give birth by 24 months of age. Therefore a generation could be expected every 2 years. In reality, biological and management variations cause losses and delays in the life cycle of animals, and decisions to buy or sell breeding livestock on the basis of economic cycles may extend generation intervals from 2 to 3 times this achievable number. For maximum productivity and standardized calving season programs each cow should give birth to a healthy calf every year on a 12-month cycle after entering the reproductive herd. Calendarizing the reproductive cycle of the breeding herd is intricately linked to excellent reproductive management, record keeping, and aggressive **culling** of animals not meeting performance expectations. Although culling rates depend on many factors such as breeding program goals, health, and the economy, annual rates of 20 to 25 percent are commonplace.

2.4.7 The Beef Industry in the Twenty-First Century

The beef industry has changed dramatically during the history of the United States. The life cycle of a beef animal depends upon geography of the production unit, feedlot versus breeding use, the economy, and goals of individual livestock producers. As consumers in the United States and the world change their financial well-being and eating habits, and as resource allocations change, the beef industry will continue to evolve its production methods to meet market demands and opportunities.

2.5 DAIRY LIFE CYCLE

Dairying has traveled a long road from the wild animals producing enough milk for their calves to the highly efficient cows housed in modern dairy barns. The way has been subjected to many influences of man and nature.

A. R. Porter, J. A. Sims, and C. F. Foreman (1965)
Dairy Cattle in American Agriculture

2.5.1 Historical Perspective of Dairy Cattle in America

Cattle arriving with the seventeenth-century colonists, as described in Section 2.4.1, were mostly multiple-purpose animals of *Bos taurus* heritage that produced milk for the new immigrants. By modern standards milk production from those cows was low, scarcely enough for calf and family. As American agriculture specialized in the nineteenth century higher-producing dairy breeds were imported from Europe and cattle were bred for a single purpose: milk production. By mid-twentieth century dairy cattle breeding efforts focused intensely on volume of milk produced. Holstein–Friesian cows surpassed all others in average total milk production and became the predominant breed in the United States. Currently Holsteins account for more than 90 percent of all cow's milk produced in the United States, dominating the dairy industry in a manner unmatched by any breed in other areas of livestock production. The focus on milk production supported by changes in breed-

ing, nutrition, and management led to a tremendous increase in average milk produced per cow in the second half of the twentieth century. Records of the National Dairy Herd Improvement Association show average Holstein milk production in 1970 was just under 15,000 lb per lactation whereas the 2000 average was over 20,000 lb, a 33 percent increase. Average U.S. milk production per cow for all breeds in 2000 was 18,204 lb. In 1996 the record world production was set by a cow in Wisconsin (70,600 lb of milk in 365 days).

Development of the dairy industry in the United States followed a unique population/geography model. Because milk is perishable it must be delivered to customers quickly or held at cool temperatures to prevent spoilage. In the early years of U.S. agriculture, without refrigerated bulk tanks and delivery trucks or high-quality rapid-transportation routes, ease and proximity for delivery were paramount. Therefore, unlike the production of meat animals, dairy cattle were traditionally located relatively close to large population centers such as New York City and Chicago. That practice continues even today with large dairy cattle populations being in or near states with high human populations such as California and New York. Table 2.4 lists the 10 states with the largest number of dairy cows, their number of dairy farms, and the average annual milk production per cow (lb). The total number of U.S. dairy farms; dairy cows; and milk production in 2000 was 105,250; 9,210,000; and 167.7 billion pounds, respectively.

Although the concept of milk production in close proximity to population base has continued, significant changes occurred late in the twentieth century. Similar to the swine industry described in Section 2.8 there has been a significant division in size of dairy operations. This division has followed a geographic trend. Individual dairy herd size has increased throughout all of the United States but size increased dramatically in certain regions such as the West Coast and the Southeast. Whereas average herd size in the United States in 1995 was about 68 cows, average herd size in California was 380.

| TABLE 2.4 | Number of Dairy Farms, Dairy Cows, and Annual Milk Production per Cow of Top Ten States, 2000 |

Rank	State	Number of Dairy Farms	Number of Dairy Cows	Annual Milk/Cow (lb)
1	California	2500	1,523,000	21,169
2	Wisconsin	21,000	1,344,000	17,306
3	New York	7900	686,000	17,386
4	Pennsylvania	10,700	617,000	18,081
5	Minnesota	8500	534,000	17,777
6	Texas	2500	348,000	16,480
7	Idaho	1100	347,000	20,816
8	Michigan	3500	300,000	19,017
9	Ohio	5500	262,000	17,027
10	New Mexico	500	250,000	20,944

The dairy industry also experienced a significant expansion of milk production in Arizona and New Mexico late in the twentieth century.

The dairy industry has been an early adopter of emerging technologies. In the mid-twentieth century automation in the milking parlor significantly changed the labor requirements and the number of cows one worker could manage. Concurrently the advent of artificial insemination techniques reduced the need to keep breeding bulls on the farm and increased dramatically the quality of genetics available to the average milk producer (Chapter 14). In the 1970s superovulation and embryo technologies facilitated rapid multiplication of the highest-quality genetics from sires and dams (Chapter 13). In the 1990s biotechnology brought commercial production of bovine growth hormone, or BST (bovine somatotropin), which via injection can increase production per cow an average of about 8 to 10 lb of milk daily. Although each of these developments experienced a period of controversy, collectively they represent the dairy industry's search for and adoption of leading-edge technologies associated with efficiencies in producing an abundance of high-quality milk.

2.5.2 Birth to Weaning

One dramatic difference between reproductive management in beef and dairy herds is calving strategy. Calving is desired on a year-round basis in typical U.S. dairy herds. A steady stream of calves ensures a steady supply of cows entering the milking herd to provide a reliable supply of milk throughout the year. But calving on a daily basis creates the need for facilities to manage parturition during winter months in colder climates. Winter season stress and the unique management of day-old calves and their dams determine that dairy calves are generally born in a calving facility as opposed to the open-pasture calving in the beef industry. Intense individual management is critical to the cow's prompt entry into the **milking string** soon after calving. The dairy cow has been selected for a high level of milk production, much more than the calf can consume. Therefore, cows commonly begin the machine-milking regime on day 1 following calving. By legal definition the milk produced the first 3 days postcalving is colostrum and cannot be added to herd milk and sold. Rather, colostrum from machine milking in the parlor can be added to the pool of milk used to feed older baby calves.

Research at the University of Arizona demonstrated that up to 40 percent of the calves studied were inefficient at nursing and did not receive adequate colostrum. Therefore, most calves are hand-fed from bottles, buckets, or esophageal tubes following birth. Hand feeding permits the manager to select higher-quality, stored colostrum from older cows for newborn calves. A mature multiple-parity cow produces colostrum with a greater concentration and range of antibodies. Colostrum from these older cows may be frozen and stored for later use. Calves are normally separated from their dam during the first day or two following birth enabling the dam to enter the milking string. Calves are moved to a calf management facility and hand-fed until weaning.

Although calves are separated from their mother at 1 day of age they continue to receive milk or **milk replacer** by bottle or bucket for 3 to 6 weeks. During the preweaning period calves receive a mixed dry starter feed and high-quality forage. Length of the liquid feeding period is determined by the environment, economics, and dry feed consumption. They are usually housed and fed individually to avoid development of behavioral vices such as attempts to nurse each other. Throughout most of dairy industry history calves in northern climates have been raised indoors in barn-type facilities. In the late twentieth century many producers moved their calf raising to individual outdoor **hutches,** which separated calves environmentally from each other and reduced the transfer of respiratory and other diseases.

As previously described for beef calves, navel dipping, identification, dehorning, and castration should be completed early in life (Section 2.4.3). Compared with the beef industry a smaller percentage of male dairy calves are kept for breeding purposes. Genetically superior preselected bull calves often result from contracted planned matings intended to produce potentially outstanding males for **bull studs.** The bull stud industry and its mating and evaluation of potential sires is highly organized (Chapter 14). A small number of elite dairy herds sell bulls for breeding purposes directly to producers. See Chapter 9 for more information on animal breeding. As a result of bull production strategies the day-old bull calf is a surplus item on most U.S. farms. These calves are commonly sold to specialized producers who raise them for veal or slaughter beef. Unlike producers buying 7-month-old beef calves, producers handling day-old dairy calves must have facilities and equipment to grow the calves through the milk-fed and early postweaning period. Dairy steers will eventually enter a typical beef feedlot environment and are managed as described in Section 2.4.5.

2.5.3 Replacements: Weaning to Calving

Month-old weaned dairy heifer calves are commonly transferred from individual housing to small groups and fed high-quality forages and a balanced grain mix. Forages remain an important base of a lean growth diet. Heifers should reach a desired size and weight to be bred by 15 months. Postbreeding heifers are fed to gain about 1.5 to 1.7 lb daily. Breeding at 15 months or earlier supports the goal of obtaining the first calf from heifers by 24 months of age. From breeding to calving the heifer continues to depend mostly on forages to provide for lean gain. During the latter stages of pregnancy, heifers begin consuming a pre**freshening** diet that includes more concentrates to prepare them for high milk production.

2.5.4 Moving to the Milk Production Herd

Ideal first-calving age for dairy heifers is 24 months or earlier. Prior to freshening the heifer is moved to a calving facility with others about to freshen. Precalving heifers and cows are changed from a predominantly forage diet to one containing forages and a significant amount of balanced **concentrate** mixture to prepare them for high levels of milk production.

Attention should be given to the prevention of **milk fever,** a life-threatening disease of dairy cattle. *Parturient* paresis (milk fever) is not an infectious disease but rather a tetany-like condition resulting from the rapid drain of calcium from blood to mammary tissue for deposition in milk. High-producing cows in general are more prone to this blood calcium disorder. Moreover, Jerseys and Guernseys are more genetically prone to the disease than other dairy breeds. Careful management of the prefreshening diet can assist in preventing the disease.

After calving, cows in good health enter the milking string. Although milking is done by hand in many countries, virtually all dairy cows in the United States are milked in a parlor equipped with semiautomated milking machines. The lactation and milking processes are described further in Chapter 15. It is important to note that sanitary conditions in the parlor and proper maintenance and use of milking equipment are critical to harvesting high-quality milk and protecting cows from mammary gland infections (**mastitis**). Most cows are milked twice daily but milking 3 times per day is also practiced. In general, the more frequently cows are milked the more milk they secrete but the increase diminishes with added milkings. Decisions relative to milking 3 times (or more) per day are related to labor availability and return on investment. Currently research teams are evaluating automated (**robot**) milking systems that allow cows to enter the milking parlor and be milked multiple times daily. Cost and effectiveness are critical issues in applying this developing technology.

Cows reach peak milk production about 45 days into their lactation. Therefore, nutrition and management in early lactation is especially important. After peaking, milk production usually follows a pattern of slow but steady decline. In effect, the mammary gland's ability to secrete milk slows as cells are lost because of the normal regression of milk-making cells, and as the secretion of secretory hormones decreases. After an appropriate shutdown (**dry period**) the cow can begin lactation again and proceed through the same profile of peak, then gradual decline in production. For maximum productivity cows should complete a 45- to 60-day dry period between lactations. Furthermore, for optimum lifetime productivity each cow should produce a calf on a 12-month cycle. These guidelines lead to a standardized annual cycle of 10 months of lactation followed by a 2-month dry period. To complete this routine cows should be bred 3 months into their lactation. Figure 15.14 illustrates typical lactation cycles (persistency).

Dairy cattle continue to mature until they reach 5 to 6 years of age when maximum productivity is commonly reached. Because of normal culling for low milk production, inefficient reproduction, and disease conditions, dairy cows leave herds, on average, at 5 to 6 years of age. Productive healthy animals have been maintained in individual herds for well over 10 years.

2.5.5 The Dairy Industry in the Twenty-First Century

The individual farm dairy cow and her milk production were important to the development of colonial America. As America grew, small dairy herds were developed close to large popula-

tion centers. As the twenty-first century begins, the dairy industry, much like most of agriculture, is destined for continued change as fewer dairies milk more cows. In 1950 a 100-cow herd was considered large in much of the United States. By 2001, 2000-plus cow herds were commonplace in several states. The future of a diverse dairy industry hinges on the ability of farms of all sizes to competitively utilize technology and access markets with high-quality milk.

2.6 GOAT LIFE CYCLE

> And there will be goats' milk enough for your food, for the food of your household, and sustenance for your maidens.
>
> **Proverbs 27:27**

2.6.1 Historical Perspective on Goats

Among the first domesticated animals, goats are believed to have descended from a species of wild goat found in Asia Minor, Persia, and nearby countries. There are dozens of biblical references to goats. A bas-relief on an ancient pharaoh's tomb shows goats feeding in the tops of trees. Goats are hardy, relatively disease-resistant, natural browsers that thrive on rough, hilly terrain with sparse pastures where a dairy cow would virtually starve. These characteristics make the goat a commonly underrated animal that supplies high-quality milk and meat for humans.

Goats provide an important source of much-needed protein and minerals—especially important nutrients deficient in human diets of many developing countries. They are ideal for families needing a dependable supply of milk but who are unable to own a milk cow. Although dairy cattle predominate in producing milk in the United States, more people consume goat's milk than cow's milk worldwide. Goats also browse a wider variety of forages than sheep or dairy cattle. Specifically, goats are more likely to select woody plants and certain shrubs and trees to browse, thereby expanding their intake of marginal quality feedstuffs relative to other domesticated ruminants.

There are approximately 600 million goats worldwide and more than 2 million in the United States where Texas ranks first in their production. It is likely that goats first came to the Americas along with sheep on the second voyage of Columbus in 1493. Goat breeds are usually divided into meat-, milk-, and fiber-producing categories. The most important products of goats worldwide include milk, cheese, meat, mohair, and leather. Although goats have long provided significant products to the human diet and the fabric industry, the first National Research Council publication, *Nutrient Requirements of Goats,* was not available until 1981. That publication noted that goats should be managed and fed differently than either sheep or dairy cattle.

Major imported milking breeds of goats in the United States include French Alpine, Oberhaslis, Saanen, and Toggenburg from Switzerland; and Nubian from England. The milking LaMancha breed was developed in the United States early in the twentieth century.

Goat's milk, like cow's milk, is an excellent source of important food nutrients for humans and when compared with milk from cows has certain unique characteristics. Goat's milk is to some degree homogenized naturally in that the fat globules are much smaller than those of cow's milk and do not rise readily to the top when milk is stored. Additionally, because of differences of proteins in the milk of cows and goats, some people with allergies to cow's milk can consume goat's milk with no adverse allergic reaction.

By legal definition "milk" in the United States comes only from cows. Therefore, dairy products made with goat's milk (e.g., fluid milk, cheese, and ice cream) must indicate the source of milk on the label.

The male goat is called a **buck,** the female a **doe.** A goat less than 1 year of age is called a **kid.**

2.6.2 Reproductive Character of Goats

Many breeds of goats, like sheep, tend to be seasonally polyestrous. In the Northern Hemisphere does begin their natural breeding season in early fall and continue through late winter to early spring. Signs of estrus are marked by increased activity, swelling of the vulva, twitching the tail, frequent urination, and **bleating.** Goats, like cattle, in full estrus (heat period) ride each other. Typical estrous cycle length in goats is similar to that of cattle (17 to 21 days), whereas gestation length is more nearly that of sheep (about 5 months). In the United States the young are usually born from late winter through spring.

2.6.3 Birth to Weaning

The number of kids born among milking breeds ranges from singles to quadruplets, although singles and twins are most common. As with the management of some ewes, does may need encouragement to accept the later born in multiple births. This is true especially when significant time has lapsed between birthings and the doe has completed the process of cleaning (shedding afterbirth) and bonding with the first- and/or intermediate-born kid. Multiple kidding may also lead to the birth of less-vigorous, smaller kids or to individuals less competitive with their stronger siblings. Because goats have only two mammary glands this can be a problem if more than two offspring must compete at nursing time. Experienced herdspersons are prepared to assist kids in obtaining colostrum and may need to "cross-foster" them (find a new mother) to more acceptable situations, such as being moved to does with a single offspring or to a doe with higher milk production.

Strong, healthy kids in commercial milking herds are weaned as early as 1 to 3 days of age, thereby freeing the doe to enter the milking string. Weaned kids are fed milk replacer and/or surplus colostrum by bottle. They may be raised in groups if careful attention is paid to individual access to bottles at feeding time. Weaning from milk commonly occurs at 1 to 2 months, providing the kids are consuming adequate amounts of dry concentrate feed.

Kids from meat-type and Angora breeds are usually left with their mothers for an extended period, similar to raising lambs. In this case milk is not the primary salable product of the doe, but rather serves as a resource to grow the young. The primary economic product with these breeds is the goat kid for meat and/or mohair production.

2.6.4 Replacements: Weaning to Kidding

After weaning most growing replacements are placed on a diet based heavily on forages, which promote lean growth. Diets are similar in composition to those fed dairy heifers, although growing goats consume feed at a higher daily proportion of body weight. Whereas a growing dairy heifer may eat dry feed equivalent to 2.5 to 3 percent of its body weight daily, a growing goat may consume from 3 to 4.5 percent of its body weight. During the milk production phase this increased feed consumption by goats continues. Lactating dairy cows ordinarily eat 3 to 5 percent of their body weight daily, whereas lactating goats may consume at a rate of up to 6 percent or higher of their body weight.

Properly managed kids attain breeding weight and good body condition by the first fall season. In general, puberty is reached at about 6 to 8 months of age. Breeding underdeveloped females can significantly impact potential milk production. Female kids born earliest in late winter or early spring are most likely to be developed sufficiently to breed late the following fall. Other does will be held over to grow and develop until their second fall before breeding. This increases significantly the input costs before providing economic returns via milk and/or their own kids.

Young males are more likely than females to attain breeding weight by their first fall and can be mated with a small number of does at that time. Goat semen can be collected, frozen, and managed using procedures similar to those employed with bull semen (Chapter 14). Because of the relatively low level of commercial utilization, goat semen is available from a limited number of specialty suppliers. Mature goats secrete strong body odors that are objectionable to many humans. Handling bucks easily transfers the odorous compounds to skin and clothing. The odor is persistent and its intensity on the handler or in the environment can be a negative factor for those wishing to own goats as pets or companion animals.

2.6.5 Milk Production of Does

The lactation period and management principles for dairy goats are similar to those of dairy cattle. Goats bred for milk production commonly produce for 10-month lactation periods and should be given a 2-month dry period between lactations. As described in Section 2.6.2, this time line is generally impacted by calendar dates because of the seasonal nature of goat reproduction, whereas season has little or no impact on dairy cow reproduction. Because of the natural fall breeding pattern goats are more likely to be bred in one season, followed by a concentrated kidding season, resulting in somewhat synchronized lactation cycles. Therefore, unlike most commercial dairies that have cows milking throughout the year, goat dairies may have a 2- or more-month dry period for the entire herd. Goats reach peak milk production at about 5 to 7 years and often remain in the milking herd 10 or more years.

High-producing milk goats are subject to the same metabolic and production-depressing disorders that affect dairy cows (e.g., milk fever and mastitis). Feeding and management strategies to help prevent these disorders in goats are similar to those described for dairy cattle (Section 2.5.4).

2.6.6 Meat Production

Goat kids entering meat and mohair production systems are weaned from their mothers at ages ranging from 2 to 4 or more months. Some kids may nurse until terminal slaughter. Earlier-weaned kids entering meat production receive a combination of grain-based concentrate feed and forage. If managed in a feedlot program they often enter the feeding program weighing 40 to 60 lb. Depending on breed size, kids are usually sold for slaughter at 100 to 125 lb live weight.

Although there are several breeds of meat goats in the United States, the Boer from Africa probably received the most attention in recent years. According to information from Oklahoma State University the growth rate of Boer goats in feedlots commonly averages 0.3 to 0.4 pounds per animal daily. Carcasses produced from meat-type goats are comparable in conformation and cuts to those of lambs. In general, goat meat is lean and has a distinctive flavor that becomes stronger with age. As with lambs, the most desirable meat comes from younger animals.

2.6.7 Mohair Production

Mohair (fiber) production among goats in the United States comes mainly from the Angora breed that was developed from Asian goats. Mohair "clip" from a well-managed goat herd ranges from 6 to 14 lb per animal, similar to the fleece weight of sheep. Mohair does not have extreme "crimp" like fine wool but has considerable strength and, like wool, retains its shape in fabric. Mohair accepts color dyes readily and once dyed has a luxurious sheen. The combination of these characteristics results in mohair being commonly used in upholstery and ornamental fabrics rather than wearing apparel. The Angora goat is most populous in Texas where about 60 percent of the approximately 2 million goats are producers of mohair.

2.6.8 Other Interests in Goats

Goats have also become companion and hobby animals. Many Americans enjoy 4-H and Future Farmers of America (FFA) projects, showing competition, and hobby farming with goats. The number of breeds is extensive and goat characteristics that attract fanciers are equally diverse.

The "Fainting Goat" is not really a breed but rather a strain of goats in the United States whose selection has resulted in its dramatic response to stress. Fainting Goats do not faint (temporarily lose consciousness) but respond to excitement or stress by stiffening muscles that lead to the inability to escape and a "falling down" syndrome that is somewhat similar to stress responses seen in other stress-prone animals. Unlike stress-prone swine, the Fainting Goat usually regains control of its locomotion in a short period of time.

Though goats are small ruminants with relatively low maintenance costs, an even smaller goat captured the attention of many goat fanciers in recent years. The Pygmy goat has grown in popularity throughout the United States as a pet and as an animal production pastime. Many Americans have adopted this breed and other small goats as hobby-farming animals or as an extension of their agricultural roots.

2.6.9 The Goat Industry in the Twenty-First Century

Goat production is an important part of agriculture in North America as well as in most other agricultural regions of the world. Because these ruminants can be raised and maintained on low-cost feeding regimens and can graze on land and terrain unsuitable for crop production; and because as browsers they are efficient in converting forages and other low-cost feedstuffs into meat, milk, and mohair; and because they are inexpensive to purchase and raise and demonstrate considerable resistance to disease, goats hold great promise as even more important animals in the service of humanity in the years ahead.

Goats are especially important sources of milk for the people of India, Africa, southern Europe, and western Asia. Moreover, because of its minute fat globules and fragile curd, goat's milk can be easily digested by infants, invalids, and most persons intolerant to cow's milk. These benefits, together with those mentioned earlier, cause us to project that the goat industry will thrive and expand globally in the twenty-first century.

2.7 SHEEP LIFE CYCLE

> In 1785, Louis XVI of France became much interested in improving the production of wool and in fostering woolen manufacturing in his country; he thenceforth asked the king of Spain for permission to import from the celebrated Spanish flocks a flock of sheep with the highest quality of fine wool.
>
> **Hilton M. Briggs and Dinus M. Briggs**
> **Modern Breeds of Livestock (1980)**
> **(Description of the foundation of the Rambouillet breed)**

2.7.1 Historical Perspective on the Sheep Industry

Sheep, like other species discussed in this chapter, arrived in the Americas with the second voyage of Columbus. They arrived shortly thereafter on the mainland of Central America. Sheep and goat flocks flourished and spread along with the explorers and missionaries throughout what was to become the southwestern United States and California. Sheep were introduced to the colonial area on the eastern American seaboard early in the seventeenth century. Modern breeds of sheep raised for wool or meat (mutton) are of the genus and species *Ovis aries*. In the

United States most emphasis has been on breeding for meat type with wool being a secondary consideration.

Sheep production in the United States became centered in states with open grazing lands as illustrated in Table 2.5. Flock size varies dramatically by state. The western grazing states have large flocks, whereas midwestern farm flocks are smaller. Lamb and mutton consumption in the United States has remained low, averaging 1 to 3 lb per person annually over the past century.

Primary emphasis in sheep breeding has been for meat production. To increase lamb crop size, increased emphasis has been placed on creating mother breeds, "ewe-type" crossbreeds that give birth to multiple lambs in spring and fall. The breed used in greatest number for its strong mothering characteristics is the Rambouillet. This breed has **fine wool** and produces a heavier, finer, more valuable fleece. Finnsheep are also used in creating ewe-type dams because they give birth to larger numbers of lambs. Rambouillet have been combined with Finnsheep and others to synthesize the crossbred mother flocks. Traditional mother breeds do not have the heavy muscling desired by the slaughter industry. Ewe-type dams are intended to be the basis of flock systems and are crossbred to meat-type rams to produce offspring for the feedlot industry. This type of breeding program is a prime example of capitalizing on strengths of selected breed types as discussed in Sections 2.2.1 and 2.2.5. Pure breeds excelling in meat characteristics and having the highest registry numbers are Suffolk, Dorset, and Hampshire. Farm income in the United States from marketing lambs in 1996 was $0.6 billion.

2.7.2 Lambing Season

Breeding seasons for sheep are generally concentrated in the fall and spring months. Most breeds of sheep produced in the United States are sexually active in the shortened day lengths of late summer through early winter. The industry terms this reproductive cycling **"seasonally polyestrous"** indicating that ewes cycle regularly during this shortened season. Sexually mature ewes begin cycling in late summer or early fall and continue cycling until late fall or early winter. Some breeds identified as **"nonseasonally polyestrous"** will breed in other months. A second spring breeding season is not uncommon for these ewes. In general, winter and spring lambing is most common in the Midwest and upper Northwest areas of the United States, whereas fall lambing is more common in the southwestern United States.

The estrous cycle of sheep (15 to 17 days) is shorter than the 19 to 21 days of other mammalian species discussed in this chapter. Pregnancy length of 148 to 152 days (5 months) for most breeds results in lambing times in late winter through spring in midwestern and northern flocks. Early lambing, especially the months of January and February, dictates greater investments in indoor facilities in higher latitudes. Later lambing (April and May) can be accomplished on pasture without extensive indoor facilities.

2.7.3 Lambing to Weaning

Management procedures include some of those described previously for beef. The lamb should receive colostrum and have its navel dipped as soon as possible following birth. Unlike beef or dairy cattle, ewes regularly give birth to twins or triplets. Multiple births in sheep often lead to a problem of abandonment of the later-born lambs. If the dam has **cleaned** and nursed her first lamb a significant length of time before giving birth to a second lamb, she may not accept the later lamb as her offspring. Shepherds devise methods to confuse the dam about the identity of her lambs so that she will accept the later-born offspring. Methods include removing both lambs to roll them in placental fluids or douse them with salt solutions to mask differing odors of the offspring. The lambs are reintroduced to the dam with the hope that she will be unable to differentiate between them and, thereby, accept both. These methods are also used to foster lambs from one dam to another in cases of death of the mother, inadequate milk production, disease or poor health, or other problems. To enhance the probability of fostering, or **grafting,** to a dam whose newborn lamb died, shepherds may skin the dead lamb and cover the lamb to be fostered with the fleece. This presents the dam with a new lamb having the odor of her own offspring.

Lambs should be identified as soon as possible within the management system of the flock owner. Vaccination systems should be defined with the flock's veterinarian. Lambs should have their tails docked as soon as possible (prior to 1 month of age). This is done at an early age to minimize stress. Docking is the process of removing the tail leaving only one or two tail vertebrae intact. The tail is removed by clamping or banding between vertebrae or by surgical removal. Docking is beneficial in later life because it prevents manure accumulation and parasite infestation of the tail. Moreover, it increases reproductive performance in replacement animals.

Castration of lambs is not an automatic management decision as it commonly is in most beef and swine operations in the United States. Because of the young slaughter age of U.S.

TABLE 2.5	**Number of Sheep, Farms, and Ranches by State, 2001**		
Rank	**State**	**Number of Sheep**	**Number of Farms and Ranches**
1	Texas*	1,100,000	6800
2	California	840,000	3000
3	Wyoming	530,000	900
4tie	Colorado	420,000	1900
4tie	South Dakota	420,000	2300
6	Utah	390,000	1500
7	Montana	360,000	1800
8	Idaho	275,000	1000
9	Iowa	270,000	4700
10	Oregon	245,000	2900
Total	**All states**	**6,915,000**	**66,000**

*Texas also had approximately 2,000,000 goats in 2000.

lambs, producers often market intact ram lambs. If lambs are to be marketed at a later age or comingled with ewe lambs they are commonly castrated before 1 month of age, usually at the time of docking.

2.7.4 Market Lambs

Lambs may be weaned early and moved to the feedlot or they may remain with their dams to nurse until marketed on an early marketing system. Typically weaning occurs from 2 to 5 months of age. Some market lambs are weaned and go directly to slaughter. Lambs weaned to the feedlot can be finished and marketed at about 3 to 4 months of age and usually weigh 110 to 130 lb. Lambs may also enter lower-energy grazing systems, which may significantly increase the age at preferred market weight. In 1996, 4.2 million sheep were slaughtered at an average weight of 128 lb to yield 269 million lb of lamb and mutton. In the United States relatively young slaughter lambs carry only about 2 lb of wool clip at slaughter. These short fleeces are used for specialty markets.

The marketing pattern reflects the seasonal nature of the breeding pattern of sheep. Commonly lambs from fall breeding programs and grazing systems reach market age and weight in the summer months. But the spring and Easter holiday market for lamb brings premium prices, so lambs born in the fall or early in the new year are often managed to benefit from this marketing opportunity.

As described in Chapter 3 maturity of the animal is a significant factor in grading meat. For lambs the age factor is critical. Age at slaughter influences intensity of "lamb flavor" in the resulting meat. To many American consumers the development of a strong flavor in older lambs is objectionable. If the lamb is less than 1 year of age its carcass will be marketed as "lamb," indicating a projection of less lamb flavor development. If the lamb is significantly older than 1 year the carcass is marketed as "mutton," which is expected to be more highly flavored and therefore commands a lower price. The USDA grader estimates age of lamb carcasses by assessing the growth plate of the lower leg at the epiphyseal–diaphyseal junction (Chapter 10). If the plate breaks easily revealing the cartilagenous surface (called **break joint**), the carcass is graded lamb. If the plate does not separate indicating the plate has ossified in the older animal, the grader removes the lower leg from the carcass at the joint above the pastern. This exposed joint (called **spool joint**) identifies the carcass as mutton. A grade in between, called yearling mutton, is used for intermediate animals with one break joint and one spool joint.

2.7.5 Replacement Lambs

Ewe lambs are raised for lean gain to enhance later reproductive and lactation performance and to avoid unnecessary feed costs. Lower-energy diets are based on good-quality stored forages or on grazing. Grain is required only when forage quality or quantity is inadequate to maintain optimum weight gain. Ewe lambs should be evaluated for their genetic potential for performance to support a well-designed replacement selection program. Effective record keeping and involvement in on-farm or formalized testing programs such as the National Sheep Improvement Program serve as the primary basis for making genetic progress in selection (Chapter 9).

Properly grown ewe lambs should be ready for breeding during their first or "yearling" fall season. Ewe lambs should weigh about 100 lb at first breeding. They may be separated from older ewes and bred later in the season to allow as much prebreeding growth as possible. Early bred ewe lambs are still growing while they are pregnant and should be monitored for adequate growth and body condition during gestation.

Ram lambs selected for breeding purposes may be evaluated in a feedlot performance trial at a **ram test station** where rams from state or regional flocks are performance evaluated under common environmental conditions. Feeding periods are commonly 4 to 5 months. Rams are evaluated on traits such as average daily gain, **feed efficiency,** estimated body composition, and structural soundness. Young rams may be used for breeding in their yearling season but should be monitored for growth and body condition during the breeding season.

2.7.6 The Ewe Flock

Breeding programs are determined by the polyestrous nature of the flock, geographic and climatic environment, and desired marketing program. In fall breeding programs mature ewes are bred earlier in the season if the production goal is to sell lambs for the higher-value spring slaughter market. The challenges created by breeding in the early fall include the need for indoor lambing facilities and more intense animal management for lambing during the winter season. Ewes that cycle in the spring may be bred for fall lambing when weather creates less stress and need for indoor facilities.

If ewes are not in good breeding condition they should be fed extra grain to improve body condition and increase ovulation rate. **Flushing** is the practice of feeding 1 to 2 lb of grain to ewes for about 2 weeks prior to breeding. A similar practice of flushing is often employed in breeding swine.

After breeding, ewes may return to forage feeding and receive grain only if body condition or forage quality is a concern. Special attention should be given to the ewe's condition as she enters the final third of pregnancy. Grain may be warranted to assist her in maintaining body condition for the last rapid gain in fetus weight and to prepare for lactation. Ewes may be shorn in late pregnancy. The shearing can involve **crutching,** which is the removal of wool in the genital and mammary area, or it may be a complete shearing. Shearing depends on management scheme, facilities, and season of lambing.

Ewes to lamb in the winter season are generally moved in groups to lambing facilities about 2 weeks prior to the projected lambing date. Supervision and grain feeding are commonly more convenient in lambing facilities. Ewes lambing in warmer seasons or out-of-doors may be concentrated in smaller paddocks or grazing areas to assist with management. After lambing the udder of the ewe should be checked for mastitis and to assure an adequate supply of milk for the newborn. Ewes and lambs from indoor lambing facilities may be comingled in groups after the ewe–lamb bonding has been established. Adequate bonding usually occurs within a few days following lambing. Ewes may be placed in groups according to number of lambs per ewe because dams of multiple lambs may need more feed to support milk production. The age to wean lambs depends on the management system as described in Section 2.7.4. After weaning their lambs ewes generally return to a grazing system to await the next breeding season.

Nonseasonally polyestrous ewes may be bred more than once per year to produce two lamb crops. Weaning must be accomplished early enough for the ewe to return to good body condition for the breeding season. It is unlikely that an individual ewe will maintain an intense, steady twice-per-year lambing cycle but will more likely produce three lamb crops in two years. With proper management ewes can be maintained in a flock for several years. Culling rates for poor performance average about 20 percent annually.

Wool is often an important product to the sheep producer. "Fine wool" breeds such as Merino and Rambouillet produce a wool of finer fiber diameter and greater weight per fleece. These fine wool fleeces have greater market value. Typical meat-type breeds have lower weight and coarser fleeces of less value. Longer wool from an annual shearing program is of greater value than shorter wool. Therefore, producers generally shear once annually often just prior to lambing.

2.7.7 The Sheep Industry in the Twenty-First Century

Sheep production is a historic part of agriculture in North America. The U.S. sheep industry continues to be concentrated in the mountainous and grazing states of the West. At the beginning of the twenty-first century western sheep flocks continue to maintain many of the traditions and methods of production of the twentieth century. The U.S. sheep industry is challenged by low per capita consumption of lamb meat and by the many alternatives to wool in the clothing and carpet industries. If concerns continue for world food supplies and population increases as expected in the twenty-first century the sheep industry may capitalize on its historical use of forages from grazing arid lands and rugged mountainous terrain to produce increased amounts of meat and fiber.

2.8 SWINE LIFE CYCLE

The general appearance of the pig should be compact and thick of body, involving shortness of head, broad back, deep body, short legs, and plenty of quality, as shown in abundance and fineness of hair and strong bone and joints.

Charles S. Plumb
Types and Breeds of Farm Animals (1906)

2.8.1 Historical Perspective on the Swine Industry

Swine were domesticated some 8000 years ago. Ancestors of modern U.S. and European hogs were the European wild hog *Sus scrofa,* whereas ancestors of Chinese breeds were *Sus*

indicus. Swine were brought to the Western Hemisphere as early as the second voyage of Columbus in 1493 and arrived on the tip of Florida in 1539. Escaped hogs from this early exploration era likely were the source of the wild hogs of the eastern United States. Swine arriving with the colonists in the seventeenth century were the source of early herds for domestic production. Swine breeds of the United States developed mainly from imports of British origin. Several breeds still carry "shire" in their name (e.g., Berkshire, Hampshire, and Yorkshire).

Swine production followed the American migration westward. In the nineteenth century the east central United States was the major swine production region, with Cincinnati acquiring the name of the "Porkopolis" because of the large number of hogs slaughtered there. For a period of time some swine were moved to slaughter in small "trail drives" similar to the movement of cattle from the western United States. During the twentieth century swine production became concentrated in the midwestern United States because of availability of their primary feed sources (corn and soybeans). The recent trend of developing large corporately owned swine units in the southeastern United States has had a significant impact on swine distribution nationally (Table 2.6).

Similar to the beef industry most swine marketed in the United States are crossbred, with pure breeds providing the genetic resource for swine improvement. The most populous pure breeds of swine (in rank order) in the United States are Yorkshire, Duroc, and Hampshire, with five other breeds providing significant but fewer numbers. Consumption of pork by Americans remained relatively steady throughout the twentieth century. For most of the century pork consumption was second only to beef but recently poultry meat consumption has exceeded that of pork and beef (Chapter 3). Worldwide, pork enjoys the highest average per capita consumption of all meats. China produces nearly one-half of the world's swine.

TABLE 2.6	Number of Swine and Operations by State, 2000		
Rank	**State**	**Millions of Swine**	**Number of Operations**
1	Iowa	15.2	12,300
2	North Carolina	9.3	3600
3	Minnesota	5.8	7300
4	Illinois	4.2	5100
5	Indiana	3.4	4400
6	Nebraska	3.0	4000
7	Missouri	2.9	3600
8	Oklahoma	2.3	2700
9	Kansas	1.6	1600
10	Ohio	1.5	5200
Total	**All states**	**59.3**	**85,760**

2.8.2 Management of Farrowing

Modern swine management units take advantage of the truly polyestrous sexual cycling of swine to produce litters from groups of sows on a year-round basis. In the mid-1900s farmers often farrowed only once or twice per year during months when field crop work was reduced. Under these limited farrowing systems sows were often sold after one to two farrowings. Contributing to the early sale of females was the sow's tendency to continue gaining weight throughout her life. When **limit feeding** programs to control sow weight became accepted the productive life of sows was extended. Farmers farrowing swine under intensive management systems often set an annual goal of 2.5 litters per sow.

The gestation length for swine is about 112 to 114 days, or the farmers' thumb rule of "3 months, 3 weeks, and 3 days." Baby pigs weigh about 3 to 3.5 lb at birth. Their immediate needs following birth include warmth and colostrum. **Neonatal** pigs do not adequately control their body temperature for several days after birth. In the wild the sow provided heat through physical contact, heat radiating from her body, as well as providing insulating materials in a "nest" not totally unlike the maternal behavior of birds (Chapter 16).

In modern swine production farrowing may still occur outdoors in farrowing huts provided with an ample supply of straw or comparable bedding for insulating the piglets. Research at Texas Tech University showed that outdoor individual-sow farrowing hutches should be large enough to allow ease of movement of the sow to avoid injury to piglets. But most pigs are born in confinement **farrowing units,** which have stalls or crates. Individual farrowing units have a separate area warmed with infrared heat lamps or a heated floor mat for the piglets. These warmed areas provide for two needs of the baby pig: a physical safety zone where they cannot be injured by the movement of their much larger mother, and radiant or contact heat to maintain their body temperature.

2.8.3 Farrowing to Weaning

Colostrum is critical as a source of antibodies and nutrients for baby pig survival. Sows and litters should be observed to determine if an adequate quantity of colostrum is being produced by the mother and that all pigs are nursing. Proper nutrient intake is critical to survival of the newborn. Survival of baby pigs weighing less than 2.5 lb at birth can be dramatically increased by hand-feeding colostrum for several feedings following birth. Litter size and subsequent piglet survival are critical factors in profitability of the swine enterprise. One interesting comparison between smaller and larger swine-producing farms is litter size per sow. The USDA published data illustrating that litter size increased as numbers of swine on the farm increased (Table 2.7). This increase may be attributed to the availability of specialists on larger farms, the farrowing environment, and other management factors.

Critical management procedures for newborn pigs include dipping the navel in an antiseptic solution to prevent infection. Procedures that improve performance but are less essential to

survival should occur the first week. These include clipping of the tips of sharp "needle teeth" with a simple side-cutter pliers to help reduce injury from biting littermates. This practice also reduces mammary irritation for the sow. Producers often dock the tail in an to attempt to avert tail biting, a common social disorder in pigs. Iron injections are given to prevent iron anemia because of the rapid growth rate in piglets and the low concentration of iron in sow's milk. Vaccination programs begin early under the direction of a veterinarian.

Castration of males intended for meat production should be done as soon as possible and before weaning. In the United States virtually all male swine fed for slaughter are **barrows.** Although it is true that intact males grow faster and leaner than castrates, the "boar odor/flavor" which emanates from cooking pork of intact males is objectionable to many consumers. Animal rights activists have had significant impact on management procedures in swine production through imposed legislation in some countries outside the United States. Animal welfare restrictions in England require that tail docking is completed under veterinary supervision and castration is banned (Chapter 7).

Pigs are weaned at a variety of ages depending on the management system. Typical U.S. systems call for weaning between 3 and 6 weeks with 3 to 4 weeks being most common. In European countries animal welfare legislation has impacted weaning management. In England weaning under 21 days is discouraged, and in Sweden piglets may not be weaned before 28 days. Research at Iowa State University on **segregated early weaning (SEW)** programs as early as 12 to 16 days of piglet age has shown improved performance because of reduced exposure to disease and increased pig survival. Piglets are weaned from sows at young ages in an attempt to break the cycle of transmission of respiratory diseases from sows to offspring. Piglets are moved to an isolated facility, often on a separate farm, to improve biological security.

To improve pig growth and prepare for weaning **creep** feeds may be provided. These feeds, as well as those used throughout the life of swine, are generally based on high-energy concentrates. Primary examples are corn and soybean meal. Creep feed formulas are often more complex than **finisher** diets. Research at Kansas State University showed a benefit from adding **lactose** (milk sugar) in creep feed to maximize performance of early weaned pigs.

TABLE 2.7	**Swine Herd Size and Number of Pigs per Litter, December 2000 to February 2001**
Herd Size	**Pigs/litter**
1–99	7.5
100–499	7.8
500–999	8.2
1000–1999	8.5
2000–4999	8.7
>5000	8.9

Source: USDA.

2.8.4 Feedlot Animals

Hogs for slaughter include **gilts** not intended for replacement use and barrows. The sexes are often segregated to form more uniform feeding and marketing groups. In the United States market hogs are usually fed *ad libitum,* which facilitates eating all the feed provided **free-choice.** Most diets are complete balanced feeds formulated to contain a balance of protein, energy, vitamins, and minerals. Swine convert feed to body weight gain at a ratio of about 3.0 to 3.5:1 with 3.25:1 being an appropriate goal for the cumulative period from weaning to slaughter. With high-energy diets swine reach acceptable market finish and weight of 230 to 260 lb at 6 or less months of age. The USDA reported average 1996 market weight was 254 lb. Marketing of properly grown hogs before 200 days of age is an appropriate goal. Finish, the amount of backfat, is as critical to marketing and profitability as are weight and age. Backfat of market hogs should be less than 1 inch at market weight. Most hogs for slaughter are sold by "direct" marketing schemes, which include packer–buyer representatives purchasing hogs on the farm or farmers delivering hogs to buying stations established by the packer. Increasingly hogs are produced and marketed under contracts established between individual farmers and meat packers. U.S. farm income from marketing of slaughter hogs in 1996 was $12.6 billion.

2.8.5 Replacement Hogs

Hogs selected for breeding purposes are separated from the general population and fed lower-energy diets, which produce leaner, slower gains. The goal is lean productive animals that can be retained in the herd for multiple farrowings. Gilts for replacement use should be evaluated for type, physical soundness, and performance at about 180 to 220 lb of body weight, when typically they are 5 to 6 months of age. The selected animals should be moved to separate facilities and placed on a **limit-fed** diet to support lean gain. Limit feeding refers to the reduction of calories while providing adequate protein, minerals, and vitamins for optimal growth. If a 250-lb gilt is permitted to eat *ad libitum* she will consume up to 10 lb of feed daily. On a limit-fed program she commonly receives 5 to 6 lb of a balanced diet.

Boars for breeding purposes are fed and managed similarly to selected gilts. Boars should be evaluated, selected, and placed on limit-feeding programs at 200 to 240 lb body weight, commonly reached in 5 to 6 months. Boars reach puberty in 5 to 8 months and may be used to breed at 8 to 9 months.

Properly managed gilts demonstrate first estrus at 6 to 7 months of age. Breeding usually occurs at about 7 to 8 months and is recommended when gilts have reached about 240 to 280 lb. Most gilts and sows are bred through the use of artificial insemination (AI). Concurrent with the monumental shift of swine production to large farm units, the use of AI increased rapidly. Indeed, during the 1980s and 1990s, the percentage of gilts and sows artificially inseminated in the U.S. increased from an estimated 10 percent in 1980 to more than 60 percent in 2001. And, while boar semen was successfully frozen for AI use in 1970, the vast majority of inseminations of swine are with fresh semen which results in a greater number of pregnancies (Chapter 14).

During most of the 112- to 114-day gestation the gilt or sow remains on a limit-fed diet of 3 to 6 lb of feed daily to avoid excess weight gain. Sows should gain about 100 lb during gestation. This 100 lb is about equally distributed to fetal gain, placenta and associated fluids, and the sow's own body weight.

Prior to farrowing, the diet formula will be changed to support the lactation period and the amount of ration will be increased to 10 or more lb of feed daily. The lactation feeding program depends on weaning strategy. An important goal during lactation is to provide for maximum milk production for piglets and minimal weight loss in the sow. After weaning the sow is returned to limit feeding and the cycle of reproduction continues.

The typically polyestrous cycling pattern of the sow allows breeding and farrowing throughout the year, although extremes in temperature and the sow's body condition may affect cycling. Sows may show estrus as early as 7 to 10 days after weaning their piglets. Sows that have been properly managed and are in good reproductive health may be bred at this time. Reproductive health is extremely important to productivity and profitability. The primary reason for culling sows is reproductive difficulty.

Sows are most productive, as measured by the number of piglets produced and piglet performance, in their third and fourth farrowing cycles. Culling for reproductive inefficiencies, disease, physical disability, or other problems often removes animals of high genetic value from the herd long before their fourth farrowing. Management programs that emphasize proper gilt development and sound sow maintenance aid in maximizing lifetime productivity and, therefore, increase swine farm profitability.

2.8.6 The Swine Industry in the Twenty-First Century

The swine industry experienced dramatic change in the late twentieth century with the growth of large corporate farm units. This resulted in a significant shift of hog populations from the Midwest to the Southeast. Current concerns for the swine industry are the control of odors and groundwater pollution from large swine units. Additionally the swine industry has received closer scrutiny from animal activists than other traditional livestock farming units. The issues of pollution and animal rights, and the industry's response to these issues, may have a large impact on swine production early in the twenty-first century (Chapter 7).

2.9 POULTRY LIFE CYCLE

> I think if required on pain of death to name instantly the most perfect thing in the universe, I should risk my fate on a bird's egg.
> **T. W. Higginson (b. 1863) as cited in**
> **The Avian Egg (1949)**
> **Alexis L. Romanoff and Anastasia J. Romanoff**

2.9.1 Historical Perspective on Poultry Production

The term *poultry* includes a great number of domestically produced birds. Most Americans visualize chickens and turkeys as poultry. Worldwide ducks, geese, guinea fowl, and others are also important sources of poultry meat and eggs. Although chickens are virtually the only source of eggs in U.S. supermarkets, duck eggs are a staple in most Asian markets. In the United States the imported Chinese pheasant has become an important species in sport hunting. Meat, leather, and feathers from ostrich and emus have become significant specialty items from farm-raised birds.

The domestic chicken, *Gallus gallus* or *Gallus domesticus,* arose from the Asian wild fowl *Gallus bankiva,* and was brought by the colonists to the Americas as early as the seventeenth century. Those same colonists found the wild turkey, *Meleagris gallopavo,* to be one of America's native species. Turkeys had been hunted as a food staple by Native Americans long before European explorers landed on the eastern shores. In the sixteenth century Spanish explorers returned to Europe with turkeys domesticated by Central Americans.

The U.S. poultry industry has completed a radical evolution in animal production. The first part of the dramatic change was in the type of poultry farm. During the first half of the twentieth century chickens were primarily produced as a source of steady cash income on a large percentage of American farms. Farm families marketed eggs on a weekly basis to local buyers and grocery stores. The cash supplemented family income between major market times for livestock and/or crops. A common term for cash used by farm families during this period was "egg money." During the mid-twentieth century larger egg and broiler producers emerged. The trend to large production units changed from the farmer with a flock of 100 to 200 hens to corporate producers with millions of poultry on farm units.

The second part of the dramatic evolution in the poultry industry was the aggressive marketing strategy that resulted in the consumption of poultry meat exceeding pork and beef (Chapter 3). As a further illustration of the change, until 1930 turkey consumption was not reported by the USDA. In that year the reported consumption of turkey per capita was 1.5 lb and chicken was 10.1 lb. Consumption for 1995 was 18 lb of turkey and 71 lb of broiler meat. This increase can be related to three major factors: the lower cost of poultry meat resulting from production efficiencies, aggressive advertising and marketing strategies of the poultry industry, and the trend of Americans to choose meats with lower fat.

The U.S. poultry industry is divided into three parts: egg layer, broiler (meat chicken), and turkey. Breeds of chickens and turkeys are no longer of commercial significance. There is essentially no dual-purpose production of chickens for meat and eggs. Hybrids (or lines) have been developed and intensely focused by breeding strategies for desired production characteristics.

The poultry industry is a model of production integration, both horizontal and vertical. Within the turkey and broiler industries a small number of companies own or control the production of the majority of birds, whereas in all parts of the poultry industry there has been a strong trend for retained ownership through levels of production. The broiler and turkey industries are more concentrated in the southeastern United States (Tables 2.8 and 2.9). The egg-production industry is dispersed throughout the United States (Table 2.10). Income from poultry in 1995 was $11.8 billion for broilers, $2.8 billion for turkeys, and $4.0 billion for eggs.

TABLE 2.8	**Number of Broilers by State, 1999**	
Rank	**State**	**Millions of Birds**
1	Georgia	1240
2	Arkansas	1196
3	Alabama	971
4	Mississippi	735
5	North Carolina	674
Total	**All states**	**8146**

TABLE 2.9	**Number of Turkeys by State, 2000**	
Rank	**State**	**Millions of Turkeys**
1	North Carolina	44
2	Minnesota	43
3	Arkansas	28
4	Virginia	25
5	Missouri	24
6	California	17
7	Indiana	14
Total	**All states**	**273**

TABLE 2.10	**Egg Production by State, 2000**	
Rank	**State**	**Millions of Eggs**
1	Ohio	8163
2	Iowa	7554
3	Pennsylvania	6313
4	California	6293
5	Indiana	6098
6	Georgia	5114
7	Texas	4423
8	Arkansas	3559
9	Minnesota	3271
10	Nebraska	2999
Total	**All states**	**84,412**

2.9.2 Hatcheries and Hatching

In commercial production fertilized eggs are artificially incubated in heated forced-air containers or rooms. Incubation length for selected species are given in Table 2.11. Eggs are placed in incubators with pointed end down; maintained at carefully controlled humidity and temperature; and are slowly, mechanically tilted from side to side to prevent the developing chick from attaching to the shell membrane. In wild or naturally incubated birds, hens provided the heat and movement of eggs as a daily routine.

TABLE 2.11	Hatching Length for Selected Birds		
Species	**Incubation Time, Days**	**Species**	**Incubation Time, Days**
Chicken	21	Ostrich	42
Turkey	28	Emu	52–56
Duck	28–35	Pheasant	24
Goose	28–34		

Eggs are usually moved from the incubator to a short-term hatching unit a few days before projected hatching. This confines hatching wastes to a limited area and reduces cleanup in incubating units. The bird has a thorny growth on its beak which assists it in breaking through the shell membranes and shell to emerge from the large (blunt end) of the egg.

Hatched birds must be kept warm by artificial means because for several weeks they do not adequately regulate their internal body temperature. The body temperature of birds is considerably higher than that of other farm animals (the body temperature of a mature layer is approximately 106°F). Birds are not fed at the hatchery. They draw much of the yolk material into their bodies through the umbilicus shortly before hatching. This yolk provides energy for the hatchling from the time of hatching until it arrives at a rearing unit.

Sex determination is critical. Egg-laying operations have no use for males. Sexing is done by various methods. Breeding programs have allowed introduction of **sex-linked** genes, which create color patterns or wing feathering to identify sexes at 1 day of age. Older methods include the use of a cloacal examination to view the rudimentary papillae of the male or a proctoscope to locate the testicles through the intestinal wall.

2.9.3 Grower Birds

Chicks and *poults* arrive in growing units as early as 1 day of age. The early growing unit is called a **brooder,** a name derived from the farmer's term for the hen's practice of setting on eggs and subsequently hovering over her chicks to keep them warm and safe during their early days. Similarly the units providing heat were named **hovers.** Brooder units are maintained with a range of temperature selections for hatchlings. At the warmest, usually the center of the unit, the temperature is maintained at or above 90°F. At the outer edges the temperature may be as low as 60°F. As the young birds grow the peak temperature is gradually reduced to about 70°F. Hovers are often used until the birds have reached 4 to 6 weeks of age.

Chicks and poults have their upper beaks clipped to decrease injury caused by the social **pecking order** and to reduce feed waste and improve feed efficiency. Chicks destined for layer units may have their combs clipped (called **dubbing**) to reduce losses to comb injury while in layer cages. Poults have their **snoods** clipped to prevent social disorder pecking by other birds attracted to the dangling snood. Beak clipping, dubbing, and desnooding may be done at the hatchery.

Birds are provided with clean water and a high-energy balanced grain mix soon after arrival at the rearing unit. Birds are observed in early life to identify any that are not eating properly. Producers may trick poults into eating by placing shiny objects too large to eat, such as cat's-eye marbles, in the feeders. Poults are curious about shiny objects and peck at them. By chance and poor aim the birds pick up feed instead and, thereby, begin eating.

2.9.4 Birds for Meat Production

Turkey poults and broiler chickens destined for meat production are fed high-energy concentrate-based diets to produce the most rapid body weight gain possible. It is common for fat to be added to the diet to increase performance. Meat-type birds are commonly raised in environmentally controlled floor-level units. Outside facilities are used in some turkey units. Confinement-reared birds may be managed under reduced light. The day lengths, as determined by artificial light, will be short to control social disorders and encourage eating and drinking during the lighted time, thereby increasing productivity.

Broiler birds are raised as intact males except for the small percentage sold in the specialty **capon** market. Capons are castrated male birds fed to a higher weight and finish, generally marketed at 5 lb or heavier. Broiler birds are fed to about 4 lb and marketed as early as 6 to 7 weeks of age. Selected strains of broiler-type birds are fed to 3 or 4 months of age and marketed as heavier **roaster** birds.

Turkey poults are commonly separated by sex for slaughter bird **grow out.** Weights of turkeys bought for Thanksgiving dinner are related to sex. Female turkeys are commonly slaughtered from 14 to 16 lb live weight, males at 25 to 30 lb. The major difference in size of wild and commercially produced birds can be attributed to modern breeding and nutrition programs. It resulted from the discovery of a useful mutation for heavy breast muscle, termed the "broad breasted bronze," and bred into the modern domestic bird. This discovery and the following paragraph illustrate that not all mutations are negative, especially when viewed in the context of domestic animal production schemes.

The wild forms of chickens, turkeys, ducks, and other meat birds were multicolored birds. Designed breeding programs have led to the propogation of white-feathered varieties of current market birds. The white-feather gene was a mutation that met a market need. Dark-feathered birds have colored pigments in the shaft of the feather. Colored pigment can stain the skin of market birds during processing. Colored pinfeathers left on the carcass are easily noticeable. The breeding of white birds eliminated both of these marketing concerns.

2.9.5 Layer Birds

Layer **pullets** are managed for rapid but lean gain to promote high egg production later in life. They are fed a high-energy concentrate diet free-choice until 9 to 12 weeks of age and then switched to a limit-fed program. The limit-fed program reduces average daily intake to about 70 percent that of free-choice, thereby creating lean slower weight gain. A short artificial day length is maintained to control activity and suppress hormonal response for egg laying. Prior to 20 weeks of age pullets are moved from the rearing facility to a laying facility. Artificial day length is then increased to 14 or more hours per day and feed supply is increased to prepare for high egg production. The average pullet is laying by 23 weeks of age; at peak production 90 percent of hens lay an egg each day. Contrary to the opinion of many, eggs are produced by hens in the absence of roosters.

Hens reach peak productivity, as measured by percent laying per day, at about 30 weeks of age. They begin by laying small less-valuable eggs but produce large-size eggs by about 44 weeks. The length of the normal laying period of commercial hens is about 12 to 14 months. At the end of this period laying productivity falls to a point where eggs produced do not pay for costs of feed and management. In many commercial operations hens are then sold for slaughter.

In some units hens will be **force molted** at the end of the first production period. Molting is accomplished by reducing feeding and returning the artificial lighting to a shortened day. During molt chickens shed a large portion of their feathers and begin to rebuild body condition. Hens may be returned to production for a second laying period although it will not last as long or be as productive as the first. However, hens begin the second laying period by producing the more valuable large eggs. Decisions for marketing or molting hens are not made on an individual bird basis but rather on the basis of entire house or unit productivity.

2.9.6 Breeder Birds

Breeder units are designed to produce fertile eggs for hatcheries, which produce the broilers, layers, ducklings, and poults that provide meat and eggs as end products. Therefore, breeder units are established to produce fertile eggs. Breeder birds are usually managed in open housing units or in lower-density caged units. In many units breeding is done naturally by males housed in the unit during the breeding period. Turkey hens are usually artificially inseminated because of concerns about injury to large birds and the inherent ineptness of natural mating by heavy males.

2.9.7 The Poultry Industry in the Twenty-First Century

The poultry industry led animal agriculture into new production and marketing strategies late in the twentieth century. Whether these changes are all positives for agriculture and consumers remains to be determined. Poultry producers were the first to become highly integrated in production systems. Integrated ownership promoted faster changes through adoption of research results and quicker responses to market demands because fewer producers controlled a much larger proportion of the industry than for any other livestock product. The poultry industry adopted aggressive, effective, unique marketing programs to position broiler and turkey meats as lower-priced, lower-fat substitutes for red meats. In the late twentieth century it was clear that "turkey was not just for Thanksgiving" in the United States. At the turn of the century the per capita consumption of poultry meats led the meat industry as the consumer's dietary meat choice in the United States (Chapter 3).

2.10 SUMMARY

Breeds of livestock and poultry have been developed, for the most part, to assure increased efficiency in producing meat, milk, eggs, and wool for humans. Exceptions include the development of certain breeds of chickens for game purposes and horses for the track and pleasure riding.

In this chapter we learned how breeds are developed for specific form and functions, how they may differ genetically, and why there are different breeds. The major breeds and life cycles of selected livestock and poultry of North America were noted. Collectively they provide animal products for humanity—the foremost function of animal agriculture in the United States.

The livestock industry underwent massive change during the twentieth century—a century that began with animals providing the majority of farm and ranch power. Needs for rendered fat as well as consumer preferences resulted in slaughter livestock with heavy fat cover and low efficiency in converting feed to body weight gain. Pure breeds of livestock and poultry were predominant in production systems. In the last third of the twentieth century dramatic changes occurred in types of livestock produced in the developed nations. By midcentury the tractor hastened the agricultural production revolution by becoming the primary source of farm power. Consumer preferences dominated changes in later years by demanding leaner meats and lower fat animal products. Changes in livestock type and production efficiency were accomplished by adjustments in breeding and nutrition management programs. Different breeds became popular, and crossbreeding became a dominant force in commercial animal production. Farmers and ranchers continue to respond to the preferences and needs of consumers as animal agriculture positions itself to provide desired and much needed, as well as competitively priced and safe, animal products in the twenty-first century.

The stage is now set to learn more about the availability, nutritional contributions, and food safety aspects related to animal products for humans in Chapter 3.

STUDY QUESTIONS

1. What is a breed of livestock or poultry?

2. Why were breeds developed?

3. What is meant by the term purebred?

4. What is a pedigree? What information does it commonly include?

5. What is a pedigree record association? Give examples of its activities and functions.

6. Name and give the color and country of origin of at least three breeds of swine, beef cattle, dairy cattle, and sheep. Name at least three breeds of horses and domestic fowl.

7. What is a dual-purpose breed? A triple-purpose breed?

8. How are new breeds of farm animals developed?

9. Is the practice of crossbreeding being followed in developing greater productivity among all farm animals? Cite examples.

10. What is meant by the term *heterosis?*

11. Cite examples of exotic breeds of cattle imported in North America in recent years. What is meant by the term *synthetic breed?*

12. Give one or more examples of the relation of form (body conformation) and function (work and/or productivity) among domestic farm animals.

13. How may differences among breeds be explained genetically?

14. Why was the imported Landrace breed crossed with established breeds of swine in the United States?

15. Name a breed of sheep that has enjoyed increased numbers in the United States in recent years.

16. When were horses first brought to America? By whom?

17. How useful and how important is crossbreeding to the poultry industry of the United States? Cite examples.

18. Give one or more factors that have favored rapid genetic progress, as measured in terms of productivity, in the breeding and development of commercial poultry in North America.

19. Outline typical life cycles of beef and dairy cattle, goats and sheep, swine, and poultry. Include both egg and meat production life-cycle aspects in chickens and turkeys.

20. What is the foremost function of animal agriculture in the United States?

ANIMAL PRODUCTS[1]

Tell me what you eat, and I will tell you what you are. The fate of nations depends upon how they are fed.

J. A. Brillat–Savarin (1755–1826)
French gastronome

3.1 HISTORY OF AVAILABILITY AND CONSUMPTION OF ANIMAL PRODUCTS

The long steps from gathering food in the wild to growing crops and from hunting wild game to the **domestication** of animals went unrecorded. These important changes began in the early Neolithic period, long before people appreciated the importance of recorded events. In the new Stone Age humans produced selected animal products: **poultry** and eggs; fish; fats from animals; and dairy foods including whole milk, sour milk, butter, and cheese. People consumed most of their food near its origin. Then someone invented the wheel, and soon ox-drawn carts were carrying animal products to towns in exchange for tools, metals, and other goods. Later, oceangoing vessels sailed with cargoes of fish, meat, butter, and cheeses.

More recently animal products have been available to a limited extent in developing nations and to a fuller extent in developed countries of the world. In general, the more highly developed nations, whose people are healthier and have a longer life expectancy than those in developing countries, are consumers of liberal amounts of animal products. The favorable effects of animal products on body development, muscle power, and general health have been noted in two African tribes that had considerable intermarriage but whose diets differed greatly. The Masai tribe consumed a diet high in protein with liberal amounts of meat and milk, whereas the Kikuyu tribe consumed a vegetarian diet of cereals (chiefly maize), legumes, sweet potatoes, and green leaves. In the latter group bone deformities, dental caries, **anemia,** pulmonary diseases, and tropical ulcers were prevalent, whereas the tribe consuming large quantities of meat and milk were 5 in taller and 50 lb heavier, had 50 percent greater muscle power, and were comparatively free from **disease.**

When the children of Japan began consuming more animal products (meat, milk, and eggs), they began growing taller and stronger and having better teeth. Of course, the importance of milk to sound teeth has been appreciated for thousands of years.

His eyes shall be red with wine, and his teeth white with milk.

Genesis 49:12

The average height of Japanese 14-year-olds increased more than 4 in since World War II. The average daily per capita Japanese intake of calcium in 1996 was approximately 300 mg compared with only 20 mg in 1946. Until this change in diet many Japanese people believed their ultimate height was merely a genetic factor. Now they realize more fully the importance of nutrition in genetic expression of adult size.

Studies have shown that children adapted to a low calcium intake respond to increases in dietary calcium with increases in linear growth. This result indicates that diets in many developing countries do not allow the full genetic growth potential of the peoples of those countries. Recent studies of **malnutrition** among preschool children in developing countries indicate that lack of nutrients during periods critical for growth and development of the central nervous system may cause irreversible cellular damage, thus inhibiting subsequent normal mental development. Malnutrition, especially protein deficiency, inhibits the development of protective anti-

[1]The authors acknowledge with appreciation the contributions to this chapter of Dr. R. T. Marshall, Department of Food Science and Nutrition, University of Missouri, Columbia.

bodies and lowers resistance to disease. Furthermore, it is well established that malnourishment during early years of life results in delayed physical maturity, even if the deficiency is temporary and a normal diet is restored later. Even more ominous is evidence that protein deficiency in infancy and early childhood results in permanent impairment of the brain. The dietary protein requirement of humans is low during the first 4 months of infancy when proteins account for only about 11 percent of the new tissue formed, but it increases sharply during the next 8 months when protein represents about 21 percent of body weight gain.

The current availability and economy of animal products have already been discussed (Sections 1.4.4 and 1.5.1), but it should be indicated here that animal products have, over the past three decades, become a good food buy relative to wages earned. For example, recent U.S. Department of Agriculture (USDA) studies showed that the average factory worker could buy considerably more animal products with an hour's pay in 1996 than in 1950, as noted below.

Animal Product	1950	1996
Frying chicken	2.5 lb	15.7 lb
Milk	8.0 qt	20.1 qt
Eggs	2.4 doz	14.0 doz
Pork	2.7 lb	6.3 lb

The above data suggest that the greatest recent strides in the efficiency of producing animal products have been made in poultry. One important reason for this is that the reproductive rate of poultry is much greater than that of beef. For example, the average cow produces only about 0.8 progeny annually, which, in terms of slaughter weight, is about 75 percent of her body weight. The average sow produces about 20 progeny per year and those pigs represent a total market weight of 10 times the sow's body weight. But, the average meat-type hen produces 150 progeny per year, or about 100 times her body weight. These differences in reproduction rate have a profound effect on the relative costs of producing different meats. They also greatly affect per capita consumption trends, which reflect retail prices. For example, in the early 1950s the retail price of chicken was 80 percent that of beef. In recent years chicken prices have averaged less than half those of beef.

The beef industry is the largest segment of U.S. agriculture constituting approximately one-fifth of all farm marketings. But with per capita consumption of beef decreasing in recent years the industry faces numerous challenges, including (1) increasing the rate of reproduction; (2) expanding the competitive advantage it holds in utilizing forage crop residues, by-products, and nonprotein nitrogen; (3) using genetic engineering to produce more useful microflora that, in turn, can provide increased potential for the conversion of nonconventional feedstuffs into usable dietary nutrients; (4) improving its production and distribution efficiency; and (5) providing consumers with a wide range of conveniently packaged and competitively priced products.

Americans have been able to spend less and less of their disposable incomes for food since the end of World War II. They spent approximately 23 percent for food from 1929 through 1949, about 14 percent during the 1969 to 1979 period, only 12 percent in 1989, and only about 10 percent at the turn of this century. Americans are consuming increasing amounts of food away from home. Considering all money spent on food, that used for "at-home eating" was 87 percent in 1929, 81 percent in 1949, but only 55 percent in 1991. Per person spending for animal products used at home amounted to $315 (47 percent of total) in 1980 and $413 (38 percent of total) in 1995. The largest change has been in expenditures for fast foods. In 1970 Americans spent $6 billion on fast foods; in 2000 they spent more than $110 billion. (Compare this with $76 billion spent on higher education and $110 billion spent on new cars in 2000.) According to the National Restaurant Association, the U.S. restaurant industry employs more than 11 million people at 840,000 locations. Their annual sales approached $400 billion in 2000 and 2001.

Farm value as a percentage of retail cost is considerably higher for animal products than it is for products of plants. In 1980 this value ranged from 51 percent for meat products to 64 percent for eggs, whereas the same statistic for products of plant origin ranged from 14 percent for cereals and baked products to 29 percent for fats and oils. By 1996 these percentages had dropped. Those for animal products ranged from 36 percent for meat and dairy products to 52 percent for eggs, whereas those for plant products ranged from 7 percent for cereals and baked products to 22 percent for fats and oils.

Egg prices have been favorable to the consumer with grade A large eggs selling at retail for $1.00 per dozen or less during the 1990s. When one considers that a dozen large eggs weighs at least 24 ounces and that the yield on breaking approaches 90 percent, it is evident that the edible portion costs about $0.75 per pound. The yolk comprises about one-third of the edible portion and contains about 50 percent solids. It yields 78 percent of the calories and nearly all of the lipid, iron, zinc, and vitamins (B_6, B_{12} and A, folic acid, pantothentic acid, and thiamine) of the whole egg. It also contains about one-half of the protein (yolks test 15 percent and whites test 10 percent) and riboflavin of the egg. As laid, yolk and albumen (white) contain about 52 percent and 12 percent solids, respectively. However, when shell eggs are stored water migrates from the white into the yolk so that commercially processed yolk contains about 44 percent solids.

Because poultry are **monogastric** their diets can be changed to modify the composition of eggs. The affected nutrients include fatty acids, fat-soluble vitamins, some of the water-soluble vitamins, iodine, fluorine, and manganese. Addition of menhaden oil, flaxseed, or flaxseed oil to hen's **rations** increases the concentration of n3 **polyunsaturated** fatty acids (see Section 3.2.2). Furthermore, enrichment of diets with oils containing high amounts of monounsaturated fatty acids increases the content of these fatty acids in the eggs produced. Limited success has been experienced in efforts to lower concentrations of cholesterol in eggs by dietary manipulation.

3.2 COMPOSITION AND COMPARATIVE NUTRITIONAL CONTRIBUTIONS OF SELECTED ANIMAL PRODUCTS

Animal products provide abundant sources of practically all known food nutrients; however, some products are especially rich in certain dietary essentials and will be noted in discussions of major classes of **nutrients.** The composition of selected animal products is given in Table 3.1.

3.2.1 Proteins

> Upon what meat doth this our Caesar feed that he is grown so great?
> **William Shakespeare (1564–1616)**
> **English dramatist and poet**

Meat, milk, and eggs are excellent sources of **protein.** The quality of proteins is determined by their ability to support growth and maintenance. Proteins are composed of amino acids, which serve as building blocks from which the body fabricates its own special protein tissues. Proteins that contain the **essential amino acids** in about the proportions needed to build body tissues are known as *complete proteins. Incomplete proteins* lack or are deficient in one or more of the essential amino acids. Animal proteins are superior to plant proteins for humans and other monogastric animals because they are better balanced in essential amino acids. Zein (a corn protein) is an example of an incomplete plant protein (Table 3.2). It is deficient in the amino acids lysine and tryptophan. Animal proteins are excellent sources of lysine, and many, especially the proteins in milk and eggs, are rich in tryptophan. The **biological values (BV)** (also referred to as net protein utilization) of animal proteins are much higher than those of plant proteins; e.g., the biological values of whole eggs and milk are 94 and 85 in contrast to whole corn or navy beans (cooked) at 60 and 38, respectively. As a result, the nutritional value of proteins of

TABLE 3.1	**Composition of Selected Animal Products**					
	Water, %	Food Energy*	Protein, %	Fat, %	CHO, %	Ash, %
Beef						
Round	66.6	197	20.2	12.3	0.0	0.9
Porterhouse	50.2	370	15.3	33.8	0.0	0.7
Rump roast	59.4	271	18.3	21.4	0.0	0.8
Hamburger	60.2	268	17.9	21.2	0.0	0.7
Lamb						
Rib chops	53.4	339	15.1	30.4	0.0	1.1
Shoulder roast	59.6	281	15.3	23.9	0.0	1.1
Pork						
Ham	56.5	308	15.9	26.6	0.0	0.7
Loin chops	57.2	298	17.1	24.9	0.0	0.9
Chicken						
Light meat	73.3	117	23.4	1.9	0.0	1.0
Dark meat	73.7	130	20.6	4.7	0.0	1.0
Turkey	64.2	218	20.1	14.7	0.0	1.0
Milk and milk products						
Whole	87.7	64	3.3	3.6	4.7	0.7
Skim	90.8	35	3.4	0.2	4.8	0.8
Nonfat, dried	3.2	362	36.2	0.8	52.0	7.9
Ice cream	60.8	202	3.6	10.8	23.9	1.0
Cottage cheese (creamed)	79.0	103	12.5	4.5	2.7	1.4
Cheddar cheese	36.8	403	24.9	33.1	1.3	3.9
Butter	15.9	717	0.9	81.1	0.1	2.1
Yogurt (plain, low-fat)	85.1	63	5.3	1.6	7.0	1.1
Eggs						
Whole	74.6	158	12.1	11.2	1.2	0.9
Whites	88.1	49	10.1	Trace	1.2	0.6
Yolks	48.8	369	16.4	32.9	0.2	1.7
Whole dried yolk	4.7	687	30.5	61.3	0.4	3.2
Fish, cod	81.2	78	17.6	0.3	0.0	1.2
Honey	17.2	304	0.3	0.0	82.3	0.2

*Calories per 100 grams (g) foodstuff.

Sources: "Composition of Foods," *USDA Handbook* 8 (1963); and "Composition of Foods: Dairy and Egg Products," *USDA Handbook* 8-1, ARS, revised November 1976.

TABLE 3.2	Percent of Selected Amino Acids of Proteins of Important Animal Foods and the Protein of Zein (Corn)					
Amino Acid	**Beef**	**Pork**	**Chicken**	**Milk**	**Eggs**	**Zein**
Arginine	6.4	6.7	6.7	4.3	6.4	1.8
Cystine	1.3	0.9	1.8	1.0	2.4	0.8
Histidine	3.3	2.6	2.0	2.6	2.1	1.2
Isoleucine*	5.2	3.8	4.1	8.5	8.0	4.3
Leucine*	7.8	6.8	6.6	11.3	9.2	23.7
Lysine*	8.6	8.0	7.5	7.5	7.2	0.0
Methionine*	2.7	1.7	1.8	3.4	4.1	2.3
Phenylalanine*	3.9	3.6	4.0	5.7	6.3	6.4
Threonine*	4.5	3.6	4.0	4.5	4.9	2.2
Tryptophan*	1.0	0.7	0.8	1.6	1.5	0.2
Tyrosine	3.0	2.5	2.5	5.3	4.5	5.9
Valine*	5.1	5.5	6.7	8.4	7.3	1.9
Total	**52.8**	**46.4**	**48.5**	**64.1**	**63.9**	**50.7**

*Essential to humans.

Source: Modified from M. L. Scott, *J. Amer. Diet. Assoc.* 35(1959):248.

TABLE 3.3	Average Measure of Protein Quality for Milk and Milk Proteins				
Type	**BV**	**PD**	**NPU**	**PER**	**PDCAAS**
Milk	91	95	86.5	3.1	1.21
Casein	77	100	76.0	2.9	1.23
Whey protein	104	100	92.0	3.6	1.15

BV: Biological value—proportion of absorbed protein that is retained in the body for maintenance and/or growth.

PD: Protein digestibility—proportion of food absorbed.

NPU: Net protein utilization—proportion of protein that is retained (calculated as BV × PD).

PER: Protein efficiency ratio—gain in body weight divided by the weight of protein consumed.

PDCAAS: Protein digestibility corrected amino acid score—the ratio of essential amino acids compared with a FAO/WHO reference pattern, then multiplied by a digestibility factor. (PDCAAS is truncated to 1.0 for **FDA** food labeling.)

Source: "Protein Quality—Supplementing the Scores," *Ingredient Insight* 3(1):2, Dairy Management, Inc., 2000.

corn, potatoes, white bread, or navy beans is substantially increased when they are consumed with animal products rich in the amino acids that are deficient in the vegetables. Research indicates that it may be possible to develop a high-lysine corn. Soy protein is recognized as beneficial in the diet, and compared with other plant-derived proteins, has a relatively good balance of essential amino acids. Methionine is the limiting essential amino acid in soy protein. Because each gram of protein in soy protein isolate supplies 14 mg of methionine and the daily recommended intake for an adult weighing 154 lb is 910 mg, 65 g of soy protein would be required. Methionine is also the limiting amino acid in milk protein with each gram supplying 25.2 mg. Therefore, the same adult could consume enough methionine in only 36 g of milk protein, the approximate amount in a little more than 1 liter of nonfat milk. For additional information see: http://www.nal.usda.gov/fnic/cgi-bin/nut_search.pl?soy+protein.

Protein quality is measured in several ways. Table 3.3 shows how milk protein and its two major protein classes (casein and whey protein) rate in terms of the five common measures of protein quality.

Thus, animal proteins might be called "supplements" for vegetable proteins. The high nutritive value of milk protein makes it of special value in the treatment of **kwashiorkor,** a type of protein malnutrition found among young children in areas where people subsist on plant foods. Nonfat dry milk (36.2 percent protein) is used in many bakery and canned foods as a protein supplement as well as a functional ingredient. It is also used to fortify many foods. The balance of essential amino acids in milk and its potential for providing daily dietary amino acids are shown in Figure 3.1. Milk—the nectar of mammals—is virtually a balanced meal in itself. Per unit of dry matter, milk is nutritionally equivalent to meat, yet when fed to animals only about 10 percent of the milk solids are recovered in the form of meat solids. Therefore, it is nutritionally wasteful to feed milk to older animals. Instead, milk produced in excess of that needed to raise young animals should be available for human consumption.

Because high-protein animal foods promote growth but do not fatten like other foods, they fit well into a reducing diet. Although excessive protein is somewhat wasteful and

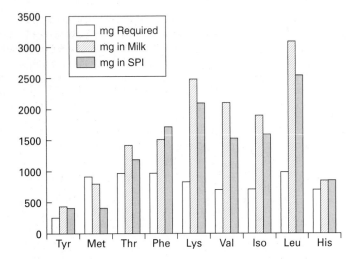

Figure 3.1 Amounts of amino acids required per day by normal adults and amounts supplied by 1 quart of nonfat milk or a like amount of soy protein isolate (SPI).

From "Recommended Dietary Allowances," Food and Nutrition Board, National Research Council, 10th ed., National Academic Press, Washington, DC, 1989; Composition of Foods, Dairy and Egg Products, USDA Handbook 8-1, 1976; USDA Nutrient Data Base (No. 16122) for std. ref., Release No. 13, November 1999; URL: www.nal.usda.gov/fnic/cgi-bin/list_nut.pl.

increases the amount of nitrogen that must be excreted by the kidneys, it serves to satisfy hunger cravings of the overweight without adding much to their fat stores. Proteins consumed in excess of normal needs are **deaminized,** and only a part of the original caloric value of the protein is retained. The balance is voided in the urine. By comparison, dietary **fats** and **carbohydrates** are utilized more fully and their metabolic end products are stored in the body as fat. It has been estimated that whereas carbohydrates yield approximately 95 percent of their potential energy to the body, only about 70 percent of the potential energy of protein is useful in meeting energy needs of the body.

3.2.2 Fats

Although few animal products are purchased for their fat content, many contribute essential fatty acids; the fat-soluble vitamins A, D, E, and K; and **precursors** of vitamins A and D. Because body fat can be produced from the carbohydrates of food, one might expect that dietary fat is not important. On the contrary, fats are essential dietary elements. Fatty foods contain certain fatty acids essential for good health that cannot be synthesized in the body. Rats soon cease to grow when fed a diet insufficient in fat but resume growth when small amounts of the essential fatty acids are added to their diet. **Saturated fatty acids** (palmitic and stearic) may be synthesized in humans from carbohydrates and stored in the body. Therefore, they are nonessential. The unsaturated (essential) fatty acids of body fat may be derived from eggs, milk, butter, and plant oils. Essential fatty acids include linoleic (two double bonds), linolenic (three double bonds), and arachidonic (four double bonds). Arachidonic acid is found only in animal tissues and milk. Any one of these fatty acids apparently can meet the body's need for essential fatty acids (Section 18.2.3). Each of these fatty acids has at least two *double bonds.* Thus, they are polyunsaturated.

The mean saturated and unsaturated fatty acid content for selected animal products is given in Table 3.4. Diet influences the composition of animal fats; e.g., lard from hogs fed a low-fat diet may have a linoleic content of only 1.2 percent in contrast with the normal of 7 percent. Animal fats have shorter-chain fatty acids than do plant fats and are more easily digested. (The **coefficient of digestibility** of milk fat is 97 for humans.)

Unsaturated fatty acids are classified into "omega" families based on the location of the first double bond counting from the methyl end of the molecule. Omega-3 (n3) unsaturated fatty acids have their first double bond at the third carbon from the methyl end while omega-6 (n6) unsaturated fatty acids have their first double bond at the sixth carbon from the methyl end. Omega-6 fatty acids predominate in most terrestrial mammals while n3 fatty acids predominate in marine fish. Omega-3 fatty

TABLE 3.4	Fatty Acids and Cholesterol in Selected Animal Products				
Food	**One Serving Amount**	**Weight, g**	**Cholesterol, mg**	**Saturated Fatty Acids, g**	**Unsaturated Fatty Acids, g**
Beef, round steak	3.5 oz	100	125	4.70	5.75
Butter	1 pat	7	20	3.25	2.17
Cheese, cheddar	1 oz	28	32	5.38	3.27
Chicken, fryer	3.5 oz	100	60–90	2.33	4.51
Eggs	1 medium	54	340	1.64	3.58
Lamb, chop	3.5 oz	100	70	17.85	12.93
Liver, beef	3.5 oz	100	320	0.87	2.14
Milk, whole	1 cup	244	33	5.24	3.79
Pork, loin or chop	3.5 oz	100	70–105	9.50	14.25
Salmon	3.5 oz	100	60	4.97	10.71
Turkey, breast meat	3.5 oz	100	8–15	1.31	3.10
Brains, beef	3.5 oz	100	2233	N/A	N/A

Source: Modified from O. B. Hayes and G. Rose, *J. Amer. Diet. Assoc.* 33(1957):27–28.

acids have been found to have anti-inflammatory effects, lessen symptoms of arthritis, and may decrease the incidence of heart disease in people. Feeding fish oils or certain plant oils (e.g., flax, linseed) containing high concentrations of n3 fatty acids to animals will increase the concentration of n3 fatty acids in their products (eggs, milk, meat). The two most important fatty acids in animal cell membranes are arachidonic acid (n6 family) and eicosapentaneoic acid (n3 family). Note location of the first double bond in each of these fatty acids in their structural formulas below:

Arachidonic acid (n6)

Eicosapentaneoic acid (n3)

3.2.3 Carbohydrates

Meat and eggs are low in carbohydrates, whereas milk solids and honey are excellent sources (Table 3.1). When the Hebrews were promised "a land that floweth with milk and honey" (Joshua 5:6), they were being offered both a necessity of life, *milk,* and a luxury, *honey.* It would be remiss not to mention some unique characteristics of these two animal products.

Honey

A jar of honey on the table was once considered a mark of wealth. The ancient Egyptians are believed to have used honey in embalming. As noted in Table 3.1, honey contains more carbohydrates (82.3 percent) than does any other animal product. A unique property of honey is its predominant sugar, fructose (levulose). This sugar is primarily responsible for the sweet taste. Honey contains a large amount of dextrose (glucose) but is low in sucrose (less than 8 percent by legal standards).

Most pure honeys granulate, or develop sugar crystals. An exception is tupelo honey, which seldom granulates because it contains a large amount of fructose in relation to dextrose. It is the dextrose that granulates, either partially or completely. Industrial users heat honey to prevent granulation.

Honey is heavier than water (the specific gravity, sp gr, of honey is 1.41), but the wax portion is lighter (sp gr 0.96). Most honeys will darken in color if stored at a temperature above 59°F.

Uses of Honey Honey is an excellent energy food because it contains simple sugars that can be used quickly by the body. Honey contains mineral salts and other materials needed by the body. It is the only form of sugar food that does not need to be refined for human use. Bakers sometimes use it in place of sugar for their products. Many cough medicines and laxatives contain honey. The wax portion is used in making candles, lipsticks, polishes, waterproofing compounds, adhesives, chewing

gum, ointments, and other materials. **Honey butter** is a popular spread and is made by beating honey and butter together.

Kinds of Honey The flavor and color of honey are influenced by the kinds of flower nectars available. Honey ranges in color from white to dark amber. Usually light-colored honeys have the mildest flavor. The most common honey plants are alfalfa and clovers. Minnesota, where clovers are common, is the leading U.S. producer of honey. California ranks second where white sage and orange blossom honeys predominate. In the East buckwheat flowers are used, whereas in the South tupelo, mesquite, sourwood, and gallberry supply nectar for honey. There are approximately 300 honeys available in the United States.

How Honey Is Made Research indicates that bees recognize honey-yielding flowers first by color and second by scent. Honey is made by bees from flower nectar. Bees sip it from the blossoms and carry it to their hives. Each worker stores the nectar it collects in a special pouch, called a *honey stomach,* inside its body. In the stomach the sugars in nectar are broken down by a process called *inversion* into two simple sugars: *fructose* and *dextrose.* After honeybees deposit nectar in the hive, they allow most of the water to evaporate and the liquid becomes thick. The nectar may contain as much as 70 percent water as collected. The workers fan the nectar with their wings, reducing the water content to about 17 percent. They also add **enzymes** that enhance the flavor. A special gland in the abdomen of the young worker produces beeswax. It is interesting that honeycomb has very thin walls (1/80 in) that can support 30 times their own weight.

The Honeybee Colony Bees are the only insects that produce food commonly consumed by humans. There are some 20,000 **species** of bees, but only honeybees make honey and wax for human use. They form a colony that includes three classes of honeybees: the queen lays eggs, the workers gather food and care for the young, and the **drones** fertilize the queen. Honeybees have a highly developed society.

The *queen* honeybee lays eggs that hatch into thousands of workers. Laying eggs is the queen's only responsibility. If two queens hatch at the same time, they fight until one stings the other to death. When the young queen makes her maiden flight she may mate with one or sometimes several drones. After mating the young queen returns to the hive and begins laying eggs 2 days later. During her reproductive surge she may lay 2000 eggs daily (more than her body weight) and more than 200,000 in a single season. With a life expectancy of about 5 years the queen may lay 1 million or more eggs in her lifetime. Bee grubs are fed a mixture of honey and pollen known as **beebread.** The length of time they receive *royal jelly* (also called *bee milk*) determines whether the female grub will develop into a queen or a worker. Queen **larvae** are reared in special large cells and receive royal jelly throughout their entire grub phase. Royal jelly enjoys a fad as a food for humans. It is a creamy substance, rich in vitamins and proteins. It is formed by a **gland** in the head of the young worker bee.

The *workers* are all females and do all the work except the laying of eggs. The worker has a long tongue for gathering nectar. It uses its hind legs to carry pollen, which is used for food. After the worker gathers as much nectar as it can carry, it takes the shortest route (hence *beeline*) back to the hive. During the busy summer season a worker usually lives about 6 weeks. A strong colony may have from 60,000 to 80,000 workers, each collecting only about 1/10 lb of honey during its entire lifetime. To make a pound of honey a bee colony may have to extract nectar from a million or more blossoms, make 37,000 trips, and travel 50,000 miles.

All female bees (queens and workers) have a **diploid** complement of **chromosomes,** whereas the males (drones) are **haploid.** Workers have barbed stingers and can sting only once because the stinger pulls out of the bee's body. A worker bee dies a few hours after losing its stinger. The queens have smooth stingers but use them only to get rid of a rival queen. Bee **venom** has been used in arthritis therapy and other disorders of humans. The favorite medicine of Hippocrates was honey. He said, "The drink to be employed should there be any pain is vinegar and honey. If there be great thirst, give water and honey."

Drones are burly, clumsy creatures. They do no work and have no sting. Drones develop from unfertilized eggs (**parthenogenesis**). The only function of a drone is to mate with a young queen. An unmated queen can lay only drone eggs; she must be fertilized in order to lay worker eggs. In a queenless colony the workers can lay eggs; however, since they are unfertilized only drones are produced. In autumn, when the honey flow is over, the workers allow the drones to starve to death. This is done because they are no longer useful to the colony and would eat too much of the stored honey.

Bees are especially beneficial in the pollination of fruits, vegetables, and seed crops. Worldwide the profit derived from pollination by bees is estimated at 25 times higher than that obtained from bee products.

> The bee helps the garden, the garden helps the bee, and people reap the harvest of both.
>
> **Anonymous**

> The bee is more honored than other animals, not because she labors, but because she labors for others.
>
> **St. John Chrysostom (c. 347–407 A.D.)**
> **Church Father and patriarch of Constantinople**

In the past century apiculture (beekeeping) has come to be considered a branch of animal science, and systematically controlled breeding, selecting, and crossing of bees, using techniques commonly employed with domestic animals, have been employed successfully. World production of honey is estimated at 1.9 billion lb. Leading honey producers include the former U.S.S.R., Australia, Argentina, western Germany, the United States, Mexico, Canada, Norway, Cuba, and China. The annual per capita consumption of honey in the United States is about 1.5 lb.

Pesticides are of special concern to beekeepers. Because bees roam in search of pollen and nectar, they may collect food from treated or drift-contaminated plants.

TABLE 3.5	Brain Size as Related to Body Weight and Lactose in Milk

Animal	Brain Weight, g	Brain Size as Percentage of Total Body Weight	Lactose in Milk, %
Human	1400	2.5	7.0
Horse	600	0.25	5.9
Elephant	5000	0.20	3.4
Whale	2050	0.0025	1.8

Source: World Book Encyclopedia, vol. 2, Field Enterprises Educational Corp., Chicago, 1965, p. 460b.

Lactose–Milk Sugar Is Unique

Milk is the only substance in nature containing the sugar known as *lactose,* and it is the principal carbohydrate or sugar present in milk. (There are only traces of other sugars.) It is amazing that the milks of all mammals, some 10,000 living species, from reptilian **monotremes** to moles, mice, monkeys, and humans, contain lactose. It must therefore be especially important or it would not occur naturally in the milk of all these species.

Early Native Americans believed that the longer the child received breast milk the longer would be his or her life, and it was not uncommon for children of Native Americans to be breast-fed until ages 7 to 9 years. It has been reported that students who were breast-fed as infants make significantly higher scores on college examinations than their bottle-fed contemporaries.[2] Why or how could this be possible? Human milk is especially rich in lactose (7 percent), which correlates with the large brain size (as related to total body mass) of humans (Table 3.5). Lactose constitutes 56 percent of the dry matter of woman's milk in contrast with 36 and 6 percent in cow's and rabbit's milk, respectively. Because lactose contains galactose, a constituent of the central nervous system (as galactosides and cerebrosides of nerve and brain tissue), milk lactose may be a "brain food," a special nutrient for the growth and development of the central nervous system of mammalian young. (The human brain reaches about 70 percent of its future adult weight by 12 months of age.) Worldwide infant nutrition could be improved by reversing the trend toward less breast-feeding. Mother's milk is a valuable natural resource.

Research studies of more than 1000 children in New Zealand, published by the American Academy of Pediatrics, found that the longer infants were breast-fed the more likely they were to have positive cognitive and academic outcomes in early childhood. In a 9-year study of 526 children in the Netherlands, those who were breast-fed for at least 3 weeks were less likely to suffer neurological abnormalities, such as coordination problems, than were those who were fed formula.

[2]J. R. Campbell, *In Touch with Students . . . A Philosophy for Teachers,* Educational Affairs Publishers, P. O. Box 248, Columbia, MO 65205, 1972.

The breasts were more skillful at compounding a feeding mixture than the hemispheres of the most learned professor's brain.

Oliver Wendell Holmes (1809–1894)
American physician and author

One of the most desirable qualities of lactose is its hygienic value. Unlike sucrose, lactose passes the **ileocecal valve** to be slowly absorbed from the intestine. Its presence in the intestine stimulates the growth of microorganisms (especially *Lactobacillus acidophilus)* that produce organic acids and synthesize many B-complex vitamins. The high acid concentration suppresses protein **putrefaction** and the growth of many **pathogenic organisms** because these organisms are sensitive to high acidity. Consumption of *L. acidophilus* via certain dairy products, such as sweet acidophilus milk, has a favorable influence during oral administration of **antibiotics** and encourages reestablishment of normal intestinal flora after therapy ceases.

Assimilation of calcium, phosphorus, magnesium, and barium is enhanced by lactose in the intestine. This unique quality of lactose also makes milk an excellent antirachitic (rickets-preventing) food even when the milk is low in vitamin D. Moreover, a woman's milk is more antirachitic than cow's milk, which may be explained by the higher level of lactose in a woman's milk (7 versus 5 percent). These unique properties of lactose would not be predicted from its chemical composition.

Lactose is involved in the antipellagric (pellagra-preventing) properties of milk. The recommended daily consumption of niacin (nicotinic acid) for an average person is 14 to 18 mg. Because a quart of milk contains only 1 mg, it would be logical to conclude that a human would have to consume 14 to 18 qt of milk daily to obtain the needed 14 to 18 mg from milk. Instead milk provides lactose, which enables microorganisms of the intestine to synthesize nicotinic acid. Thus milk is, in fact, an excellent antipellagric food. Additionally, milk contains tryptophan, and it is recognized that 1 mg of niacin is derived from each 60 mg of dietary tryptophan. A quart of milk contains approximately 540 mg of tryptophan, equivalent to about 9 mg of niacin.

Although lactose and sucrose have the same empirical formula, they differ in many respects. Lactose is about one-fifth as sweet as sucrose and is less quickly metabolized to acids than other sugars. This makes lactose less irritating to the stomach and intestinal mucosa than the more quickly metabolized sugars. Thus milk is valuable in diets used in the treatment of ulcers of the stomach and duodenum.

Lactose may have some value in reducing diets because of its less-rapid absorption compared with the more quickly metabolized carbohydrates, and its ability to reduce fat deposition in the body. In one study **carcasses** of animals fed high-lactose diets were found to contain only 58 percent as much fat as those that received high-glucose diets.

The relatively low solubility of lactose in water, about 18 g/100 ml, causes it to tend to crystallize in dairy foods that are low in moisture and high in lactose. These products include frozen desserts, condensed and dry milks, and whey. Because of the low solubility of lactose, special practices are followed in the manufacture and storage of these products so their qualities are preserved. For example, as the water is frozen out of an ice cream mix a thick syrup forms. Lactose may crystallize in this syrup if the temperature is high enough to make the product soft for an extended time. The crystals cause the consumer to feel sandlike particles in the mouth as the ice cream is eaten.[3]

Some persons suffer from insufficient *lactase* in their intestine and are unable to digest all of the lactose consumed. The symptoms of lactose intolerance include a bloated feeling resulting from gas production in the colon, and diarrhea in severe cases. Lactase splits the 12-carbon disaccharide, lactose, into two 6-carbon monosaccharides: glucose and galactose. Both are easily absorbed through the intestinal walls. However, intact lactose is not readily absorbed and is moved through the small intestine to the colon where fermentative bacteria metabolize it making acid and gas. If the concentration of lactose in the colon is very high the high osmotic pressure within the colon attracts water causing loose stools. Persons of African and Asian descent are more prone than Caucasians to be lactase-deficient. However, the NHANES study published in the May 2000 issue of the *Journal of the American Dietetic Association* showed that African-American girls who had not been drinking milk adapted within 21 days to intakes of sufficient dairy foods to supply 1300 milligrams of calcium per day, the amount recommended for teenagers.

The NHANES study showed that typical African-American girls consume only about 750 milligrams of calcium per day, far below that needed to form strong bones and teeth. The National Dairy Council recommends that persons who believe they suffer from lactose maldigestion begin with smaller portions of lactose-containing dairy foods and slowly increase the serving size. Milk should be consumed with other foods so it passes from the stomach slowly. Natural cheeses have had all or most of their lactose fermented by ripening bacteria. Cultured milk products have had their lactose partially digested, and they contain desirable bacteria that carry lactase to the intestine.

3.2.4 Minerals

Animal products provide abundant sources of essential dietary minerals. The total ash content of selected animal products is given in Table 3.1. Data on the content of six essential dietary minerals in animal products are presented in Table 3.6. Milk, milk products, and fish are excellent sources of dietary calcium and phosphorus. Dairy products, beef, chicken, **lamb,** liver, pork, and turkey are good sources of phosphorus. Liver, beef, and whole eggs are rich in iron, whereas milk is deficient. Interestingly, most newborn mammals are provided with liver stores of iron, copper, and manganese to last them beyond the normal nursing period. (The pig is not.) Animal products are good **supplements** to plant foods such as cereals, potatoes, and white bread, which are deficient in calcium. Phosphorus, sodium, potassium, and magnesium are other minerals important to human health. Diets rich in animal products that provide sufficient calcium are likely to supply enough phosphorus as well.

[3]R. T. Marshall and W. S. Arbuckle, 1996 *Ice Cream,* Chapman & Hall, N.Y., N.Y.

TABLE 3.6	Mineral Content of Selected Animal Products, mg/100 g, Edible Portion					

Animal Product	Calcium, mg	Phosphorus, mg	Iron, mg	Sodium, mg	Potassium, mg	Magnesium, mg
Recommended daily allowance (RDA), adult humans	800	800	10.0 (15.0, women)	500	2000	350 (280, women)
Beef, good grade	10	152	2.5	65	355	18
Cheese, cheddar	750	478	1.0	700	82	45
Chicken, white meat	11	265	1.3	64	411	19
Eggs, whole	54	205	2.3	122	129	11
Honey	5	6	0.5	5	51	3
Ice cream	146	115	0.1	63	181	14
Lamb, choice grade	10	147	1.2	70	290	23
Liver, beef	11	476	8.8	184	380	18
Milk						
Whole	118	93	Trace	50	144	13
Skim	121	95	Trace	52	145	14
Human	33	14	0.1	16	51	4
Nonfat, dry	1308	1016	0.6	532	1745	143
Pork, ham	10	236	3.0	65	390	17
Salmon, pink	196	286	0.8	387	361	30
Turkey, cooked	8	251	1.8	130	367	28

Sources: National Academy of Science–National Research Council, Food and Nutrition Board, Washington, DC, 1989; "Composition of Foods," *USDA Handbook* 8 (1963); and "Nutritive Value of American Foods," *USDA Handbook* 456 (1975).

Milk is the best nutritional source of calcium, not only because of its richness in calcium but also because of the favorable Ca/P ratio. Furthermore, research published in the *American Journal of Clinical Nutrition* showed that calcium was absorbed from cow's milk better than from calcium-fortified soy beverages. Approximately 60 percent more calcium was needed in soy beverages than is present in milk to provide comparable absorbed amounts of calcium. Of special interest is the parallel between milk composition and the maturing speed of newborn mammals among species (Table 3.7). Nature provided milk having a relatively low protein and mineral content for humans, who mature slowly. Contrast this with the milk of the cow, which contains a much larger amount of minerals and protein. Bovine offspring grow rapidly and thus need large quantities of calcium and phosphorus for bone development and protein for muscle development. Sow's milk is even richer in minerals than that of the cow, which correlates with the very fast early growth of the newborn pig. The same is true of dogs and rabbits.

An interesting and important fact regarding the nutrition of the young is that nature designed the milk of a given species specifically for the growth and development of the offspring of that particular species. A German scientist, Bunge, found that dog's milk had an ash content of exactly the same composition as ash of the newborn puppy. The mineral content of the milk was therefore perfectly adapted for the construction of new puppy tissue. However, it was quite different in composition from woman's, cow's, or other milk. Only in the case of iron is the quantity lower than corresponds to the composition of the offspring, but this is offset by the fact that the body is richer in

iron at birth than it is at any other period of life. The caseins (milk proteins) of different milks are different in chemical behavior. Moreover, the renninlike enzymes (Section 19.7.2) of the stomach are adapted to cause coagulation of the casein produced by the female of the same species.

Furthermore, the concentration of constituents in milk important to normal growth and development (especially minerals and protein) is dependent on the rapidity with which an animal grows. It is interesting that a higher percentage of milk ash is absorbed by infants receiving women's milk than by those fed cow's milk. In one study babies were found to absorb about 80 percent of the mineral from their mother's milk and only 60 percent from cow's milk. Adults were observed to absorb about 53 percent of the ash of cow's milk.

TABLE 3.7	Milk Composition and Growth Rates of Selected Animals		
	Milk Composition, %		Time, in Days, for the Newborn
Animal	Protein	Ash	to Double Its Weight
Woman	1.6	0.2	180
Mare	2.2	0.4	60
Cow	3.3	0.7	47
Goat	3.7	0.8	19
Sow	4.9	0.9	18
Dog	7.1	1.3	8
Rabbit	14.4	2.5	6

3.2.5 Vitamins

Animal products vary in their vitamin content. In supplying dietary needs for both fat-soluble and water-soluble vitamins, liver is likely the best-balanced animal product source of vitamins. Aside from thiamine (B_1), ascorbic acid (C), and vitamin D, a serving of beef liver (100 g) would provide the approximate daily vitamin needs for the average human as recommended by the National Research Council. Meat is an excellent dietary source of several B vitamins. The vitamin content of fresh pork reflects the amount of vitamins in the animal's ration. This ration effect is not observed in the meat of ruminants presumably because of the synthesis of B vitamins by rumen microorganisms (Section 19.3.2). Only animal products provide reliable natural sources of vitamin B_{12}, which is essential in small amounts for the proper growth of humans, production of red blood cells, and functioning of the central nervous system.

Cheese, eggs, milk, and certain organ meats offer good sources of vitamin A. Unless fortified, all animal products except eggs are poor sources of vitamin D. Furthermore, liver is the only animal product with an appreciable amount of vitamin C. Ham, liver, eggs, and milk are good sources of thiamine (B_1), whereas liver, eggs, and dairy products that contain the aqueous portion of milk provide liberal amounts of riboflavin (B_2). Dietary niacin may be readily obtained from meat and milk (niacin equivalent, Section 3.2.3). Oranges and other citrus fruits are the most reliable daily source of vitamin C.

3.3 PURCHASING FOOD NUTRIENTS VIA ANIMAL PRODUCTS

North Americans consume liberal amounts of animal products. These foods provide approximately two-thirds of our dietary protein, calcium, phosphorus, and riboflavin; about one-half of our fat and niacin; and about 40 percent of our iron, vitamin A, and thiamine (Figure 3.2). Animal products provide only 6.2 percent of our dietary carbohydrates, but of more interest to weight-conscious persons is the fact that only one-third of our food energy is derived from animal products. Moreover, animal products are high in protein as related to total calories and therefore fit well into reducing diets. In addition, the nutritious balance of most food nutrients present in animal products make them a valuable purchase for the consumer.

For the economy-minded consumer who wishes to purchase food nutrients on a least-cost basis, costs of various food nutrients can be computed in several ways; e.g., a pound of cottage cheese with 62 g of protein at $1.30 could provide all the needed dietary protein daily (56 g for an adult male) for about $1.20, whereas if one chose porterhouse steak at $6.80/lb to provide that same amount of protein (at 61 g protein per lb) it would cost about $6.24.

3.4 TRENDS IN PER CAPITA CONSUMPTION OF ANIMAL PRODUCTS

North Americans today consume about the same amount of protein, phosphorus, and total calories; fewer carbohydrates; and more fat, calcium, iron, vitamins A and C, thiamine, riboflavin, and niacin than in 1910. The increased fat is in the form of vegetable fats and oils.

When the dietary nutrients furnished by major food groups are considered, the amount of protein being obtained from animal products is found to be increasing and the amount of fat to be decreasing.

One of the hardest things to sell weight-conscious shoppers is animal fat. They prefer to do their cooking in vegetable oil and want their steak lean. Through selective breeding and feeding meats having more protein with less fat and fewer calories are now available (Figure 3.3). Consumption of lard and butter in 1997 was about one-fifth that in 1950 (Figure 3.4). Consumption of beef increased until 1970 and has decreased since then in the United States. The consumption of **lamb, mutton,**

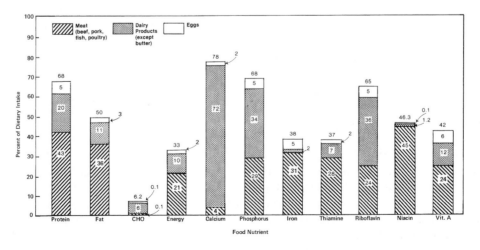

Figure 3.2 Contributions of animal products to the human dietary intake of selected food nutrients in the United States.

Adapted from National Food Review, USDA, ERS, NFR-13.

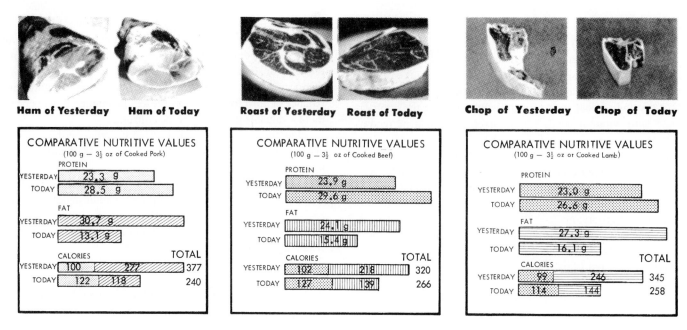

Figure 3.3 Comparative nutritive values of old- and new-type meats.
National Live Stock and Meat Board.

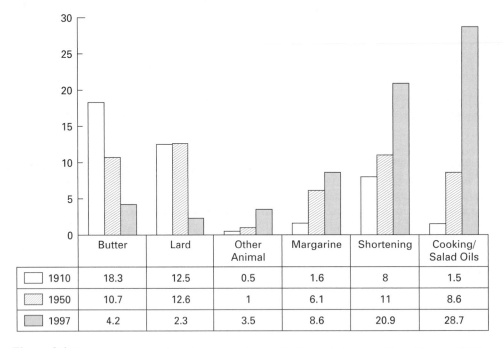

	Butter	Lard	Other Animal	Margarine	Shortening	Cooking/ Salad Oils
1910	18.3	12.5	0.5	1.6	8	1.5
1950	10.7	12.6	1	6.1	11	8.6
1997	4.2	2.3	3.5	8.6	20.9	28.7

Figure 3.4 Per capita consumption of animal and vegetable fats and oils in the United States—1910, 1950, and 1997.
Adapted from USDA Economic Research Service statistics.

and **veal** has decreased and that of pork has had a modest increase during the past half century (Table 3.8). Total meat consumption increased about 40 percent during the past five decades primarily because of increased poultry consumption.

From the 1970s to the 1990s total pounds of meat consumed increased about 8 percent from 177 to 191 lb per person, but the proportions of beef and pork dropped from 71 to 59 percent (16 lb less red meat), whereas that of poultry and fish com-

bined increased from about 27 to over 41 percent (28 lb more poultry and 2 lb more fish) (Table 3.8).

These changes in consumption resulted in lowered intake of saturated fat and cholesterol by about one-third. Concurrently, the average amount of fat in processed ground beef dropped approximately one-third. Several studies have shown that more than 25 percent of ground beef sold at retail is at least 86 percent lean and 10 percent or more is at least 91 percent lean.

TABLE 3.8	Per Capita Annual Average Consumption of Meat and Eggs in the United States by 10-Year Periods—1950–1999

	Per Capita Annual Averages				
Item	**1950–1959**	**1960–1969**	**1970–1979**	**1980–1989**	**1990–1999**
Total meats, lb	133	162	177	183	191
Red meats	102	123	129	122	113
Beef	53	69	81	72	64
Pork	41	48	45	48	48
Veal and Lamb	9	6	4	2	2
Poultry	20	28	35	47	63
Chicken	16	23	28	37	49
Turkey	4	5	7	10	14
Fish and Shellfish	11	11	13	14	15
Eggs, number	373	320	285	257	238

Note: Totals may not add due to rounding.

Source: Adapted from USDA Economic Research Service statistics.

Consumption of fish has held steady (modest increases during the past 30 years); however, there have been fluctuations in the amount of fresh, frozen, and canned fish consumed.

Food consumption patterns of Americans have changed significantly. The Economic Research Service of the USDA reported that compared with the 1970s each American in the 1990s consumed an average of 77 lb more commercially grown vegetables, 62 lb more grain products, 54 lb more fruit, 28 lb more poultry, 7 gallons more milk lower in fat than whole milk, 16 lb less red meat,

47 fewer eggs and 12 gallons less whole milk. Overall milk consumption per person in all forms in the United States amounted to about 591 lb during the 1990s, up from a low of 548 lb in the 1970s (Table 3.9). Consumption of natural cheeses more than doubled between 1970 and 2000. This accounted for much of the gain in total milk utilization. On a volume share basis cheeses sold in 1999 were in the forms of chunks/loaves, 33 percent; slices, 32 percent; shredded/crumbled, 21 percent; spreads/snacks, 7 percent; grated, 3 percent; all other forms, 4 percent.

TABLE 3.9	Per Capita Annual Average Consumption of Products of Milk in the United States by 10-Year Periods—1950–1999

		Per Capita Annual Averages				
Item	**Unit**	**1950–1959**	**1960–1969**	**1970–1979**	**1980–1989**	**1990–1999**
All milk products*	lb	700	619	548	575	591
Cheese‡	lb	7.7	9.5	14.4	21.5	27.0
Cottage cheese	lb	3.9	4.7	4.9	4.1	2.9
Frozen desserts	lb	22.8	27.4	27.8	27.4	29.1
Ice cream	lb	18.0	18.3	17.7	17.7	16.1
Low-fat ice cream	lb	2.7	6.3	7.6	7.3	7.5
Sherbet	lb	1.3	1.5	1.5	1.3	1.3
Other	lb	1.0	1.5	1.0	1.2	4.3
Nonfat dry milk	lb	4.9	5.9	4.1	2.4	3.1
Dry whey	lb	0.2	0.6	2.1	3.3	3.6
Condensed and evaporated	lb	21.4	15.7	9.4	7.5	7.6
Cream products	$^{1}/_{2}$ pt	18.0	13.3	10.1	12.8	15.8
Yogurt	$^{1}/_{2}$ pt	0.1	0.7	3.2	6.5	8.5
Whole milk	gal	33.3	28.8	21.7	14.3	9.2
Lower-fat milk	gal	2.9	3.7	8.1	12.2	15.5

***Milk-equivalent** basis, milk fat basis; includes butter. Individual items are on a product-weight basis.

‡Natural equivalent of natural cheese and cheese products; excludes full-skim American, cottage, pot, and baker's cheeses.

Source: Adapted from USDA Economic Research Service statistics.

Eggs are consumed in two forms, shell eggs and egg products. Shell eggs are used as produced by the hen. Egg products are made from whole egg, egg white, or egg yolk and are used primarily by food manufacturers and the food service industry. The yolk, white, or whole egg must be pasteurized and may be preserved by freezing, drying, or by lowering the water activity (a_w) by addition of salt or sugar. Products made from these bulk ingredients include pastas, confections, and mixes for baking. Total annual consumption of eggs in all forms dropped by about 1.2 eggs per year between the 1950s and the early 1990s (from 285 to 238). Numbers of eggs used during the late 1990s increased slightly from 234 to 245 per person with 174 being shell eggs. The highest per capita egg consumption in the United States was 402 in 1945.

3.5 ATHEROSCLEROSIS

Heart disease is the nation's number one killer accounting for more than one-half of all deaths in the United States today, contrasted with only 20 percent in 1900. **Atherosclerosis** is the underlying cause of coronary heart disease (CHD). **Cholesterol, phospholipids,** fats, iron, and proteins are deposited within the inner lining of the blood vessel walls. These compounds form a mass (**atheroma**) that protrudes into the **lumen** of blood vessels, and later calcium may be deposited in the atheroma causing "hardening of the arteries," or the disease called *atherosclerosis.*

After decades of nationwide publicity and public concern, cholesterol has become a dietary ingredient feared by many. Because of the cholesterol in meat, milk, and eggs there have been those who associate these foods with atherosclerosis and advise against their consumption.

Studies of cholesterol **metabolism** in humans have demonstrated that the liver synthesizes about 0.8 (0.5 to 2) g daily, or approximately twice as much cholesterol as is consumed in the average daily diet. Synthesis from an important metabolic intermediate (acetyl coenzyme A) and degradation of cholesterol go on simultaneously in the animal body. Cholesterol is an essential part of the structure of cell membranes. It is also the starting material from which the body makes its own supply of sex and adrenal hormones (Chapter 11). The body also can convert cholesterol into vitamin D, which is essential for proper calcification of bones and teeth. Approximately 80 percent of the cholesterol metabolized is transformed into various bile acids. In healthy individuals the body tends to maintain **plasma** cholesterol concentration by compensating for dietary intake through adjustment of synthesis and also through degradation and excretion of cholesterol and its products.

Of special interest is the fact that **herbivorous** animals (especially the cow) eat little or no cholesterol, yet their blood **serum** cholesterol level approximates that of humans. Elephants eat no animal fats, but in a recent study of the hearts and aortas of 415 elephants it was found that 72 percent of the aortas and 27 percent of the coronary arteries had visible atherosclerotic lesions similar to those in human atherosclerosis.

Atherosclerosis has also been observed in red deer on an island near Scotland, in African buffalo, cattle, sheep, goats, and caribou (all ruminants).

Why do some people have elevated blood cholesterol levels? There are two main reasons. First, a relatively small number of people suffer from such diseases as hypothyroidism and nephrosis, and these individuals almost invariably have elevated blood cholesterol. But for the majority, the explanation is hereditary, just as it is for diabetes and high blood pressure.

Nevertheless, a popular hypothesis in the United States, commonly accepted as fact, is that diet is responsible for increased blood cholesterol and consequently for the prevalence of coronary heart disease. The reasoning seems logical. It is known that CHD results from the accumulation of fatty material on the walls of the coronary arteries. Furthermore, this material is largely cholesterol. It is generally assumed that the cholesterol of circulating blood settles onto the arterial walls much like silt from a stream of running water. It is contended further that the amount of blood cholesterol reflects the amount of saturated fat and cholesterol in the diet.

The above thinking is supported by the fact that CHD is rarely seen in some areas of the world, especially in the developing countries of Africa and Asia. Moreover, blood cholesterol concentration is low in the people of those areas and their diets are low in saturated fat. Their main source of nourishment is whole-grain food with a high fiber content.

On the other hand, there are as many areas where people consume extremely large amounts of saturated fat and their blood cholesterol levels are nevertheless exceedingly low. Examples include the tribes of northern Kenya where the main dietary substances are cow's milk and meat with animal fat providing 60 percent of the total caloric intake, yet blood cholesterol levels are about one-half of those in people of the United States and CHD is virtually unheard of. Interesting as well, most farmers in the Swiss Alps live primarily on dairy products yet they too have low blood cholesterol and a low incidence of death due to heart disease.

The three most important modifiable risk factors for coronary heart disease are cigarette smoking, high blood pressure, and elevated levels of blood cholesterol. Other factors include obesity, diabetes mellitus, lack of physical exercise, and genetic predisposition.

In recent years blood lipid–protein agglomerates, called lipoproteins, have been associated with risk of CHD. Based on relative weight of each molecule these lipoproteins are divided into four classes: chylomicrons, very low-density lipoproteins (VLDL), low-density lipoproteins (LDL), and high-density lipoproteins (HDL). An elevated concentration of LDL in the blood is associated with increased CHD. In contrast, a high concentration of HDL is desirable. Although some change in the amount and type of lipoprotein can be affected by diet, heredity plays the dominant role.

Although high intakes of total and saturated fat are associated with elevated blood total and LDL cholesterol concentrations, the direct relation to CHD is difficult to prove. In 1990 saturated fatty acids in the typical American diet were contributed primarily from four sources: meat, poultry, and fish, 38 percent; fats and oils, 33 percent; dairy foods, 21 percent; and other sources, 8 percent. Saturated fatty acid intake is a strong predictor of concentrations of cholesterol in the blood. However, the offending acids are lauric, myristic, and palmitic acids C 12, C 14 and C 16, respectively—but not stearic (C 18). Short- and medium-chain fatty acids do not significantly increase levels of plasma cholesterol.

3.5.1 Are Animal Fats at Fault?

Cholesterol intake from animal fats was at first thought to be responsible for the genesis of atherosclerosis, but this theory has not been proved scientifically. Records on trends in per capita consumption of animal fats (Figure 3.4) are pertinent and of special interest. During the 87-year period from 1910 to 1997 annual per capita consumption of butter and lard, the principal animal fats, decreased nearly 80 percent (30.8 to 6.5 lb), whereas consumption of vegetable fats and oils increased 434 percent (11.1 to 59.3 lb) during the same period.

Most research studies indicate that persons with very high blood cholesterol levels have relatively more atherosclerosis and a greater risk of heart attack than those with normal levels. However, so far research does not indicate that decreasing the intake of dietary cholesterol will significantly lower blood cholesterol and reduce the risk of CHD. Thus, high blood cholesterol level may be a symptom of the disease but not the cause itself.

Although much of the research related to dietary influences on blood cholesterol levels has been directed toward lipid components of the diet, recent studies indicate that the type of protein may influence blood cholesterol levels. For example, in rabbits used as test animals soy protein diets were associated with lowered plasma cholesterol, and extracted egg diets with high-plasma cholesterol concentrations.

Eggs are high in cholesterol. Yet nature designed the egg to nourish a developing embryo. The blood cholesterol concentration of pregnant animals is about 50 percent above normal. Moreover, nature designed milk to nourish the newborn and milk contains cholesterol. This leads one to conclude that nature had a purpose in providing cholesterol.

Although eggs contain high levels of cholesterol, research at the University of Missouri at Columbia and at the University of California School of Medicine and Public Health at Los Angeles showed that the consumption of one or two eggs daily caused no significant difference in average blood cholesterol levels between the test group (one or two eggs daily) and the control group (no eggs), when persons with normal blood cholesterol levels were used as research subjects.

Boiled shrimp has considerably more cholesterol than most meats from farm animals (e.g., the cholesterol content per ounce of boiled shrimp is about 43 mg compared with 25 mg for chicken, 26 mg for beef and pork, and 28 mg for lamb and veal).

3.5.2 Evidence That Cholesterol Alone Does Not Cause Heart Disease

The United States heads the list among countries in number of deaths from CHD, but stands sixteenth in per capita consumption of milk fat. Moreover, in many countries having greater life expectancy than the United States (Netherlands, Sweden, Norway, Canada, Denmark, and New Zealand), the people consume liberal amounts of animal products. The Navajo Indians consume considerably more saturated fat than their fellow Americans, yet they have little CHD. The Masai, an east African tribe, consume about 7 liters of milk per capita daily and saturated fatty acids constitute about 60 percent of their diet. Yet they rarely develop atherosclerosis. It would seem then that the condemnation of saturated fat as the hypercholesteremic agent in animal fats is an excellent example of guilt by association.

Increasing numbers of medical and nutritional scientists are concluding that heart disease is caused by a mix of factors and that the indictment of animal fats was not only premature but also probably incorrect. Increased incidence of atherosclerosis appears to parallel prosperity. Heredity, obesity, physical inactivity, hypertension (multiple stresses such as the strain or pressure of modern living), too many cocktails, excessive smoking, and other socioeconomic factors appear to be related to heart disease.

The above facts notwithstanding, research still does not permit unequivocal conclusions as to the possible relations between diet and atherosclerosis. Considerable evidence suggests that elevated levels of blood cholesterol and triglycerides are associated with accelerated rates of atherogenesis and CHD. Cholesterol and triglycerides are transported in the blood with specific aggregates of lipids and proteins called lipoproteins. Triglycerides are present mainly in a very low-density (90 percent lipid) lipoprotein fraction (VLDL), whereas cholesterol is carried mainly in the low-density (75 percent lipid) lipoprotein fraction (LDL), with some also present in the high-density (50 percent lipid) lipoprotein fraction (HDL). Elevated blood concentration of LDL has been associated with an increased incidence of CHD, and intake of high levels of saturated fats tends to increase blood LDL and cholesterol levels. Physical exercise, dietary modifications, abstinence from smoking, and weight control may help change the LDL/HDL ratio to favor a lower CHD risk.

With so much at stake in terms of human health and longevity, as well as much of animal agriculture and associated businesses, further long-term comprehensive, carefully controlled investigations of possible causative or contributory dietary factors of multifactorial CHD are urgently needed.

3.6 FORTIFICATION OF ANIMAL PRODUCTS

The purposes of fortifying animal foods are to improve their nutritional qualities and to provide a more uniform product throughout the year. The vitamin A potency of milk, for example, may vary on average as much as threefold in the United

States from its high in May through September, when carotene (the precursor of vitamin A) is abundant in grasses and forages, to its low in November through April, when forages are often low in carotene.

Fortification of foods with vitamins was first reported in 1924 when it was found that antirachitic properties (vitamin D) could be imparted to selected foods by irradiation with ultraviolet light. Liquid vitamin A concentrates (carotene) were introduced in about 1930, and pure crystalline vitamin D became available in 1932. From 1932 to 1943 industry utilized advances in science to fortify milk so that it became the first and only food to provide humans with the minimum daily vitamin requirements, with the exception of vitamin C.

3.6.1 Legal Aspects of Fortifying Animal Products

Certain states restrict the amount and kind of vitamin fortification. Milk is the only animal food approved and recommended for vitamin fortification jointly by the American Medical Association's Council on Foods and Nutrition and the National Research Council's Food and Nutrition Board.

Labeling

Milk being sold in the United States must have a statement on the container indicating the name and quantity of the product, its grade, and the name and address of its manufacturer. Federal Food and Drug Administration (FDA) regulations require nutritional labeling. The declaration of nutritional information on the label must contain the following: (1) serving size; (2) servings per container; (3) on a per serving size basis the total calories, calories from fat, the grams each of fat, protein, total carbohydrate, dietary fiber, and sugars; and (4) the percentages of the Reference Daily Intake (given as percent daily value) for all of above-listed nutrients plus vitamin A, vitamin C, thiamine, riboflavin, niacin, calcium, and iron. In determining the percentage daily value for protein, the protein digestibility–amino acid score is important. The reference protein is casein (the major protein of milk). Fatty acid composition and cholesterol content, sodium concentration, and contents of other nutrients may be declared. Percent daily values are based on a 2000 calorie diet for adults or for the target group of consumers.

A food is deemed to be an imitation and is subject to federal labeling controls if it is a substitute for and resembles another food but is nutritionally inferior to the food imitated. If it is not nutritionally inferior and bears a common or usual name that is not false or misleading, it need not be labeled as imitation.

3.6.2 Fortifying with Vitamins

Vitamins A and D being fat-soluble are not present in nonfat milk. Federal standards require addition of enough vitamin A to provide 2000 IU per quart in any liquid milk product from which fat has been removed. The daily vitamin A requirement for a human is 5000 IU. Addition is optional for dried nonfat and low-fat milks. The addition of vitamin D to milk is optional, but nearly all processors add 400 IU per qt. Fortifica-

tion with vitamins A and D adds about $0.0005 to the cost of a quart of milk. The fortification of nonfat dry milk with vitamins A and D is approved for overseas shipment.

The recommended daily dosage of vitamin D for humans is 200 to 400 IU. Because few foods contain appreciable amounts of vitamin D, it is the vitamin most likely to be deficient in an otherwise balanced diet. (Only a minimal amount of vitamin D is stored in the body. Therefore, it should be consumed daily.) Vitamin D is required for the **absorption** and metabolism of calcium and phosphorus and is therefore commonly added to fortify milk, which has an abundant supply of these two important dietary minerals. The most commonly used method of Vitamin D fortification of milk is to add it in the form of **irradiated ergosterol** (vitamin D_2, a plant **sterol**).

Dangers of Vitamin D

Excessive doses of vitamin D cause losses of calcium and phosphorus from tissues, reversing the effects of normal doses. Vitamin D poisoning may cause nausea, **diuresis,** headache, **asthenia,** loss of appetite, and low retention of calcium and phosphorus. Dr. Cooke of Johns Hopkins University found that excessive vitamin D intake by pregnant women can lead to birth defects and mental abnormalities in infants.

3.6.3 Fortifying with Nonfat Milk Solids

In recent years United States consumers have increased their use of nonfat and low-fat milk and milk products. To add nutritive value (especially protein, lactose, and minerals), as well as flavor and body, to these low-fat products some milk processors add 1 to 2 percent nonfat dry milk solids to fluid skim milk prior to pasteurization. Others are using sophisticated equipment to remove water from milk, thereby providing nonfat milk solids content equivalent to the total solids of low-fat milk or whole milk.

Any milk labeled "with added milk solids not fat" must contain at least 10 percent milk-derived nonfat solids, and the ratio of protein to the total nonfat solids of the food and the protein efficiency ratio of all protein present must not be decreased as a result of adding such ingredients.

3.6.4 Milk Toning

In view of the limited world supply of milk and protein many private and government-supported programs involve shipment of nonfat dry milk to countries and territories abroad. Much of this skim milk powder (with an appropriate amount of pure filtered water) is added to the native milk, such as high-fat buffalo milk, to *tone* it to 3 percent milk fat. Further reduction of the fat level to 1.5 percent is called *double toning*. Toning is practiced regularly in India.

3.6.5 Natural Foods

The words "natural," "no preservatives," and "no artificial additives" are embraced by many consumers, whereas anything that smacks of chemical treatment is viewed with alarm. However, concerns about additives and preservatives are, in some cases, ill-founded and overdrawn. A natural product, just because it is "natural," is not necessarily safe for human consumption. Few people would question the nutritional wisdom of consuming many, if not most, foods in their natural state. However, certain plants used as foods contain naturally occurring toxicants. They usually occur in minute amounts that the human body can tolerate safely. Yet if subjected to the same FDA testing as food additives those "natural" plants would not be approved for human use. Various state and federal agencies have as their main mission the monitoring of foods and food additives to assure the consuming public of a safe and wholesome food supply. This includes extensive, rigorous experimental tests of the safety of food additives.

3.7 PRESERVATION OF ANIMAL PRODUCTS

To preserve animal products is to enhance their keeping qualities and thereby increase their shelf life. Reasons for preserving animal products include (1) meeting the constant demand for meat, milk, and eggs; (2) the seasonal supply of certain animal products; (3) the necessity of transporting from producer to consumer; and (4) the demand for an abundant and safe food supply.

Until early in the twentieth century, when transportation and refrigeration provided ways of moving and preserving humanity's food supply, many food items were available on only a seasonal basis. Milk, eggs, and poultry meats were available the year-round but cattle, swine, and sheep could be slaughtered only in cold weather so that the meat could be used, smoked, salted, or fried down before it spoiled. The principal meat for the farm family was pork. Cattle served as a cash crop because they could be driven to and sold for slaughter in nearby cities. About Thanksgiving time each farmer would butcher a number of hogs in accordance with family size. Hams, shoulders, and side meat were packed in salt for about 6 weeks and then removed and further cured with hickory smoke (or that of another hardwood). Each farmstead had a smokehouse. Sausage was made from trimmings of the shoulders and hams plus the hearts, livers, and other bits of meat. The women fried down the sausage, backbones, and spareribs. (They fried the meat, packed it in earthenware crocks, and covered it with melted lard rendered from the fatter parts.) This process preserved the meat for at least a year.

3.7.1 Methods of Preservation Applicable to Animal Products

The foremost concern in preserving animal products is to control microorganisms that cause spoilage or illness. One or more of the following methods may be employed.

High Temperature

In 1809 a Frenchman, Nicholas Appert, was awarded 12,000 French francs for his work leading to the publication of *The Book for All Households on the Art of Preserving Animal and Vegetable Substances for Many Years.* His studies on canning foods were prompted by an offer from the French government for a method of preserving foods for the country's armies that were often far away during the expeditions of Napoleon.

In 1864 Louis Pasteur, another Frenchman, invented the process of pasteurization, or killing bacteria in fluids with heat. This process is widely used to preserve animal products, especially milk products. These thermal treatments are the most widely used method of *killing* spoilage and potentially pathogenic microorganisms in animal products. They are employed in pasteurization and canning.

Low Temperature

Refrigeration is the most widely used method employed with animal foods to *inhibit the growth* of microorganisms without killing them. The lower the temperature the lower the rate of growth. Most bacteria grow well at normal room temperatures. The single cells of bacteria multiply by dividing into two daughter cells. Typical rates of division at 72°F (22°C) are every 30 to 60 minutes (min), depending on nutrients available and genetic capabilities. However, some bacteria can divide in as few as 10 min. Time required for a cell to divide is called generation time (GT). When GT is 30 min, one cell can produce 1000 cells in approximately 5 h. The GT for most bacteria that spoil animal products is extended to at least 12 h when the temperature is kept at 40°F (4.4°C). This means that time for numbers to increase 1000-fold is extended to some 5 days. Typical maximal temperatures of storage of fresh animal products are 45°F (7°C) for shell eggs and liquid pasteurized egg products; 40°F for fluid milk products, butter, and processed meats; and 35°F (2°C) for fresh meats. Some cured meats, for example salt-cured hams, can be stored at room temperature without spoilage if they are protected from growth of molds. As observed later, the low availability of water in salt- and sugar-cured meats prevents growth of most microorganisms.

Removal or Destruction of Microorganisms

The process for removing bacteria from liquid (e.g., milk) by means of centrifugal force is called *bactofugation.* The bactofuge utilizes differences in specific gravity between bacteria and the constituents of milk. The specific gravity of bacteria varies between 1.07 and 1.13, as compared with 1.032 and 1.036 for whole and skim milk, respectively. Specific gravity is the ratio of density of a substance to the density of water at equal temperatures. If the coefficient of expansion of the two substances varies greatly with temperature, so will specific gravity.

Dehydration

Sun-drying is the oldest and least expensive type of **dehydration** and is still used for certain foods. Air-drying of meat has been practiced for centuries. Marco Polo reported early in the 1300s that every Tartar Mongol warrior carried 10 lb of dried meat as part of his food rations. Mechanical dryers were first mentioned by Marco Polo in the fourteenth century, but these were not used commercially until the twentieth century. American troops in the Revolutionary War ate dried beef, sometimes called **jerky.** This consisted of thin strips of beef or game that were dehydrated by hanging in the sun to air-dry. An adaptation of jerked beef, known as *pemmican,* has been used by Native Americans and Arctic explorers. More recently cooked meat has been dehydrated in hot-air dryers. Milk and certain poultry products may be spray-dried or dried with drum dryers having heated rollers. Dried eggs were imported from China for decades, but not until 1927 were eggs dried commercially in the United States. The common method of drying milk and eggs is by spray drying.

Chemical Preservatives

Curing agents for meat include a mixture of sodium chloride, **sodium nitrite, sodium nitrate (color fixative)** and sugar. Sorbitol is added to assist in meat preservation. Sorbic acid and propionates are often incorporated into packaging materials to inhibit the growth of molds inside the package. Sorbic acid migrates from the wrapper into the product. Other chemical preservatives include ascorbic acid (ascorbate, an **antioxidant**) and certain phosphates (e.g., tripolyphosphate). Unfortunately many chemicals in concentrations that adversely affect growth of microorganisms may be harmful to humans.

Irradiation

Meats can be preserved by means of **radiation** doses that *sterilize* (kill all organisms) or *pasteurize* (kill most organisms). Radiation can be used in combination with other methods of preservation such as chemical additives or heat. Secondary reactions associated with radiation present problems when the process is incorrectly done. These reactions include changes in flavor, color, odor, and texture and the loss of some nutrients. Radioactive particles strike and kill microorganisms in or on food without appreciably raising the temperature of the product. The process is therefore referred to as *cold pasteurization* or *sterilization.*

Approval for irradiation of chicken was given by FDA in 1982 and for red meats in 1997. Low doses of radiation can kill at least 99.9 percent of *Salmonella* in poultry and an even higher percentage of *Escherichia coli* 0157:H7 in ground beef. The most promising application of ionizing radiation to foods is in treating ground beef to kill these toxin-producing *E. coli.* This bacterium, a significant contaminant of ground beef, is killed by cooking to 160°F. However, failures to cook meat adequately have led to outbreaks of hemorrhagic colitis and several deaths among children.

Much research on the application of ionizing radiation to food was done in the 1950s, but passage in 1958 of the Food Additives Amendment to the Food, Drug and Cosmetic Act delayed commercialization of irradiation for three decades. This amendment classified radiation of food as food additives. Thus the amendment required an authorizing regulation prescribing safe conditions of use and a premarket review with acceptance by the FDA. Irradiated foods were first produced for sale in the United States in 1992. The cost of irradiating poultry sold in Chicago in 1993 was $0.02 per pound.

In approving the use of ionizing radiation for foods the FDA considers nutritional adequacy of the treated food plus safety from harmful radiological, microbiological, or toxicological effects. DNA base damage and breaks in single- and double-stranded DNA are the major effects of irradiation.

Approved sources of ionizing radiation are gamma rays (produced from cobalt-60 and cesium-137), machine-generated X rays, and electrons that are generated by **linear accelerators.** X rays and electrons are produced by machines that can be turned on and off, but gamma rays are constantly emitted from their source, which must be constantly shielded. The approved low-treatment doses do not cause the food to become radioactive.

Amounts of radiation energy are measured in **grays** where 1 *gray* (Gy) equals 1 joule per kilogram. Depending on the dose of radiation applied and the moisture in the food (high moisture equals high effectiveness), foods can be mildly treated to kill insects and to prevent sprouting of tubers (*radicidized*), moderately treated to kill pathogenic microorganisms (*radurized*), or highly treated to kill all microorganisms (*radappertized*). Freezing of foods increases the amount of radiation needed to achieve the desired effect but decreases the undesired effects on flavor and color that sometimes occur. For example, the maximum doses permitted for raw meat are 4.5 kGy and 7.0 kGy for unfrozen and frozen forms, respectively. Packaging used on treated foods must be approved for exposure to irradiation. Minimization of air around the product is important to minimize loss of vitamins. FDA requires that affected vitamins in treated foods not be significant in the overall diet. For example, pork is a major source of thiamine, the most radiation-sensitive water-soluble vitamin, but only 2.3 percent of the thiamine in American diets would be lost if all the pork in the United States were irradiated.

Environmental Preservation

Maintaining conditions unfavorable to the growth of microorganisms, for example, using sealed or evacuated containers and substituting a gas, such as nitrogen or carbon dioxide for oxygen, in sealed and impermeable containers helps preserve animal products. The food industry refers to the practice as modified atmosphere packaging (MAP).

Low pH (High Acidity)

Low **pH** is an important method of preserving many cultured milk, cheese, and meat products. It is normally accomplished by fermentation. Following slaughter, muscle glycogen is fermented to lactic acid, reducing the pH of beef muscle from near neutral to about 5.7. This lower pH retards bacterial growth. Biologically produced lactic acid enhances flavor and gives stability to fermented sausages, Lebanon bologna, and dry summer sausages. Fermentation in meats is accomplished through the addition of a fermentable sugar and a starter (**culture**). Similarly, a culture (controlled bacterial population) may be added to certain milk products to develop lactic acid and desirable flavorful substances such as diacetyl. Sour cream and cottage cheese are also frequently produced by direct acidification with a food-grade acid or with glucono-delta-lactone. This lactone slowly hydrolyzes producing gluconic acid. Artificial flavors must be added to directly acidified milk products to make them taste similar to their cultured counterparts.

Many chemical preservatives are more effective at a low pH; for example, the minimum inhibitory concentration of sodium nitrite, an additive to meat products, is reduced over 40-fold when the pH is reduced from 6.9 to 5.

3.7.2 Animal Food Safety

Illnesses caused by the seven dominant causative agents of food-borne illnesses in the United States were estimated to cost from $6.6 to 34.1 billion in 1996. The two approaches used in making these estimates were called the "human capital approach" and the "labor market approach." The wide range in cost is due to many uncertainties. For example, many such illnesses are not reported making it necessary to estimate numbers of affected persons and the consequent effects on various cost factors. Nevertheless, it is certain that the cost is high. Furthermore, most cases of food-borne illness can be traced to animal foods. This is true because the offending microorganisms are often associated with the intestinal tracts of animals.

Campylobacteriosis, the most frequent form of food-borne diarrhea, is caused by *Campylobacterium jejuni* or *C. coli*. The infection is mostly associated with undercooked poultry, but other meats and raw milk have been implicated. In a small number of the cases patients develop Guillain-Barré (G-B) syndrome, a neurological disease that is characterized by rapid onset, various degrees of numbness, pain, progressive weakness, or paralysis over 1 to 4 weeks. Almost all G-B patients are hospitalized and roughly 20 percent are left disabled.

Another prevalent food-borne disease is salmonellosis. This intestinal infection is caused by numerous different serotypes of "nontyphoid *Salmonella*," bacteria that are associated with the intestinal tracts of humans, animals, and fowl. Up to 4 million cases are thought to occur annually in the United States. Most cases are mild, but death can occur among infants, the elderly, and immunocompromised persons. The average annual number of outbreaks of food-borne disease reported during 1993 to 1997 was 550. (An outbreak is defined as an occurrence of multiple cases of illness resulting from the same cause.) *Salmonella* serotype Enteritis accounted for the largest number of outbreaks, cases, and deaths. Most of these outbreaks were attributed to eating inadequately prepared eggs or foods containing them.

Incidences of disease from enterohemorrhagic *Escherichia coli* and *Listeria monocytogenes* are relatively low, but the symptoms can be severe. Some *E. coli* can cause hemorrhagic colitis and hemolytic uremic syndrome, a severe life-threatening illness especially among young children. Listeriosis usually produces mild symptoms in adults but may lead to stillbirths among pregnant women and to mental retardation among infants. The most common manifestation of listeriosis is meningitis, which produces high fever, severe headache, neck stiffness, and nausea, as well as serious and sometimes fatal infections in those with weak immune systems (infants, the frail or elderly, and persons with chronic disease, HIV infection, or persons taking chemotherapy).

Some 30 percent of live birds and 60 percent of processed raw poultry have been found contaminated with salmonellae, campylobacteria, and/or listeriae. Each of these bacteria is also frequently present in raw meat or milk. Usually the interior of the freshly laid shell egg is sterile, but one serotype of salmonellae, *Salmonella enteritidis,* is able to infect the hen's ovary. Because the yolk is formed in the ovary it can become contaminated with these infecting bacteria. Subsequent thermal abuse of the egg can cause multiplication of the bacteria. For this reason cooling and storage temperatures for eggs were recently lowered to 45°F (7°C). Eggs can be pasteurized in the shell by immersion in hot liquid under carefully controlled conditions that prevent gelation of the white. Such eggs retain the functionality of raw eggs except that whites are slower to whip.

The Food Safety and Inspection Service (FSIS) is the agency within the USDA responsible for ensuring the safety, wholesomeness, and accurate labeling of meat, poultry, and egg products. This agency issued a new regulation in July of 1996 entitled Pathogen Reduction: Hazard Analysis and Critical Control Point (HACCP) Systems. This regulation, in part, sets pathogen reduction performance standards for *Salmonella* that slaughter plants and plants producing raw ground products must meet. Large plants, which are federally inspected establishments employing 500 or more employees, became subject to the *Salmonella* testing requirements in January 1998. Medium and small plants were affected in the same month of 1999 and 2000, respectively.

In 1996 FSIS started the process of moving from a system in which each carcass was observed for obvious abnormalities such as visible lesions and filth to a system that relies more on preventing contamination. Four essential elements of this system are (1) all state and federally inspected plants that slaughter or process meat or poultry must have an approved HACCP program, (2) these plants must all develop written sanitation standard operating procedures (SSOP), (3) FSIS tests for *Salmonella* on raw meat and poultry products to verify that pathogen reduction standards are being met for *Salmonella,* and (4) slaughter plants test for generic *E. coli* on carcasses to verify that the process is under control with respect to preventing and removing fecal contamination.

The process for developing a HACCP plan calls for (1) identifying all potential hazards likely to cause illness or injury (quality is not the primary focus) that exist in the operation; (2) locating steps in the process where there is an opportunity to prevent, eliminate, or reduce a food safety hazard to an acceptable level; (3) setting limits on the process that will assure meeting safety criteria; (4) monitoring planned observations or measurements to assess whether a critical control point is under control and to produce an accurate record for the future; and (5) establishing corrective actions to be taken in the event of a possible failure. Each plan must be validated by collection and evaluation of scientific and technical information to determine whether the plan, when properly implemented, will control the hazards.

In March 1999 the agency released a report that summarized the *Salmonella* prevalence in broilers, swine, ground beef, and ground turkey among large plants during the first year of the testing program. *Salmonella* prevalence in 199 large plants for each of these four products was lower after the first year of

TABLE 3.10	Prevalence of *Salmonella* in Samples from 199 Large Meat and Poultry Processing Plants Examined in Baseline Studies (Pre-HACCP) and Again after Implementation of HACCP Practices in 1999

Product Class	Pre-HACCP, %	Post-HACCP, %
Ground turkey	49.9	36.4
Broilers	20.0	10.9
Swine	8.7	6.5
Ground beef	7.5	4.8

Source: HACCP Implementation: First Year *Salmonella* Test Results: www.fsis.usda.gov/OPHS/salmdata.htm.

HACCP implementation than in baseline studies conducted before HACCP implementation (Table 3.10). It is unlikely that all of these reductions are solely attributable to the implementation of HACCP. For these four product classes combined, 100 of the 114 plants (88 percent) with complete data sets met their respective *Salmonella* performance standards.

Characteristics of the United States food safety system include the separation of powers among the executive, congressional, and judicial branches of government, plus transparent science-based decision-making and public participation. Congress (the legislative branch) enacts statutes designed to ensure the safety of food. Congress also authorizes executive branch agencies to implement statutes, and they may do so by developing and enforcing regulations. When enforcement actions, regulations, or policies lead to disputes, the judicial branch is charged to render impartial decisions.

The FDA and USDA website at http://www.fsis.usda.gov/OA/codex/system.htm describes the food safety system as:

> . . . based on strong, flexible, and science-based federal and state laws and industry's legal responsibility to produce safe foods. Federal, state, and local authorities have complementary and interdependent food safety roles in regulating food and food processing facilities. The system is guided by the following principles: (1) only safe and wholesome foods may be marketed; (2) regulatory decision-making in food safety is science-based; (3) the government has enforcement responsibility; (4) manufacturers, distributors, importers and others are expected to comply and are liable if they do not; and (5) the regulatory process is transparent and accessible to the public. U.S. food safety statutes, regulations, and policies are risk-based and have precautionary approaches embedded in them.

To promote food safety internationally and to harmonize regulations United States agencies interact with the Codex Alimentarius Commission, the World Health Organization (WHO), and the Food and Agriculture Organization (FAO) of the United Nations. Principal federal regulatory organizations responsible for providing consumer protection are the Department of Health and Human Services' (HHS) Food and Drug Administration, the USDA's Food Safety and Inspection Service (FSIS) and Animal and Plant Health Inspection Service (APHIS), and the Environmental Protection Agency (EPA).

The web page cited above describes the roles of the principal agencies as follows:

> The FDA is charged with protecting consumers against impure, unsafe, and fraudulently labeled food other than in areas regulated by FSIS. FSIS has the responsibility for ensuring that meat, poultry, and egg products are safe, wholesome, and accurately labeled. EPA's mission includes protecting public health and the environment from risks posed by pesticides and promoting safer means of pest management. No food or feed item may be marketed legally in the U.S. if it contains a food additive or drug residue not permitted by FDA or a pesticide residue without an EPA tolerance or if the residue is in excess of an established tolerance. APHIS' primary role in the U.S. food safety network of agencies is to protect against plant and animal pests and diseases. FDA, APHIS, FSIS, and EPA also use existing food safety and environmental laws to regulate plants, animals, and foods that are the results of biotechnology.

In recent years the federal government has focused on risks associated with microbial pathogens and on reducing those risks through a comprehensive, farm-to-table approach to food safety. This emphasis is based on the principle that multiple protective steps are required throughout the farm-to-table chain. In earlier years the agencies concentrated on physical examination of animals and their carcasses plus managing chemical hazards from the food supply by regulation of additives, drugs, pesticides, and other chemical and physical hazards to human health. The new approach recognizes that safety concerns from biological hazards differ from those presented by chemicals, and that microorganisms are invisible to the naked eye.

3.7.3 Providing Safe Honey

The chief sources of microorganisms in honey are the nectar of flowers and the honeybee. The high sugar content of honey produces high osmotic pressure that restricts the growth of most microorganisms (especially bacteria). Because the pH of honey is low (3.2 to 4.2), the growth of osmophilic yeast presents the major microbiological problem. If the moisture content of honey is held to 21 percent or less, the problem of yeast growth is greatly reduced. Honey is pasteurized at 160 to 170°F for 5 min and promptly cooled to 90°F. This treatment kills the yeasts and prevents fermentation of honey.

3.7.4 Providing Safe Meat

In the early days pork was considered "unclean." This idea may have resulted from the observation that pork meat could cause illness. In 1847 Joseph Leidy discovered the nematode **parasite** *Trichinella spiralis* in pork when cutting a cold ham sandwich. This discovery led to the conclusion that humans might become infected by eating raw pork. All meat intended for interstate trade must be slaughtered and processed in plants approved by the FSIS.

Trichinosis

Trichinosis is caused by the small roundworm *T. spiralis,* which is killed when meat is (1) heated to an internal temperature of 137°F with no holding time, (2) frozen 20 or more continuous days at 0°F or below, or (3) smoked at 80°F for 40 h or more, followed by 10 days in a drying room at 45°F.

Trichinella larvae are rendered incapable of completing their life cycle in the new host when infected carcasses are exposed to gamma irradiation at a dosage of 0.3 to 1 kGy.

Curing Meat

Methods of curing meat include (1) dry cure, in which the curing agent is rubbed into the meat as in bacon, ham, and beef trimmings intended for sausage making; (2) pickle cure, in which the meat is immersed in a solution of ingredients (e.g., 15 percent NaCl); (3) injection cure, in which the internal injection of curing agents (e.g., 24 percent solution of NaCl) results in a rapid and uniform distribution of the cure throughout the tissues (the curing agent is introduced by a single-needle injection into the vascular system of hams and by multiple-needle injections in smaller cuts, such as bacon, **jowl,** and shoulder); and (4) direct addition, in which the curing agent is mixed and ground in the meat, as in sausage making.

With rapid processing techniques and improved refrigeration, the practice of curing meat as a means of preservation per se has decreased. The principal reason for curing meat today is to impart various flavors and to prevent the germination of spores of Clostridium botulinum, the spore-forming bacterium that produces the world's most lethal toxin. Vegetative cells of this bacterium normally are killed during the processing of meat, but spores are not killed. Because this is an anaerobic bacterium it does not grow in the presence of air. However, when meat is packaged in oxygen-impermeable containers, the pH is above 4.6 and the water activity above 0.93, it is imperative that the temperature be kept below 45°F.

Smoking

Smoking and heat processing of meat products are accomplished concurrently. The use of oak and/or hickory sawdust in smoking imparts a characteristic flavor. The combination of heat and smoke aids in preservation by reducing the bacterial population and by causing surface dehydration. Smoked poultry meat is frequently held 6 to 8 h at a smokehouse temperature of 170°F and then at 185°F until the internal temperature of the meat reaches 160°F. Few red meat foods are processed in which smoking constitutes an important role in preservation against bacterial spoilage. "Ready-to-eat" hams have been heated to an internal temperature of 137°F, whereas smoked hams are commonly processed at lower temperatures.

Canning

As noted in Section 3.7.1, Nicholas Appert is credited with the invention of canning for the military. The practice of thermal processing after packing in containers is followed today for preserving most canned meats including hams, sausages, and fish products.

Preservation and Stability of Meat Nutrients

Most vitamins of meat are relatively stable to processing. However, thiamine is partially destroyed in the course of curing, smoking, cooking, canning, heat dehydration, and irradiation.

Meat Tenderizers

Tenderness is one of the most important features of high-quality meat. It depends largely on the condition of muscle fibers, which in turn consist of protein fibers interwoven and grouped together by a delicate sheath called the *sarcolemma.* Certain **hydrolytic** enzymes may be used to break the sarcolemma, resulting in greater tenderness.

The use of enzymes to help tenderize meat is not new. At least 800 years ago Mexican Indians tenderized meat by wrapping it in papaya leaves. Early explorers found this practice among Native Americans and South Sea islanders also.

Several proteases of animal, plant, and microbial origin are used as meat tenderizers: trypsin (from the pancreas), bromelin (from pineapple), ficin (from figs), papain (from papaya), and *Aspergillus* fungal protease. Cooking temperature is of great importance since enzymatic action must occur before the enzyme is denatured at 140 to 185°F (60 to 85°C). Enzymes used to tenderize meat are inactivated during later stages of cooking. Hams are also tenderized by proteases. Commonly a heat-sensitive protease is added through the pickling solution in combination with phosphate salt. The enzyme is inactivated by heating the ham at 140 to 158°F (60 to 70°C).

Proteases are also used to tenderize animal casings for sausages and other processed foods and to prepare hydrolysates (products of hydrolysis) of meat, milk, fish, and plant proteins for use in manufacturing special diet foods, condiments, and animal feeds.

Certain proteolytic enzymes aid in skinning fish by selectively liquefying gelatin under the skin. Old roosters and tom turkeys may be tenderized by injecting a protease solution intravenously about 5 min before the birds are killed. Another enzyme, glucose oxidase, combined with catalase is used to ferment glucose of eggs prior to drying to prevent the browning of egg powder.

Research in Australia indicates that meat may be tenderized by high-pressure treatment at elevated temperatures. The process seems to act like an accelerated aging.

Poultry Meat

All poultry meat must be inspected by USDA personnel for wholesomeness and processed under sanitary conditions. In 1970 USDA inspection programs of red meat and poultry meat were combined.

Broilers are usually marketed as fresh (but refrigerated) poultry. The **shelf life** of broilers in the display case and home refrigerator is directly proportional to the number of bacteria on the meat surface after processing. Therefore, sanitary processing, rapid chilling, and subsequent refrigeration are important adjuncts to maximal shelf life. Any salmonellae that may be present on poultry meats are destroyed during proper cooking;

however, there is some risk of cross-contamination of foods that will not be cooked when they touch surfaces of poultry or surfaces contaminated by it.

Processing considerations applicable to broilers are also important for turkeys. However, most turkey meat is frozen, which minimizes bacterial spoilage. The problem of lipid oxidation and the resultant off-odors has been largely overcome by using packaging films that are virtually impermeable to oxygen.

More than one-third of all poultry meat marketed in the United States enters "further processed" items such as pot pies, frozen dinners, canned poultry rolls, and frankfurters. Bony parts (necks and backs) are deboned mechanically so meat losses are minimized. Up to 15 percent of this meat emulsion can be combined with red meat in the manufacture of sausages and wieners. There has been good market acceptance of 100 percent turkey and poultry "franks."

Pasteurization of Deboned Poultry Meat Rapid development of further processed poultry products and the strong demand for animal protein have accelerated developments in mechanically deboned poultry meat (MDPM)—a finely ground product that has been processed through an automatic deboner—obtained from turkey frames, poultry backs and necks, or the entire carcass of hens that have completed commercial egg laying. The deboner grinds the meat and bones and, through a series of screens, separates the edible portion from the nonedible portion (bone and gristle).

Because of the finely ground nature of MDPM, and because it is an excellent medium for bacterial growth, it deteriorates quickly if not handled and/or stored properly. Spoilage caused by microbial growth is the major problem in nonfrozen MDPM, whereas oxidative rancidity of the lipid portion is a major cause of deterioration of frozen MDPM. Researchers at the University of Missouri showed that freshly ground raw chicken packaged to restrict air entry with bacterial populations of about 1,000,000 per gram spoiled within 7 days when held at 40°F (4°C). Samples of the same meat stored in the same way had estimated spoilage times of 27, 70, and >96 days after exposure to high isostatic pressures at 408, 616 or 888 mega Pascals, respectively. This treatment results in only small increases in temperature. Ultrahigh pressure processing is still in the developmental stage. The major problem is development of a continuous process with high throughput.

Research at the Pennsylvania State University demonstrated that the storage life of MDPM can be extended by heat pasteurization, up to 6 min at temperatures ranging from 138°F (59°C) to 160°F (71°C), which reduces the bacterial load in meat.

Because MDPM is frequently used in frankfurters and other meat products that depend on meat protein to bind water, and because high pasteurization temperatures result in denaturization of the protein in MDPM, it is important that the times and temperatures used in pasteurizing MDPM be controlled closely.

3.7.5 Providing Safe Poultry Products

Refrigeration, pasteurization, freezing, and drying are important processes in providing consumers with safe poultry products. Of course, healthy birds and sanitary practices are basic in this regard.

Shell Eggs

It may be the cock that crows, but it is the hen that lays the eggs.
Lady Margaret Thatcher (b. 1925)
Former Prime Minister (1979–1990), United Kingdom

Most eggs are sterile when laid, but the shell surfaces immediately become contaminated with microorganisms. Less contamination occurs when eggs are produced in cages than in nests. Although most contaminating microorganisms have limited **public health** significance, *Salmonella* may be present in cracked eggs or infrequently in eggs from hens with infected ovaries, causing food poisoning in humans. Eggs should be cooled to 45°F soon after being laid so quality is preserved and bacterial growth is impeded. Shell eggs are washed and sanitized in mechanical washers before reaching consumers. Cartoned shell eggs should be refrigerated in the store and home.

Nature provides shell eggs with protection from invasion by microorganisms. The thin layer of protein on the shell's surface provides protection until it dries and cracks or is removed in cleaning. When this occurs motile bacteria can invade pores in the shell. However, once they reach the shell's fibrous membranes they encounter physical and chemical resistance. The membrane contains **lysozyme,** an enzyme that can *lyse* many bacteria, especially **Gram-positive** ones. The white, being highly viscous, presents a physical barrier to invasion by bacteria and to diffusion of oxygen that many bacteria need for growth. The high pH of albumen (9.6) is unfavorable to many bacteria. Also contained in egg white are lysozyme and proteins that bind certain essential nutrients: ovotransferrin and conalbumin bind iron and other minerals, avidin binds biotin, ovoflavoprotein binds riboflavin. Once an invading bacterium reaches the yolk it is able to grow well if temperature is favorable. It is primarily the **Gram-negative** bacteria that cause spoilage of eggs.

Pasteurization (thermostabilization) of shell eggs can be accomplished by heating at 130°F for 15 min in water and oil. Pasteurization kills bacteria, stabilizes egg whites, and pasteurizes the shell thereby prolonging shelf life. The oil prevents moisture loss and retains CO_2 thereby maintaining the character of the albumen. Commercially a fine mist of mineral oil is sprayed on the large end of shell eggs before packaging to minimize loss of moisture and CO_2.

Egg Products

In 1998 more than 67 billion eggs were produced in the United States, and nearly 20 billion were broken for use in egg products. Some egg products are refrigerated for formulations where fresh eggs might be used, others are frozen or dried. Since 1971 all egg products must be processed under USDA inspection. (Inspection is not required for facilities where eggs

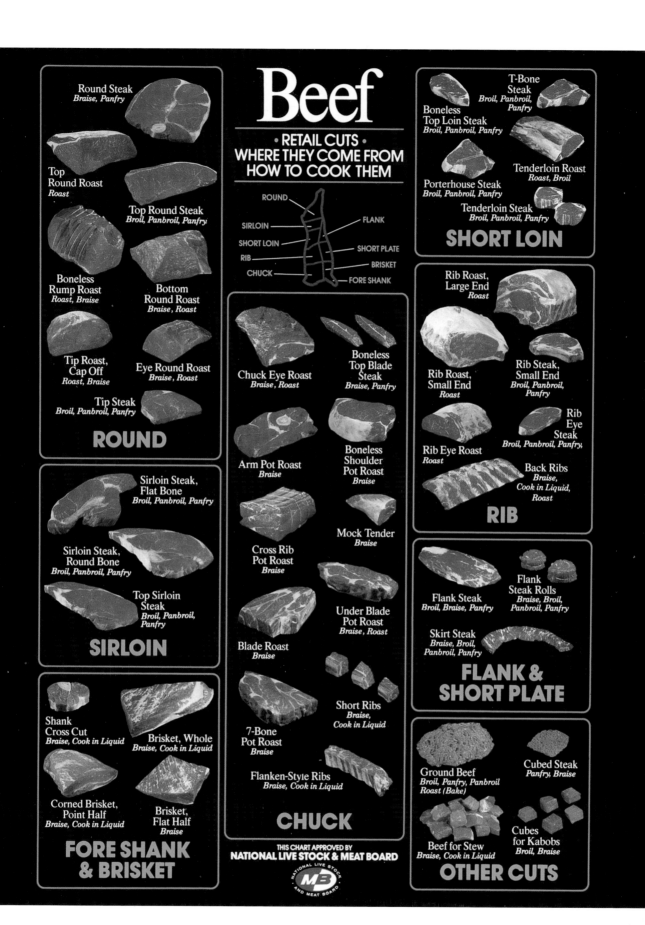

Color Plate 3A Courtesy of the National Cattlemen's Beef Association © 1986.

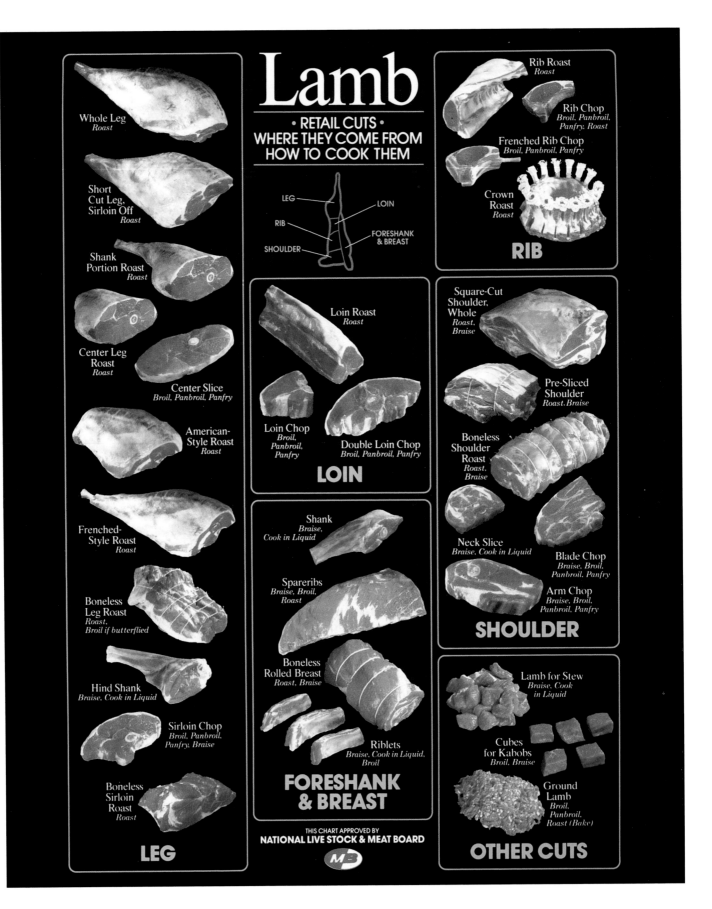

Lamb

· RETAIL CUTS ·
WHERE THEY COME FROM
HOW TO COOK THEM

LEG
LOIN
RIB
FORESHANK & BREAST
SHOULDER

Whole Leg
Roast

Short Cut Leg, Sirloin Off
Roast

Shank Portion Roast
Roast

Center Leg Roast
Roast

Center Slice
Broil, Panbroil, Panfry

American-Style Roast
Roast

Frenched-Style Roast
Roast

Boneless Leg Roast
Roast, Broil if butterflied

Hind Shank
Braise, Cook in Liquid

Sirloin Chop
Broil, Panbroil, Panfry, Braise

Boneless Sirloin Roast
Roast

LEG

Loin Roast
Roast

Loin Chop
Broil, Panbroil, Panfry

Double Loin Chop
Broil, Panbroil, Panfry

LOIN

Shank
Braise, Cook in Liquid

Spareribs
Braise, Broil, Roast

Boneless Rolled Breast
Roast, Braise

Riblets
Braise, Cook in Liquid, Broil

FORESHANK & BREAST

THIS CHART APPROVED BY
NATIONAL LIVE STOCK & MEAT BOARD

Rib Roast
Roast

Rib Chop
Broil, Panbroil, Panfry, Roast

Frenched Rib Chop
Broil, Panbroil, Panfry

Crown Roast
Roast

RIB

Square-Cut Shoulder, Whole
Roast, Braise

Pre-Sliced Shoulder
Roast, Braise

Boneless Shoulder Roast
Roast, Braise

Neck Slice
Braise, Cook in Liquid

Blade Chop
Braise, Broil, Panbroil, Panfry

Arm Chop
Braise, Broil, Panbroil, Panfry

SHOULDER

Lamb for Stew
Braise, Cook in Liquid

Cubes for Kabobs
Broil, Braise

Ground Lamb
Broil, Panbroil, Roast (Bake)

OTHER CUTS

Color Plate 3B Courtesy of the National Live Stock and Meat Board © 1986.

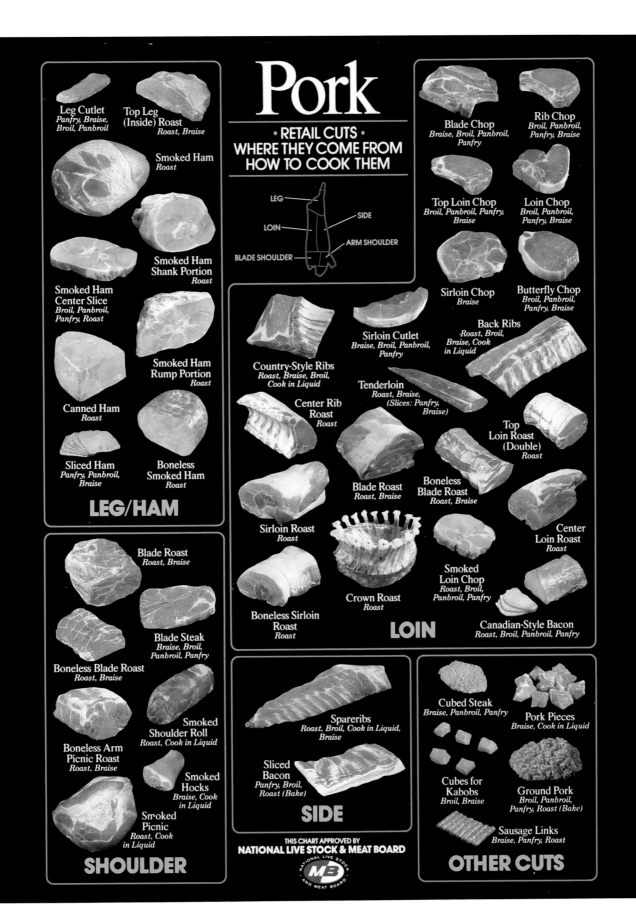

Pork

· RETAIL CUTS ·
WHERE THEY COME FROM
HOW TO COOK THEM

LEG
LOIN
SIDE
ARM SHOULDER
BLADE SHOULDER

LEG/HAM

Leg Cutlet
Panfry, Braise, Broil, Panbroil

Top Leg (Inside) Roast
Roast, Braise

Smoked Ham
Roast

Smoked Ham Shank Portion
Roast

Smoked Ham Center Slice
Broil, Panbroil, Panfry, Roast

Smoked Ham Rump Portion
Roast

Canned Ham
Roast

Sliced Ham
Panfry, Panbroil, Braise

Boneless Smoked Ham
Roast

SHOULDER

Blade Roast
Roast, Braise

Blade Steak
Braise, Broil, Panbroil, Panfry

Boneless Blade Roast
Roast, Braise

Smoked Shoulder Roll
Roast, Cook in Liquid

Boneless Arm Picnic Roast
Roast, Braise

Smoked Hocks
Braise, Cook in Liquid

Smoked Picnic
Roast, Cook in Liquid

LOIN

Blade Chop
Braise, Broil, Panbroil, Panfry

Rib Chop
Broil, Panbroil, Panfry, Braise

Top Loin Chop
Broil, Panbroil, Panfry, Braise

Loin Chop
Broil, Panbroil, Panfry, Braise

Sirloin Chop
Braise

Butterfly Chop
Broil, Panbroil, Panfry, Braise

Country-Style Ribs
Roast, Braise, Broil, Cook in Liquid

Sirloin Cutlet
Braise, Broil, Panbroil, Panfry

Back Ribs
Roast, Broil, Braise, Cook in Liquid

Center Rib Roast
Roast

Tenderloin
Roast, Braise, (Slices: Panfry, Braise)

Top Loin Roast (Double)
Roast

Sirloin Roast
Roast

Blade Roast
Roast, Braise

Boneless Blade Roast
Roast, Braise

Center Loin Roast
Roast

Crown Roast
Roast

Smoked Loin Chop
Roast, Broil, Panbroil, Panfry

Boneless Sirloin Roast
Roast

Canadian-Style Bacon
Roast, Broil, Panbroil, Panfry

SIDE

Spareribs
Roast, Broil, Cook in Liquid, Braise

Sliced Bacon
Panfry, Broil, Roast (Bake)

OTHER CUTS

Cubed Steak
Braise, Panbroil, Panfry

Pork Pieces
Braise, Cook in Liquid

Cubes for Kabobs
Broil, Braise

Ground Pork
Broil, Panbroil, Panfry, Roast (Bake)

Sausage Links
Braise, Panfry, Roast

THIS CHART APPROVED BY
NATIONAL LIVE STOCK & MEAT BOARD

Color Plate 3C Courtesy of the National Pork Producers Council © 1986.

Veal

• RETAIL CUTS •
WHERE THEY COME FROM
HOW TO COOK THEM

LEG (ROUND)
SIRLOIN
LOIN
RIB
SHOULDER
FORESHANK & BREAST

Rib Roast
Roast

Boneless Rib Roast
Roast

Crown Roast
Roast

Boneless Rib Chop
Braise, Panfry, Broil

Rib Chop
Braise, Panfry, Broil

Short Ribs
Braise, Cook in Liquid

RIB

Blade Roast
Braise, Roast

Arm Roast
Braise, Roast

Blade Steak
Braise, Panfry

Arm Steak
Braise, Panfry

Boneless Shoulder Arm Roast
Braise, Roast

Boneless Shoulder Eye Roast
Braise, Roast

SHOULDER

Boneless Rump Roast
Braise, Roast

Round Steak
Braise, Panfry

Top Round Steak
Braise, Panfry

Leg Cutlet
Braise, Panfry, Broil

LEG (ROUND)

Breast
Braise, Roast

Boneless Breast Roast
Braise, Roast

Cross Cut Shank
Braise, Cook in Liquid

Riblet
Braise, Cook in Liquid

Shank
Braise, Cook in Liquid

FORESHANK & BREAST

THIS CHART APPROVED BY
NATIONAL LIVE STOCK & MEAT BOARD

Loin Roast
Roast

Boneless Loin Roast
Roast

Loin Chop
Braise, Panfry, Broil

Kidney Chop
Braise, Panfry

Top Loin Chop
Braise, Panfry, Broil

Butterfly Chop
Braise, Panfry, Broil

LOIN

Sirloin Roast
Roast

Boneless Sirloin Roast
Roast

Sirloin Steak
Braise, Panfry, Broil

Top Sirloin Steak
Braise, Panfry, Broil

SIRLOIN

Veal for Stew
Braise, Cook in Liquid

Ground Veal
Panfry, Broil

Cubes for Kabobs
Braise

Cubed Steak
Braise, Panfry

OTHER CUTS

Color Plate 3D Courtesy of the National Cattlemen's Beef Association © 1986.

are cooked or incorporated into other foods.) Moreover, egg products are pasteurized to protect consumers against salmonellae. Because egg products vary in composition several pasteurization conditions are employed. For example, liquid whole egg is pasteurized at 140°F for 3.5 min. Salted (10 percent) egg yolks require higher temperatures to effect pasteurization. (As high as 154°F is recommended.) Dried egg white is pasteurized by holding at 130°F for 7 days. Glucose is removed from egg whites before drying to prevent browning caused by the interaction of certain amino acids with this reducing sugar. However, glucose-free corn syrup or sucrose may be added to some egg products to preserve whipping properties and improve storage stability. Quality assurance personnel routinely check all lots of egg products for the presence of salmonellae prior to shipment.

Spray drying is the most common method of dehydrating eggs. Stabilized egg white powder (the glucose is removed) has virtually unlimited shelf life. Dried products containing egg yolk are subject to lipid oxidation and often are refrigerated if stored for extended periods.

Both pasteurization and drying affect the foaming properties (reduce the cake volume) of egg white and whole egg. Freezing causes egg yolk to thicken (gelatinize). These changes are minimized by the addition of corn syrup, sugar, or salt prior to pasteurization. The nutritive value of egg products is not affected significantly by pasteurization or by freezing. However, minor nutrient loss may occur during dehydration and prolonged storage.

3.7.6 Providing Safe Milk Products

Milk and certain milk products are pasteurized to protect against pathogens. The temperatures and times appropriate to the *holding* and *high-temperature-short-time* (HTST) methods of milk pasteurization are 145°F for 30 min and 161°F for 15 seconds (sec), respectively. Because of the protection afforded bacteria by the higher-fat and milk-solids content, cream, ice cream, and certain other milk products are pasteurized at higher temperatures. These times and temperatures were established to provide a safe margin of time and/or temperature in the destruction of *Brucella abortus* (the bacterium that causes brucellosis), which may be shed into milk from an infected cow and transmitted to humans causing undulant fever; *Mycobacterium bovis,* which may be transmitted through milk and cause tuberculosis in humans; and *Coxiella burnetii,* which also may be transmitted through milk and cause Q fever in humans (Sections 22.3.2 and 22.3.3). Extended-shelf-life (ESL) products are being produced using a combination of ultrapasteurization (UP: 280°F = 138°C for 2 sec) and aseptic filling (without contamination) of containers.

Pasteurization might be termed the final protective measure for the consuming public. It is the major critical control point in the production of most milk products. All cows are routinely tested for tuberculosis and brucellosis. **Reactor** animals must be disposed of (Section 22.3.3). Close control is maintained over milk production, handling, and processing methods to prevent contamination with microorganisms and to prevent

multiplication of those that may get in. Bacteria normally found in well-protected milk are nonpathogenic and most are readily killed by pasteurization. Pasteurization does not significantly affect the vitamin content of milk (there is a slight reduction of vitamin C).

A substantial amount of milk is preserved and marketed in sealed cans as condensed or evaporated milk. These products have about 60 percent of the water removed. Sweetened condensed milk has approximately 45 percent sugar added, which aids in preservation (changes the **osmotic pressure**) as well as adding sweetness and food solids.

Vitamin Loss in Storage

Milk or milk products sealed in airtight, dark containers and stored for extended periods of time show little or no loss in vitamin potency. However, milk bottled in clear glass and exposed to sunlight for only 1 h may lose 30 percent of its vitamin A, a considerable amount of vitamin C, and may develop an oxidized flavor. Paper containers have essentially eliminated this problem in the United States; however, an oxidized flavor may develop in milk in translucent plastic containers exposed to a strong source of fluorescent light. The fluorescent-light-induced reaction destroys part of the riboflavin and vitamin C of milk. An orange filter covering the fluorescent light or orange pigment in the plastic bottle will prevent passage of the offending wavelength, which is between 400 and 500 μm.

3.7.7 Recent Trends in the Preservation of Animal Products

Advances in equipment design and fabrication have made more economical and desirable several new methods of preserving animal products, particularly dehydration. Cooperation between equipment manufacturers and health officials leading to the adoption of the 3-A sanitary standards for the manufacture of dairy equipment has been a major contribution. E3-A standards apply to egg-processing equipment.

Agglomeration

By a process of holding milk or whey in a nearly dry state and then completing the drying, agglomerates of powder can be produced that have excellent sinkability, dispersibility, and solubility properties. The result is "instant" powders. New products include combinations of milk with fruit or vegetable components.

Freeze-Drying (Lyophilization)

Commercial application of freeze-drying (**lyophilization**) to animal products is of recent origin; however, the Incas of South America dried potatoes by this process many centuries ago. In the winter months they spread their potatoes on the ground and let them freeze overnight. The next day's sun warmed the potatoes, but it did not thaw them because the temperature was cold at that altitude. Moisture in the ice crystals left the potatoes and sublimed into the air as water vapor. This evaporation was possible because of the high altitude and low air pressure (low vapor pressure). This type of drying is called *freeze-drying*. It simply means that the food is dried while it is frozen (just as a snowbank will sometimes disappear without melting). Today a vacuum (pressure of less than 1 millimeter of mercury, mmHg) is substituted for the low air pressure of the high mountains. Freeze-drying is a **sublimation** process that removes moisture from frozen products without appreciably changing their shape, color, or taste. *Sublimation* means that *ice in the food goes directly from solid to vapor, bypassing the liquid phase.* This process is based on Le Châtelier's second principle, which states that *a given mass of a substance occupies a larger volume as a gas than as a solid, and that if the outside pressure is below the vapor pressure, the equilibrium will shift from the solid to the gaseous state.*

The first lyophilized foods were marketed in about 1960. Today steaks, roast beef, pork chops, ham, poultry meats, scrambled eggs, cream cheese, and even entire meals are available in this form. Many "space foods" have been lyophilized so that quality, convenience, and safety from spoilage are ensured. Dried animal foods are presently used by the armed services, campers, food processors (as ingredients), restaurants, vending machines, hospitals, and consumers.

The foremost limiting factor in the widespread acceptance and use of lyophilized animal foods is cost. Poultry meat is expected to be the animal product lyophilized to the greatest extent during the next decade followed by red meats, dairy products, and fish.

Lyophilization is especially desirable for the preservation of egg proteins because coagulation is reduced to a minimum by freeze-drying. The high temperatures required for killing bacteria will coagulate egg albumen.

3.8 Antibiotics and Antibiotic Resistance in the Production of Foods from Animals

Millions of pounds of antibiotics are fed in subtherapeutic doses to U. S. livestock annually. This practice began in 1951. Current estimates are that nearly all poultry, 90 percent of the swine, and 70 percent of the cattle receive antibiotics in their feed or water. The limit on concentration of antibiotics in animal feed is 200 g per ton of feed.

Antimicrobials are used in animals to promote growth, prevent disease, and, therapeutically, treat bacterial infections. The prophylactic use of antibiotics is especially needed when animals are reared in confinement. As antibiotics stabilize the health of farm animals energy is diverted from building the immune system for fighting infections to building animal tissues. Increases in the estimated value of these animals to producers is billions of dollars annually because feed efficiencies increase by 10 to 20 percent. Moreover, the mammary glands of today's high-producing dairy cow are susceptible to invasion and infection by several different microorganisms. These infections often require treatment with antibiotics. Treatment during the dry period is usually more effective in eliminating an infection than is treatment during lactation. Furthermore, risks that antibiotics will get into milk are minimized by application during the dry period. The economic value of antibiotics for treatment of mastitis is illustrated by a culling rate of approximately 30 percent in cows of treated "traditional herds" versus nearly 50 percent in "organic herds" in which antibiotic therapy is not practiced.

When animals are healthy the probability they will pass pathogenic bacteria to consumers of animal products is minimized. Proper antibiotic use leads to healthy animals. However, this gain in efficiency and disease prevention may come at the price of development of resistant bacteria within treated animals. This has been the subject of much debate for several years. The World Health Organization concluded in 1997, "*Antimicrobial use leads to selection of resistant forms of bacteria in the ecosystem of use. This will occur with all uses including treatment, prophylaxis, and growth promotion.*" Three adverse consequences of selecting resistant bacteria in animals were noted: (1) increased prevalence with increased risk of transfer to humans via direct contact or consumption of food or water, (2) transfer of resistance genes to human bacteria, and (3) more human infections that are refractive to antibiotic therapy.

The risk to humans is that the reservoir of resistance genes in bacteria from farm animals will increase and that resistant pathogens will be passed to humans and cause disease. If resistant pathogens are present in animal products there is some risk that they will not be killed during food preparation and will be consumed. If the consumer does not have sufficient immunity to the invading bacteria, disease may result. Because passage of resistant pathogens to humans would be a stepwise process, blocking the process at any step up to establishment of disease will prevent illness. This emphasizes the importance of careful food preparation.

In the early days of feeding antibiotics to food animals the drugs of choice were limited to penicillin G and the tetracycline antibiotics, Aureomycin® and Terramycin®. However, these also were the antibiotics of choice to treat human infections. Today many more antibiotics and derivatives of the earlier ones have been discovered. This permits selection of some antibiotics to use as feed additives for animals. However, development of resistant bacteria in the animal population presents a formidable challenge to scientists to develop replacement drugs for use in treating infections.

Bacteria vary in their natural resistance to antibiotics. For example, Gram-positive bacteria are generally sensitive to penicillin G because its mode of action is to attach to the bacterial cell wall and to lyse (disintegrate) the cell. Gram-negative bacteria have a protective layer of lipid on their cell surface and so are protected from attachment of penicillin molecules. Streptomycin, on the other hand, is more effective against Gram-negative than Gram-positive bacteria.

There are many resistance factors. Among them is the production of enzymes that destroy antibiotic molecules. The best known of these is penicillinase or β-lactamase, the enzyme that hydrolyzes the β-lactam ring structure of penicillin G rendering the antibiotic inactive. Most strains of *Staphylococcus aureus,* a bacterium that was generally sensitive to penicillin in early days of its use, have now acquired the genes to produce this enzyme and have become penicillin resistant.

Exposure of bacteria to marginally lethal doses of antibiotics tends to kill many cells but to permit growth of a subset of the population that is resistant. Long-term use of a single drug may, therefore, favor colonization with the more-resistant bacteria. Complicating the issue is the potential for transmission of resistance genes, R-factor, from one bacterium to another. The resistance gene does not have to exist in a pathogenic cell to be passed to a drug-sensitive pathogen. It can be contained in any bacterium capable of transferring genes with a pathogen. Fortunately, passage from a cell of one species to a cell of another is generally limited to bacteria of similar genetic makeup. These resistance genes may become incorporated in the chromosomal DNA or may reside in extrachromosomal DNA known as plasmid DNA.

Another factor to consider is the potential for bacteria to mutate, that is to change their genetic makeup rather quickly. Bacteria grow rapidly, sometimes duplicating themselves in as little as 10 min. At the more usual generation time of 30 min, a single cell can grow to 1 million within 10 h and to 1 billion within 15 h. If the rate of nonlethal mutation is 10^{-7}, 100 mutants would be produced among these cells.

Whether resistance is natural, transferred, or a result of mutation, presence of an antibiotic that kills much of the flora provides opportunities for resistant cells to grow far beyond their normal capacity: the fittest survive and rise to prominence.

The subject of acquired bacterial resistance to antibiotics today is focused on three pathogens common to animals and humans: *Salmonella, Campylobacter* and enteropathogenic *Escherichia coli.* Elderly, infants, and immunocompromised persons are at greatest risk of infection with these bacteria.

In 1996 the FDA established the National Antimicrobial Resistance Monitoring System (NARMS) for veterinary isolates of *Salmonella.* It calls for tests of minimal inhibitory concentrations (MICs) for 17 drugs. The first report was issued in April 1998. *Salmonella* were isolated from cattle, swine, chickens, turkeys, horses, and a few from exotic animals and pets. Among the more than 2000 strains tested *resistant isolates* ranged up to 27 percent. Isolates from turkeys showed the most and those from cattle the least resistance. Antimicrobials to which there was the greatest resistance were tetracycline, 27.4 percent; sulfamethoxazole, 17.7 percent; streptomycin, 17.6 percent; ampicillim, 12 percent; ticarcillin, 11.6 percent; and kanamycin, 10.1 percent.

In January 1997 U.S. government–sponsored activities were initiated and expanded to improve the safety of the U.S. food supply. The FDA's Center for Veterinary Medicine (FDA/CVM) is involved in surveillance and research activities of the Food Safety Initiative (FSI).

In 1976 the European Community banned feeding to animals any antibiotic used in human medicine. According to the Animal Health Institute this ban on "preventive use" of antibi-

otics brought no measurable reduction in the incidence of antibiotic-resistant infections in humans in England. In the 1990s avoparcin was fed to animals in some European countries while vancomycin (a similar drug) was used sparingly in hospitals. Avoparcin was not fed to animals in the United States but was used widely in hospitals. Consequently, vancomycin-resistant enterococci (VRE) have been prevalent in U.S. hospitals but sparse in hospitals of Europe. Conversely, significant numbers of VRE have been isolated from human feces and chicken carcasses in Europe, but not from feces of humans in communities or from animals or meat products in the United States.

Considerable information about vancomycin and avoparcin has been gained from experiences in Denmark. The Danish Veterinary Laboratory concluded that "use of the glycopeptide avoparcin as a growth promoter had created in food animals a major reservoir of *Enterococcus faecium,* that contain the high level glycopeptide resistance determinant vanA, located on the Tn 1546 transposon." This transposon can be transferred among enterococci in the intestinal tract and from *Enterococcus* to *Staphylococcus aureus.* Vancomycin is often the only antibiotic clinically effective against *E. faecium.* The ban on feeding avoparcin in Denmark resulted in a drop in occurrence of VRE in poultry from 82 percent to 12 percent at slaughter, with a similar decrease in Germany. The human carrier rate dropped from 12 percent to 3 percent over 3 years.

One of the newest classes of antibiotics developed for treating infections in people and animals is the fluoroquinolones. Fluoroquinolones kill bacteria by inhibiting bacterial DNA synthesis (specifically the enzyme DNA gyrase). FDA gave approval in 1995 for sarafloxacin, a fluoroquinolone, to be used in drinking water for the prophylactic treatment of *E. coli* in poultry. A second fluoroquinolone, enrofloxacin, was approved for use in chickens and turkeys in 1996. However, shortly after the approval of these drugs in poultry, an increased incidence of fluoroquinolone-resistant *Campylobacter* infections was noted in humans. In October 2000 the FDA/CVM proposed to withdraw approval of the use of enrofloxacin in poultry. The manufacturers of sarafloxacin voluntarily withdrew production of this drug for use in poultry.

Although there continues to be much concern about buildup and transfer of resistance because of nontherapeutic use of antibiotics in food animals, it is important to consider the following conclusions from the 1998 report of the National Research Council Institute of Medicine entitled "The Use of Drugs in Food Animals: Benefits and Risks": (1) the use of drugs in the food-animal industry does not appear to constitute an immediate public health concern, (2) the incidence of human disease linked to use of antibiotics in food animals is very low, and (3) constant vigilance in monitoring trends in antibiotic resistance in farm animals and humans is strongly encouraged.

In addition to the potential for antibiotic-resistant organisms to arise because of on-farm use of antimicrobial drugs there has been a consistent potential for human illness due to consumption of antibiotic-contaminated animal products. The danger is primarily to people with allergies to drugs. Approximately 2 percent of the U.S. population is allergic to penicillin. Consumption of even minute quantities of penicillin can lead to severe reactions (**anaphylaxis**) and even death of penicillin-allergic individuals. Lack of controls and improper use of antibiotics to treat farm animals in the early days led to a high incidence of antibiotic-contaminated milk. Of the milk samples tested in the United States in 1962 and in Britain in 1963, 12 percent and 11 percent, respectively, were found to contain microbial growth inhibitors, but by 1988 less than 1 percent of all animal products tested in the United States contained actionable concentrations of inhibitors.

Milk-producing animals suffering from mammary gland infections are treated by infusions of antibiotics directly into the infected glands. Milk from all quarters of treated cows must be withheld from sale, usually for 3 days. The antibiotic(s) chosen should be determined by tests of antibiotic sensitivity of the infecting bacteria.

Each tanker load of milk delivered to dairy plants in the United States is tested for presence of antibiotics. Any positive test must be reported to regulatory authorities who will be responsible for retesting to confirm the finding. All samples from individual producers are tested and a positive test stops sale of milk from the offending farm until the cause is determined and corrected. Dairy farmers have a special concern about increases in resistance among mastitis bacteria because the more resistant a bacterium the higher the dose needed to treat an infection by it, and the resultant longer withholding time. Furthermore, this increases the risk that antibiotics may get into the milk supply and calls for tighter rules in the use of antibiotics for treating infected cows.

3.9 GRADING ANIMAL PRODUCTS

Quality grades of meat, milk products, and eggs are based on chemical, physical, and sensory characteristics. Although they are related to safety, they primarily reflect quality. Governmental agencies determine quality grades based on criteria established jointly by the producing industry and the responsible component of the Agricultural Marketing Service (AMS) of the USDA.

3.9.1 Egg Grades and Grading

Grade is determined by examining the external and internal quality of the egg at the time it is packed for sale as a fresh egg. Grades AA, A, and B reflect the cleanliness, soundness, texture, and shape of the shell plus firmness of the white and yolk, size of the air cell, and freedom from blood spots caused by minor hemorrhaging during the laying process. All eggs must be clean, but a small amount of stain is permitted on grade B eggs. Color of the shell is a breed-specific characteristic and does not relate to egg quality.

Interiors of eggs are observed in a process called "candling" in which they are rolled over a strong light while an experienced grader watches for defects such as cracked shells and blood spots. Also observed are size of air cell, shape of the yolk, and viscosity of white. The latter is reflected in the ability of the white to keep the yolk centered as the egg is twirled. The air-filled pocket at the large end of the egg is the "air cell." When first laid the air cell is very small or nonexistent. However, as the egg cools from about 105°F to ambient temperature, the internal contents shrink more than does the shell. This causes the two shell membranes to separate and allows air and CO_2 to accumulate in the cell. As eggs age CO_2 evolves and the pH of the albumen drops from 9.3 to as low as 7.4. Air cell depth of eggs in grades AA and A should not exceed 1/8 and 3/16 in, respectively.

Supplemental to candling is the breaking of randomly selected eggs onto a level surface for measurement with a micrometer of height of the yolk and thick portion of the white. The content of a grade AA egg covers a small area while the yolk stands tall and is surrounded by much thick white and little thin white. The grade B egg has an enlarged and flattened yolk and the white appears weak and watery. The grade A egg has characteristics in between those of grades AA and B. Color of the yolk reflects the type of feed consumed by the hen. Feeds, such as alfalfa meal and yellow corn containing high amounts of xanthophylls, produce dark yellow yolks. Excessive amounts of cottonseed meal in rations produce an olive-brown mottling color of yolk because of an interaction of gossypol of cottonseed with iron in the yolk.

3.9.2 Sizing Eggs

Size classification is independent of grade. Size is determined by weight per dozen eggs (weights of individual eggs may vary within limits.). In descending order egg sizes are jumbo, extra large, large, medium, small, and peewee. The most common sizes laid by hens are medium, large, and extra large. Size classes differ by 3 ounces per dozen with a dozen large eggs weighing at least 24 ounces. When a dozen large eggs cost $1.00, the equivalent value of medium and extra large sizes differs about $0.12 per dozen while that of small and jumbo sizes differs by about $0.25. (Subtract if size is smaller and add if size is larger.)

3.9.3 Grading Milk Products

Although fluid milk products bear the grade A label this grade reflects the sanitary care taken in production rather than being the traditional indicator of quality at the time of packaging. Regulations for the production of grade A milk are recommended by the FDA in conjunction with the Interstate Milk Shipper's organization. States and municipalities have adopted these regulations and enforce them so milk can move freely among states without regulators in receiving states having to inspect facilities and test milk and its products of shipping states. This practice is called "reciprocity." In brief, the regulations define milk and its products, and they set standards for numbers of bacteria and somatic cells, storage temperatures, freedom from adulterants, acceptable design and construction of equipment, and the pasteurization process. Inspections and tests are conducted periodically and without notice to the producing entity, whether it be at the farm or the processing plant. Permits granted to each producer or processor are suspended or revoked when regulations are violated.

Grading manufactured dairy foods, on the other hand, is quality oriented and done voluntarily under contract by the manufacturer of butter, cheese, or dry milk products with the AMS. Grading is done at the producing facility. In addition to the quality attributes observed, the agency requires raw milk used to make these products be produced under sanitary conditions and that standards for content of bacteria, somatic cells, and adulterants be applied. These standards are similar to those applied to grade A milk except the bacterial count is permitted to be higher. Furthermore, requirements for construction of facilities at the farm are less rigorous. States adopt the USDA's "Recommended Requirements for Milk for Manufacturing Purposes and its Production and Processing." To be approved for grading service a plant must be shown by inspection to meet the requirements of USDA's "General Specifications for Approved Dairy Plants." This inspection informs management about the quality of raw material, adequacy of sanitation practices, condition of the plant and equipment, and suitability of processing procedures.

Grades are based on nationally uniform standards developed by the USDA Dairy Programs' experts in cooperation with industry representatives. Sellers can request grading services to assure that products meet specific grade or contract requirements and have good keeping-quality properties. Buyers can request grading services to assure that products have uniform high quality. Those wishing to use the services must request, qualify for, and pay a fee commensurate with the cost of providing them. Two examples of such services are the grade label program for butter and Cheddar cheese and the acceptance service for volume buyers. Under the grade label program consumer packages bear an official identification indicating the U.S. grade. As a part of the acceptance service the user has all deliveries examined by the USDA to certify that they meet specifications.

Milk products sold to the federal government are inspected by AMS dairy graders. The stocks of government-owned dairy products are inspected periodically to ensure that quality has not deteriorated during storage.

Almost all dairy products can be graded but the service is used most widely for butter, Cheddar cheese, instant nonfat dry milk, and regular nonfat dry milk. Inspectors also grade other cheeses, dry whey, dry buttermilk, and dried and condensed milk. The following descriptions of grades of various products are from the AMS web page under the following URL: http://www.ams.usda.gov/dairy/grade.htm.

There are three grades for butter: U.S. Grades AA, A, and B. The ratings are assigned on the basis of flavor, body, and color. The quality of the cream from which the butter is made determines the flavor factor in assigning grades.

There are four grades for Cheddar cheese: U.S. Grades AA, A, B, and C. As with butter, all grades may be used in the wholesale trade but only the top grade is used at the retail level. To rate top grade cheese must have a consistently fine Cheddar

flavor. In addition, there are grades for Swiss cheese, Emmentaler, Colby, Monterey (Monterey Jack), and bulk American cheese for manufacturing.

If instant nonfat dry milk meets the standard for quality it may carry the U.S. Extra grade shield. This means that laboratory tests show that it possesses a sweet and pleasing flavor, a natural color, and satisfactory solubility. USDA inspectors also check the instant milk for other quality factors such as moisture, fat, bacteria, scorched particles, and acidity.

Dry buttermilk and regular nonfat dry milk, which are sold in bulk to producers of ice cream, bakery products, and some processed meat processors, can be graded either U.S. Extra or U.S. Standard. The lower-grade "Standard" may be the result of excess moisture or scorched particles from the drying process or other quality factors.

The grades of U.S. Extra and U.S. Standard for dry whole milk are based on quality factors like those for other dry dairy products. Grade requirements for dry whole milk also include a maximum bacteria content. Bacteria limits are designed to ensure a safe product that has good keeping quality.

Dry whey—a co-product in the making of natural cheese— is tested for flavor, appearance, amount of milkfat, and moisture. It must have a good, sweet taste to earn the U.S. Extra grade. Whey of this top quality is desired by manufacturers because it is used as an ingredient in other foods.

For cottage cheese, processed cheese, cream cheese, or any other dairy product for which no U.S. grade standards have been established there is a USDA program for official quality approval. Such products may earn the "Quality Approved" rating, which is based on a USDA inspection of the product and the plant where the product was made. The product must be wholesome and measure up to a specific level of quality to earn the rating. The "Quality Approved" shield may be used on retail packages.

3.9.4 Grading Meat Products

The Meat Grading and Certification branch of the AMS uses university-researched, USDA-developed, and industry-recognized standards. Grading determines the quality and yield of carcasses. Quality grades vary depending on the species.

Table 3.11 shows the species graded, the quality grades of each, and yield grades where used. Retail cut charts of beef, veal, pork, and lamb are depicted in Color Plates 3A to 3D.

Official U.S. Standards for Grades have been published for the following: vealers and slaughter calves, feeder cattle, slaughter cattle, feeder pigs, slaughter swine, slaughter lambs, yearlings, and sheep, and for carcasses of veal, calves, beef, pork, lamb, yearling mutton, and mutton. To view these see the following web page: http://www.ams.usda.gov/lsg/stand/st-pubs.htm#Official.

3.9.5 Quality Standards for Beef

There are five yield grades for all classes of beef with yield grade 1 representing the highest degree of cutability. The yield grade of a beef carcass is determined by considering four characteristics: (1) amount of external fat; (2) amounts of kidney,

TABLE 3.11	United States Grades of Meat by Species	
Species	**Applicable Quality Grades**	**Yield Grades**
Beef	Prime, Choice, Select, Standard, Commercial, Utility, Cutter, and Canner	1–5
Lamb and yearling mutton	Prime, Choice, Good, Utility, and Cull	1–5
Mutton	Choice, Good, Utility, and Cull	1–5
Veal and calf	Prime, Choice, Good, Standard, and Utility	N/A
Barrows and gilts	U.S. No. 1, 2, 3, 4, 5, and Utility	N/A
Sows	U.S. No. 1, 2, 3, Medium, and Cull	N/A

Source: MGC web page at http://www.ams.usda.gov/lsg/mgc/grade.htm.

pelvic, and heart fat; (3) area of the ribeye muscle; and (4) carcass weight.

Carcasses are split in half down the back and one or both sides are separated into forequarter and hindquarter portions for grading. The purpose is to expose the lean muscular structure of the ribeye and the distribution of lean and fat between the twelfth and thirteenth ribs. For steer, heifer, and cow beef, quality of the lean is evaluated based on marbling and firmness as observed at the cut surfaces in relation to evidences of maturity. Maturity is determined by evaluating the size, shape, and ossification of bones and cartilage, especially the chine bones (backbones), and the color of the lean flesh as well as size and shape of rib bones. Except for bullock carcasses, to which only the five maturity classes are applicable, grading standards set five levels of maturity and seven levels of marbling ranging from practically devoid to slightly abundant.

The USDA also offers the following services for beef:

Beef Carcass Data Service (BCDS) facilitates the flow of data on carcass quality and yield, thus on carcass value, to cattle producers and feeders. BCDS provides producers and feeders a valuable management tool to use in their selection and feeding programs.

Beef Carcass Information Service (BCIS) provides information to feeders and others who buy or sell cattle on a formula or carcass grade and yield basis. This information is provided on a lot rather than an individual animal basis.

Beef Carcass Evaluation Service (BCES) is normally used by collegiate and university researchers to evaluate quality factors such as texture—firmness and color of lean, texture of marbling, bone and muscle maturity—and yield grade factors including ribeye tracings. Other data collection programs have been developed for specific customers to assist in genetic selection and value-based marketing systems.

3.9.6 Quality Standards for Swine

There are five classes of pork carcasses, comparable to the same five classes of slaughter hogs: barrow, gilt, sow, stag, and boar. Standards for grading of carcasses apply to barrows, gilts, and sows but not to stags and boars. Grades for barrow and gilt carcasses are based on (1) characteristics of the lean and fat and (2) expected yield of the four lean cuts: ham, loin, shoulder, and Boston butt. The levels of quality are "acceptable" and "unacceptable." Quality of the lean is determined by viewing the cut surface of the loin at the tenth rib. To be acceptable this surface must be slightly firm, have a slight amount of marbling, and be grayish pink to moderately dark red in color. Bellies must be thick enough (at least 0.6 in at any point) to produce bacon of acceptable quality. Expected yields of and yield grades for "acceptable" chilled carcasses are:

Grade	Yield
U.S. No. 1	60.4 percent or more
U.S. No. 2	57.4 to 60.3 percent
U.S. No. 3	54.4 to 57.3 percent
U.S. No. 4	Less than 54.4 percent

The above yields are based on use of cutting and trimming procedures of the USDA.

Grades of barrow or gilt carcasses are determined by considering the backfat thickness over the last rib and the thickness of that muscle in relation to skeletal size. Pork carcasses with average fatness have 1.1 to 1.2 in of fat over the last rib. Each 0.1-in change in backfat thickness over the last rib changes the grade by 40 percent of a grade. Barrows and gilts have three degrees of muscling, namely, thick (superior), average, and thin (inferior). The mathematical equation for determining barrow and gilt carcass grade is:

Carcass grade =
$$(4.0 \times \text{in of backfat thickness})(1.0 \times \text{muscling score})$$

The following is an example of the rules applied:

U.S. No. 1 barrow or gilt carcass: Carcasses in this grade have an acceptable quality of lean and belly thickness and a high expected yield (60.4 percent and over) of four lean cuts. They must have less than average backfat thickness over the last rib with average muscling, or average backfat thickness over the last rib with thick muscling.

3.10 THE FUTURE OF ANIMAL PRODUCTS

With the world population increasing at the annual rate of approximately 86 million, and the nutritional level of a large percentage of the populace presently below desired standards, the projected need for animal products presents both a challenge and an opportunity to all involved in activities related to their care and production. Present international trends toward increased consumption of animal products will continue as the economic conditions of the world's peoples become more favorable. Efficiencies of converting feed nutrients into animal products will continue to improve. This will result in a reduction in cost to consumers (relative to wages earned and per capita real income) and an increase in the consumption of animal products. These trends support an optimistic outlook for the producer, processor, distributor, and consumer of animal products. We expect increased nationwide, indeed worldwide, emphasis on human nutrition. This has important implications for animal products, because the very dietary essentials so frequently deficient in human diets—protein, calcium, riboflavin, and iron—are found in abundance in many animal products.

In a recent USDA study of 14,500 U.S. men, women, and children the nutrients most commonly found below recommended dietary allowances were calcium and iron. (They averaged more than 30 percent below recommended allowances set by the Food and Nutrition Board of the National Academy of Sciences–National Research Council.) Iron in the diets of infants and children under 3 years of age was about 50 percent below recommended amounts. Diets of adolescent girls and of women were below recommended levels of calcium, iron, and thiamine and for some age groups below recommended levels of vitamin A and riboflavin. Older men had diets low in calcium, vitamin A, riboflavin, and vitamin C.

The above study clearly indicates a great need for additional consumer education in human nutrition. In this statement it is assumed, of course, that more persons in the United States are malnourished because of nutritional ignorance and misinformation than because of poverty, although isolated instances of the latter also prevail. Nonetheless, increased intake of animal products could alleviate the deficiencies observed.

The future of animal products is interwoven with the costs of production, processing, and distribution. Recent trends have been toward increased processing, longer storage, and further shipment of foods. Yet these activities require additional energy. Increases in energy costs could impede the trend toward more convenience foods.

The prospects of increased food formulation and the so-called **nutraceutical** foods may significantly affect future consumption of animal products. Several supermarket chains are selling hamburger mixed with vegetable proteins made from soybeans. These mixtures usually have a lower fat content than regular hamburger and cost less. The mixture tends to bind and hold more moisture than does pure ground beef. The competition between soybean products and animal fats, as well as between soybean products and certain meat products, in supermarkets is of great interest to livestock producers (and to soybean growers) and will be important to the future of animal products. Researchers at Utah State University reported using textured whey protein as an additive to ground beef for making burgers. Compared with soy protein used at the same 40 percent replacement rate, those containing whey protein were superior in flavor and texture.

In the end, of course, decisions of consumers will govern the demand for and future of animal products.

> Unless our product has consumer demand, we could well be sitting on a bushel of gold and starve to death.
>
> **John A. Moser**

If livestock producers continue to increase production efficiency, breed leaner-type meat animals, and market milk having a higher protein/fat ratio, the future looks bright; however, failure to meet consumer demands will encourage the increased use of synthetic animal products. Changes in food habits within the United States are frequently brought about by advertising and promotion. Thus, producers of animal products may enhance demand for their products through increased advertising. The automobile industry spends approximately 10 times more on the research and development of new models than the entire food industry invests in new product development.

3.10.1 New Animal Products

Technological advances expected to expand the use of present and new dairy foods include development of extended-shelf-life products, new formulations of yogurt and other dairy products containing healthful bacteria (*Bifidobacterium, Lactobacillus reuteri* and others) plus fructooligosaccharides (FOS) to make these friendly bacteria grow well in the human intestine. More flavored cheeses will come to market as the invention of Utah State University researchers that injects flavorings into prepared cheeses is used. Whey protein concentrates will be used increasingly. Body builders will select whey proteins because they contain the highest concentrations of the branched chain amino acids leucine, isoleucine, and valine of any protein source. Specialty proteins will be marketed. Individuals with phenylketouria can get peptides devoid of phenylalanine by selecting concentrates of glycomacropeptide, the portion of kappa-casein that is liberated by the enzyme chymosin or rennin during the coagulation of casein in cheese making. Lactoferrin is likely to be used in pharmaceutical products because of its ability to inhibit certain bacteria.

Prepackaged, portion-controlled, fully prepared foods are perhaps the ultimate in *convenience* and are sought by many American consumers. Many fresh meats have been converted into frozen form and others are packaged in a vacuum or controlled atmosphere to improve storage life. Sausage products and luncheon loaves (table-ready meats) will increase. There is an increasing trend toward ready mixes, bake (heat) and serve, and instant types of animal products that can be prepared by microwave heating. New types of ripened low-fat cheeses are being developed and marketed. Instant milk and cheeses, dry butter, smoked cheeses, frozen whipped cream, and cottage whey sherbet are of relatively recent origin. More animal food products will follow.

Restructured Meat Products

Sales of restructured meat products have increased. Meat products commonly referred to as restructured include flaked and formed, sliced and formed, and sectioned and formed. Such products, texturewise, have characteristics somewhere between those of ground meats and intact muscle cuts. They have the advantages of (1) resembling the preferred whole muscle products such as cutlets, chops, and steaks and having similar storage stabilities; (2) being less expensive than whole muscle products; (3) having controlled fat content; and (4) better utilizing secondary carcass parts. Restructured meat products received high acceptability in consumer tests.

Processed Meats

The per capita consumption of processed meats increased from 42 lb in 1930 to 61 lb in 1980. By 1995, 60 to 70 percent of the pork and 15 percent of the beef and poultry in the United States were sold in processed forms. These trends are expected to continue.

Frozen Scrambled Egg Mixes

The traditional scrambled egg is a mixture of yolk and white with milk frequently added. This simple product is satisfactory for immediate consumption. However, many uses of such a product are more demanding, and special mixes have been developed for these purposes.

Institutions such as dormitories and cafeterias require a longer holding period between cooking and consumption. The two most common problems with holding scrambled eggs on the steam table are (1) green discoloration, which is caused by a reaction of hydrogen sulfide formed with iron during cooking to form the green ferrous sulfide; and (2) weeping (properly termed *syneresis*), which is the separation of liquid from the coagulated egg protein after cooking.

The addition of selected ingredients improves scrambled egg mixes for institutional uses. Scrambled eggs made from these mixes have improved color and textural characteristics. Each ingredient added provides a useful function. For example, the most common additions include (1) nonfat dry milk, which improves the nutritional quality by supplying additional calcium and phosphates to provide better color retention; (2) vegetable oil, which produces a softer scrambled egg coagulum; (3) gums, which reduce weeping by binding water; and (4) citric acid and/or phosphates, which reduce the pH and chelate (bind) iron to prevent greening.

Cooked-Frozen-Thawed-Reheated Egg Products

Omelets, quiches, scrambled eggs, and soufflés are examples of frozen egg products included in this category. They all contain cooked egg white. Unfortunately, when cooked egg white is frozen the gel-like structure is destroyed by the formation of ice crystals and the liquid separates from the coagulum. To demonstrate this freeze damage the reader is encouraged to freeze and thaw a hard-cooked egg. Note that the egg white is laid down in layers. Two methods are used to overcome this textural defect. First, freeze-stable starches can be added to the raw egg white to bind the watery liquid. This is the technique used in the "long egg" produced by the Ralston Purina Company. This "egg" is about 1 ft long, about 1.5 in in diameter, and has a yolk center. A

second method is to rapidly freeze the product at a low temperature by using liquid nitrogen, carbon dioxide, or Freon. The technique of fast freezing retards ice crystal growth and reduces freeze-thaw damage.

Hard-Cooked-Peeled Eggs

Hard-cooked eggs were probably the first processed egg product consumed by humans. For example, a forest fire could have cooked the eggs. Cooking an egg improves the flavor and aroma compared with those of a raw egg.

Since about 1970 hard-cooked-peeled eggs have been marketed commercially. Some institutions use only these prepeeled, hard-cooked eggs. They are commonly distributed in 5-gallon polyethylene buckets. The shelf life under refrigeration is commonly 1 to 2 months.

Normal cooking procedures are used to cook the eggs in the shell. If they are cooked too long at a high temperature the surface of the yolk will turn green. This results from the reaction of hydrogen sulfide (produced in the white during cooking) with iron from the yolk to form iron sulfide. After the eggs are peeled they are packed in a solution containing sodium benzoate and citric acid. To users of this product there is a savings in labor and energy and no loss because of incomplete or ineffective peeling.

3.10.2 Synthetic Animal Products

During the past half century the dairy industry lost much of its butter market to the more economical product, margarine. More recently **filled milk** and **imitation milk** have been marketed in several states. Filled milk has nonfat milk solids as the nonfat base and vegetable oil as a source of fat. Imitation milks are produced from vegetable protein or sodium caseinate, vegetable fat, corn syrup solids (or other sugars), stabilizers, and emulsifiers. Most imitation milks have been lower in protein, minerals, and unsaturated fatty acids and higher in carbohydrates and **saturates** than cow's milk. More synthetic foods that compete with animal products are certain to follow. Yet it should be noted that the interrelationships existing among food nutrients are an important consideration. The nutritional merits of milk as a whole, for example, are greater than the total individual values of its components. The same is true of most other animal products.

The use of meat substitutes is not new. The protein mainstay of the German army during World War II was not meat but rather "brattling," a soybean–skim milk sausage. Moreover, peanut and cottonseed meals (after the oil had been extracted) were used in Germany during World War II as a meat substitute in the form of sausage. More recently considerable interest has been generated in the production of synthetic meats. The technology of spinning fibers from protein powder and texturizing it by extrusion makes it possible for simulated meats to be produced from vegetable proteins. Thus far these products are made mainly in dry or frozen form and have a longer shelf life than chilled meats.

Figure 3.5 Synthetic turkey loaf of flavored spun soybean protein. *Worthington Foods, Inc., Worthington, OH.*

Isolated soybean protein has been fabricated into meatlike cuts (Figure 3.5) and is currently available for consumption. Soy concentrates are used in certain manufactured products not requiring the texture of meat and where a deviation in flavor would not likely be noticed. It is estimated that more than 50 million people in the United States do not eat meat some or any of the time for reasons of religion, doctor's recommendations, or personal preference. Eating nonmeat foods resembling frankfurters, hamburgers, pork sausage, dried beef and roast beef, fried chicken, turkey loaf, and other items may appeal to this group. Current regulations allow the addition of soy protein concentrate up to 3.5 percent in properly labeled sausage products.

The retail cost per pound of protein has not generally favored soy-based meat analogues over natural animal products. Soy flour is higher in crude protein than meat but it is not utilized as well in the body and has a somewhat less nutritionally desirable amino acid profile than meat protein. However, the pressing need for additional protein throughout the world has stimulated research in the production of amino acids from organic molecules in the hope that quality protein can be produced economically on a mass basis.

3.10.3 Pesticides and Animal Products

Prior to the twentieth century the consumer knew the producer personally. If anything was wrong (sour milk, cracked eggs, or other questionable standards of quality) the difference was quickly resolved or future sales were lost. Today the public expects (and rightly so) agriculture to provide foods that are wholesome and safe. This means food is free from harmful pesticide residues, disease agents, or other toxic substances.

Pesticide chemicals have played a significant role in increasing agricultural productivity both in the United States and worldwide. About 5 lb of pesticide is applied per person annually for pest control in the United States. Uptake of pesticides by animals, leading to residues in animal products, can result either from direct application of the pesticide to an animal or from the animal ingesting feed carrying pesticide residues.

Concentrations of pesticides and other agricultural chemicals in animal products (and in other foods) are monitored closely by governmental agencies. Based on research findings to date, present constraints on tolerable amounts of these organic chemicals in foods are more than adequate to assure consumers of safe and wholesome animal products.

3.11 SUMMARY

The nutritive value of food is a positive factor in determining the health and quality of life from conception until death. Provisions for adequate nutrition in early years contribute to a more productive and enjoyable life in later ones. Most animal products are rich in high-quality protein and relatively low in calories. Animal products also make important contributions to humanity's needs for fats, carbohydrates, minerals, and vitamins. These dietary essentials are available through animal products at a reasonable cost to consumers.

For many centuries domesticated farm animals have made a major contribution to human welfare as sources of food, clothing, transportation, power, soil fertility, fuel, and pleasure. The most important of these is food. Foods from animals are the most nearly complete foods nutritionally. In the United States they provide more than half the dietary protein, calcium, phosphorus, and several vitamins. Animal proteins are of high quality because they contain the amino acids required for human nutrition in proportions similar to those in the human body. Meat and poultry are important sources of iron.

As "natural" foods, animal products effectively balance and supplement plant foods. Meat, milk, and egg proteins are high in biological value and provide essential amino acids often deficient in plant foods. Additionally, milk and milk products are rich in calcium and riboflavin; meats are rich in iron, niacin, and nicotinic acid. Milk and eggs also are important sources of vitamin A. Moreover, animal foods contribute to palatability and variety in the diet of humans.

Improved methods of food preservation increase the availability of animal products, and fortification ensures a more uniform and nutritious product composition. Mixtures of foods from animals and plants, especially fabricated foods, are likely to become more important in the diets of Americans and will provide opportunities for the development of new foods and new menus.

Modern production, processing, packaging, and distribution practices, coupled with close supervision from regulatory agencies, assure today's consumers of having animal products that are safe in terms of disease.

STUDY QUESTIONS

1. What is the relation of the consumption of animal products to the following characteristics of a nation, its people, or both: (a) health, (b) life expectancy, (c) progressiveness?

2. Cite an example of how diet influenced the physical stature of a nation's people.

3. Why are animal proteins superior to vegetable proteins for humans and other monogastric animals?

4. Corn is especially deficient in which 2 *essential* amino acids?

5. What are the *essential* fatty acids for humans? Are they found in animal products?

6. Are animal products good sources of carbohydrates?

7. Why does honey taste so sweet? Is it heavier or lighter than water?

8. What are several uses of honey?

9. What state leads in the production of honey?

10. What is the function of (a) the queen bee, (b) workers, and (c) drones?

11. Of what importance is royal jelly to the honeybee colony?

12. How much honey does the average worker bee gather in its lifetime?

13. How many times can the female bee sting a human?

14. What happens to the drones during the winter months?

15. Considering your answer to question 14, how can you explain the presence of drones in the spring?

16. What is lactose? What plant products contain lactose?

17. Relate the percentage of lactose in the milk of a given species to the brain size of that species (as related to total body mass).

18. Discuss the unique properties of lactose.

19. What is the relation of lactose to calcium, phosphorus, magnesium, and barium assimilation in the body? Does milk contain significant amounts of these minerals?

20. Does woman's milk or cow's milk contain the most lactose? Relate this to the prevention and therapy of rickets.

21. Is milk a good source of niacin? Explain the antipellagric effect of milk as a food for humans.

22. Compare lactose with sucrose in sweetness.

23. What is a placebo? (See Glossary.)

24. List several animal products that are good sources of calcium and/or phosphorus.

25. Is milk a good source of iron? Of copper?

26. Does cow's milk contain more or less mineral (especially calcium and phosphorus) than woman's milk? What correlation was made between the ash content of milk and the growth rate of a given species?

27. Animal products are somewhat deficient in which vitamins?

28. Do animal products have a favorable protein/calorie ratio for those concerned with their body weight?

29. Which animal products are economical sources of protein?

30. Do American consumers ingest more or fewer calories now than in 1910? What about the consumption of other major nutrients?

31. What is the trend in American consumption for (a) beef, (b) fish, (c) eggs, (d) poultry meats, (e) nonfat milks, (f) butter, (g) total animal fats, (h) vegetable fats?

32. Does the body synthesize cholesterol? Is it needed in the body?

33. Do people live longer in the United States than in all other countries of the world?

34. Why do food processors fortify animal products?

35. Can milk be economically fortified with vitamins A and D?

36. Is nonfortified skim milk a good source of the fat-soluble vitamins? Why?

37. Is an excessive intake of vitamin D beneficial? Harmful?

38. What is milk toning?

39. Why are people concerned with preserving animal products?

40. What is the effect of high and low temperatures on bacterial growth?

41. What is bactofugation?

42. Is dehydration a recent innovation in the preservation of animal products?

43. What is a limitation of the use of chemical preservatives in animal products?

44. What is cold sterilization?

45. What natural component of honey restricts the growth of most microorganisms?

46. What are the times and temperatures used to effect the pasteurization of honey?

47. How may pork be rendered safe for human consumption (safe from trichinosis)?

48. Why is meat smoked?

49. Who is credited with the invention of canning?

50. Are most vitamins of meat destroyed in processing?

51. Why should eggs be heat treated?

52. What are the times and temperatures used to pasteurize eggs?

53. What is thermostabilization of eggs?

54. What are the times and temperatures used to pasteurize milk?

55. Why is milk pasteurized? Does pasteurization destroy the vitamins of milk?

56. What is the effect of sunlight on the flavor and vitamin A potency of milk?

57. How are instant animal product powders manufactured?

58. What is lyophilization? Sublimation? Are you using the Glossary?

59. For what are lyophilized animal products being used?

60. What is the primary limiting factor in the expanded use of lyophilized animal products? Which ones are likely to increase the most in use during the next decade?

61. Which agency within the U.S. Department of Agriculture is responsible for ensuring the safety, wholesomeness, and accurate labeling of meat, poultry, and egg products?

62. Of what significance was the implementation of HACCP practices in the assurance of safe animal products?

63. Does the use of irradiation in selected animal products cause the food to be radioactive? Discuss.

64. What factors might influence the expanded use of animal products?

65. What are some trends in new animal products?

66. Do you think synthetic foods are a threat to animal products?

67. What is mechanically deboned poultry meat? Why is it pasteurized?

68. What problem may arise from the use of a strong source of fluorescent light in refrigeration cases in which milk is being marketed in translucent plastic containers?

69. Cite advantages of the marketing of restructured meat products. Cite examples of such products.

70. Are "natural" foods always safe for human consumption? Discuss.

4

COMPANION ANIMALS

Karen L. Campbell[1]

Near this spot are deposited the remains of one who possessed beauty without vanity, strength without insolence, courage without ferocity, and all the virtues of man, without his vices. This praise, which would be unmeaning flattery if inscribed over human ashes, is but a just tribute to the memory of Boatswain, a dog.

Lord Byron (1788–1824)
English poet

An Overview

Those who have experienced firsthand the personal pleasure associated with looking squarely into the affectionate eyes of their cuddly cats and devoted dogs know why companion animal biology is increasingly being included in the study of animal science. This chapter provides an overview of the most pertinent topics related to people's most common, lovable pets—cats and dogs. These topics include the domestication of cats and dogs; major breeds and characteristics of cats and dogs; contributions, including therapeutic uses, of companion animals to humans; companion animal–human bonding; the behavior, care and training, feeding and management, and health and disease aspects of cats and dogs; as well as career opportunities associated with companion animals. Please enjoy!

4.1 INTRODUCTION

Approximately two-thirds of American families and more than one-half of all households in English-speaking countries share their homes with one or more companion animals. Relationships between companion animals and people are long, lasting, and often intense. For many people, living with a companion animal is living with another family member. The person is greeted at the door by the animal when he/she arrives home; there is a companion with whom to share the couch and television; there is an animal friend to shop, feed, and care for and thereby add to the paced, circular rhythm of family life. The

animal provides a means for fun, pleasure, exercise, and in some cases physical protection. Household pets teach children responsibility and respect for life. Pets provide mental security; dogs and cats do not reject their owners.

Animals serve as a catalyst for establishing human contact and interaction. A dog on a leash serves much the same sociability function as a baby in a carriage inviting attention, comment, and admiration. The politician who includes a cocker spaniel in the family portrait makes a politically astute gesture. Folklore is replete with tales of heroic animals—dogs who saved families from household fires and rescued drowning children. Media mythology of Lassie and Rin-Tin-Tin has immortalized the exploits of dogs. Assistance dogs for the blind, deaf, and physically, mentally, or emotionally handicapped have improved the quality of life for countless people. Dogs also work as police dogs, tracking dogs, herding dogs (Figure 4.1), water rescue dogs, sled dogs, guard dogs, and mascots (e.g., McGruff, the

[1]Dr. Karen L. Campbell is Professor of Veterinary Clinical Medicine; Section Head, Dermatology and Specialty Medicine, College of Veterinary Medicine, University of Illinois at Urbana–Champaign. She holds B.S. and D.V.M. degrees from the University of Missouri–Columbia, and the M.S. degree in Veterinary Clinical Pathology from the University of Georgia–Athens. She is a Diplomate in the American College of Veterinary Internal Medicine and a Diplomate in the American College of Veterinary Dermatology.

Figure 4.1 Shetland Sheepdog herding sheep.
Courtesy of Jane Rothert, Urbana, IL.

Figure 4.2 Companion animals and their owners commonly display strong bonds.
Courtesy of Kathy L. Wall, Mexico, MO.

crime-prevention dog; Sparky, the fire-protection dog; the Georgia Bulldog; Texas A&M's Revelry; and the Washington Huskies).

Webster's Collegiate Dictionary defines a pet as "a domesticated animal kept for pleasure rather than utility." The term *companion animal* connotes one is frequently in the company of, associates with, or accompanies the animal. The factor that distinguishes a companion animal from other pets is the relationship with the owner. Relationships with companion animals are often similar to relationships with human companions. The precise relationship depends upon the personality of the owner and what he/she wishes to derive from the partnership. People often pick pets who mirror their personality, appearance, and behavior.

An important factor contributing to the popularity of dogs and cats as companion animals is their power of nonverbal expression. Both species possess a rich repertoire of postures, gestures, facial signals, and sounds used to communicate with their owners. Many of these signals are interpreted by people as expressions of attachment and affection. The animal conveys to the owner a feeling of being valued and needed (Figure 4.2). Because they cannot speak companion animals are incapable of lying to, deceiving, or criticizing their owners. Thus they combine many benefits of human relationships with few of the threats. Except for livestock and poultry, companion animals represent the largest and most influential category of domestic animals in modern society.

> The animal does not judge but offers a feeling of intense loyalty . . . it is not demanding . . . it does not expose its master to the ugly strain of constant criticism. It provides the owner with a chance to feel important.
>
> **A. Siegal**

In a Michigan survey of children through grade 8, approximately 90 percent owned one or more animals. The order of animal preference based on ownership was dogs, cats, rabbits, hamsters, fish, birds, horses, and other large animals (calves, goats, sheep, pigs). There are dozens of pets that deserve discussion in a chapter devoted to topics related to companion animals. However, because of space limitations we have concentrated our discussion on the most conventional companion animals—cats and dogs—clearly the two most common companion animals in the United States.

Wild animals are not recommended as pets. Many wild animals may seem cute and cuddly as babies but as adults prove to be unmanageable, unpredictable, and dangerous. Love and concern for these animals sometimes prompts people to attempt to keep them as pets—a decision that often proves unpleasant for both people and animals. Therefore, children and older people alike are encouraged to adopt only domesticated animals as pets.

A 1996 national survey estimated that 58.2 million U.S. households (58.9 percent) owned a companion animal. Animals considered as companion animals in the study included dogs, cats, birds, horses, hamsters, guinea pigs, gerbils, rabbits, and fish. Approximately 31.6 million U.S. households owned a dog. The total number of dogs was estimated at 52.9 million (mean of 1.68 per dog-owning household).

Cats have increased steadily in popularity as companion animals (Figure 4.3). They are especially popular as apartment pets because of their ease of litter training and adaptability to a totally indoor lifestyle. In 1996 approximately 27.0 million U.S. households (27.3 percent) owned a cat. The total number of cats owned was estimated at 59.0 million (mean of 2.19 per cat-owning household).

Figure 4.3 Cat and its owner sharing a precious moment.

Other pets owned by U.S. households in 1996 included 12.6 million birds, 4.0 million horses, 55.6 million fish, 4.9 million rabbits, 1.9 million hamsters, 764,000 gerbils, and several million other household pets of various species. Many households own more than one type of pet. Approximately 42.0 percent of dog owners also owned a cat and approximately 48.6 percent of cat owners also owned a dog.

Demographic factors associated with pet ownership include household size and life stage, income, and home ownership. Families with children comprise 40 percent of all U.S. households and 50 percent of them have pets. Singles comprise 26 percent of U.S. households but only 17 percent of them own pets. Families with incomes above $25,000 comprise 65 percent of pet-owning households and 61 percent of the total population. Seventy-six percent of companion animal owners own their homes compared with 73 percent of all U.S. households. On the other hand, only 18 percent of pet owners rent homes compared with 21 percent of all U.S. households.

The feeding and care of companion animals contribute to multibillion-dollar pet-related industries. The production of dog food is a $10.1 billion, 7-million ton industry in the United States with substantial additional tonnage manufactured in Japan and European countries using ingredients purchased from the United States. Cat owners in the United States annually spend approximately $2 billion on cat food and $300 million on cat litter. The number of veterinary visits per household and the expenditure for each visit have increased in recent years. Veterinary expenditures were approximately $8.0 billion for dogs and $4.0 billion for cats in 1998. Additional billions are spent annually on pet care products ranging from shampoos and flea control products to toys, beds, leashes, and fences. Pet-related industries also include pet training facilities, grooming facilities, boarding kennels, cat and dog shows, obedience trials, herding trials, sled dog races, dog racing, service dog organizations, therapy dog organizations, humane societies, animal behavioral therapists, and many others.

4.2 HISTORY AND DOMESTICATION OF DOGS AND CATS

Pets upgrade the quality of life, bring us closer to nature, provide companionship and emphasize the fact that animals must be accepted as desirable participants in society.

Boris Levinson
American child psychologist

The dog was the first animal species to be domesticated and universally has been the most popular pet. It is among the most intelligent of domestic species and is also the most expressive. People and dogs understand each other so well because both are naturally playful, gregariously sociable species that instinctively defend their territories and hunt for food. Initial stages of domestication of the dog began at least 12,000 years ago at the end of the Pleistocene Period and the Paleolithic Age. The earliest records of dog remains come from the late Paleolithic cave in Palegawra, Iraq, dated 12,000 B.C.

Most paleontologists agree that dogs were domesticated from the gray wolf, *Canis lupis*. Wolves may originally have been simply camp followers, living in close contact with people as an easy means of obtaining food. It is likely that hunters brought young wolves into their camps to entertain the children. The territorial nature of wolves likely made them valuable as camp guards. Wolves may also have joined hunters in the chase of game for food. Because of the wolf's adaptable nature taming soon led to controlled breeding and domestication. By the time of the New Kingdom in Egypt (1570 B.C.) several varieties of dogs existed. Most were hunting or battle dogs, some had straight tails, others curly; some had prick ears, others had pendulant ears; some were spotted, others solid colored. From that time onward the insatiable desire of humans to improve on nature has led to the development of many varieties and breeds of dogs.

Through their association with humans, domestic dogs have developed a large repertoire of expressions they use in relating to people. They employ many movements of submissiveness that can be seen in the wolf in greeting his/her master, obviously accepting him/her as the pack leader. Dogs added tail wagging, a trait not found in wild canines. Many skills learned by domestic dogs are a development and sophistication of instincts associated with survival in the wild. It is natural for dogs to carry things in their mouths. This habit has been used by people as a basis for training dogs to retrieve. The guarding of its territory developed into the guarding of homes and people. Some breeds work by sight, others by scent. These traits have been refined in developing the sporting and working breeds of dogs. The relationship between dogs and people continues to evolve and develop to the benefit of both species.

Cats were first domesticated in Egypt approximately 6000 years ago. It is probable that the African bush cat, *Felis lybica,* took up residence in Egyptian grain storehouses to prey upon mice and rats that fed on stored grains. The Egyptians likely brought kittens into their houses for children to pet and play with. The inner mysticism of the cat—the obscure wisdom and secrecy that seems to dwell in its eyes—attracted the attention of peasants, priests and nobles alike. The cat's role became diversified. Domestic cats trained to hunt birds were popular as partners of Egyptian aristocrats indulging in this sport. Early Egyptians adopted cats as objects for deification as exemplified by the cat-headed goddess Bast. The term "puss" was likely derived from the cat's identification with Bast (also known as Pasht, Ubastet, and Bubastis), goddess of the home and of cats (stealthy and a lover of pleasure). Phoenician trading vessels brought cats into Europe about 900 B.C. Cats were in great demand in Europe as killers of flea- and lice-ridden disease-spreading rodents. During the Middle Ages cats became associated with witches and warlocks and linked with Satan worship and demonology. Superstitious people consider black cats to be familiars of Lucifer. Thousands of cats were killed by religious zealots during the Dark Ages. Killing cats likely contributed to the spread of bubonic plague (Black Death) throughout Europe in the fourteenth century.

Fortunes of the cat as a domestic creature have swung back and forth from one extreme to another. Through it all the cat has not lost its mystique or its role as a controller of rodent populations. In the burst of creative thinking that followed the French Revolution and the Enlightenment artists, novelists, poets, painters, and composers discovered the aesthetic delights of the cat: its grace, agility, beauty, charm, keen mind, and humorous idiosyncrasies. There is no evidence that Native Americans kept cats. Domestic cats were introduced to the Americas by European ships carrying them as pets and defenders of foodstuffs. To this day there is still a polarity of feeling about cats. In general, a person either loves cats with a passion or hates them with equal fervor. Ailurophobia, an abnormal fear of cats, affects many people. The victim in ailurophobia is not the cat, which has survived persecution by people for more than 2000 years, but rather the cat hater who is deprived of the pleasure, comfort, thrill, stimulation, amusement, and sheer delight a cat can give.

4.3 BREEDS AND CHARACTERISTICS OF COMPANION ANIMALS

For centuries people have controlled the breeding of animals to accentuate or diminish certain characteristics. Dogs inherited a range of characteristics and behaviors from wolves. By Roman times these characteristics had been refined into various categories of dogs including those used as herding dogs, warrior dogs, sight and scent hunters, terriers, and companions. The American Kennel Club (AKC) was established in 1884 as a nonprofit organization devoted to the advancement of purebred

dogs. The AKC classifies purebred dogs in seven categories and a miscellaneous group (Table 4.1). The seven categories are (1) sporting dogs, (2) hounds, (3) working dogs, (4) terriers, (5) toy breeds, (6) nonsporting dogs, and (7) herding dogs. The miscellaneous group includes rare breeds, which have active enthusiasts but are not yet well-established enough to be admitted into the *AKC Stud Book.* Approximately one-fourth of the dogs licensed in the United States are registered with the AKC. The AKC currently registers dogs of 147 breeds. Approximately 1.5 million new dogs are registered annually. Other active purebred dog registries in the United States include the United Kennel Club and the States Kennel Club. Color photographs of 24 of the most commonly registered breeds are shown in Color Plates 4B and 4C.

Dogs have been said to have a "plastic germ plasm." Their type is easily changed by selective breeding. It is estimated that there are more than 400 dog breeds in the world. Breeds vary in popularity from one decade to the next. The 20 most often registered breeds in 2000 are listed in Table 4.2. Interest in purebred AKC-registered dogs has increased because of the promotion associated with dog shows, agility trials, obedience trials, field trials, working trials, herding trials, and similar events open only to AKC-registered dogs. The wide variety of dogs gives owners an opportunity to select the breed best adapted to their personal preferences and objectives in dog ownership.

There has been less selective breeding of cats than of other domesticated animals. For long periods of time cats were primarily valued for their role in the control of rodent populations. Twentieth-century Americans were the "new Egyptians" and rediscovered the cat as a valued companion animal. There are approximately 70 established breeds of cats (Table 4.3). Breeds of cats are divided into three types: natural, human-developed, and spontaneous mutation. *Natural breeds* such as the Persian, Russian Blue, and Turkish Angora evolved naturally in different locales and were refined and stabilized by cat fanciers through selective breeding. *Human-developed* breeds are hybrids such as the Exotic Shorthair and Bombay, created by cat breeders combining existing breeds. *Spontaneous mutations* such as the Manx and Scottish Fold resulted from spontaneous changes in genetic codes.

There are seven major feline stud-book organizations in the United States: American Cat Fancier's Association; American Cat Association; Cat Fancier's Association, Inc.; Cat Fancier's Federation; Crown Cat Fancier's Federation, Inc.; National Cat Fancier's Association, Inc.; and United Cat Federation. Some breeds are accepted in all associations, others are recognized by only a few. Breed standards may vary from one association to another. Twelve major breeds of cats in the United States are depicted in Color Plate 4A.

Cats are classified as having one of three basic body types: (1) foreign (or oriental), (2) cobby, or (3) moderate (modified). The *foreign* (oriental type) is typified by the Siamese as ideally a sleek, svelte, slim, lithe, elegant fine-boned cat. *Cobby* refers to a heavy, low-lying, short-legged,

TABLE 4.1	Breeds by Groups as Recognized in 2000 by the American Kennel Club (AKC)

Group I: Sporting Dogs
Brittany
Pointer
Pointer, German Shorthaired
Pointer, German Wirehaired
Retriever, Chesapeake Bay
Retriever, Curly-Coated
Retriever, Flat-Coated
Retriever, Golden
Retriever, Labrador
Setter, English
Setter, Gordon
Setter, Irish
Spaniel, American Water
Spaniel, Clumber
Spaniel, Cocker
Spaniel, English Cocker
Spaniel, English Springer
Spaniel, Field
Spaniel, Irish Water
Spaniel, Sussex
Spaniel, Welsh Springer
Vizsla
Weimaraner
Wirehaired Pointing Griffon

Group II: Hounds
Afghan Hound
Basenji
Basset Hound
Beagle
Black and Tan Coonhound
Bloodhound
Borzoi
Dachshund
Foxhound, American
Foxhound, English
Greyhound
Harrier
Ibizan Hound
Irish Wolfhound
Norwegian Elkhound

Otterhound
Petit Basset Griffon Vendeen
Pharaoh Hound
Rhodesian Ridgeback
Saluki
Scottish Deerhound
Whippet

Group III: Working Dogs
Akita
Alaskan Malamute
Bernese Mountain Dog
Boxer
Bullmastiff
Doberman Pinscher
Giant Schnauzer
Great Dane
Great Pyrenees
Greater Swiss Mountain Dog
Komondor
Kuvasz
Mastiff
Newfoundland
Portuguese Water Dog
Rottweiler
Saint Bernard
Samoyed
Siberian Husky
Standard Schnauzer

Group IV: Terriers
Airedale Terrier
American Staffordshire Terrier
Australian Terrier
Bedlington Terrier
Border Terrier
Bull Terrier
Cairn Terrier
Dandie Dinmont Terrier
Fox Terrier, Smooth
Fox Terrier, Wire
Irish Terrier

Jack Russell Terrier
Kerry Blue Terrier
Lakeland Terrier
Manchester Terrier (Standard)
Miniature Bull Terrier
Miniature Schnauzer
Norfolk Terrier
Norwich Terrier
Scottish Terrier
Sealyham Terrier
Skye Terrier
Soft Coated Wheaton
Staffordshire Bull Terrier
Welsh Terrier
West Highland White Terrier

Group V: Toys
Affenpinscher
Brussels Griffon
Cavalier King Charles Spaniel
Chihuahua
Chinese Crested
English Toy Spaniel
Havanese
Italian Greyhound
Japanese Chin
Maltese
Manchester Terrier
Miniature Pinscher
Papillon
Pekingese
Pomeranian
Pug
Shih Tzu
Silky Terrier
Yorkshire Terrier

Group VI: Nonsporting Dogs
American Eskimo Dog
Bichon Frise
Boston Terrier
Bulldog

Chinese Shar-Pei
Chow Chow
Dalmatian
Finnish Spitz
French Bulldog
Keeshond
Lhasa Apso
Löwchen
Poodle
Schipperke
Shiba Inu
Tibetan Spaniel
Tibetan Terrier

Group VII: Herding Dogs
Australian Cattle Dog
Australian Shepherd
Bearded Collie
Belgian Malinois
Belgian Sheepdog
Belgian Tervuren
Border Collie
Bouviers des Flandres
Briard
Canaan Dog
Collie
German Shepherd Dog
Old English Sheepdog
Puli
Shetland Sheepdog
Welsh Corgi, Cardigan
Welsh Corgi, Pembroke

Miscellaneous Breeds
Polish Lowland Sheepdog
Plott Hound
Spinoni Italiani

compact broad-chested body typified by the American Short-hair, Burmese, Chartreux, British Shorthair, and Persian. *Moderate* or *modified* "intermediate" breeds include the Russian Blue and Abyssinian. Feline coat types include the hairless cats (e.g., Sphynx), shorthair cats (e.g., Siamese, Bombay, Burmese, Oriental Shorthair, Chartreux, Manx, Russian Blue, and American Shorthair), wirehair coats (e.g., American Wirehair), curly coat (e.g. Cornish Rex and Devon Rex), and longhair cats (e.g., Persians, Turkish Angora, Somali, Maine Coon Cat, Ragdoll, Himalayan, Birman, Balinese). Some breeds prefer solitude, whereas others prefer human interaction. Certain breeds are notably active, others sleep much of the time. For many people selecting a breed is a matter of aesthetic preference and economic considerations. Prospective cat owners should choose a cat on the basis of personal preferences, home, and lifestyle.

(a) Abyssinian

(b) American Bobtail

(c) American Shorthair

(d) Bengal

(e) Birman

(f) British Shorthair

(g) Chartreux

(h) Cornish Rex

(i) Korat

(j) Maine Coon

(k) Persian

(l) Siamese

Color Plate 4A Major Breeds of Cats in the United States.

Courtesy of (*a*) Barb Martinec (also *c, f,* and *h*), (*b*) Allan Paul (also *e, g,* and *i*), (*d*) Cassie Hale, (*j*) Alice Pursell, (*k*) Judie Ohnemus, and (*l*) Jackie Wyandt.

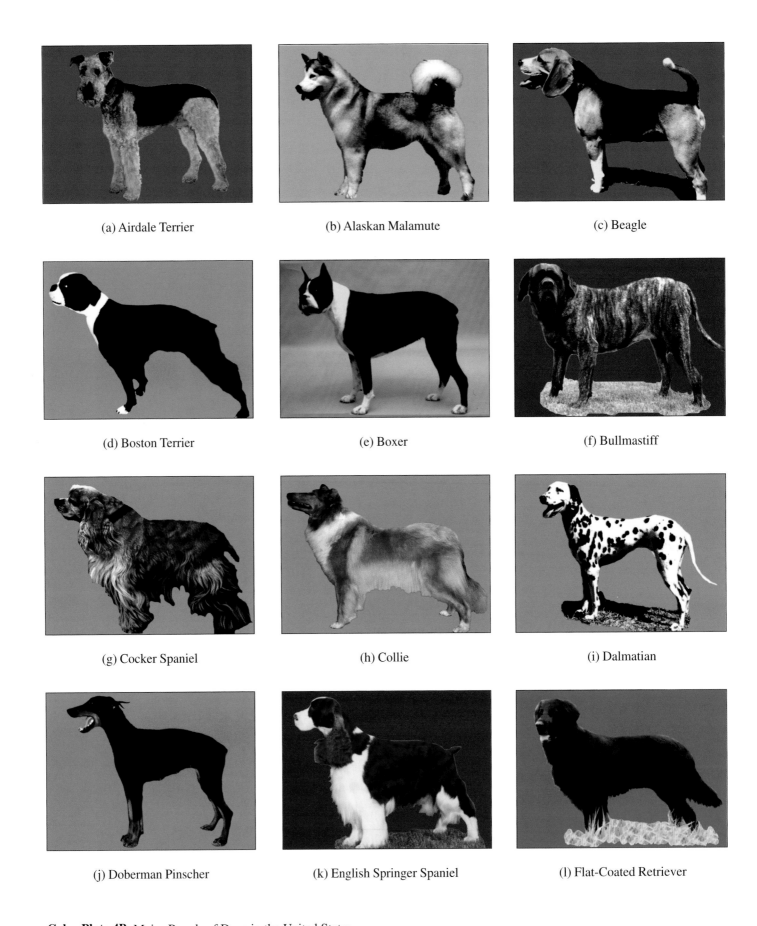

(a) Airdale Terrier

(b) Alaskan Malamute

(c) Beagle

(d) Boston Terrier

(e) Boxer

(f) Bullmastiff

(g) Cocker Spaniel

(h) Collie

(i) Dalmatian

(j) Doberman Pinscher

(k) English Springer Spaniel

(l) Flat-Coated Retriever

Color Plate 4B Major Breeds of Dogs in the United States.

Courtesy of (*a*) Karen Campbell (also *h* and *j*), (*b*) Robin Haggard, (*c*) Lisa Klopp, (*d*) Allan Paul (also *g*), (*e*) Sue Kamienski, (*f*) Sandra Grable, (*i*) Linda Klippert, (*k*) Dawn Morin, and (*l*) Pat Norris.

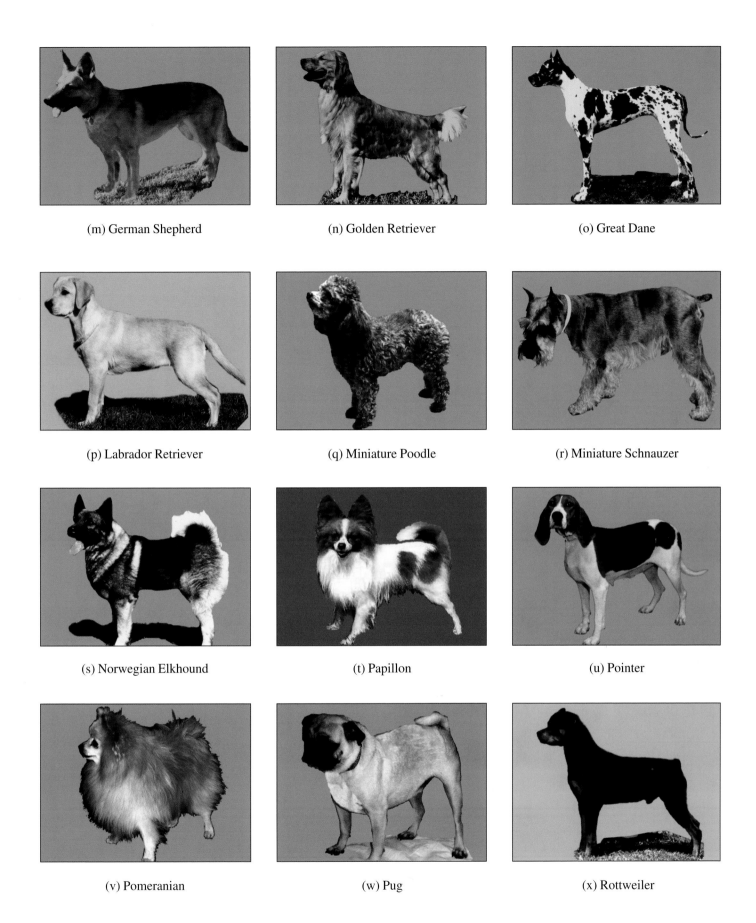

(m) German Shepherd

(n) Golden Retriever

(o) Great Dane

(p) Labrador Retriever

(q) Miniature Poodle

(r) Miniature Schnauzer

(s) Norwegian Elkhound

(t) Papillon

(u) Pointer

(v) Pomeranian

(w) Pug

(x) Rottweiler

Color Plate 4C Major Breeds of Dogs in the United States (continued).

Courtesy of (*m*) Jan Sears, (*n*) Allan Paul (also *s*), (*o*) Drs. Thomas Burke and Jane Schumann, (*p*) Kim Kensell, (*q*) Karen Campbell (also *r, u,* and *v*), (*t*) John Larkin, (*w*) Jennifer Matousek, and (*x*) Jennifer Cameron and Harriet Stockanes.

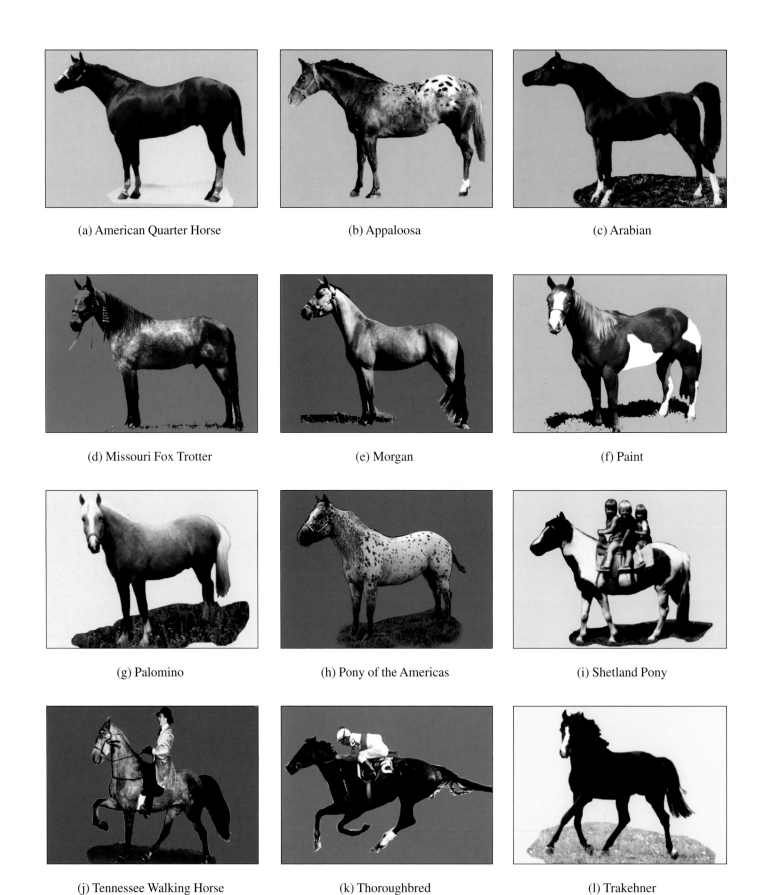

Color Plate 4D Major Breeds of Horses in the United States.

Courtesy of (a) American Quarter Horse Association, *(b)* Pam Staley, *(c)* Dr. Richard Fredrickson and Lyn Freel, Crystal Castle Arabians, *(d)* Steve and Jackie Wyandt (also *e*), *(f)* University of Illinois *(g)* Karen Campbell (also *i*), *(h)* Douglas Hutchens, *(j)* Ann Johnson, *(k)* The New York Racing Association, Inc., and *(l)* Ted Lock, University of Illinois.

4.4 CONTRIBUTIONS OF COMPANION ANIMALS TO HUMANS

Pets have kept many of us out of mental institutions. The family pet—dog, cat, bird, or hamster—is a cure for worry that is worth a fortune. Pets relieve anxiety. They act as emotional stabilizers because of the way they unconditionally accept you.

Wesley Young

Only recently have we begun to understand and appreciate the various ways in which companion animals contribute to human lives. Once viewed as misplaced sentimentality, the human–companion animal bond is now recognized as a powerful cable woven from a wide variety of strands each possessing great individual strength. Indeed, we now recognize the value of companion animals for physical health, mental health, emotional health, and human social health.

A traditional picture of "the good life" depicts a family gathered about the fireplace at night while winter winds blow snow and cold outside. The family dog is in the picture lying at length on the warm floor, its back near the fire. This picture conveys a sense of shelter and warmth, of protection and family bonds. The dog has an honored place in many homes. It is a special companion who awakens the household each morning, fetches slippers and papers, barks at strangers, plays with children when they come home, and takes the warmest spot by the fireside.

Stuffed animals and animal toys are part of a child's early environment. The first books given and read to children are often about animals. Animals also dominate television programs designed for children, who find it easy to identify with animals and their characteristics. Children talk to their toys, animate them, personify them, and attribute a variety of feelings and behaviors to them. Live animals have a distinct advantage over inanimate objects in that they can initiate interactions, return affection, and participate directly in activities of children. Children unconsciously view their pets as an extension of themselves and treat their pets as they want to be treated (Figure 4.4). Petting an animal helps children feel both caring and cared for.

The entire life cycle, activities, and habits of animals are readily observable making them excellent educational tools. Children frequently learn about life, death, reproduction, and biological processes by first observing animals. They beg their parents for a pet and are often told they must help take care of the animal. Children given jobs such as exercising, brushing, feeding, and cleaning up after their pets learn important lessons in assuming responsibility for the care of another living creature. Children participating in 4-H projects, cat or dog shows, obedience training, field trials, or herding work learn to discipline themselves as well as their animals and learn valuable lessons in dealing with the wins and inevitable losses associated with competition. The adoration of a pet helps children cope with stresses and attacks on his/her self-esteem from peers.

Animals . . . bolster the pet owner's morale and remind him that he is, in fact, a special and unique individual.

K. M. G. Keddle

TABLE 4.2	Twenty Most Often Registered AKC Breeds in 2000
1. Labrador Retrievers	11. Rottweilers
2. Golden Retrievers	12. Pomeranians
3. German Shepherd Dogs	13. Miniature Schnauzers
4. Dachshunds	14. Cocker Spaniels
5. Beagles	15. Pugs
6. Poodles	16. Shetland Sheepdogs
7. Yorkshire Terriers	17. Miniature Pinschers
8. Chihuahua	18. Boston Terriers
9. Boxers	19. Siberian Huskies
10. Shih Tzu	20. Maltese

TABLE 4.3	Cat Breeds as Listed by *Cat Fanciers* Website			
Abyssinian	British Angora	Devon Rex	Munchkin	Siamese
American Bobtail	Brazilian Shorthair	Egyptian May	Nebelung	Siberian
American Curl	British Shorthair	European Shorthair	Norwegian Forest Cat	Singapura
American Longhair	Burmese	Exotic Shorthair	Ocicat	Snoeshoe
American Shorthair	Burmilla	Havana Brown	Ojos Azules	Sokoke
American Wirehair	California Spangled Cat	Himalayan	Oriental Longhair	Somali
Angora	Chantilly/Tiffany	Honeybear	Oriental Shorthair	Sphynx
Asian Shorthair	Chartreux	Japanese Bobtail	Persian	Spotted Mist
Asian Semi-Longhair	Chausie	Javanese	Pixie Bob	Sterling
Australian Mist	Cerubim	Korat	Ragamuffin	Tiffanie
Balinese	Colorpoint Longhair	LaPerm	Ragdoll	Tonkinese
Bengal	Colorpoint Shorthair	Maine Coon	Russian Blue	Turkish Angora
Birman	Cornish Rex	Malayan	Scottish Fold	Turkish Van
Bombay	Cymric	Manx	Selkirk Rex	York Chocolate

Figure 4.4 Child and her dog prepare to kiss as a mark of mutual affection.
Courtesy of Mike Volpe, San Francisco, CA.

Companion animals decrease the loneliness and depression of their owners by providing a source of companionship, an impetus for nurturance, and a source of meaningful daily activities. They decrease owners' anxiety by (1) providing a relaxing focus for attention, (2) permitting the exchange of affectionate touch, and (3) providing a feeling of safety. Companion animals contribute to human longevity by being a stimulus to exercise and keeping the human companion busy with feeding, grooming, and other pet-related activities. Pet owners participating in cat and dog shows, obedience trials, field trials, tracking, herding, agility training, and other pet-related competitions have many opportunities to form friendships with other pet owners and to develop a hobby they can enjoy throughout the various stages of their lives. Medical studies have verified the benefits of pet ownership. Pet owners have lower blood pressure and increased longevity in comparison with their nonpet-owning peers.

Dogs also provide invaluable services in guarding homes and businesses, working with police and law enforcement agencies, sniffing out narcotics, participating in search and rescue operations, hunting with people by tracking prey, pointing out and retrieving game, herding livestock, and as service dogs for the handicapped and therapy dogs for the lonely.

Following the September 11, 2001, terrorist attacks on the New York City World Trade Center, more than 350 canine search specialists worked 12-h shifts to locate victims. Some of these dogs came from as far away as Europe to assist in the efforts. Search-and-rescue dogs are rigorously trained to meet standards of performance set by the Federal Emergency Management Agency. A dog's sense of smell is more than a thousand times as sensitive as that of a human, thereby enabling it to detect scents through concrete. In addition to locating victims, dogs carried cameras strapped to their sides through tunnels in the debris to provide engineers with information on the stability

of an area prior to sending in human workers. Search and rescue dogs were handled by members of fire departments and emergency medical technicians. Veterinary assistance sites were set up to provide fluids and treats to the dogs and also clean their ears, eyes, and feet after each work shift.

4.5 THERAPEUTIC USES OF COMPANION ANIMALS

> Animals are such agreeable friends; they ask no questions, pass no criticism. They are devoted and unjudging, content to love and be loved.
>
> **George Eliot (1819–1880)**
> **English novelist**

Pets provide people with many therapeutic benefits including companionship, love, humor, play, exercise, a sense of power, outlets for displacement, projection, and nurturance. Petting and talking to animals has been shown to reduce stress, promote feelings of reverie and comfort, and to enhance longevity and physical health.

The best medicines are frequently those perceived by patients to make them feel better. Animals often fit this perception. As previously mentioned, the presence of a dog has been shown to lower blood pressure in both children and adults. People who are anxious and withdrawn, and who will not talk freely with a psychiatrist or psychologist, will often open up to a dog or cat in the presence of the therapist. They feel safe in the presence of the dog or cat, probably because the animal will not be critical or reject them. Children who have been abused frequently respond to the presence of an animal and develop deep feelings of love, care, and empathy for the animal. Companion animals can replace absent parents and siblings and provide children with opportunities to play out their fantasies, express feelings, and act out conflicts and dreams. Companion animals can enhance the ego development and social skills of regressive individuals. They help diffuse hostilities and provide a shared source of social interests in families whose members are struggling against pathological patterns of relating to one another.

Recently there has been an increased interest in the way animals can be used therapeutically to improve the physical and emotional health of the elderly. A survey of nursing homes and retirement communities in the United States revealed that nearly one-half of them are using animal-facilitated visitation programs (Figure 4.5). The elderly frequently suffer from a loss of relatives and friends and may withdraw from active participation in human affairs. Objects and animals that provided security in early life may assume a greater importance in later life. Pets can be important love objects and can be loved without fear of rejection. The self-esteem of senior citizens can be restored by the assurance that animals love them. Too often the only physical contact for institutionalized patients is from nurses and doctors in painful injections and unpleasant examinations. The opportunity to touch and talk to an animal can help reverse the negative effects of institutional life. It is important

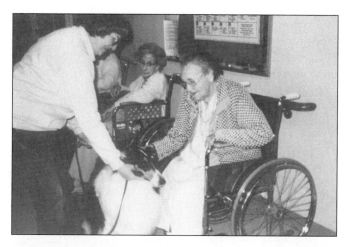

Figure 4.5 Nursing home residents enjoy regular interaction with therapy dogs.

Courtesy of Marilyn Campbell, Georgetown, IL.

that more elderly have the opportunity to experience the joys derived from having an animal companion—a special friend they can like who likes them.

There are more than 24.1 million people in the United States with a disabling handicap. Of these, more than 1.6 million have a disability involving blindness or serious visual impairment, more than 15.0 million have a disability involving loss or lack of physical mobility, and more than 1.6 million people have a disability involving deafness or serious hearing impairment.

Approximately 8000 blind people in the United States utilize guide dogs. There are 14 schools in the country providing guide dogs for the blind. The best known is The Seeing Eye Inc. of Morristown, New Jersey, which was founded in 1929. The schools set strict standards for both human and canine trainees. To be accepted for training a person must have medical proof of legal blindness and otherwise good health, mobility with the ability to walk independently, sufficient manual dexterity to harness the dog and enough physical strength to control it, references to show the applicant is able and anxious to venture outside a home, and prove that he/she is responsible enough to care for a dog. Dogs must be physically sound, have excellent temperaments, not be afraid of loud noises or traffic, be able to ignore distractions, and have the initiative to do guiding work. Three of the most common breeds used as guide dogs are Golden Retrievers, Labrador Retrievers, and German Shepherds. These breeds are enthusiastic and eager to work, tractable, possess endurance, are easy to groom, have even temperaments, and are a comfortable size for most adults to walk with using a harness. Most guide schools breed their own pups and use volunteer families to raise the puppies from weaning until they are ready to start formal training at around 14 months of age. Approximately one-half of dogs selected for training are able to finish the programs and become guide dogs. Those not completing the training program are offered for adoption to the foster families that raised them.

The use of dogs as service animals has expanded in recent years. Dogs for the Deaf is a national Hearing Dog training and placement service that began in 1978 to rescue unwanted dogs from animal shelters and train them to become the ears for the deaf and hearing impaired. Through programs such as Dogs for the Deaf and Paws with a Cause, dogs are trained to alert their partners to sounds vital to independent living: a smoke alarm, a telephone, a knock at the door, a baby's cry, and other important sounds the hearing-impaired human is commonly unable to hear. The Delta Society Pet Partners Program and other groups are training dogs to help the physically disabled in wheelchairs by pulling the persons for long distances and up inclines and ramps, turning the switches of lights and appliances on and off, and opening doors. Dogs are also being used as cotherapists and aides in pet-assisted therapy programs for the convalescing, the retarded and autistic, and the mentally and emotionally handicapped. Rehabilitation centers have found that pets provide motivation to patients to improve their strength, coordination, and mobility. Activities vary from remembering the names of animals to petting and brushing animals, gripping and throwing balls, and walking with animals. Patients often develop renewed interest in physical therapy and recreational pursuits after regular pet sessions. Pet-assisted therapy is moving beyond "pet-a-pup" programs into a serious, legitimate health profession with special training and formal certification of participants.

Companion animals have also been used to improve the behavior and outlook of inmates in penal institutions. There is an innovative program called "Friends for Folks" at the Lexington Correction Facility in Oklahoma that trains dogs for the elderly, disabled, and others needing or wanting a trained dog as a companion. Working cooperatively with faculty of the College of Veterinary Medicine at Oklahoma State University, the inmate housing unit that has dogs in residence is one with reduced tension and brims with enthusiasm in meeting its objectives. Inmates embrace the feeling that they are helping others and take great delight in their ability to win over a dog considered unteachable, untrainable, or mean. Indeed, inmates take personal pride in their accomplishments, gain self-respect from fellow prisoners, and learn self-discipline as well. Additionally, although displaying affection for fellow inmates is considered to be an unacceptable behavior, inmates feel perfectly comfortable and enjoy hugging and petting their dog trainee. Dogs are selected from the Last Chance Shelter by the program director who attempts to match a dog's personality with that of the inmate trainer. When training is completed the director matches dogs with those awaiting a pet companion. The program enjoys a three-way benefit: inmate, dog, and recipient.

An example of the "Friends for Folks" program benefit is a senior citizen who had obtained one of the trained dogs. She suffered a stroke and was in a coma for an extended period of time. It appeared she would not recover. Knowing of the special closeness and bonding between the woman and her favorite dog her son asked the attending physician to permit the pet dog to visit its comatose owner. Within days the woman recovered and independently resumed her lifestyle. Her son wrote to the program's director stating, "There is no doubt that the dog saved my mother's life." Funds for the program are donated privately. It receives no state or federal funding.

4.6 HUMAN–COMPANION ANIMAL BONDING

> The great pleasure of a dog is that you may make a fool of yourself with him, and not only will he not scold you, but he will make a fool of himself too.
>
> **Samuel Butler (1835–1902)**
> **English novelist and satirist**

The relationship between humans and companion animals is one of *symbiosis,* an intimate living together of two dissimilar organisms in a mutually beneficial relationship. For centuries people and animals have lived together in valuable caring relationships. As long as animals are treated compassionately they enjoy a loving bond with people. People and animals communicate in subtle ways and often seem to anticipate each other's feelings and needs (Figure 4.6). Dogs are social animals that thrive in a group where there is a well-established hierarchy. The dog should become the lowest-ranking member of the family in which it lives. When the dog sees its master as the leader of its pack it gives the same loyalty to its master as to a pack leader. Dogs seek attention and physical contact from their owners and, in return, provide them with companionship and affection.

Cats do not have the same pack instinct and are less dependent upon their owners as masters and leaders. However, most cats become dependent upon their owners for sustenance, comfort, and love and offer their own affection in return (Figure 4.7). A common example of a cat's affection toward its owner is an offering of its catch to the owner as if to say, "I bring you this fine, fat mouse because you are good to me and love me, and I am, in return, fond of you."

The human-animal bond is especially strong with children, the elderly, and in families where members are vulnerable to loneliness following a traumatic life change such as death of a family member, divorce, or departure of children from home. Bonding between the elderly and their companion animals is particularly strong. Pets give the elderly someone to talk to, take care of, nurture, and love. Research suggests that companion animals may permit the elderly to live independently in their own homes longer, experience better health, and reduce their dependence on drugs. This role of companion animals in improving the quality of life for the elderly and in decreasing their health care costs will become increasingly important as the world's population ages. The present population of people over 65 years of age in the United States is projected to double by the year 2030 and represent more than 20 percent of the total population. Cost-saving strategies will be needed to reduce the fiscal burden of health care costs. Additional research is needed to determine how companion animals can best contribute to cost reductions in health care.

Figure 4.6 Scott Rowland "sings" along with his dog Isaac during a fund-raiser for a local humane society.
Photograph by Chris Birks, Crystal Lake, IL.

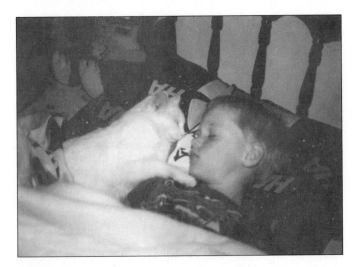

Figure 4.7 Boy and his favorite pet share a catnap.
Photograph courtesy of Sherry Behrens.

4.7 COMPANION ANIMAL BEHAVIOR

. . . we have become aware of the inherent emotional value of pets to human beings . . . as noncompetitive, nonjudgmental companions.

H. A. Nieburg
American author
from *Pet Loss: A Thoughtful Guide for Adults and Children*

Dogs descended from pack animals that were hunters and scavengers. They have excellent sight that enables them to see the slightest movement of potential prey; a sense of smell to scent game, territory odors, and emotional states in other animals; acute hearing to distinguish sounds over great distances; a refined sense of touch that provides pleasure in contact as when pack animals play together, huddle together for warmth, or lick and paw each other as a means of signaling rank within the pack. As pack animals dogs have a subtle and sophisticated range of communication methods largely based on body language and odors. In the absence of fellow canines, dogs look upon humans as other members of their pack. Separation anxiety results in howling, barking, pacing, and destroying objects as signs of the stress of being left alone. Urine scents are left in strategic places as markers to other dogs providing information about the sex and state of mind of the marker.

A dog exhibiting dominant behavior will maintain direct eye contact, hold its ears erect and alert, hold its tail high, brace its legs to allow the dog to lean forward, and have its muscles taut and ready for action. Ritual submissive behavior involves flattening its ears, lying down exposing its belly, tucking its tail, and raising a leg as a signal to more dominant animals that it is not a threat. The play-bow is a body language signal to invite other dogs or humans to chase or have a rough-and-tumble game with it. Dogs wag their tails to indicate friendliness and happiness; it is also a sign of submissiveness toward a dominant member of its pack. Understanding the meanings of a dog's body language can help prevent situations in which the animal shows aggression toward a person. For example, a person who behaves submissively toward a dog by crouching, avoiding eye contact, and showing fear may arouse the dog's sense of dominance and result in an attack. Direct eye contact with a dog exhibiting dominant behavior may be perceived as a challenge by the dog. The owner of a dog exhibiting dominant behavior toward the person should receive advice from an experienced dog trainer to guide him/her in establishing the human as the dominant member of "the pack."

In contrast to the friendliness prompting a dog to wag its tail, a cat flicking its tail is expressing displeasure of some sort. A cat knocked onto its back with its belly exposed does not show the submissive behavior of a dog in the same position; the cat in this position draws its hindquarters to its chest and rakes its razor-sharp claws into an opponent's belly while holding the attacker with its front feet and gripping the opponent's neck with its teeth. These examples illustrate the diversity of body language among various species. Cats often augment their body language with vocal sounds that convey a definite meaning. An angry cat will flatten its ears back against its skull, arch its back, hold its tail straight out, curl its lips, and spit and snarl at its opponent. Dilated pupils, a fluffed-up coat and tail, and an arched back signify fear.

Cats use soft physical touch such as a pat to the face to show affection toward their owners. Cats also express pleasure by "kneading," working their forepaws back and forth on their owner's lap or on a bed or pile of clothes with great determination as if they were kneading bread dough. Cats purr when content, when playing, and sometimes when hurt. They do not purr when hunting or actively involved in pursuit. There are many theories on the source of a cat's purr. Some believe the sound is produced by false vocal cords, others believe the sound is produced by vibrations in the walls of major blood vessels in the cat's chest. Cats have many intonations of their purrs and meows to express a variety of feelings. During kittenhood cats learn four basic games: play-fighting, prey-pouncing, bird-swatting, and fish-scooping. Many owners enjoy joining in these games with their kittens and continue to play the games during the adult life of their cats.

Cats are considered to be a "nonsocial" species because they do not live in groups as adults. In rural settings cats have territories that can be as large as 3 acres per female and 30 acres per male. Cats may share a "core area" but most of their hunting and other activity occurs within exclusive territorial boundaries. Crowding of cats creates territorial disputes and results in problems such as fighting and urine marking.

By nature cats are nocturnal animals that sleep during the day and hunt at night. Owners can help cats readjust their schedules by playing with them early in the evening. Fifteen or more minutes of playtime will help consume the cat's play energy and decrease "midnight frenzies." A typical outdoor cat will spend 40 percent of its time sleeping (includes numerous "catnaps"), 20 percent of its time traveling, 14 percent hunting, 2 percent eating, 12 percent grooming, and 12 percent resting.

Of all God's creatures, there is only one that cannot be made the slave of the lash. That one is the cat. If man could be crossed with the cat, it would improve man, but it would deteriorate the cat.

Mark Twain (1835–1910)
American writer and humorist

4.8 CARE AND TRAINING OF COMPANION ANIMALS

Man's utilization of animals cannot be taken for granted, but must be tempered with compassion, care, and concern. It is of little value that we serve the good on one hand if we do injustice on the other. In the further bonding of humans and animals in new and beneficial ways, not only can the lives of humans be enriched and made whole, but those animals which we use can be afforded genuine love, companionship, and care.

John A. Hoyt

Training and caring for animals provides both children and adults opportunities for developing self-esteem, empathy, responsibility, self-control, and the satisfaction of completing an important job. Ownership of a companion animal carries a dual responsibility—to meet basic needs of the pet and also to ensure its well-being in the community and in its relations with other animals, family members, friends, and neighbors.

Companion animals should be trained from an early age to be sociable and even-tempered. Early socialization will ensure that the animal is relaxed and comfortable in the presence of other animals and strangers. Many dog-training clubs offer puppy socialization classes to help novice owners train their puppies to be good canine citizens. Dogs should be trained to sit quietly in a group of other dogs without showing signs of aggression and to walk on a leash beside their owners in the "heeling position" and to not pull or drag a person. "Sit," "down," "stay," and "come" are basic commands that all dogs should be taught to obey. All dogs require daily exercise. The amount needed will vary depending on the dog's breed, age, and state of health. Dogs denied mental and physical activity may spend their energy in destructive behavior. Training a dog to retrieve objects is an excellent way to expend a dog's energy in a short period of time. Playing "fetch" with a dog helps reinforce the owner's authority, because the dog is dependent upon the owner to throw the object. Dogs provided with toys can exercise their minds and senses in their owner's absence. Chewing a toy can be a satisfying way of filling a dog's time when left alone.

Kittens that are handled by humans early in life will be socialized and more adaptable to interacting with humans. The most important characteristic to look for in a cat is its temperament. The ideal kitten or cat will be playful yet have a calm, stable disposition. A friendly outgoing kitten will respond to something dangled in front of it such as a string or peacock feather, or to something rolled such as a ball or empty spool. Play and exercise are important for a kitten to develop intelligence, dexterity, and social skills for interacting with humans and other animals.

Daily grooming of dogs and cats helps keep their coats clean, shining, and free of mats. Time spent in grooming helps develop a closeness between the owner and animal and becomes a time that both look forward to each day. Daily brushing to remove loose hairs is a must for longhaired cats to prevent formation of hairballs in the cat's stomach from self-grooming. Daily grooming enables the owner to detect skin disorders or other physical abnormalities that might otherwise go undetected. Dogs and indoor cats should have their nails trimmed monthly. Cats should also have access to a scratching post made of wood or carpet-covered wood. The mouths of dogs and cats should be checked monthly for tartar buildup, infected gums, and broken teeth. Dental problems require treatment from a veterinarian to prevent serious infections. Toothbrushes, finger cots with bristles, and "doggie toothpaste" are available for owners to use in minimizing the accumulation of tartar on their pet's teeth. Human toothpaste should not be used for pets because it may irritate the stomach and/or result in fluoride poisoning if consumed by the pet.

All dogs and cats should have an annual examination by a veterinarian (see Section 4.10). Pets should wear identity and license tags. Methods for permanent identification of an animal include tattoos and microchip implants. Dogs showing aggression toward strangers should be taught to wear a basket-type muzzle while in public places such as a veterinarian's office.

Dogs and cats are naturally clean animals that do not willingly soil their sleeping areas. Cats are naturally mimics and often learn to use a litter box as kittens by imitating their dam. If litter training is needed the kitten or cat should be taken to the litter box after eating or drinking, left there sufficient time for elimination, and then praised lavishly. Similarly, a puppy should be taken outdoors or to a newspaper for elimination after eating, drinking, playing, or napping. Sniffing the ground is often a sign that the puppy is looking for a place to eliminate; it should be scooped up and taken to the appropriate place and then praised lavishly. When the puppy is left alone it should be kept in a crate or similar confined area to encourage it to wait until the owner returns and takes it to another location for elimination. Many local ordinances require owners to "clean up" after their dogs. A plastic bag or "pooper-scooper" can be used to pick up feces for disposal. Canine feces are aesthetically unpleasant and can transmit internal parasites and thus should be disposed of in a responsible manner.

Millions of healthy dogs and cats are put to death each year in the United States for lack of people to adopt them. This problem is largely the result of allowing dogs and cats to breed without consideration of the availability of homes for the newborn. The decision to breed an animal should carry with it the responsibility of ensuring that the offspring are placed in loving homes. The decision to spay or neuter a dog or cat has health benefits for the animal and also prevents potential cruelty, starvation, and death for unwanted offspring. For male dogs neutering can help reduce fighting and aggression and prevent the development of prostatic disease. For male cats neutering can help decrease fighting, roaming and territorial aggression, urine spraying, and odor. For female cats and dogs early spaying reduces the risk of mammary cancer, eliminates the risk of uterine infections and tumors, and eliminates the nuisances of dealing with heat cycles when female dogs "spot" blood and attract stray male dogs in the neighborhood, and when female cats become extremely vocal with frantic rolling, restlessness, and amorous behavior. An ideal time to spay or neuter a dog or cat is around 6 months of age. Maximum protection from mammary cancer is obtained by spaying prior to the first estrous cycle.

4.9 FEEDING AND MANAGING COMPANION ANIMALS

Proper nutrition is essential for good health, performance, and long life of pets. Dietary management must consider the changing nutritional needs of the animal throughout its life during times of health and disease. A newborn puppy doubles its birth weight in about 8 days. To meet the nutritional needs of her offspring, the milk of a bitch is rich in protein (7.1 percent) and minerals (1.3 percent). (For comparison, a human infant requires approximately 6 months to double its birth weight and the protein and mineral content of human breast milk are 1.6 percent and 0.2 percent, respectively.) The puppy's high-nutrient needs continue during its growth period until full growth is reached between 12 and 18 months of age. Puppies between 2 and 9 months need 50 to 100 Kcal per lb of body weight compared with 30 to 45 Kcal per lb for active adult house dogs. It is easy to guarantee a well-balanced diet for a dog by feeding it a commercial food designed for its current life stage.

Commercial pet foods are available in canned, semimoist, and dry formulations. The major difference between these formulations is in their water content. Canned foods are typically 75 to 82 percent water, dry and semimoist foods are approximately 10 and 25 percent water, respectively. Many people incorrectly assume that canned foods are meat-based and dry foods are cereal-based. In fact, many canned food are primarily cereal grains shaped as "imitation meatballs" and many dry foods contain a high proportion of meat and/or poultry products. Commercial pet foods produced by reputable companies have undergone rigorous feeding trials and are complete and balanced to meet the nutritional needs of the targeted life stage(s) of the pet. The choice of dry versus canned formulations is one of personal preference for the owner. Canned foods are typically more expensive on a nutrient basis; however, some owners derive pleasure from feeding their pets canned food that closely resembles human food but is balanced to meet the pet's nutrient requirements.

Dry dog foods have the advantage of providing greater chewing exercise and help slow the accumulation of tartar by scraping the teeth. Dogs are not obligatory carnivores and meats need not comprise more than 25 percent of their total diet. A pet owner wishing to feed a homemade diet should consult with the pet's veterinarian and/or an animal nutritionist to ensure that the homemade diet is nutritionally complete and balanced. Some foods enjoyed by people are not good for dogs. These include chocolate (can cause convulsions and death) and onions (can damage a dog's red blood cells).

Cats are considered true carnivores. They have an absolute requirement for nutrients such as the fatty acid *arachidonic acid* and the amino acid *taurine,* which are found primarily in products of animal origin. As hunters cats consume the entire body of their prey including visceral organs and intestines filled with partially digested plant matter. A 2-month-old kitten requires approximately 125 Kcal daily per lb of body weight. The requirement decreases to 65 Kcal per lb of body weight by 8 months of age and to 40 to 45 Kcal per lb of body weight after 1 year of age. It may be as low as 30 Kcal per lb for a sedentary adult cat living indoors. The nutrient requirements of cats are notably different from those of dogs. The cat, compared with the dog, requires a higher protein diet (which must include taurine) and a higher fat diet (which must include arachidonic acid).

Feeding cats a food that is nutritionally adequate for dogs can result in blindness and heart disease in cats because taurine may not be present. The mineral content of the diet is also critical in cats. A diet high in magnesium may lead to development of a potentially fatal urinary tract obstruction called feline urinary syndrome (FUS) in cats (primarily males). Phosphoric acid helps lower cat urine pH (increases acidity) and helps prevent FUS. Cats fed diets containing large amounts of fish, especially red meat tuna, can develop steatitis, a painful and sometimes fatal inflammation of their body fat associated with vitamin E deficiency. Thus it is important that cats be fed a diet formulated specifically to meet their unique nutrient requirements.

Excellent commercial cat foods are available in dry, semimoist, and canned formulations. Dry foods are frequently less expensive on a nutrient delivery basis and have the advantage of helping decrease tartar accumulation on teeth. Because of lower mineral and higher water content, canned foods or dry foods designed specifically to counteract FUS deposition may be preferable for cats with a history of urinary tract obstruction. Some cats are finicky eaters and may refuse to eat certain diets. To minimize "finickiness" this author recommends routinely feeding cats both dry and canned cat foods.

Water is the most essential component of nutrition for the maintenance of life. It constitutes approximately 70 percent of an adult animal's body. Newborn puppies have 82 percent of their body weight as water. Living cells require a constant supply of water. Symptoms of dehydration are apparent following a 5 percent loss in total body water, a 10 percent loss results in severe morbidity, and a 15 percent loss is incompatible with life. Dogs and cats should have access to fresh, clean water at all times.

Many dogs and cats enjoy having their own beds such as a basket or airline crate lined with a washable mattress. Other cats and dogs may prefer to sleep in a quiet corner, under furniture, in a closet, a drawer, or with their owners. Even if not used for a bed, an airline crate is a good investment for most pet owners. Dogs and cats that are accustomed to being confined in an airline crate will experience minimal stress when placed in one for transport to a veterinary clinic or for safety during periods of turmoil within the house (e.g., during a noisy party). Often dogs and cats elect to use the security of their cages during any family stress. A crate is highly recommended for confinement during the house training of dogs.

Litter boxes are essential for indoor cats. Cats are reluctant to soil their sleeping area and naturally use litter boxes with almost no training because they prefer to urinate and defecate in loose material. There are many types of litter ranging from absorbent clays to sawdust, wood shavings, ground corn cobs, shredded paper, pelleted newspaper, sand, peat moss, and synthetic materials. No one litter seems to work best for all cats. The most obvious way to keep the contents of the litter box from becoming odorous is to keep it clean. The litter should be shifted daily and changed as often as necessary. Baking soda is commonly used to neutralize litter box ammonia and helps during cleaning. Ammonia should not be used because it worsens odor. Clorox should not be used because the combination of chlorine and ammonia from the urine produces a poisonous gas. Some people train their cats to use a toilet. This is convenient but requires patience on the owner's part and cooperation by the cat. The cat is first trained to use a litter box, the litter box is then moved to the toilet seat, the amount of litter placed in the pan is gradually reduced and a slit is cut in the pan to allow urine and feces to fall through. Eventually the cat may learn to balance itself on the sides of the seat so the pan can be removed.

4.10 HEALTH AND DISEASE ASPECTS OF COMPANION ANIMALS

Dogs and cats can suffer from almost all the physical diseases and ailments that afflict humans. In addition, each species has unique infectious diseases and hereditary disorders that may affect them. The increasing awareness of the importance of the human–companion animal bond has resulted in research into the cause and treatment of most diseases. Modern veterinary care provides an important array of preventive vaccines, diagnostic aids, and treatments to enable animals to live long, healthy lives.

Preventive medicine for dogs includes the provision of adequate exercise, health-sustaining foods, routine stool checks for intestinal parasites, heartworm preventative medication, and regular vaccinations against diseases such as distemper, hepatitis, leptospirosis, parvovirus, parainfluenza virus, adenovirus, and rabies (some rabies vaccines provide a 3-year immunity; other vaccines must be given annually for maximum protection). Other vaccines may be given on an individual basis (e.g., coronavirus and bordetella for dogs going to boarding kennels or dog shows). During the annual examination veterinarians check for signs of dental, heart, and ear diseases; cancer; external parasites (such as fleas, lice, and mange mites); and will evaluate a blood test for heartworms and other diseases as deemed appropriate for the individual animal (e.g., checking for diabetes or kidney or liver disease).

Preventive care for cats is similar to that of dogs. Vaccinations routinely recommended for cats will include panleukopenia, rhinotracheitis, calcivirus, and rabies. Other vaccines may be given on an individual basis (e.g., feline leukemia virus, chlamydia, feline infectious peritonitis) as deemed appropriate by the veterinarian and owner. Cats with recurring illnesses will be checked for infection with feline immunodeficiency virus and/or feline leukemia virus, two incurable viruses producing symptoms in cats similar to those of AIDS in humans (human immunodeficiency virus).

Zoonoses are diseases that can be transmitted to humans from animals (Chapter 22). The list of potential zoonotic diseases is long. However, in actuality the incidence of zoonotic diseases acquired from companion animals is low.

Systemic Diseases

Rabies is the most dangerous zoonotic disease because it is almost always fatal. The virus is spread in saliva and usually transmitted through bites. All dogs and cats should be immunized against rabies. Tuberculosis is a bacterial infection that can be spread from one species to another, often through bacteria present in sputum. Fortunately tuberculosis is rare in cats and dogs. Animals diagnosed with tuberculosis are routinely destroyed. *Toxocara* are roundworms commonly present as intestinal parasites of puppies and kittens. Humans (especially children) ingesting roundworm larvae can develop a condition known as visceral larval migrans (VLM) from larval worms migrating in the person's body. Migration of larval worms into the eye can produce blindness. During previous centuries almost all puppies and kittens harbored roundworms and required treatment with an anthelmintic (dewormer). With modern anthelmintics few "confined-to-the-home" dogs or cats have worms and the risk of VLM is low in areas where pets receive preventive health care. However, dogs and cats should have routine stool checks and be dewormed if intestinal parasites are found.

Toxoplasmosis is a systemic protozoal disease that can affect all species of animals, birds, and humans. Infected cats may develop diarrhea, pneumonia, neurological signs, and liver failure. Some infected cats do not show signs of disease but may still shed oocysts in their feces, which can infect people. The major danger in people is in transplacental infection causing birth defects in children. To minimize the risk of infection pregnant women should never touch cat feces—another member of the family should be responsible for changing the cat's litter box. Blood titers and/or cat fecal examination for *Toxoplasma* oocysts are used to screen for exposure to and infection with toxoplasmosis. Antibiotic treatment and/or coccidiostats are required to help eliminate infections. Leptospirosis is a bacterial infection that, depending upon the strain of *Leptospira,* can sometimes cause severe kidney and/or liver disease. The bacteria are passed in urine of infected animals. Many species including dogs, rodents, cattle, and people are susceptible to leptospiral infection. Vaccination can help prevent this infection in dogs; antibiotics are useful in treatment.

Intestinal Diseases

Campylobacteriosis and *Salmonellosis* are bacterial diseases that can cause abdominal cramps and diarrhea in humans and animals. Infections can be spread through bacteria in feces. Fecal swabs and cultures are used to diagnose these infections; antibiotics are effective in treatment. *Giardia* is a common protozoa that can cause diarrhea in animals and humans. The organism is passed in feces. Water from streams, particularly streams contaminated with beaver and other animal feces, is the usual source of infection. The organism can be diagnosed by fecal examinations and treated with antibiotics such as metronidazole.

Skin Diseases

Sarcoptes spp. and *Cheyletiella* spp. are mange mites that are transmissible among species. These mange mites produce itching and papular eruptions in animals and people. Parasiticides for use on dogs and cats are required to kill these mites. Occasionally the premises (e.g., house) must be treated. Another common mite species affecting dogs and cats is *Demodex* spp. (*Demodex canis* and *Demodex felis,* respectively). These mites are highly species specific and not transmissible to humans. Dogs and cats can be affected by lice. Lice are highly species specific and those affecting animals are not transmissible to humans. Fleas can affect both animals and humans causing itching and papular rashes. The adult flea spends most of its time on the pet. Flea eggs and immature fleas are found in the pet's environment. Successful control of a flea infestation for indoor pets often requires treatment of the environment (including daily vacuuming and daily vacuum bag disposal) in addition to treating all animals in the household.

Ringworm infections are fungi that cause circular areas of hair loss in animals and circular scaly skin lesions in people. Ringworm infections are diagnosed by culturing hairs adjacent to lesions in the dog or cat. Antifungal treatments are effective in clearing the lesions. The environment can harbor spores on infected hairs that were shed by animals, thus environment cleaning is required to prevent a reoccurrence of the infection. Bacteria such as *Pasteurella* are normal inhabitants of the mouth of dogs and cats. Bite wounds frequently result in purulent wounds and abscesses. People bitten by a dog or cat should wash the area and seek medical advice concerning prophylactic antibiotics, tetanus boosters, and rabies quarantine of the animal.

Allergies

Allergies are not communicable diseases. People developing allergies to animals have a hereditary predisposition to form a type of antibody known as immunoglobulin E (IgE) to foreign proteins. Allergic reactions to dogs and cats commonly result from proteins in the saliva or dander of animals. Frequent grooming and bathing of the animal, both done outside the home by a nonallergic person, can help minimize the concentration of the proteins causing the allergic reaction. People with severe allergic symptoms should consult a human allergist for advice regarding hyposensitization, adjuvant measures (such as control of pollens, molds, and house dust), or the necessity of removing animals from the household. Similarly, many dogs and cats develop allergies to environmental and dietary substances. Clinical symptoms of allergies in dogs and cats include excessive licking, scratching, and chewing with hair loss and secondary skin infections. These symptoms may be alleviated by use of oral antihistamines or by identification of the allergy cause by a veterinary allergist or dermatologist who will recommend specific treatment for the allergic pet.

Pet Health Care

Because dogs and cats cannot seek veterinary care on their own, their owners must notice signs of illness and take appropriate action. Changes in an animal's behavior, appetite, drinking, eliminations, habits, or appearance should prompt the seeking of advice from a veterinarian. Just as in human medicine, veterinarians now utilize sophisticated tests including endoscopy, ultrasonography, and computerized tomography in the diagnosis of health problems in pets. Veterinary specialists can provide sophisticated care for animals suffering from complicated health disorders.

Signs of aging appear at different rates in different breeds of dogs and cats. As dogs and cats become older their hearing and vision may deteriorate, their skin loses elasticity, their hair coat becomes thinner and drier, muscle tone decreases, and signs of arthritis may develop. As a general rule smaller breeds of dogs age slower than larger ones. The typical life span of small breeds such as a toy poodle is 14 to 17 years, that of middle-sized breeds such as a collie or German shepherd is 12 to 14 years, and that of larger-sized breeds such as a Great Dane or Newfoundland is only 7 to 11 years. The typical life span of a household cat used to be 12 to 14 years, but with today's improved nutrition and veterinary care the average life expectancy of a domestic cat is now about 17 years with occasional reports of cats living into their twenties and thirties. Older animals benefit from being fed high-quality, easily digested foods in sufficient quantity to meet their nutritional needs without excesses to burden the liver and kidneys; regular dental care to prevent dental infections; regular grooming to keep skin and hair coat clean and free of infections; and gentle massages to help keep limbs and joints mobile.

With regular care dogs and cats are able to maintain a good quality of life in their "golden years." When the time comes that the quality of life is no longer acceptable owners may consider euthanasia, which derives from Greek and means *death with peace*. Euthanasia is painless for the animal. It is simply an overdose of an anesthetic. However, with regard to a pet the decision to choose euthanasia may be the most difficult one an owner will ever make. This subject is replete with controversy and confusion. In circumstances in which a companion animal is terminally ill, carries a disease dangerous to humans, suffers from one or more ailments that cause it pain, or has experienced an overwhelming injury from which no successful recovery can be expected, euthanasia performed by a professional expert may be the most humane solution. A veterinarian should be consulted. The veterinarian can help the owner judge the seriousness of the animal's condition and the realities of the situation, and has the proper equipment, drugs, and professional expertise to perform euthanasia in the most humane manner possible.

4.11 CAREER OPPORTUNITIES ASSOCIATED WITH COMPANION ANIMALS

> His amiable eyes
> Are very friendly, very wise;
> Like Buddha, grave and fat,
> He sits, regardless of applause,
> and thinking, as he kneads his paws,
> what fun it is to be a cat!
>
> **Christopher D. Morley (1890–1957)**
> **American novelist and poet**

The enthusiasm and unbridled enjoyment of life shown by companion animals extends to those who work and play with them. Professions involving interactions with companion animals provide workers with opportunities to genuinely help people through their contributions to the human–companion animal bond. The ability to improve the quality of life of a pet and, therefore, its owner provides companion animal workers with a sense of accomplishment and personal satisfaction.

The first requirement for the human-animal bond is a suitable animal. For many people the breeding and raising of pure-bred dogs and cats is a hobby. For others it becomes a profession. The responsible breeder will only breed dogs or cats that (1) have excellent temperaments, (2) are good representatives of their respective breed, (3) are in excellent health, and (4) will produce offspring for which there is a demand and good homes can be found. To promote their breed and create a demand for dogs and cats, many hobby and professional breeders participate in dog and cat shows. Career opportunities associated with dog and cat shows include professional judges, trainers, groomers, handlers (persons who exhibit the animals in show rings), photographers, and show coordinators.

Many puppies and kittens are bought directly from the breeder. This is ideal because it minimizes stress for the animal and the prospective owner can obtain information on the background of the puppy or kitten (e.g., amount of socialization given, health history), observe the animal's interaction with littermates, and note the temperament of the mother. Other puppies and kittens are sold through dealers and pet shops. These are less desirable because the animals typically are poorly socialized and less attention has been given to pet health and longevity characteristics. The business of supplying pet accessories and equipment is large and growing. Pet shops sell toys, treats, over-the-counter medications, grooming equipment, feeding bowls, cages, carriers, kennels and their fittings, shampoos, cosmetics, personalized cat doors, cat condos, scratching posts, cushions, beds, collars, leashes, and even "doggie sanitary pads." The pet food industry has also expanded rapidly in recent years. A growing proportion of pet foods are now sold through specialty pet stores and veterinary offices (rather than through grocery or discount stores) with the seller providing advice to owners in the selection of foods to meet the nutritional needs of their pets. Career opportunities associated with the pet food industry include workers in nutrition, development, quality assurance, research, and sales.

Owners take great pride in the appearance of their pets. Professional groomers can offer a variety of clipping styles for dogs such as poodles and also provide skin and coat care for all breeds of dogs and cats. Many so-called "poodle parlors" are operated in conjunction with boarding kennels. The business of boarding companion animals has increased substantially in recent years, paralleling the increased number of owners traveling on business and taking extended holidays. Animals being imported from or exported to foreign countries may require quarantine in an approved kennel for variable lengths of time. In addition to traditional boarding kennels, pet-sitters are in great demand to care for pets in their home while owners are away.

Unfortunately many animals are abandoned and/or destroyed because of behavioral problems. Professional trainers and animal behaviorists can help owners learn techniques for correcting behavioral problems and thereby save the animal. The owner of a well-trained dog derives great pleasure and pride in the performance of his/her animal. Professional trainers can help owners teach dogs to "be good citizens" and are also important in the training of working dogs for hunting, herding, pulling, field trials, and obedience trial competitions as well as training Schutzhund dogs (personal protection), search and rescue dogs, hearing-ear dogs, seeing-eye dogs, drug-sniffing dogs, police dogs, and therapy dogs. Animal-assisted therapy is a rapidly developing area in which trained professionals use dogs to assist with physical therapy and with providing other services to physically, emotionally, or mentally handicapped individuals. Working with people having disabilities heightens one's respect for them and one's understanding of their problems. People attracted to pet-assisted therapy not only have the pleasure of working with animals but also the personal satisfaction derived from helping other people.

Half a century ago few veterinarians were engaged in companion animal practice. Today nearly three-quarters of the veterinarians in private practice in the United States are involved in providing health care to companion animals. Rapid advances in medical knowledge have opened many new important areas of specialization in veterinary medicine. The American Veterinary Medical Association currently has specialty boards in the areas of anesthesiology, animal behavior, cardiology, clinical pharmacology, dentistry, dermatology, emergency medicine and critical care, imaging (radiology), internal medicine, laboratory animal medicine, microbiology, nutrition, oncology, ophthalmology, pathology, preventive medicine, surgery, theriogenology, and zoological medicine. Sophisticated tests such as magnetic resonance imaging (MRI), computed tomography (CT scans), nuclear scans, endoscopic exams, and ultrasonography are available as aids in the diagnosis of diseases affecting companion animals.

The opportunities to *make a difference* in the lives of animals and their owners have never been greater than they are for today's veterinarian. Indeed, few people have the opportunity in their lives to genuinely help as many people as the small-animal practitioner. There are approximately 26,235 veterinarians practicing small-animal medicine in 13,736 practices in the United States. They attend to more than 157 million pet visits annually. That number, multiplied by an average of 1.7 persons accompanying each pet to the practitioner's clinic, is a total of approximately 267 million human contacts annually by the veterinary profession. Another opportunity in animal health care is a career as a veterinary technician or assistant.

4.12 SUMMARY

Companion animals can and do contribute immensely to the lives of people in all life stages and circumstances. Happily, animals accompany many of us on our journey through life. We enjoy, appreciate, and benefit from them for the many reasons mentioned in this chapter and they in turn need, enjoy, and appreciate us. It is impossible to assign a monetary value to the social, societal, and personal benefits derived from companion animals. Indeed, such things are simply priceless. Many readers of this chapter have personally experienced the closeness and one-on-one bonding that commonly exists and the affection that is frequently exchanged between people and their favorite animal friends. All of life is a learning experience, and companion animals add substantial depth and enrichment to our lives.

We should note that limited space has precluded our including more than a small fraction of the interesting information that has been published on topics related to companion animals. Interested readers can find dozens of books and hundreds of articles in popular magazines and professional journals available through bookstores, libraries, and on the Internet. Selecting and obtaining a companion animal involves assuming the responsibility for another living creature—a responsibility that continues for the rest of the life of the animal and/or owner. Understanding and undertaking this responsibility should include studying the best ways to make the life of the animal as long and healthy as possible. Acquiring new information is exciting and rewarding. It is in that spirit that we should all read and research further. Enjoy it!

STUDY QUESTIONS

1. What proportion of American families own one or more pets?

2. Give 6 or more ways in which dogs serve people.

3. What is meant by the term, *companion animal*?

4. Why are companion animals commonly considered members of the family?

5. Are wild animals recommended as pets? Why?

6. What are the most common companion animals in the United States?

7. Of what economic significance are the dog and cat food industries in the United States?

8. Name 10 or more pet-related industries in the United States.

9. What was the first animal species to be domesticated? When and where?

10. When and where were cats domesticated?

11. How many breeds of dogs are recognized by the American Kennel Club (AKC)? List the 7 categories of purebred dogs classified by the AKC. Cite 3 or more breeds in each category.

12. There are approximately how many established breeds of cats? Name the 3 types into which breeds of cats are divided.

13. Cite 10 or more ways in which companion animals contribute to the lives and well being of humans.

14. Cite 6 or more therapeutic uses of companion animals.

15. Name the 3 most common breeds used as guide dogs. What traits make them a good choice for that purpose?

16. What is meant by "pet-assisted therapy"?

17. Explain the symbiotic relationship between humans and companion animals.

18. Why is the human-animal bond especially strong with children and the elderly?

19. How may companion animals contribute to cost reductions in the health care of an aging population?

20. What traits enable dogs to be good hunters?

21. What is the meaning of a dog wagging its tail? Does this hold true in cats, as well?

22. Do cats purr at all times? Approximately what percentage of a cat's day is spent sleeping and resting?

23. What responsibilities are incurred in owning companion animals?

24. Why is daily grooming of companion animals important?

25. What is the commonly recommended age to spay or neuter a dog or cat?

26. How many days are required for a newborn puppy to double its birth weight? How long is required for a human infant to double its birth weight? Are these periods of time related to the protein and mineral content of milk of the bitch and human breast milk? Discuss.

27. Discuss the available forms of commercially prepared pet foods. What are the major differences in their water content?

28. Are dogs and cats true carnivores? Discuss the significance of this in formulating cat and dog foods.

29. Discuss special dietary needs of cats and dogs.

30. Do cats and dogs suffer from any of the physical diseases and ailments that afflict humans?

31. Discuss 8 or more preventive medicine measures for dogs and cats.

32. Of what significance are *zoonoses* to humans?

33. Cite examples of: (a) systemic diseases, (b) intestinal diseases, (c) skin diseases, and (d) allergies in cats and dogs. Name those having the zoonotic potential to affect human health.

34. Is the average life span the same in all sizes of dogs? Explain and give examples.

35. Which group of companion animals commonly lives the longest—cats or dogs? Is size of dogs commonly related to their life span?

36. What is meant by the term euthanasia? Cite circumstances that can lead to consideration of performing euthanasia.

37. Give 7 or more career opportunities associated with the care and management of cats and dogs.

38. Is the veterinary profession expanding or contracting with regard to small animal medicine in the United States? Explain.

39. How many specialty boards are provided professional oversight by the American Veterinary Medicine Association? Name them.

40. Where can curious canine and feline admirers find additional information pertaining to their favorite companion animals?

5 HORSES[1]

There is nothing as good for the inside of a man as the outside of a horse.

William Penn Adair (Will) Rogers (1879–1935)
American actor, author, and humorist

5.1 INTRODUCTION

The pathways of history are paved with the bones of horses. Archaeological records indicate that the horse was first **domesticated** approximately 5000 years ago (Chapter 1). The domesticated horse immediately took a leading role in the destiny of the human race. No other animal has so inspired the human imagination. History reports demon horses, angel horses, ghost horses, sea horses, headless horses, and others. The **centaur** of the early Greeks (a weird combination of a horse's lower body with a man's upper body and head) probably resulted from impressions formed by the first encounter of Greek warriors with the enemy mounted on horses. It is said that Native Americans also believed that horse and rider were one being when they first met mounted Spanish soldiers.

One of the first uses made of horses was that of helping conduct wars. As early as 2000 B.C. the Assyrians used horses to draw bronze chariots that carried armored soldiers. Soldiers of ancient Egypt and Greece rode horses in their conquests. Ghenghis Khan led his army of over 700,000 horsemen in the conquest of China. In addition to riding horses in battle his soldiers used them to round up large **herds** of goats and sheep. England had its own breeds of horses before the Roman invasion, but the Romans also brought horses with them so that interbreeding with British breeds likely occurred. In the age of knighthood in the British Isles each knight used horses in defending his king, religion, and

women. Horses were also used as pack animals for weapons, armor, and baggage during war.

Columbus brought five **mares** and two **stallions** to the West Indies in 1493. Spanish colonists later brought additional horses to the New World. Cortés took horses with him when he invaded Mexico in 1519. These were the first horses on the American continent since prehistoric days. Cortés later imported hundreds more to use in fighting the Aztecs. Some of the horses escaped as did some of those belonging to early explorers and missionaries who journeyed into the southwestern United States. The horses that escaped reproduced rapidly because of almost ideal conditions and by 1580 wild horses were so plentiful that Native Americans were capturing, breaking, and riding them. These **ponies** were called **cayuses** (named after the Cayuse now living in Oregon).

Horses changed the lives of Native Americans who learned to depend on the speed of horses to help hunt and kill wild buffalo, a major source of meat. They made ropes of twisted horsehair and used horsehides to make clothing, beds, tents, saddles, and moccasins. In many tribes the wealth of a man came to be judged by how many horses he owned. When a chief died his favorite horse was killed and placed in his grave with many of his other belongings, because of the belief that the chief's horse would go with him to the "hereafter."

Later, when cowboys used these wild horses to **work** (cut out, round up, move) cattle on the range, they were called **mustangs** (from the Spanish *mustengo* meaning "wild"). A **bronco** (**bronc**) was the name given a wild mustang that had not been tamed and broken. The cattle industry in the southwestern United States would not have thrived as it did without the tough, durable cow horse. Even today there are parts of the

[1]The authors acknowledge with appreciation the contributions to this chapter of Dr. David W. Freeman, Department of Animal Science, Oklahoma State University, Stillwater; Dr. Laurie M. Lawrence, Department of Animal Science, University of Kentucky, Lexington; and Dr. Gary D. Potter, Professor, Equine Program Leader, Department of Animal Science, Texas A & M Unvierstiy, College Station.

range where cattle cannot be worked without horses. The cowboy and his pony are as close as were the Native American and his pony. The horse also proved to be a hero of the Pony Express.

The development of farming in the grain belt of the United States depended on **draft** horses to pull farm implements. The 1900 Census of the United States Department of Agriculture (USDA) found 79 percent of all farms had horses. Later horses were replaced in large part by tractors. Horses gradually decreased in numbers during the mid-twentieth century. By 1992 the Census of Agriculture found only 18 percent of farms in the United States had horses. Recently horse numbers have increased significantly because of the popularity of light horses for shows, sports, racing, and pleasure riding. The USDA estimates that horses consume an amount of feed equivalent to 40 percent of that used by the poultry industry. The value of live horses exported from the United States in 1995 was $285 million compared with $86 million for live cattle exported that year.

A year-long study commissioned by the American Horse Council in 1999 found 6.9 million horses in the United States. These horses are owned by 1.9 million people; however, the impact of the horse industry is far greater than indicated by the number of horse owners. Indeed, over 7.1 million Americans are involved in the industry as horse owners, trainers, service providers, employees, riders, and volunteers. Tens of millions more Americans participate as spectators at horse races and shows. Approximately 725,000 horses are involved in the racing industry, 1,974,000 horses are used in showing, 2,970,000 are enjoyed as pleasure horses, and 1,262,800 are used in other activities including farm and ranch work, rodeo, polo, police work, and as therapy horses. The number of full-time jobs associated with horses exceeds 1.4 million. The total economic impact of the equine industry in the United States topped $112 billion in 1999.

Horses are also used as teaching animals in 4-H projects, schools and colleges, and in handicap riding programs. Additionally horses are used for the production of certain **vaccines** and **hormones;** for example, Jumbo, a **bay gelding,** provided enough blood over an 11-year period to produce tetanus **antitoxin** and pneumonia **antiserum** inoculations for 210,000 people. Pregnant mare serum gonadotropin (PMSG), also called equine chorionic gonadatropin (eCG), is harvested from the blood serum of mares between 40 and 80 days of pregnancy for use in certain hormonal therapies in veterinary medicine. One such therapy, P.G. 600® (400 IU PMSG and 200 IU human chorionic gonadatropin, hCG), is used to induce estrus in weaned sows and prepubertal gilts. Another product, Folligon® (containing 400 to 1000 IU PMSG per injection), is used to control fertility problems and for estrus synchronization in domestic animals including cattle and sheep. Currently Folligon is not approved for use in the United States, but it is approved for use in over 50 countries including Canada. Estrogens obtained from the urine of pregnant mares are widely used in estrogen replacement therapy to prevent osteoporosis in postmenopausal women.

Most horse owners do not live on farms; nevertheless, they are careful to maintain the health and well-being of their prized animals. It is to this group that this chapter is especially dedicated.

5.2 CHARACTERISTICS AND TYPES OF HORSES

The horse is a member of the order Perissadactyla (odd-toed hoofed mammals) and family Equidae. This family includes all living horses, donkeys, zebras, and **onagers,** as well as their many extinct ancestors. All living Equidae are members of the genus *Equus,* which includes seven **species:** domestic horses (*E. caballus*) and their closest wild relatives (*E. caballus przewalski*); the asses, domestic and wild; the onagers, wild asses of southwestern Asia; and three kinds of zebras.

The major external parts of the domestic horse are shown in Chapter 10. The skeleton of the horse is presented in Figure 5.1. A knowledge of the skeletal parts is useful in describing the location of certain abnormalities in horses.

5.2.1 Size and Type

Domesticated horses vary greatly in height and weight. American Miniature Horses, the smallest, weigh less than 300 lb. Ponies commonly weigh 300 to 800 lb; draft horses may weigh a ton (2000 lb) or more at maturity.

The height of horses is measured in **hands** from the ground to the highest point of the *withers,* the ridge between the shoulder bones. A hand is the average width of a man's hand, 4 in. In the formal "English" horse society, mature horses measuring less than 58 in (14.5 hands) in height are called *ponies.* Certain draft horses attain a height of 20 hands, whereas most horses of the riding and racing breeds are from 14 to 16.2 hands high. Racing horses weigh from 800 to 1100 lb and most riding horses from 1000 to 1400 lb.

Humans have developed the natural qualities of horses by **selective** matings. For example, speed and power have been vastly increased. **Breeds** for many other purposes such as riding and working livestock have been developed. More than 111 breeds of domestic horses are currently recognized.

Horses are generally divided into three types: *light horses,* which have small bones, thin legs, and weigh approximately 900 to 1200 lb at maturity; *heavy horses,* which have large bones, thick and sturdy legs, and weigh 1600 lb or more at maturity; and *ponies,* which usually weigh less than 800 lb at maturity. Each of these types includes several breeds and there may be more than one type within the same breed (e.g., there are Hackney *ponies* and *heavy* Hackneys). Horse enthusiasts often refer to horses as being "hot bloods" (which includes the majority of sport horse breeds and saddle horse breeds), "warm bloods" (includes many breeds used for dressage and jumping competitions), and "cold bloods" (draft horse breeds). Names of selected breeds of horses were presented in Table 2.1. The names and general usages of various breeds are given in Table 5.1. Twelve leading breeds of horses are shown on Color Plate 4D.

5.2.2 Gait

The way a horse walks or runs is referred to as its **gait.** In describing gaits consideration is given to speed, beat, and leg movement. Gaits overlap somewhat so that it is difficult to

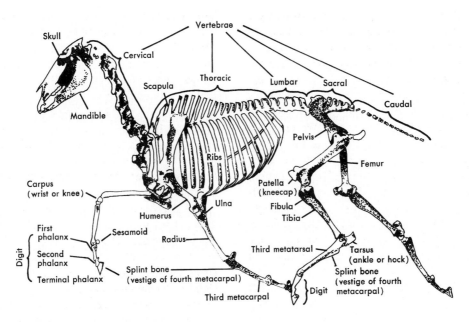

Figure 5.1 The skeletal system of a horse.
Courtesy American Museum of Natural History, New York.

TABLE 5.1	Selected Breeds of Horses Categorized by Form and Function

Saddle and sport horses	Light harness horses (roadsters)	Warmblood breeds	Wildhorses
American Saddlebred Horse	Morgan	Hanoverian	Ass
Appaloosa*	Standardbred (American trotting horse)	Holsteiner	Forest horse
Arabian		Trakehner	Przewalski's horse
Buckskin*		Warmblood (American, Australian, Dutch, and Swedish)	Zebra
Lipizzan (Spanish Riding School horses)	Draft horses	Westphalian	
Missouri Fox Trotter	Belgian		
Morgan	Clydesdale		
Mustang	Friesian	Ponies	
Paint*	Percheron	American Miniature Horse	
Palomino*	Shire	Connemara (good jumper)	
Paso Fino	Suffolk Punch	Dartmoor (strong and surefooted)	
Pinto*		Hackney	
Quarter Horse (highest registration)	Heavy harness horses (coach horses)	Haflinger (good jumper)	
	Cleveland Bay	Pony of the Americas (small Appaloosa)	
Tennessee Walking Horse	French Coach (Normand)	Shetland	
Thoroughbred	German Coach (Oldenburger)	Welsh	
	Hackney		

*Color breeds.

describe them clearly. The term "lead" is used to refer to which foreleg of the horse is the farthest ahead ("leading"). If a horse is on the left lead its left leg is the farthest forward; this is the correct lead for a horse circling counterclockwise. A horse circling clockwise should be on its right lead with its right forefoot landing the farthest forward.

Horses have four natural gaits: *walk, trot,* **canter** or **lope,** and *gallop. Walk* is the slowest gait (approximately 4 miles per hour, mph) with four beats (e.g., right hind, right fore, left hind, left fore). At the walk the horse always has two or three hooves on the ground making this a very steady and comfortable gait for both horse and rider. The *trot* is a two-beat gait in which each diagonal pair of legs (**near** hind and **off** fore, followed by off hind and near fore) hits the ground concurrently. This gait has a period of suspension when the horse springs from one diagonal to the other. During this transition ("spring") all four feet are off the ground. The trotting speed is approximately 9 mph. Riders often post to make the trot more comfortable for the horse and rider. To post the rider rises out of the saddle for one beat and then sits down in the saddle for the next beat. English riders should post on the outside diagonal, this means the rider is in the up (out of the saddle) position when the horse's outside foreleg (the one closest to the outside of the ring) is the farthest forward. The *canter* (English) or *lope* (Western) is a comfortable, three-beat rhythmic riding gait at 10 to 12 mph. There are three hoofbeats to each stride of the canter. The steps of the right-lead canter are (1) left hindfoot, (2) right hindfoot and left forefoot, and (3) right forefoot. The steps of the left-lead canter are (1) right hindfoot, (2) left hindfoot and right forefoot, and (3) left forefoot. Many people consider the canter to be a slower version of the gallop. A galloping horse may move at speeds of 40 or more mph. The *gallop* is a four-beat gait in which the two hind feet give a two-beat. The steps of the right-lead gallop are (1) left hindfoot, (2) right hindfoot, (3) left forefoot, and (4) right forefoot. The steps of the left-lead gallop are (1) right hindfoot, (2) left hindfoot, (3) right forefoot, and (4) left forefoot.

Horses may be trained to use four artificial gaits: pace, slow gait, fox trot, and rack. These gaits are natural in some breeds of horses. When a horse *paces* it moves the legs on the same side of the body at the same time, and with some pacers there is a movement from side to side as well as up and down. The natural pacing ability of a horse is influenced greatly by **heredity.** The **pace** is an uncomfortable riding gait and is avoided by those using cow ponies on the range. The **slow gait** is a high, methodical, showy, stepping pace done very slowly and with restrained speed. A slight variation of the slow gait is the *running walk*. This is a four-beat gait seen most commonly in the Tennessee Walking Horse in which each foot hits the ground separately and never in combination with one of the others. It is quite comfortable for the rider. The **fox trot** is a broken trot in which the hindfoot reaches the ground slightly ahead of the diagonal forefoot. This gives the appearance of walking in front and trotting behind. It is a medium-speed four-beat gait that is easy on both horse and rider. The **rack** is a rapid four-beat gait free from any lateral or pacing motion in which the legs move in pairs, but not quite simultaneously, so that each foot is lifted and put down separately. The rear foot strikes the ground a fraction of a second before the front on the same side. An experienced person can recognize this gait by ear if blindfolded. It is normally found only in the American Saddlebred **five-gaited horse.**

5.2.3 How Horses Perceive and React to Their Surroundings

Horses are quite perceptive to sights, sounds, movement, touch, and smell. They have long survived on their ability to quickly recognize and respond to threats of danger.

The eyesight of horses is keen in both daylight and darkness. Horses are believed to be capable of distinguishing between their masters and other people at a distance of one-half mile. Their oval-shaped eyes are set at the sides of their head and move independently. Although horses focus on objects in front of their body similar to how people see (binocular vision), eye placement along the sides of the head also allows them to focus on two different fields of vision concurrently (monocular vision). Horses have areas along their body where vision is restricted. There is a blind spot in front of and under the head. The horse also has marginal sight along the side of the hips and cannot see directly behind its body unless it turns its head and neck to the area. Some horsepersons use *blinkers* (blinders that fit alongside the eyes) to block a horse's side and rear vision.

Horses have a well-developed sense of hearing. They will direct their ears toward a sound, and handlers can observe ear placement to determine where a horse is directing its attention.

The sense of touch is also highly developed in horses. Usual sensitive areas are the head, ears, flank, and lower legs. Gentling techniques such as brushing and rubbing are done to increase the horse's acceptance to handling and training.

Wild horses have learned to react quickly to danger. Being a prey specie, their main defense is to flee from danger. Their other defense is to fight. Domestic horses show this same "fight-or-flight response" when they interact with people, other horses, and their surroundings. Understanding how horses perceive and react to their environment helps horse owners train and handle these prized companions safely.

5.2.4 Coat Color

The **coats** of horses are many colors and combinations of colors. Solid body colors include black, brown, **bay** (reddish brown), buckskin (golden), liver chestnut (dark reddish brown), light chestnut or **sorrel** (red shades of chestnut or yellowish brown), **palomino** (golden), gray, and white. Many gray horses are born black but gradually turn lighter with age. Some gray horses (e.g., the **Lipizzans** of Austria) are white by maturity. Bays and buckskins have black *points* (legs, mane, and tail). Palominos have white manes and tails. Chestnuts vary the most in color.

Some horses have mixed colors and these are given various names. **Piebald** indicates a black coat with white spots; **skewbald** is any color, except black, spotted with white (any piebald or skewbald horse may also be called a **pinto**); *pied* is a solid-colored coat with a few small spots; **tobiano** is a white horse with large spots of color, the white crosses the topline; **overo** is a colored horse with jagged white markings originating on the side or belly and spreading upward but rarely crossing the topline; **roan** is a solid color (red or black) with white hairs growing throughout the coat.

Horses may be grouped according to *color type*. The terms *pinto* and *paint* color types are often used synonymously. Pinto horse registries place emphasis on coat color pattern for qualification purposes in registration. As such, many different breeds may be represented in pinto registries. Paint horses are, however, restricted for registration to horses with registered paint, Quarter Horse, or **Thoroughbred** breeding. Palominos have golden coats and light blond or silvery manes and tails. **Appaloosas** typically have dark brown or black leopard spots on a colored background. They have also been called "raindrop" horses because of their spots. The Appaloosa color pattern in horses is variable with at least six basic patterns: (1) frost, (2) leopard, (3) marble, (4) snowflake, (5) spotted blanket, and (6) white blanket. The skin is mottled around the **genitalia** and sometimes around the eyes and nostrils. The **sclera** of the eye is white, the hooves of an Appaloosa are striped. True *albinos* have a white or pale-colored coat and pink eyes. In nearly all mammals except the horse the gene for albinism is recessive. However, in horses the gene for albinism is believed to be **lethal,** so that apparently no horses are *true albinos*. Pseudo (false) albinos may have a light skin and hair coat, but their eyes are colored. **Cremellos** are horses that are such a pale cream that they appear white. These horses have blue eyes.

Markings differ considerably among horses. Selected markings are depicted in Figure 5.2. Short descriptions of these markings are as follows:

Snip is any marking, usually vertical, between the two nostrils.

Star is any marking on the forehead.

Strip is a long vertical marking running down the entire length of the face from the forehead to the nasal peak.

Blaze is a broader, more open strip.

Star and strip is a marking on the forehead with a strip to the nasal peak; the strip does not have to be an extension of the star.

Star, strip, and snip are markings on the forehead with a narrow extension to the nasal peak and opening up again between the nostrils.

Bald is a broad blaze; it can extend out and around the eyes and down to the upper lip and around the nostrils.

Coronet is any narrow marking around the coronet above the hoof.

Half pastern is a marking that includes only half the pastern above the coronet.

Lightening mark is an irregular white marking on the leg that does not contact the hoof.

Pastern is a marking that includes the entire pastern.

Sock is a marking that extends around the leg from the coronet halfway up the **cannon bone,** halfway to the knee on the foreleg, or halfway to the hock on the back leg.

Stocking is a full marking almost to the knee on the foreleg and almost to the hock on the hind leg; it is an extended sock.

Suggested modes of inheritance of the many coat colors are presented in Tables 5.2 and 5.3. Not all geneticists agree on these. Changes will be made as additional information is gained from research.

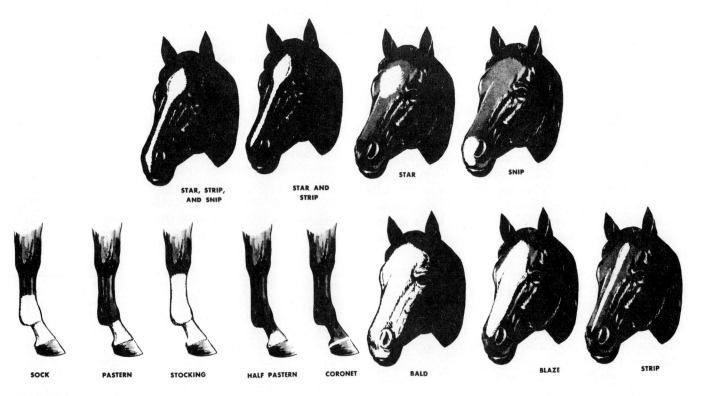

STAR, STRIP, AND SNIP **STAR AND STRIP** **STAR** **SNIP**

SOCK **PASTERN** **STOCKING** **HALF PASTERN** **CORONET** **BALD** **BLAZE** **STRIP**

Figure 5.2 Common markings of horses.
Courtesy American Quarter Horse Association.

TABLE 5.2	Genes Thought to be Involved in Determining Coat Colors in Horses

Gene(s)*	Coat Color(s)
AA, Aa	A is a pattern determiner, making color appear in the mane and tail, causing black mane and tail, or white or light mane and tail of the sorrel and palomino.
aa	Uniform color, mane and tail same color as body.
BB, Bb	Black pigment in hair, skin, and eyes.
bb	Brown pigment in hair, skin, and eyes.
c^{cr}	Sometimes called the palomino gene. It dilutes bay to buckskin, chestnut or sorrel to palomino when heterozygous and to glass-eyed whites when homozygous, regardless of the background genotype for color. It has no effect on black or mouse colors.
D	D is a dominant dilution gene separate and apart from the c^{cr}, gene. It converts black into mouse, bay into yellow dun with dark mane and tail, and chestnut or sorrel into yellow dun with dun mane and tail.
DD	Extremely dilute color (almost white); D is **incompletely dominant** to d.
Dd	Moderately dilute color (palomino).
dd	Allows color with no dilution.
EE, Ee	Full extension of black or brown pigment throughout the coat except as modified by the A gene.
E^d	Dominant black color, nonfading, Shetland type, completely covers the effect of the A gene. Thus E^d is **epistatic** to A.
ee	Red and yellow range in the body coat.
RR, Rr	**Roan.** (Also referred to as the Rn gene.)
rr	Nonroan.
GG, Gg	Progressive graying; any color at birth but grays with age.
gg	Does not cause a progressive graying of the hair.
SS, Ss	**Piebald** and **skewbald** spotting.
ss	Nonspotted, solid color.
FF, Ff	Face and legs solid color.
ff	White face and legs; ranges from extreme to star in forehead.
Ww	White (WW not seen—probably lethal; W is epistatic to all other colors.
ww	Allows pigmentation with other color genes.
W^{ap}	Appaloosa with blanket.

*None of these genes appears to be linked on the same **chromosome.**

5.3 SELECTION OF HORSES

The fact that horse breeders have been able to develop many types and breeds of horses is sufficient evidence to conclude that their form, structure, and function are strongly inherited, just as has been shown with food-producing animals.

The improvement of horses through the application of proven breeding methods involves finding/identifying superior individuals and then mating them with the expectation that they will produce superior **offspring.** Superior individuals may be located and identified in several ways. These include close observation of the type and performance of an individual and its close relatives. **Type** refers to the physical form and structure that should best fit individuals for a particular function. To know what is the most desirable form and structure, one must become familiar with the parts of a horse (Figure 10.5, page 182) and the characteristics of importance for specific purposes. One must also be able to evaluate and compare body conformation (type). For this reason a guide for evaluating conformation is presented in Section 5.3.1. Performance involves proper functioning of the horse. Thoroughbreds are **selected** for speed; saddle horses for various gaits; draft horses for the ability to pull heavy loads; and show horses for their ability to compete in performance events, dressage, or halter (conformation) classes. To be successful in finding superior breeding stock one must have a definite performance goal in mind.

The selection of breeding stock can be based on the animal's own type and performance, often called *individuality.* Three kinds of records of relatives may also aid in selection of breeding stock: those of an individual's ancestors in its **pedigree,** those of its **progeny** or descendants (if any), and those of its **collateral relatives** (equivalent to human brothers, sisters, uncles, aunts, and cousins). The latter are not related to the individual as ancestors or descendants (Table 9.5, page 170).

A rule of thumb for improving horses through breeding methods is to select superior individuals that have superior relatives. The closer the relationship between an individual and its relatives, the more attention should be given to the type and performance of those relatives. Most of the major horse breed associations now publish this information.

5.3.1 Evaluating Body Conformation of Horses

The usefulness and, to a large extent, the longevity of a horse are dependent on its conformation. Sound, strong feet and legs, for example, are essentials in all types of horses that serve humanity. Because conformation tends to denote service and utility, a few selected desirable and undesirable traits are presented in Figure 5.3 on p. 104.

5.4 CARE AND MANAGEMENT OF BREEDING HORSES

The reproductive rate in horses is often lower than that of other farm mammals because horses have a comparatively longer **gestation** period, are older when they first produce young, and commonly produce only one **foal** at **parturition.** However, mares have a long reproductive life, often reproducing until they are 20 or more years of age. Fertility rates among mares on most horse farms exceed 80 percent, depending on the level of management and care. Rates are usually higher when horses are on pasture than when confined. (This assumes that the stallion:mares ratio on pasture is within the stallion's ability to impregnate the mares.)

Spring is the usual breeding and **foaling** season for horses, although some foals are produced during other seasons. Spring foaling is advantageous because of the warm weather and abundance of green grass for both mother and foal. These induce fast growth and development of the young.

TABLE 5.3 | Suggested Genotypes of Horses of Different Coat Colors

	Gene Pairs (or Alleles)									
Color	A, a, a^t	B, b	W, w	C, c^cr	D, d	E^d, E, e	R, r	G, g	S, s	Comments
Bay										
Normal	AA or Aa	BB or Bb	ww	CC	dd	EE or Ee	rr	gg	ss	Bay, black mane and tail
Red	AA or Aa	BB or Bb	ww	CC	dd	ee	rr	gg	ss	Red, black mane and tail
Black	AA or Aa	BB or Bb	ww	CC	dd	E^dE^d or E^de	rr	gg	ss	Dominant black, will not fade
Normal	aa	Bb or Bb	ww	CC	dd	EE or Ee	rr	gg	ss	Black (fades brown in sun)
Smoky	aa	BB or Bb	ww	CC	dd	ee	rr	gg	ss	Black (fades red in sun)
Dilute	aa	BB or bb	ww	CC	Dd	Ee	rr	gg	ss	Dirty black; sometimes called good black
Chestnut	AA or Aa	bb	ww	CC	dd	EE or Ee	rr	gg	ss	Dark-brown mane and tail
Liver chestnut	aa	bb	ww	CC	dd	EE	rr	gg	ss	Uniform brown body, mane and tail
Seal brown	a^ta^t	BB or Bb	ww	CC	dd	EE or Ee	rr	gg	ss	Very dark brown
Sorrel	AA or Aa	bb	ww	CC	dd	ee	rr	gg	ss	Light mane and tail
Sorrel, uniform	aa	bb	ww	CC	dd	ee	rr	gg	ss	Uniform color body, mane, and tail
White			WW or Ww							WW is lethal, not seen
Gray			ww					GG or Gg		Foal colored at birth becomes gray with age
Piebald and skewbald (see Sec 5.2.4)			ww						SS or Ss	Spotted any color with white, depending on color of parents
Roan										
Red	AA or Aa	BB or Bb	ww	CC	dd	EE or Ee	Rr	gg	ss	Bay and white with dark mane and tail
Blue	aa	BB or Bb	ww	CC	dd	EE or Ee	Rr	gg	ss	Blue (black and white hairs)
Strawberry	AA or Aa	bb	ww	CC	dd	ee	Rr	gg	ss	Chestnut or sorrel and white with light mane and tail
Buckskin	AA or Aa	BB or Bb	ww	$c^{cr}C$	dd	EE or Ee	rr	gg	ss	Sooty cream, black mane and tail
	AA or Aa	BB or Bb	ww	$c^{cr}C$	dd	ee	rr	gg	ss	Clear cream, black mane and tail
Palomino	AA or Aa	bb	ww	$c^{cr}C$	dd	EE or Ee	rr	gg	ss	Sooty type, white mane and tail
	AA or Aa	bb	ww	$c^{cr}C$	dd	ee	rr	gg	ss	Clear type, white mane and tail
	aa	bb	ww	$c^{cr}C$	dd	EE or Ee	rr	gg	ss	Sooty type, mane and tail same color
	aa	bb	ww	$c^{cr}C$	dd	ee	rr	gg	ss	Clear type, mane and tail same color
Pseudo albino	AA or Aa	bb	ww	$c^{cr}c^{cr}$	dd	EE or Ee	rr	gg	ss	Pink skin, blue eyes, sooty pale cream
	AA or Aa	bb	ww	$c^{cr}c^{cr}$	dd	ee	rr	gg	ss	Pink skin, blue eyes, clear pale cream
	aa	bb	ww	$c^{cr}c^{cr}$	dd	EE or Ee	rr	gg	ss	Cremello
	aa	bb	ww	$c^{cr}c^{cr}$		ee	rr	gg	ss	Cremello
Mouse, grulla*	AA or Aa	BB or Bb	ww	CC	Dd	EE, Ee, or ee	rr	gg	ss	Sooty black
Dun*	AA or Aa	BB or Bb	ww	CC	Dd	EE, Ee, or ee	rr	gg	ss	Yellow dun with dark mane and tail
Dun*	aa	bb	ww	CC	Dd	ee	rr	gg	ss	Yellow dun with dark mane and tail

*The actual genotype and phenotypes are somewhat controversial at present.

Source: Courtesy Dr. M. G. Neuffer, University of Missouri–Columbia and Dr. Laurie M. Lawrence, University of Kentucky–Lexington.

5.4.1 The Stallion

Although the stallion attains **puberty** in about 1 year he is seldom used for breeding before 2 years of age. Even then he is used on a limited number of mares because he has not fully matured in his ability to produce semen. Because of this most breeders do not use a stallion for breeding purposes until he is at least 3 years of age. Older, mature, highly fertile stallions may be hand-mated (closely scheduled and supervised natural matings) with 75 to 100 mares during a long breeding season. Usually breeding is limited to once daily. However, if the occasion demands, two well-spaced services per day are possible for a short period of time. Proper **nutrition** and exercise are important in maintaining good physical **condition** in stallions to be used for breeding purposes.

Artificial insemination of horses has been practiced for many years (Chapter 14), and is used on most farms if breed associations permit it. In 1988 the Hamilton–Thorne Company of Massachusetts developed a specialized container for the transport of horse semen. This container was developed to lower the temperature of the semen by 32.5°F (0.3°C) every minute until the temperature stabilizes at 39° to 43°F (4° to 6°C). This container can be used to store and ship semen for use within 3 days following collection. For storage beyond 3 days horse semen is diluted with a nutrient-rich freezing extender and frozen in straws placed in liquid nitrogen (temperature −321°F, or −196°C). The majority of breed associations, with the exception of the Thoroughbred Jockey Club, allow the use of transported chilled or frozen semen.

The volume of semen per **ejaculate** from the stallion is from 50 to 150 cc (the average for Thoroughbreds is 50 to 60 cc), and the normal **sperm** concentration is 30 to 800 million per cubic centimeter. Thus the total number of sperm per ejaculate is in the billions. Sperm production varies with an individual, its age, and **environmental** conditions.

5.4.2 The Mare

Fillies (young females) reach sexual maturity between 10 and 18 months of age, but they do not reach physical maturity (mature body size) until age 3 to 5 years. If kept as **broodmares** they are not usually bred until they are at least 3 years old. Saddle or racing mares are not commonly bred until they are older.

The usual estrous season begins in the spring and continues through the summer. As such, mares are called long-day breeders because most mares do not cycle (become anestrous) during seasons of short day length (late fall to early spring).

Once cycling, the length of the estrous cycle in normal mature mares averages 19 to 21 days. The length of **estrus** varies from 2 to 9 days, with an average of 5 to 6. **Ovulation** most frequently occurs 1 or 2 days before the end of estrus (Chapter 13). In some mares ovulation occurs even though they do not show outward signs of estrus (silent estrus). Commonly only one egg is released from the **ovary,** although twin ovulations may occur (up to 20 percent) depending on breed and season. Twin ovulations seldom result in delivery of two live foals, thus breeding farms often screen mares using ultrasonography for twin pregnancies and destroy one of the fetuses.

Because thin mares have lower reproductive efficiency, mares should enter the breeding season with fat cover over the ribs and evidence of fat on the tailhead, neck, and behind the shoulder (a minimum moderate to fleshy condition). Mares in healthy physical condition at breeding time are more likely to conceive. Proper feeding, health care, and exercise are vital. Mares usually obtain sufficient exercise if allowed to run in large pastures. When confined, however, they should be exercised regularly.

One can usually determine if a mare is in **heat** by observing her acceptance to a stallion. The stallion is restrained by a handler or pen. He is allowed to tease the mare with a barrier such as a rail, stall, or teasing pen between them. The mare may be bred if she is receptive to the stallion. The optimum time to breed mares, if only a single service is given, is late in the heat period because this is when most ovulations occur. To increase conception one may breed mares on alternate days during estrus if the stallion is not being used too frequently on other mares.

During estrus quiet mares will stand for service, whereas nervous ones may have to be **twitched** or **hobbled** to prevent movement and the possibility of their injuring (kicking) the stallion. The use of strong breeding hobbles or a properly constructed breeding stall is recommended. Occasionally mares may be tranquilized. Most mares come into heat 5 to 10 days after foaling (foal heat). However, breeding at this time is recommended only for mares free of **infection.** Uterine infections are indicated by the birth of a diseased or dead foal and a retained **placenta.** Mares with uterine infections should be treated and not bred until they return to a healthy condition.

Pregnancy is indicated by the cessation of the usual estrous periods. If a stallion is kept on the same farm it is desirable to check mares for estrus on alternate days, beginning the fifteenth day after breeding. Pregnancy may be diagnosed by means of rectal palpation, ultrasound, or blood tests. Ultrasonography and palpation can detect the presence of an embryo as early as a few weeks after conception. Blood tests used to detect heightened levels of hormones present during pregnancy can be used between 40 and 80 days of pregnancy.

The gestation periods of mares average approximately 340 days but vary between 335 and 345 days. A mare shows several signs of approaching parturition. Her **udder,** often enlarging for several days, becomes distended with milk, which may drip from the teats 24 to 48 h before foaling. Beads of wax form on the teat canal in some mares just before foaling. The muscles around the tail head relax and the lips of the vulva become enlarged. Mare restlessness is indicated by successive lying down and rising.

Foaling usually occurs while the mare is lying down. Shortly before parturition the fetal membranes (water bag) appear. They rupture and soon the front feet and nose of the foal appear at the vulva. The normal and usual presentation of the foal is front legs first and outstretched, with the head and chin resting between the legs. The back of the young should be near the top of the mare's pelvis. Any other presentation is abnormal and increases the probability that the mare will require assistance for completion

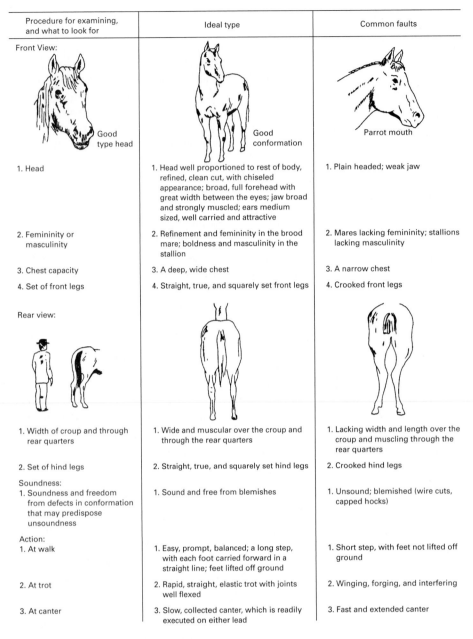

Procedure for examining, and what to look for	Ideal type	Common faults
Front View:	Good type head	Parrot mouth
	Good conformation	
1. Head	1. Head well proportioned to rest of body, refined, clean cut, with chiseled appearance; broad, full forehead with great width between the eyes; jaw broad and strongly muscled; ears medium sized, well carried and attractive	1. Plain headed; weak jaw
2. Femininity or masculinity	2. Refinement and femininity in the brood mare; boldness and masculinity in the stallion	2. Mares lacking femininity; stallions lacking masculinity
3. Chest capacity	3. A deep, wide chest	3. A narrow chest
4. Set of front legs	4. Straight, true, and squarely set front legs	4. Crooked front legs
Rear view:		
1. Width of croup and through rear quarters	1. Wide and muscular over the croup and through the rear quarters	1. Lacking width and length over the croup and muscling through the rear quarters
2. Set of hind legs	2. Straight, true, and squarely set hind legs	2. Crooked hind legs
Soundness: 1. Soundness and freedom from defects in conformation that may predispose unsoundness	1. Sound and free from blemishes	1. Unsound; blemished (wire cuts, capped hocks)
Action: 1. At walk	1. Easy, prompt, balanced; a long step, with each foot carried forward in a straight line; feet lifted off ground	1. Short step, with feet not lifted off ground
2. At trot	2. Rapid, straight, elastic trot with joints well flexed	2. Winging, forging, and interfering
3. At canter	3. Slow, collected canter, which is readily executed on either lead	3. Fast and extended canter

Figure 5.3 Guide for evaluating the conformation of light horses.
Courtesy Dale Foster, Albers Milling Company. From "Selecting, Feeding, and Showing Horses," Albers Research Bull., 1966.

continued on page 105

of the birth process. If labor is too difficult and prolonged (normal delivery time is 10 to 15 min), a veterinarian should be called. It is interesting that sizable breeding fees (often several thousand dollars in **purebred** horses) charged for selective matings are guaranteed by the result of a live, nursing foal. When foaling appears normal it is better not to disturb the mare, although assistance from trained personnel may be given after the withers of the foal have passed the rim of the pelvis. Pulling the legs downward and outward toward the mare's hocks when contractions occur is often helpful. Once the fetal membranes have ruptured normal delivery requires only a few minutes, and delays (especially more than 45 min) may result in the death of the foal.

One should make certain that the foal is breathing after delivery and that all fetal membranes are removed from the mouth and nostrils. If the foal is not breathing, blowing into its mouth, tickling the nostril with a straw, working the ribs in artificial respiration, or lifting the foal and dropping it gently to the ground will often initiate the breathing process.

The **afterbirth** (placenta) is usually expelled within 2 to 6 h postfoaling. If it is retained longer than 6 h, a veterinarian should be called. Even when the afterbirth is expelled normally, the hindquarters and tail of the mare should be washed and **disinfected** and stall litter or other debris removed. This aids in preventing uterine infections.

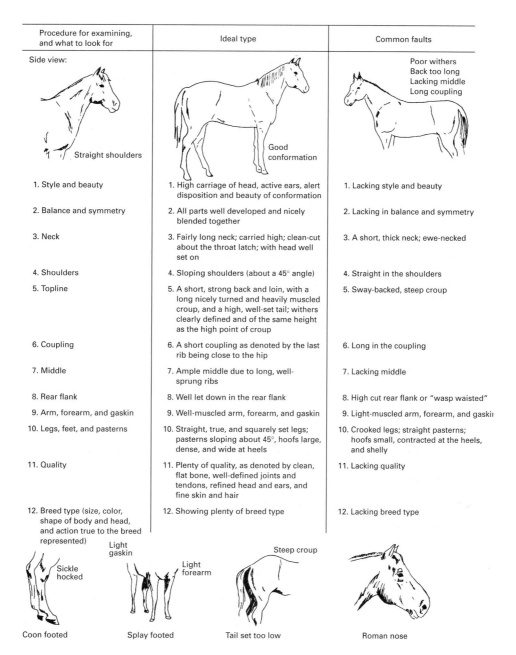

Procedure for examining, and what to look for	Ideal type	Common faults
Side view:		Poor withers Back too long Lacking middle Long coupling
Straight shoulders	Good conformation	
1. Style and beauty	1. High carriage of head, active ears, alert disposition and beauty of conformation	1. Lacking style and beauty
2. Balance and symmetry	2. All parts well developed and nicely blended together	2. Lacking in balance and symmetry
3. Neck	3. Fairly long neck; carried high; clean-cut about the throat latch; with head well set on	3. A short, thick neck; ewe-necked
4. Shoulders	4. Sloping shoulders (about a 45° angle)	4. Straight in the shoulders
5. Topline	5. A short, strong back and loin, with a long nicely turned and heavily muscled croup, and a high, well-set tail; withers clearly defined and of the same height as the high point of croup	5. Sway-backed, steep croup
6. Coupling	6. A short coupling as denoted by the last rib being close to the hip	6. Long in the coupling
7. Middle	7. Ample middle due to long, well-sprung ribs	7. Lacking middle
8. Rear flank	8. Well let down in the rear flank	8. High cut rear flank or "wasp waisted"
9. Arm, forearm, and gaskin	9. Well-muscled arm, forearm, and gaskin	9. Light-muscled arm, forearm, and gaskin
10. Legs, feet, and pasterns	10. Straight, true, and squarely set legs; pasterns sloping about 45°, hoofs large, dense, and wide at heels	10. Crooked legs; straight pasterns; hoofs small, contracted at the heels, and shelly
11. Quality	11. Plenty of quality, as denoted by clean, flat bone, well-defined joints and tendons, refined head and ears, and fine skin and hair	11. Lacking quality
12. Breed type (size, color, shape of body and head, and action true to the breed represented)	12. Showing plenty of breed type	12. Lacking breed type

Light gaskin

Sickle hocked

Light forearm

Steep croup

Coon footed   Splay footed   Tail set too low   Roman nose

Figure 5.3 *continued*

After foaling the mare should have free access to clean water. The quantity of grain may be reduced about one-half normal **ration** after foaling to minimize **colic.** This amount should be gradually increased for the first few days at which time the mare can be fed levels she received before foaling. Feed should be increased after this time because nutritional needs increase with lactation.

5.4.3 The Foal

A disease caused by navel infection, *navel ill* or *joint ill,* is serious in new foals. It is characterized by lameness and swollen joints. The navel cord should not be cut. Rather it should be allowed to dry up and drop off naturally. The navel cord may be disinfected with iodine or dusted with a drying powder recommended by a veterinarian 2 or 3 times daily until it dries and drops off.

It is imperative that within 12 to 24 h of birth the foal gets **colostrum,** which is rich in vitamins and **antibodies** and is a natural laxative. Antibodies do not pass from the blood of the mare to that of the foal. Thus the newborn foal without colostrum may succumb to infections. Because of the importance of colostrum to the foal, the mare should not be milked before foaling.

A strong, healthy foal should be standing one-half h after birth. The foal should not be encouraged to stand right away but rather should be given ample time to stand on its own. Some foals nurse almost immediately, others may need assistance. A weak foal may be fed colostrum with a bottle and nipple. Owners should watch for constipation and diarrhea in the foal the first few days after birth and treat if necessary.

Most farms pasture mares and foals in groups. Thoroughbred foals are often born in foaling stalls. Many foals begin nibbling hay and grain by 2 to 3 weeks of age. The foal should be encouraged to eat grain as soon as possible so dietary transition changes at weaning will be made easier. Foals are usually weaned at 4 to 6 months of age by being separated from their mothers. Weaning systems vary among farms. Some abruptly remove mares from foals; others practice gradual weaning where foals and mares are allowed to stall next to each other for several days before removal from sight or sound. Foals weaned in pairs and foals eating grain before weaning have less stress at weaning.

Horse **colts** may be **castrated** when only a few days of age, but most horse owners prefer waiting until they are approximately 1 year or older believing the horse will develop more musculature as a result of androgens produced by the testicles. In general, the older the horse at castration the greater the danger of complications. Springtime—before extremely hot weather, flies, and danger of screwworms (these have been eradicated from most parts of the United States)—is an ideal time for castration. Colts can then be turned out on pasture where there is minimum exposure to infectious microorganisms.

5.4.4 The Young Horse

Yearlings and 2-year-olds will develop rapidly and possess sounder feet and legs if they are on pasture rather than continually confined in close quarters. Periodic handling and grooming, together with leading and training, are important. If properly trained during the nursing period, there will be less training or breaking (except riding) needed when the young horses are older. It is desirable to teach the colt to lead as soon as possible, at latest by 1 month of age. Additionally, handling and care of the feet at 1 month is a good practice. It is also particularly important to control **parasites** (external and internal) in young horses.

5.4.5 The Working Horse

Horses perform many tasks, and as such, the need for and types of conditioning and training programs vary. Nutritional requirements will change with varying workloads.

5.5 NUTRITION OF HORSES

Feed is an expensive item in raising and maintaining horses. Rations should be adequate in both quantity and quality, yet formulated as economically as possible. The best diet for horses depends on the availability and relative costs of feeds.

Horses are nonruminant herbivores. Protein, fat, vitamins, and minerals are digested and absorbed in the foregut (stomach and small intestines). Plant fibers are digested and absorbed in the hindgut (cecum and large intestines). The cecum is a large fermentation vat containing billions of bacteria, which produce enzymes that break fiber down into volatile fatty acids and carbohydrates that the horse can absorb and use for energy. These bacteria also digest protein and synthesize vitamins (e.g., B-complex).

Horses require the same nutrients as other **livestock.** These include carbohydrates, fats, proteins, minerals, vitamins, and water. Grain, dry **forage,** and pasture supply most of these nutrients. Nutrient requirements of horses depend on age, size, and amount of work performed.[2] Horses used for breeding have additional requirements.

5.5.1 Concentrated Feeds

Most horses are fed a mixed-grain ration to **supplement** hay or pasture. Oats, corn, barley, and milo are mixed with various high-fiber feeds, minerals, vitamins, molasses, and other ingredients. Grains may be fed whole, coarsely processed, pelleted, or extruded. Oats are the most common single grain fed to horses in many parts of the United States, especially in the Midwest where they are readily available. They are **palatable** and provide sufficient fiber to prevent too much carbohydrate being fed at a single feeding (which may cause **colic** and/or **founder**). However, like most grains, oats have a high-phosphorus/low-calcium ratio, which may cause a nutritional imbalance if not supplemented with calcium.

Corn is a low-fiber high-carbohydrate feed. It is high in caloric density and is an excellent source of energy for hardworking horses. However, it is usually combined with a higher-fiber grain such as oats to lessen the chance of feeding too much carbohydrate at one feeding. It is low in protein, which must be supplied by good-quality hay or a small amount of protein supplement.

Grain sorghums can be fed to horses after being coarsely processed or incorporated into pellets. Whole wheat is not a commonly recommended feed for horses; however, when competitively priced it may be fed in limited quantities with other grains. Molasses is frequently added to horse rations because it increases palatability and decreases dustiness. Wheat bran is a desirable feed for horses because of its bulk, palatability, nutrient content, and laxative properties.

Linseed meal, used in the early 1900s, has been replaced in most horse diets by soybean meal, which is a higher-quality protein. Protein supplementation is usually only necessary for growing horses and pregnant or lactating mares.

Flaxseed meal is rich in omega-3 (n3) fatty acids, which will quickly put a "bloom" (glossy shine) to the hair coat of horses. Omega-3 fatty acids also improve the absorption of minerals (e.g., zinc), decrease inflammation in joints (a treatment for arthritis), lessen symptoms associated with allergies (itching and pulmonary disease), and can strengthen the function of horse immune systems.

---

[2]For additional information, see *Nutrient Requirements of Horses,* 5th ed., National Academy of Sciences–National Research Council, Washington, DC, 1989.

Beet pulp is used in many "complete feed" formulations. Beet pulp is a highly fermentable fiber and a good source of energy. Beet pulp also promotes good health by favoring the growth of beneficial bacteria in the intestinal tract.

Corn oil is used as a high-energy food for horses needing extra calories to maintain weight because of old age or hard work. Up to 2 cups of corn oil can be safely added to grain twice daily. Extruded grains and complete feeds (containing pelleted forage in addition to grains) are also easy to digest and are used in diets formulated for older horses.

5.5.2 Harvested Forages

Bright-green grass hay is the most common dry forage fed to horses. Timothy, Bermuda grass, and orchard grass hays are popular because they are commonly free of dusts and molds. Because grass hay is low in protein it should be fed with a **concentrate** or protein supplement so that a balanced ration is provided. Prairie and well-cured oat hays are approximately equal to timothy hay as forage for horses. Bright-green and properly cured **legume** hays such as alfalfa, red clover, and lespedeza are also excellent for horses. Legumes are high in calcium and attention should be given to assuring an overall Ca:P ratio between 1:1 to 2:1 by feeding in combination with grass hays or grains higher in phosphorus. It is important that they be free of dust, molds, and insects.

Good-quality **silage** that is finely chopped and free of decay and molds may be used to replace about one-half hay in the diet. Corn silage is preferred but grass silage, legume **haylage,** or milo silage can be fed. Silage should be added to the ration gradually. Most silages are too bulky for working horses, which require a high-energy diet, but they are satisfactory for less-active ones.

Pelleted feeds for horses are popular because they are convenient to handle and individual feedstuffs are not sorted by the horse. Extruded feeds have become popular as a way of adding fat to the ration. Pellets may be somewhat more expensive than a diet consisting of farm-produced hay and grain but are more digestible and may increase longevity in horses with poor teeth.

5.5.3 Pasture

This is the natural feed for horses. It more nearly provides the nutritional needs of horses than any other feed. Mature horses doing little or no work can obtain all their nutrients from grass. Good pasture (improved and properly fertilized) is almost indispensable for broodmares, foals, and young horses. The Bluegrass Region of Kentucky is noted as a center of Thoroughbred breeding, largely because of its excellent pastures.

5.5.4 Salt and Minerals

The working horse has an additional dietary need for salt (NaCl) because it may lose 75 g or more daily through sweat and urine. A horse's need for salt may be met by having block salt available at all times. Granular salt may be included in the diet at the rate of 0.5 to 2.0 percent of the daily ration.

Pregnant mares, **lactating** mares, foals, and young growing horses require additional minerals, the most noted being calcium and phosphorus. Milk production and the growth of bones increase these requirements. Mature horses, other than broodmares, require no supplemental minerals other than salt while on good-quality pasture or when fed high-quality hay, a portion of which is legume. Horses do not self-regulate their mineral needs so free-choice mixes of calcium, phosphorus, and other major minerals are not generally effective. However, the common practice of providing trace mineralized salt (NaCl with minerals needed in very comparatively small amounts) free-choice in block form has special merit for exercising horses, growing horses, and horses in production. To ensure adequate intake most commercially prepared grain mixes supplement minerals well above minimal recommended levels. The forced feeding of trace (micro-) minerals can be detrimental (Chapter 21).

5.5.5 Vitamins

Horses receiving good pasture or high-quality hay (including legumes) do not need supplemental vitamins. Those confined inside for long periods and fed low-quality roughages are more likely to develop vitamin deficiencies. The vitamin requirements for such horses may be satisfied with suitable commercial supplements. Excess vitamins contribute nothing to the horse's health and are an added expense.

B-complex vitamins are synthesized in the large intestine and cecum of healthy horses. However, a need for supplemental dietary thiamine has been demonstrated when horses consume certain plants that contain thiaminase, which impairs the utilization of thiamine (Section 22.6), and a low riboflavin intake appears to contribute to periodic ophthalmia (moon blindness). Green plants and high-quality legume hays are good sources of all B-complex vitamins and also of vitamin A. The latter is essential for good health of epithelial tissues, especially those of the eyes and skin. Vitamin A is also necessary for proper bone growth and maintenance (Chapter 20). Vitamin D requirements of horses are usually satisfied when horses are exposed to sunlight. Research identifying relationships of vitamin E with increasing immune response of animals is ongoing; however, vitamin E is especially important in equine athletes.

Like all supplements, vitamins should be added when needed to increase levels in the regular daily diet. Because most if not all vitamin needs are supplied in fresh forage, supplementation may not be necessary in most horses with access to pasture and freshly cured hays and forages. Nonetheless, because research on vitamins is complicated and virtually nonexistent with horses, vitamins are routinely supplemented to commercially prepared rations. Although vitamin deficiencies are rare, or entirely undocumented in many instances, they are the most commonly supplemented nutrient. Recommendations are for horse owners to use care when using vitamin supplements because vitamin toxicities can occur.

5.5.6 Water

Horses should be provided with fresh, clean drinking water at all times. Horses commonly drink 10 to 12 gal of water daily. Additional water is needed when they are working or when the weather is hot. Horses not having access to water free-choice should be watered at regular intervals and at least twice daily. An inadequate water intake may cause **impaction** and subsequent colic.

5.5.7 Feeding Horses

Horses vary widely in the amount of feed required to maintain weight and condition. Some are **easy keepers;** others are **hard keepers.** Defective teeth and/or parasite **infestations** increase feed requirements.

Many commercially prepared rations provide feeding-level guidelines for different types of horses. A rule of thumb for feeding a horse doing light work is 1 lb of grain and 1 lb of hay daily per 100-lb body weight. An idle horse usually requires only about one-half this amount of grain. If a horse is ridden or driven hard the amount of hay may remain constant, but the grain should be increased to 1.5 percent of body weight. A stallion used heavily for breeding purposes should be fed approximately the same amount as a horse doing heavy work. Broodmares when nursing a foal may do well on high-quality pasture, although most housing and pasture conditions require grain supplementation for them to maintain body condition. A good-quality feed should be available in a creep (a structure barring entrance of larger horses) to foals 2 or more months of age. In winter broodmares should be fed small amounts of grain in addition to dry forage.

Young horses (**weanlings** to 2 years of age) should be fed sufficient amounts to maintain moderate body condition while growing at steady, moderate rates. They may gain satisfactorily if fed one-half lb grain per 100-lb body weight daily when on pasture or when fed good-quality hay. This amount should be increased to 1 lb if pasture or hay is of inferior quality.

5.6 TRAINING AND GROOMING HORSES

Proper training of horses is important. Early training of a foal usually results in a gentler and better-dispositioned individual. Lessons should be given one at a time and in proper sequence.

Place a halter on the foal at 2 to 3 weeks of age and allow the foal 2 or 3 days to become accustomed to it. The foal may then be handled near its dam for 15 to 30 min daily. Brushing and grooming a foal helps it to accept people. One can train the foal to lead by first leading it with its mother and then by itself. Because early learning experiences greatly affect the long-term behavior of horses, foals should be handled with consistent cues that enforce acceptable and unacceptable behavior. Patience, consistency, and stepwise training are of great importance when handling foals.

Breaking a young horse to a saddle or harness may be accomplished during the winter before it is a 2-year-old. (Inter-estingly, Thoroughbred horses have the same official birthday, January 1.) Many expert "bronc riders" break horses that have not been handled as colts and are wild when brought in from the range. Each horse breaker has methods he or she prefers, and it is surprising how quickly young horses can be broken and trained. One cowboy who could scarcely write his name was outstanding in breaking and training cow horses. When asked how he got so much out of a horse he replied, "You just have to be smarter than the horse."

Horses that are to be shown must be well mannered and well groomed. The coat can be conditioned by proper feeding and grooming. A balanced diet fed in adequate amounts helps ensure a pleasing appearance. The best grooming is useless if a horse is underfed and **unthrifty.**

A currycomb or rubber brush should be used to remove mud and soil from the coat. Otherwise vigorous rubbing with a stiff brush followed by rubbing with a soft one is recommended. Large stables often use a vacuum to remove loose hairs and dirt. The vacuum prevents dirt from being redeposited on the coat. Rubbing the coat briskly with a cloth will make the hair shine. It has been said that the best way to make a horse's coat shine is to apply plenty of "elbow grease." The groom should attend to the entire body (including the face, belly, and legs). It is recommended that a brush be used with care on the mane and tail of show horses because it may remove considerable hair. Liberal use of coat conditioner and hand picking will minimize breaking of hairs.

Daily washing is not commonly recommended because it removes oils that give the coat its natural sheen and glossy appearance. However, it is sometimes necessary to wash certain parts of the body if the hair is stained. White legs and ankles and light manes of show horses are usually washed before showing. Cornstarch or baby powder can be used to enhance white markings. Most people wash a horse after a hard workout. Racing horses are usually washed following a race. Methods used to groom, bathe, and cool down horses vary with different trainers and environmental conditions. However, all horses should be cooled down slowly and be given a rubdown following a ride.

Blanketing of show and pleasure horses that are kept in a stable helps keep their coats clean, smooth, and short. This is especially beneficial in cold weather. The shedding of hair is influenced by day length. (Horses shed and maintain shorter hair coats when day length is longer.) Hair length can be kept short in the winter by using artificial lighting programs. These programs extend the natural length of daylight with artificial lighting to equal the length of light during summer days. Horses look neater if the long hair on the ears, jaws, and back of the fetlocks is kept trimmed. Some of the mane can also be trimmed at the point where the headstall of the bridle is normally located (commonly called the "bridle path," at the poll). Some owners of pleasure horses prefer to roach (clip) the mane completely. It is recommended that those contemplating the showing of horses investigate fully the methods used by the most successful show winners.

5.7 Common Defects and Unsoundness in Horses

Defects and/or unsoundness in horses may involve both anatomy and physiology. They may be due to environment or heredity or the interaction of the two.

5.7.1 Defects of Movement

The normal movement of a horse going forward is for the legs to move parallel to an imaginary centerline drawn in the direction of travel. A defect in movement is one that deviates from this in any way.

Cross-firing is "scuffing" of the inside of the diagonal forefeet and hindfeet and is usually found only in pacers and trotters.

Dwelling is a noticeable pause in the foot's flights, making it appear that the stride was completed before the foot reached the ground. This is most noticeable in trick-trained horses.

Interfering is a condition in which the fetlock or cannon is struck by the opposite foot. It is observed most frequently in base-narrow, toe-wide, or splayfooted horses.

Lameness is an indication of a structural or functional disorder in which the individual favors one or more feet while standing or walking. It is also called *claudication.* Pressure between the foot and ground is eased, and the head bobs up as the affected front foot contacts the ground and bobs down as the affected hind foot contacts the ground.

Paddling involves throwing the forefeet outward from the body as they are lifted and before they reach the ground again. This is most common in toe-narrow or **pigeon-toed** horses.

Pointing is a more than usual extension of the stride with little or no flexion (joint movement).

Pounding is a heavier-than-normal contact of the feet with the ground (observed especially in Thoroughbreds, hackneys, and some draft horses).

Rolling is an excessive **lateral** motion of the shoulder (usually observed in horses with protruding shoulders).

Scalping is a condition in which the toe of a forefoot hits a hindfoot at the top of the hairline.

Speedy cutting is contact between the inside of the diagonal fore and hind pastern. It may be seen in fast-trotting horses.

Stringhalt is excessive and involuntary flexing (bending) of the hocks during progression (moving ahead) and may affect one or both rear limbs. It is more noticeable when the horse is turning.

Trappy is the term for a short, quick, and choppy **stride,** which is most common in horses with short, straight pasterns and shoulders.

Winding, or *rope-walking,* is a twisting of the striding leg around and in front of the supporting leg, as with a person walking a rope.

Winging, an inward motion, is particularly noticeable in splayfooted horses.

5.7.2 Blemishes and Unsoundness

Blemishes include abnormalities such as cuts and splints that usually do not affect the performance or service of horses. However, *unsoundness* includes defects that may impair performance or service. Although blemishes seldom disqualify horses from show-ring competition, unsoundness (particularly hereditary conditions) does. Points of the body where some of these abnormalities occur are shown in Figure 5.4. Those interested in buying and breeding horses should become familiar with these defects and seek the advice of a veterinarian who will do a prepurchase (soundness) evaluation.

Definitions for selected examples of blemishes and unsoundness include the following:

Blindness is either impaired eyesight or complete blindness.

Blood spavin is enlargement of the hock due to an enlargement of the superficial saphenous vein on the anteromedian aspect (inside) of the hock.

Bog spavin is the enlargement of the hock joint capsule due to an accumulation of fluid.

Bone spavin (or *jack spavin*) is a bony growth on the hock.

Bowed tendons are inflamed, enlarged tendons behind the cannon bones of the legs.

Calf-kneed is the opposite of knee-sprung, so that the horse stands with its knees too far backward.

Capped elbow is the enlargement of the elbow (bursa).

Capped hock is a swelling on the point of the hock.

Cocked ankles is a condition in which the legs (usually rear) are bent forward at the fetlocks in a cocked position.

Contracted heels is a contraction or drawing in of the hoof at the heel.

Corns are bruises of soft tissue under the horny part of the hoof's sole.

Curb is an enlargement at the plantar aspect of the fibular tarsal bone due to an inflammation and thickening of the plantar **ligament;** in acute cases lameness results.

Developmental bone diseases in horses include osteochondritis dessicans (OCD), osteochondrosis (diseased **cartilage**), and other forms of physitis (joint abnormalities). These may produce lameness, joint swellings, and limb deformities in horses under 3 years of age.

Fistulas are inflammations or infections in the withers accompanied by passages or sinuses through the tissues to the skin's surface.

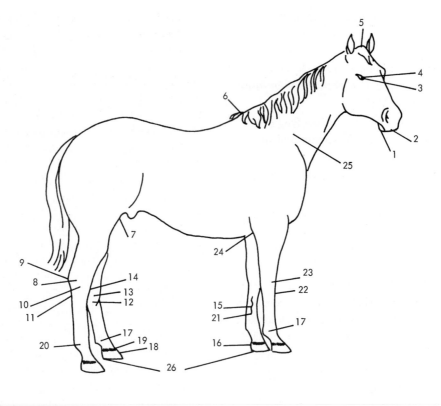

1. Undershot jaw
2. Parrot mouth
3. Blindness
4. Moon blindness
5. Poll evil
6. Fistulous withers
7. Stifled
8. Thoroughpin
9. Capped hock
10. Stringhalt
11. Curb
12. Bone spavin or jack spavin
13. Bog spavin
14. Blood spavin
15. Bowed tendons
16. Sidebones
17. Cocked ankles
18. Quittor
19. Ring bone
20. Windpuffs
21. Splints
22. Knee sprung
23. Calf kneed
24. Capped elbow
25. Sweeney
26. Contracted feet, corns, founder, thrush, quarter or sand crack, scratches or grease heel

General: heaves, hernia, labored breathing, roaring, thick wind

Figure 5.4 Location of common points of unsoundness in horses.
Adapted from USDA Farmers' Bull. *2127, 1962.*

Founder (*laminitis*) is an inflammation of the hoof laminae.

Knee-sprung (*bucked-kneed*) is a condition in which the knees protrude too far forward.

Moon blindness (*periodic ophthalmia*) is a periodic cloudy or inflamed condition of the eye that appears and disappears at intervals of approximately 1 month; it was given this name because people thought it was related to moon cycles. It may be caused by infections (bacterial, viral, and parasitic), trauma, allergies, and rarely by nutritional deficiencies.

Navicular disease is joint inflammation followed by degeneration of the navicular bone in the front feet.

Parrot mouth is a condition in which the lower jaw is shorter than the upper jaw.

Poll evil is a swelling on the top (poll) of a horse's head.

Quarter (or *sand*) *crack* is a vertical split in the inside of the horny wall of a hoof that extends from the coronet downward.

Quittor is an open draining sore at the coronet (hoof's head).

Ringbone is a bony growth on the pastern bone (usually of a forefoot).

Scratches (or *grease heel*) is inflammation of the **posterior** surfaces of the fetlocks; it occurs more frequently in the hind legs.

Sidebones is the hardening, or ossification, of the lateral cartilage of the third phalanx.

Splints is an inflammation of the ligament between the large and small metacarpal bones most frequently affecting the forelimbs of young horses. Splints are most commonly found on the medial aspect of the limb and may be a result of excessive strain or trauma.

Stifled is an upward fixation of the patella (kneecap), which locks the hindleg in extension.

Stringhalt is an involuntary flexion of the hock, which is most noticeable when the horse is rested.

Sweeney is the **atrophy,** or shrinking, of the shoulder muscles because of a nerve injury; it results in a depression of the shoulder.

Thoroughpin is a swelling just above the hock on both sides of the leg; it involves the tarsal sheath that encloses the deep digital flexor **tendon.**

Thrush is a disease of the horny layers of the **frog** of a foot; it is usually accompanied by a discharge and an unpleasant odor.

Undershot jaw is a condition in which the lower jaw is longer than the upper jaw.

Wind puffs are soft enlargements on the fetlock due to excessive production of **synovial** fluid.

5.7.3 Stable Vices and Other Bad Habits

Horses in confinement often develop habits that may be annoying to their owners and even harmful to themselves. Most stable vices (bad habits) result from boredom, others are forms of obsessive-compulsive behaviors.

5.7.3.1 Vices

Bolting is a horse's habit of eating its concentrate feed very rapidly and without chewing it properly.

Cribbing refers to a condition in which the horse bites the manger or another object while sucking in air. It sometimes causes a **bloated** condition and the horse may be more likely to develop colic. Cribbing may be remedied by buckling a strap snugly around a horse's neck in such a way that it compresses the **larynx** when the neck is flexed or extended. If properly placed such a strap will cause no discomfort at other times.

Halter pulling refers to the habit of pulling back on the halter rope or bridle reins when tied. The halter and/or strap may be broken allowing the horse to run away.

Kicking or striking the stall or other objects with their rear feet is a habit developed by some horses. They apparently do this for the satisfaction derived from striking something.

Tail rubbing is a habit of persistently rubbing the tail against the side of a stall or on another object. This occasionally results from parasites such as pinworms.

Weaving is a condition in which a horse weaves, or sways from side to side, while standing in a stall.

5.7.3.2 Bad Habits

Draft horses may develop the habit of *balking* when hitched to heavy loads. Thus they refuse to push their shoulders against the collar. Such a habit sometimes develops because of a poorly fitting collar, which may cause a sore on their necks or shoulders.

A horse may develop the habit of *rearing* on its hind legs when an attempt is made to lead it. Other horses may rear when

mounted. Some horses develop the habit of *striking out* with the forefeet when they are being led and/or when they rear up on their hindlegs. Other horses develop a habit of *shying* when ridden and may even unseat their riders. Draft horses sometimes develop the dangerous habit of *running away* when hitched to a wagon or other moving vehicle. Riding horses may be "cold-jawed," a habit of taking the bit in their teeth and running at top speed. Pulling on the reins, and other methods commonly used in stopping them, are often unsuccessful. Once these runaway habits develop it is very difficult to correct them.

Another undesirable habit horses develop is that of playing "hard to catch" when their owners try to drive them into a stable or catch and bridle them in a pasture or pen. Other horses object to being harnessed or saddled. Gentle horses may develop the habit of expanding their chest cavity (breathing deeply, commonly called *swelling up*) when the saddle cinch is tied or buckled. They exhale air after the cinch is fastened and thereby cause the saddle to be loosened. Some horses, probably because of the way they are first broken and handled, develop the habit of *bucking*. This habit marks some horses as unsuitable for riding; others seem gentle but may "break in two" and throw their riders when least expected.

Most vices and bad habits of horses are difficult to correct once they are developed. Preventing the behavior is much more desirable than trying to correct it. Older horses, however, can be broken of many undesirable habits and taught new ones. It should be remembered that, because of their size and swiftness, horses must be handled with respect and taught to respect people.

5.7.4 Care and Trimming of the Feet

> For want of a nail the shoe was lost,
> For want of a shoe the horse was lost,
> For want of a horse the rider was lost,
> For want of a rider the battle was lost.
> **George Herbert (1593–1633)**
> **English poet**

Most horses are now kept for either riding or driving purposes. These uses involve movements of various kinds and proper movement is dependent on healthy, sound feet. In fact, many movement defects among horses are related directly or indirectly to improper care and trimming of the feet. Therefore, horse owners should care for the hooves of their horses properly and use the services of a professional when appropriate. Tools most often used to trim the feet include hoof nippers, hoof knife, and rasp. Properly and improperly trimmed hooves are shown in Figure 5.5.

Proper care of feet starts with the young foal. An unshapely hoof causes uneven wear and may result in unsound legs. Faulty legs can often be corrected by proper trimming of the hooves; however, it is more desirable to keep hooves properly trimmed as a **prophylactic** measure against such unsoundness. The foal's feet should be observed while it is standing on a hard surface and then when in action, both at a walk and at a trot.

Most necessary foot trimming of young foals can be accomplished with a rasp because the feet seldom need more than a little smoothing and rounding. Even if little shaping is necessary,

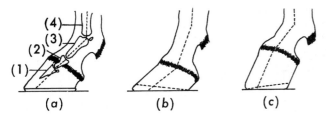

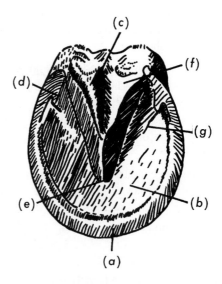

Figure 5.5 Trimming horse feet. (a) Properly trimmed hoof with normal foot axis. (1) Coffin bone, (2) short pastern bone, (3) long pastern bone, (4) cannon bone. (b) Toe too long, which breaks the foot axis backward. Horizontal dotted line shows how hoof should be trimmed to restore normal posture. (c) Heel too long, which breaks the foot axis forward. Horizontal dotted line shows how trimming will restore the correct posture.
USDA.

Figure 5.6 Parts of a horse's foot. (a) Bearing edge and hoof wall; (b) sole; (c) median furrow of the frog; (d) lateral furrow of the frog; (e) apex of the frog; (f) branch of the frog; (g) bar.
USDA.

handling the foal's feet makes shoeing easier at maturity. Regular hoof care in the foal and young horse will favor proper development of the legs and correct movement. It is seldom necessary to put shoes on young foals, although light plates are sometimes used on young show horses. Some people who raise horses on the range prefer to run them in a rocky area or where the ground is hard and firm. They believe that horses raised under such conditions have harder and more durable feet later in life. Mares and foals are never shod under such conditions.

The hoof of a normal healthy horse grows about 1/2 in monthly. If the ground is mostly soft and the hoof is not trimmed the wall will break off and wear unevenly. Under such conditions the hoof should be trimmed monthly. Nippers may be used to trim the horn and the wall may be leveled with a rasp. If the hoof is uneven it may be necessary to correct this over a period of time with several trimmings. The frog of a foot should also be carefully trimmed by removing the ragged edges so that dirt or other extraneous matter will not accumulate in the crevices (Figure 5.6). Because it serves as a cushion for the hoof, the frog should not be trimmed excessively. The sole should be trimmed little, if at all. It is recommended that the outside walls of the hoof not be rasped because this removes the outer protective layer that prevents **evaporation.** (Evaporation of water from the hoof causes the hoof wall to become dry and brittle.) The bars (either of the recurved ends of the wall of a horse's hoof where they curve in to the sole) should never be trimmed below the level of the walls. Frequently the novice trims these excessively, which allows the wall to collapse inwardly and thereby pinches the lateral cartilage.

Shoes are a necessity when horses pull loads or are ridden on hard surfaces. Otherwise the hooves wear down, the feet become tender, and lameness results. Cow horses worked on the range, especially on rocky ground, must be shod. Shoes should be the proper size for each horse. They are shaped to fit the contour of the hooves. When nailing on a shoe one should take care not to drive the nail into the "quick," or sensitive, part of the hoof. When nailed through the hoof the sharp ends of nails should be twisted off and the remaining blunt ends tapped lightly to bend them downward and thereby hold the shoe firmly in place. Shoes are commonly replaced or reset at inter-

vals of 4 to 8 weeks. Those worn too long may allow the hoof to grow out of proportion and thereby throw a walking or running horse off balance. Corrective shoeing to relieve an abnormality requires skill.

5.8 DETERMINING THE AGE OF HORSES

A mature male horse has 40 teeth, whereas a mature female has only 36. Foals of both sexes have 24. Frequently a small pointed tooth, sometimes called a "wolf tooth," may appear in front of each first molar in the upper jaw thus increasing tooth numbers by two. Because these teeth may interfere with the bit they are often removed at a young age. The mature male (usually not the female) commonly has **tushes,** or pointed teeth, located between the incisors and molars. The appearance of teeth is used to determine a horse's age, which is important when one is purchasing horses. Age is determined by noting the time of appearance, shape, and degree of wear of temporary and permanent teeth. Temporary (milk) teeth are easily distinguished from permanent teeth because they are smaller and whiter. Even the novice can recognize a **smooth-mouthed** horse (normally considered a horse 10 or more years old). A description of the teeth of horses at various ages is given in Figure 5.7.

5.9 DISEASE AND PARASITE CONTROL ASPECTS OF HORSES

The best possible breeding, feeding, and training will be of no avail if a horse is sick and unhealthy. Important selected characteristic body functions of healthy horses are described in Section 5.9.1. Variations from these, indicating illness or disease, are discussed briefly in Sections 5.9.2 to 5.9.5.

Temporary incisors to 10 days of age; first or central upper and lower temporary incisors appear.

Temporary incisors at 1 1/2 years; intermediate temporary incisors show wear.

Temporary incisors at 4 to 6 weeks of age; second or intermediate upper and lower temporary incisors appear.

Temporary incisors at 2 years; all show wear.

Temporary incisors at 6 to 10 months; third or corner upper and lower temporary incisors appear.

Incisors at 4 years; permanent incisors replace temporary centrals and intermediates; temporary corner incisors remain.

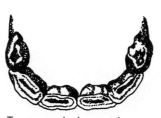

Temporary incisors at 1 year; crowns of central temporary incisors show wear.

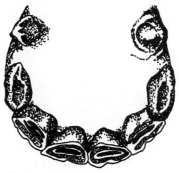

Incisors at 5 years; all permanent; cups in all incisors.

Figure 5.7 The teeth of horses at various ages.
Redrawn from "Breeding and Raising Horses," USDA Agr. Handbook No. 394, 1972.

continued on page 114

5.9.1 Selected Normal Body Functions

The pulse rate of a horse at rest varies between 36 and 42 beats per min. During extreme exertion it may increase to more than 200 beats per min, but when the horse is rested the pulse rate usually returns to normal in a few minutes. Pulse rate is higher in young than in mature horses. It may be determined by placing the fingertips over the external maxillary artery that comes around the **ventral** surface of the **mandible** (jawbone) immediately in front of the large masseter (cheek) muscle.

The respiration rate of horses at rest varies between 9 and 12 per min. It may increase to 100 or more per min during running or vigorous work. It usually returns to normal within a few minutes after the physical exertion stops. Breathing should be free, easy, and noiseless without a whistling or roaring sound. One may determine respiration rate by observing movements of the flanks and/or nostrils.

The normal body temperature of a horse at rest is about 100°F. High **ambient temperatures,** excitement, and exercise

Incisors at 6 years; cups worn out of lower central incisors.

Incisors at 9 years; cups worn out of upper central incisors; dental star on upper central and intermediate pairs.

Incisors at 7 years; cups also worn out of lower intermediate incisors.

Incisors at 10 years; cups also worn out of upper intermediate incisors, and dental star is present in all incisors.

Incisors at 8 years; cups worn out of all lower incisors, and dental star (dark line in front of cup) appears on lower central and intermediate pairs.

Incisors at 11 or 12 years; cups worn in all incisors (smooth mouthed), and dental star approaches center of cups.

Characteristic shape of lower incisors at 18 years.

Figure 5.7 *continued*

increase body temperature, which may be determined by inserting a thermometer in the rectum and allowing it to remain for at least 3 min.

5.9.2 Disease Symptoms

The frequency of defecation and appearance of its feces are indicative of a horse's condition. Normally horses **defecate** 8 to 10 times daily. Hard feces may be caused by consumption of feed that is too dry, an inadequate water intake (especially in winter), and/or inadequate exercise. Soft, watery feces often result from consumption of large quantities of green watery grass, too much bran, or an intestinal irritation. Slimy, strongly odoriferous droppings suggest that the feed is too concentrated, intestines are irritated, or something has caused a shift from normal in the intestinal bacterial **flora.** If the droppings contain large amounts of whole grain this indicates either that the horse eats too fast, without properly chewing, or that it cannot chew grain properly because of defective teeth.

Indications of disease are a loss of appetite, high body temperature (above 101°F), fast respiration and heart rates, profuse sweating, stiffness, nasal discharge, coughing, diarrhea, constipation, pawing, rolling, groaning, and lameness. When such symptoms are observed a veterinarian should be consulted.

Studies of horses treated in the Veterinary Clinic at the University of Missouri–Columbia indicate that wire cuts are the number one reason for veterinary treatment. Thus barbed-wire fences are undesirable for restricting horses, especially in small lots. Braided electric wire, woven wire, and board or vinyl fences are preferred.

5.9.3 Nutritional Diseases and Maladies

Selected nutritional **diseases** and maladies (abnormal conditions) of horses and their probable causes, symptoms, and methods of prevention and treatment are given in Table 5.4. Poor management is the general cause of most noncommunicable horse diseases. Because horse owners can prevent many of these diseases, it is well for them to become acquainted with such information.

TABLE 5.4	**Selected Nutritional Diseases of Horses**					
Disease	**Cause**	**Symptoms (and Age or Group Most Affected)**	**Distribution and Loses**	**Treatment**	**Control and Eradication**	**Prevention**
Anemia, nutritional	Commonly an iron deficiency but may be caused by a deficiency of copper, cobalt, and/or certain vitamins.	Loss of appetite, progressive emaciation, and death.	Worldwide. Losses consist of retarded growth.	Provide dietary sources of the nutrient or nutrients lacking.	Balanced diet.	Feed adequate levels of iron, copper, cobalt, and vitamins.
Azoturia (Monday morning disease, blackwater)	Associated with faulty carbohydrate metabolism and with work following a period of idleness in the stall on full diets.	Profuse sweating, abdominal distress, wine-colored urine, stiff gait, reluctance to move, and lameness.	Worldwide, but the disease is seldom seen in horses at pasture and rarely in horses at constant work.	Absolute rest and quiet. While awaiting the veterinarian, apply heated cloths or blankets or hot-water bottles to the swollen and hardened muscles.	Azoturia is noncontagious. When trouble is encountered, decrease the diet and increase exercise on idle days.	Restrict the diet and provide daily exercise when animals are idle, or turn idle horses out to pasture.
Calcium deficiency	Inadequate dietary calcium intake.	Calcium deficiency symptoms are fragile bones, reproductive failures, and decreased milk production.	Calcium-deficient areas have been reported in parts of Florida, Louisiana, Nebraska, Virginia, and West Virginia.	Select natural feeds that contain sufficient quantities of calcium and phosphorus.	Increase the calcium and phosphorus content of feeds through fertilizing the soils.	Feed balanced diets and allow animals free access to a phosphorus and calcium supplement.
Colic	Improper feeding, working, or watering; internal parasites.	Excruciating pain; depending on the type of colic, other symptoms are distended abdomen, increased intestinal rumbling, violent rolling and kicking.	Worldwide.	Call a veterinarian. To avoid danger of inflicting self-injury, place the animal in a large, well-bedded stable or take it for a slow walk.	Proper feeding, working, and watering.	Adequate diet with minimal dust. Control internal parasites.
Developmental orthopedic bone disease (osteochondritis dissecans, OCD)	Overfeeding with excesses in calcium, protein, and/or energy, improper calcium:phosphorus balance.	Young horses that will mature to a height over 15 hands are most often affected; lameness, joint disease, bog spavin, and joint effusion are common signs; the shoulder, hock, and stifle joints are most often affected.	Worldwide. Unsoundness is a common sequelae.	Some cases require surgery to remove abnormal bone and/or cartilage. Correct Ca:P balance; supplement diet with copper and zinc; avoid overfeeding.	Feed balanced diets. Avoid overfeeding.	Avoid overfeeding young horses. Analyze mineral content of hay and grain, ideal Ca:P ratio is 1.8:1. Copper (25 mg/kg) and zinc (65 mg/kg) should also be in the diet and in a ratio of approximately 1:4.

Continued on next pages.

TABLE 5.4 *continued*

Disease	Cause	Symptoms (and Age or Group Most Affected)	Distribution and Loses	Treatment	Control and Eradication	Prevention
Founder (laminitis)	Overeating, overdrinking, or inflammation of the uterus following parturition. Also intestinal inflammation.	Extreme pain, fever (103 to 106°F), and reluctance to move. If neglected, chronic laminitis develops.	Worldwide. Actual death losses from founder are not very great.	Pending arrival of the veterinarian, the attendant should stand the animal's feet in a cold-water bath.	Alleviate the causes, namely, overeating, overdrinking, and/or inflammation of the uterus following parturition.	Avoid overeating and overdrinking (especially when hot).
Heaves	Often associated with the feeding of damaged, dusty, or moldy hay. Often follows severe respiratory infections.	Difficulty in expiring air, resulting in a jerking of flanks (double flank action) and coughing. The nostrils are often slightly dilated.	Worldwide. Losses are negligible, however performance is impaired.	Affected animals are less bothered if used only at light work, if the hay is sprinkled lightly with water at feeding, or if the entire ration is pelleted.	See Prevention. Bronchodialators may be prescribed.	Avoid the use of damaged and/or dusty feeds.
Iodine deficiency (goiter)	A failure of the body to obtain sufficient iodine from which the thyroid glands can form thyroxine.	Foals may be weak.	Northwestern United States and the Great Lakes region.	Once the iodine deficiency symptoms appear no treatment is entirely satisfactory.	At the first signs of iodine deficiency, an iodized salt should be fed.	In iodine-deficient areas, feed iodized salt containing 0.01% potassium iodine.
Osteomalacia	Lack of vitamin D. Inadequate intake of calcium and phosphorus. Incorrect ratio of calcium and phosphorus.	Phosphorus deficiency symptoms are depraved appetite (gnawing on bones, wood, or other objects or eating dirt), lack of appetite, stiffness of joints, failure to breed regularly, decreased milk production.	Southwestern United States is classified as a phosphorus-deficient area.	Select natural feeds that contain sufficient quantities of calcium and phosphorus.	Increase the calcium and phosphorus content of feeds through fertilizing the soils.	Feed balanced diets and allow animals free access to a phosphorus and calcium supplement.
Periodic ophthalmia (moon blindness)	One cause is deficiency of riboflavin. Leptospiral microorganisms may also be involved. Other bacterial infections, viruses, parasites, and trauma can cause signs.	Periods of cloudy vision, in one or both eyes, which may last for a few days to a week or two and then clear (recurs at intervals).	In many parts of the world. In the United States, it occurs most frequently in the states east of the Mississippi River.	Antibiotics administered promptly are helpful.	If symptoms of moon blindness are observed, immediately change to greener hay or grass or add riboflavin to the diet at the rate of 40 mg per horse daily. Consult with a veterinary ophthalmologist.	Feed high-riboflavin green grass or well-cured green leafy hays, or add riboflavin to the diet at the rate of 40 mg per horse daily.
Rickets	Lack of either calcium, phosphorus, or vitamin D and/or incorrect ratio of the two minerals.	Enlargement of the knee and hock joints; and the animal may exhibit great pain when moving about. Irregular bulges (beaded ribs) at juncture of ribs with breastbone, and bowed legs.	Worldwide. It is seldom fatal.	If the disease has not advanced too far, treatment may be successful by supplying adequate amounts of vitamin D, calcium, and phosphorus, and/or adjusting the ratio of calcium to phosphorus.	See Prevention.	Provide sufficient calcium, phosphorus, and vitamin D and a correct ratio of the two minerals.

Continued on next page.

TABLE 5.4	*continued*

Disease	Cause	Symptoms (and Age or Group Most Affected)	Distribution and Loses	Treatment	Control and Eradication	Prevention
Salt deficiency	Lack of salt (sodium chloride).	Loss of appetite, retarded growth, loss of weight, a rough coat, lowered production of milk.	Worldwide.	Salt-starved animals should be gradually accustomed to salt, slowly increasing the hand-fed allowance.	See Prevention and Treatment.	Provide salt-free choice.
Urinary calculi (gravel, stones, water belly)	Unknown, but incidence is higher when there is a high potassium intake, an incorrect Ca/P ratio, or a high proportion of beet pulp or grain sorghum.	Frequent attempts to urinate, dribbling or stoppage of the urine. Usually only males affected.	Affected animals seldom recover completely.	Once calculi develop, dietary treatment appears to be of little value. Smooth-muscle relaxants may allow passage of calculi if used before rupture. May require urethrostomy (surgery).		Good feed and management appear to lessen the incidence. 1 to 3% salt in the concentrate diet may help (using the higher levels in the winter).
Vitamin A deficiency (night blindness)	Vitamin A deficiency.	Night blindness, the first symptom of vitamin A deficiency, is characterized by impaired vision.	Worldwide.	Treatment consists of correcting the dietary deficiencies.	See Prevention and Treatment.	Provide carotene (vitamin A) through green, leafy hays; silage; or lush, green pastures. Synthetic preparations of vitamin A are recommended for winter feeding.

5.9.4 Communicable Diseases

Most diseases of horses are caused by bacteria, **viruses, fungi,** or **parasites.** As with noncommunicable diseases, the application of preventive measures is more desirable than trying to treat a disease once symptoms appear. Preventive measures include daily observation and care, cleanliness, disinfection of premises, isolation and/or **quarantine** of sick animals, and **vaccination** when appropriate. Close cooperation with veterinarians is important in planning preventive measures and is essential for proper treatment of diseased animals.

Encephalomyelitis ("sleeping sickness") is a viral disease transmitted by mosquitoes. It affects horses, rodents, birds, and humans. There are at least three strains ("Eastern," "Western," and "Venezuelan"). Signs of disease include loss of appetite, staggering, paralysis, and death. Vaccines are available and should be given each year prior to the mosquito season.

Equine infectious anemia (EIA) ("swamp fever") is a contagious disease that affects horses worldwide. It is caused by a virus closely related to the human immunodeficiency virus (HIV). EIA is transmitted by bloodsucking insects and contaminated needles. There are three clinical forms: (1) the acute form causes depression, fever, and incoordination; (2) the second phase is characterized by weight loss, recurring fevers,

weakness, and anemia. Horses may die during the first two phases or may survive and enter (3) the final or chronic stage where they may appear clinically normal, but are life-long carriers of the virus and become ill again (relapse) if stressed. A blood test was developed by Dr. Leroy Coggins in the 1960s to detect antibodies to EIA in the blood. A negative Coggins test is required prior to transportation, sale, or showing of horses in most states. If a horse tests positive it must be quarantined or destroyed. There is no vaccine for this disease.

Influenza is an acute, contagious, viral disease affecting the upper respiratory tract of horses. Similar to human colds, there are many strains of this virus. Several vaccines are available with protection lasting only a few months following vaccination.

Potomac horse fever is a seasonal disease occurring in summer months. It is caused by the rickettsial organism *Ehrlichia risticii.* This disease is characterized by high fever, diarrhea, depression, weight loss, and severe founder (laminitis). Many affected horses die. A vaccine is available and should be given to horses in endemic areas.

Protozoal myelitis is a neurological disease caused by the protozoa *Sarcocystis neurona.* Clinical signs associated with this disease include weakness, incoordination, and muscle atrophy. The protozoa are spread in the feces of opossums (often contaminants in hay and/or stored grains).

Rabies produces fatal neurological disease. Horses, like other mammals, are susceptible to rabies. A vaccine is approved for annual administration to horses.

Rhinopneumonitis is a viral disease that can produce respiratory disease, abortion, and paralysis. There are two strains of **rhinovirus:** equine herpes virus-1 (abortion, paralysis) and equine herpes virus-4 (respiratory disease). Vaccines are available for both strains.

Strangles is a contagious bacterial disease caused by *Streptococcus equi*. It affects the upper respiratory tract causing inflammation and abscessation of the lymph nodes in the upper neck and throat region. A killed bacterium is available for use in horses at high risk of exposure to this bacteria.

Tetanus ("lockjaw") is an acute disease caused by toxins produced by the bacterium *Clostridium tetani*. The bacteria enter wounds (especially puncture wounds) and produce toxins which cause tetany (prolonged muscle contractions). Annual vaccination is recommended for all horses. Tetanus antitoxin should be given to non-vaccinated horses following any type of injury to the skin or feet.

West Nile Encephalitis was first diagnosed in horses in the United States in 1999. Mosquitoes have spread the virus throughout the eastern states and into the Midwest. Clinical signs include fever, weakness, incoordination, convulsions, and death. A vaccine is available for use in horses (Section 22.3.1).

5.9.5 Common Horse Parasites

Parasitic infestations may cause the appearance of **acute** symptoms, but they are more likely to result in a gradual and progressive **insidious** unthriftiness. Parasitic infestations are associated with general unthriftiness or weakness, **emaciation,** rough hair coat, and frequent attacks of colic. They are most damaging to foals and young horses and often result in slow growth and development. Heavily parasitized older horses perform inefficiently and have increased nutrient requirements.

Most parasite eggs laid internally pass in the feces and are deposited on pasture or in stalls and lots. They develop to the infective stage on the ground and may reenter a horse's body with grass, other feed, or water. Preventive measures include breaking the **life cycle** at some vulnerable point. (Horses are commonly dewormed at 2-month intervals.)

The female botfly attaches her eggs to hairs on a horse's chin, nose, or legs. Eggs hatch in 7 to 10 days and are taken into the mouth when the horse licks or bites the hair. The **larvae** then burrow into the mucous membranes of the tongue, where they remain approximately 2 weeks before emerging and passing to the stomach. They remain attached to the stomach walls for 8 to 10 months or until they fully develop. The larvae are then passed with the feces, **pupate** in the ground for 20 to 70 days, and finally emerge as adult flies (Chapter 23). Large numbers of bots cause irritation and injury to the stomach walls, and sometimes severe colic develops.

Ascarids, or large roundworms, infect horses and when fully grown may attain a length of 12 in. They locate mostly in the upper portion of the small intestine but sometimes locate in the stomach and cecum. Eggs pass from the body in feces and, in the infective stage, may reenter the body when swallowed by the horse. Mature horses apparently build up a resistance to ascarids; consequently these parasites are primarily a threat to the health of young horses. Heavy infestations with ascarids may even be fatal to young horses, although they more frequently cause unthriftiness, rough hair coat, and slow inefficient weight gains. Cleanliness of quarters and pasture rotation are recommended preventive practices. Several available drugs are active against ascarids.

Without deworming, large strongyles (bloodworms, or palisade worms) are the most important parasites of horses. These bloodsucking worms attach themselves to the colon and cecum. Their greatest damage is caused by the loss of blood, intestinal injury, and the migration of larvae through many body organs. Migrating larvae cause many problems depending on where they move and lodge. Many "colics" in adult horses are caused by restriction of the blood supply to the large bowels by parasite damage to the arteries. Small strongyles are also an important internal parasite in horses. Serious infestations cause general unthriftiness, **anemia,** and weakness. The life cycle of strongyles includes the discharge of eggs into the feces with larvae entering the body when the horse **grazes.** Prevention includes keeping quarters clean and rotating pastures. Drugs are available for control of large and small strongyles, but should be chosen after consultation with a veterinarian.

Other species of intestinal parasites that may cause problems for horses include lungworms, pinworms (may cause tail rubbing), strongyloides (threadworms), and tapeworms. These rarely cause serious disease. If they become a problem a veterinarian can assist with diagnoses and treatment.

Various species of lice and ticks also parasitize horses and can be effectively controlled through spraying with recommended **insecticides.**

5.10 SUMMARY

> Where in this wide world can one find nobility without pride, friendship without envy, or beauty without vanity? Here, where grace is laced with muscle, and strength by gentleness confined. He serves without servility; he has fought without enmity. There is nothing so powerful, nothing less violent, there is nothing so quick, nothing more patient. England's past has been borne on his back. All our history is his industry. We are his heirs, he is our inheritance. The HORSE.
>
> **Sir Ronald Duncan (1914–1982)**

Because horses were first domesticated approximately 5000 years ago they have been of great value and service to humanity. Horses probably were first used to carry soldiers into battle, but they soon found a role in transportation and farming (especially for draft purposes). The introduction of horses into America had a significant effect on the lives of Native Americans.

They rode horses to capture and kill buffalo for food and made clothing and other items from horsehair and horsehide.

Tractors, trucks, and other power-driven vehicles have almost replaced horses for draft purposes in the United States, and for a period of time the future of horses appeared bleak. However, the number of horses increased decidedly in the last half of the twentieth century. This resulted from an increased demand for horses for pleasure, recreation, and sporting events.

Machines will never replace horses for many purposes. The beauty of horses and the wide variety of sizes, colors, and gaits make them popular for riding, driving, and showing. The horse project is one of this nation's largest 4-H youth projects. Horse racing is the world's most popular sport (Chapter 1). In some areas range cows are still worked with horses, and this will probably always be true.

The breeding, feeding, managing, training, and showing of horses require patience and skill, but in this day of missiles, rockets and spaceships the horse offers a pleasurable and desirable means of recreation and relaxation to those seeking a change of pace.

STUDY QUESTIONS

1. What was probably the first major use of the horse by humans?

2. How and why did the horse gain a foothold in the New World?

3. Why did the wealth of an American Indian chief come to be measured in terms of the number of horses he owned?

4. Why did horse numbers decline on farms a few decades ago? Why have they increased in recent years?

5. What are some of the most important uses of horses at the present time?

6. List and describe the 3 major types of horses.

7. What are the 4 natural gaits of a horse? The 4 artificial gaits?

8. How does the trot differ from the pace?

9. What are some distinguishing features of the eyes of horses?

10. How many different pairs of genes are known to affect coat color?

11. Why does mating a palomino mare to a palomino stallion fail to always produce a palomino foal?

12. List some of the common faults of conformation in the light horse.

13. Define filly, gelding, stallion, foal, colt, and mare.

14. Outline the recommended care that should be given a foal at parturition and shortly thereafter.

15. What are the major nutrient requirements of horses?

16. Describe some important factors to consider in training and grooming horses.

17. Name and describe some of the common defects and types of unsoundness often found in horses.

18. What are some stable vices and other bad habits of horses? How may they be corrected?

19. Describe some important points to consider in trimming horses' hooves.

20. How many teeth does a horse have? What are wolf teeth?

21. Name the major parts of a horse's foot.

22. What is moon blindness in horses? How did it get its name?

23. What are some important communicable diseases of horses?

24. What are some common horse parasites? Describe a treatment for each.

25. What food can be used as a high-energy supplement for horses with dental disease?

6

AQUACULTURE

Carolyn L. Orr and James H. Tidwell[1]

Give fish to hungry people and they will be nourished for a day, but teach them how to grow and catch fish, and they will be fed the rest of their lives.

Chinese proverb

6.1 INTRODUCTION

Although there are thousands of years of experience supporting **aquaculture** in China and other countries, it is a relatively young industry in the United States. It is also one of the fastest growing segments of animal agriculture. Aquacultural output has grown at a rate of more than 10 percent a year since 1990, compared with only 1.6 percent for captured fisheries and 3 percent for livestock meat production (**FAO,** 2000). While only 13 million tons of fish were produced in North America in 1990, there were over 31 million tons produced in 1998. Concur-

rently, the value of this production doubled from $490 million to nearly $1 billion (**USDA,** 1998).

Human consumption of fish and related foods is increasing each year. Demand for fish products is greater than the ability of the oceans to supply them. A reduction in wild-caught fish coupled with increased consumption by American consumers has led to increased profitability, improvements in technology required for captive production, and a rapid expansion of farmed aquatic species. As a result aquaculture is becoming a major alternative crop globally and in many areas of the United States. Figure 6.1

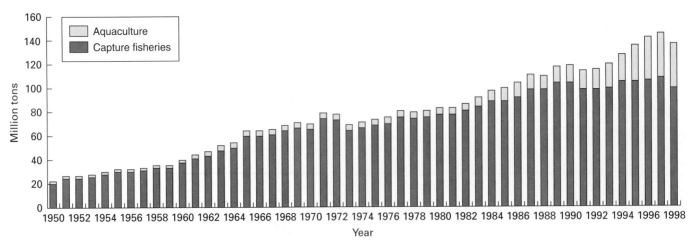

Figure 6.1 Total world fish production.
FAO, Rome

[1]Dr. Carolyn L. Orr, P.A.S., is Professor and Chair, Department of Agriculture, Berea College, Berea, Kentucky; and Dr. James H. Tidwell is Professor of Aquaculture, Kentucky State University, Frankfort, and President of the World Aquaculture Society.

illustrates the increasing percentage of the total world fish production supplied by aquaculture in relation to the oceanic catch. Because the oceanic fish catch is no longer increasing significantly, additional production depends on growing more fish through aquaculture.

Aquaculture ensures the availability of fish for human consumption and increases the efficiency of production through improved environmental control. It is limited, however, by gaps in current scientific knowledge and technology. These gaps resulted from our having invested minimally heretofore in salt- and freshwater food production research compared with research investments made in food production from land. Approximately 75 percent of the earth's surface is covered by water, yet humans harvest only about 1 percent of their total food supply from this source. Expanding aquaculture holds tremendous opportunity to provide a greatly increased supply of nutritious human food so urgently needed in much of the developing world (Chapter 1). In contrast to meat production, which is concentrated in industrial countries, approximately 85 percent of fish farming occurs in developing countries.

6.2 AQUACULTURE DEFINED

Aquaculture is the rearing of aquatic organisms under controlled or semicontrolled conditions. In simple terms, aquaculture is underwater agriculture. **Mariculture** refers to marine aquaculture, which uses saltwater rather than freshwater animals. Some species are raised in brackish water, a term referring to the water where salt- and freshwater merge. There is no analogous term for freshwater or brackish-water aquaculture. Nearly 59 percent of the 1998 aquaculture production was freshwater culture. Mariculture and brackish water accounted for 35 and 6 percent of production, respectively (Figure 6.2).

Whereas aquaculture refers to the cultivation of plants and animals, fish are the primary aquacrop in North America. Production is for one of five purposes: food, bait, feed, ornamen-

tals, or sport. Food fish and aquatic species produced for human consumption include trout, shrimp, and catfish. The ability of aquaculture to efficiently supply quality food for a growing population is the most important reason for its expansion. Small fish such as the fathead minnow are grown for fishers to use as baitfish. The marine species menhaden and others are used as feed ingredients for both captive fish rations and livestock feed formulation. Fish produced for pets in aquariums are referred to as ornamentals. A wide variety of species are produced for private ownership ranging from angelfish to goldfish. The growing popularity of pay lakes, as well as the need for pond and stream stocking, has seen an increase in sport fish produced and sold in the United States. Predominant species of sport fish include trout and catfish. Today aquaculture can refer to cultured sea species; crops produced in streams, ponds, or lakes; and even intensive tank culture in buildings (Figure 6.3).

6.3 HISTORY OF AQUACULTURE

Over 4000 years ago the Chinese, through trial and error, developed a systems approach to harvesting the nutritional riches of water. They found that different species feed in different natural niches. Using this information they could raise multiple species

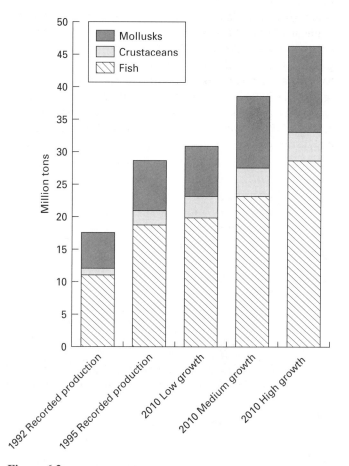

Figure 6.3 Projected aquaculture growth.
FAO, Rome.

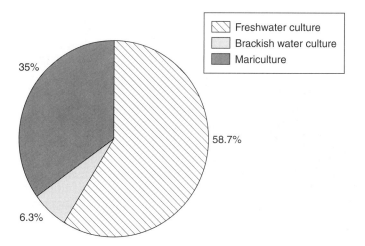

Figure 6.2 Contribution to U.S. aquaculture by production environment.
USDA, 1998.

in the same body of water. The animals of particular use to the ancient Chinese were complementary varieties of carp. The common carp (*Cyprinus carpio*) is a **benthic** or bottom feeder that consumes **invertebrates** from the bottoms of ponds. The black carp (*Mylopharyngodon piceus*) eats snails and small shellfish. Adding the "cattle of the pond world," a grass carp like *Ctenopharyngodon idellus,* utilizes vegetation in the middle layers of the pond. Finally, the addition of filter feeders including the Big Head Carp (*Aristichthys nobilis*) and Silver Carp (*Hypothalmichthys molitrix*) utilizes the microscopic animals and plants, **zooplankton** and **phytoplankton,** present in ponds. These free-swimming carp feeders are termed **pelagic** feeders.

Combining the different species, specialized by their feeding niches, better utilizes the various microenvironments of ponds. As the free swimmers feed and defecate, the excreta go to the pond bottom where it is eaten by common carp and **macroinvertebrate** bottom dwellers. The excreta also supply nutrients for zooplankton and phytoplankton growth providing feed for the free swimmers. This allows nutrients to continue to cycle in a manner that promotes efficient fish growth.

For thousands of years carp species provided the Chinese with a sustainable harvest. The Chinese discovered that the use of manure as fertilizer could enhance the productivity of a pond. In the 1800s **aquaculturists** in eastern European countries also cultured the common carp. Interestingly, the common carp is the most cultured species in the world. Only in the United States has the common carp failed to evolve as a desirable food fish.

Aquaculture was launched in the United States when the federal government began hatching and stocking programs to restock salmon lost because of the damming of spawning rivers in the Northwest. The damming resulted from the drive to produce electrical power for an expanding human population. Many rivers that became homes to hydroelectric dams were important spawning grounds for **anadromous** salmon species. Numbers reduced dramatically and many fishers lost their livelihoods when salmon could no longer go upstream to spawn.

In an attempt to increase salmon numbers in the wild, the federal government began hatching programs that were basically "put-and-take" systems. Eggs were collected, hatched in captivity, and fingerlings released. This program, although successful in increasing wild-caught numbers, was actually more successful in providing initial information and research for the science of aquaculture. Hatching and restocking into the wild required development of the same technology as that needed by commercial hatcheries. Early **salmonid** work came almost exclusively from government hatching centers. Today commercial salmon landings (fishing) account for over 600 million lb annually. This makes salmon the major marine fish (mariculture) industry of the United States with more than half the production coming from public or private hatcheries (Figure 6.4).

Another spin-off of governmental spawning programs was development of trout production in flowing-water systems. This system involves stocking hatchlings in concrete or earthen raceways approximately 3 ft deep. Productivity is limited by the flow rate of water, averaging 10 to 40 lb of fish per gal per min of flow. Idaho is the leading trout producing state with

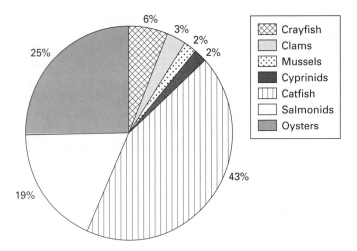

Figure 6.4 United States production of major aquaculture species.

Figure 6.5 Aerial view of aquaculture ponds in the southeastern United States.
Photo courtesy of R. Durborow.

approximately 67 million lb of trout marketed annually. The requirement for large volumes of fresh cold water for raising trout in confinement continues to be a limiting factor in the commercial profitability of this species. In his 1868 book titled *American Fish Culture,* Thaddeus Norris wrote ". . . farmers are more favorably situated for the whole routine of breeding and growing trout than persons of any other occupation."

In the 1950s the U.S. government supported programs of farm pond construction and stocking. Although more than 2 million ponds were constructed they have demonstrated little importance or priority for food production. In the 1960s, ponds were constructed specifically for catfish in the southern United States where production increased geometrically from the 1970s through today (Figure 6.5). Figure 6.4 shows catfish production ranks first among aquaculture species produced in North America. Important early catfish production and research, particularly nutrition research, occurred in Kansas. Then as a result of the availability of land, favorable water temperature, and climate matching catfish-production needs, much of the actual production of catfish moved to the Mississippi Delta region.

The federal research laboratory in Stuttgart, Arkansas, helped promote catfish by providing information and support. Regional government aquaculture research centers have helped bolster growth and profitability of the industry. Aquaculture production in North America involves diverse farming systems in diverse areas. It is most concentrated in the southern region of the United States (Figure 6.6). Today the U.S. catfish industry is the largest aquaculture industry in the world employing more than 25,000 people and generating over $1 billion annually. More than 178,000 acres of catfish ponds on 1300 farms produce approximately 100 million metric tons per year. Currently 80 percent of the catfish production occurs in the state of Mississippi with 105,000 acres concentrated on 400 farms.

Many other commercial species are also centered in the Southeast. Louisiana is the leading producer of crawfish and Arkansas leads in baitfish production. Florida is the leading producer of ornamental fish. Increasingly there are niches being found for hybrid striped bass, freshwater prawns, redfish, and tilapia throughout the country with tilapia and striped bass production growing significantly. Production of the American cupped oyster is predominately from the United States (Figure 6.4). Other species that have potential include alligators, bullfrogs, and turtles.

6.4 CURRENT TRENDS IN GLOBAL FISH CONSUMPTION

> We remember the fish we used to eat in Egypt.
>
> **Numbers 11:5**

Although fish consumption in the United States is relatively low compared with much of the world, it is increasing. Much of this increased consumption results from the perceived health advantages of eating fish, its greater availability, and competi-

tive pricing. The meat of fish is similar to other meats in nutritive value. It is an excellent source of protein, n3 fatty acids, B vitamins (especially nicotinic acid), and vitamins A and D. United States consumption of fish increased from 3700 million lbs in 1990 to over 4100 million lbs in 2000 (edible meat weight). This is an average increase of almost 1 lb per person annually in 10 years. Although demand for fish and seafood is increasing, the harvest of wild fish is static or decreasing worldwide. FAO data indicate that about 35 percent of the 200 major wild fishery resources are showing declining yields, about 25 percent are plateauing at a high exploitation level, and 40 percent are still "developing." There are none at undeveloped levels. This indicates that about 60 percent of the major wild fish resources are either declining or plateauing.

The production of the ocean is not limitless. Because commercial fishers are good at their trade and the level of technology available increases continuously, the species people find most desirable are being harvested in greater numbers than they can reproduce. Additionally, the effects of pollution on wild species, particularly the pollution of **estuaries** that serve as nurseries, result in a reduction of the population of wild species. The United States has typically relied on imported fish to meet consumer demands. This has resulted in a trade deficit in the fish and seafood industry.

In 1999 U.S. imports of fish were valued at $9 billion, exports at only $2.8 billion. As wild-caught supply decreases and the trade deficit in fish products increases, the value and economic sustainability of U.S. farm-raised fish is expected to increase. An increase in U.S. aquaculture production could assist in lowering the international trade deficit. The import/export $6.2 billion deficit in edible fish and seafood was the largest deficit of any natural resource other than oil according to the National Marine Fisheries Service. Shrimp leads the list of imported species. In the next 15 years the Fisheries Service projects the annual global demand for fish and seafood will increase more than 50 percent to 120 million metric tons.

It is thought that much of the future growth in aquaculture will be in new species and geographic areas. The rapid growth of the catfish industry has moderated. Depletion of groundwater resources has become a limiting factor to its continued growth and expansion in the Mississippi Delta region. As a result new production facilities will likely develop in other areas and may well utilize other aquaculture species. Examples include the hybrid striped bass industry developing on the eastern seaboard of the United States and tank tilapia (St. Peter's fish) production occurring nationwide.

6.5 CHARACTERISTICS OF FISH

There are several anatomical structures of fish that are not part of other animal species discussed in this textbook. One obvious difference is the presence of gills (Figure 6.7), which function in several ways. Important is the fact that they are responsible for the intake of oxygen. However, fish also use their gills to excrete ammonia. Gills are small **filamentous** organs over

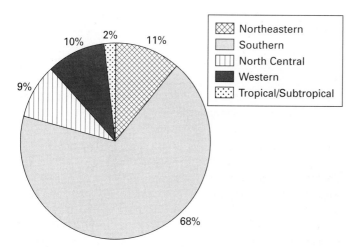

Figure 6.6 Percentage of aquaculture producers by region in the United States.
USDA, 1998.

which water passes when the fish swims. Oxygen is absorbed and carbon dioxide and nitrogenous wastes in the form of ammonia are released. Because fish excrete ammonia passively through their gills directly into their environment they need not burn energy in storing, detoxifying, transporting, or excreting waste products through urine. This represents an important energy savings. However, it also results in fish living in intimate contact with their waste products. What they excrete in water is breathed back in if the water quality is not sufficient to dilute it. Species can be categorized by their requirements for oxygen, water temperature, pH, and salinity.

The body of fish can be divided into the head, trunk, and tail (Figure 6.8). The head, which terminates at the end of the gill covers (operculum), contains the nose, snout, nostrils, eyes, and mouth. The top or dorsal part of the head is termed the nape. The **ventral** part is the thorax. Eye size is related to whether the fish must see to find food. Catfish use the sense of smell for finding food and have smaller eyes than trout, which are sight-finding predators. Having no external ears fish sense vibrations through ear bones in their heads. Olfactory sacs, which provide fish with the sense of smell, open through the **nares** located between the eyes and the snout. Catfish have taste buds all over their bodies. Indeed, they have more taste buds than any other animal. In most fish the major sense organ is not the eyes or nose but rather the **lateral line.** It is comprised of a series of pores running the length of the trunk. Fish sense their environment through changes in pressure against pores in the lateral line.

In addition to the lateral line the trunk contains the dorsal fin, a pair of **pectoral** fins, and pelvic fins. The tail, which begins at the anus and ends with the far tip of the **caudal** fin, also has another fin, the anal fin, just behind the anus. Some species including the salmon and trout have an **adipose** fin on the tail. All fins are supported by **soft rays** whereas others also contain spines in front of the rays.

Fish exist in a buoyant thermally changing environment. This provides major advantages in terms of energy requirements in that fish do not burn calories fighting gravity. Yet it requires fish to have some manner of controlling their buoyancy and swimming level in water. This is achieved through the swim bladder, an internal "balloon" that fish inflate as needed to maintain position in the water column. Fish move through water by moving separate muscle segments.

Fish are **poikilotherms** (cold-blooded), which is a "double-edged sword" in their production. It is advantageous as long as the ambient temperature of their water is optimum for growth. Like mammals, the biochemical enzymes that control fish metabolism have optimum operating ranges. When the water environment is within these ranges fish do not use dietary energy to maintain body temperature, another savings in terms of their maintenance energy requirement. However, when their environmental temperature is outside the optimum range, their body temperature will be the same as that environment and many of their biochemical reactions will slow. Thus the rate of metabolism and growth of a fish varies with water temperature.

Physiologically these differences represent an energy savings and a resultant increased efficiency of feed utilization. Because of their unique biological mechanisms—including coldbloodedness, reduced energy expenditure for waste excretion, and movement—fish, as a category, have lower energy needs than other animals. The result is that fish convert feed to weight gain extremely efficiently, particularly if their environment is managed properly.

The digestive system of the fish contains a mouth, esophagus, stomach, intestine, and anus. Most digestion occurs in the stomach but the size and shape of the system depends on the natural diet. Fish species show differences in their choice of diet. Grass carp are herbivorous compared with omnivorous catfish and carnivorous hybrid striped bass, trout, and salmon. Fish that feed on aquatic vegetation have a small stomach with a long intestine, whereas a short intestine and large stomach characterize carnivorous species. Even within these categories there are differences in nutrient needs. Protein is one of the most critical nutrient needs of fish. Fish generally require 25 to 40 percent protein in their diets with at least 10 essential amino acids. Much of the protein needed is supplied by aquaculturists in the form of fish meal.

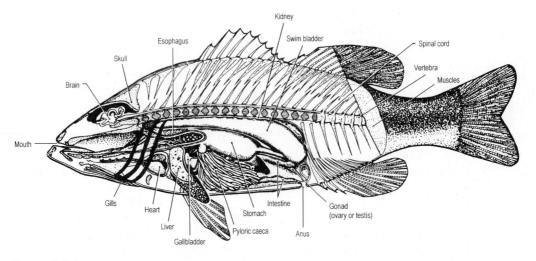

Figure 6.7 Diagram of fish anatomy.
Courtesy of George Luker, Langston University, OK. Adapted from Karl F. Lager, et. al., Ichthyology, Wiley, Inc., NY, 1977.

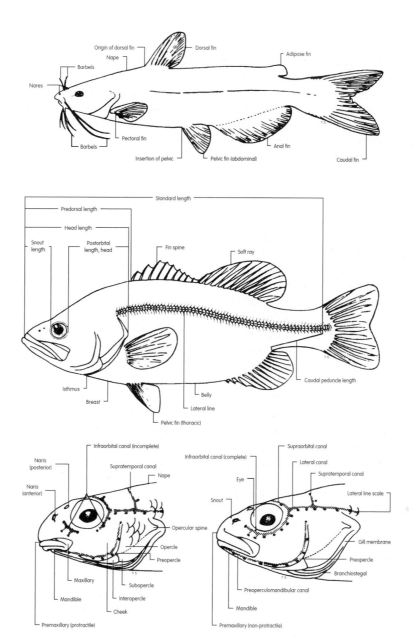

Figure 6.8 Fish schematics.

Courtesy of George Luker, Langston University, OK. Adapted from Rudolph J. Miller and Henry W. Robinson, The Fishes of Oklahoma, *Oklahoma State University Press, Stillwater, 1973.*

Some carnivorous species (e.g., salmonids) require a diet as high as 45 to 50 percent fish meal. Other animal by-products such as blood meal can supply some of this protein. Marine fishes in general seem to have a higher requirement of dietary fish meal than freshwater fishes. Catfish require about 4 percent fish meal for optimal growth. Apparently there is a genetic correlation between feed requirements and geographic areas. Moreover, warm-water fish tend to need less protein, particularly fish meal, than cold-water fish.

Other dietary differences involve the need for quantities of n3 (omega-3) fatty acids. Marine fish have a dietary requirement for these fatty acids whereas most freshwater fish do not. Additional dietary fat usually comes from animal sources. Cold-water fish such as trout require relatively high levels of unsaturated fats, whereas warm-water catfish and tilapia utilize monounsaturated and saturated animal fats in relatively low levels in their diets. The quantity of dietary energy derived from carbohydrates is also related to the type of fish. Herbivorous fish require a higher level of dietary carbohydrates and are better able to digest them than catfish or trout. The latter require less than 10 percent carbohydrates in their diet. Fish are able to absorb some minerals from their water environment. Therefore, it is important to know the characteristics of the water to determine the mineral supplementation required in a fish's diet. Vitamin deficiencies resulting in poor growth and skin lesions are common in captive fish, so their diets are commonly supplemented with both fat-soluble and water-soluble vitamins.

6.6 WATER QUALITY

The most important aspect of fish production is managing the quality of their water environment. Any discussion of water environment must begin with dissolved oxygen because it is the first factor limiting intensive production of fish. Oxygen is a much rarer commodity in aquatic environments than on land. Gaseous atmosphere is comprised of 18 percent oxygen, whereas dissolved oxygen in water is 0.0001 percent by weight. This is reported as parts per million (ppm). The solubility of this oxygen changes as the temperature changes. Cold water holds more oxygen than warm water. Dissolved oxygen levels need to be at 5 ppm or greater for proper growth of fish. Most water needs dissolved oxygen added to it for intensive fish production. This process is termed **aeration.**

Many people believe oxygen in a pond diffuses from surface air to the water. Although this occurs, the amount of dissolved oxygen derived this way is limited. It is difficult for oxygen to diffuse into water. Most oxygen available in a pond is derived from photosynthesis by phytoplankton. Maintenance of phytoplankton populations and their production of oxygen are important in intensifying pond fish production to commercially viable levels. Without aeration, production is limited to 500 to 1500 lb of fish per surface acre of pond. Aeration can increase this level to 1500 to 4000 lb of fish per acre. Thus, the advent of mechanical aeration has made commercial pond production of fish economically viable.

Oxygen can be depleted from water through **biological oxygen demand (BOD).** This term refers to the use of oxygen in the water by natural processes. Some of these processes include the decay of weeds, leaves, feed, or other organic matter. In addition, because natural aeration depends on photosynthesis, cloudy days and dark nights combined with high stocking rates can deplete oxygen supplies.

The second limiting factor in commercial fish production is the accumulation of nitrogenous waste products. Fish passively excrete ammonia into water, which can reenter fish through their gills. Total ammonia exists in a dynamic equilibrium between the ionized (nontoxic) and un-ionized (toxic) forms. Temperature and pH of water are major influences in determining which way this balance goes, with an increase in temperature and pH shifting the equilibrium to the un-ionized form. The ammonia in fish blood cannot diffuse into the water when water ammonia levels increase. High water ammonia levels increase the blood ammonia levels and can cause fish to cease growing and even die.

Maintaining an adequate pond environment depends on bacterial nitrification, a nutrient-cycling process that occurs naturally to remove ammonia from water. Detoxification and nitrification are the same. As ammonia (NH_3) is converted to nitrite (NO_2) it reduces the toxicity to fish; then nitrite is converted to nitrate (NO_3), which reduces toxicity still further. The chemical reaction is written as $NH_3 \rightarrow NO_2 \rightarrow NO_3$ and occurs only in the presence of *Nitrosomonas* and *Nitrobacter* bacteria. These changes require adequate oxygen and optimum temperature and pH to function efficiently. The resulting nitrate is harmless and is largely released as gas or used as fertilizer by phytoplankton.

6.7 PRODUCTION METHODS

Effective and efficient production measures must address specific biological requirements for each species including providing oxygen and nutrients, removing waste products, and maintaining optimum temperatures. The most commonly used production systems include ponds, raceways, cages, and recirculating tanks (Figure 6.9).

The use of each of these systems provides differing production intensity levels. The more intensive a production system, the greater the density of animals and the higher the level of technology and environmental control required. The greater the intensity of production the more knowledgeable and attentive the producer must be, culminating in greater attention to financial investment and returns as well. Some species are better adapted to different environmental temperatures and cultural conditions than are others. Certain species have compatible requirements and can be raised together, a production method referred to as **polyculture.**

Ponds are the most common system used for raising fish. Generally constructed with earthen dams or levees, they may range in size from 1 to 50 acres. Average catfish pond size in Mississippi is 20 surface acres, including the levees (Figure 6.10). These ponds are 4 to 5 ft deep and are a compromise between the economies of scale for construction and fish feeding, while having a manageable unit for aerating and harvesting. Rectangular ponds are easier to manage. The pond bottom should be smooth and free of holes for ease of seine harvesting. High clay soils make the best ponds. With aeration, a well-designed pond can produce 4000 to 5000 lb of catfish per acre.

Some nontraditional species of pond fish are raised in 5- to 10-acre drainable, earthen ponds. These ponds have freshwater requirements of 40 to 50 gal per min per surface acre, typically

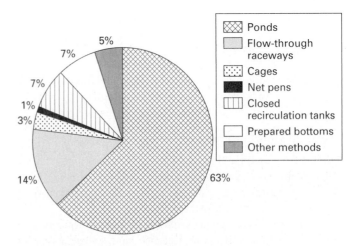

Figure 6.9 Methods of aquaculture production used in the United States.
USDA.

Figure 6.10 Catfish ponds in Mississippi.

Photo courtesy of J. H. Tidwell.

Figure 6.11 Caged fish production.

Photo courtesy of Kentucky State University, Frankfort.

Figure 6.12 Trout production by raceway.

Photo courtesy of J. H. Tidwell.

from a well. Watershed ponds—those that follow the contour of the landscape and catch the water runoff from a land area—are generally deeper with less surface area and are therefore less productive per acre. Watershed ponds may be harder to harvest because of the configuration of the bottom. Ponds are most efficiently harvested by seine so harvesting is easiest in ponds specifically constructed for fish production. **Corral seines** are used where fish are trained to feed in certain areas of the pond, thus facilitating seining in the feeding areas. When fish cannot be seined directly or corral seined, they must be raised in a cage (Figure 6.11).

Cage producers typically use existing waters that cannot be harvested by other means. Cage production allows harvest and protects species from predators in ponds, lakes, rivers, oceans, and estuaries. When maintained in a cage they are easily harvested. Cage production has disadvantages, however, including localized water quality deterioration and increased disease transmission because of more intensive confinement of the fish. Because fish are confined they must be fed more frequently. Other difficulties include the inability of fish to react to water quality problems and poaching. Cage sizes vary from 1 m^3, capable of holding 400 lb of fish, to $4 \times 4 \times 8$- or $8 \times 8 \times 12$-ft cages, or even large "net" pens. Under cage conditions ponds can support up to 2000 lb of fish per acre.

Raceways are concrete tanks built in a terraced fashion so water runs from one tank into the next (Figure 6.12). This is a commercially feasible and economically proven technology, provided sufficient running water is available. Greater rates of water flow provide for more intensive production. Twenty to forty pounds of trout can be raised per gallon per minute of water flow. This flow-through system utilizes high-quality oxygenated water and high fish densities with the water conveyed onto successive raceways. Thus water quality deteriorates as it flows from one raceway to the next with oxygen level decreasing and ammonia levels increasing. The ammonia is mostly in nontoxic ionized form because it is a cold-water system. The water is reoxygenated to some extent as it drops from one raceway to the next in the terraced design. The water can be used through as many as eight sequential raceways before its ammonia level becomes too high for reuse. Closed raceways will treat the water for reuse, whereas open raceways dispose of this water, but it may require treatment before being released into a running stream.

The use of recirculating aquaculture tank systems is a new technology currently being tested for commercial feasibility (Figure 6.13). A great deal of research is underway in the upper Midwest where the tanks are built under roof to maintain favorable water temperatures.

Large tanks are designed so water flows through the tank continuously. As water leaves the bottom of the tank it carries solids, which are removed by mechanical filtration. Water then passes through **biofiltration** systems where bacterial nitrification by *Nitrosomonas* bacteria at an intensified level convert ammonia to nitrite, then to nitrate. A biofiltration unit uses plastic balls or similar material that provide a large amount of surface area. Populations of *Nitrosomonas* bacteria grow on this

Figure 6.13 Recirculating tank aquaculture.
Photo courtesy of Kentucky State University, Frankfort.

surface area and when water is moved across these bacteria the ammonia is converted by bacterial enzymes to NO_2 and then to NO_3. Thus toxicity of the ammonia is reduced by bacteria. This is similar to what happens in nature. The water is aerated so that water leaving the biofiltration unit can be recycled through tanks in a closed system. This requires intensification of both management and technology in an artificial environment.

Advantages of the recirculating system include the ability to maintain temperature and water quality at optimum levels for each fish species. Using these systems tropical fish can be raised in areas like Chicago. (Interestingly, about 1 billion lb of seafood and freshwater fish are consumed annually in the Midwest, yet the region produces less than 2 percent of that amount according to the North Central Regional Aquaculture Center at Michigan State University.)

6.8 Aquaculture Production Cycles

Catfish lead in aquaculture production in United States. A typical production schedule for catfish would begin when 4- to 6-year-old **broodfish** (4 to 8 lb in weight) are stocked in spawning ponds in early spring with two females per male. These broodfish are selected genetically much as cows in the beef seedstock industry (Chapter 9). They are stocked at 800 to 1000 lb per acre. Spawning containers like milk cans that imitate hollow logs are submerged into ponds after the water temperature exceeds 70°F. Containers are checked for eggs several times weekly (Figure 6.14). Eggs are transferred to artificial hatcheries where the water is kept moving to simulate actions of the male fish in nature.

Hatching occurs in 6 to 10 days with the shorter time corresponding to warmer water temperatures. At hatching the young fish, termed **fry,** have attached egg sacs that provide nutrition for a few days. This is similar to the situation with newly hatched chicks (Chapter 16). When this nutritional store is gone, fry begin eating. They are generally stocked in **fingerling** ponds at 50,000 to 150,000 per acre depending on desired fin-

gerling size. Following hatching, the fry are fed a finely ground high-protein diet for 3 to 4 weeks. After fry begin to feed actively they are gradually switched to floating feed pellets. Once fingerlings reach a length of 2 to 6 in, they are transferred to growing ponds either in the fall or the following spring.

Fingerlings are fed floating pellets containing 32 percent protein. Catfish stocked in the spring commonly weigh 1 to 10 lb by fall. Fish are harvested by seine and transported live to a processing plant. They are held in live wells until entering the processing line. The desired food fish is 1 to 4 lb. Catfish have a dressing percentage of 55 percent, meaning they produce 55 percent of their live weight as marketable food product. Fish go from swimming to a processed product in a matter of minutes.

Trout are a cold-water fish and have been cultured longer than any other species in the United States. They grow best in water with a temperature range of 50 to 68°F. Trout production is most successful in highly oxygenated flowing water. The 2- to 5-in fingerlings are stocked at rates of 300 to 800 per acre. Wild trout are carnivorous and captive trout continue to require the high-protein diet associated with the eating of insects, fish, and fish eggs. Protein levels of 42 percent in a diet fed twice daily provide for maximum gain rates. Approximately 2 years are required to produce marketable trout from fertile eggs.

Tilapia are a new species to U.S. producers, but have been cultured for 4000 years in Africa. Also known as *St. Peter's fish,* tilapia cannot be released into the wild in the United States because of fear they will take over lakes and streams. Tilapia are disease-resistant fish that grow best in shallow fertile ponds with warm water, although they are tolerant of poor water quality and temperatures to 50°F. They will feed on aquatic vegetation, insects, and commercial feeds which average 32 percent protein. Marketable weights of 2 lb can be achieved 1 year from hatching.

The exact process of producing ornamental fish is species dependent but requires a source of water, water containers, heaters, and water filtration equipment. If city water is used it must be allowed to age for a minimum of 24 h or be chemically

Figure 6.14 Harvesting catfish eggs.
Photo courtesy of J. H. Tidwell.

treated to remove the chlorine added at the water treatment plant. Goldfish, guppies, and barbs are easy-to-grow ornamentals.

6.9 AQUACULTURE, THE ENVIRONMENT, AND ANIMAL WELL-BEING

It is important that aquaculturists pay particular attention to protecting the environment, conserving natural resources, and preserving the well-being of animals. Water is of great concern with respect to both source of fresh and disposal of used water. The aquaculture industry faces challenges from some environmental groups concerning the possible contamination of the environment by effluents, excess food, or **chemotherapeutics.** Producers must comply with environmental safeguards and minimize any possible adverse effect on the environment. Access to rivers and streams does not give producers the right to alter either the course or quality of water. Many state and federal regulations apply to both the removal and discharge of water as well as the disposal of fish waste and mortalities. Processing plant waste disposal requires even greater planning.

With growth of the aquaculture industry it is important that producers become knowledgeable of fish characteristics and requirements to ensure that principles of animal well-being are followed (Chapter 7). Fish kept as pets need an environment that is conducive to their growth and well-being. Species must not be released into environments to which they are not adapted. Water temperature and quality, stocking rate, and feed requirements are all important issues in ensuring favorable animal state of being. The introduction of exotic species into natural bodies of water is of concern to fish and wildlife experts. Exotic species often have tremendous impacts on the ecology and species of natural populations. In many states it is illegal to release such species as tilapia and grass carp, species originally used to control aquatic vegetation but found to threaten native species.

6.10 MARKETING

In early commercial fish production, prior to the initiation of commercial processing, most marketing was done through live haulers or pay lakes. Live haulers are "middlemen" who transport fish to pay lakes or for pond stocking. Pay lakes are owned by individuals who have a highly stocked lake that is open to public fishing for a fee. Similar to the "pick-your-own" strategy in fruit and vegetable marketing, this is a "catch-your-own" process that has worked well as a family activity, especially near urban centers. In areas where commercial processing has not developed, production can be marketed through farmers markets and direct sales of fresh fish.

Processors provide the lowest return to producers, yet they provide a steady market and quality assurance for consumers. Most processors taste-test the fish prior to harvesting to ensure product quality. The product form of fish has shifted from approximately 60 percent of the catch being sold as whole and 40 percent as fillets to more than 60 percent of the market in the

form of fillets. This follows consumer trends toward convenience items, as is the case for other meat products (Chapter 3). Shelf life of fresh-caught catfish is 3 to 4 days in the refrigerated section. This may be lengthened to 10 or 11 days when kept on ice and up to 20 days when ice packed. Frozen catfish has a shelf life of up to 12 months.

Compared with fish and shellfish products from wild capture, aquacultured products have numerous advantages including supply, size, and more consistent quality assurance characteristics. Product can be sized for target markets and timed to produce greater returns. Aquaculturists can assure the integrity of their products particularly in relation to the environmental impacts of pollution and pathogens. Specimens can be taste-tested weeks before harvest, thus the processing plant can determine quality of their product before it is processed. Applying scientific knowledge of genetics and nutrition, producers can modify fat content of fish including the n3 fatty acid content of their product.

6.11 PROCESSING OF AQUACULTURE PRODUCTS

Preparing the products of aquaculture for consumers requires specialized processing plants. "Dressing" fish is similar to processing livestock. It follows stunning by an electric shock and includes removal of the head, internal organs, skin, and/or scales. Shellfish require even more specialized processing, which includes deheading, peeling, and deveining. All species require frequent washing and inspecting prior to fabricating (cutting) and packaging. Growth of the value-added segment providing seasoned, precooked, or smoked consumer-friendly items adds additional jobs to the basic processing procedure.

6.12 SUMMARY

Aquaculture is one of the fastest-growing segments of animal agriculture. The importance and financial magnitude of aquaculture production in the United States has more than doubled during the last 15 years. Greater demand for fish products and reduced wild catches have fueled this increase in captive fish production. The raising of aquatic organisms under controlled systems includes species as diverse as trout, catfish, shrimp, and alligators. Chinese aquaculturists developed sustainable production systems 4000 years ago using fish that had different feeding niches. Today specialized producers harvest, process, and freeze thousands of pounds of boneless fish fillets for the consumer market.

Fish biology provides greater efficiency in producing meat than red meat animals through reduced energy expended for movement in a weightless environment, their being cold-blooded, and less energy expended for waste excretion. Maintenance of excellent water quality with sufficient dissolved oxygen and low concentrations of nitrogenous waste products are key aspects of profitable aquaculture production.

The food cycles in water systems of the world depend on microscopic animals that feed upon the phytoplankton and are in turn fed upon by larger organisms, ultimately yielding animal protein for direct human and domestic animal consumption. This cycle offers vast potential for improving the nutritional well-being of young and older humans alike in both developed and developing countries. With a higher priority for research related to fish and other aquatic life, the production of food in various bodies of water holds great promise in improving the nutrition of humanity on a global basis.

STUDY QUESTIONS

1. Name one or more countries that pioneered the development of aquaculture.

2. Is aquaculture a decreasing or an increasing segment of animal agriculture in the United States? Why? Discuss.

3. Define aquaculture. What is mariculture?

4. What was the motive for the U.S. government becoming interested in fish hatching and stocking programs?

5. Where did catfish production research begin in the United States? What U.S. region currently produces the most catfish?

6. How does the U.S. catfish industry rank as a global aquaculture industry?

7. Which state leads the nation in the production of ornamental fish? In crawfish? In baitfish production?

8. Discuss the nutritional contributions of fish and related foods to the diets of humans.

9. Is the per capita consumption of fish decreasing or increasing in the United States?

10. Does the United States export more or less fish than it imports? Is the global demand for fish increasing or decreasing?

11. What is the most obvious anatomical difference between fish and other food-producing animals?

12. How do fish excrete ammonia? Is a buildup of ammonia in water of any consequence in fish farming? Why?

13. Do fish use their eyes as the exclusive means of finding food? Discuss.

14. Of what significance is the fact that fish are poikilotherms? Discuss.

15. Do all fish have a dietary need for n3 fatty acids? Any difference between marine and freshwater fish in this regard?

16. Discuss the importance of water quality in the production of fish.

17. Of what significance is biological oxygen demand in aquaculture?

18. What are the most commonly used production systems in raising fish? Which system is used the most?

19. How many pounds of catfish can be produced per acre in a well-designed pond?

20. What is an important advantage of the recirculating system in fish production? Discuss.

21. Outline a typical production cycle for catfish. For trout.

22. Why cannot tilapia fish be released in the wild in the United States?

23. Are there any special environmental considerations that aquaculturists should take into account? Discuss.

24. How are most freshwater fish marketed?

25. Compared with fish and shellfish products from wild capture, cite some advantages held by aquacultured products.

26. Discuss product quality assurance in the marketing of aquaculture products.

27. Does aquaculture hold any special promise in providing improved diets for humans in the developing countries? Discuss.

28. Did reading this chapter give you an expanded perspective as to the potential importance and benefits of aquaculture domestically and globally? Discuss.

STATE OF BEING OF DOMESTIC ANIMALS[1]

Stanley E. Curtis[2]

We need another and a wiser and perhaps a more mystical concept of animals. Remote from universal nature, and living by complicated artifice, man in civilization surveys the creature through the glass of his knowledge and sees thereby a feather magnified and the whole image in distortion. We patronize them for their incompleteness, for their tragic fate of having taken form so far below ourselves. And therein we err, and greatly err. For the animal shall not be measured by man. In a world older and more complete than ours they move finished and complete, gifted with extensions of the senses we have lost or never attained, living by voices we shall never hear. They are not brethren, they are not underlings; they are other nations, caught with ourselves in the net of life and time, fellow prisoners of the splendour and travail of the earth.

Henry Beston (1888–1968)
Author, The Outermost House, 1928

7.1 INTRODUCTION

Since Ruth Harrison wrote *Animal Machines* (1964), which called **"factory farming"** methods cruel to animals, the **state of being** of agricultural animals has been prominent on public agendas across North America, western Europe, Australia, and New Zealand. The farm animal–welfare issue emerged in the United States during the late 1970s. Specific areas of concern are much the same now as they were then. This complex issue continues to stir debate among members of a pluralistic society who have special sympathy for the extent to which animals experience well-being and those who have—through agriculture, biology, economics, philosophy, the law, public policy, or politics—a business, professional, or scientific interest in animals.

According to surveys conducted in the mid-1980s by several agricultural organizations, most Americans and some animal protection groups support the use of animals as food and

believe farmers generally treat their animals in humane fashion. Nevertheless, many individuals also support governmental regulation to ensure humane treatment of animals in production agriculture.

Debate around the world has not resolved the issue. Although several European nations, commissions, and councils have attempted to deal with the farm animal–welfare issue for more than three decades, it remains prominent on public agendas there, as in North America. During this same period, agricultural groups in the United States have developed voluntary guidelines for the care and treatment of various species. But the debate continues. The essential role of animals in the world food-production enterprise and the global catastrophe that would ensue if that role were to cease still oblige all stakeholders to continue searching for thorough understanding of the well-being of agricultural animals and an inclusive resolution to the farm animal–welfare issue.

The purpose of this chapter is threefold:

- To outline and provide examples of ethical, moral, economic, legal, and policy aspects of public issues concerning the state of being of agricultural animals both on and off farms and ranches.
- To describe scientific approaches to assessing the state of being of agricultural animals.
- To identify areas in which additional scientific insight would help ensure that animals actually and usually experience well-being in agricultural production systems.

[1]Much of this chapter is based on the Executive Summary of Task Force Report No. 130, *The Well-Being of Agricultural Animals,* published in 1997 by the Council for Agricultural Science and Technology (CAST). Members of the multidisciplinary task force are acknowledged at the end of this chapter. Dr. Curtis chaired that task force, and is responsible for the changes and additional materials included in this chapter.
[2]Dr. Stanley E. Curtis is Professor of Animal Sciences, University of Illinois at Urbana–Champaign. A distinguished alumnus of Purdue University and recognized authority and frequent speaker and writer on applied ecology and ethology, state of being, and care of kept animals, he has received numerous awards and honors for research, teaching, and industry outreach. He has presided over the American Society of Animal Science, Association for Assessment and Accreditation of Laboratory Animal Care International, and Federation of Animal Science Societies.

7.2 ORIGINS OF THE DEBATE

7.2.1 Philosophical Aspects

Welfare and Rights: The Distinction

Some describe those advocating moderate change in agricultural systems as *animal-welfare advocates* and those seeking greater change as *animal-rights advocates.* In scholarly works, however, **rights** more often are analyzed instead in terms of claims made by one party against another. There are informal, moral, and legal rights.

Any rights claim depends on some context of validation. One intervening on behalf of a victimized animal makes (on behalf of that animal) a rights claim validated by custom, ethics, and even law. This claim need not imply, however, either support for an animal's right not to be used for human food or validation of other claims made by animal-rights groups.

Rights also may assert the priority of an individual interest over the interests of a group and even the common good. But political examination of "trade-offs" (costs and benefits) is constrained by rights. For example, trade-offs that would violate individual rights are "trumped" by valid rights claims. Contentious public policy issues often revolve around where the line between trade-offs and trumps should be drawn, and **welfare** and rights can be opposing concepts in a variety of ethical, legal, and political controversies.

An example of a current controversial animal rights issue can be found in policies pertaining to the slaughter of "downer animals." The 1999 National Cow and Bull Audit indicated that the percentage of non-ambulatory dairy cows arriving at slaughter plants increased from 1.1 percent in 1994 to 1.5 percent in 1999. Animal rights activists insist that downer cows should undergo immediate euthanasia. A compromise to fulfill humane care while supporting the common good of salvaging edible meat and income for the producer is to give downer cows priority for handling. The Humane Slaughter Act prohibits the dragging of downed or crippled livestock. The animals are transported using slide boards to an area where they can be examined by an inspector and moved to slaughter as quickly as possible. Animal welfare advocates provide cattle producers and haulers with guidelines to prevent crippled and downer cattle (use of non-slip flooring, nutritional advice to prevent laminitis and **milk fever,** breeding of heifers to bulls which sire low birth weight calves to prevent calving paralysis).

Ethics and Morals

> We serve, in never fulfilled repayment, the animals that
> accompany humankind.
>
> **Petrus C. Boutens (1870–1943)**
> **Dutch poet**
> **In Entryway to Veterinary College**
> **University of Utrecht, Netherlands**

Two distinct approaches to showing how a rights claim might be validated depend on different kinds of ethical principles. The **utilitarian strategy** considers an action or a policy justified in light of the sum of its consequences for all affected parties, whereas the **rights strategy** states that individuals have certain rights that above all must be protected, and the morality of an act is to be judged according to whether it respects others' rights. As illustrated above, the Humane Slaughter Act is an example of a utilitarian strategy for the handling of downer animals. Proponents of the rights strategy would forbid the slaughter of downer animals. Philosophical tension exists between utilitarian and rights philosophers. Major leaders of activist political organizations also differ over which philosophical principles best justify their initiatives, upon which they generally agree.

Major Figures

Hundreds of essays have been published on the ethical basis for the reform of animal agriculture, as well as on the moral basis for relationships between humans and other animals generally. The works and opinions of four authors—R.G. Frey, Peter Singer, Tom Regan, and Bernard Rollin—will be reviewed briefly.

Probably only R. G. Frey has developed a significant reputation for defending status quo attitudes toward animals, and even he has been somewhat critical of recent trends in animal agriculture (in *Issues in Agricultural Bioethics,* T. B. Mepham and colleagues [eds.], 1995). Future work on the moral status of animals and on the ethics of production methods in agriculture undoubtedly will integrate ethical theory with scientific approaches to the study of animal behavior and cognition.

Peter Singer established his basic approach, an extension of the utilitarian philosophy in 1973. He argued that the common practice of limiting the evaluation of the consequences of one's actions to the effect on people nearby in space and time is morally arbitrary and should be abandoned in favor of a more comprehensive assessment. With respect to animal welfare, this entailed his critique of speciesism—that is, of arbitrarily favoring the interests of human beings over those of nonhuman animals.

Singer's thinking on animals was extended in his book *Animal Liberation* (1975), now in its second edition (1990). This work laid the foundation both for much of the political movement on behalf of sweeping reform in the use of animals and for **sentience** views in the debate over animal welfare and animal rights. The **sentient experiences** of pain and suffering form the basis for extending moral consideration to an entity or organism.

Timberlake claimed in 1980 that a fundamental problem with Singer's work is the lack of an adequate definition of **suffering. Animal suffering** has become a term so emotionally charged that attempts to analyze it critically have been rejected in favor of immediate action to stop it, whatever *it* may be.

Singer's writings do not necessitate vegetarianism, but do open the door for carefully regulated production of animals for food.

Tom Regan became well known with his 1983 book, *The Case for Animal Rights.* Regan accepts the sentience criterion proposed by Singer, but argues that any organism possessing consciousness is "the subject of life," and as such is entitled to the strong protection of its individual interests associated with a trumps view. Regan's basic position thus is a general rejection of the utilitarian trades view. Regan requires radical changes not only in animal production, but also in the very use of animals by humans.

Bernard Rollin, in his 1981 book, *Animal Rights and Human Morality,* presented the argument that people already do implicitly recognize animal rights in their speech and conduct. In more recent work, Rollin has argued that the implicit social consensus on the moral standing of animals requires more explicit and diligent attention to how animals fare in production, transport, and slaughter situations.

7.2.2 Economic and Policy Aspects

Further Examination

Certain intensive animal-production methods have improved production efficiency, but at times they have put egg, meat, milk, and wool producers in defensive positions. Animal protection activists and other critics have branded certain methods as "factory farming." Yet, in the opinion of some agriculturalists, the modifications called for by some activists would increase production cost while not necessarily improving animal state of being. Most animal-rights advocates call for the outright end of raising animals for food and coproducts useful to humans.

Choices

From a public policy perspective, there exist at least four alternative approaches to the animal state-of-being controversy:

- Enact laws requiring modifications in controversial production methods.
- Enact laws requiring modifications in production methods that have been documented to cause animal suffering.
- Encourage food-animal production systems that are intensive and at the same time engender animal well-being.
- Allow consumers to choose among foods labeled as being from animals kept in a range of intensive and extensive production systems. The Free Farmed Certification Program is a voluntary service available to producers of animals raised for food to certify that the animals were treated according to animal welfare standards developed by the American Humane Association.

Changes in how agricultural animals are treated could affect what is eaten and worn by humans and what medicines remain available for use by physicians and veterinarians. Recent legislation in certain European nations outlawed some production systems. This led to the collapse of affected agricultural sectors as well as to the importation of foods originating in other nations' production systems similar to those forbidden by law. In the meantime, many such laws have been modified or rescinded, because citizens of those nations came to recognize domestic economic realities and the inevitability of undesirable events cascading from those laws. For example, the raising of veal calves in crates is banned in the United Kingdom, Sweden, and Germany. However veal is a popular meat in other European countries including Belgium, France, Holland, and Italy. Consequently the ban on raising calves in crates, the system which produces the most desirable light-colored meat, is controversial due to its effects on exports and imports. (Countries which ban the use of veal crates cannot produce or export veal with the "desired" light colored meat and may instead import this meat for use in restaurants within the country.)

Economics

Intensive production tends to have economic advantages, but evaluation must be made carefully, and generalizations should not be made. One threat to U.S. animal agriculture would be requirements that animal products in international trade be produced under conditions specified by laws based on the standards of certain animal protection groups. Also, if animal agriculture were partly or wholly discontinued, hundreds of thousands of farm families and communities worldwide would be devastated. For example, the Animal Welfare Institute's criteria for humane on-farm husbandry of pigs specifies that each farm must be a family farm with housing for animals designed to allow the animals to behave naturally with continuous access to pens bedded with straw or chopped corn stover or pasture. The reality is that the increased cost of labor and facilities required to meet these criteria would require major increases in pork prices to be economically feasible. In addition, it can be argued that modern swine units, which house sows in crates with guard rails and heating lamps (or mats), have improved the well-being of piglets by preventing crushing and **hypothermia.**

Countermeasures

Animal agriculture organizations are resisting the efforts of animal protection activists that would restrict or destroy the industries and institutions dependent on traditional uses of animals. The animal industries have major political advantages over organizations and philosophies that would disrupt the national economic base and standard of living.

Trends

Power clusters within food and agriculture as well as elsewhere provide public support and facilitate legislative action to maintain and to protect various animal sectors in the American economy. How animals are treated in the future, however, will depend on public attitudes and ethical and social values, all of which are changing. Will producers of food animals sense the need for increasing personal responsibility for humane care and shift their actions closer to the positions of reformists? Public pressure for increased regulation will persist, but placing animal issues in the forefront for legislative action will be a formidable task; other issues currently rank higher on legislative agendas at both state and federal levels. Still, judging from the European experience, U.S. animal producers should expect ethical values to influence change in animal care practices.

7.2.3 Legal Aspects

Existing Framework

Existing legislation in the United States deals very little with the treatment of livestock and poultry. Concerns related to any ethical rights of agricultural animals generally have not been recognized as legal rights.

Production Practices

Public calls for regulation of agricultural–animal care practices have been more successful in Europe than in the United States, although standards differ markedly among the nations there, and trans-European legislation has been slow in coming. For example, in England castration of pigs is banned, as is weaning of piglets prior to 21 days. Sweden bans the weaning of piglets prior to 28 days of age. However research at Iowa State University has demonstrated that piglets weaned at 12 to 16 days and moved to segregated housing units have improved performance due to less exposure to respiratory diseases (Chapter 2). An added benefit to early weaning programs is increased reproductive efficiency due to earlier rebreeding of the sows. Americans dislike the "boar odor/flavor" in meat from intact males and a ban on castration could significantly decrease pork consumption in the United States.

In the United States, most states have anticruelty legislation to prohibit gross mistreatment of animals, although they are generally criticized as being relatively ineffective, often because of alleged apathetic enforcement. Some state legislation excludes agricultural animals altogether, whereas application of other statutes is limited to practices other than those customary in farming. Still other legislation applies only to clearly unjustifiable actions or practices. At the federal level, only limited legislation exists related to the humane treatment of animals, and none is specifically focused on animals residing on-farm.

7.3 SCIENTIFIC ASSESSMENT OF THE STATE OF BEING OF AGRICULTURAL ANIMALS

7.3.1 Needs for Scientific Assessment

. . . whose devotion to his animals is second only to his love of God and family . . . who cheerfully braves personal discomfort to make sure his livestock suffer not . . . his coming is greeted with demonstrations of pleasure, and his going with evident disappointment . . . may his kind multiply and replenish the earth.

Herbert W. Mumford (1871–1938)
Dean, College of Agriculture, University of Illinois
(1922–1938)
"A Tribute to the Stockman," *Breeder's Gazette,* **1915**

Although the issue of agricultural animal state of being probably will be resolved politically, for several reasons its scientific assessment is nonetheless needed. Specific recommendations of the CAST Task Force on the Well-Being of Agricultural Animals were:

- Producers should adopt scientifically based practices.
- Voluntary animal care guidelines published by producer organizations have been based on scientific assessment of husbandry practices and should be followed.
- Education of the general citizenry should be based on scientific assessment.

- The Congress of the United States should continue to consider scientific assessment and opinion seriously when addressing specific issues.
- The public should consider requesting scientific assessments of (1) the presence of—and, therefore, the *actual need* to alleviate—animal suffering and (2) the degree to which proposed alternative practices would indeed alleviate any suffering.
- Future designs of animal accommodations and care practices should reflect the results of scientific assessment.

7.3.2 Approaches to Scientific Assessment of Animal State of Being

Underlying Assumptions

Farm animal ecologists and ethologists hold several common assumptions about animal state of being that are based on science. The following points about agricultural animals are some emerging ones.

- Humans have the right to use animals in agricultural production and are morally obligated to treat them appropriately.
- The undomesticated progenitors of agricultural animals were unusual creatures.

[I]n an evolutionary sense, domesticated animals chose us as much as we chose them.

Stephen Budiansky
Writer and editor
The Covenant of the Wild:
Why Animals Chose Domestication **(1992)**

- Agricultural animals have been molded by genetic selection, and so they have specific environmental sensitivities, tolerances, and needs (Chapter 9).
- They can experience cruelty of two fundamental kinds— abuse and neglect—and perhaps of a third sort— deprivation of opportunities to express internally motivated behaviors; yet it is difficult to enforce well-being regulations except to minimize gross abuse.
- Their productive and reproductive functions are sensitive to stressors, which can diminish performance (Chapter 13). For example, dairy cows are susceptible to environmental heat stress because of the large amounts of feed they must consume to meet the energy requirements of high milk production. The digestion of food generates a large heat load (Section 17.7) and so heat stress occurs when the cow's total heat load is higher than her ability to dissipate the heat. Signs of heat stress include reduced feed intake, lower milk production, decreased fertility, failure to demonstrate signs of estrus, early embryonic deaths, lower calf birth weights, and cow death. To combat heat stress, cows must have access to shade, air movement, ample

drinking water, and/or other effective means of supporting body-temperature regulation.

- Production systems unsupportive of animals have resulted from careless building designs. For example, confinement buildings for housing swine contain gutters and manure pits which are a potential source of hazardous pit gases (e.g., ammonia and hydrogen sulfide). A functioning ventilation system is needed to exhaust these pit gases from the building before they can accumulate in the animal zone. A well-functioning ventilation system improves performance of the animals and the health of animals and human workers in the buildings.
- Agricultural animals are confronted by stress ranging from eustress (good stress) to distress; their state of being still is difficult if not impossible to define precisely in practical terms; and, as with any creature, it should not be expected that they will experience well-being continuously.
- They are endowed with a variety of useful and effective adaptive traits—behavioral, immunological, physiological, and anatomical.
- Physiological changes in responses to stressors may indicate that they have reached a prepathological state. For example, pigs that are stressed by excitement or heat have difficulty dissipating body heat and may develop malignant hyperthermia or porcine stress syndrome followed by prostration and sudden death. To prevent this, pigs should be handled quietly and sprayed with large droplets of water to help keep them cool during hot weather (do not spray cold water on a pig that is already down but rather wet the floor around it and allow the pig to cool by evaporation).
- Agricultural animals probably have internally motivated behaviors, which should be considered behavioral needs that should be accommodated by the environment (Chapter 24). For example, research has shown that "environmental enrichment," achieved by hanging strips of cloth in pens for the pigs to play with, results in pigs which are calmer and easier to drive through pens and chutes.
- Their immune systems are influenced by stressors (often negatively), and these in turn influence other responses to stressors.
- They perceive various physical and psychological stressors consciously, although little is known yet about any feelings connected with these perceptions.
- They can experience diminished state of being because of either acute or chronic stressors. For example, lameness in cattle may be acute following nutritional stress from grain overload resulting in acute laminitis, or lameness may be due to chronic stress resulting from slippery manure-covered ground, poor hoof trimming, or infectious diseases such as hairy warts.
- Residing in a wide range of conditions provided by a variety of agricultural systems, animals can experience an ethically acceptable state of being.

Defining State of Being

The quality of an animal environment cannot be evaluated by only interpreting measures either of that environment or of the economics of the animal-based enterprise. Such evaluation should be based on the animal's **response category** in the environment of interest. Only if an environment actually supports an appropriate state of being in an animal may it be considered satisfactory. This approach emphasizes **performance criteria** of environmental quality and de-emphasizes **design criteria.** In other words, environmental quality is determined on the basis of the state of being of the animal residing in the environment of interest (performance specification), not whether or not the environment fulfills certain design specifications.

Because neither **environmental stimuli** nor *animal responses* are discontinuous, lines between *ranges of response category* still are nebulous, and the ranges themselves—from very well to extremely ill—will overlap (Table 7.1).

- In the *range of basal regulation 0,* an animal is easily and readily coping, and only *ordinary* physiological and psychological regulatory mechanisms are operating. This happens in an environment with which the animal is generally in harmony, one it finds agreeable, favorable. The animal's state of being is *very well* or *well.*
- In the *range of response category 1,* an animal's composite adaptive responses to multiple stimuli are such that the animal is still handily coping. But in doing so, it must invoke *extraordinary* regulatory processes—responses in addition to those that operate within the basal range. Still, such environmental conditions are indefinitely tolerable, and **well-being** is engendered. Even in the range of response category 1, however, significant amounts of internal resources (especially metabolizable energy) must be reallocated to supporting bodily and psychic integrity. These resources must be diverted away from supporting performance processes, and so performance rate usually will decrease.
- In the *range of response category 2,* the animal usually can cope, but perhaps with some difficulty, via adaptive responses (*low-2*). Sometimes internal constancy may be compromised, and then the responses constitute **minor stress responses** (*high-2*). Such environmental conditions are tolerable to marginally tolerable; **stress** and **fair-being** sometimes ensue; and—if environmental quality continues to deteriorate—energy expenditure rate for maintenance will rise to progressively higher levels, so performance will progressively fall.
- In the *range of response category 3,* multiple stimuli are such that the subject animal cannot cope. Its limit of adaptability—at the **intolerable-** or *unsustainable-response level*—has been exceeded; frank **stress responses** ensue; and the animal is said to be experiencing **ill-being.** Here, the animal
  - May be *overloaded* by **multiple stimuli.**
  - Usually has extremely high **maintenance needs** to support regulatory mechanisms, and hence performance

TABLE 7.1	**Ranges of Response Categories**						
Range of Response Category	Nature of Response	Description of Animal	State of Being	Resource Expenditure	Fitness Index	Change in Animal Performance	Performance Index
Basal regulation 0	Normal physiological regulation; no response effected	In harmony with environment	Very well	Basal	10	Basal	10
Category 1	Adaptive responses	Readily coping by invoking homeokinetic mechanisms	Well	Small increment (+)	8–9	Small decrement (-⇓<5%)	8–9
Category 2	Adaptive responses; some stress responses	Coping, but with some difficulty	Well to fair	Medium increment (+ +)	6–8	Medium decrement (--⇓5% to ⇓15%)	6–8
Category 3	Stress responses	Not coping; collapse will ensue, eventually death if not mitigated	Ill	Large increment (+ + +)	4–6	Large decrement (---⇓>15%)	4–6
Category 4	Stress responses, if any	Quickly overwhelmed; quickly succumbs if not mitigated	Very ill or dead	—	0–3		

Source: Stanley E. Curtis, et al., *Environmental Aspects of Animal Care,* Iowa State University Press, Ames, IA, 2002.

usually will be markedly reduced (paradoxically, though, the animal's appetite often is reduced).

- Usually progresses to *breakdown* or *collapse,* which, if not mitigated, soon will lead to *death.*

• In the *range of response category 4,* either a single stimulus or multiple stimuli impinge(s) so abruptly and so severely that the subject animal is immediately overwhelmed and quickly succumbs, often experiencing **extreme ill-being**—but usually only for a very short period—before it dies.

Of course, it is unrealistic to expect that any animal should continuously experience well-being. Instead, *the goal should be that a kept animal will experience well-being most of the time; fair-being some of the time; ill-being only rarely.*

For example, environmental temperature can influence basal regulations at levels 0–4. In a thermally neutral environment (basal regulation 0) an animal does not expend any extra energy in temperature regulation. Under mild temperature stress (response category 1), an animal may spend energy seeking shade from heat or shelter from cold winds, thereby decreasing its food consumption and productivity, but readily maintaining good health. Under moderate temperature stress (response category 2), an overheated animal will drink more and seek cool damp areas while cold animals will huddle together. These animals will have significant decreases in productivity. Under more severe temperature stress (response category 3), cattle will have repro-

ductive losses in addition to production losses. Under severe temperature stress (response category 4), animals are unable to compensate and will succumb to hyperthermia or hypothermia.

Proposals for Assessing State of Being

No scientific consensus has emerged regarding the definition of state of being in agricultural animals. Until there is convergence in this respect, the sort of strides called for by some in ensuring that animals usually experience well-being are unlikely to happen. Yet, in the opinions of experts, state of being always will be difficult if not impossible to define precisely.

Perhaps the difficulty in assessing state of being has more to do with its complexity than with anything else. In turn, this complexity may have much to do with the current level of our scientific ignorance. Agreement on approaches to scientifically assessing the state of being of animals in agricultural settings remains wanting. Neither is there consensus as to meaningful indicators of well-being, which are essential for making prudent change. Nevertheless, several proposals have emerged, and they need not be mutually exclusive.

The First Attempt to Assess Well-Being

The report of a special committee to the British Parliament in 1965 constituted the first attempt at addressing these matters. The report states:

• *Welfare* refers to "both physical and mental well-being."

- Its assessment must involve "scientific evidence available concerning the feelings of the animals that can be derived from their structure and functions and also from their behavior."
- There are sound reasons for assuming that sensations and emotional states are substantial in animals and should not be disregarded.

That report also established "five freedoms": "An animal should at least be able without difficulty, to turn around, groom itself, get up, lie down and stretch its limbs." In the main, subsequent attempts to establish meaningful assessment of the overall state of being of an individual animal have sprung from that framework constructed some 40 years ago.

Other Proposals for Assessing Well-Being

Numerous approaches to assessing the state of being of agricultural animals have emphasized one or more of the following:

- Behavioral and cognitive indicators.
- Anatomical, physiological, and immunological indicators.
- Fitness and agricultural performance indicators.
- Multiple indicators.

Each of these approaches has merit, and any of them may serve as a basis for consensus. As a group, these approaches recognize (1) that there are differences between acute and chronic incidents of anxiety, frustration, discomfort, and pain; and (2) that the state of being of an animal involves biological systems that may change over the life of an individual as well as over the natural history of a population. Moreover, the approaches generally advocate multiple categories of indicators of state of being, demonstrate awareness of the human-animal interface, and acknowledge the ongoing nature of domestication.

Off-Farm Experiences

The well-being of an animal may be compromised more in its off-farm than in its on-farm experiences. One of the most important determinants of an animal's off-farm experience is the attitude of management personnel in the succession of firms typically responsible for animal care and handling between rearing and slaughter: at markets, during transportation, or in a holding area at the abattoir.

Livestock handling systems have been designed to promote animal well-being by minimizing stresses to animals going through these systems. Key principles for well-designed handling systems include (1) crowd pens must always be level; (2) single file chutes should lead from the crowd pen to the restraint area; (3) if the system includes a ramp, it should be located within the single file chute; (4) an animal standing in the crowd pen should be able to see 2 to 3 body lengths up the single file chute before it curves; (5) the sides of the chute should be solid to prohibit the animals from seeing through them; and (6) the flooring should be non-slip. Following these principles will facilitate the movement of animals through the chutes with minimal stress.

7.4 SCIENTIFIC ASSESSMENT OF THE CURRENT STATUS OF ANIMAL STATE OF BEING

7.4.1 Overview

Designers and operators of animal agricultural systems are constrained nowadays by insufficient knowledge. Small design differences (design specifications) can cause major differences in how effectively animal needs are fulfilled (performance specifications). An animal accommodation may be

- Designed and operated well, evidently supporting an ethically adequate degree of animal well-being.
- Designed well but operated incompletely or apathetically—hence not supporting adequate animal well-being.
- Designed so poorly that deficiencies are insurmountable regardless of the ethical concern or technical competence of animal care personnel.

In animal agriculture in the United States, traditional business priorities still often prevail over emerging ethical considerations. In view of the nature of our economic system, politics, and governance, this probably will be the case so long as assessment of animal state of being is based on absent or inadequate science or lack of interest. Today, again, the problem is primarily that of inadequate science. The current situation of business priorities receiving more weight than ethical considerations probably will remain the norm until meaningful scientific assessment of animal state of being becomes realized. For example, milk producers recognize that heat stress not only endangers the well-being of cows but also results in significant economic losses from decreased milk production, increased illnesses, and reproductive losses. It is cost-effective to provide dairy cows with access to shade, air movement, ample drinking water, and at times with sprinkler systems and changes in diet to improve the well-being of the cows during periods of hot weather.

7.4.2 Welfare Plateau

On any animal farm, achieving continuously the highest possible level of animal well-being is still infeasible and perhaps inadvisable. D. C. Hardwick proposed that an ethically acceptable state of being is not limited to one ideal set of circumstances, but rather that it exists over a range of conditions provided by a variety of agricultural production systems. This range of acceptable environments comprises what has been called the "welfare plateau" (Figure 7.1). On this plateau, an improvement in the animals' environment can increase total well-being (if only slightly); however, a production system located even at the lowest point on the welfare plateau still would meet the minimum standards for ethical acceptability with respect to animal well-being.

The concept of the welfare plateau is important for the design and operation of facilities for agricultural animals. A brief working description of this concept follows. Note that as environmental conditions increase in quality toward the welfare plateau (in the upward direction), the cost of production generally increases, as well, leading to a dilemma for the producer.

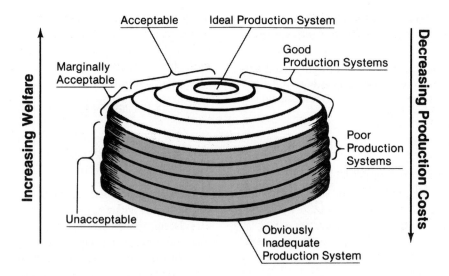

Figure 7.1 Hardwick's notion of the welfare plateau.
After D. C. Hardwick in S. E. Curtis, "Animal Well-Being and Animal Care," The Veterinary Clinics of North America: Food Animal Practice: Farm Animal Behavior, 3(2) July 1987:369–382.

1. At the lower limit of the welfare plateau stand production systems that are marginally acceptable in terms of animal well-being.

2. Below these stand systems that more or less fail to fulfill the animals' basic needs sufficiently well to be considered ethically acceptable.

3. In this range of ethically unacceptable environments, small environmental improvements result in economic returns more than adequate to pay for the improvements (i.e., there is a positive overall relationship between animal state of being and the economic vigor of an agricultural enterprise).

4. The animal producer often finds that, in marginally acceptable production systems, the law of diminishing economic returns is in effect; it is certainly in effect *on* the welfare plateau.

5. Therefore, the animal production systems adopted most widely tend to reflect a compromise between animal state of being and economics, and thus tend to be the marginally acceptable ones with respect to animal well-being. Producers tend to provide their animals an existence as free of suffering as possible, and must adopt production systems that are optimal for prevailing economic conditions. Thus, they face the reality of denying animals the greatest possible well-being for the sake of the economic viability of their businesses. This fact means that the most widely adopted compromise production systems are located either in the upper part of the marginally acceptable range or on the lower part of the welfare plateau itself.

7.4.3 Needed Scientific Insight

Animal scientists and other agricultural stakeholders generally agree on the prioritization of researchable questions regarding these matters. Agreement has resulted from four consensus initiatives in recent years. Members of the CAST task force[1] gen-

erally subscribed to this emerging consensus, which can be summarized as follows.

General Research Areas Identified by Discussants in the Food Animal Integrated Research (FAIR) '95 Process

Six research areas were identified:

- Bioethics and conflict resolution.
- Responses of individual animals to the production environment.
- Stress.
- Social behavior and space requirements.
- Cognition.
- Alternative production practices and systems.

Overall Research Objectives on which Consensus Emerged in the FAIR '95 Process

Two research objectives identified during these deliberations were

- To determine scientific measures of well-being in food-producing animals.
- To develop short-term production practices and long-term management systems based on scientific research findings about animal well-being.

[1]Members of the CAST task force were: Stanley E. Curtis (Animal Science), University of Illinois, and Frank H. Baker (Animal Science, deceased), Winrock International, cochairs; Jake W. Looney (Law), University of Arkansas; Paul B. Siegel (Poultry Science), Virginia Polytechnic Institute and State University; Paul B. Thompson (Philosophy), Purdue University; Jack L. Albright (Dairy Science), Purdue University; James V. Craig (Poultry Science), Kansas State University; Temple Grandin (Animal Science), Colorado State University; Harold D. Guither (Agricultural Economics), University of Illinois; John J. McGlone (Animal Science), Texas Tech University; James D. McKean (Veterinary Medicine), Iowa State University; Gerald L. Riskowski (Agricultural Engineering), University of Illinois; W. Ray Stricklin (Animal Science), University of Maryland; and Carolyn L. Stull (Veterinary Medicine), University of California–Davis.

The full report is available from CAST, 4420 West Lincoln Way, Ames, IA 50014-3447.

Contributions by the Workgroup at the 1993 Purdue University/USDA Food Animal Well-Being Conference and Workshop

To develop measures of the state of being of agricultural animals, three priority research areas were identified:

- Adaptations and adaptiveness.
- Social behavior and space requirements.
- Cognition and motivation.

7.4.4 Interim Recommended Approach

Taking advantage of the multitude of approaches and disciplines now involved seems most reasonable at present. Decades probably will be required for adequate scientific data to be generated, however. Meanwhile, one rational approach to establishing provisional multifactorial indices of well-being in agricultural animals would involve

- Assembling a multidisciplinary team of several scientists specifically knowledgeable and experienced.
- Asking the team to assemble a worldwide database of reliable information of all kinds bearing on matters of farm animal state of being.
- Asking the team to employ appropriate multivariate parametric and nonparametric statistical analytical methods to elucidate and to determine multifactorial indices of state of being in agricultural animals.

7.5 SUMMARY

The state of being of agricultural animals residing in modern production systems is a public issue in North America and many other parts of the world. It is a complex issue that stirs much debate. In the United States surveys have shown that most people believe farmers treat their animals humanely, but most still would like to see governmental regulation of animal treatment on farms and ranches. Although several farm organizations have developed voluntary guidelines for humane treatment of farm animals, the issue remains unresolved and the debate continues. The debate has philosophical, economic, public policy, and legal origins. Scientific assessment of the situation will ultimately be needed if the issue is to be resolved.

Unfortunately, although designs of animal accommodations and care practices should reflect the results of scientific assessment of the state of being of agricultural animals, this work is in its infancy. Fortunately, consensus among scientists from a variety of specialties as to how best to assess an animal's state of being is beginning to emerge. For one thing, it is unrealistic to expect that any animal should continuously experience well-being; instead, the goal should be that an animal will experience well-being most of the time; fair-being some of the time; ill-being only rarely.

As a group, the emerging consensus approaches to assessing an animal's state of being recognize that (1) there are differences between acute and chronic incidents of anxiety, frustration, discomfort, and pain; and (2) the state of being of an animal involves biological systems that may change over the life of an individual as well as over the natural history of a population or a series of generations. They also generally advocate multiple categories of indicators of state of being, demonstrate the human-animal interface, and acknowledge the ongoing nature of domestication.

STUDY QUESTIONS

1. What significant input did Ruth Harrison make to the issue and public awareness of the state of being of agricultural animals? When?

2. Define "state of being" of agricultural animals.

3. Differentiate between *animal-welfare advocates* and *animal-rights advocates*.

4. How does the "utilitarian strategy" differ from the "rights strategy" in its approach to showing how an animal rights claim might be validated?

5. Discuss briefly views related to the ethical basis for reforming animal agriculture advanced by authors R. G. Frey, Peter Singer, Tom Regan, and Bernard Rollin. Generally, what moral basis for relationships between humans and other animals did each author put forward?

6. Cite four alternative approaches to dealing with the agricultural animal state-of-being controversy.

7. What impact did legislation outlawing certain production practices have on agricultural sectors in European nations? Should economic aspects be considered in determining which animal production systems are employed?

8. Do agricultural animals have legal ethical rights? Should they have?

9. Regarding the issue of agricultural animal state of being, what recommendations were made by the broad-based 14-member CAST Task Force on the Well-Being of Agricultural Animals? Were their recommendations objective? Discuss.

10. Cite six or more common assumptions of farm animal ecologists and ethologists regarding animal state of being that are based on science.

11. Review Table 7.1 (as needed) to discuss the "range of response" categories of varying stress levels animals may experience when exposed to different environmental conditions. How do the various stimuli affect the animal's immune system, appetite, maintenance needs, performance, and general well-being?

12. Does this chapter's author believe it is realistic to expect animals to continuously experience a maximal level of well-being? What should be the goal of those responsible for providing proper care to agricultural animals in this regard?

13. What were the "five freedoms" of animals established by the special committee to the British Parliament in 1965?

14. Discuss differences between "on-farm" and "off-farm" experiences of agricultural animals.

15. Is the pool of scientific knowledge regarding evaluation of the current status of animal state of being adequate? Discuss.

16. Why is the concept of the *welfare plateau* important for the design and operation of facilities for agricultural animals? Discuss.

17. Cite six or more research areas needing further study to determine appropriate production practices for acceptable multifactorial indices of state of being in agricultural animals.

18. Did reading this chapter provide you with any new thoughts regarding the issue of animal state of being? Discuss.

FUNDAMENTAL PRINCIPLES OF GENETICS[1]

If a man leaves children behind him, it is as if he did not die.

Moroccan proverb

8.1 INTRODUCTION

Ongoing discoveries in molecular genetics show clearly that **genes,** the determiners of **heredity,** play an important part in all the biochemical reactions in an animal's body. Because these reactions are necessary for proper form and function of the body, the importance of heredity in animal production and human welfare can readily be seen. The purpose of this chapter is to outline and discuss the basic principles of animal genetics.

8.2 THE CELL THEORY OF INHERITANCE

The **cell** theory states that all plants and animals are made of small building blocks called *cells.* Each plant cell is bounded by a tough outer cell wall and an inner membrane, and each animal cell is bounded by a membrane. Both plant and animal cells usually have a **nucleus.** Between the nucleus and the cell wall is the cellular material called **cytoplasm,** which contains certain components necessary for the function of that particular cell (Figure 8.1).

All cells originate from other cells through the process of cell division, and many of these new cells are capable of growing and dividing to produce other cells like themselves. Within the nucleus are **chromosomes,** which carry the hereditary material called *genes.* In body cells, these genes occur in pairs on chromosomes that also occur in pairs and are similar in their size, shape, and proportions. A pair of such chromosomes is known as

homologous chromosomes (*homo* meaning "alike" or "equal" and *logous* meaning "proportion"). The number of pairs of homologous chromosomes in body cells is constant in normal individuals within a **species.** Different species, however, often have a different number of chromosomes, as shown in Table 8.1. In animals, the body cells possess one pair of chromosomes known as **sex chromosomes,** because they are related to the sex of the individual. In mammals they are referred to as the X and Y chromosomes, the X chromosome being at least two to three

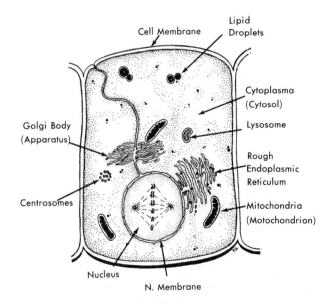

Figure 8.1 Present-day conception of a cell.
Courtesy of Dr. Robert Godke, Louisiana State University.

---

[1]The authors acknowledge with appreciation the contributions to this chapter of Dr. David S. Buchanan, Professor of Animal Science, Oklahoma State University, Stillwater.

times larger than the Y. The X chromosome carries much more genetic material (more genes) than the Y chromosome. In such species, the male is XY (heterogametic) whereas the female is XX (homogametic) (Figure 8.2). In chickens and turkeys, the male is ZZ (homogametic) and the female is ZW (heterogametic), which means the female has a W chromosome much smaller than the Z. This situation is the opposite of that found in mammals.

All chromosomes in body cells other than the sex chromosomes are known as **autosomes.** Thus, in humans, there are 22 pairs of autosomes and one pair of sex chromosomes, a total of 23 pairs. For each pair of chromosomes an individual possesses in the body cells, he/she received one chromosome from his/her father and the other from his/her mother. This means each parent normally contributes one-half the chromosome number to each **offspring.** Each offspring, when it becomes sexually mature and reproduces, will transmit only one of each pair of

chromosomes to any one of its offspring. Which one it transmits will be determined by the law of probability.

Pairs of chromosomes in body cells are referred to as the 2n number, or **diploid** (*di* meaning "two") number. In **gametes,** or sex cells, where only one of each pair of chromosomes is found, these are referred to as the 1n, or **haploid,** number of chromosomes.

8.3 CHROMOSOMAL ABNORMALITIES

Chromosomal abnormalities or *chromosomal aberrations,* can cause drastic changes in the appearance (**phenotype**) of an individual, or they may even cause its death. Death can occur shortly after the fertilization of the **ovum** by the **spermatozoon,** during **intra**uterine life, or even some time after the individual is born. Several types of chromosomal aberrations have been described in humans. A few have been described in animals other than humans. A substantial proportion of spontaneous abortions (miscarriages) are caused by chromosomal aberrations.

Nondisjunction is one general form of chromosome aberration. This term means that the homologous chromosomes fail to separate when the sex cell is formed and that both members of a particular pair of homologous chromosomes go into the sex cell rather than the normal one-half of each pair. Two general forms of nondisjunction have been observed. The first, known as **aneuploidy,** results when an individual has the basic 2n number of chromosomes but also has one or more chromosomes duplicated or missing. Thus the individual possesses either one (or more than one) more chromosome than is normal, as in humans with Down syndrome, or one (or more than one) fewer chromosomes than the normal number. An extra or missing chromosome often causes abnormal conditions in animals if the chromosome is small, but is likely to cause death if it is one of the larger chromosomes. For example, Down syndrome results from an extra copy of the smallest human chromosome and such individuals have a good survival rate. Extra copies of larger human chromosomes are usually lethal either during gestation or

TABLE 8.1	Characteristic Numbers of Chromosomes in Selected Animals
Animal	**Chromosome Number (2n)**
Donkey	62
Horse	64
Mule	63
Swine	38
Sheep	54
Cattle	60
Human	46
Mink	30
Dog	78
Lion	38
Domestic cat	38
Bengal tiger	38
Chicken	78

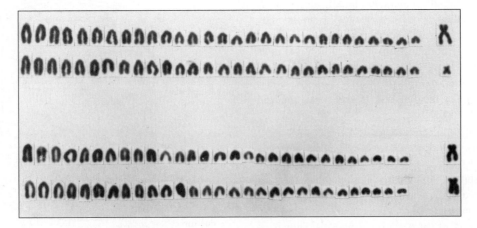

Figure 8.2 The chromosome **karyotype** of cattle. Males possess X and Y chromosomes (top); females possess two X chromosomes (bottom). These are actual photographs of chromosomes at the metaphase, where each chromosome appears doubled.
Courtesy of Missouri Agr. Expt. Station.

soon after birth. A second form of nondisjunction is known as **polyploidy, or euploidy.** This term means that entire sets of chromosomes are duplicated, giving, instead of the normal $2n$ number in body cells, a $3n$, $4n$, and so on, number. Polyploidy is common in plants but rare in higher animals. One example of polyploidy in animals is the triploid "grass carp" used by pond owners to control aquatic weeds.

Translocation is another form of chromosomal aberration. This term includes those instances in which a chromosome is broken and the smaller piece attaches itself to another chromosome not homologous to it. This condition has been reported in humans, cattle, and swine and other species of animals.

A chromosomal aberration is called a **duplication** when a portion of a chromosome attaches itself to the chromosome to which it is homologous. Duplication is probably due to the fact that the homologous chromosomes that *synapse* (come together) in **meiosis** do not completely separate at the reductional division, leaving a portion of one chromosome attached to its homologous mate.

Deletion refers to a chromosomal abnormality in which a portion of a chromosome is lacking. Large deletions may be lethal because the portion of a chromosome that normally carries genes that perform a function vital to the life of an individual is lacking.

Another form of chromosomal abnormality is known as an **inversion.** This and other examples are illustrated in Figure 8.3. An *inversion* means that *a portion of a chromosome has been rearranged so genes occur on the chromosome in a new or inverse order from their original sequence.*

8.4 CELL DIVISION

An animal's body consists of millions of building blocks called *cells*. Cells multiply or increase in numbers by undergoing division. Two general kinds of cell division are known: **mitosis** and **meiosis.** Mitosis refers to the kind of cell division in which each cell divides and forms two cells, both of which possess two

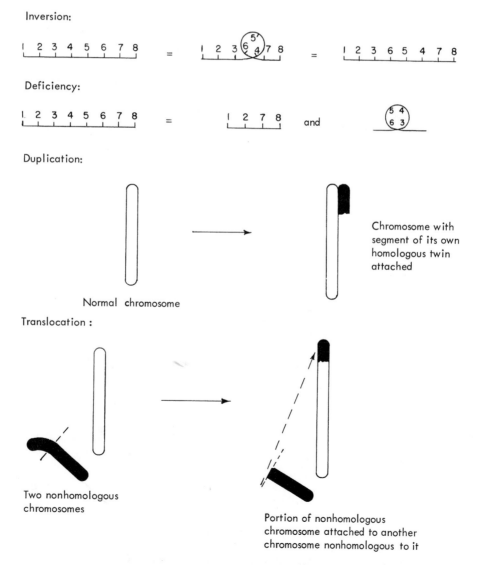

Figure 8.3 Illustration of various kinds of chromosome abnormalities in animals.

complete sets of chromosomes just like those found in the mother cell. Meiosis refers to a type of cell division in the sex cells, the sperm and the egg, in which the chromosome number is reduced by one-half from the diploid (2n) to the haploid (1n) number. The reduction is necessary so that the union of the sperm and egg in fertilization will restore the diploid number of chromosomes normally found in body cells but will not exceed this number. It also ensures that one-half the chromosomes of the **progeny** are contributed by each parent.

8.4.1 Mitosis

Mitosis has several phases: prophase, metaphase, anaphase, telophase, and interphase. Only two pairs of homologous chromosomes will be used to illustrate this type of cell division, although there are many more pairs than this in the body cells of animals.

The **prophase** is the beginning of mitosis. In early prophase, the chromosomes begin to shorten and thicken, and each chromosome appears as a double strand of identical **chromatids,** or sister chromosomes, connected together by means of a centromere (Figure 8.4). During midprophase the nuclear membrane disappears, and in late prophase the spindle fibers form and the centromeres become attached to a spindle fiber.

During **metaphase,** the chromosomes (chromatids) line up on the equator of the spindle fibers, the mitotic centers (**centrioles**) appearing at each pole.

The **anaphase** is characterized by each centromere separating into two centromeres, each with a chromatid. Each **centromere** then moves toward the mitotic center, dragging its chromatid with it.

In the **telophase** the nucleus is re-formed, and the nuclear membrane reappears around each mitotic center, resulting in

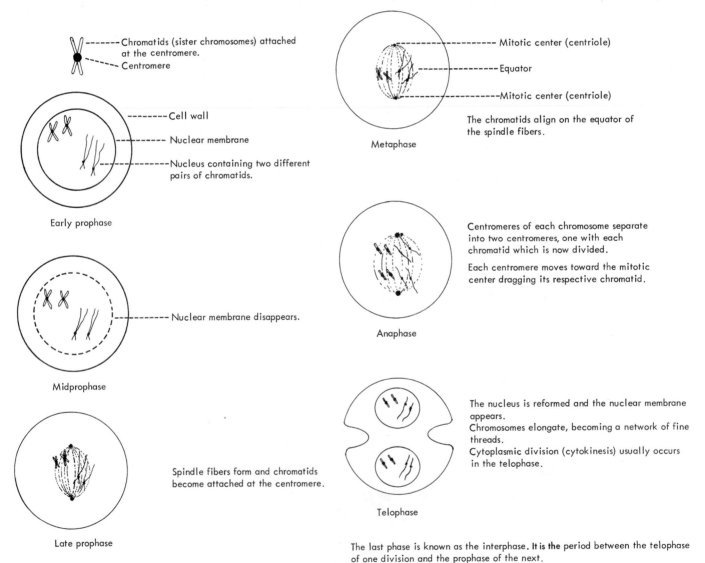

Figure 8.4 The different stages in mitosis. Chromosomes are not visible in stained slides in all stages, but they are presented here as if they were in order to illustrate the duplication of cells and chromosome numbers.

two new nuclei. The chromosomes grow longer and disperse into a network of fine threads. This phase of mitosis is completed when the cytoplasm divides (**cytokinesis**) forming two new cells that possess the same chromosome complement as the original mother cell. The period of time between the telophase of one cell division and the prophase of the next is known as the **interphase.**

8.4.2 Meiosis

Meiosis is similar to mitosis in some respects but different in others of importance. Meiosis is the type of cell division that forms the gametes (sex cells) with the haploid ($1n$) number of chromosomes. It involves a **first** and a **second meiotic division.**

In the prophase of the first meiotic division (Figure 8.5), the chromosomes shorten and thicken, each chromosome appearing as two chromatids connected by a centromere, as in mitosis. A distinguishing feature of meiosis, however, is the pairing up (**synapsis**) of two homologous chromosomes, giving the appearance of four chromatids. These synapsed pairs are known as **tetrads.** Synapsis seldom, if ever, occurs in mitosis. During synapsis in meiosis, the homologous chromosomes may exchange parts, which is known as **crossing over.** The spindle fibers then appear, as in mitosis, and the synapsed chromosomes become attached to the spindle fibers at the centromeres. The chromosomes again align themselves at the equator of the spindle fibers during the metaphase.

In the anaphase of the first meiotic division, each tetrad separates into two **dyads** (a pair of chromatids connected by a centromere). In the telophase, the nucleus and cytoplasm reform, resulting in the production of two new cells, each one of which contains only one chromosome of each original homologous chromosome pair. Thus it is a reductional division.

In the prophase of the second meiotic division, each chromosome in the nucleus is present in the haploid state. Each chromosome appears, however, as a pair of sister chromatids connected at the centromere. These sister chromatids align themselves at the equator of the spindle fibers in the metaphase and then separate in the anaphase, the two chromatids going to

FIRST MEIOTIC DIVISION

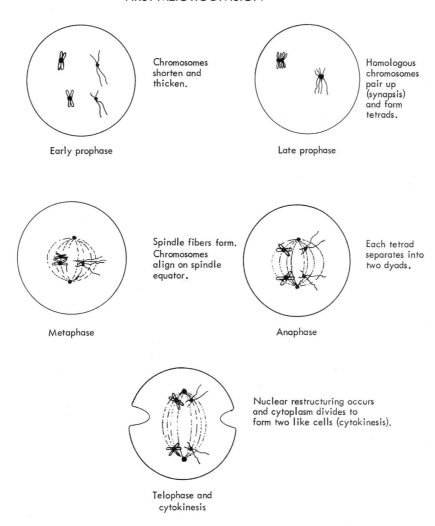

Figure 8.5 Different stages within the nucleus of the first meiotic division. (Only the nucleus is shown.)

SECOND MEIOTIC DIVISION
(nucleus only)

Chromosomes are haploid.

Prophase

Spindle fibers form and chromosomes align on equator.

Metaphase

Each chromatid divides and moves to each mitotic center or centriole.

Anaphase

Two nuclei are formed and cytoplasm divides (cytokinesis).

Telophase

Figure 8.5 *continued*

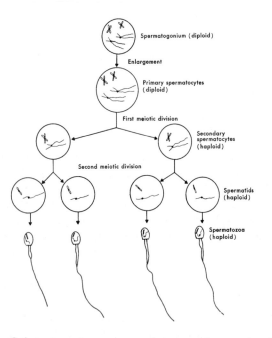

Figure 8.6 Outline of spermatogenesis (using only two pairs of homologous chromosomes and showing the nucleus only).

opposite poles of the spindle fibers. In the telophase, the nucleus re-forms and the cytoplasm divides, forming two cells that each have the haploid number of chromosomes.

The process of meiosis in the male **gonads** produces four spermatozoa from each spermatogonium, as is shown in Figure 8.6. Each oocyte, in meiosis, produces an ovum and three polar bodies, as shown in Figure 8.7.

8.5 THE GENE AND HOW IT FUNCTIONS

There have been many new discoveries about the gene and its functions. Genetics is a rapidly growing field with new information coming at an overwhelming rate. A key element in the development of genetics is the Human Genome Project. This project was initiated in an attempt to locate all the genes in humans. Scientists have collaborated on a global basis to find genes which affect human health and well-being. Similar efforts are underway to examine the genome of other species including cattle, swine, sheep, horses, dogs, and mice (Section 8.16).

8.5.1 The Gene

Many years ago a gene was often defined as *the smallest unit of inheritance.* This definition probably was used because the actual chemical composition of the gene was unknown. Research has shown that the gene is a portion of a **DNA** (deoxyribonucleic acid) **molecule.** Much information is now available on the chemistry and functions of genes. Indeed, the amount of knowledge about DNA, how it works and how it can be manipulated, is growing rapidly.

The DNA molecule may be described as the backbone of the chromosome, analogous to the vertebral column, which is the backbone of the vertebrate animal body. The DNA molecule resembles a long, twisted ladder in which the two strands (sides) are joined together by rungs (Figure 8.8). Each strand is called a *polymer (poly* meaning "many," *mer* meaning "part") because it is composed of many repeated units, called *nucleotides.*

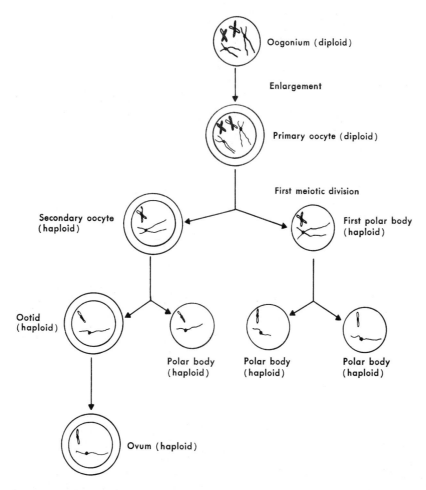

Figure 8.7 Outline of oogenesis (using only two pairs of homologous chromosomes).

A **nucleotide** is composed of a nitrogenous base (either a purine or a pyrimidine) linked to a sugar, which is in turn linked to a phosphoric acid molecule (Figure 8.9). The sugar in **RNA** (ribonucleic acid) is ribose whereas in DNA it is deoxyribose. The various chemical constituents of RNA and DNA molecules are shown in Figures 8.10 and 8.11. A schematic arrangement of several nucleotides in a single strand of DNA is shown in Figure 8.12. The two single strands (nucleotides) of a double-strand molecule of DNA are joined together by hydrogen bonds between the bases. Adenine (A) always joins to thymine (T), and guanine (G) always joins to cytosine (C), as shown in Figure 8.13. The average gene (cistron) is thought to consist of a portion of a double-strand DNA molecule containing about 600 consecutive base pairs.

8.5.2 Functions of the Gene

The functions of the gene are to replicate itself when new cells are produced and to send the code to the cytoplasm to build certain proteins.

It is now known that during replication, the two strands of the DNA molecule separate, as shown in Figure 8.14. Each strand then serves as a template, or mold, for synthesizing the missing part. Thus, if the sequence of bases in the single strand is T, G, C, G, it follows that the sequence of bases in the missing strand should be A, C, G, C, because A and T always go together, as do G and C. This base pairing allows the DNA molecule to copy the double-strand DNA molecule accurately, regardless of the number of times each cell divides to form new cells.

DNA, the substance of the gene, is found almost entirely in the cell nucleus. Proteins, however, are synthesized in the cytoplasm by ribosomes, which link various amino acids together in a way instructed by the gene. Because DNA cannot leave the chromosome, it synthesizes messenger RNA (mRNA), another nucleic acid, which carries instructions from DNA to the cytoplasm as to the kind of protein to build. Besides differing from DNA in the sugar molecules it contains, RNA also differs from DNA in containing the nitrogenous base uracil instead of thymine. Thus uracil is found in RNA but not in DNA, whereas thymine is found in DNA but not in RNA.

Proteins are organic compounds that are found in all body cells. They consist of the 20 naturally occurring amino acids

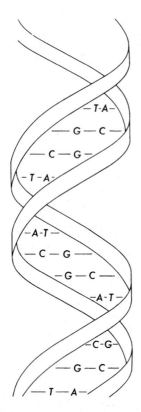

Figure 8.8 Diagram of a portion of the DNA molecule. The letters A, C, T, and G represent certain bases, or cross-links, connecting the two strands. The code sent by the gene to the cytoplasm depends on the kind and arrangement of these bases.

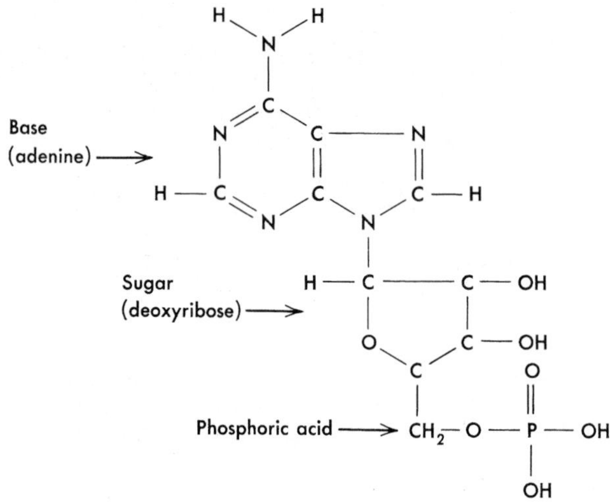

Figure 8.9 A nucleotide, adenylic acid. It consists of a combination of the base adenine, the sugar deoxyribose, and phosphoric acid.

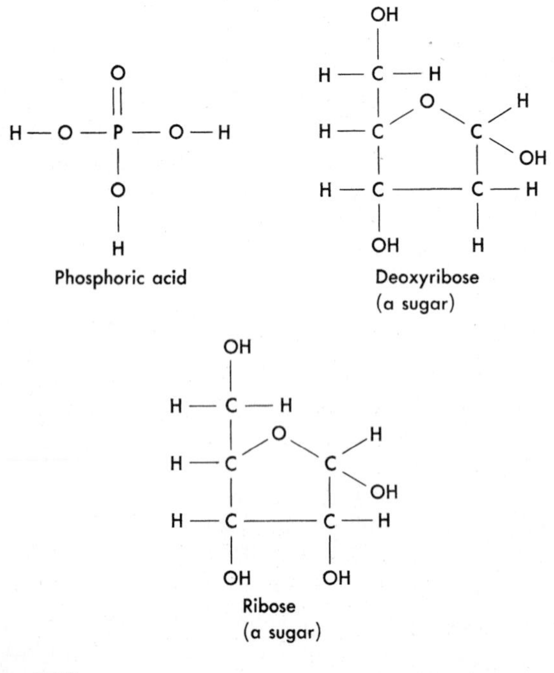

Base (adenine) ⟶

Sugar (deoxyribose) ⟶

Phosphoric acid ⟶

Figure 8.10 Chemical formula of phosphoric acid and sugar molecules found in DNA and RNA.

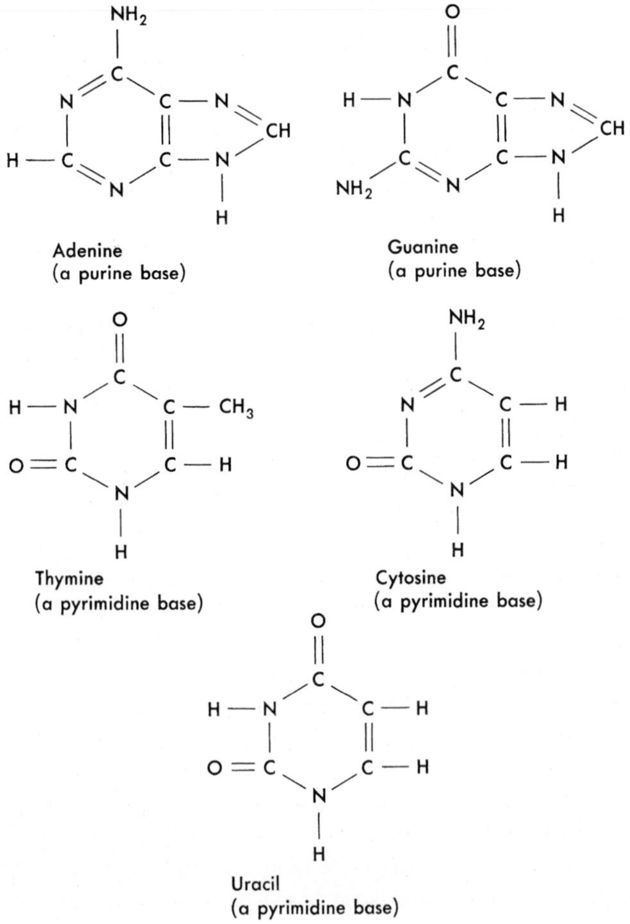

Phosphoric acid

Deoxyribose (a sugar)

Ribose (a sugar)

Adenine (a purine base)

Guanine (a purine base)

Thymine (a pyrimidine base)

Cytosine (a pyrimidine base)

Uracil (a pyrimidine base)

Figure 8.11 Purine and pyrimidine bases found in RNA and DNA.

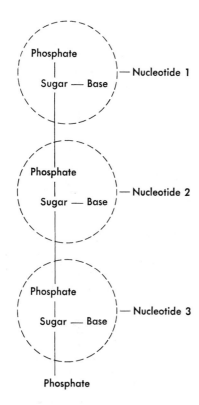

Figure 8.12 Schematic arrangement of a series of nucleotides (phosphate, sugar, base) in a nucleic acid. This represents a single strand of nucleic acid with three nucleotides.

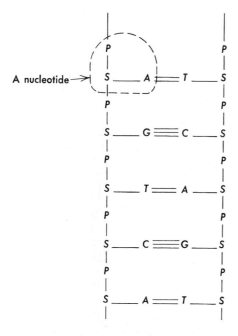

Figure 8.13 Schematic representation of a double-strand DNA molecule. P, phosphoric acid; S, sugar deoxyribose; A, base adenine; G, base guanine; T, base thymine; and C, base cytosine. Base A always pairs with base T and G with C. A nucleotide is the combination of P + S + a base, as shown by the dotted circle.

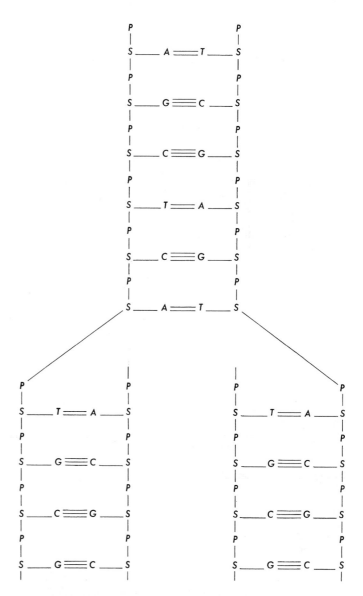

Figure 8.14 Illustration of how the double-strand molecule exactly duplicates itself. P, phosphoric acid; S, sugar deoxyribose; A, base adenine; T, base thymine; G, base guanine; and C, base cytosine.

listed in Table 8.2. These amino acids, joined together, form thousands of different proteins. The specific protein produced depends on the kind, number, and arrangement of amino acids in the protein molecule, just as a specific English word depends on different kinds, numbers, and arrangements of the 26 letters in the English alphabet.

8.5.3 The Genetic Code

Genetic information is stored in the DNA molecule in the form of a triplet code, called a *codon*. A codon consists of *a sequence of three (triplet) nitrogenous bases that code only one specific amino acid*. Because the DNA molecule contains a total of four different bases, and any combination of three specifies only one certain amino acid, the total possible combinations of these four bases in groups of three would number 4^3, or 64. These 64 groups are more than enough possible combinations to specify the 20 amino acids. In fact, more than one combination of three nitrogenous bases can specify the same amino acid, as shown in Table 8.3.

The gene is thought to carry the code for the formation of a single polypeptide chain of a protein, as well as instructions to start and stop the sequence of amino acids in that polypeptide chain. Part of one of the two strands of the DNA molecule (gene) transcribes the code, or message, for a protein into a long single strand of mRNA. This message is then carried to the cytoplasm.

In the ribosomes, mRNA forms the template for a sequence of particular amino acids. Completion of the message carried by mRNA is the work of another group of molecules, called *transfer RNA* (tRNA). Each tRNA molecule specifies only one of the 20 amino acids. Each group of tRNA molecules also recognizes a specific enzyme that joins it to a specified amino acid. Transfer RNA molecules also contain a series of three bases, called

anticodons, and each anticodon seeks out and complements a codon on the mRNA molecule, carrying its particular amino acid with it, as shown in Figure 8.15. This ensures the correct order of amino acids as specified by the DNA. The specific amino acids are then combined into a protein molecule. Two bases in the codon pair with two bases in the anticodon, with adenine (A) pairing with uracil (U) and guanine (G) with cytosine (C). The third base in the codon, corresponding to the first base in the anticodon, has a certain amount of uncertainty in it pairing (the "wobble" theory). It can be seen from this discussion that protein synthesis is complex and some details are not completely understood.

8.5.4 Control of Gene Function

Each body cell that has a nucleus contains a sample of all the genes an individual possesses. Not all these genes function at any given time. Thus there must be something within the cell that controls the function of some genes.

The present theory of gene regulation proposes that two principal kinds of genes exist: the *structural* gene, which functions to synthesize specific proteins and enzymes, and *control* genes (regulator and operator genes), which are not directly concerned with protein synthesis but regulate the activity of the structural genes and the amount of protein synthesized. The *operator* gene is thought to be on the same chromosome as structural genes and adjacent to them. Each operator gene may control several structural genes in the way of an off-on switch. The *regulator* gene, on the other hand, may be on a different chromosome from the structural gene. The regulator gene is thought to produce substances that may combine with the proteins produced by the structural genes to form other substances called *repressors*. The repressor substance is thought

TABLE 8.2	**Amino Acids Utilized in Protein Synthesis in the Animal Body**	
Alanine	Glycine	Proline
Arginine	Histidine	Serine
Asparagine	Isoleucine	Threonine
Aspartic acid	Leucine	Tryptophan
Cysteine	Lysine	Tyrosine
Glutamic acid	Methionine	Valine
Glutamine	Phenylalanine	

TABLE 8.3	**Messenger Codons and the Amino Acids They Are Believed to Specify**							
UUU	Phenylalanine	CUU	Leucine	AUU	Isoleucine	GUU	Valine
UUC	Phenylalanine	CUC	Leucine	AUC	Isoleucine	GUC	Valine
UUA	Leucine	CUA	Leucine	AUA	Isoleucine	GUA	Valine
UUG	Leucine	CUG	Leucine	AUG	Methionine	GUG	Valine
UCU	Serine	CCU	Proline	ACU	Threonine	GCU	Alanine
UCC	Serine	CCC	Proline	ACC	Threonine	GCC	Alanine
UCA	Serine	CCA	Proline	ACA	Threonine	GCA	Alanine
UCG	Serine	CCG	Proline	ACG	Threonine	GCG	Alanine
UAU	Tyrosine	CAU	Histidine	AAU	Asparagine	GAU	Aspartic acid
UAC	Tyrosine	CAC	Histidine	AAC	Asparagine	GAC	Aspartic acid
UAA*	End	CAA	Glutamine	AAA	Lysine	GAA	Glutamic acid
UAG*	End	CAG	Glutamine	AAG	Lysine	GAG	Glutamic acid
UGU	Cysteine	CGU	Arginine	AGU	Serine	GGU	Glycine
UGC	Cysteine	CGC	Arginine	AGC	Serine	GGC	Glycine
UGA*	End	CGA	Arginine	AGA	Arginine	GGA	Glycine
UGG	Tryptophan	CGG	Arginine	AGG	Arginine	GGG	Glycine

*UAA, UAG, and UGA signal the end of polypeptide chains.

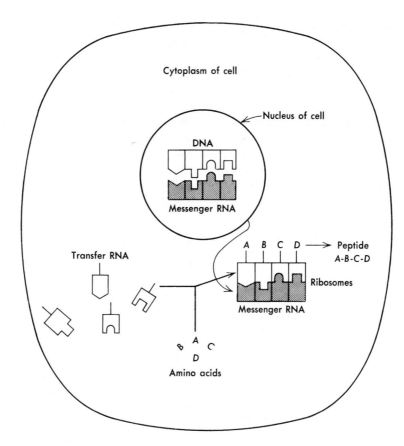

Figure 8.15 Diagram showing how proteins are built in the cell. Messenger RNA is synthesized by DNA in the nucleus and then moves to the cytoplasm (ribosomes), where transfer RNA moves amino acids to fit the mold of the mRNA. Amino acids are then arranged together in a sequence, as instructed by the DNA. Ribosomal RNA is also present in the cytoplasm, but its function is not known.

to proceed to the operator gene and to keep it turned off, thereby inhibiting enzyme formation by the structural gene. If the repressor substance is blocked by a metabolite (inducer), the operator gene then turns on the structural gene that produces gene products. This complete gene complex is called the *operon.*

Some genes modify the phenotypic expression of other genes. This is referred to as epistasis. An example of epistasis is the inheritance of horns and scurs in cattle. Scurs are small bony masses in the horn pits of **polled** cattle. Scurs are only present in animals that are polled. It does not matter whether a horned animal has the genes for scurs, a horned animal cannot also have scurs. Horns and scurs are controlled by different gene pairs, but the scur gene can only express itself if the animal is polled.

Various endocrine secretions are known to affect the phenotypic expression of certain genes. The male hormone appears to be involved in a group of sex-influenced traits that are dominant in males and recessive in females. Certainly, genes for milk or egg production do not express themselves in males, although males carry genes for these traits. Hereditary baldness

in the human male is associated in some way with the male hormone testosterone, although the genes that produce hair on a man's head appear to function for a period of time, commonly until the man is between 20 and 30 years of age, when he may become bald. The relations between certain endocrine secretions and some traits in livestock are also of interest. Physiological age at maturity differs in breeds of livestock, showing that genes are involved in regulating the onset of maturity.

8.6 GENES AND EMBRYOLOGICAL DEVELOPMENT

The fertilized egg divides into two cells, then four, then eight, and so forth, for a considerable length of time. At first, cells in the new embryo are the same, mother and daughter cells being identical. Later, mother cells produce daughter cells that are different, thus forming the different body tissues and organs. Why they begin to produce cells unlike themselves in certain stages of embryonic development is not completely understood.

Genes act throughout the embryonic period to cause formation of various body organs and tissues. Evidence for this is suggested by the fact that different species of animals vary in their rate of embryological development in the uterus. These differences must be controlled by genes. Also, many **lethal genes** (killer genes) are known to have their effects during various stages of embryonic development. Some defects of the embryo are known to be due to environmental factors that resemble inherited defects. Viral infections and certain substances (e.g., thalidomide) taken by the pregnant mother at a specific stage of pregnancy cause defects in the developing fetus similar to genetic defects, probably because they block cell division.[2] Such defects may be more extensive in farm animals than is usually recognized.

8.7 BIOTECHNOLOGY

The potential of biotechnology is indeed enormous. Biotechnology includes all technologies that pertain to molecular manipulation of living material. An important subset of biotechnology is the area of genetic engineering. Genetic engineering has been called one of the four most significant scientific breakthroughs of the twentieth century, on a par with unlocking the atom, escaping the earth's gravity, and the computer revolution.

Genetic engineering is a frequently used, yet loosely interpreted, term. Here, the term means the use of unconventional methods of modifying the animal genome, basically, transferring a gene from one individual to another. Genetic engineering approaches that have been proposed for animal improvement include genetic transformation using isolated DNA, haploid animal production through cell cultures, transfer of organelles from one animal to another, fusion of protoplasts of different species, cellular selection, and variant production by animal regeneration from protoplasts or the fusion of protoplasts. The possibility of selecting for many desired traits at the cellular level holds considerable promise as a useful tool in the genetic improvement of animals.

The technique of inserting genes into other cells (recombinant DNA or genetic transformation technology) has enabled scientists to code genes for desirable compounds and insert them into microorganisms, thus facilitating large-scale fermentation production of the compounds.

8.7.1 Microbe Engineering

The results of recombinant DNA technology, one form of genetic engineering, has made, and will continue to make, many important contributions to humans and domesticated animals in areas such as aging, cancer, immunology, and vaccines.

A segment of DNA that codes for a particular protein can be inserted into the genetic material of bacteria or other microbes such as yeast, indeed even into animal cells grown in tissue cultures. These new or altered forms of life are then multiplied into billions of cells that produce the chemical for which they have been engineered. Already, several biological products are being produced in this way, including human and cattle growth hormone, insulin, and interferon. Development of a wide range of vaccines is underway using these techniques.

The most commonly used "factory" for manufacturing these products is the bacterium *Escherichia coli,* which is one of the organisms studied most by microbiologists. It is precisely this background of knowledge about the organism that enables scientists to remove a circle of DNA called a plasmid from *E. coli,* cut the circle by using specific enzymes, and splice segments of DNA from another organism into the gap in the circle (hence the term gene-splicing). When the reconstructed plasmid is reinserted into the bacterium, it produces the protein product for which it was coded.

Strains of bacteria have been developed that consume petroleum and are helpful in cleaning oil spills. The dairy industry has used somatotropin (growth hormone) from recombinant sources as an injection to increase milk production. Perhaps cellulose-degrading genes could be introduced from certain bacteria into *E. coli,* which normally inhabits the digestive tract of animals. This could be helpful to some animals in more effectively utilizing cellulose, which is not digested by the enzymes normally present in single-stomached animals (see Chapter 19).

Another approach to genetic engineering is to chemically synthesize pieces of DNA from the genetic code for the desired amino acids and insert the synthesized genes into a plasmid. This approach was used, in part, for producing pure insulin.

Genetic engineering offers many advantages over other methods of producing vaccines. One of the most important is safety. To produce a conventional vaccine, the disease organism must be grown in large numbers, then inactivated. Both procedures can be dangerous and/or difficult. For example, in countries where the incidence of foot-and-mouth disease is controlled by vaccinating cattle, occasional outbreaks have been caused by the escape of the virus from the vaccine-production laboratories or by the inoculation of animals with improperly inactivated vaccines. Strict guidelines for genetic engineering studies have been developed. These are designed to help ensure the continued safety of such research. In addition to safety, genetically engineered vaccines have the advantage of being less expensive than vaccines produced from whole virus.

8.7.2 Transgenesis

Recombinant DNA technology is receiving widespread attention and research funding. The introduction of mammalian genes for various proteins, for example, insulin, into bacteria has resulted in the bacteria manufacturing the protein as a by-product. This procedure avoids the costly and time-consuming one of purifying a protein from an animal tissue source. Also of interest is being able to insert genes with unique functions directly into the developing embryo. This process is called *transgenesis.* Transgenesis in mammals was first accomplished

---

[2]The technique of amniocentesis (examining the fluid taken from the womb of a pregnant woman) can reveal any of some 70 possible genetic disorders in a woman's fetus.

with the insertion of extra copies of genes for growth hormone directly into mouse embryos in the early 1980s. This research was successful in that some of the mice with extra copies of the growth hormone gene passed those extra copies to their off-spring. It may follow that transgenic farm animals with extra copies of growth hormone genes may grow faster, be leaner, and/or produce more milk. This new potential of science opens a wide range of possibilities for introducing superior production and conformation traits and disease resistance.

8.7.3 Applications of Genetic Engineering in Plants

Green plants function as solar energy machines: they convert sunlight and plant nutrients into food, fiber, oil, and biomass (energy) for an ever-expanding world population that currently exceeds 6 billion people. Crop plants are the heart of a vast food and agricultural system that represents one of humanity's most precious natural resources. Recent discoveries in gene-splicing raise the exciting possibility that virtually any gene from any source can be engineered into plants.

Traditionally, plant breeders have combined the desirable genes or traits from two different parents into one descendant through selective matings. However, breeding has been possible only between closely related plants. Corn, for example, will not cross with wheat. The technologies being developed through genetic engineering may ultimately make it possible for the plant breeder to take genes from any plant species and insert (splice) them into any other plant.

There have been several exciting examples of genetic engineering in crop species. Resistance to either herbicides or insects is a critically important characteristic that appears to have promise for applications from molecular genetics. There are several crops, including soybeans and sugar beets, for which seeds are marketed that have inserted genes for resistance to the herbicide Roundup®. This greatly improves weed control in fields because the herbicide can be applied without damaging the crop. Transgenic corn and cotton are being sold that have genes from the microorganism *Bacillus thurigensis*. This bacterium conveys resistance to the corn borer and boll weevil. The FlavrSavr® is a genetically engineered tomato that has a longer shelf life. The genetic modification allows it to remain on the vine until ripening occurs, rather than being picked while green.

8.7.4 Cloning

A clone is a genetic identical. In this sense, identical twins are clones of each other. Until recently, development of twins (or more) naturally or through embryo splitting was the only way to have clones in farm animals. Clones were developed through embryo splitting by dividing the embryo when it was still early in development. Once differentiation of tissues began, clones could not be developed. This was the standard discussion of cloning in farm animals until 1997. In the spring of 1997, Dr. Ian Wilmut of the Roslin Institute in Scotland announced the successful cloning of an adult ewe. His research group joined cells from the mammary gland of an adult ewe with an egg cell

from which the contents of the nucleus had been removed. This cell developed and the resulting lamb became the first reported successful birth of a mammal as a clone of an adult. They named it "Dolly." Dolly possessed the same genetic material as the adult which donated the mammary gland cell. The difficult part of this process was not the manipulation of the DNA. Rather, the most challenging and monumental scientific breakthrough was making the mammary gland cell revert to an undifferentiated state. Recall we noted earlier in this chapter that an embryo will start developing specialized cells which will become the various body tissues. Once this begins, the genes that do not pertain to that specialized tissue no longer have a function. The developers of Dolly were able to manipulate the cell in such a way that specialized development was arrested and the cell was ready to start developing from the beginning.

It might seem advantageous to use cloning to make many copies of superior individuals. When the technology for cloning is refined, this may be used to assist in genetic improvement. If a producer could accurately identify the best high-producing dairy cow in his/her herd, it would be nice to make many copies of her. The technology to accomplish such a feat is probably some time away. There is also a problem with accurately identifying the best cow in the herd. She may have the highest production because she has the best genetics or because she has been exposed to the best environment and management. Environmental differences among cows will keep even genetically identical cows from performing identically.

A more logical early use for cloning may be in the development of pharmaceuticals. For example, a scientist may insert a human gene (through transgenesis) into a sheep or pig so that it produces a pharmaceutical product in its milk. People could obtain this drug simply by drinking the milk. Normal reproduction of such an animal would yield some offspring with the gene, others without it. However, if clones were produced, each one would have the gene and would produce the product. Another example could be in developing lines of pigs for organ transplants. Genes could be inserted into pigs which would provide certain characteristics of a human immune system, thereby reducing the frequency of organ rejection. Clones of these pigs could be developed as sources of organs rather than waiting for a human donor to die.

8.7.5 Marker-Assisted Selection

Improvement of livestock normally proceeds by choosing the best young animals to be parents of the next generation. The best animal can be determined by its own performance and/or the performance of relatives. New developments in molecular genetics have opened the possibility of selection based upon the genes themselves.

A genetic marker is a segment of DNA that can be identified through the techniques of molecular genetics. If a marker is within, or very near, a gene that has an influence on an important trait, that marker can be used to assist in the selection process. This is called "marker-assisted selection." A young animal could be evaluated for several genetic markers and these evaluations could be combined with regular performance data to make selection more accurate.

8.8 SEGREGATION AND RECOMBINATION OF GENES

Genes occur in pairs in the body (**somatic**) cells. This statement is true except for **sex-linked** traits, the genes carried on the portion of the X chromosome that is not homologous to the Y chromosome. Another relatively rare exception might occur in cases where the body cells of an individual possess one or three chromosomes rather than the usual homologous pair. Of each pair of genes an individual possesses, one gene comes from the father and one from the mother. The individual will transmit one gene or the other, but not both, to any one offspring.

Mendel's law of **segregation and recombination of genes** states that when two genes are paired in body cells, they segregate independently of each other in the gametes. This may be illustrated by the following example:

	Father		Mother
P_1 generation	AA	$\times$	aa
F_1 generation		All Aa	
F_2 generation	Aa	$\times$	Aa
	AA aA		Aa aa

This example shows that in the F_1 (first filial) **generation** the genes A and a were paired. They are called **allelomorphs** (or **alleles**) because they are carried at the same locus (location) on the same pair of homologous chromosomes and affect the same trait but in a different way. When they produced sex cells, the F_1 individuals put gene A into their gametes about 50 percent of the time and gene a about 50 percent. Thus, even though they were paired in the F_1 individuals, they did not stay together in the gametes but rather segregated independently. They also recombined in pairs at random, giving a $1AA:2Aa:1aa$ ratio in the F_2 (second filial) **generation.**

Linkage of two or more genes prevents segregation of the different pairs of genes in gametes. *Linkage* means that *two or more pairs of genes are carried on the same homologous chromosomes,* as shown below:

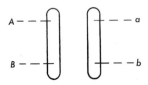

Linkage means that genes A and B would segregate together in the gametes, as would genes a and b. An exception to this, however, occurs when homologous chromosomes

come together (synapse) in the process of meiosis. Sometimes during synapsis the homologous chromosomes exchange parts, so that A and b, as well as a and B, may get together, as shown in the following diagram. This is known as *crossing over* and occurs more frequently in genes carried farther apart on homologous chromosomes than in those carried closer together.

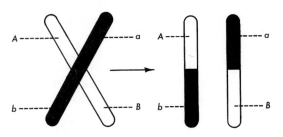

8.9 LAWS OF PROBABILITY AND ANIMAL BREEDING

The laws of probability apply to the segregation and recombination of genes. For example, an individual of **genotype** Aa will transmit gene A to approximately one-half its offspring. The same is true for gene a. This leads to the assumption that the probability that an individual of genotype Aa will transmit either gene A or gene a to a particular offspring is one-half, or 0.5. The probability that an individual of genotype AA will transmit gene A to any particular offspring is 1, whereas the probability that it will transmit gene a to its offspring is zero. These probabilities are valid provided no **mutations** of genes A and a occur.

The laws of probability also apply to the segregation of two or more pairs of genes in the gametes of an individual, provided each pair of genes is carried on a different pair of homologous chromosomes. The law involved states that the probability of two or more independent events occurring together is the product of each separate probability. To illustrate, what is the probability that an individual of genotype $AaBb$ will transmit genes AB to a single offspring? If the two pairs of genes A or a and B or b are carried on different pairs of homologous chromosomes, they segregate independently of each other according to the laws of probability. The segregation of each pair would be a single independent event. Thus the probability that individual $AaBb$ will transmit gene A in a gamete is 0.5, and the probability that it will transmit gene B in a gamete is also 0.5. The probability that both genes A and B will be transmitted through the same gamete is 0.5×0.5, or 0.25. Obviously, an individual of genotype $AABB$ could not transmit combinations of genes such as aB, Ab, or ab through the same gamete unless a mutation occurred, and so the probability of this transmission is zero. The probability of the segregation of a certain combination of genes from an individual whose genotype possesses many different pairs of alleles can be calculated in the same way as long as each pair of allelic genes is on a different pair of homologous chromosomes. For example, the probability that an individual of genotype *AaBbCcDd* will transmit genes *ABCD* in the same gamete is $0.5 \times 0.5 \times 0.5 \times 0.5$, or 0.0625.

Laws of probability can be applied to the recombination of genes in the **zygote**. For example, if two individuals that are of genotypes *Aa* are mated, the probability of offspring of the four possible genotypes would be

$$AA = 0.5 \times 0.5, \text{ or } 0.25$$
$$Aa = 0.5 \times 0.5, \text{ or } 0.25$$
$$aA = 0.5 \times 0.5, \text{ or } 0.25$$
$$aa = 0.5 \times 0.5, \text{ or } 0.25$$

\} 0.5

This recombination gives the familiar 1:2:1 ratio between the cross of two individuals that are heterozygous for one pair of alleles.

The laws of probability can also be used to calculate the possibility of a recombination of two or more genes in the zygote. For example, the probability that the mating of two individuals of genotypes *AaBb* will produce an offspring of genotype *AABB* is 0.5×0.5, or 0.25, for genes *AA;* and 0.5×0.5, or 0.25, for genes *BB;* or 0.25×0.25, or 0.0625, for genes *AABB*. The probability of certain other combinations of genes in the offspring of such parents can be calculated in a similar way.

The laws of probability can be expanded to cover many other aspects of gene combinations in gametes and zygotes. For example, the probability that parents, both of whom are of genotype *Aa*, will produce a homozygous recessive *aa* offspring is 1 out of 4, or 0.25 (0.5×0.5). The probability of four offspring, all of which are of genotype *aa*, from such parents, would be 0.0039 ($0.25 \times 0.25 \times 0.25 \times 0.25$).

8.10 MUTATIONS

Genes possess the remarkable property of duplicating themselves precisely generation after generation. Occasionally, however, mistakes are made in the gene-duplicating process and a new gene (allele) is born. Basically, most if not all mutations result in a change in the code sent by the gene to ribosomes by means of mRNA to form a particular protein. If the wrong code is sent, a different protein is formed. The missing protein or the new protein may cause a defect or a new genetic trait to appear. Differences that can be seen or measured between individuals are due to an accumulation of different mutations (old or new) within populations. These mutations are responsible for differences in coat color, size, shape, behavior, and other traits in various species. These differences among individuals are the raw material with which the animal breeder has to work.

8.11 PHENOTYPIC EXPRESSION OF GENES (NONADDITIVE)

The genotype of an individual refers to *its actual genetic makeup.* For example, individuals that are *AA, Aa,* or *aa* repre-

sent three different genotypes. *Phenotype* refers to *those differences in individuals that can be measured by means of the senses,* such as black or white, tall or short, horns or no horns (polled).

Genes almost always segregate in gametes and recombine in zygotes in the same general way as long as they are carried on different chromosomes. However, even though these genes segregate and recombine in the same way, they can express themselves in the phenotype in many different ways. This is what is meant by different kinds of phenotypic expression of genes.

8.11.1 Dominance and Recessiveness

A gene is said to be **dominant** when it covers or hides the expression of its allele. For example, in Angus cattle the gene for black *B* is dominant to the gene for red *b,* its own allele, because an individual of genotype *Bb* is black. With two alleles such as *B* and *b,* there can be three genotypes and two phenotypes, as follows:

Three Genotypes	Two Phenotypes
BB	Black
Bb	Black
bb	Red

Three genotypes exist because the two genes have combined in pairs in three different ways. Two phenotypes (black and red) exist because both the *BB* and the *Bb* individuals are black and the *bb* individuals are red. The two black genotypes are quite different in their breeding ability, however. The homozygous black individuals *BB* will always produce black offspring, even if mated to red *bb* individuals. This is illustrated as follows:

Kinds of Matings	Offspring	
	Genotype	Phenotype
BB × BB	*BB*	Black
BB × Bb	1*BB*:1*Bb*	Black
BB × bb	*Bb*	Black

The heterozygous black *Bb* individuals do not always breed true, however. This is illustrated as follows:

Kinds of Matings	Offspring	
	Genotype	Phenotype
Bb × BB	1*BB*:1*Bb*	Black
Bb × Bb	1*BB*:2*Bb*:1*bb*	3 black to 1 red
Bb × bb	1*Bb*:1*bb*	1 black to 1 red

If red *bb* individuals are mated together, they should always produce red offspring. However, they do not breed true when mated to black individuals, as illustrated below.

Kinds of Matings	Offspring	
	Genotype	Phenotype
bb × *BB*	*Bb*	Black
bb × *Bb*	1*bb*:1*Bb*	1 red to 1 black
bb × *bb*	*bb*	Red

From the foregoing discussion, one can see that it is relatively easy to develop a purebreeding strain that is **recessive,** or red *bb,* in Angus. One merely has to mate recessives *bb* to recessives *bb* to get recessives *bb*. Developing a pure dominant or black strain is more difficult, however, because one cannot tell by observation whether individuals are homozygous dominant *BB* or heterozygous *Bb*. Special progeny tests must be conducted to distinguish between the two dominant genotypes. One test is to mate the black individual of the dominant phenotype (either *BB* or *Bb;* one would not know which) with homozygous recessive *bb* individuals. One red or homozygous recessive *bb* offspring from such a mating proves the black individual heterozygous *Bb*. If five matings are made to red *bb* individuals without a single red offspring being produced, the black individual is **homozygous** black *BB,* and this would be correct at the 95 percent level of probability.

Sometimes homozygous recessive *bb* individuals are not available for mating. In such a case, known **heterozygotes** *Bb* can be used for progeny tests. They are known heterozygotes *Bb* if they produce one homozygous recessive *bb* offspring or if one of their parents is homozygous recessive *bb*. One homozygous recessive offspring *bb* from such a mating proves the individual heterozygous *Bb*. Eleven such matings without a homozygous recessive offspring *bb* being produced proves the dominant, or black, individual homozygous *BB* at the 95 percent level of probability, and 17 such matings are proof at the 99 percent level.

The same principles of progeny testing used in these examples apply in all cases in which one pair of alleles, dominant and recessive, is concerned.

8.11.2 Lack of Dominance

Sometimes neither of two alleles is dominant to the other, and if both are present, each expresses itself in the phenotype. Coat color in Shorthorn cattle is a typical example of such inheritance. A Shorthorn is red if it has two genes for red (*RR*) and white if it has two genes for white (*WW*). A Shorthorn is roan, or a mixture of both red and white, if it has one gene for red and one for white (*RW*). In this case there are three genotypes and three phenotypes, and there is little or no difficulty in distinguishing between them.

8.11.3 Partial Dominance

Some genes are only partially, and not completely, dominant to their alleles. An example of this type of phenotypic expression of genes is the comprest gene in Hereford cattle. Two comprest genes (*CC*) cause the individual to be a dwarf. Two normal genes (*cc*) cause the individual to be of normal size or phenotype. One comprest gene and one normal gene (*Cc*) cause the individual to be a comprest, ranking midway between the normal and the dwarf in size or phenotype.

8.11.4 Overdominance

Overdominance refers to *a phenotypic expression of genes in which the heterozygote (a^1a^2) is superior in phenotype to either homozygote (a^1a^1 or a^2a^2)*. Many examples are known to illustrate overdominance. One of the better-known examples is sickle-cell anemia, found in many African blacks. In certain regions of Africa, the disease malaria is prevalent. This disease kills many individuals homozygous for normal adult **hemoglobin** (genotype *AA*). Those individuals homozygous for the sickle-cell gene *SS* die of **anemia.** No *AS* individuals die of anemia, and few of them die of malaria. The genotype *AS* resists, in some way, damage to the red blood cell by the malaria organism. Thus the heterozygote *AS* individual is superior in survival to either *AA* or *SS* individuals in this **environment.** In the United States, where there is little or no malaria, the *AA* individuals would survive as well as the *AS* individuals, and there would be no heterozygote advantage.

Although heterozygotes are superior when overdominance is involved, they never breed true when mated together. They can always be produced by mating the two homozygotes together (a^1a^1 and $a^2 a^2$), provided both of these genotypes are available for mating. Overdominance probably has its greatest effect on traits related to physical fitness, the heterozygote being superior to either homozygote.

8.11.5 Epistasis

Epistasis may be defined as a type of phenotypic expression of genes due to the interaction of two or more pairs of genes that are not alleles. Epistasis may be distinguished from overdominance in that the latter is due to an interaction between genes that are alleles. An example of epistasis is as follows:

a^t is the gene for tricolor in collie dogs. It is recessive to A^t, the sable and white gene.

M is the gene for merle in collies, giving a dilution effect; *m* is the gene for nonmerle.

A collie that has the genetic makeup $a^t a^t mm$ is tricolored, or black, tan, and white. A collie with the genetic makeup $a^t a^t Mm$ is a blue merle, whereas one with the genetic makeup $a^t a^t MM$ is a white merle. Thus the gene *M* influences the way the tricolor gene a^t expresses itself phenotypically even though the *M* and a^t genes are not alleles. Epistasis may affect the phenotype in different ways through various interactions between nonallelic genes.

Dominance and recessiveness, overdominance and epistasis are all types of nonadditive phenotypic expression of genes. A second general type of phenotypic expression of genes is known as *additive.*

8.12 PHENOTYPIC EXPRESSION OF GENES (ADDITIVE)

An additive type of gene action means that *the effect of each gene that contributes something to the phenotype of an individual for a certain trait adds to the phenotypic effect of another gene that contributes something to the same phenotype.* An example of additive gene action is skin color inheritance in humans. It has been proposed that two different pairs of genes affect skin color in whites and blacks, although this may be an oversimplification because more than two pairs of genes may be involved. The proposed type of inheritance suggests that white persons are of genetic makeup *aabb,* and that when either an *A* or a *B* gene is present, each causes a darker skin color. Thus the possible genotypes and phenotypes would be:

Genotype	Phenotype
aabb	White
Aabb or *aaBb*	Light
AAbb, aaBB, or *AaBb*	Medium
AABb or *AaBB*	Dark
AABB	Black

Each contributing gene *A* or *B* makes the skin color darker, whereas neutral genes, *a* and *b,* do not affect skin color. Here it is assumed that genes *A* and *B* always contribute the same amount to skin color, but this may not be absolutely true. If they did differ slightly, it would not affect the validity of the example to any appreciable extent.

Both additive and nonadditive types of gene action affect many economic traits in animals. These traits include rate of gain, **feed efficiency,** lactation, and egg laying as well as other traits. Many pairs of genes may affect such traits. It is possible that some traits may be almost entirely affected by nonadditive gene action whereas other traits may be affected almost entirely by an additive type of gene action. Still other traits may be affected by both additive and nonadditive gene action. How to determine what type or types of gene action affect various traits is discussed in Chapter 9.

8.13 SEX-LINKED INHERITANCE

Sex-linked inheritance refers to inheritance due to genes carried on the nonhomologous portion of X chromosomes. Sex-linked genes are usually recessive in their phenotypic expression.

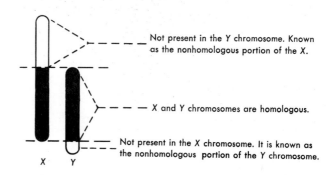

Figure 8.16 Diagram of the pairs of sex chromosomes in the male which has one X and one Y chromosome. The female of such an animal has two X chromosomes.

It was pointed out earlier that each animal species possesses a characteristic number of chromosomes and that the different species have different numbers of chromosomes (Table 8.1). For example, humans possess 23 pairs whereas cattle possess 30 pairs. In farm animals and humans, there is one set of sex chromosomes known as the X and Y chromosomes (except poultry, in which the sex chromosomes are called Z and W).

The X chromosome in humans and farm animals is much larger and longer than the Y chromosome. This is illustrated in Figure 8.16. In the male, there is a portion of the X chromosome that does not pair with the Y, or, genetically speaking, that is not homologous (of like proportion) to the Y. Genes carried on this portion of the X chromosome are said to be *sex-linked.* Because only one X chromosome is present in the male, a gene on the nonhomologous portion of the X chromosome will express itself in the phenotype even if it is a recessive gene. In the female, where two X chromosomes are present, two recessive genes are required to cause a recessive trait to appear.

As is shown in Figure 8.16, a portion of the Y chromosome probably has no counterpart, or homologous portion, in the X. Genes carried there are transmitted only from fathers to sons. This type of inheritance is known as *holandric inheritance.*

Many sex-linked recessive traits, such as **hemophilia** and red–green color blindness, are known in humans, but very few are known in farm animals, possibly because sufficient studies have not been made in farm animals to determine the inheritance of such traits.

8.14 SEX-INFLUENCED INHERITANCE

Certain genes carried on the autosomes are present in body cells in pairs, but the sex of an individual determines whether the trait is dominant (in males) or recessive (in females). The gene for **scurs** (S_c) in European cattle appears to be inherited in this way. The inheritance of scurs in polled European breeds of cattle seems to be as follows:

	Phenotype in	
Genotype	Male	Female
S_cS_c	Scurred	Scurred
S_cS_n	Scurred	Not scurred
S_nS_n	Not scurred	Not scurred

Hereditary baldness in humans and horns in some breeds of sheep also appear to be sex-influenced traits, being dominant in males and recessive in females.

8.15 SEX-LIMITED TRAITS

Certain traits, such as lactation in cattle and egg laying in poultry, are affected by many genes but are expressed in only one sex. Thus there is something about the sex of the individual that determines whether or not the traits appear phenotypically. The appearance of these traits is related to the production of sex hormones.

Even though certain traits are limited to one sex, both sexes possess and transmit genes for such traits to their offspring. This means that both parents should possess superior genes for such traits if their offspring are to be superior. Because dairy bulls do not lactate and **roosters** do not lay eggs, it becomes a more complicated problem in males than in females to determine the kind of genes they possess for these traits. Estimates of their genetic merit must be based on the performance of female relatives such as their dam, sisters, or daughters (Chapter 9).

Single gene effects, such as cock-feathering or hen-feathering in chickens, are also limited in their phenotypic expression to only one sex. Such traits also appear to be related to interactions of genes and sex hormones.

8.16 GENOME PROJECT

During the late 1980s a research effort called the Human Genome Project was initiated. The goal of the Human Genome Project is to fully describe all the DNA (every gene) in every chromosome in humans. This ongoing project is being conducted at several government, university, and private laboratories in the United Stated and in other countries. Tremendous progress has been made in identifying several genes on each chromosome. Concurrently, numerous research projects are being conducted to map the genome in cattle, swine, sheep, horses, dogs, and mice. Each of these research initiatives operate in conjunction with each other and the

Human Genome Project. The projects assist and complement one another because many genes are arranged in similar patterns in different species. If two genes are close to each other in one species they will frequently be close to each other in other species. Closely related species have more genetic similarities than species that are more distantly related. For example, there are more similarities between cattle and sheep than between cattle and humans.

Completion of the Human Genome Project may have numerous profound effects. Many human diseases have a genetic component, and knowledge of the location of genes that influence a particular disease will help understand how to diagnose, prevent, or treat the disease. In farm animals, knowledge of the locations of genes will provide important information to assist in selecting superior breeding stock (marker-assisted selection).

8.17 SUMMARY

Cells are the building blocks of which the body is made, and each individual possesses trillions of them. The main parts of the cell include the nucleus and cytoplasm. The nucleus is the heart and brain of the cell and contains the chromosomes. Each species of animal possesses the characteristic number of chromosomes for that species. Chromosomal abnormalities sometimes occur and usually result in an abnormal phenotype or death of the individual. Chromosomal abnormalities include variations from normal in structure and numbers.

Genes, the determiners of heredity, are carried on chromosomes. The gene is a portion of a DNA molecule and has the ability to replicate itself when new cells are formed. It also sends the code for protein structure and synthesis to ribosomes in the cytoplasm by means of messenger RNA. The genetic code and protein synthesis are complex, and functions of the gene in protein synthesis are a recent discovery.

Genes may express themselves in the phenotype in two general ways, known as additive and nonadditive phenotypic expressions. Genes also segregate into the gametes (sperm and ovum) and recombine in the zygote (the new individual resulting from the union of a sperm and an ovum) according to certain laws of probability.

A promising field of scientific endeavor is genetic engineering. The ability to locate genes on chromosomes, transfer them between animals within and among species, and clone individual animals with characteristics that have value is blossoming. This will provide many opportunities to assist in disease recognition and treatment in humans and in the development of animals that can provide important benefits to humanity.

STUDY QUESTIONS

1. Wild squirrels, which are usually brownish-gray in color, sometimes produce offspring that are albinos (white). When these white squirrels are mated, they always produce white offspring. What type of inheritance is involved in the production of albino offspring? Why do the albino squirrels breed true? Why are most wild squirrels of a brownish-gray color and not white? How did the albino characteristic originate?

2. The black gene *B* in Angus cattle is dominant to the red gene *b*. Breeders of black Angus cattle have never kept red animals for breeding. Why do red calves still appear in some matings of black Angus parents? How would you start a new breed of red Angus cattle?

3. How would you design a program to eliminate the red gene *b* from the black Angus breed?

4. The comprest gene *C* in Hereford cattle is only partially dominant to the normal gene *c*. Therefore normal individuals for this gene are *cc*, comprests are *Cc*, and comprest dwarfs are *CC*. How would you start a pure strain of comprest Herefords?

5. White coat color in some breeds of swine is dominant to black coat color, and the mule-footed condition is dominant to the normal split-hooved condition. The offspring of a certain white, mule-footed boar are always white and mule-footed, regardless of the phenotype of the sows to which he is mated. What is the probable genotype of the boar if white is *W*, black is *w*, the mule-footed condition is *M*, and the normal foot condition is *m*?

6. Another boar is white and mule-footed, but when mated to black sows, about one-half the offspring are black, and when mated to normal-footed sows, about one-half the offspring have normal feet. What is a possible genotype of this white, mule-footed boar?

7. Solid color in Angus cattle is dominant (*S*) to white spotting (*s*), found in Holsteins. What would be the expected color of a calf from an Angus bull mated to a Holstein cow?

8. Some breeds of chickens are white because of the presence of a dominant epistatic pigment inhibitor (*I*), whereas other breeds are white because of the presence of the recessive allele for no color (*c*) in the homozygous condition. The two pairs of genes are on separate pairs of homologous chromosomes. If a white male of genotype *IICC* is mated to white females that are *iicc*, what would be the phenotype of the F_1 offspring? If the F_1 individuals are mated among themselves, what would be the phenotypes of the F_2 offspring?

9. A certain kind of hemophilia in humans is a sex-linked recessive trait. A man who had hemophilia married a woman who did not have the disease but whose father did. What kinds of sons, genetically, could they have for this trait? What kinds of daughters could be expected?

10. How many sons of the hemophiliac father mentioned in question 9 would have hemophilia if their mother had not been a carrier of the recessive gene for hemophilia? How many of his daughters would have hemophilia in such a case? How many would be carriers of the gene for this disease?

11. Assume that you are a genetic consultant. A man comes to you with this problem: He works for an atomic energy plant, where he is exposed to radiation. Before he started working at the plant, he had three sons and two daughters who were normal in every way. Two years after he started working for this company, he had a fourth son who had hemophilia (sex-linked). He wants to sue the company for damages, insisting that it was a mutation from radiation that caused him to produce the son with hemophilia. What advice would you give him?

12. The A, B, and O blood groups in humans can be used at times to determine parentage cases in lawsuits. Both genes *A* and *B* are dominant to a gene for *O*, but they are not dominant to each other. Which of the following combinations would be possible?

Blood Groups		
Child	**Father**	**Mother**
(1) Group O	Group A	Group B
(2) Group O	Group AB	Group B
(3) Group A	Group B	Group A
(4) Group B	Group AB	Group AB
(5) Group AB	Group AB	Group O
(6) Group A	Group O	Group A

13. Assume, again, that you are a genetic consultant. A girl comes to you with a problem seeking advice. She is Rh negative. She has three boyfriends of whom she is about equally fond. One is Rh negative; a second is Rh positive and both his parents were Rh positive; and a third is Rh positive, but his mother was Rh negative (Chapter 10). From a genetic standpoint, which boy would you recommend she choose for a husband? Why?

14. One hundred Native Americans in New Mexico were blood-tested, and it was found that 59 were of blood group M, 33 were MN, and 8 were NN. What was the frequency of the *M* and the *N* gene in this group of Native Americans? (See Section 9.4 before answering questions 14, 15, and 16.)

15. Of 1200 Native Americans who were blood-typed, assume 16 percent were of blood type N. What was the frequency of the *N* and the *M* gene in this population? How many should be blood group M and blood group MN? (M is not dominant to N, and vice versa.)

16. Suppose you bought 20 Angus cows that were homozygous black *BB* and a bull that was a carrier of the red gene *Bb*. If all cows produced a calf during the first calving, what would be the probable frequency of the red gene in the entire herd of 20 cows, 1 bull, and 20 calves?

17. A couple has just married. They plan to have four children and would like to have four boys. What is the probability their desire will be fulfilled if they have four children?

18. Joe Jones has a family of nine boys and no girls. (a) What is the probability that in families of nine all will be one sex? (b) What is the probability that the tenth child in this family will be a girl?

19. Henry Smith's father was not bald, but his mother was. What is the probability that he will be bald when he reaches 40 years of age?

20. A brown-eyed girl of blood type MN marries a boy who is brown-eyed and is of blood type N. The boy is not color blind, but the father of the girl was. Their first child is a girl of blood type N with blue eyes and normal vision. What is the probability that their second child will be a boy of blood type MN who is color blind and who has brown eyes? (The red–green color-blind gene is a sex-linked recessive.)

21. Andalusian chickens are of three types: black, splashed-white, and blue. The black and splashed-white breed true, but when mated together they always produce blue. When the blue Andalusians are mated together, they do not breed true but produce one black, two blue, and one splashed-white. Explain why blue individuals do not breed true. How could blue Andalusians always be produced?

22. Several different kinds of combs in chickens are inherited. A single-combed individual is of genotype *rrpp*, a rose-combed is *R-pp*, a pea-combed is *rrP-*, and the combination of *R* and *P* in the genotype produces a walnut comb. What are the F₁ genotypes and phenotypes if a single-combed rooster is mated to rose-combed hens of genotype *RRpp*? What would be the phenotypic ratio in the F₂?

23. If rose-combed and pea-combed individuals, when mated, always produced walnut-combed chicks, what would be the genotype of the parents? What would be the genotypic and phenotypic ratios of the F₂ offspring, starting with the rose-combed and pea-combed parents in this question?

24. What kind of phenotypic expression of genes is necessary for the production of walnut combs in chickens? Explain. Would it be possible to develop a purebreeding strain of walnut-combed chickens? Explain.

25. In poultry, barring *B* is a sex-linked dominant trait and black *b* is its recessive allele. If barred *BB* males are crossed with black females, what would be the genotypic and phenotypic ratios of the offspring? If heterozygous barred *Bb* males are mated with black (*bw*) females, what would be the genotypes and phenotypes of the offspring? Remember that the female in chickens is *ZW* in chromosome composition.

26. In question 25, if barred females (*Bw*) were mated with black males, what would be the genotypes and phenotypes of the offspring? What would be the sex of the barred chicks? What would be the sex of the black chicks?

ANSWERS TO STUDY QUESTIONS

1. The albino gene *a* is a recessive gene. For an albino to be produced, it has to have genotype *aa*. Albino parents *aa* should always produce albino offspring *aa*. Most squirrels are brownish-gray because the gene for this color is dominant *A* to the albino gene *a*. Only a few albino genes *a* are present in the population. The albino gene probably arose by a mutation of gene *A* to *a*.

2. Because many black Angus still carry the red gene *b*, when black parents of genotype *Bb* are mated, one-fourth of their offspring should be red, or *bb*. A new breed of red Angus could be developed by mating red *bb* to red *bb*.

3. Discard all red Angus *bb* when they appear and do not use black parents of a red Angus calf for breeding because they are carriers of the red gene (genotype *Bb*). Progeny-test bulls to determine if they are *BB* or *Bb*. This can be done by mating the bull in question to red *bb* cows. One red offspring proves the bull *Bb*. Five black and no red offspring out of red cows prove him *BB* at the 95 percent level of probability. Seven black and no red offspring from seven red cows prove him *BB* at the 99 percent level. Other progeny tests are to mate him to known carriers (*Bb*) of the red gene, or to at least 23 of his own unselected daughters. Actually to eliminate the red gene, both cows and bulls must be tested and proved *BB*. This would be difficult to do in cows.

4. It could not be done because comprests (*Cc*) do not breed true.

5. The genotype of the boar is probably *WWMM*.

6. The genotype of this boar is probably *WwMm*.

7. Solid black of genotype *Ss*.

8. The F_1 offspring would all be white. The F_2 offspring would be 13 white to 3 colored.

9. The woman is normal but a carrier of the gene for hemophilia. One-half of the sons would be normal and one-half would have hemophilia. Also, one-half of the daughters would be normal and one-half would have hemophilia. However, the normal daughters would be carriers of the recessive gene for hemophilia.

10. None. None of the daughters would have hemophilia, but all would be carriers of the gene for hemophilia.

11. The radiation was not at fault. The father transmits his X chromosome only to his daughters. The sons always receive the X chromosome from their mother.

12. (1) is possible if the parents are *Aa* and *Ba*. (2) is not possible because the child is O, or genotype *aa*, and has to receive the *a* gene from both parents. The *AB* father does not possess the *a* gene. (3) is possible, (4) is possible, (5) is not possible, and (6) is possible.

13. If she chose the Rh negative/Rh negative boy, she would have no trouble with erythroblastosis fetalis

in her children. However, all her daughters would be Rh negative/Rh negative, and she would just delay the problem for another generation. The Rh positive/Rh positive boy with Rh positive parents would be the greatest risk, genetically.

14. The frequency of the *M* gene is 0.755; that of the *N* gene is 0.245.

15. The frequency of the *M* gene is 0.6, and that of the *N* gene is 0.4; 432 should be of blood group M and 576 of blood group MN.

16. Frequency of the red gene *b* would be 0.134.

17. 1 out of 16.

18. (a) 1 out of 512 families of 9. (b) 1 out of 2.

19. Probability is 1. His mother is of genotype *BaBa*, and he would receive a *Ba* gene from her. Because baldness is dominant in the male and recessive in the female; one *Ba* gene would be enough to cause him to be bald.

20. 3 out of 32.

21. The blue individuals would be heterozygous. They could always be produced by mating black and splashed-white individuals.

22. The F_1 genotypes would all be *Rrpp*, and the phenotypes would all be rose-combed. The F_2 genotypes would be one *RRpp*, two *Rrpp*, one *rrpp*. The phenotypes would be

three rose-combed to one single-combed.

23. The rose-combed individuals would be *RRpp* and the pea-combed *rrPP*. The F_2 genotypes would be one *RRPP*, two *RRPp*, one *RRpp*, two *RrPP*, four *RrPp*, two *Rrpp*, one *rrPP*, two *rrPp*, one *rrpp*. The phenotypic ratio in the F_2 would be nine walnut-combed, three rose-combed, three pea-combed, and one single-combed.

24. Epistasis, because it is the interaction between nonallelic genes. Yes, it is possible to develop a purebreeding strain of walnut-combed chickens by finding those that are *RRPP* and mating them. The difficulty of developing such a strain would be very great because one would have to progeny-test to find those that were *RRPP*.

25. Both males and females would be barred, with the males *Bb* and the females *Bw*. Heterozygous barred *Bb* males mated to black *bw* females would give one *Bb* male that was barred, one *bb* male that was black, one *Bw* female that was barred, and one *bw* female that was black.

26. Genotypes = one *Bb* to one *bw*; phenotypes = one black to one barred. The barred chicks would be males, and the black chicks females.

PRINCIPLES OF SELECTING AND MATING FARM ANIMALS[1]

Though reason is progressive, instinct is stationary. Five thousand years have added no improvement to the hive of the bee or the house of the beaver.

Charles C. Colton (1780–1832)

9.1 INTRODUCTION

Many pairs of **genes** affect the expression of economic traits in farm animals. Economic traits include **fertility,** rate and efficiency of body weight gains, milk and egg production, and **carcass** quality. These traits determine profit or loss in commercial livestock and poultry production. Because many genes affect these economic traits, there is usually no sharp distinction between **phenotypes** but rather a continuous variation from one extreme to another. This type of inheritance is referred to as **quantitative inheritance.** Conversely, **qualitative inheritance** refers to such traits as coat color or horns (or **polledness**), in which only one or a relatively few pairs of genes are involved and there is a sharp and distinct difference between phenotypes. Different selecting and mating schemes are required for these two general types of inheritance.

9.2 PHENOTYPIC VARIATIONS IN QUANTITATIVE TRAITS

In a large population of animals, a quantitative trait, such as rate of gain in the feedlot, varies from individuals that make slow gains to others that gain rapidly. The rate of gain for a majority of individuals, however, will tend to cluster around an average figure somewhere between the two extremes. If data for individuals making different rates of gain are graphically plotted, there will be a tendency for the rates of the entire group to fall into a curve resembling the shape of a bell, as shown in Figure 9.1. This bell-shaped curve is called the "normal" distribution. The phenotypic variation and shape of the resulting curve are expected within a large group of individuals. Variation is the raw material with which animal breeders must work to improve their **livestock.** Without this variation there would be no hope of improvement through the application of breeding methods.

Qualitative inheritance involves genetic principles that deal specifically with the individual. This is because there are distinct differences between phenotypes and because few genes are involved. *Quantitative* inheritance, conversely, involves the individual less and the total population more. Thus principles of genetics of populations are important for improving quantitative traits such as those of economic importance in livestock and poultry.

9.3 STATISTICAL EVALUATION OF QUANTITATIVE TRAITS

Statistical methods for describing the phenotypic variation of quantitative traits have been devised. The base, or starting point, is the average (mean) of the entire population being studied. This is determined by adding the phenotypic value of a trait for all the individuals in a population and dividing this sum by the total number of values in the population. If the letter X represents the phenotypic value for each individual, the Greek symbol Σ the sum of all values when added, and n the number of values in the population, the mean of that population may be described by the formula

$$\text{Mean} = \frac{\Sigma X}{n}$$

[1]The authors acknowledge with appreciation the contributions to this chapter of Dr. David S. Buchanan, Professor of Animal Science, Oklahoma State University, Stillwater.

Within a population, quantitative traits vary widely between two extremes, a low value and a high value. This is called the *range*. A more useful statistical description of variation is the *variance,* which is an average of the squared deviations from the mean. The variance may be determined, as shown in Table 9.1, by calculating the phenotypic mean of the population and then subtracting each phenotypic value from the mean. Each deviation from the mean is then squared to remove the negative signs and to give more weight to extreme values. The sum of the total squared deviations is then divided by $n - 1$, the value n representing the total number of phenotypic values in the population. The answer thus derived is the variance.

The variance may be determined by a short-cut method, using a calculator and the formula

$$\text{Variance} = \frac{\Sigma X^2 - \Sigma (X)^2/n}{n - 1}$$

where X is each phenotypic value in the population, Σ the Greek symbol that means to sum all X or X^2 items, and n the total number of items. By using special and appropriate statistical techniques, the variance can be separated into several portions, giving an estimate of the proportion of various factors causing phenotypic variation in a trait for that particular population.

Sometimes the phenotypes of a population for a quantitative trait are described by using the mean plus or minus (±) the *standard deviation* (abbreviated SD). The standard deviation is the square root of the variance. The mean ± 1 standard deviation should include approximately 68 percent of the phenotypic values for the trait in question in the described population. The mean ± 2 or 3 standard deviations should include 95 or 99 percent, respectively, of all phenotypic values in any respective population.

9.4 FREQUENCY OF GENES IN A POPULATION

Genetic differences among populations, such as families, races, breeds, and even species, are largely due to differences in gene frequencies. The *frequency* of a gene refers to *how rare or abundant a particular gene is in a population as compared with*

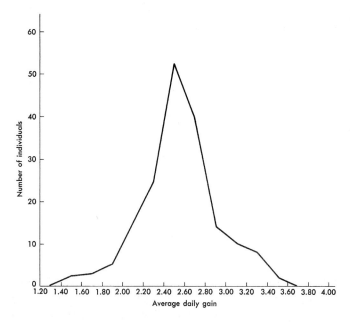

Figure 9.1 Frequency distribution curve of the average daily gain of 176 beef bulls **full-fed** for 140 days. Rate of gain is a quantitative trait in beef cattle in which there is a continuous variation in the phenotype from one extreme to another, with no sharp distinction between phenotypes.

TABLE 9.1	Calculation of the Variance and the Standard Deviation for Average Daily Gains of 10 Bulls on a Performance Test			
No. of Bull	**Average Daily Gain (ADG)**	**Mean for All Bulls**	**Deviation from Mean**	**Square Deviation from Mean**
1	3.20	2.61	0.59	0.348
2	2.70	2.61	0.09	0.008
3	1.80	2.61	–0.81	0.656
4	2.40	2.61	–0.21	0.044
5	2.60	2.61	–0.01	0.000
6	2.90	2.61	0.29	0.084
7	3.00	2.61	0.39	0.152
8	2.20	2.61	–0.41	0.168
9	2.50	2.61	–0.11	0.012
10	2.80	2.61	0.19	0.036
Total (sum)	26.10		0.00	1.508

Mean for all bulls = $\dfrac{26.10}{10}$ = 261

Variance = $\dfrac{1.508}{n - 1} = \dfrac{1.508}{9}$ = 0.1675

Standard deviation = $\sqrt{\text{variance}} = \sqrt{0.1675}$ = 0.4092

its own **allele** *(or alleles).* In most examples used to illustrate genetic crosses, the frequency of the two alleles in each generation is kept equal. For example, the cross of a red *RR* Shorthorn **bull** with white *WW* Shorthorn **cows** should give the following ratios:

P$_1$	*RR* red	×	*WW* white
F$_1$		All *RW* roan	
F$_2$		1*RR* + 2*RW* + 1*WW*	

If there were 20 roan individuals *RW* in the **F$_1$ generation,** there would be a total of 40 red *R* and white *W* genes, because genes occur in pairs in body cells. Of these 40 genes, 20 would be red *R* and 20 would be white *W*. Thus the frequency of the red gene *R* would be 20/40, or 0.5, and the frequency of the white gene *W* would also be 20/40, or 0.5. Gene frequencies are expressed as fractions of 1.

In the **F$_2$ generation,** there is a 1:2:1 ratio when the parents are both *RW*. Therefore, a total of eight red *R* and white *W* genes can be assumed. Of the eight genes, 4/8 would be red *R* with a frequency of 0.5 and 4/8 would be white *W* with the same frequency. Thus the frequencies of the red and the white genes are equal in the F$_2$ generation. Identical gene frequencies between parents and offspring are always expected, even if the gene frequencies are not 0.5 *R* and 0.5 *W*.

In populations of animals, the mating of particular **genotypes** is not controlled, because individuals of any one genotype could mate, by chance, with individuals of any one of the other genotypes. As a result, the frequencies of the two alleles in the offspring of a single mating may not be equal.

To illustrate the calculation of gene frequencies, it will be assumed that in a herd of 100 Shorthorns, 49 are red *RR,* 42 are **roan** *RW,* and 9 are white *WW*. This is not a 1:2:1 ratio as in the previous example (F$_2$). The reason the genotypic ratio in this example herd is not a 1:2:1 ratio is that there are many more red *R* than white *W* genes in this population. The frequency of the red and white genes in this **herd** of 100 Shorthorns can be calculated as follows:

Number of Animals of Different Colors	Genotype	Total No. of Genes in Herd	
		Red	White
49 red	*RR*	98	0
42 roan	*RW*	42	42
9 white	*WW*	0	18
Total number of genes		140	60

Thus the frequency of the red gene *R* would be 140/200, or 0.7; the frequency of the white gene *W* would be 60/200, or 0.3.

In populations of animals the frequencies of two alleles are seldom equal, and in some populations one gene may be present at a much higher frequency than its contrasting allele. Obvi-

ously, if all individuals in a certain population are **homozygous** for a particular gene, the frequency of that gene is 1 and the frequency of the contrasting allele is 0. Genetic variation at a particular **locus** is present when the frequencies of the alleles are less than 1 and more than 0.

The examples previously used to explain the concept of gene frequencies included traits, such as coat color in Shorthorns, in which **dominance** was not complete and genotypes could be determined by observing phenotypes. The frequency of genes in a population can also be estimated for traits in which dominance is complete and phenotypes of the homozygous dominant and the **heterozygous** genotypes are the same. However, there are two prerequisites for calculating the frequency of alleles when dominance is complete: (1) matings must be at **random** (no selection) for at least one generation, and (2) the accurate proportion of the homozygous recessive individuals born within the population concerned must be determined. With this information available, the frequency of the **recessive gene** can be calculated by merely extracting the square root of the proportion of homozygous recessive individuals in a particular generation of **progeny.** The frequency of the dominant allele may be calculated by subtracting the frequency of the **recessive** allele from 1.

To illustrate, assume that in a large herd of black Angus cattle, 1 out of each 100 calves born was red. The black gene *B* is dominant in Angus, and the red gene *b* is recessive. Thus the frequency of the red gene *b* can be calculated by taking the square root of the frequency of the red calves born in the herd. This would be the square root of 1/100 (1/10, or 0.1). The frequency of the black gene *B* would be 1 – 0.1, or 0.9. Knowing the frequency of both the black *B* and the red *b* genes, the proportions of black individuals that are homozygous and heterozygous in the calves can be calculated. The proportion of homozygous black individuals would be the frequency of the black gene (0.9) times the frequency of the black gene (0.9), or 0.81. The proportion of heterozygous individuals in this population would be twice the frequency of the red gene (0.1) times the frequency of the black gene (0.9), or 0.18 (2 × 0.1 × 0.9 = 0.18).

The above calculations can be made by applying the Hardy–Weinberg law. This law describes the situation that occurs in large populations, where the frequency of one of two alleles is equal to *p* and the frequency of the other is equal to *q, and* the sum of the frequencies *p* + *q* equals 1. If matings are at random, the offspring of the three genotypes will occur in a definite ratio, or will be in equilibrium in the next generation at the frequencies of p^2, 2*pq*, and q^2. Under these conditions, we can calculate the frequencies of many different pairs of allelic genes in large populations.

As stated previously, two alleles may not occur in equal frequencies in populations. Four factors are known to be responsible for these differences: (1) mixture of populations (migration), (2) **mutations,** (3) **selection,** and (4) genetic drift. These are commonly called "the four forces that change gene frequency." If genetic improvement is to be made in a population (herd, breed, etc.) these four forces must be examined.

9.4.1 Mixture of Populations (Migration)

To illustrate the meaning of this factor, assume there is a herd of 20 white Shorthorn cows. The frequency of the white gene W in this group of cows would be 1, and the frequency of the red gene R would be 0. If a red bull RR were mated to these white cows and produced 20 calves, all these calves would be roan RW, and the frequency of the white gene W in the calves would be 0.5 as compared with 1 in their dams.

The same change in frequency of a particular gene would occur if one population in which the frequency of some particular gene was high was mixed and mated with another population in which the frequency of that same gene was low. The frequency of the gene in the offspring of the mixed population would be somewhere between the frequencies of the gene in the two parent populations, but it is not likely to be the same as that of either parent population. Migration can be an extremely powerful force for causing genetic improvement. Purchasing new breeding stock for a herd and importing new breeds into a country are examples of migration. If a large number of animals which are markedly different from the original population are imported migration can have a dramatic effect. This can be accomplished by actions as simple as purchasing a new herd sire.

9.4.2 Mutations

A mutation, as defined in Chapter 8, is a chemical change in a gene. Genes and their alleles mutate at different rates. Some seldom mutate, whereas others may mutate often. The difference between alleles in their mutation rate may cause a change in gene frequencies in a population over a long period of time. For example, if in a large population the frequency of gene A were 1, the frequency of its allele a would be 0. However, gene A may mutate to a twice as rapidly as gene a mutates to A. Therefore, in this population, there would be a tendency over an extended period of time for the frequency of gene A to decrease and that of gene a to increase, until there were twice as many a genes as there were A genes in that population. When this occurred, the total number of new mutations of A and a would be approximately equal. The frequencies of these two alleles would remain in equilibrium in the population and would not change unless there were forces other than mutation rate involved.

Examination of mutation rates as described in the previous paragraph reveals the fact that recurrent mutations occur at such low rates that changes in gene frequency due to recurrent mutations would rarely have an observable effect on a herd or a breed. This might lead to the conclusion that mutations are not important considerations in livestock improvement. History reveals differently. Characteristics like dwarfism in cattle, stress syndrome in swine and spider syndrome in sheep are ample evidence that mutations can have a dramatic effect. Dwarfism in Angus and Herefords was a substantial problem in the late 1940s and 1950s. However, there is evidence that the mutations that led to dwarfism occurred more than 50 years earlier. The mutated gene remained in the population at very low levels for many years until selection (or chance) led to an increase in the frequency of the mutated gene, which resulted in it being expressed phenotypically.

9.4.3 Selection

Selection can be a potent force in changing the frequency of a gene, or genes, in a population. This is illustrated by the example

| F_1 | Aa | $\times$ | Aa |
| F_2 | | $1AA + 2Aa + 1aa$ | |

The frequency of the recessive allele a in the F_2 is 0.5. If the homozygous recessive individuals aa were culled (selected against), the frequency of the recessive allele a would be 0.33 instead of 0.5. Conversely, the frequency of the dominant allele A would have been increased to 0.67 by culling the aa individuals. Thus the main genetic effect of selection, if selection is effective, is to change gene frequencies. Selection, like migration, can be used by a manager to cause a directed change in gene frequency in a population. Although its effect may not be as dramatic as the effect of migration, controlled genetic change through selection can improve a population in a predictable, balanced manner.

9.4.4 Genetic Drift

Genetic drift is attributed to the sampling nature of inheritance and means that the frequency of a particular gene in a small population may be quite different than in the larger population from which the small population was derived. Suppose, for example, the frequency of genes A and a is 0.5 in a large population. If a small group of individuals leaves this larger population (migrates) and settles in another area where they become isolated and interbreed among themselves, the frequencies of alleles A and a in this small group could be 0.7 and 0.3, or any other frequencies, depending on chance. Therefore the frequencies of genes A and a in the new isolated population could be quite different than in the original population from which they came.

Genetic drift may occur in small herds or flocks but is unlikely to have much effect in large populations. A zoo is a place in which genetic drift may play a role because zoo populations are almost always small.

9.5 CAUSES OF PHENOTYPIC VARIATION

Basically, there are three general causes of phenotypic variation in quantitative traits: (1) **heredity,** (2) **environment,** and (3) the joint action of heredity and environment.

Hereditary differences among individuals are those due to genotype, or genetic makeup, of individuals. The genotype of an individual is fixed at conception, when the **spermatozoon** and **egg** unite in the process of **fertilization.** An individual dies with the same genotype with which it was born (except for possible mutations). The phenotype of the individual, however, is subject to change and does change quite often throughout life.

The genotype of an individual determines what it can and will transmit to its offspring. Genotype also affects phenotype for a particular trait, not only because of the kinds of genes the individual possesses, but also because of the way in which certain genes act in different combinations within the individual.

Environmental differences are those caused by anything other than heredity. These include differences due to disease, *nutrition,* management, accidents, and other factors. Superiority or inferiority due to environment will not be transmitted from parents to their **offspring.** Therefore it becomes important to the animal breeder to determine whether heredity or environment or both cause an animal to be superior.

For some traits, the interaction between heredity and environment may be an important source of phenotypic variation. What this interaction really means is that two genotypes may perform quite similarly in one environment but quite differently in another. For example, in the semiarid regions of the southwestern United States, **steers** from the British breeds and steers from Brahman × British crosses will gain at about the same rate when fed at the same place and on the same ration in winter months when it is cool. In hot summer months, however, the Brahman **crossbred** steers will usually gain faster than the **purebred** British steers because they have the genetic ability to withstand hot dry weather. In the above example, **breeds** were the genotypes and season was the environment. Another example of a genotype by environment interaction would be evidenced when sires rank differently on the basis of offspring raised in different parts of the country. Some sires may produce outstanding offspring in the Midwest but only mediocre offspring on the Gulf Coast. The reverse may be true for other sires.

9.5.1 Heritability Estimates

Heritability estimates indicate the proportion of total phenotypic variation that is attributable to heredity. These may be recorded in percentages. A heritability estimate of 40 percent means that 40 percent of the total phenotypic variation is due to heredity. If the heritability estimate is subtracted from 100 percent, the difference is an estimate of the total phenotypic variation that is attributable to environment. Heritability estimates for various classes of livestock are shown in Table 9.2.

The lower the heritability for a trait, the slower the progress one can expect in improving the trait by choosing superior parents. Also, the lower the heritability, the larger the proportion of total phenotypic variation that is due to environment. Traits with low heritability could be improved in the offspring by crossing different families, inbred lines, or breeds.

9.6 SELECTION

Selection may be defined as *causing or permitting certain individuals within a population to produce the next generation.* The selection that the animal breeder practices is usually referred to as *artificial* selection; that which nature does through the survival of the fittest is known as *natural* selection. Both artificial and natural selection are at work in herds of livestock. Artificial selection by animal breeders is an attempt to increase the proportion of desirable genes in a population for a particular purpose. However, natural selection is still in effect. Artificial selection that attempts to negate the effects of natural selection is probably destined to be ineffective.

9.7 SELECTION FOR DIFFERENT KINDS OF GENE ACTION

Quantitative characters involve many different genes, which may express themselves in many ways in the phenotype. The phenotypic expression of genes may be divided into two general types, known as *nonadditive* and *additive.* Nonadditive include dominance and recessiveness, **overdominance,** and **epistasis,** as well as any other type of expression in which the phenotypic effects of genes are not of the additive type. The purpose here is to discuss methods of selection for the different kinds of additive and nonadditive gene action.

TABLE 9.2	**Heritability Estimates for Selected Traits in Cattle,* Sheep, Swine, and Poultry**			
	Percent Heritability			
Trait	**Cattle**	**Sheep**	**Swine**	**Poultry**
Fertility	0–15	0–15	0–15	0–15
Ovulation rate	—	—	27–37	
Age at puberty	—	20–30	28–38	
Conception rate	12–22			
Testes size	43–53	—	30–40	
Birth difficulty	10–20	—		
Number born	—	5–15	4–14	
Number of young weaned	10–15	10–15	10–15	
Birth weight	26–36	10–20		
Weight of young at weaning	19–29	20–30	12–22	
Postweaning weight	26–36	35–45	26–36	
Food conversion	27–37	20–30	25–35	
Milk production, lb	26–36			
Milk fat, lb	25–35			
Milk protein, lb	26–36			
Percent lean cuts	40–50	—	43–53	
Fat thickness	39–49	25–35	44–54	
Marbling score	33–43			
Eye muscle area	37–47	45–55	42–52	
Feed efficiency	—	—	—	20–25
Total egg production	—	—	—	20–30
Age at sexual maturity	—	—	—	30–40
Egg shape	—	—	—	35–40
Body weight	—	—	—	35–45
Egg weight	—	—	—	35–45
Eggshell color	—	—	—	40–45
Viability	—	—	—	5–10

*See also Table 15.3.

9.7.1 Selection for Dominant or Recessive Alleles

Selection against a dominant gene requires that all animals showing the dominant phenotype be culled. This assumes **penetrance** of the gene is complete. If penetrance of the gene is incomplete (not 100 percent), it will be impossible to discard all individuals that possess the dominant gene, because its presence will not always be observed in the phenotype.

Selection for a dominant gene means that individuals showing the presence of the gene by their phenotype are kept for breeding purposes. The homozygous recessive individuals are discarded. Effective selection for a dominant gene on the basis of phenotype is difficult, because it is impossible to distinguish between homozygous dominant *DD* and heterozygous dominant *Dd*. Thus some dominant phenotypes are selected that carry the recessive gene. **Progeny tests** must be conducted to find which individuals of the dominant genotype are heterozygous *Dd,* and these must be discarded.

Selection for a recessive gene is relatively simple if penetrance of the gene is complete. One merely keeps individuals showing the recessive trait, and then these are mated. Recessives mated to recessives should produce all homozygous recessive offspring, although mutations and rare interactions with other genes may occasionally affect the phenotype of recessive individuals.

Selection against a recessive gene requires that all individuals showing the recessive trait be discarded. Heterozygous individuals *Dd* that are of the dominant phenotype but are carriers of the recessive gene must be detected by means of progeny tests, and they must also be discarded. Thus selection for a dominant gene and selection against a recessive gene present the same problem; that is, the distinction between homozygous dominant *DD* and heterozygous dominant *Dd* phenotypes must be considered in both.

Several kinds of matings can be made to determine if an individual is of the dominant phenotype but a carrier of the recessive gene. The kind and number of matings to prove an individual homozygous dominant *DD* at the 95 and 99 percent levels of probability are given in Table 9.3. Only one recessive offspring from any of these matings proves the individual being tested to be heterozygous dominant. Conversely, one cannot be positive that an individual tested is homozygous dominant, although the degree of confidence increases as the number of offspring from matings with the tester animals produces offspring only of the dominant phenotype.

9.7.2 Selection When There Is Overdominance

Overdominance is a type of gene action in which heterozygous individuals are superior in merit to homozygous ones. Usually, the superiority is in vigor or performance. Many pairs of genes may be involved for an economic trait, but selection for this type of phenotypic expression of genes will be illustrated here with only two pairs. Assume there are individuals of the following genotypes and that heterozygous individuals are superior:

Individual Number	Genotype
1	$A^1A^1B^2B^2$
2	$A^1A^1B^1B^2$
3	$A^1A^2B^1B^2$
4	$A^2A^2B^2B^2$
5	$A^2A^2B^1B^2$
6	$A^2A^2B^1B^1$

Because heterozygous individuals are superior when overdominance is involved, individual 3 should be selected because it is heterozygous for two pairs of genes. When individuals of this genotype are mated, some offspring will be of quality equal to that of the parent, but many will be inferior. The average performance of offspring from several such matings would be considerably below that of heterozygous parents. Thus selection of superior phenotypic individuals would result in selecting those most heterozygous, and they would not breed true. The average of their offspring would tend to regress to a point below the average of the heterozygous parents. From this example, it can be seen that selection on the basis of an individual's merit would be disappointing.

Although heterozygous individuals will not breed true when mated together, the mating of the right genotypes should always give offspring that are heterozygous. For example:

Parent 1	×	Parent 2
$A^1A^1B^1B^1$		$A^2A^2B^2B^2$
All offspring would be		
$A^1A^2B^1B^2$		

TABLE 9.3	**Number and Kinds of Matings Required to Prove an Individual Homozygous Dominant *DD* at the 95 and 99 Percent Levels of Probability**

Animal of Dominant Phenotype but Genotype Unknown Can be Mated with	Number of Matings Producing No Recessive Offspring	
	95% Level of Probability*	99% Level of Probability*
Homozygous recessive individuals	5	7
Known heterozygotes	11	16
Unselected different daughters of a known heterozygous sire	23	35
A sire mated to his own unselected daughters (all different daughters)†	23	35

*Probability level at which the individual is proved homozygous dominant *DD*.
†Tests for any recessive gene the sire might be carrying and not only for a specific recessive gene.

Double-heterozygous offspring would also be produced if each of the parents were homozygous for the opposite alleles $(A^1A^1B^2B^2 \times A^2A^2B^1B^1)$. Once heterozygous offspring are produced, however, one must return to the mating of homozygous parents (homozygous for different alleles) to produce all $A^1A^2B^1B^2$ offspring. Selection for overdominance, then, is selection among different lines, families, or breeds to find those that combine the best to give the most heterozygous offspring. This is done in the production of hybrid corn. Inbred lines of corn are formed by inbreeding to make all individuals within the lines homozygous for all genes. The many lines formed by inbreeding are then crossed in various combinations to find those that combine best to produce the most vigorous (heterozygous) offspring. Those that combine the best are probably homozygous in opposite ways. For example, line 1 might be $A^1A^1B^1B^1C^1C^1$, whereas line 2 might be $A^2A^2B^2B^2C^2C^2$. Crossing these two lines would produce offspring heterozygous for the three pairs of genes. In actual practice the **linecross** offspring may be made heterozygous for many pairs of genes, but there is no way of knowing for which pairs or how many.

9.7.3 Selection for Epistasis

Epistasis is a type of gene action in which one pair of genes may affect the phenotypic expression of another pair of genes that are not their alleles. These genes may vary widely in the way they interact with each other. In spite of this variation, the method of selection would be similar for all types of epistasis. To illustrate selection for epistasis, let it be assumed that two pairs of alleles *(Aa and Bb)* are involved and that a combination of genes A and B in the genotype gives the desirable phenotypic effect. The following gene combinations of these two pairs of alleles would be possible:

Individual Number	Genotype
1	AABB
2	AABb
3	AAbb
4	aaBB
5	aaBb
6	aabb
7	AaBB
8	AaBb
9	Aabb

The desired combination of A with B would be found in genotypes 1, 2, 7, and 8, and selection for superior individuals would result in retention of individuals of these genotypes for breeding purposes. Obviously, the goal would be to save only those individuals that are *AABB,* because, when mated together, they would produce only *AABB* offspring, or would breed true. Selection on the basis of the individual's phenotype, however, would probably cause us to keep genotypes 1, 2, 7, and 8 in equal numbers, but all except genotype 1 would fail to breed

true when mated together. Thus it is possible to establish a pure line that would breed true for a certain **epistatic** effect, but this is not highly probable in practice. It must be kept in mind that many pairs of alleles (more than two) may affect various economic traits in farm animals. This leads to the conclusion that selection for superior merit of the individual (selection on the basis of individuality) is not very effective when epistasis is involved.

Probably the most effective way to select for an epistatic effect would be to form **families** (inbred lines or breeds) and test them to find those that combine the best in crosses to give superior offspring. Those combining the most satisfactorily would be retained and crossed to produce superior progeny. The original lines (families or breeds) would be kept pure and crossed again and again to produce superior crossbred or linecross offspring.

Selection among families or lines is therefore probably the best method of selection when overdominance and/or epistasis is important.

9.7.4 Selection When There Is Additive Gene Action

Additive gene action is a type of phenotypic expression of genes in which the phenotypic effect of one gene adds to the phenotypic effect of another.

Selection for additive gene action can be on the basis of individual performance, although attention to the merit of the individual's close relatives should help make selection more accurate. To illustrate why additive gene action is selected for on the basis of individuality, the following examples will be used. Assume that genes A and B add two points to the superiority for a trait when the residual genotype *aabb* is given a value of 20 points. The value of each genotype would be

Genotype	Value of Genotype, Points
aabb	20
aabB	22
aaBB	24
aABB	26
AABB	28

If these individuals were selected in a comparable environment so that phenotypic variations due to environment were minimized, the individuals most likely to be selected would be those of genotype *AABB,* with a phenotypic score of 28 points. If such individuals were selected and mated, they should produce superior offspring. It is true that to a certain extent superiority or inferiority due to environment would be confused with that due to heredity, but selecting and mating the best to the best for several generations (mass selection) should improve the overall merit of individuals in the population.

Superiority for a trait when additive gene action is important, therefore, is dependent on the kind and number of desirable genes (positive additive genes) an individual possesses. It does not depend on a certain combination of alleles or nonallelic genes.

TABLE 9.4	How to Determine if an Economic Trait Is Affected Largely by Additive or Nonadditive Gene Action or Both

	Type of Gene Action Indicated		
	Additive	**Nonadditive**	**Both**
Heritability estimate	High	Low	Medium
Crossbreeding effect	None	Large	Low–medium
Inbreeding effect	None	Large	Low–medium

9.7.5 How to Determine if Additive or Nonadditive Gene Action Affects a Quantitative Trait

Qualitative traits are affected by relatively few genes, and there is a sharp distinction between phenotypes. Thus it is not too difficult to determine the type of phenotypic expression of genes involved when appropriate progeny-testing procedures are applied. Quantitative traits, however, may be expressed through the action of many pairs of genes, and there is no sharp distinction between phenotypes. Moreover, environment may also significantly influence the phenotypic expression of quantitative traits. This raises the important question of how to determine whether additive or nonadditive gene action has the greater influence on a particular economic trait. Information summarized in Table 9.4 shows, in a general way, how this can be accomplished.

Heritability estimates for various traits are summarized in Table 9.2; **heterosis** effects (crossing) for various traits are summarized in Table 9.8. **Inbreeding** effects are usually the opposite of **crossing** effects. Thus, if crossing improves a trait, inbreeding should have a detrimental effect.

Selection in farm animals is usually concerned with several traits. Some traits may be largely affected by *additive* gene action, others by *nonadditive* gene action. Some traits may be affected by both, but in varying degrees. How can a selection program be designed if this is true? A manager must determine which traits have an effect on the economic efficiency of the herd or flock and attempt to improve those traits that are most important. Although all important traits should receive some selection emphasis, producers can also take advantage of **crossbreeding** to enhance performance in lowly heritable traits.

9.8 SELECTION OF SUPERIOR BREEDING STOCK

Superior breeding animals are those that possess a large proportion of superior genes for a desirable trait, or traits, in a homozygous state. The identification of superior breeding stock may be based on the phenotype of the individual and/or on that of its close relatives, especially if accurate records are available.

9.8.1 Individuality

Attention should be given to the type and performance of the individual when possible. An individual's own merit is most important when the trait or traits sought are medium to highly heritable. Such traits are determined largely by additive gene action, and individuals that are superior in a group tested are more likely to be superior because of their genetic makeup. The selection procedure in such a case becomes one of finding the best potential parents followed by mating in an optimum manner.

Traits of low heritability are those largely affected by environment and nonadditive gene action. Superior individuals for such traits may be disappointing in their performance as parents. As mentioned previously, traits of low heritability require the formation of families, inbred lines, or breeds. These must be tested in crosses to find those that combine best when crossed to produce superior linecross or crossbred progeny.

The main error encountered when selection is based on individuality is that effects due to heredity and environment are often confused. Confusion of environmental and genetic effects can be avoided by comparing the individuals being considered for breeding purposes at the same location, during the same time, and when they are fed and managed alike. These precautions will not eliminate all variation due to environment, but they will reduce it.

When many individuals are compared under standardized environmental conditions, those that are superior in performance are more likely to be superior because of their genetic makeup. Therefore they should prove to be superior parents. The principle involved is that when the proportion of the total phenotypic variation attributable to environment is reduced, more of the remaining variation is due to heredity. This may be illustrated by the example

$$\text{Percent of variation due to heredity} = \frac{\text{variation due to heredity}}{\text{variation due to environment} + \text{variation due to heredity}} \times 100$$

If the variation due to heredity is 6 and that due to environment is 6, obviously the proportion of variation due to heredity is

$$\frac{6}{6+6} \times 100, \text{ or } 50 \text{ percent}$$

If the amount of variation due to environment is reduced to 4 by appropriate methods, the proportion of total variation due to heredity is

$$\frac{6}{4+6} \times 100, \text{ or } 60 \text{ percent}$$

Rate and efficiency of gains in most livestock are moderately to highly heritable and therefore are affected mostly by additive genes. Performance testing of a group of individuals includes determining rate and efficiency of gains from a period of time shortly after weaning until they near market weight. To obtain efficiency of gain records on an individual, it is necessary to feed each animal and to record the amounts of feed consumed and its gain. From these records, the amount of feed required to make a pound of gain can be calculated. Because of the time and expense required to feed each animal individually, feed-efficiency records are seldom determined in a performance test. One can easily determine rate of gain for each animal within a group by merely obtaining the weights of each animal at the time it goes on full-feed and then again when the full-feeding period is completed. Pounds of gain made during the feeding period divided by the number of days fed gives average daily gain (**ADG**).

Performance testing as described above is done mostly with beef cattle and swine and to a lesser extent with sheep.

TABLE 9.5	**Probable Percentage of Genes (Relationship) That an Individual Has in Common with His/Her Collateral Relatives* above the Average of the Population**

Kind of Relative	Percent Relationship
Identical twins	100.00
Fraternal twins[†]	50.00
Full brothers or sisters (*full siblings*)	50.00
Half brothers or sisters (*half siblings*)	25.00
First cousins	12.50
Double first cousins	25.00
Second cousins	3.13
Third cousins	0.78
Aunts or uncles	25.00

*Collateral relatives are those not related as ancestors or descendants. In these examples it is assumed that none of the relatives are inbred.

[†]It is possible for fraternal twins to be from two different sires. They would be related by 25.00 percent.

TABLE 9.6	**Probable Percentage of Genes (Relationship) That an Individual Has in Common with His/Her Ancestors above the Average of the Population**

	Percent Relationship	
Kind of Ancestor	Not Corrected for Inbreeding	Corrected for Inbreeding
No inbreeding involved		
Parent	50.00	
Grandparent	25.00	
Great-grandparent	12.50	
Great-great-grandparent	6.25	
Inbreeding involved		
Both a parent and a grandparent*	75.00	67.08
Double grandparent	50.00	47.14
Double great-grandparent[†]	25.00	24.62
Triple great-grandparent	37.50	36.38
Quadruple great-grandparent	50.00	47.14

*Individual resulting from the mating of a sire to his daughter or a dam to her son.

[†]The same individual is a grandparent of both the sire and the dam. If a double great-grandparent is in only the sire's or dam's pedigree, the relationship need not be corrected for inbreeding and would be 25.00 percent.

9.8.2 Records of Relatives

The performance of close relatives of an individual also helps determine if an individual is genetically superior. The closer the relationship, the more attention should be given to the merits of those relatives. The degree of relationship among various individuals is shown in Tables 9.5 and 9.6.

An individual has three kinds of relatives: (1) ancestors in its **pedigree;** (2) its descendants, or progeny; and (3) its collateral relatives, which are those not related to it as ancestors or descendants. **Collateral relatives** include brothers, sisters, cousins, uncles, and aunts.

Selection for traits that can be measured only after the death of the individual, or in only one sex, often requires that criteria be based on the performance of relatives rather than the individual. The carcass quality of some farm animals is difficult to measure accurately in the live animal. Use of the backfat probe and ultrasound, however, has helped increase the accuracy of selection for carcass quality in the live animal (Figure 9.2). Selection for carcass quality in swine can also be based on the carcass quality of **littermates** after slaughter. Littermates

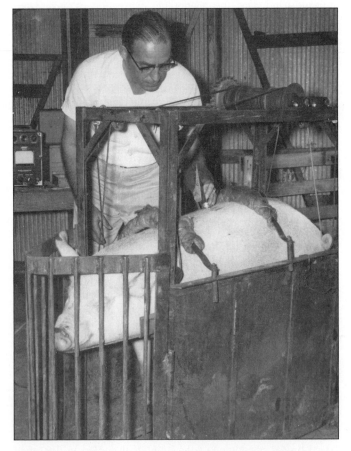

Figure 9.2 Measuring fat-cover and loin-eye area in the live hog by means of the sonoray, an ultrasonic instrument. The instrument sends out sound waves that go through tissue without harming the animal. The time needed for the sound waves to pass through and bounce back from different tissues varies with tissue thickness. The measurement is quite accurate.

Courtesy of Missouri Agr. Expt. Station.

TABLE 9.7	Selection Differentials among Boars and Gilts in a Selection Experiment for Backfat Thinness			
Generation of Selection	No. of Boars Selected per Generation	Selection Differential for Boars, mm	No. of Gilts Selected per Generation	Selection Differential for Gilts, mm
P_1	7	2.9	30	2.8
F_1	7	3.3	20	2.0
F_2	6	2.7	26	0.8
F_3	7	2.8	28	1.4
F_4	6	2.2	25	2.6
Total and average	33	2.8	129	1.9

Source: Missouri Agr. Expt. Station.

that are full siblings (brothers and sisters) are 50 percent related. When single births predominate, as with beef cattle, usually the closest relatives that can be slaughtered are half siblings. Their relationship is 25 percent (Table 9.5).

Selection among dairy cattle for improvements in milk production requires intense selection among bulls. This necessitates measurement of performance of relatives because milk production is a sex-limited (measured in only one gender) trait.

Much more selection pressure can be applied to bulls because, through the use of artificial insemination, one bull can be mated with thousands of cows annually. Conversely, a cow commonly produces only 5 to 10 calves during her lifetime (and many cows produce fewer than 5).[2] Moreover, most females must be raised and relatively few culled if herd size is to be maintained. This situation results in far less intensity of selection among dairy females than among males. However, selection pressure for milk production should be applied through rigid culling of low producers when possible.

Notwithstanding the fact that bulls do not secrete milk, they do possess and transmit genes for milk production. A bull's genetic worth must be determined from the production records of closely related females, especially his daughters.[3]

9.9 PREDICTING THE AMOUNT OF PROGRESS POSSIBLE THROUGH SELECTION

The preceding discussion dealt with how to select for quantitative and qualitative traits. In this section factors useful in predicting selection progress will be discussed.

9.9.1 For One Generation

The amount of progress made through selection for a particular trait in one generation is equal to the heritability of the trait times the selection differential. The formula for this calculation is

Genetic progress = heritability × selection differential

Heritability estimates for various traits are usually obtained from experimental reports (Table 9.2). Selection differentials, however, must be calculated using data from animals where selection is practiced.

The *selection differential* may be defined as *the average superiority of those selected as parents over the average of the population from which they were selected.* The selection differential is sometimes referred to as the *reach.* To illustrate the calculation of expected genetic progress through selection, let us assume that in a herd of cattle the average weaning weight of all calves adjusted to a "bull-calf basis" was 400 lb. From these calves, bulls that weighed 500 lb were selected for breeding, and the selected heifers weighed 450 lb. The selection differential for the bull calves would be 500 − 400, or 100 lb. For **heifers,** the selection differential would be 450 − 400, or 50 lb. The selection differential for both parents combined would be 100 lb (bulls) + 50 lb (heifers) ÷ 2, or 75 lb. If the heritability of weaning weight is 25 percent, the expected genetic improvement (genetic progress) in the offspring of the selected parents would be 25 percent of 75 lb, or 18.75 lb.

The magnitude of the selection differential is dependent on the number of individuals that can be culled. Table 9.7 presents selection differentials for boars and gilts in a selection experiment for backfat thinness in swine at the Missouri Agricultural Experiment Station. The selection differential for **boars** was 2.8 mm, compared with 1.9 mm for gilts. Thus the selection differential for boars was larger because fewer of them were kept for breeding purposes.

The more traits selected for, the smaller the selection differential for any individual trait. This is because it is more difficult to find an individual that is excellent for two or more traits than it is to find an individual that is excellent for only one. Selection among farm animals should focus on traits of economic importance. If several traits are economically important, each should receive attention in the selection program.

9.9.2 For Several Years

Genetic progress in selection over a period of years depends on (1) the degree of heritability of the trait selected, (2) the size of the selection differential, and (3) the length of the generation interval. This may be expressed in the equation

[2]The technology associated with superovulation and embryo transfer will enable the animal breeder to raise many more offspring from cows (see Chapter 13).
[3]Even when traits are highly heritable, as are rate and efficiency of gain in beef cattle, selection on the basis of individuality may be strengthened by selecting individuals with good-performing relatives. Therefore, when records on relatives are available, they are useful in all methods of selection.

$$\text{Annual genetic progress} = \frac{\text{selection differential} \times \text{heritability for trait}}{\text{length of generation interval, years}}$$

Generation interval may be defined as *the average age of parents when their offspring are born.* It averages about 2 to 3 years in swine, 4 to 6 years in cattle, 30 to 35 years in humans. For some insects and small laboratory animals, the generation interval may be only a few days or a few weeks in length. Because sows and boars may be sold after the first litter is produced, the generation interval for swine could be as short as 1 year. Progeny testing and the use of its results as a basis for selection may double, or even more than double, the length of the generation interval. This decreases the amount of progress made by selection in a period of several years.

9.10 GENETIC CORRELATIONS

Genetic correlations may be determined mathematically, or they may be determined by selection for only one trait. If success is achieved in improving the single trait selected for, possible changes can also be observed in unselected traits. A genetic correlation means that two or more traits are affected by many of the same genes. Genetic correlations between two traits may influence the amount of progress made in selection for one of them.

Because most economic traits in farm animals are influenced by many genes, it is reasonable to expect that some of the same genes may affect more than one trait. This is probably because the same genes affect a certain physiological process that determines to a certain extent the expression of two or more traits. Many instances are known in which a single major gene may cause the appearance of two different traits in animals. This is known as **pleiotropy.**

Genetic correlations may be either positive or negative. Traits may be positively correlated in such a way that improvement is made in both traits, regardless of which one receives the selection emphasis. For example, selection for increased yearling weight in cattle will also tend to increase weaning weight. However, positive correlations are not always beneficial. For example, that same selection for yearling weight will also tend to increase birth weight and, as a result, calving difficulty. Negative correlations may also be either desirable or undesirable. An example of a favorable negative genetic correlation exists between average daily gain and days to market weight in swine. Selection for high average daily gain will decrease the length of time required for pigs to reach market weight. Conversely, selection for increased milk production in dairy cattle tends to decrease fat and protein percentage, which is an example of an unfavorable negative genetic correlation.

9.11 NATIONAL PERFORMANCE PROGRAMS

Organizations of people with an interest in performance testing in livestock have been developed that promote the use of performance records and make recommendations for performance programs. These organizations include the Beef Improvement Federation, National Swine Improvement Federation, National Sheep Industry Program, and Dairy Herd Improvement Association. These groups have been invaluable in popularizing the use of performance records and, by extension, the improvement of efficiency of livestock production.

9.11.1 Dairy Cattle

Among large food animal industries, the dairy industry has made effective use of performance records for the longest period of time. The Dairy Herd Improvement Association, in cooperation with the United States Department of Agriculture, has effectively used performance records to make amazing improvements in milk production for more than 50 years. A standardized length of lactation (305 days) was adopted for measurement of milk production and adjustment factors were developed so that known sources of environmental variation could receive proper correction. With proper use of adjustment factors, milk production of cows could be compared fairly across different ages of cows and months of the year. Routine methods for obtaining milk production records and for obtaining information about components of milk production (protein, fat, solids–not–fat) were developed so that an extremely large proportion of dairy producers obtain, and use, performance information for making decisions related to selection and mating.

9.11.2. Beef Cattle

The Beef Improvement Federation was initiated in the early 1970s as an organization of scientists, extension specialists, test station managers, and producers with an interest in performance testing. It meets annually with a symposium format to advance ideas related to performance testing and, periodically, publishes a document titled "Guidelines for Uniform Beef Improvement Programs." These guidelines contain the latest information about recommended performance programs. Adjustment factors, to account for age of dam, are presented for birth and weaning weights. Equations for calculating adjusted 205 day weaning weight and adjusted yearling (365 day) or long-yearling (452 or 550 day basis) weights are given. There are also recommendations for appropriate assessment of frame score (hip height), carcass merit, and reproductive traits. Appreciation of the importance of performance testing beef cattle has increased immeasurably since the Beef Improvement Federation was formed.

9.11.3 Swine

The National Swine Improvement Federation was formed soon after the Beef Improvement Federation with similar goals and organizational structure. It meets annually and periodically publishes "Guidelines for Uniform Swine Improvement Programs." There are equations and adjustment factors for performance programs that include litter size, litter weight at 21 days (to measure maternal ability), postweaning average daily gain, age at 250 lb and feed efficiency. There are also recommendations to measure backfat thickness and loin eye area adjusted to market weight (250 lb). The guidelines also present selection indexes to combine information from several traits. These indexes include a maternal index that combines litter size, litter weight and age and backfat thickness at 250 lb. A paternal index contains age and backfat thickness at 250 lb. These indexes can be used by swine producers to improve general genetic merit.

9.11.4 Sheep

The sheep industry has been slower to adopt widespread performance testing programs. However, the National Sheep Improvement Program was developed to make recommendations concerning performance testing procedures. Adjustment factors and equations for proper evaluation of weights at various ages, optimum evaluation of reproductive performance, and carcass merit are also discussed.

9.12 NATIONAL GENETIC EVALUATION

A major difficulty in assessing the genetic merit of livestock is the problem of making fair comparisons among animals from different herds. Differences in environment and management can be so large that performance varies widely among herds, even among animals of similar genetic merit. Livestock shows were, in part, used to compare animals among herds but, realistically, differences among animals that have been brought together for a short period of time from different herds are largely due to differences in environment or management. Test stations bring animals together from different herds to be evaluated in a common environment for an extended time period and contribute to the ability to compare animals among herds, although traits are limited to those measured postweaning. In addition, test station capacity is too limited to have a major effect on genetic change in breeds of livestock.

National performance programs were originally developed with the goal of providing information to livestock owners for use in making within-herd selection decisions. However, the databases grew rapidly and managers of the programs realized that the data could also be used for comparing animals among herds. Several breeds have more than 1 million animals with performance information. The advent of artificial insemination helped make this possible. To have fair genetic comparisons among animals from different herds, there must be genetic ties among the herds. The simplest example of genetic ties can be illustrated as follows:

Sire 1 450	Sire 2 470	Sire 3 460
Sire A 460	Sire A 450	Sire A 430
Herd 1	Herd 2	Herd 3

In this example Sire A is referred to as the "Reference Sire." It is used in all three herds as a means of comparison. The other sires are used in only one herd each. They can be compared with each other through within-herd comparisons with the Reference Sire. Sire 1 has calves that are 10 lb lighter than those by the Reference Sire, whereas Sire 2 and Sire 3 have calves that are 20 and 30 lb heavier, respectively. Because Sire 3 has the largest advantage over the Reference Sire, it should be ranked first.

When artificial insemination is used widely, several sires are used in numerous herds. It is unnecessary to have sires used in every herd but examination of dairy, beef, or swine records reveals that there are many sires that are used in a large number of herds. These sires act like the Reference Sire in the above example to provide ties among herds. Additionally, ties are established through relationships between parents used in different herds.

Data from performance programs are analyzed by use of an animal model. In the animal model, a prediction of genetic merit can be made for every animal in the database, including all parents. This prediction of genetic merit is referred to as the Expected Progeny Difference (EPD). In dairy cattle the EPD is called the Predicted Transmitting Ability (PTA). Confidence in the EPD is presented as the Accuracy (*Reliability* for dairy cattle). Information from each animal's own performance, siblings, progeny, and every other type of relative is combined to calculate a single EPD for each trait. The animal model has several advantages over previous analysis methods. Because all relatives are used to calculate the EPD, accuracy is maximized. EPDs can be obtained for any individual, male or female, of any age. Bias from nonrandom mating (some sires mated to better dams) is also eliminated with the use of the animal model.

The most visible result of national genetic evaluation is the Sire Summary. Sire Summaries are available in printed form and on the World Wide Web. Typically, Sire Summaries list EPDs or PTAs for several hundred sires with the most available information (highest accuracies). Dairy Sire Summaries include PTAs for milk production, protein, fat, solids-not-fat and type. Beef Sire Summaries are published by individual breed associations and list EPDs for birth weight, weaning weight, yearling weight, and milk. In addition, some breeds include other growth traits, certain carcass traits, or some characteristics that pertain to reproduction. Swine Sire Summaries are prepared through the STAGES (Swine Testing and Genetic Evaluation System) program and include EPDs for litter size, litter weight, age at market weight, and backfat thickness. The swine Sire Summaries also have EPDs for maternal and paternal indexes. The National Sheep Improvement Program has started presenting sire summary information, as well.

9.13 MATING SYSTEMS FOR LIVESTOCK IMPROVEMENT

The preceding discussion was directed toward selection principles and their use in animal improvement. Once superior animals are identified and selected, it is necessary to devise mating systems that will maximize efficiency of production. The mating systems that may be used are inbreeding, linebreeding, linecrossing, and crossbreeding.

9.13.1 Inbreeding

Inbreeding may be defined as *the production of progeny by parents that are more closely related than the average of the population from which they came.* A base (average of the population) to determine where inbreeding begins is necessary because, even though they are not referred to as relatives, all individuals within the same species have many genes in common. Humans recognize relatives as those individuals that have the same ancestor (or ancestors) in the pedigree of both father and mother. Individuals are typically considered to be relatives if they have a common ancestor within the last four or five generations.

The major genetic effect of inbreeding is to increase the number of pairs of genes that are homozygous in an inbred population. These genes are made more nearly homozygous by inbreeding regardless of whether they express themselves phenotypically as additive or nonadditive. Probably the most practical effect of inbreeding is that it pairs recessive genes, which are usually detrimental in their effect, and because of this pairing, they express themselves in the phenotype.

Inbreeding does *not* create recessive genes. If recessive genes are covered up by dominant genes in the original noninbred population, they will be paired by inbreeding. The more intense the inbreeding, the more pairs of genes will be made homozygous. Likewise, if few or no recessive genes are present in the original population, they will not be paired by inbreeding, and the results are not so detrimental.

Inbreeding has been used in laboratory animals to study numerous parameters among strains (e.g., disease resistance; immune responses; endocrine functions; and response to drugs, cancer, and variations in dietary requirements), as well as to study coat color patterns and body conformations (Figure 9.3).

Increased homozygosity due to inbreeding increases breeding purity. In the following example, individual 1 could transmit only one combination of genes (*ABCD*) to its offspring through the gametes. Individual 2 could also transmit only one combination of genes to its offspring (*abcd*). Individual 1, which is homozygous dominant for four pairs of genes, would cause all offspring to resemble it regardless of the genotypes of the other individuals to whom it is mated. Individual 1 is said to be **prepotent** because it is homozygous dominant for these four pairs of genes. Individual 2 is not prepotent because it transmits only recessive alleles. Thus, if individual 2 is mated to individuals possessing the dominant alleles, the offspring will resemble the other parent rather than individual 2. Individual 3 is heterozygous for four pairs of genes, and it could transmit 16 different combinations of genes to its offspring through its gametes. Therefore it would not breed true.

	Genotype	Genes in Gametes
Individual 1	*AABBCCDD*	*ABCD*
Individual 2	*aabbccdd*	*abcd*
Individual 3	*AaBbCcDd*	16 possible combinations of genes

Increased homozygosity due to inbreeding also tends to fix recessive genes in a small inbred population. These recessive genes may be present at high frequencies before their presence is detected and their genetic nature is understood. For example, an autosomal recessive gene for hemophilia (failure of blood to clot) was discovered in a small inbred line of Poland China swine at the Missouri Agricultural Experimental Station. Many of the homozygous recessive boars bled to death when **castrated,** and some sows bled to death at **parturition** before the

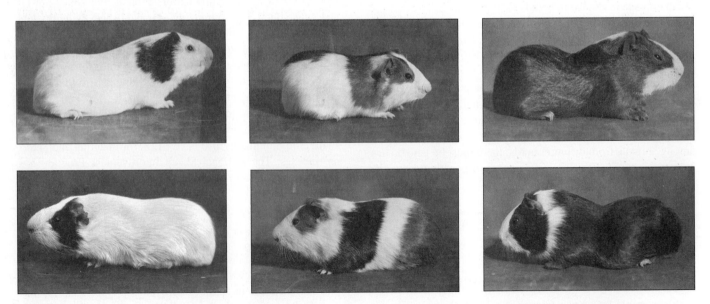

Figure 9.3 Illustration of varying coat color patterns and conformation in closely inbred strains of guinea pigs, males (top) and females (bottom), in generations 18, 12, and 13 (left to right, respectively) of brother–sister matings. Note the relative degree of uniformity of the left, middle, and right individual strain pairs.

Courtesy of Dr. Sewall Wright, Emeritus Professor of Genetics, University of Wisconsin, Madison.

cause of death was determined. So many individuals within this small line were either homozygous recessive or phenotypically normal carriers of the gene that when the genetic cause of the defect was discovered, the line had to be abandoned for breeding purposes. The recessive gene was retained at the Missouri Station, however, mainly for studies of blood-clotting mechanisms.

Increased homozygosity due to inbreeding, when combined with selection, can be used to eliminate detrimental recessive genes from an inbred population. When inbreeding is practiced and recessive defects are paired and uncovered, the incidence of the recessive gene may be greatly reduced by discarding the individuals and their close relatives. However, a large inbred line is necessary before such rigid culling can be practiced, and the breeder must know what to look for as inbreeding progresses.

Increased inbreeding, accompanied by selection for certain performance traits such as **type,** conformation, and rate of gain, is a good method of establishing families within a breed that are quite different in their phenotype. The major phenotypic effects of inbreeding are to uncover recessive defects and to cause a decline in traits related to physical fitness. These traits include vigor at all ages, but especially from shortly after birth to weaning. They also include fertility.

The main reason for inbreeding is to produce inbred lines that may be used for crossing purposes to take advantage of heterosis. A classic commercial example of this is found in the production of hybrid seed corn. More than 99 percent of all corn produced in the United States today comes from planting hybrid seed that was produced by means of a systematic combination of certain inbred lines. The system used in hybrid seed corn production is also being used in the production of other crops, such as hybrid tomatoes and hybrid sorghum. A similar system is also used in producing certain kinds of commercial poultry.

9.13.2 Linebreeding

Linebreeding is *a form of inbreeding in which an attempt is made to concentrate the inheritance of one outstanding ances-*

tor in the pedigree. An example of linebreeding is shown in Figure 9.4. Other kinds of matings can be used to illustrate linebreeding, but all have as their objective the concentration of the inheritance of some desirable ancestor in the pedigree. Homozygosity (inbreeding) usually does not increase as rapidly when linebreeding is practiced, as is true of some forms of inbreeding, because usually only matings more distant than first cousins are made. Even though close matings are avoided, homozygosity will inevitably build up with the loss of performance that normally results from inbreeding.

9.13.3 Linecrossing and Crossbreeding

Linecrossing refers to *mating unrelated families within the same breed.* **Crossbreeding** refers to *mating individuals from different breeds.* Crossbreeding is more intense in its genotypic and phenotypic effects than linecrossing, but these effects are similar in both types of breeding.

The main genetic effect of linecrossing and crossbreeding is opposite to that of inbreeding. They result in increased heterozygosity, or decreased homozygosity, within the population in which they are practiced. These systems of mating, because they increase heterozygosity, tend to cover up recessive genes, decrease breeding purity, and eliminate families in just one generation. Several cycles of crossbreeding may result in greater variability in the crossbred population than is found in purebreds for traits such as color and, possibly, body size and conformation. The first cross (the F_1) between two breeds (P_1, P_2), however, may be more uniform than purebreds for certain performance traits.

The main phenotypic effect of linecrossing and crossbreeding is to cause an improvement in traits related to physical fitness. This is the opposite of the effect of inbreeding. *The increased vigor of crossbreds as compared with the average of the purebred parents that make the cross is known as heterosis.* This is of great value in the commercial production of plants and animals.

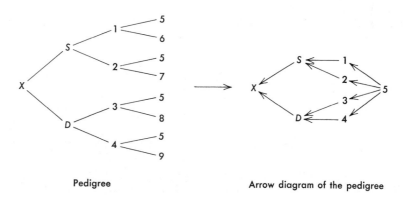

Pedigree   Arrow diagram of the pedigree

Figure 9.4 A pedigree and its arrow diagram illustrating linebreeding. Note that the only common ancestor of the sire S and dam D is individual 5 and that there are four pathways connecting individuals X and 5 through the sire and dam. Hence it is called *linebreeding.*

Heterosis vigor may be defined as the superiority of crossbred offspring over the average of the purebreds used to make the cross. This may be expressed as follows:

$$\text{Purebred average} = P = \frac{P_1 + P_2}{2}$$

$$\text{Percent heterosis} = \frac{\text{crossbred average} - \text{purebred average}}{\text{purebred average}} \times 100$$

$$= \frac{F_1 - P}{P} \times 100 = \frac{F_1 - (P_1 + P_2)/2}{(P_1 + P_2)/2} \times 100$$

Estimates of the amount of heterosis expressed by certain traits in farm animals are given in Table 9.8.

Heterosis is due to nonadditive gene action, such as dominance, overdominance, or epistasis, but it is not due to additive gene action. In general, the farther apart the parent breeds are in relationship, the more heterosis is expressed for those traits showing heterosis. Theoretically, the reason for this is that one breed may be mostly homozygous for one allele (such as *DD),* whereas the other breed may be mostly homozygous for the other allele *(dd).* This could be true for several pairs of alleles; therefore, the offspring from such crosses would be heterozygous for more gene pairs than would crosses of more closely related breeds or individuals.

TABLE 9.8	Estimates of Amount of Heterosis Expressed for Certain Traits in Farm Animals When a Linecross or Crossbred System of Mating Is Followed

	Amount of Heterosis			
Trait	**Cattle**	**Swine**	**Sheep**	**Poultry**
Conception rate	Some*	Some*	Some	Some
Birthrate	Some	Medium	Some	
Survival—birth to weaning	Large	Large	Large	
Survival—weaning to market	Some	Some	Some	
Weaning weight	Some	Some	Some	
Postweaning growth rate	Little	Little	Little	
Postweaning feed efficiency	Little	Little	Little	
Milk production	Some	Some	Some	
Carcass†	Little	Little	Little	Little
Egg production	—	—	—	Medium
Fertility and hatchability	—	—	—	Medium

*In a few crosses there is evidence that crossbreeding gives a lower conception rate than does purebreeding. Perhaps, if true, it is due to incompatibility of sperm and egg or of embryo and mother.

†Crossbreds may be slightly fatter than the purebred parents.

Crossbreeding is the preferred system of mating for commercial production of many classes of livestock. This is particularly true of swine, because more pigs survive from birth to weaning and because they may be heavier at weaning when produced in a **three-breed** cross. Crossbred sows may wean 40 to 45 percent heavier *litters* (reflecting heterosis) than purebreds. Heterosis probably approximates 20 to 25 percent for offspring produced per ewe for sheep. Crossbred cows, on the average, have heavier calves at weaning than purebreds, and more crossbred than purebred calves survive to weaning. The increased weight of calf weaned per crossbred cow in the herd is probably near 20 percent for beef cattle.

Several systems of crossbreeding may be used in the commercial production of livestock. These have been used more widely on a practical scale in swine production than with sheep and cattle.

The *single cross* refers to *crossing any two breeds.* The parents are purebreds, whereas the offspring are crossbreds. Heterosis from this system of mating is limited to crossbred progeny. Such a system gives a considerable amount of heterosis, but for best results it is necessary to use females from a breed that is known for its high prolificacy, good milking qualities, and mothering ability. Over 90 percent of the commercial **broilers** produced today come from the single cross of white Plymouth Rock females, rated fair in meatiness and good in egg production, with White Cornish males (Chapter 2). Individuals of the Cornish breed are large-breasted and meaty individuals, but they are poor in egg production. The crossbred offspring combine the desirable characteristics of both parent breeds in their growth and body shape, plus some heterosis for physical fitness. A similar scheme is now being followed in some regions of the United States in commercial swine production.

The **rotation** is another crossbreeding system that may be used for commercial production of livestock. As an example, suppose Angus × Hereford females were crossed. The offspring of such a mating would be 1/2 Angus and 1/2 Hereford. Females of this cross could be kept for breeding and mated to an unrelated Angus bull. The calves produced would be 3/4 Angus and 1/4 Hereford. Heifers from this cross could then be mated to an unrelated Hereford bull, giving calves that were 5/8 Hereford and 3/8 Angus. This rotation system of mating, using crossbred females mated in alternate generations to unrelated Angus and Hereford bulls, could be followed for many generations. Such a crossbreeding system has the advantage of using the more productive crossbred mothers, but some heterosis may be lost in later generations, as compared with the first **backcross** between these two breeds.

The three-breed cross is a practical approach that maximizes heterosis in the crossbred offspring and the crossbred mother. An example of the three-breed cross would be the use of a Duroc boar on crossbred Hampshire × Yorkshire females. A rotation of purebred boars from the three breeds on subsequent generations of selected crossbred gilts could be continued indefinitely. Or a fourth breed could be used if desired. After the three-breed cross, one is merely attempting to retain as much of the heterosis obtained as possible.

Regardless of the crossbreeding system used, only the best crossbred females and the best purebred males should be used. The reason for this is that heterosis, for traits that reflect it, adds to the average merit of the parents. For example, assume that heterosis for litter size at weaning is 20 percent, and the litter size at weaning in each of three breeds is as follows:

Breed A	7.0
Breed B	8.0
Breed C	9.0

If the heterosis is nearly the same for any combination of breeds (it may not be in actual practice), one would want to cross breeds B and C because the litter size of 20 percent from heterosis would add to the average of these two breeds (8.5),

giving a total of 10.2 pigs as compared with 9.0 for the cross of breeds A and B.

9.14 SUMMARY

Quantitative inheritance, which involves many pairs of genes, affects most economic traits in farm animals. Phenotypic variation is the raw material with which the animal breeder must work to improve livestock and poultry. Methods of estimating the proportion of total phenotypic variation due to heredity have been developed, and from this knowledge, principles of improvement of livestock through breeding have evolved.

Tools the animal breeder has available to use in molding the raw material are selection, inbreeding, linebreeding, linecrossing, and crossbreeding. Proper use of these tools has greatly increased the efficiency of livestock production. More systematic use of these tools in the future should add to the income of the livestock producer.

STUDY QUESTIONS

1. In a herd of cattle the average weaning weight of all calves is 400 lb. Heifers weighing 450 lb and bulls weighing 500 lb are selected for breeding purposes. What is the selection differential for those animals selected for breeding?

2. For the situation in question 1, if only average heifers were selected for breeding, what would the selection differential be?

3. In one study, the correlation between husbands and wives for IQ was higher than the correlation between parents and children. What is a possible explanation for this?

4. In an experiment with chickens, selection was practiced only for longer shanks. Selection was effective, and shank length increased as the number of generations of selection increased. Because selection was based on individuality, or for individuals with longer shanks, what type of gene action was probably involved?

5. What are some possible explanations for different races of people with different skin colors?

6. What is responsible for the snowshoe hare in Alaska turning white in winter and brown in summer? Why do rabbits in the United States not change colors in winter and summer, as do snowshoe hares?

7. Assume that, in beef cattle, you have 10 heifers with the following 205-day weaning weights: 560, 550, 530, 510, 500, 450, 420, 410, 400, and 390 lb. If you saved five of the heaviest heifers for your herd, what would be their selection differential if the average heifers in this herd weighed 400 lb? If you saved three of the top heifers, what would their selection differential be? Because the three selected heifers had a larger selection differential than the five heifers, what important principle does this illustrate?

8. Calculate the generation interval in your family (it is the average age of your parents when all children were born).

9. Compare the generation interval of question 8 with those of four of your friends. From these data, what would you conclude about the length of the generation interval in humans?

10. In a herd of cattle in which the average daily gain in the feedlot for all calves is 2 lb per animal, heifers that gain 2.2 and bulls that gain 3 lb are kept for breeding. If the offspring gain 2.3 lb per animal daily, what is the heritability of rate of gain from these data?

11. Backfat thickness at 200 lb in an entire pig crop is 1.5 in. Boars measuring 1 in in backfat and gilts measuring 1.3 in are retained for breeding. If backfat thickness in swine is 45 percent heritable, what would be the expected backfat thickness in the progeny of these parents?

12. In question 11, why was the backfat thinner in the selected boars than in the selected gilts?

13. Is the following statement true or false? "All humans are inbred." Explain your answer.

14. Mating a sire to 23 of his own unselected daughters is one system of progeny testing for the presence of a recessive gene in his genotype. What are the advantages and disadvantages of this kind of a progeny test?

15. Calculate the mean, variance, and standard deviation for the heifers in question 7.
16. Is weaning weight in beef cattle a quantitative or a qualitative trait? Why?
17. Although black Angus breeders have never kept red Angus for breeding, about 1 of 200 calves born among black Angus parents is red. Why does the red gene persist in the black Angus breed?
18. Outline a program that would help eliminate the red gene from the black Angus breed.
19. How would you start a pure breed of red Angus?
20. Assume you have a herd of red Angus and one of your cows produced a black Angus calf. How would you explain this?

ANSWERS TO STUDY QUESTIONS

1. The selection differential is 75 lb.
2. The selection differential for heifers is 50 lb.
3. It appears that there is a good correlation between the IQ of the husband and wife because a person with a high IQ tends to marry another with a high IQ. The IQ of parents is similar because of both hereditary and environmental effects. The correlation between IQ of parents and their offspring is mostly due to heredity.
4. Probably an additive type of gene action.
5. Mutations have been responsible for the genes for different skin color. Natural selection in a particular part of a country might have been effective in developing a certain skin type. In other words, one skin color might have caused better survival than another skin color in a given area.
6. Snowshoe hares in Alaska possess genes for changing coat color in different seasons. The ability to change coat colors probably arose from new mutations and increased in frequency, because the ability to do this made them better fit for survival in Alaska. Rabbits farther south probably do not change coat colors with seasons because they would be at a disadvantage when snow was not on the ground, because a white coat would make them more vulnerable to predators.

In fact there probably has been selection against the ability to change coat color in the Midwest and farther south in the United States.

7. The selection differential of the five heaviest heifers would be 130 lb. The selection differential for the three heaviest would be about 147 lb. This illustrates the general principle that the fewer animals kept for replacement, the larger the selection differential, and vice versa.
8. Answer depends on the individual.
9. Answer depends on the individuals.
10. About 50 percent.
11. Approximately 1.34 in in the progeny.
12. Because fewer boars are kept for breeding and the selection intensity would be greater than in gilts.
13. False. Inbreeding refers to an increase in the number of homozygous gene pairs when relatives are mated as compared with the average of the population (or base population).
14. The main advantage is that such a progeny test is for any recessive gene the sire might be carrying, rather than a specific one. One disadvantage is that the offspring would be at least 25 percent inbred when from a father-daughter mating. This would result in a decline of vigor in the inbred offspring. Another disadvantage

would be that the sire might be so old by the time the progeny test was completed that he might be out of **service** or dead.

15. The mean is 472 lb; the variance, 4262.22 lb; the standard deviation, 65.29 lb.
16. It is a quantitative trait because it is affected by many genes, and there is no sharp distinction between phenotypes with measurements for the trait in the population falling into a normal frequency distribution curve.
17. Perhaps there has not been enough emphasis on identifying and culling black carriers of the red gene *Bb*. Another remote possibility might be that selection favors heterozygous red *Bb* individuals.
18. Discard all red individuals and their parents. Progeny-test both cows and bulls to determine those that are homozygous black *BB* and keep only these for breeding. Homozygous black *BB* individuals would not produce red offspring unless a new mutation from black to red occurred.
19. Mate red individuals with red individuals.
20. The red gene might have mutated from red to black. Only one black gene is required to make an individual black. Also, one should check the records and make certain both parents were actually red.

ANATOMY AND PHYSIOLOGY OF FARM ANIMALS

We are so accustomed to investigating the functions of different organs or parts of the body that we do not see the individual as a living whole.

A. Stewart Paton (1903–1988)
South African writer

10.1 INTRODUCTION

Although there is considerable variation among **species** within the animal kingdom, domestic farm animals show many similarities in form (anatomy) and function (**physiology**). In many instances, humans have altered the form and function of domestic animals for their own needs. In this chapter significant anatomical and physiological characteristics of farm animals are discussed. A knowledge of these is important because it enables one to describe animals more clearly in judging and selecting for breeding purposes and to properly apply husbandry and veterinary practices.

Because of the complexity of the bodies of humans and animals, scientists divide anatomy into several branches. *Gross anatomy* is the study of structures that can be seen by the unaided eye. *Microscopic anatomy,* or *histology,* is the study of tissues by means of microscopes. *Comparative anatomy* compares the body structure of different species. **Embryology** is the study of the body as it develops **in utero** in farm mammals and in the fertilized eggs of **poultry.** A knowledge of body structure is essential in understanding the function of **morphology** in maintaining health and resisting disease. *Physiology* is the branch of biology related to studies of body-organ function, individually and collectively, in systems.

The anatomy and physiology of the reproductive system are presented in Chapter 13. The **endocrine** system is discussed in Chapter 11. The physiology of growth, **lactation,** egg laying, and digestion are discussed in Chapters 12, 15, 16, and 19, respectively. Certain aspects of environmental physiology are presented in Chapter 17.

10.2 EXTERNAL BODY PARTS

Various external body parts of selected farm animals are shown in Figures 10.1 to 10.6. Other terms, not given in the figures, are also used to describe and identify the location of body parts of an animal. These include dorsal, ventral, caudal (posterior), and cranial (anterior). **Dorsal** refers to the top or back side of an individual; **ventral** refers to the belly or underside. **Cranial** (anterior) means toward the front, whereas **caudal** (posterior) means toward the rear.

Skin is the exterior covering of the body and is continuous with the exterior membranes of the respiratory, **urogenital,** and digestive tracts. Hair, wool, horns, feathers, and hooves are considered to be modified appendages of skin. The skin functions in many ways, such as protecting against infections, regulating temperature, and helping the individual respond to its **environment** by means of **sensory** nerves. The skin also contains **glands** of secretion and excretion (Figure 10.7).

The skin consists of two layers, the **epidermis** and the **dermis.** The epidermis is the outer layer of epithelial cells. The inner layer, known as the *dermis,* or corium, is composed of a connective tissue network and includes blood **vessels, lymph** vessels, nerves, glands, hair follicles, and muscle fibers. The principal glands of the dermis are sweat glands (sudoriferous) and sebaceous glands. The latter secrete oily substances that lubricate the hair and skin. The dermis also contains special types of glands, such as the modified sweat glands found in the snout of swine. Even mammary glands are considered to be highly modified glands of skin. Skin thickness varies with species, **breed,** sex, and body location (e.g., eyelids are thin, whereas the sole of the foot is thick).

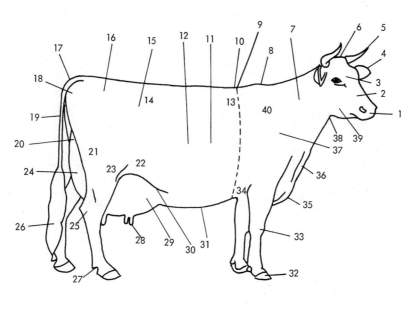

1. Muzzle	11. Barrel	21. Thigh	31. Mammary veins
2. Bridge of nose	12. Ribs	22. Flank	32. Hoof
3. Forehead	13. Back	23. Stifle	33. Knee
4. Ear	14. Loin	24. Rear udder	34. Point of elbow
5. Horn	15. Thurl	25. Hock	35. Brisket
6. Poll	16. Rump	26. Switch	36. Dewlap
7. Neck	17. Tailhead	27. Dewclaw	37. Point of shoulder
8. Withers	18. Pinbone	28. Teats	38. Throat
9. Heart girth	19. Tail	29. Fore udder	39. Jaw
10. Crops	20. Rear udder attachment	30. Fore udder attachment	40. Shoulder blade

Figure 10.1 External parts of a dairy cow.

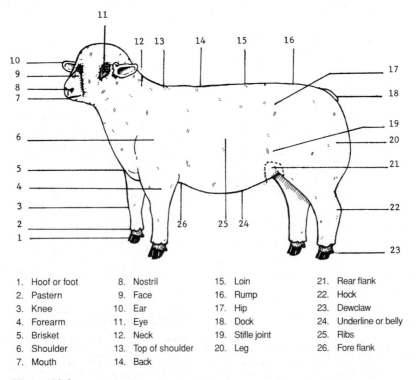

1. Hoof or foot	8. Nostril	15. Loin	21. Rear flank
2. Pastern	9. Face	16. Rump	22. Hock
3. Knee	10. Ear	17. Hip	23. Dewclaw
4. Forearm	11. Eye	18. Dock	24. Underline or belly
5. Brisket	12. Neck	19. Stifle joint	25. Ribs
6. Shoulder	13. Top of shoulder	20. Leg	26. Fore flank
7. Mouth	14. Back		

Figure 10.2 External parts of a sheep.

Courtesy of Dr. Robert Godke, Louisiana State University.

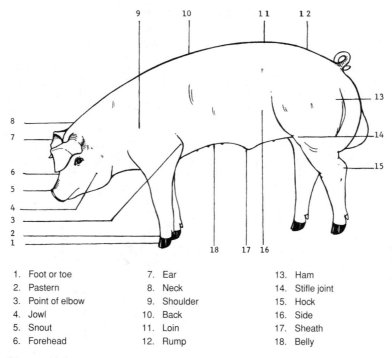

1. Foot or toe	7. Ear	13. Ham
2. Pastern	8. Neck	14. Stifle joint
3. Point of elbow	9. Shoulder	15. Hock
4. Jowl	10. Back	16. Side
5. Snout	11. Loin	17. Sheath
6. Forehead	12. Rump	18. Belly

Figure 10.3 External parts of a market hog.
Courtesy of Dr. Robert Godke, Louisiana State University.

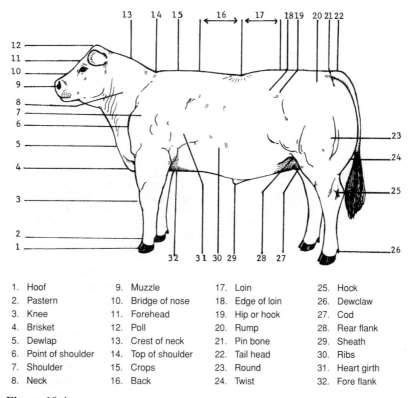

1. Hoof	9. Muzzle	17. Loin	25. Hock
2. Pastern	10. Bridge of nose	18. Edge of loin	26. Dewclaw
3. Knee	11. Forehead	19. Hip or hook	27. Cod
4. Brisket	12. Poll	20. Rump	28. Rear flank
5. Dewlap	13. Crest of neck	21. Pin bone	29. Sheath
6. Point of shoulder	14. Top of shoulder	22. Tail head	30. Ribs
7. Shoulder	15. Crops	23. Round	31. Heart girth
8. Neck	16. Back	24. Twist	32. Fore flank

Figure 10.4 External parts of a steer.
Courtesy of Dr. Robert Godke, Louisiana State University.

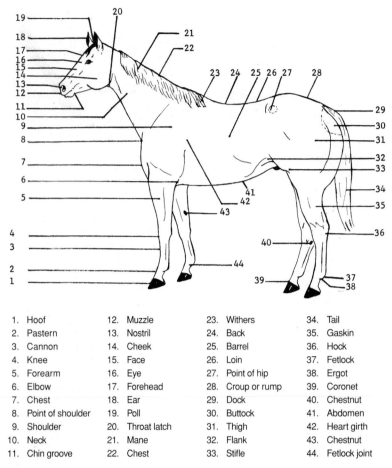

Figure 10.5 External parts of a horse.
Courtesy of Dr. Jack L. Kreider, Louisiana State University.

1.	Hoof	12.	Muzzle	23.	Withers	34.	Tail
2.	Pastern	13.	Nostril	24.	Back	35.	Gaskin
3.	Cannon	14.	Cheek	25.	Barrel	36.	Hock
4.	Knee	15.	Face	26.	Loin	37.	Fetlock
5.	Forearm	16.	Eye	27.	Point of hip	38.	Ergot
6.	Elbow	17.	Forehead	28.	Croup or rump	39.	Coronet
7.	Chest	18.	Ear	29.	Dock	40.	Chestnut
8.	Point of shoulder	19.	Poll	30.	Buttock	41.	Abdomen
9.	Shoulder	20.	Throat latch	31.	Thigh	42.	Heart girth
10.	Neck	21.	Mane	32.	Flank	43.	Chestnut
11.	Chin groove	22.	Chest	33.	Stifle	44.	Fetlock joint

Hair is the coat covering of cattle, goats, horses, and swine, whereas wool is the covering of sheep. Wool differs from hair by being finer-textured, soft, and curly. In temperate or cold climates, the hair coat of animals grows long for the winter and is shed in the spring. The wool of sheep is usually shorn (clipped off) by the caretaker in the spring, before the onset of hot weather. This allows the sheep to be more comfortable in the summer months and also yields a product with economic value. The wool usually grows out again before the following winter, thus protecting the sheep from cold temperatures. Wool growth increases during the spring and summer as an adaptive mechanism for insulation against radiation heat gain in hot, arid climates. *Feathers* are the body covering of poultry. Hens shed their feathers (**molt**) in late summer and grow new ones in time to provide a body covering in the winter. It is interesting that hair, wool, and feathers are proteinaceous in composition and are produced through the action of **genes** (Chapter 8).

Cattle, goats, sheep, and swine have cloven (split or divided) hooves; horses do not. A breed of swine known as the *mule-foot* (describes hoof appearance) does not have cloven hooves. Mule-foot in swine is genetically **dominant** over the normal cloven hoof. A similar heritable condition of cattle in which the hoof is not cloven has been described.

10.3 THE SKELETAL SYSTEM

The bony tissues that form the body's framework constitute the skeletal system. This system includes the long bones of legs, ribs, vertebrae, skull, and other bony processes. In farm animals and other mammals the skeletal systems are internal (endoskeleton) and basically alike. A giraffe's neck has the same number of bones as a mouse's neck; it is merely the length and size of the bones that vary. In many species, such as insects, the skeleton is outside the body (exoskeleton) as discussed in Chapter 23.

The outer layers of bony tissues are filled with mineral deposits, mainly in the form of calcium and phosphorus (Figure 10.8). The inner core is a soft tissue known as *bone marrow*. Some of this tissue is yellow and consists mostly of fat; it is called *yellow marrow*. The red tissue of bone marrow is known as *red marrow* and functions in blood cell and platelet formation.

The skeletal tissue increases in size and length as an animal grows. Growth of bones occurs at the ends in the region of the **cartilage** between the **epiphysis** (the end) and **diaphysis** (the shaft). This cartilage, known as *epiphysial–diaphysial cartilage,* gradually becomes calcified and is replaced by bony material. Once the cartilage is completely replaced by bone,

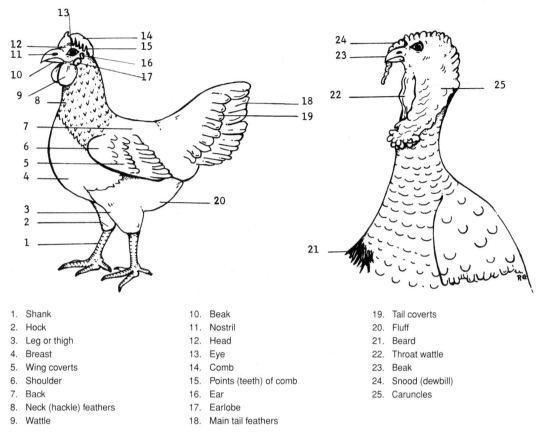

1.	Shank	10.	Beak	19.	Tail coverts
2.	Hock	11.	Nostril	20.	Fluff
3.	Leg or thigh	12.	Head	21.	Beard
4.	Breast	13.	Eye	22.	Throat wattle
5.	Wing coverts	14.	Comb	23.	Beak
6.	Shoulder	15.	Points (teeth) of comb	24.	Snood (dewbill)
7.	Back	16.	Ear	25.	Caruncles
8.	Neck (hackle) feathers	17.	Earlobe		
9.	Wattle	18.	Main tail feathers		

Figure 10.6 Parts of a chicken and turkey.
Courtesy of Dr. John A. Sims, Iowa State University, redrawn by Dr. Robert Godke, Louisiana State University.

further bone growth ceases. **Calcification** of the epiphysial–diaphysial cartilage at the ends of long bones occurs at physical maturity. Further growth in body proportions (except fattening) then ceases (Chapter 12). It is interesting that bone is continually being reabsorbed and replaced in mature animals, although no net bone growth occurs. The diameter of long bones also increases as they grow and lengthen. This increase in size results from production of new bony tissue by the **periosteum** that surrounds the cortex of bone. As new bone is deposited, portions of the deeper, inner bone are removed, in turn increasing the size of the marrow cavity.

Hormones, **vitamins,** and other **nutrients** can affect proper bone growth. The mineral tissues of bone can be depleted in cases of nutritional deficiencies (Chapters 20 and 21), and they may become fragile and/or distorted. The skeletal systems of the cow and chicken are shown in Figures 10.9 and 10.10. The skeletal systems of all farm **mammals** are similar, but only that of the cow is shown. Cattle normally have 13 pairs of ribs; however, some occasionally have 14 pairs.

Bony tissues are fairly rigid and become more brittle with age. For this reason, they are sometimes fractured or broken, but the body has the power to mend or repair the break. Normal mending of bones occurs when both ends of the fracture are brought together and held securely in place by a splint, bone

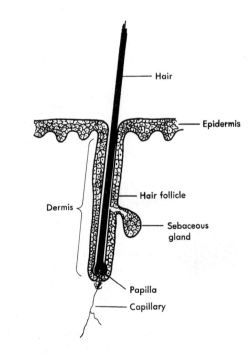

Figure 10.7 Diagram of hair follicle of the skin.

pin, or other device. The break is first filled by a fibrin clot, which is then invaded by capillaries to furnish a blood supply. The break is later filled with true bony tissue.

The skeletal system functions in body support. This is especially true of long bones. The skeletal system also provides attachment sites and leverage for muscular movement. Movement is made possible by several kinds of joints. *Ball-and-socket joints* allow movement in all directions. An example is the shoulder joint, where the forearm is attached. *Hinge joints* allow certain body parts (e.g., the elbows) to move in two direc-

tions. *Pivot joints* are found in the neck and allow the head to be turned in more than one direction. *Gliding joints* as found in vertebrae allow the body to be flexible so that it can be bent forward, backward, or in several directions. Bones are joined at the joints by **ligaments** outside the joint capsule. Within the joint capsule is **synovial** fluid, which lubricates the joints and allows them to move freely and without friction.

Another important function of the skeletal system is protection of vital body organs. The rib cage protects the heart, lungs, and some abdominal organs. It also aids in breathing. The skull protects the brain. The importance of this is illustrated when a brain concussion (jarring of the brain by a blow) occurs. Such a concussion can cause unconsciousness and even permanent damage to this vital organ. The spinal column protects the spinal cord from injury. Such protection is essential for the life and well-being of an individual. Damage to the spinal cord by a serious injury may result in paralysis of body parts **distal** to the injury. This paralysis is often permanent because nerves cannot regenerate.

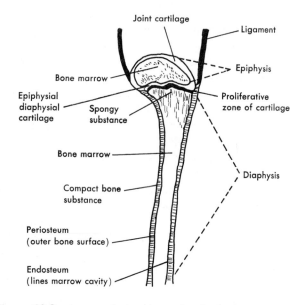

Figure 10.8 Diagram of a long bone, showing various parts.

10.4 THE MUSCULAR SYSTEM

The external muscular system of the cow is shown in Figure 10.11. Muscle consists largely of **protein** and is the lean portion of the **carcass** in meat animals. Muscle cells usually occur in bundles or sheets, although they may occasionally occur as individual cells scattered throughout the tissue.

Muscles are commonly classified as voluntary or involuntary, that is, under or not under the control of the individual's will. Voluntary muscles function when the animal calls on

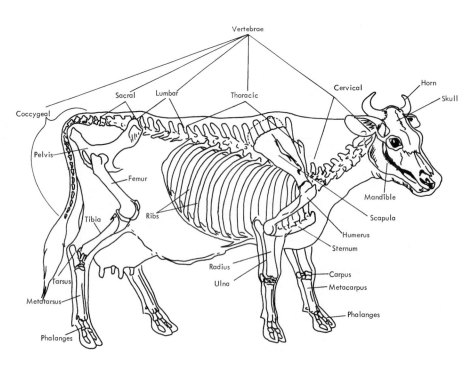

Figure 10.9 Skeleton of a cow.
Courtesy of Dr. P. D. Garrett and Dr. D. E Rodabaugh, University of Missouri.

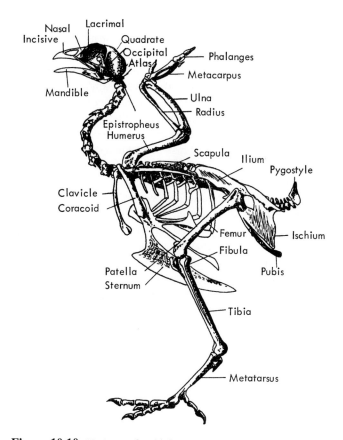

Figure 10.10 Skeleton of a chicken.

Adapted from M. C. Nesheim, R. E. Austic, and L. E. Card, Poultry Production, 12th ed., Lea & Febiger, Philadelphia, 1979.

them. Muscles are also classified as *smooth* (unstriated) or *striated* (striped). When viewed microscopically, striated muscles show cross striations (stripes or streaks), whereas smooth muscles do not. These stripes result from the proteins *myosin* and *actin*. Voluntary muscles are striated; involuntary muscles may be either smooth or striated.

Skeletal muscles are voluntary. They are connected to bones by **tendons,** which consist of dense connective tissue. Muscles are usually attached to two bones and provide power for movement of various body parts. Skeletal muscles are also classified to indicate the type of movement they produce. *Extensor* muscles cause body parts to straighten; *flexor* muscles cause them to bend. *Abductor* muscles cause body parts to move away from a plane through the body, and *adductor* muscles draw them toward the body plane. Each voluntary muscle fiber is controlled by a branch of a voluntary nerve, or motor neuron. One motor neuron and its many branches that supply muscle fibers are collectively called a *motor unit.*

Involuntary (smooth) muscles are found in the walls of organs of many body systems, such as the digestive, urogenital, and respiratory systems. Moreover, involuntary muscles are located in many secretory organs and function to expel or force out their secretions. (An exception is the endocrine glands where secretion is controlled by specific factors circulating in the blood or in the local environment.) Within the digestive tract, involuntary muscles provide impetus for movement of **ingesta.** In the female reproductive tract, involuntary muscles are responsible for uterine motility, which moves sperm up the tract to meet the descending **ovum.** They also aid in concert with voluntary muscles to expel the **fetus** at **parturition.** Involuntary muscles increase or decrease the diameter of blood vessels, thus regulating or controlling the amount of blood flowing to a particular body area. Contractions of involuntary smooth

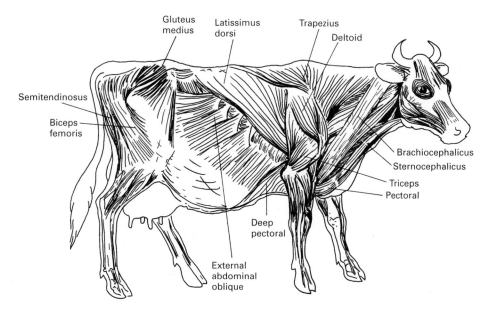

Figure 10.11 Selected muscles of the cow.

Courtesy of Dr. P. D. Garrett and Dr. D. E. Rodabaugh.

muscles are regulated by the autonomic nervous system. The hormones *epinephrine* (adrenalin) and *norepinephrine* (noradrenalin) stimulate autonomic nerves to cause relaxation and contraction, respectively, of the involuntary smooth muscles.

Cardiac muscle is a unique and specialized type of striated muscle which forms the four chambers of the heart. Although the heart beats spontaneously (involuntary), its rate and force of contraction can be altered by the autonomic nervous system.

The physiology of muscle action has received considerable attention from research workers, yet it is not completely understood. Several years ago, muscle contraction was thought to result from energy released on the breakdown of proteins. If this were true, it would mean that following work, there should be an increased excretion of the end products of protein **metabolism** from the body. Research studies failed to show this. Rather, the energy utilized for muscle action came largely from nonprotein sources. However, if food intake during work was insufficient and the animal lost weight, both proteins and body fats were broken down and used as energy sources.

Details of the chemical reactions involved in muscle contraction are complex. Therefore only a simple summary will be given. The chief source of energy used for muscle contraction is now thought to be adenosine triphosphate (ATP). When a muscle is stimulated, ATP breaks down into adenosine diphosphate (ADP) and a phosphate radical, releasing energy, which is used in the contraction process. The breakdown of muscle glycogen to lactic acid serves to generate more ATP, thus providing a cycle for production of energy that can be utilized during muscle contraction.

Another step in the chemistry of muscle action includes the breakdown of phosphocreatine to creatine and phosphoric acid. This reaction also supplies the energy and phosphoric acid needed to generate ATP. Phosphocreatine is resynthesized from phosphoric acid and creatine, using energy available when ATP is in excess, and thus serves as a storage battery. About one-fifth of the lactic acid produced from glycogen is oxidized to carbon dioxide and water. Energy released by this **oxidation** is used by the liver to resynthesize glycogen from the remaining four-fifths of the lactic acid.

Restoration of the original chemical condition of the muscle to that before contraction results in loss of part of the glycogen by means of oxidation to lactic acid by **aerobic** activity. Thus glycogen must eventually be replaced by some means for normal contraction to occur. Energy for muscle contraction comes from an **anaerobic** reaction, using energy released by lactic acid formation. (Muscle glycogen is converted anaerobically into lactic acid.) This reaction cannot continue indefinitely, because there is an accumulation of lactic acid and a lack of oxygen to oxidize it (to combine it with oxygen). The deficit of oxygen needed to replace that used aerobically is called the *oxygen debt*. This debt must be repaid before there can be a normal resumption of muscular activity. Through the temporary anaerobic recharging of the muscle, nature has provided animals with a means of expending several times more muscular activity in a short time than would be possible if oxygen had to be supplied concurrently with muscle action. Studies at the

Missouri Agricultural Experiment Station showed that horses and human males, during intense muscular work, may expend energy at 100 times the resting state. Such activity cannot be maintained long because of the onset of fatigue. It would seem that individuals who can expend a greater amount of energy before their oxygen debt halts muscular activity would be those most likely to survive in environments that require rapid and prolonged muscular activity. An expenditure of energy 6 to 8 times greater than that of rest may be endured during prolonged hard work. Under these conditions oxidation keeps pace with energy expenditure.

Certain minerals are also involved in muscle activity. Calcium **ions** function as an **enzyme**-activating factor necessary for muscle contraction (Chapters 20 and 21). One theory states that calcium ions initiate the enzyme activation of myosin filaments in muscle. Relaxation of muscle is believed to result from the binding of calcium, which makes it unavailable in the myosin area. Magnesium ions can activate the calcium-binding enzyme. Thus magnesium ions are also involved in muscle relaxation.

10.5 THE CIRCULATORY SYSTEM

> All history shows the power of blood over circumstances, as agriculture shows the power of the seeds over the soil.
>
> **Edwin Perry Whipple (1819–1886)**

The circulatory system of animals consists of the heart, veins, capillaries, arteries, lymph vessels, and lymph **nodes.** The bone marrow functions in production of blood cells. The spleen serves as a reservoir for storage of blood cells and also removes damaged and old blood cells from circulation.

10.5.1 Anatomy and Physiology

The *heart* is located in the thoracic cavity between the lobes of the lungs (Figure 10.12). The mammalian heart is a marvelous and complicated organ. It is the pump that circulates blood to all body parts. Failure to perform this function (heart failure) terminates life. It has been estimated that the heart of a person 70 years old has beaten at least 2.5 trillion times and has pumped more than 435,000 tons of blood. A typical human heart pumps blood through an estimated 60,000 miles of blood vessels daily (enough blood to fill a 4000-gal tank car).

Arteries are blood vessels that carry oxygen-rich blood from the heart to various body tissues. Arterial walls are thick and contain heavy muscle layers that can withstand the blood pressure resulting from the heart's beating. *Veins* are blood vessels that return blood from throughout the body to the heart and that convey it from the lungs, where carbon dioxide and oxygen are exchanged, to the heart. Veins have comparatively thin walls (compared with those of arteries), which are collapsible. In places, veins contain valves that aid the flow of blood to the heart.

Capillaries are minute blood vessels that lie between the terminal arteries and arterioles and the beginnings of the venules and veins. They are very tiny and thin-walled (one cell thick) and are widely distributed in all body tissues. Transfer of

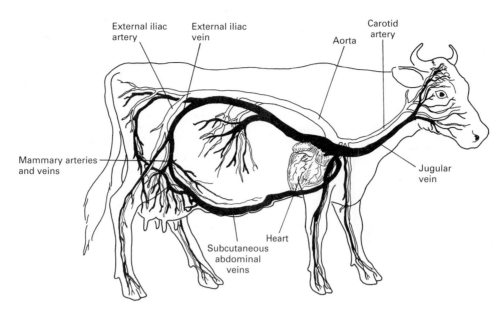

Figure 10.12 Systemic blood circulatory system of the cow (veins are dark, arteries are light).
Courtesy of Dr. P. D. Garrett and Dr. D. E. Rodabaugh.

nutrients from blood to tissues, and waste products from tissues to blood, occurs in capillaries. It is interesting that every pound of excess fat contains an estimated 200 miles of capillaries. This partly explains how being overweight can overwork the heart and contribute to eventual heart failure.

Lymph vessels are accessories to the body's circulatory system. They originate in tissue spaces and converge to form larger ducts as they pass through lymph nodes. Many such lymph ducts converge into a single duct, or vessel, which empties into the large blood veins of the circulatory system. Lymph nodes filter foreign substances from lymph, preventing their passage into the bloodstream. They also produce **lymphocytes,** the white blood cells responsible for immune responses (immunity to infections and recognition of foreign materials and tissues).

Contraction of the heart circulates blood through the circulatory system. The heart controls its own beat through the action of the sinoatrial node, called the *pacemaker,* which is located in the right atrium. The heartbeat is also influenced by accelerator and inhibitor nerves. In humans, faulty pacemakers can be replaced by an electronic timer that controls the heartbeat. Such an operation-and-replacement therapy has been employed successfully for many years.

Contraction of the heart begins in the right atrium and quickly spreads to the left atrium. Blood is then forced into the ventricles, which contract, closing the atrioventricular (AV) valves. This forces the blood into the major arteries (Figure 10.13). The heart itself requires a rich blood supply. It represents only about 0.5 percent of the body weight of humans but requires approximately 5 percent of the total blood supply.

The *systemic circulatory system* (somatic circulation) refers to the heart and vessels that move oxygenated arterial blood to all body regions and return venous (unoxygenated) blood to the heart. Systemic circulation is complex and depends on particular tissue requirements (e.g., digestion versus exercise). Arterial blood in the systemic circulation is of a bright red color, but after passing to the veins, it becomes dark or brownish red. The *portal system,* a part of the systemic circulatory system, conveys venous blood from the stomach, pancreas, small intestine, and spleen to the liver, where it passes through a second capillary bed for the filtering of toxins and other harmful substances before returning to the heart.

The vascular system that circulates blood through the lungs, in which it is oxygenated, is known as the *pulmonary system.* This system differs from the systemic circulatory system in that veins, rather than arteries, carry oxygenated blood. The pulmonary artery receives unoxygenated blood from the right ventricle of the heart and divides after leaving the heart, sending one branch to the right lobe and another to the left lobe of the lungs. Each branch of the pulmonary artery subdivides many times, first forming *arterioles,* and finally capillaries of the lungs, where oxygen is taken into the blood and carbon dioxide is exhausted from the blood. Once blood is forced through these capillaries, it passes into *small veins (venules),* which converge to form *pulmonary veins.* These veins return oxygenated blood to the left atrium of the heart, where the pulmonary circulation ends.

The pressure required to pump equal volumes of blood through the systemic circulation is much greater than that required in pulmonary circulation. Consequently, the heart muscles of the left ventricle develop a much greater muscle mass than those of the right ventricle (Figure 10.13).

10.5.2 Blood Composition

Approximately 50 to 65 percent of the total blood volume consists of **plasma.** The corpuscles are suspended in blood plasma. Plasma is the golden-straw-colored liquid that remains after

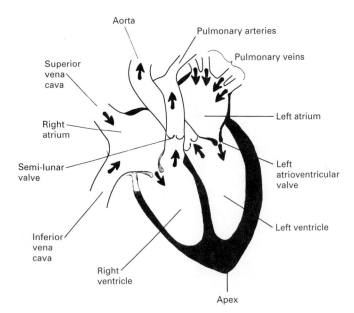

Figure 10.13 Diagram depicting the four chambers of a mammalian heart. The arrows indicate the direction of blood flow. Deoxygenated blood enters the right atrium through the vena cava. Blood is sent from the right ventricle to the lungs via the pulmonary arteries. Oxygenated blood is returned to the left atrium via the pulmonary veins and is pumped through the aorta to the general circulation by the left ventricle.

Drawn by F. E. Staten, University of Illinois.

Species	Milliliters of Blood per Pound of Body Weight	Millions of Erythrocytes per Cubic Millimeter of Blood	Thousands of Leukocytes per Cubic Millimeter of Blood
Cattle	26.1	6–10	5–12
Chickens	25.0	2–3	20–35
Dogs	42.7	5–8	8–15
Goats	32.0	13–18	6–14
Horses	41.0	6–10	5–11
Sheep	30.0	9–11	5–10
Swine	30.0	6–7	15–25

TABLE 10.1 | **Average Values for Certain Blood Components in Selected Animals***

*Many factors, such as age and stress, affect values of blood components.

cells have been removed. To obtain plasma, a blood sample is treated to prevent clotting and let stand until the cells settle to the bottom of the container. (Plasma may also be obtained by centrifuging whole blood.) Plasma usually contains approximately 90 percent water and 10 percent solids. The solids are composed of inorganic salts and organic substances such as **antibodies,** hormones, vitamins, enzymes, proteins, and glucose (blood sugar). The cellular, or nonplasma, portion of blood contains erythrocytes (red blood cells), leukocytes (white blood cells), and platelets. Erythrocyte and leukocyte values for selected species are presented in Table 10.1. **Serum** is the fluid remaining after blood has clotted. Serum, then, is plasma from which the fibrinogen (a clotting factor) has been separated in the clotting process.

Red blood cells contain **hemoglobin,** which has the important function of transporting oxygen from the lungs to various body tissues. Red blood cells also have **antigens** (on their surface), which are responsible for the various blood types of different species.

10.5.3 Hemoglobins

Hemoglobin is responsible for giving red blood cells their characteristic red color. Chemically, hemoglobin is an organic compound consisting of heme (an iron–porphyrin complex) and globin (a protein). Hemoglobin readily absorbs oxygen from air in the lungs, forming oxyhemoglobin. Oxygen is held rather loosely by hemoglobin and is readily given to tissues

within the body. Red blood cells in an adult mammal are formed in the red bone marrow. By the time they reach the bloodstream, their **nucleus** has been lost. In birds, the nucleus remains throughout the cell's life. The life of the red blood cells in the circulatory system usually varies from 90 to 120 days, after which the cells are removed by the spleen. **Anemia** is a condition in which the number of red blood cells, or the amount of hemoglobin, is reduced to a subnormal level. It may be caused by loss of blood from a wound or by **infestations** of bloodsucking **parasites.** Anemia may also result from a subnormal rate of red blood cell production because of faulty nutrition (Chapters 20 and 21) or from a shorter-than-normal life of red blood cells (the half life is only 60 days in sickle-cell anemia in humans).

Many types of hemoglobins in humans have been described. These are caused by gene **mutations.** When a new gene mutation occurs, a different protein (globin) is produced because of a change in the code sent from the genetic material (**DNA**) to the ribosomes in the **cytoplasm** of cells for building a particular protein (Chapter 8). At least two types of hemoglobin are normally found in the red blood cells of animals. *Fetal* hemoglobin is present in the red blood cells of young at birth, but is soon replaced by *adult* hemoglobin. Research has shown that humans and animals sometimes possess abnormal types of hemoglobin. An abnormal type, known as *sickle-cell hemoglobin,* is found mostly in the African-American race and has been valuable in studies designed to determine how genes function at the cellular level.

10.5.4 Blood Types

Red blood cells of humans and animals possess certain gene-determined antigens on their surfaces. Antibodies against certain antigens may also occur in blood serum (Section 22.2.5). Antibodies for a particular red blood cell antigen are not found in the blood of the same individual because the reaction between the two would **agglutinate** the red blood cells and cause death. Such a reaction may be responsible for some early embryonic death losses in humans and animals.

The first blood group discovered was the A, B, and O series in humans. At least three genes (*A, B,* and *a*) are involved. Genes *A* and *B* are dominant to *a,* which is the gene for blood group O. However, *A* and *B* are not dominant to each other. Gene *A* produces antigen A; gene *B,* antigen B; and group O individuals possess neither antigen A nor B in their red blood cells. Individuals belonging to blood group A can be of **geno-types** *Aa* or *AA;* those of blood group B, of genotypes *Ba* or *BB;* those of blood group AB, of genotype *AB;* and those of blood group O, of genotype *aa.* Although the antibody for a particular antigen does not occur together with it in the same individual, the opposite antibody does occur. For example, individuals of group A possess antibody B in their serum, those of group B possess antibody A, those of group AB have neither antibody, and those of group O possess both antibodies A and B.

A knowledge of the A, B, and O blood-group series has been valuable in matching types of blood for successful whole-blood transfusions (Figure 10.14), for use in certain medical and legal problems, in parentage disputes, and in selected genetic studies.

The discovery of the Rh blood group and its importance in humans has been of widespread interest. It received its name because the antigen was discovered first in the rhesus monkey (hence Rh). About 85 percent of the white population in the United States possess this antigen. They belong to the Rh positive blood group. Research has shown, however, that the Rh positive group consists of several different alleles (genes). The other 15 percent belong to the Rh negative blood group. It has been found that the Rh negative gene is recessive to all Rh positive genes. Thus an Rh positive person can be of genotypes Rh positive Rh positive or Rh positive Rh negative, whereas an Rh negative individual can have only the Rh negative Rh negative genotype.

The antibody against the Rh positive antigen does not occur naturally, but an Rh negative person receiving a whole-blood transfusion of Rh positive blood can build antibodies against the antigen. The first transfusion of this kind would cause no complications; however, subsequent transfusions would cause agglutination of the recipient's red blood cells, possibly causing death.

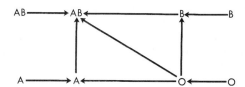

Figure 10.14 Schematic diagram showing compatibility and incompatibility among blood types for whole-blood transfusions. The key to transfusions possibly depends on whether or not incoming cells are agglutinated. The direction of the arrows indicates which type can receive whole-blood transfusions from the other types. Blood type O is a *universal donor* and blood type AB is a *universal recipient.*

The Rh positive antigen–antibody reaction was later found to be responsible for the condition known as *erythroblastosis fetalis* in human infants at birth. This condition results in jaundice, anemia, and often death. It occurs only when an Rh negative woman marries an Rh positive man and has Rh positive children. The Rh positive antigenic substance in some way crosses the placenta from the blood of the baby to that of the mother. The Rh negative mother builds antibodies against the Rh positive antigen from her child. These antibodies then cross the placenta from the blood of the mother to that of the child. When the mother produces enough antibody, many of the baby's cells are destroyed. Two-way blood transfusions at birth often save such a child. Usually, the Rh negative mother can have several Rh positive children before she builds enough antibody against the Rh positive antigen to cause serious difficulty. However, an Rh negative mother who has received a whole-blood transfusion with Rh positive blood may have enough antibody to affect her first Rh positive child. Rh negative children are not affected. When a father is Rh positive Rh negative and the mother is Rh negative Rh negative, approximately one-half of their children are Rh negative Rh negative. If the father is Rh positive Rh positive, all the children are Rh positive Rh negative.

A method for preventing erythroblastosis fetalis in human infants has been developed. This method is based on the discovery that an Rh negative mother of blood type O (genotype *aa*) always possesses antibodies A and B in her blood serum. If her baby carries blood antigens A or B on its red blood cells and some of these get into the mother's bloodstream at birth, they will be destroyed by means of the antibody–antigen reaction. This reaction also destroys Rh positive–bearing blood cells at the same time and before they have an opportunity to stimulate Rh positive antibody production by the Rh negative mother. As an experiment, Rh negative mothers with newborn Rh positive babies were injected with serum that contained antibodies against the Rh positive cells of their children. This serum did destroy a high proportion of the Rh positive–bearing cells, but it was found that the treatment later stimulated Rh positive antibody production. Additional experimentation led to the successful use of anti-Rh gamma globulin, which coats the antigen of the Rh positive cell so that it does not come in contact with antibody-forming cells. This treatment inhibits production of antibodies against the Rh positive antigens if given within 72 h after the birth of the first Rh positive baby. This protects the second child from erythroblastosis fetalis. Injections must be repeated with each successive Rh positive baby the Rh negative mother has thereafter. The treatment is ineffective if given to Rh negative mothers who have already given birth to Rh positive babies and therefore possess Rh positive antibodies. It will, however, greatly benefit Rh negative women who start families in the future.

Many blood types and groups have been discovered in farm animals. Some appear to cause diseases similar to the erythroblastosis fetalis in humans, but the antibody in animals is usually transferred from mother to young through **colostrum.** The placental membrane of nonprimate mammals is less permeable than that of **primates** and does not allow antibodies to pass through to the young within the uterus.

Hemolytic disease of the newborn has been reported in cattle, horses, pigs, dogs, and cats. The newborns are healthy at birth but begin to show signs of acute anemia and icterus (yellow discoloration of eyes, skin, and mucous membranes resulting from destruction of red blood cells) within 12 to 24 h after nursing. Treatment involves giving the newborn a blood transfusion from a compatible donor and feeding the newborn a milk replacer for 72 h while colostrum (containing the destructive antibodies) is milked from the dam and discarded.

Hemolytic disease can be prevented by selecting a sire that is of the same blood type as the dam. If this is impossible, the newborn's red cells should be tested (cross matched) with the dam's serum and shown to be negative for clumping (agglutination) prior to allowing the newborn to nurse colostrum.

Blood-typing of farm animals has several uses on a practical basis. It may be used to identify individuals and their parents. (Blood-typing is required of bulls used in commercial artificial insemination.) Blood types appear to be associated with performance in some instances, and the mating of individuals having certain blood types may result in superior performance. Considerable blood-typing has been done in poultry in relation to the hatchability of eggs and egg production. In swine, blood-typing has been used to identify individuals that are genetically susceptible to porcine stress syndrome (PSS), a condition that can affect survival under stress conditions and can also cause pale, soft, exudative (PSE) carcasses to be produced.

All proteins in the blood of animals are genetically determined, and certain tests for their presence can be used for identification purposes. Those commonly used are albumins, haptoglobins, prealbumins, and transferrins.

10.5.5 White Blood Cells (Leukocytes)

Leukocytes include basophils, eosinophils, lymphocytes, monocytes, and neutrophils. They differ from erythrocytes in possessing a nucleus during their residence in the circulatory system. Neutrophils and lymphocytes commonly represent 85 to 90 percent of the leukocytes in farm mammals. The total number of these two types of cells in farm mammals is approximately equal, although the proportion varies somewhat among species. The biological significance of these different proportions, if any, is not fully understood. Temporary stress in farm mammals results in a highly significant increase of neutrophils in proportion to lymphocytes. The proportion returns to the original level a few hours following relaxation or the removal of stress. The change in proportions of these leukocytes during periods of stress results from the interaction of a hormone from the anterior pituitary gland (ACTH—adrenocorticotropin hormone) and adrenal cortical hormones (glucocorticoids).

Neutrophils are produced in bone marrow. They fight disease by migrating to the point of infection, engulfing (**phagocytizing**) bacteria, and destroying them. Leukocytes generally increase in number during bacterial infections, but viral infections cause their number to decrease (leukopenia). Thus a differential leukocyte count (a count used to determine the relative percentages of neutrophils, eosinophils, basophils, monocytes, and lymphocytes in circulation) may be used to diagnose disease. Eosinophils are so named because they contain granules that turn red when stained with eosin (a red dye). Eosinophils normally constitute less than 5 percent of the total leukocyte count in farm animals. An increase from normal in eosinophils is indicative of an allergic reaction within the body or some type of parasitism.

Lymphocytes are produced in the lymph nodes, spleen, thymus, and other lymphoid tissues. They fight disease by producing and releasing antibodies when infections occur. Lymphocytes are involved in the production of antibodies necessary for the development of immunity to certain diseases. Lymphocytes are also important in recognition and destruction of foreign materials in the body and tumor cells. Lymphocytes undergo division readily when placed in a suitable culture outside the body. Dividing lymphocytes clearly show individual **chromosomes** during the **metaphase** stage of division. Culturing of lyphocytes is widely used in studies of chromosome numbers in many animal species.

Blood platelets (thrombocytes) are formed by large cells known as *megakaryocytes* in bone marrow. A cubic millimeter of blood contains 200,000 to 600,000 platelets. Platelets form a plug at the site of injury to a blood vessel and are involved in blood clotting.

10.6 THE DIGESTIVE SYSTEM

The digestive system consists of a tube that extends the full length of the body. It includes the mouth, pharynx, esophagus, stomach (or stomach compartments in **ruminants**), small intestine, cecum, large intestine, and rectum (Figures 10.15 and 19.1).

Various modifications of the digestive system are found in different species. Ruminants (cows, goats, and sheep) have four stomach compartments. These include the **rumen (paunch)**, **reticulum** (honeycomb), **omasum** (manyplies), and **abomasum** (true stomach). Poultry possess a crop, proventriculus (true stomach), and ventriculus (gizzard), as shown in Figure 10.16 (see also Chapter 19). Horses have a very large cecum, which serves as a primary site for the digestion of forages.

Accessory organs of the digestive system include the liver, pancreas, and salivary glands. The salivary glands, located in the throat region, secrete saliva into the mouth and pharynx. Saliva moistens and lubricates food to facilitate swallowing. Saliva also aids in initiating the digestive process. The liver filters blood, stores certain nutrients absorbed from the intestine, and secretes bile, which aids in fat digestion. The pancreas is located adjacent to the liver and secretes enzymes necessary for the digestion of foods and insulin, which regulates blood glucose concentration and metabolism (Table 19.2). A more complete discussion of the anatomy and physiology of the digestive system is presented in Chapter 19.

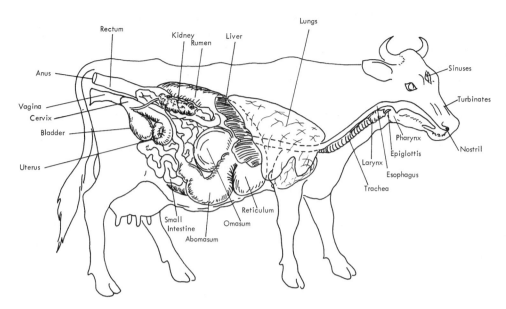

Figure 10.15 Selected internal parts of the cow.
Courtesy of Dr. P. D. Garrett and Dr. D. E. Rodabaugh.

10.7 THE RESPIRATORY SYSTEM

The respiratory system of animals includes the lungs and passageways through which air is brought into and exhaled from the lungs. These passageways include the nostrils, nasal cavity, **pharynx, larynx,** and **trachea** (Figures 10.15 and 10.17).

The *nostrils* are two in number and are the external openings of the respiratory tract. Each nostril leads to a nasal cavity, which is separated from the mouth by the hard and soft palates. Both food and air pass through the *pharynx,* but its structure is such that air cannot be inspired at the same time food is being swallowed. The passage of food and air through the pharynx is controlled by a valvelike structure, called the *epiglottis.*

In mammals, the *larynx* is known as the *voice box.* It also controls the inspiration and expiration of air. Moreover, it prevents the inhalation of foreign objects into the lungs. The *trachea* is a continuation of the larynx. It consists of adjacent rings of cartilage, resembling the vacuum hose seen on many electric floor sweepers. These rings are rigid and prevent collapse of the trachea, so that it always remains open. The trachea continues as a single tube to the base of the heart, where it divides into two tubes, called *primary bronchi.* In birds, this is the area of the voice box syrinx. One bronchus passes into each lobe of the lungs. The primary bronchi branch into still smaller bronchi and finally into very small tubes, known as *bronchioles,* in the lungs. The ends, or terminal branches, of bronchioles are the *respiratory bronchioles,* which in turn open into alveolar ducts that lead to the final and smallest portions of the respiratory passageway, called *alveoli* (Figure 10.17). Birds differ from mammals in having relatively nonexpanding lungs and accessory air cavities in the body cavity and major long bones. The accessory air cavities decrease body density and aid in making birds air-mobile. Chickens have nine air sacs (four pairs

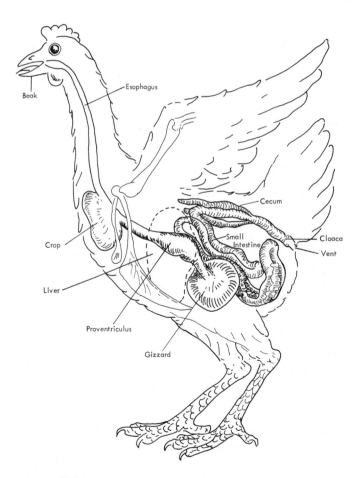

Figure 10.16 Digestive system of the chicken.
Courtesy of Dr. P. D. Garrett and Dr. D. E. Rodabaugh.

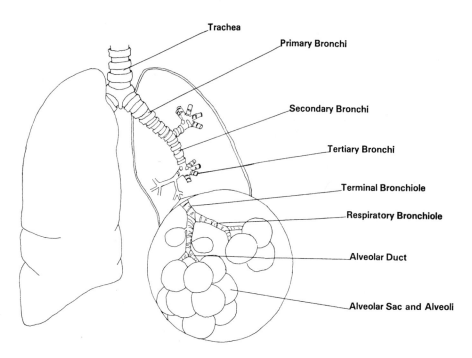

Figure 10.17 Diagram of the mammalian lung. Enlarged area depicts the terminal respiratory structures.

Drawn by F. E. Staten, University of Illinois.

and a single) and foramen (small cavities or perforations in bones) through which some respiratory action occurs. These air sacs also increase their buoyancy (ability to float on water and fly in air).

Each lobe of the mammalian lung consists of elastic, spongy material that is greatly expanded when filled with air. When expanded to full capacity, the lungs completely fill the space available in the thoracic cavity.

The primary function of the respiratory system is an exchange of gases with the atmosphere. Oxygen is the gas absorbed by the lungs from inhaled air, and carbon dioxide is the gas exhaled from the lungs. If carbon monoxide is inhaled, it unites with the iron of hemoglobin (the oxygen-carrying component of blood) and does not dissociate, forming stable compounds called *carboxyhemoglobins* that cannot carry oxygen. As a result, the animal dies from lack of oxygen or, in a sense, of suffocation. Other chemicals, such as nitrates, chlorates, cyanide, and prussic acid, are also poisonous to animals because they interfere with internal respiration or with the normal utilization of oxygen by tissues (Section 22.6).

Air is brought into mammalian lungs (inspired) by contraction of the diaphragm and enlarging of the thoracic cavity, thereby creating a vacuum. This action causes the lungs to enlarge, drawing air into them. A partial vacuum exists within the thoracic cavity under normal conditions. If this vacuum is destroyed by a puncture, the lung collapses. Air is forced out of the lungs (expired) by a decrease in the vacuum in the thoracic cavity. This is caused by relaxation of the diaphragm, contraction of the muscles that decrease thoracic cavity volume, and retraction of the elastic fibers within the alveoli.

The rate of breathing is controlled by a group of nerve cells in the medulla of the brain. This region, known as the *respiratory center,* regulates inspiration and expiration of air by the lungs. It is influenced by the carbon dioxide content of blood, body temperature, and other brain centers.

Oxygen passes from the alveoli in the lungs to the red blood cells of the circulatory system by means of simple diffusion. A reverse diffusion occurs, resulting in the discharge of carbon dioxide. Oxygen and carbon dioxide in body tissues are exchanged in a similar manner.

10.8 THE NERVOUS SYSTEM

Normal animals have the ability to adjust or to react to many stimuli in their internal and external environments. Their ability to do this depends basically on their nervous system. Thus the nervous system coordinates physical activities of the body and provides basic pathways for the action of all senses (hearing, sight, smell, taste, and touch).

The nervous system may be grossly divided into two major parts: the central nervous system (brain and spinal cord) and the peripheral nervous system (somatic and autonomic nerves).

Nerve cells are called *neurons.* Neurons have a single long fiber, called an **axon,** and several branched threads, called *dendrites.* The dendrites receive stimuli from other nerves or from a receptor organ, such as a sense organ. The impulse passes from the dendrite through the cell body to the axon. The axon conducts impulses to the dendrite of another neuron or to an *effector organ,* such as a muscle cell. The place where axons and dendrites come together is called a *synapse*. For an impulse to pass from a receptor organ to the brain or from the brain to an effector organ, it must travel over numerous neurons, crossing several synapses. Some impulses such as those involved in leg reflexes do not pass through the brain.

A nerve may be either a single neuron (axon) or a bundle of several neurons (axons). Numerous nerve fibers bound together are called a *nerve trunk*. Some axons, or nerve fibers, are covered by a fat-containing *myelin*, or *medullary, sheath*. A small bundle of cell bodies gathered together outside the brain and spinal cord is called a *ganglion*. Nerve cells that *receive* stimuli are called *sensory*, or *afferent, neurons*. These neurons carry impulses from the sense organs to the central nervous system. Nerve cells that carry impulses from the brain or other nerve centers to muscles or glands are called *motor*, or *efferent, neurons*.

10.8.1 Gross Anatomy of the Brain

The *brain* consists of the cerebrum, cerebellum, pons, and medulla oblongata (Figure 10.18).

The *cerebrum* is the largest component of the brain. Its surface area consists of numerous folds and is wrinkled in appearance. It serves as the decision-making center of the brain and controls such mental activities as voluntary muscle control and interpretations of various sensations, for example, hearing, seeing, and tasting. It is also involved in reasoning. The cerebrum is usually large relative to body size in animals of higher intelligence. Destruction of the cerebrum does not cause a complete cessation of body functions.

The *cerebellum* functions as a coordinator of the brain's other centers and is a mediator between them and the body. It functions as a coordinator of muscular activity in eating, vocalizing, running, and walking. Damage to the cerebellum results in incoordination, which interferes with voluntary muscular action, but does not cause paralysis. Farm animals have a large cerebellum, although elephants and whales probably have the largest relative to body size.

The *pons* and *medulla oblongata* control reflex actions such as breathing, swallowing, vomiting, and blinking of eyelids. They usually act independently of the cerebrum and cerebellum.

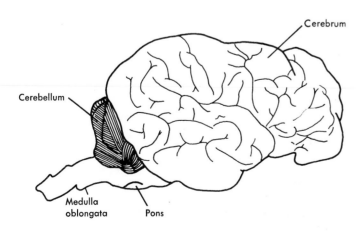

Figure 10.18 Gross anatomy of the mammalian brain.

10.8.2 The Spinal Cord

The spinal cord is located in the center of the vertebral column and is the main line through which messages are transmitted to and from the brain to various body parts. It is a continuation of the *medulla oblongata*. The spinal cord is segmented (divided), and each segment gives rise to a pair of spinal nerves. The spinal cord receives sensory, or afferent, nerve fibers, which transmit impulses from different parts of the body via the **dorsal** roots of the spinal nerves. They also yield efferent, or motor, nerve fibers, which transmit impulses from the brain and spinal cord to various body parts through the ventral roots of the spinal nerves.

10.8.3 The Peripheral Nervous System
Somatic Nerves

This system includes all nervous structures outside the brain and spinal cord. Peripheral nerves consist of bundles of nerve fibers that convey impulses from the brain and spinal cord to muscles (motor nerves) or carry stimuli from the external body to the spinal cord or brain (sensory nerves).

Autonomic Nerves

This is a relatively independent system of nerves positioned between the spinal cord and brain on one side and the various glands, blood vessels, viscera, heart, and smooth muscles of other body regions on the other. This system provides for semi-automatic regulation of involuntary functions subject to central influence. Whereas the peripheral nervous system is associated with what are usually known as somatic (body) structures, the autonomic nervous system is associated with **visceral** organs.

10.9 THE URINARY SYSTEM

The urinary system consists of the kidneys, ureters, bladder, and urethra. Major parts of the urinary system are shown in Figure 10.19.

The *kidneys* are paired and located ventral to the lumbar vertebrae in the abdominal cavity. Each kidney is closely attached to the abdominal wall by a **fascia,** vessels, and the **peritoneum.** The kidneys of the cow and hen are lobulated. The right kidney of the horse is heart-shaped and the left is bean-shaped. The kidneys of swine and sheep are bean-shaped. The gross structure of the kidney consists of an outer tissue layer, called the renal *cortex,* and an inner portion, called the renal *medulla.* The medulla connects with the expanded beginning of the ureter at the renal pelvis. Microscopically, the cortex of the kidney consists of thousands of tiny connections of vascular and urinary tubules. Blood is filtered from the capillary bed, called the *glomerulus,* into the blind sac of the urinary tubule, called *Bowman's capsule* (Figure 10.20). Together these unions of vascular and urinary tubules are called *Malpighian bodies.* These tubules join together in the medulla, forming larger and larger tubes, which eventually empty into the central cavity (pelvis) of the kidney, continuous with the ureter. Blood is conveyed to the kidney through the renal artery and distributed to the glomeruli.

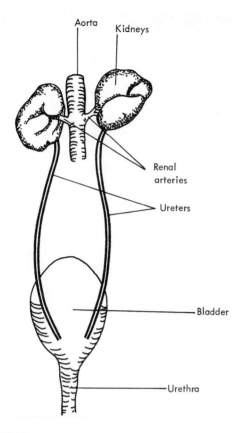

Figure 10.19 Major components of the mammalian urinary system.

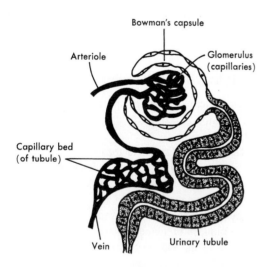

Figure 10.20 Secretion unit of the kidney. Filtration proceeds from the glomeruli capillaries into Bowman's capsule and then down the urinary tubule.

The *ureter* is a single tube leading from each kidney to the bladder. The *bladder* is a very elastic sac that stores urine until it is voided. Chickens do not have a urinary bladder, but rather the ureters end in the *cloaca,* and feces and urine are voided together. The *urethra* is a highly elastic tube that leads from the bladder to and through the penis in the male (Figure 13.2) and to the vagina in the female (Figure 13.7).

One function of kidneys is to filter waste products from blood. These are voided from the kidneys in urine that flows steadily into the ureters and then into the bladder. Urine normally contains various mineral salts; urea from protein metabolism; uric acid from metabolism of a modified protein, called nucleoprotein; creatinine from metabolism of protein and muscle; and many other materials. The end product of protein and nucleoprotein metabolism in poultry is uric acid. Most animals have enzymes that convert (oxidize) uric acid into allantoic acid, which is excreted in the urine. The exceptions are primates, Dalmatian dogs, birds, and reptiles (Figure 10.21). Uric acid precipitates at very low concentrations and can form urate deposits in soft tissues and bone joints, thereby causing the condition called *gout.* The kidneys also function to regulate blood composition and to maintain the normal internal conditions (**homeostasis**) necessary for maintenance of life. If both kidneys cease functioning, toxic wastes and electrolyte imbalances result in death of the animal.

10.10 SUMMARY

In anatomy and physiology, farm animals are similar in many respects. All have skeletal, muscular, circulatory, digestive, respiratory, reproductive, nervous, and urinary systems. Each of these systems performs the same general function within each individual in the many different species. Abnormalities of anatomy and physiology may be of an inherited or an environmental origin. These can greatly impair proper functioning and morphology of the body and may even cause death.

The blueprint for proper development and function of each individual is carried within the genetic material (DNA) of the fertilized egg. These blueprints are followed closely in the anatomical and physiological development of an individual. However, they can be modified to a certain extent by factors within the environment, such as nutrition and disease. The blueprint in the DNA calls for a pig to be a pig and a calf to be a calf. This part of the blueprint is followed to completion without exception.

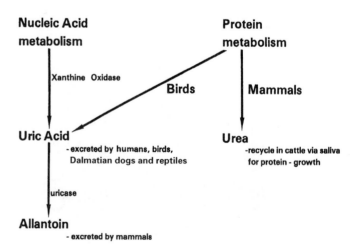

Figure 10.21 Urinary end products of nitrogen and nucleic acid metabolism in birds and mammals.
Courtesy of Dr. P. C. Harrison, University of Illinois, at Urbana–Champaign.

STUDY QUESTIONS

1. Define (a) gross anatomy, (b) histology, (c) comparative anatomy, (d) embryology, and (e) physiology.

2. What is meant by (a) dorsal, (b) ventral, (c) cranial, and (d) caudal?

3. Name the two layers of skin and the major glands of skin.

4. What is the coat covering for sheep? For horses? For poultry?

5. Which farm animals do not possess cleft (split) hooves?

6. Distinguish between an endoskeleton and an exoskeleton.

7. What substances in a ration may affect bone growth?

8. Name the different kinds of joints found in an animal's body.

9. Distinguish between voluntary and involuntary muscles.

10. What are the major functions of the skeletal system?

11. What classifications are there for muscle other than voluntary and involuntary?

12. Describe a motor unit.

13. Describe the chemical reactions involved in muscle action.

14. Define the term *oxygen debt*.

15. Name the parts of the circulatory system, and give their functions.

16. In what way does the pulmonary circulatory system differ from the systemic circulatory system in its function?

17. Distinguish between blood serum and blood plasma.

18. Describe anemia in animals.

19. In humans, what blood type is a universal donor? A universal recipient?

20. Explain how the disease erythroblastosis fetalis may occur in some human infants.

21. What kinds of leukocytes are found in blood, and where are they produced?

22. Name the parts of the digestive system.

23. Name the four stomach compartments in the ruminant.

24. What are the accessory organs of the digestive system?

25. Name the parts of the respiratory system.

26. Explain how air is brought into and forced out of the lungs.

27. What are the major portions of the brain? What is a function of each?

28. What is included in the peripheral nervous system? The autonomic nervous system?

29. Outline the gross anatomy of the kidney.

30. What are the main functions of the kidney?

31. What are the end products of protein and nucleoprotein metabolism in birds and farm mammals?

11

THE APPLICATION OF ENDOCRINOLOGY TO SELECTED ANIMALS AND HUMANS[1]

Vital mechanisms have one object, preserving constant the internal environment.

Claude Bernard (1813–1878)
French physiologist

11.1 INTRODUCTION

An animal's body is a complex and delicately balanced machine. It is made of many cells and organs that perform specific functions necessary for the maintenance of life. The proper function of these parts is dependent to a great extent on the action and interaction of chemical substances secreted by endocrine glands. Failure of cells and organs of the body to function properly may cause inefficient performance or sickness and death of an individual. Because so many important bodily functions are dependent on endocrine glands, an understanding of their secretions and how they function is basic to the study of animal science.

11.2 THE SCIENCE OF ENDOCRINOLOGY

Endocrinology is the science that deals with the study of endocrine glands and their secretions. **Endocrine** glands are located in different regions of the body and contain cells that secrete chemical substances known as *hormones* (Figure 11.1). Endocrine glands are ductless and release their secretions directly into the bloodstream. For this reason they are referred to as *glands of internal secretion*. **Exocrine** *glands* differ from endocrine glands in that their secretions are released into ducts that lead to the body surface or to cavities and their surfaces within the body. The pancreas is unique because it is both an exocrine and an endocrine gland, secreting pancreatic juice (exocrine) and insulin (endocrine).

Endocrine glands of the body secrete and release small amounts of their hormones into the bloodstream. Tiny quantities of these hormones have powerful effects on the body. For example, 1 μg (1/1,000,000 g) of oxytocin will cause the immediate ejection of milk from the mammary glands of a lactating woman.

Once hormones are in the bloodstream, they are carried to various regions of the body where they cause target organs to perform specific functions. The term **hormone** describes their action; it means "to stir up" or "to stimulate." Except for small amounts that may be held in the endocrine organs themselves, hormones are not stored in the body. Therefore in cases of endocrine deficiency, administration of repeated small doses, rather than giving large doses at infrequent intervals, are normally required to correct the deficiency.

Hormones have an important effect on body processes such as growth and fattening (Chapter 12), reproduction (Chapter 13), lactation (Chapter 15), and egg laying (Chapter 16). A high rate of performance in all animals depends on the proper level and balance of these chemical substances at their points of activity. Hormones are believed to instruct cells how to use metabolic energy (e.g., in growth, work, or fat deposition).

11.3 ENDOCRINE GLANDS AND THEIR SECRETIONS

The **anterior** pituitary gland looks like a cherry with the stem attached to the base of the brain and is known as the "master gland" of the body. It is called this because it directly or indi-

[1]The authors acknowledge with appreciation the contributions to this chapter of Dr. D. J. Kesler, Department of Animal Sciences, University of Illinois at Urbana–Champaign.

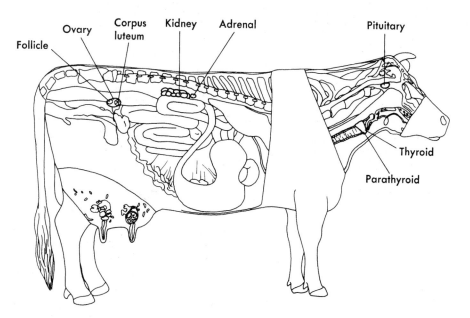

Figure 11.1 Endocrine glands of the cow. The same endocrine glands are present in females of other mammalian species. They are also present in the male, except the male gonad is the testicle, whereas the female gonad is the ovary. The endocrine structures in the testicle are interstitial cells, whereas those of the ovary are the follicles and corpora lutea.
Missouri Agr. Expt. Sta. Bull. *795.*

rectly controls the secretion of many hormones (Table 11.1). Removal of the pituitary gland by surgical means (**hypophysectomy**) affects the function of many other endocrine glands. For example, in a hypophysectomized animal the adrenal cortices, **gonads** (ovaries and testes), and thyroid glands regress and do not secrete their hormones. Thus the effects of hypophysectomy are (1) arrested growth; (2) **atrophy** of the gonads and indirectly of the accessory sex organs; (3) suppression of lactation; (4) atrophy of the thyroid glands and adrenal cortices, and probably degeneration in the parathyroids; (5) hypoglycemia, increased sensitivity to insulin, and reduction in liver and muscle glycogen; (6) lowered metabolic rate; and (7) diminished resistance to infection and shock. Therefore one may readily understand why many describe the pituitary gland as the most important gland in the body. Indeed, without it mammals do not reproduce, lactate, or carry out metabolic processes essential for life and well-being.

The pituitary consists of two distinct parts, the anterior and posterior pituitary glands, and it is located in a bony depression at the base of the brain (Figures 11.2 and 11.3).

Located directly above the pituitary gland is the hypothalamus. (Major structures of the cow brain are depicted in Figure 11.3.) The hypothalamus is a very small area of the brain that regulates and coordinates hormonal activity. It regulates the release of hormones from the anterior pituitary gland and produces hormones that are stored and released from the posterior pituitary gland. Releasing or inhibiting hormones produced by the hypothalamus are carried to the anterior pituitary gland by small blood vessels known as the hypothalamic–hypophyseal portal blood vessels. Oxytocin and vasopressin, however, are produced by nerve cells in the hypothalamus and travel by way of nerve cells to the posterior pituitary gland where they are stored. Additional functions of the hypothalamus include regulation of body temperature, appetite, and thirst.

The anatomy and physiology of other glands of internal secretion and the function of their various hormones that deal with reproduction, growth, and milk and egg production will be discussed in detail in subsequent chapters.

11.4 THE CHEMICAL NATURE OF HORMONES

Chemically hormones fall into five general classes: (1) **proteins** and glycoproteins, (2) peptides, (3) amines, (4) steroids, and (5) nonsteroidal lipids. Proteins, glycoproteins, and peptides are composed of amino acids. The general chemical structure of amino acids is as follows:

$$R - \underset{\underset{NH_2}{|}}{\overset{\overset{H}{|}}{C}} - COOH$$

The letter R is called the *side chain* of amino acids and consists of various combinations of carbon and hydrogen and in some cases nitrogen, oxygen, and/or sulfur. In the amino acid glycine, the side chain R is replaced by hydrogen (H) whereas in alanine it is replaced by CH_3, a methyl group. The dividing line between proteins and peptides is arbitrary with proteins

TABLE 11.1	Hormones Secreted by the Endocrine Glands and Their Major Functions

Endocrine Gland	Hormone Secreted	Major Physiological Function
Hypothalamus	Gonadotropin-releasing hormone (GnRH)	Stimulates release of LH and FSH
	Corticotropin-releasing hormone (CRH)	Stimulates release of ACTH
	Thyrotropin-releasing hormone (TRH)	Stimulates release of TSH
	Growth hormone–releasing hormone (GHRH)	Stimulates release of growth hormone
	Growth hormone–inhibiting hormone (somatostatin)	Inhibits release of growth hormone
	Prolactin-releasing hormone (PRH)	Stimulates release of prolactin
	Prolactin-inhibiting hormone (PIH)	Inhibits release of prolactin
	Oxytocin*	Causes ejection of milk, expulsion of eggs in hens, and uterine contractions
	Vasopression (antidiuretic)*	Causes constriction of the peripheral blood vessels and water resorption in the kidney tubules
Anterior pituitary	Growth hormone (GH or somatotropin)	Promotes growth of tissues and bone matrix of the body
	Adrenocorticotropin (ACTH)	Stimulates secretion of steroids (especially glucocorticoids) from the adrenal cortex
	Thyrotropin or thyroid-stimulating hormone (TSH)	Stimulates thyroid gland to secrete thyroxine
	Prolactin (Prl)	Initiates lactation and induces maternal behavior
	Gonadotropic hormones	
	Follicle-stimulating hormone (FSH)	Stimulates follicle development in the female and sperm production in the male
	Luteinizing hormone (LH)	Causes maturation of follicles, ovulation, and maintenance of the corpus luteum in the female. Causes testosterone production by the interstitial cells of the testes in the male
Thyroid	Thyroxine, triiodothyronine	Increases metabolic rate
	Calcitonin	Lowers the concentration of calcium in the blood and promotes incorporation of calcium into bone
Parathyroid	Parathyroid hormone	Maintains or increases the level of blood calcium and phosphorus
Adrenal glands		
Cortex (shell)	Glucocorticoids	Mobilizes energy, increases blood glucose level, has an antistress action
	Mineralocorticoids	Maintains salt and water balance in the body
Medulla (core)	Epinephrine (adrenalin)	Stimulates the heart muscles and the rate and strength of their contraction
	Norepinephrine	Stimulates smooth muscles and glands and maintains blood pressure
Ovaries		
Follicles	Estrogens	Causes growth of reproductive tract and mammary duct system
Corpus luteum	Progesterone	Prepares reproductive tract for pregnancy, maintains pregnancy, and causes development of mammary lobule–alveolar system
	Relaxin**	Causes relaxation of ligaments and cartilage in the pelvis, which assists in parturition
Uterus	PGF$_{2\alpha}$	Causes luteolysis and a decrease in progesterone
Testes	Androgens (testosterone)	Causes maturation of sperm; promotes development of male accessory sex glands and secondary sex characteristics
Pancreas (islets of Langerhans)	Insulin	Lowers blood glucose
	Glucagon	Raises blood glucose
Placenta (in some species)	Gonadotropins (pregnant mare serum gonadotropin also called equine chorionic gonadotropin, human chorionic gonadotropin), estrogens, and progesterone	Promotes the maintenance of pregnancy
	Relaxin**	Causes relaxation of ligaments and cartilage in the pelvis, which assists in parturition
Pineal	Melatonin	Controls seasonal breeding

*Oxytocin and rasopression are produced by the hypothalamus, and stored in and released from the posterior pituitary.

**Relaxin is produced by the corpus luteum in swine but is produced by the placenta in most other species.

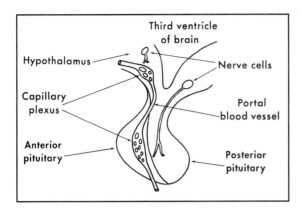

Figure 11.2 Diagram of the hypothalamus and the pituitary gland.

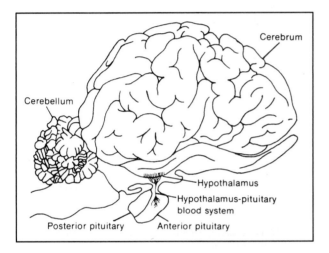

Figure 11.3 Major structures of the cow brain.

From T. R. Troxel and D. J. Kesler, "The Bovine Estrous Cycle Dynamics and Control," Univ. of Illinois Coop. Ext. Cir. *1205, 1982.*

having a molecular weight greater than 10,000. Glycoproteins are proteins that are glycosylated; they are conjugated to one or more carbohydrate residues. Glycosylation generally increases the half-life of a molecule and allows it to remain in circulation and be biologically active for a longer period of time. Insulin and ADH are polypeptide hormones. Examples of glycoprotein hormones include FSH and LH.

Epinephrine (secreted by the adrenal medulla), thyroxine (secreted by the thyroid), and melatonin (secreted by the pineal gland) are amines (Figure 11.4). These hormones are derived from the amino acids tyrosine and tryptophan.

Steroid hormones (Figure 11.5) are secreted by the ovaries, testes, adrenal cortices, and certain fetal membranes. Several kinds of estrogens and androgens are found in body fluids; they are all steroid in nature and closely allied in chemical structure. Cholesterol is the precursor for biosynthesis of all steroid hormones. Synthetic compounds have been synthesized in laboratories and used for treating hormone deficiencies and in animal production (Sections 11.8.3 and 18.6).

Although there are a large number of nonsteroid lipids, one that has significant biological function is prostaglandin $F_{2\alpha}$ ($PGF_{2\alpha}$). In the cow $PGF_{2\alpha}$ is synthesized by the uterine endometrium and is a long-chain fatty acid (Figure 11.6). Prostaglandin $F_{2\alpha}$ from the uterus is transported in the blood by a counter-current mechanism to the ovary where it causes luteolysis. Arachidonic acid is the precursor for biosynthesis of prostaglandins. Prostaglandins were first identified in semen and seminal tissues. The substances were given the name *prostaglandins* because it was thought that they were secreted by the prostate gland. Later it was discovered that these substances came from the seminal vesicles and are synthesized by many other body tissues as well. In many tissues prostaglandins act locally rather than being transported via the bloodstream. Aspirin blocks the enzymes involved in the production of prostaglandins.

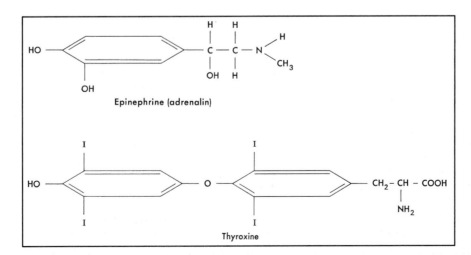

Figure 11.4 Chemical structures of epinephrine and thyroxine, which are examples of hormones that are amines.

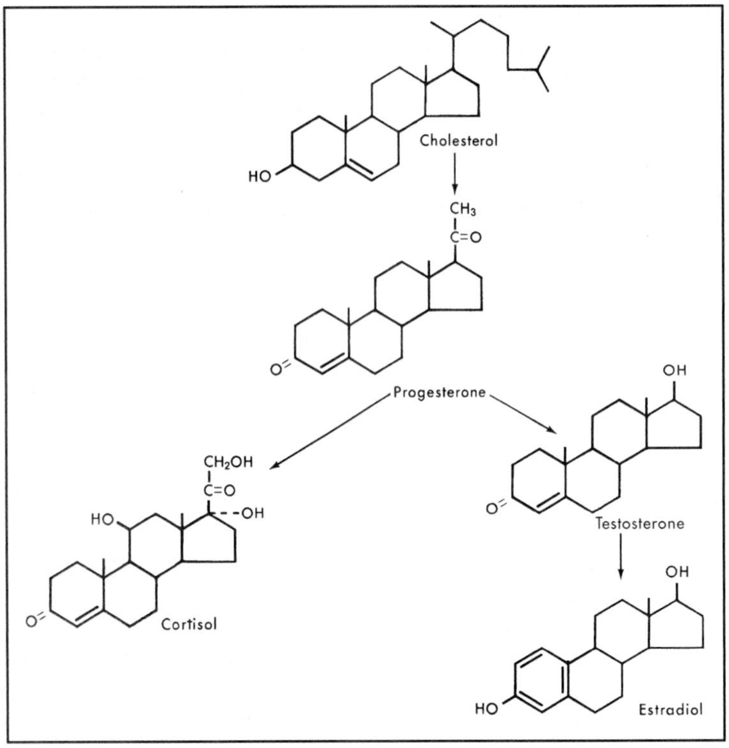

Figure 11.5 Biosynthetic pathways and chemical structures of selected steroid hormones.

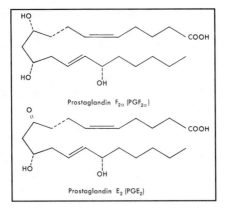

Figure 11.6 Chemical structures of selected prostaglandins. Note their configurational similarities.

11.5 FUNCTIONS OF HORMONES

Very small amounts of hormones are required to perform a particular function. Hormones affect growth, body shape, and the way in which the body uses food and, perhaps most important, they allow the body to make proper adjustments to changes in the environment. When overdoses are administered to animals the effects may be detrimental. For this reason proper dosage level for treatment is critical. In addition, two individuals may respond differently to the same hormone dosage. This is possible because two or more individuals differ in the amount of a hormone they secrete. Moreover, a given hormone may act on different tissues in different species. For example, prolactin stimulates the mammary gland to secrete milk and causes the pigeon's crop gland to **proliferate** and secrete "pigeon milk."

Many hormones are quickly deactivated when administered to animals and may disappear from the blood in only a few hours. Even though the hormone has disappeared from the bloodstream its effect may not become apparent until a few hours or days after it is administered.

11.6 MECHANISM OF HORMONE ACTION

After hormones are produced by the endocrine glands they are released into and travel through the circulatory system. In the circulatory system they are in very low concentrations, generally in nanogram (10^{-9} g) or picogram (10^{-12} g) amounts. Because they are in such low concentrations they do not reach target organs merely by chance. Hormones present in circulating blood adhere to receptors that have an affinity for specific hormones. Hormone receptors are located in the cell membrane or within the cell and bind with specific hormones. Only a specific hormone will interact with a specific receptor. For example, receptors on the interstitial cells in the testes will interact only with luteinizing hormone (LH). Once the hormone has interacted with the receptor, the action that hormone elicits will be carried out within the cell. For example, once LH has interacted with the interstitial cell receptors in testes, the synthesis of testosterone is initiated.

The intracellular events involved after the receptor has been activated vary according to cellular function. In some cases enzyme systems may be activated whereas in other cases messenger RNA synthesis may be initiated.

11.7 REGULATION OF HORMONE SECRETION

The activity of endocrine glands is carefully balanced. If a gland becomes either too active or not active enough illness results. For good health and development the glands must work together as a unit. Secretion of hormones by endocrine glands may be regulated in several ways.

11.7.1 Feedback Mechanism

An interaction between two hormones may regulate the secretion rate of both. This interaction is known as a *feedback mechanism.* For example, the production of LH by the anterior pituitary gland stimulates the testes to secrete testosterone. As more and more testosterone is produced by the testes it enters the bloodstream and is carried to the hypothalamus and the anterior pituitary gland where it causes a decreased production of LH. Thus when testosterone concentration is high, LH production is decreased whereas when testosterone secretion is low, LH production is increased.

11.7.2 Chemical Homeostasis

Hormone secretions may be controlled by the blood level of chemicals (not a hormone) on which the hormone acts. For example, a high concentration of blood calcium causes a low secretion rate of parathyroid hormone, the hormone secreted by the parathyroid gland. A low concentration of blood calcium permits release of the parathyroid hormone. This interaction between calcium and parathyroid hormone keeps calcium at a fairly constant concentration in blood. Keeping calcium concentration constant in the bloodstream is especially important in lactating animals. Another example of chemical regulation of hormone secretion is the role of blood glucose in the control of insulin secretion. High levels of blood glucose trigger secretion of insulin by the pancreas. Low levels of blood glucose inhibit insulin secretion and trigger the release of other hormones that act to increase blood glucose concentration (e.g., epinephrine, cortisol, and glucagon).

11.8 PRACTICAL USES OF NATURAL AND SYNTHETIC HORMONES

The discovery of hormones and the elucidation of their major functions in the body have led to their use in the control or regulation of certain body functions and in practical livestock production.

11.8.1 Reproduction in Farm Animals and Humans

It was estimated in 2000 that over 100 million women of reproductive age throughout the world were taking compounds orally in the form of pills as a means of birth control. Many others receive steroids via sustained-release injections or subcutaneous implants for the same purpose. A wide variety of steroid substances are extremely effective in suppressing the production of gonadotropic hormones and thus in blocking ovulation. The most widely used product contains a combination of a synthetic progesterone and a synthetic estrogen; however, synthetic progestins are being used effectively alone.

Progestins have been used experimentally to inhibit and synchronize estrus and ovulation in farm mammals. When administered to a group of females for a period of several days they inhibit ovarian follicle development and ovulation. When these compounds are removed follicular growth is resumed and most females in the group will come into estrus and ovulate within a period of 3 to 5 days. Hence estrus and ovulation are caused to occur in a group of females at about the same time (synchronized). $PGF_{2\alpha}$ will also effectively synchronize estrus and ovulation in cattle, horses, and sheep. $PGF_{2\alpha}$ acts by destroying the corpus luteum and thereby shuts down the secretion of progesterone.

The ovulation rate in farm mammals may be increased (superovulation) by injecting gonadotropic-hormone preparations during the follicular phase of the estrous cycle. Twinning has been induced in cattle by treatment with gonadotropins, and the treatment with gonadotropins of certain women of low fertility has resulted in multiple births. Gonadotropins are also used to superovulate females for embryo transfer (see Chapter 13).

The administration of various hormones to male animals has had little or no value for increasing sperm production, although certain chemicals prevent testes development and sperm production. For example, estrogenic compounds can be used to caponize (chemically castrate) male chickens, although this method is not used on a practical basis.

Gonadotropin-releasing hormone (GnRH) is approved by the **FDA** for the treatment of cystic ovarian disease. Of cows with ovarian cysts treated with GnRH, 80 percent will reestablish ovarian cycles and become pregnant. GnRH has also been used experimentally to induce ovulation in **anestrous** cows. If GnRH is administered to dairy cows 2 weeks postpartum, they have a lower probability of developing ovarian cysts. GnRH has been used in combination with progestins and $PGF_{2\alpha}$ to synchronize estrus and ovulation. Experimentally GnRH has also been used to increase conception rates in repeat breeder cows. For cows that have had difficulty in becoming pregnant, an injection of GnRH at estrus improves conception rates significantly. Therefore GnRH has numerous therapeutic uses in farm mammals.

Testosterone can be administered to female farm mammals to induce male sexual behavior. These testosterone-treated females act like males and can be used as aids in estrous detection.

11.8.2 Mammary Growth and Lactation

Many hormones are known to be involved in the growth of mammary glands and their secretion of milk (Chapter 15). Administration of estrogen and progesterone for 7 days will initiate lactation in about 70 percent of cows treated. Use of these compounds for inducing lactation has not been approved by the FDA.

11.8.3 Growth and Fattening

Estrogens and synthetic estrogen compounds have been shown to increase the rate and efficiency of weight gains in **ruminants** (cattle, goats, and sheep). A synthetic progesterone, melengestrol acetate (MGA), is also used to increase the rate and efficiency of gain in cattle. The primary mechanism by which MGA increases growth rate and gain is by causing the persistence of ovarian follicles and the continued synthesis of estradiol by those follicles.

11.8.4 Hormones and Public Health

Synthetic hormones are known to affect reproductive and other body processes. If residues of these hormones are present in meat and other foods consumed by humans and animals they may pose a health hazard. Sensitive methods have been developed for detecting minute residues in tissues. In general, residues of these compounds in animals do not appear to be a **public health** hazard. Great care is taken to prevent residues of these hormones from being present in human food. They are under constant surveillance by the FDA.

The relation of hormones to the occurrence of cancer has been given considerable attention by research workers. Estrogens have been investigated most extensively. It is possible that estrogens may cause cancer in humans when given in large doses for prolonged periods of time, although it appears unlikely that this hazard would be encountered in humans under normal conditions.

11.9 SUMMARY

A remarkable **homeostatic** interrelation exists between hormones and other factors affecting animal health (including appetite, diet, **nutrition, stress,** and **environmental** responses). Caution should be taken with hormone supplements to avoid disturbing the hormonal balance. As noted by Brody,[2] the following quotation is an excellent summary of this balance concept.

> The most important concept in endocrinology which emerges from the feverish activity of the past decade is the principle of endocrine balance. Discovered and rediscovered by several investigators, the paradoxical truth has dawned finally that man or beast may suffer less from the loss of several glands than from losing a single one. For each of the precious juices the other secretions supply a partial antidote, so that health and personality may be preserved, delicately poised. Henceforth, the practicing consultant and laboratory investigator, or both, must think in terms of integrated hormonal effects.[3]

STUDY QUESTIONS

1. How do endocrine and exocrine glands differ?

2. What is the master endocrine gland of the body? Where is it located?

3. List several functions of the hypothalamus.

4. List the hormones secreted by the anterior pituitary gland and give their main functions.

5. What hormones are produced by the adrenal glands? What hormones are produced by the placenta in some species? Which species?

6. List the five general chemical classes of hormones.

7. What is meant by the feedback mechanism in hormone control?

8. What is the role of the hypothalamus in the release of hormones from the anterior pituitary gland?

9. What are prostaglandins? Give some of their functions.

10. List some uses of hormones in humans and farm animals.

11. Which hormones may be involved in the growth of mammary glands? In the secretion of milk?

12. Which synthetic hormones may be used to increase the rate and efficiency of weight gain in beef animals?

13. In what ways could hormones be a hazard to public health? What precautions are taken to prevent them from being a health hazard?

14. If hormones were not well balanced within the animal body, what consequences might be observed?

[2]S. Brody, *Bioenergetics and Growth,* Reinhold Publishing, NY, 1945.
[3]W. T. Slater, *The Endocrine Function of Iodine,* Harvard University Press, Cambridge, MA, 1940.

THE PHYSIOLOGY OF GROWTH AND SENESCENCE[1]

If we call the weight of the human egg one, then the weight of the individual produced from the egg is about 16 billion times that of the egg itself.

G. H. Parker

12.1 INTRODUCTION

All humans begin life as a single cell (the fertilized egg). This single cell is less than 0.00143 inch in diameter and is so small that its weight cannot be determined accurately. Through the process of *cell division* and **differentiation** this cell develops into a mature individual weighing between 150 and 200 lb (mature human males) and possessing as many as 250 trillion cells. Few cells in the mature body resemble the original fertilized egg because, through the processes of differentiation and **morphogenesis,** changes occur that form all the organs and tissues characteristic of humans. Growth results from physiological processes that cause one cell to develop into a many-celled individual.

A knowledge of the factors responsible for body growth is of fundamental importance to students of animal science. A certain amount of growth must occur before animals can convert their foods into animal products and provide services useful to humanity. Rapidly growing animals are generally the most efficient in converting food into weight gains. However, proper production practices are essential to maximize the efficiency of growth and the quantity and quality of products and services yielded by various farm animals. Few aspects of animal science are more intriguing than the development of a completely new animal originating from one cell. This development proceeds in such a precisely controlled manner that all the intricate organization of cells, tissues, organs, and systems characterizing the functioning adult develops with rarely a flaw.

12.2 THE PHENOMENON OF GROWTH

If nature were our banker, she would not add the interest to the principal every year; rather would the interest be added to the capital continuously from moment to moment.

J. W. Mellor

All living things (plant and animal) grow. A giant redwood in California grows from a seed measuring 0.0625 inch in diameter to become a tree often 300 or more ft tall. A microscope is required to see a whale egg, yet a mature whale may be more than 100 ft long and weigh over 160,000 lb. In the lower animal forms, striking changes occur between the time a single cell starts growing and an adult individual emerges. The single cell of certain **insects** grows first into a wormlike **larva,** which in turn enters a **pupal,** or quiet, stage before the adult insect finally emerges (Chapter 23).

In the higher vertebrates (animals with backbones), the tiny cell steadily develops into a form that at birth resembles the adult. Although there are significant differences in the proportion of various body parts, the **neonate** resembles the adult. Growth is the normal process of increase in size produced by the accumulation of tissues that are similar in composition to that of the original tissue or organ. True growth involves an increase in the size and weight of structural tissues such as muscles, bones, heart, and brain and of all other body tissues (except **adipose** tissue) and organs. From a chemical standpoint, true growth is an increase in the amount of **protein** and mineral matter accumulated in the body. Increases in weight due to fat deposition or to the accumulation of water are not true growth.

All parts of an animal's body grow in an orderly manner. The legs and arms in humans grow in proportion to the height

[1]The authors acknowledge with appreciation the contributions to this chapter of Dr. Elisabeth Huff-Lonergan, Department of Animal Science, Iowa State University, Ames.

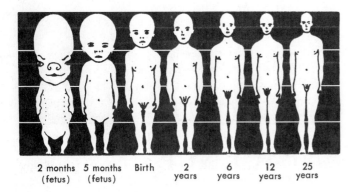

Figure 12.1 Changes in the relative size of body components during fetal and postnatal growth in humans.
S. Brody, Bioenergetics and Growth, *Reinhold Publishing, New York, 1945.*

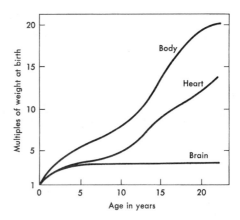

Figure 12.2 Graph depicting differences in relative growth rates of body, heart, and brain tissues in humans.
Modified from D'A. W. Thompson, On Growth and Form, *Cambridge University Press, New York, 1942; and W. T. Keeton,* Biological Science, *3d ed., Norton, New York, 1980.*

and length of the body. Following birth the head, however, grows proportionately much more slowly than the limbs (Figure 12.1). Nevertheless there is wide variation in size and body proportions within a **species.** For example, a Clydesdale (**draft** horse) is many times larger than a Shetland **pony.** Even within a **breed** there is considerable variation in size and weight among individuals.

Although the various body parts grow in an orderly manner the body does not grow as a unit because various tissues grow at different rates from birth to maturity. For example, muscle, heart, and total body tissues may show a 45- to 50-fold, 12- to 15-fold, and 20- to 22-fold increase, respectively, in size and weight compared with the brain, which may show only a 3- to 4-fold increase (Figure 12.2). The head and brain grow fast early in the lives of vertebrates. Furthermore their legs may increase in length faster than the balance of the body. For example, **foals** have long legs relative to other body components. Certain parts of an animal's body continue growing after others have stopped. The incisor teeth of rats continue to grow throughout their lives. Some animals have shells that restrict their growth beyond a certain size. Others (e.g., crabs) shed their shells and construct larger ones.

In animal production growth is usually measured as an increase in body weight. Although this is generally accepted as an adequate measure, technically it is inaccurate. It is possible to feed a young animal in such a way that its total body weight does not increase while the skeleton and internal organs may show considerable growth.

Regeneration is a special feature of growth. When a part of an animal's body is injured, the lost tissue may be (1) replaced by the same kind of tissue or (2) replaced by another kind, called *scar tissue.* Sponges are capable of almost complete regeneration. Starfish can grow whole new bodies from a single ray if a small amount of the center portion remains. A salamander that loses a leg will grow a new one. Worms can grow new heads. Several kinds of long-tailed lizards break off their tails to escape the grip of an enemy and they soon grow new tails. In higher animals the power of regeneration is not as remarkable. When a dog's tail is cut off in the lawn mower or a bear's paw is lost in a trap new parts do not grow. Similarly a cow, chicken, or horse cannot grow a new leg to replace a severed one. Most

higher animals can regenerate hair (feathers), nails (hooves), skin (hide), blood cells, liver, and certain other tissues.

12.2.1 Growth and Development of Humans

During the first stages of growth the slowly developing human is called an *embryo.* At 2 months it is only about 1.5 in long but has the form of a human being. Although all its parts are well formed the head is quite large compared with the trunk and limbs (Figure 12.1). Beginning with the third month the fully developed embryo is called a **fetus.** At 7 months it weighs about 2 lb and is approximately 15 in long. Two months later, or just prior to **parturition,** the fetus commonly weighs 6 to 8 lb and is from 19 to 21 in long. Thus the body is growing very rapidly at birth. The weight at birth may be influenced by (1) the mother's **nutrition** during pregnancy, (2) the number of previous pregnancies, (3) sex, (4) race, and (5) **environment.** Each successive pregnancy leads to an average increase of about 75 g (2.65 oz) in the weight of the child at delivery.

The rapid growth rate continues for about 2 years and then gradually declines until adolescence. For approximately 2 years during the adolescent period, the growth rate of boys and girls is accelerated. This is commonly referred to as the *adolescent spurt* in height growth. Boys usually exhibit this spurt of growth between the ages of 13 and 15 and may grow 4 to 12 inches in height. Girls normally begin the adolescent growth spurt about 2 years earlier and their maximum growth rate is somewhat less than boys. Men are taller than women chiefly because of this difference in adolescent growth spurt. Until that period is reached, the average heights of boys and girls are about equal. The tendency to be tall or short is governed largely by **heredity.** Thus tall parents usually have tall children, short parents have short children.

When People Stop Growing

Most humans stop growing between the ages of 18 and 30. (A person is usually tallest at age 26.) After attaining maximum height a person begins an exceedingly slow decrease, which is usually not noticeable until the person reaches old age. It is caused by a thinning of the pads of **cartilage** that grow between the bones of the vertebral column (backbone).

12.3 THE CELL IS THE UNIT OF GROWTH

The living cell is to biology what the electron and proton are to physics. The two theories are independent exemplifications of the same idea of "atomism."

A. N. Whitehead (1861–1947)
English mathematician and philosopher

Growth results from an increase in both size and number of body cells. An increase in the number of cells is called **hyperplasia;** an increase in cell size is called **hypertrophy.** *Accretionary growth* is an increase in size resulting from accumulation of noncellular material such as albumin in blood and calcium in bone. When cells undergo hypertrophy it is usually the protein structures in the **cytoplasm,** and not the **nucleus,** that increases in size. The process by which a single cell grows from within is also called growth by *intussusception,* meaning to swell or enlarge. It is interesting that the cell size of a mouse is about the same as an elephant, although the elephant, of course, has many more cells than the mouse.

The most significant aspect of hyperplasia is the increase in **DNA.** Because most cells have only one nucleus, and each nucleus contains the same content of DNA, the amount of DNA present in tissue can be used as an index of the number of cells in that tissue. During hypertrophy cell size increases, but the DNA content remains constant. In most cells the increase in cell size is accompanied by an increase in cellular protein. Therefore, the protein:DNA ratio can be a useful indicator of growth. The maximum protein:DNA ratio varies with tissue, species, age, and other factors. Under optimal environmental conditions and nutritional parameters a maximum protein:DNA ratio is obtained; hence, if the animal is to become larger in size it must synthesize additional DNA. During a mild nutritional setback such as a brief period of starvation the amount of tissue protein will decrease, whereas the content of DNA remains constant. Then, when proper feeding of the animal is resumed, the content of cellular protein quickly increases to the previous protein:DNA ratio. This rapid rate of growth after the resumption of adequate nutrition is called *compensatory growth.*

Hyperplasia and hypertrophy of all cells in the body occur during embryonic life. In the adult animal, however, three different types of cells are found. *Permanent cells* cease dividing early in prenatal life. Their numbers remain relatively constant thereafter. Muscle and nerve cells are examples of this type. *Stable cells* continue to divide and increase in number during the major portion of the growth period, but division ceases and the number becomes fixed in the adult. Most body organs such as the lungs and kidneys contain cells that fit this category. A third group of cells is known as **labile** (apt to change) cells. Epithelial and epidermal tissues fit this category. They continue to divide and differentiate throughout the animal's life in response to the life-sustaining need to replace worn-out tissue and cells.

Some body cells are capable of undergoing hypertrophy during different stages of life. For example, muscle tissues greatly enlarge when extra demands are placed on them by bodily exercise. Other cells are also capable of hyperplasia but do not undergo this process unless there is a special demand for it. For example, in the healing of wounds capillaries present in the surrounding tissue bud from endothelial cells, with new growth occurring in the direction of the wound. As growth progresses the capillaries become hollow, forming channels through which blood circulates. This growth continues until the raw surface of the wound is covered. Along with the new capillaries comes a procession of **fibroblasts,** which produce connective tissue in the wounded tissue. They form fibers that heal the wound. As another example, nerve cells cannot divide after they are formed in the developing fetus. Therefore, when a nerve cell is severely damaged it will not be replaced by a new cell. However, the axis cylinder (fiber) can grow from the nerve cell end and find its way into the tissue formerly supplied by the severed nerve. This explains how some aspects of normal nerve tissue function may be restored.

Studies of cell division using radioactive **isotopes** show that only about 3 percent of cells within the adult human body are capable of dividing and repairing tissues. A normal dividing cell produces two daughter cells, one of which is capable of dividing whereas the other is not. The latter perishes leaving no descendants. Through this method of division the total cell number is approximately constant, thereby allowing normal tissues to maintain their size. In some cases, as an individual ages, there may be a slight reduction of cell numbers in some tissues. Cancerous tissue provides an example of what happens when the normal cell division process is deregulated. Cancerous cells divide no more rapidly than normal cells, but they increase much more rapidly in numbers because there are fewer restraints and/or controls on cell division. What controls and eventually stops the division of normal cells is not clearly understood.

12.3.1 Growth and Development of Muscle, Fat, and Bone

The development of three different types of tissue will be examined briefly in this section. Skeletal muscle is an unusual tissue in that individual muscle cells are long and thin and contain thousands of nuclei. Muscle cells are formed by a distinct sequence of events (Figure 12.3).

During fetal development single nucleated cells that are predestined to become muscle cells divide many times until another type of single nucleated cell, called a *myoblast,* is formed. Myoblasts do not divide but instead fuse together forming a *myotube* (Figure 12.3), which is an immature muscle cell containing the nuclei from the fused myoblast. Once the nuclei are inside the myotube they are unable to divide further. The tremendous growth of muscle that occurs following birth is primarily due to an enlargement in size of the existing cells (hypertrophy). Even though few new muscle cells are added postnatally, there is evidence that the number of nuclei can

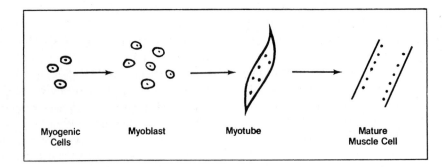

Figure 12.3 The sequence of events accompanying the formation of muscle cells.
Courtesy of Dr. P. J. Bechtel and the University of Illinois at Urbana–Champaign.

increase in existing muscle cells after an animal is born. These additional nuclei are necessary to support the synthesis of protein needed as cells increase in size. These new nuclei originate from specialized mononucleated cells associated with muscle cells known as *satellite* cells. In postnatal muscle these cells are located just outside the muscle cell membrane (sarcolemma). Unlike mature muscle cells these satellite cells retain the ability to proliferate, differentiate, and fuse with muscle cells. By fusing with muscle cells, satellite cells provide additional DNA (in the added nuclei). This added DNA ultimately provides more capacity for the growing muscle cells to create additional protein needed for hypertrophy.

Fat consists of adipose cells and supporting connective tissue. The mature adipose cell is called an *adipocyte.* These large cells contain a droplet of lipid, which is surrounded by other cellular components (nucleus, mitochondria, etc.). The mature adipose cell is derived from an immature fat cell called an *adipoblast.* The adipoblast contains a number of tiny lipid droplets. The mature adipose cell is commonly 50 to 100 times larger than the adipoblast. Adipose tissue is very dynamic and its size increases or decreases in accordance with the animal's plane of nutrition, as well as with other factors. Once an adipose cell has been formed it can be reduced to a very small size, as in response to starvation. However, the cell will fill rapidly with lipid materials under the appropriate conditions. There are two types of adipose (fat) tissue, which are referred to as *white fat* and *brown fat.* Most body adipose tissue is white and functions as a depot of stored energy. Brown adipose tissue is found primarily in newborn animals, or cold-adapted and hibernating animals. Brown adipose tissue is very active metabolically and can be used to maintain body temperature, which is especially important in neonatal animals.

Bones grow during both the prenatal and postnatal periods. Bone of a mature animal is composed of approximately 50 percent mineral (primarily hydroxyapatite, $[Ca_{10}(PO_4)_6(OH)_2]$) and 50 percent organic material and water on a weight basis. Volume-wise, mineral constitutes only about 25 percent of bone. Most of the organic material in bone is collagen with a small amount of cellular material. Bone is formed by the interaction of three different cell types: chondrocytes, osteoblasts, and osteoclasts. *Chondrocytes* are cells that produce cartilage, *osteoblasts* produce bone collagen and other bone components,

and *osteoclasts* break down bone during the process of resorption. Bones grow in length by ossification at the epiphysial plate (see Chapter 10). Once the epiphysial cartilage has been completely ossified, bones stop growing in length. Bones can grow in width, and can also undergo some degree of remodeling through the process of readsorption of existing bone by osteoclasts and the formation of new bone by osteoblasts.

12.4 PERIODS OF GROWTH

Growth in farm **mammals** may be divided into two main periods: (1) prenatal and (2) postnatal.

12.4.1 Prenatal Growth

The prenatal period of growth occurs between the time ovum is fertilized and young are born. Only 21 days are required for the fertilized hen egg (a mass of yolk and albumen) to develop into a fully functioning baby **chick.** Prenatal development takes much longer in farm mammals.

The new individual arises from the union of a **sperm** and an ovum, forming one cell[2] with the **diploid** number of **chromosomes.** One-half of each chromosome pair comes from the father and one-half from the mother. The single fertilized cell divides to form two, then four, then eight, and so on, cells, following a pattern in which each mother cell produces daughter cells, both of which are capable of dividing to form two new cells (Chapter 8). When they are separate the ova and sperm do not have the power to grow and reproduce themselves and eventually die. After combining during the process of **fertilization** they have the ability to divide and thus increase cell number.

Somatic cells formed by cell division have a diminishing power to grow by means of hyperplasia. The power to reproduce is lost in many highly differentiated cell types. For example, the mature red blood cell of mammals loses its nucleus before it enters the bloodstream, but its cytoplasm becomes filled with **hemoglobin,** which transports oxygen from the lungs to the various body parts. The red blood cell dies after

---

[2]It is interesting that the eggs of the Texas armadillo regularly divide into four parts, giving rise to four young. In some parasitic insects one egg often forms as many as 2000 separate embryos.

about 90 to 120 days in the bloodstream, but bone marrow continues to produce replacements. Skin cells arise from the germinal layer of skin. **Keratin** accumulates in their cytoplasm; they lose their nucleus and ability to divide. The end result of this differentiation is cellular death. Thus, in the body when certain cells perform a function in which their ability to divide is lost, other cells must divide and increase in numbers to replace those that are destroyed. This is the process used to create new liver and skin cells. A problem occurs when a differentiated cell such as a nerve or brain cell is destroyed, because there is no mechanism to create new cells of these types. The result is an irreversible diminishing in numbers of such cells, and the organism then loses some of its ability to regulate itself.

The rate of cell division in newly fertilized ovum varies among species. The number of cells in a fertilized rabbit ovum is approximately 32 within 40 to 50 h after mating. The fertilized ovum of swine is in the 2- to 4-cell stage at that same postmating time.

The developing fertilized ovum soon goes through a process of differentiation (growth and organization of cells into specific structures) in which the mother cell produces different kinds of daughter cells that eventually form the brain, heart, and other specialized tissues. This process is irreversible in that these differentiated cells cannot be transformed back into egg cells. More and more organs are formed as ovum development proceeds until all are complete, then further differentiation ceases. The organization of various dividing cells into special organs with a particular makeup is known as *morphogenesis,* or **organogenesis.** During organogenesis differentiated cells synthesize new structural proteins and **enzymes** that are specific for the tissue being formed. How cells are instructed to differentiate into organ cells is not known, but the process has been repeated billions and billions of times. Body organs are formed in an orderly sequence. The head portion of the embryo develops before the tail, and the cephalochord (head of spinal column) develops before other organs.

Prenatal growth in farm mammals proceeds at varying rates among different species. The pig, for example, spends only 110 to 115 days within the mother's uterus, compared with 335 to 345 days for the foal (baby horse). All farm mammals are born at a fairly comparable degree of maturity in spite of the great variation among species in the length of time spent in the uterus. However, the degree of maturity at birth varies greatly among animal species other than farm mammals. For example, the opossum is born 12 days after **conception.** The young are very immature at birth; and since there are no fetal membranes, the baby opossums must find their way from the birth canal to their mother's pouch where they attach their mouths to teats and continue their development. The number of baby opossums born greatly exceeds the number of teats the mother possesses. Those not finding a teat perish. This is a form of natural selection for the more vigorous individuals.

The baby rat is born 21 to 22 days after conception. At birth it is quite helpless; it has no hair, its eyes are closed, and it is unable to walk. The guinea pig, on the other hand, is born about 67 to 68 days after its parents mate. The young are quite mature

at birth; they possess a coat of hair, their eyes are open, and they can walk and run almost immediately (Figure 12.4).

The human infant is carried **in utero** about 9 months and is helpless at birth, even though its eyes are readily opened and the senses soon begin to function. Babies cannot walk until they are nearly 1 year of age. In early prenatal life the human female exceeds the male in biological growth rate. Thus a full-term male is somewhat more immature than a full-term female at birth. Moreover, males are more susceptible to death losses. Thus sex **hormones** appear to influence the degree of maturity of the human infant at birth.

Genes transmitted by the **sire** limit the birth size of **progeny** of large mothers, but in small mothers the uterine environment is more important in limiting the size of the progeny at birth than are the genes transmitted by the sire. For example, a small Shetland **mare** mated to a large Shire **stallion** produces a small foal at birth. Conversely, when the mother is a large Shire mare and the father a small Shetland stallion the foal is much larger. Maternal effects on size tend to diminish with advancing age of the offspring, although differences often persist for many months. The persistency of maternal effects in later life depends on the time in life influences have had their effect. The earlier in life they occur the longer they persist. This is probably also true of the effects of other environmental factors that limit growth.

Growth in the number of muscle cells an individual possesses slows at the beginning of the fetal stage. Thereafter increased muscle size depends primarily on an increase in size (hypertrophy) of individual muscle cells. Skeletal size may be modified to a certain extent by environmental factors, but it cannot exceed the limit set by the individual's genetic constitution.

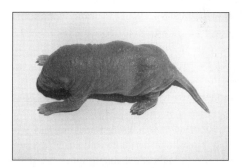

Figure 12.4 Note the marked difference in the degree of maturity at birth in the rat (top) and guinea pig (bottom).
Courtesy of Missouri Agr. Expt. Station.

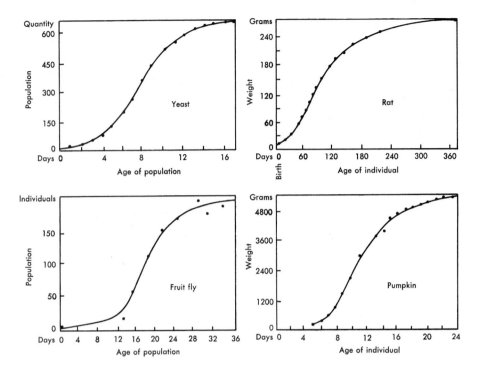

Figure 12.5 Typical S-shaped growth curves of unicellular and multicellular plants and animals. Note their similarity.
Redrawn from Missouri Agr. Expt. Sta. Res. Bull. *97.*

12.4.2 Postnatal Growth (Growth after Birth)

Growth usually begins slowly then goes through a more rapid period, and then slows down again or may even stop. This pattern yields the characteristic sigmoid (S-shape) growth curve. The general course of growth after birth in all species of farm mammals is somewhat similar. If the body weight of an animal is plotted on the *Y* axis (ordinate) of a graph and its age to maturity on the *X* axis (abscissa), the curve resembles the letter S (Figure 12.5). However, the bend in the growth curve is about one-third the distance from the baseline to the top of the curve. There are apparently few exceptions to the sigmoid form of growth curve. (Some organs such as the **gonads** and mammary glands are cyclic in growth.) It applies to the growth of populations of either unicellular or multicellular plants and animals. These curves are similar because the multicellular individual represents a population of cells that is similar to a colony of single-celled individuals.

The growth curve of farm mammals after birth may be divided into two principal segments. The first is represented by an increasingly steep slope that extends from the beginning of growth until one-third to one-half of the mature weight is reached. The length of this segment depends on the species and the influence of environmental factors that may affect growth. The second segment extends from the end of the first segment to maturity, or to the end of the growth period. The general shape of the curve is molded by two opposing forces, a *growth-accelerating force* and a *growth-retarding force* (Figure 12.6). The growth rate at any given time depends on the magnitude of each of these opposing forces.

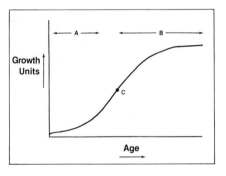

Figure 12.6 Standard postnatal growth curve: (A) growth-accelerating phase, (B) growth-retarding phase, and (C) point of inflection.
Courtesy of Dr. P. J. Bechtel and the University of Illinois at Urbana–Champaign.

The growth-accelerating force is present in body cells or in the individual cells constituting a population of single-celled individuals. The growth-retarding force is found in the environment surrounding the cells or the individuals in a population. This force involves (1) exhaustion of the supply of essential nutrients, (2) the accumulation of waste products, which results in a toxic environment, and/or (3) space limitations. Results from studies of cell cultures appear to verify the importance of these factors. When a piece of tissue was cut from a living chicken's heart and placed in a culture medium where nutrients were supplied and waste products continuously removed, the tissue lived for many years, doubling its mass every 24 h. In the live intact chick, heart cells soon reached the point where cell division

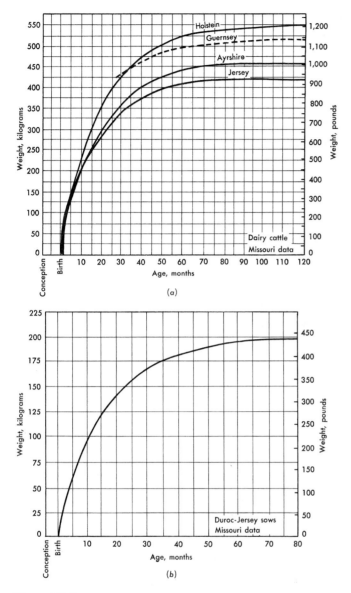

Figure 12.7 Average changes in body weight observed with advancing age of (a) four breeds of dairy cattle; (b) Duroc sows. *Redrawn from* Missouri Agr. Res. Bull. 96.

(growth) ceased entirely. This experiment suggests that cells are potentially capable of infinite division under ideal conditions.

The point in the growth curve where the growth rate ceases to increase and from which it begins to decrease is referred to as the point of **inflection** (Figure 12.6). This point, according to Brody,[3] generally indicates the (1) period of maximum velocity of growth, (2) age of **puberty,** (3) beginning of the period of increasing specific **mortality,** and (4) period of reference for the determination of age equivalents in different animals and populations. The point of inflection occurs at about 14 years in humans, 5 months in cattle, and 2 months in sheep. Factors limiting the maximum size of a single multicellular individual

include (1) its genetic constitution, (2) its **nutrient** supply, (3) the effects of certain **diseases,** and (4) the hormone supply, which may also be related to an individual's genetic constitution. A fifth factor, body-surface:volume ratio, may limit extremely large size. As size increases, animal volume increases much more rapidly than surface area. Within a species an extremely large size eventually makes it difficult to obtain enough food to sustain the animal and to eliminate body waste products efficiently.

Growth rate from birth to weaning may be greatly influenced by the amount of milk secreted by the mother and by the individual's health. Total growth during this period also depends on the individual's inherent growth potential. Growth of body parts after birth proceeds at different rates in most farm mammals. Shortly after birth skeletal parts generally grow more rapidly than other tissues in proportion to total body weight. This rapid rate of skeletal growth decreases by weaning time. Maximum muscular development occurs somewhat later than maximum skeletal growth and is occurring at an increasing rate by weaning time. Body weight increases that occur after both skeletal and muscular tissues have passed their maximum rates of increase are due primarily to increased fat deposition. These body weight increases occur much later than the usual weaning time, commonly near the usual market age and weight.

Inadequate nutrition delays peak muscle growth and slows the rate of fat deposition, whereas proper nutrition hastens the occurrence of peak rates of both. Early-maturing animals fatten at a younger age than late-maturing ones. Breeds within a species vary in their **maturation** rate as do individuals within a breed. Typical changes in body weight with advancing age for selected breeds of dairy cattle and swine are shown in Figure 12.7.

12.5 HORMONAL CONTROL OF GROWTH

Secretions of several **endocrine** glands are known to have a vital effect on growth in most animal species. Research in which the endocrine glands were surgically removed or where injections of extracted hormones were administered to young growing animals has clearly demonstrated hormonal effects. Some naturally occurring abnormalities of endocrine secretions have also provided information pertaining to these glands.

12.5.1 Growth Hormone

The anterior pituitary gland, located in the skull at the base of the brain, secretes a hormone closely related to an individual's growth (Table 12.1). It is present in the pituitaries of all vertebrates and is known as *growth hormone* (GH) or *somatotropin* (STH). Growth hormone is a protein and is secreted by acidophilic cells (cells having an affinity for acid dyes). Highly purified preparations of active GH have been obtained from the pituitaries of many species including a pure crystalline form from cattle and humans. The chemical structure of GH from several species of animals has been determined and commercial preparation, utilizing the principles of genetic engineering, has become a timely reality (Chapter 11).

[3]*Missouri Agr. Expt. Sta. Res. Bull. 97.*

TABLE 12.1	Growth Hormone Content of the Pituitary Glands of Holstein Heifers of Different Ages

Age, Weeks	Milligrams of GH per Gram of Anterior Pituitary
16	11.67
32	10.90
48	8.18
64	6.98
80	7.73

Source: D. T. Armstrong and W. Hansel, *J. Animal Sci.* 14(1955):1242.

Human growth hormone (HGH) consists of a long chain of 188 amino acid residues bridged at two points by sulfhydryl bonds. GH is species-specific so that preparations usually are not effective unless prepared and injected within the same species. Thus the human pituitary dwarf must be treated with HGH to be **therapeutically** affected.

Removal of the anterior pituitary gland in young developing animals causes growth to cease and body weight to decrease. The daily injection of anterior pituitary extracts will promote resumption of growth in such animals. Long-term injections of GH preparations in young normal animals produce extremely large individuals (giants). This has been demonstrated in rats and dogs. Secretion of excess HGH in young humans has been credited with the production of giants 7 to 10 ft tall. If excessive GH is administered to mature individuals a condition called **acromegaly** results. It is a condition in which parts of the body such as the head, hands, and feet become enlarged. Growth hormone does not appear to have a major effect on growth during fetal life. When rat, rabbit, and swine fetuses are **decapitated** they continue to grow at a normal rate. However, GH is present in the anterior pituitary glands of fetal mice, chickens, and pigs and it appears to be capable of stimulating growth when administered to postnatal animals. Removal of the maternal pituitary gland has little or no effect on fetal growth, perhaps because in many species GH from the mother cannot pass from her blood (through the placenta) to that of the young. Different breeds of cattle grow at varying rates before birth. In a Missouri study **purebred** Angus, Hereford, and Charolais calves weighed 63, 71, and 82 lb, respectively, at birth. Part of this variation in birth weight was due to differences in maternal size, but some of the variation was postulated to be due to differences in GH secretion or the response of the tissues to GH (Chapter 11).

Even though growth finally ceases in mature animals GH is present in their pituitaries (Table 12.1), although probably in reduced amounts. Perhaps skeletal growth ceases because the tissues are no longer responsive to GH. In mature animals the **epiphysial** grooves of the long bones are closed, preventing further growth. A selection experiment at the Illinois Agricultural Experiment Station was effective in developing two genetically different lines of the Hampshire breed of swine. At 180 days of age one had a heavy weight, the other a light weight. Pigs from the *heavier* line had significantly greater amounts of GH per unit of anterior pituitary tissue at all ages than those from the *light* line. Similar results were reported by the Cornell Agricultural Experiment Station for fast- and slow-growing Holstein **heifers.**

Hypophysectomy causes skeletal growth to cease resulting in dwarfism. The administration of GH causes resumption of skeletal growth. This is proof that GH is involved in skeletal size, especially in long-bone growth.

Growth hormone affects amino acid **metabolism.** This effect is apparent in the growth of young in which muscle mass increases require the synthesis of protein and thus an increased bodily retention of dietary nitrogen. GH apparently stimulates the synthesis of proteins from amino acids. How it does this is not fully understood. Growth hormone is thought to act directly on certain tissues; other effects are believed to be indirect such as the stimulation by GH of the secretion of *somatomedins,* which then act directly on a target tissue.

Growth hormone reduces the amount of fat stored in the body. Following GH injections there is usually an increase of free fatty acids in the blood, and it is thought that this rise is due to the release of fatty acids from adipose tissue. Some research suggests that GH enhances the conversion of carbon from fatty acids into carbon that may be incorporated into proteins. Growth hormone also has varied effects on **carbohydrate** metabolism, some of which are not fully understood.

Probably one of the best indicators of the functions of GH is illustrated when GH preparations are administered to normal growing animals. The treatment results in an increased proportion of **carcass** protein and water and a reduction in the proportion of fat. The carcass content of a treated animal is similar to that of a very young animal. Increases in body size are due not to fat deposition but to an increase in tissue size.

12.5.2 Thyroid Hormone

The thyroid hormone of blood is almost entirely L-thyroxine, although there are traces of triiodothyronine. Production and release of thyroid hormones is controlled by thyroid-stimulating hormone (TSH) from the anterior pituitary gland (Chapter 11).

Thyroxine contains four molecules of iodine and triiodothyronine contains three molecules of iodine. Thus iodine is essential for the synthesis of thyroid hormones. A dietary deficiency of iodine results in increased TSH secretion and hypertrophy of the thyroid gland (goiter). Despite its enlargement, without iodine the thyroid gland cannot produce thyroid hormones and the animal soon becomes deficient in thyroid hormones (hypothyroid).

Thyroidectomy (removal of the thyroid glands) and subsequent thyroid hormone deficiency is followed by disturbances in metabolism, development, and growth. The main physiological effect of thyroxine is to increase energy production and oxygen consumption of most body tissues. Thyroxine secretion is essential for metabolic functions involved in normal growth. A lack or deficiency of thyroxine in the young results in a condition called *cretinism,* which usually includes dwarfism and idiocy. Naturally occurring congenital hypothyroidism is due to an autosomal recessive defect in thyroglobulin synthesis. Breeds that have been reported with this defect in thyroglobulin synthesis include Merino sheep, Afrikander cattle, and Saanen–Dwarf crossbred goats. Affected animals are born weak and usually die within a few weeks. The hair coat of the affected newborns is short and fuzzy. The neonates are unthrifty with a tendency to bloat, retarded growth, and high susceptibility to infections.

Low thyroid activity (**hypo**thyroidism) is usually accompanied by (1) reduced food intake, (2) low blood sugar concentration, (3) less liver glycogen storage, and (4) lowered nitrogen retention. Fat deposition may be increased. An overactive thyroid (**hyper**thyroidism) increases metabolic rate and, if severe, may cause **muscle catabolism** accompanied by weight loss. Several inexpensive thyroid-active substances have been developed, and many of them have been tested to determine if they have an economic value (1) as growth stimulants in farm animals or (2) for increasing **lactation** and egg laying. Other substances have been developed that suppress thyroid secretion. In general neither thyroid-activating nor thyroid-suppressing compounds appear to have a useful application in livestock production (see discussion on use of iodinated casein and thiouracil in Chapter 18). Thyroid hormone replacement therapy is effective in treating individuals suffering from hypothyroidism. Thyroid-suppressing compounds are used in treating animals with hyperthyroidism.

12.5.3 Androgens

These male hormones are secreted by the interstitial cells of the testes and to a lesser extent by the adrenal glands. Androgens stimulate growth in young farm animals as shown by size differences in males and females, the males being significantly heavier at almost any age. Males also have better rates of body weight gain and **feed efficiency** than either females or castrated males. **Castration** of males slows growth to some extent and accelerates the fattening process in cattle, sheep, and swine. The faster growth of noncastrated males appears to be due to a greater development of muscular and skeletal tissue. It has been shown in many animal species that androgens favor tissue nitrogen retention, which means there is a greater amount of protein synthesized and utilized and/or a lesser amount of protein **catabolized** in the presence of these hormones.

Research with laboratory animals suggests that small amounts of testosterone are secreted by the testes and adrenals of prepubertal males. These small amounts appear to stimulate growth by functioning with GH to give a maximum growth response. At puberty, however, or when large doses of testosterone are administered, the male hormone appears to block the growth hormone effect on the epiphysial cartilage. When the epiphyses fuse to the long bones linear growth ceases. It is known that as an animal matures this fusion occurs and long-bone growth ceases. Perhaps the larger amounts of androgens secreted after puberty directly or indirectly enhance this epiphysial closure, resulting in the final mature height of the individual.

Anabolic steroids are synthetic hormones with growth-promoting effects. These hormones have been widely used in the beef industry to promote muscle growth, weight gain, improve feed conversion ratio (lb feed/lb weight gain), and improve carcass quality by increasing the lean meat content. Anabolic steroids are implanted into the middle of an animal's ear when it enters a feedlot. Implanted cattle grow 15 to 20 percent faster than untreated animals. Approximately 63 percent of all cattle and 90 percent of feedlot cattle in the United States are treated with hormone implants (2001 data). Steers respond best to implants containing a combination of synthetic androgens plus estrogens. Heifers respond well to androgens alone (because they naturally secrete estrogens). There is no significant effect on the animals in terms of disease, discomfort, health, or behavior. Some favorable effects on the environment result from reduced nitrogen and waste excretion (because of improved feed efficiency). By the time the implanted cattle are slaughtered, the concentration of hormones in their meat is at a level similar to untreated cattle and much lower than concentrations found naturally in other foods consumed by people. (A person would have to eat over 13 lb of meat from a treated steer to equal the amount of natural hormones in one chicken egg. A bowl of split pea soup has more than 9 times as much naturally occurring estrogen as a 5 ounce portion of meat from an implanted steer.) The FDA has researched the safety of growth-promoting hormones since the 1950s and regulates the use of these hormones in the United States. The European Union currently bans the importation of meat from cattle treated with growth-promoting hormones. There is no scientific evidence of any adverse health consequences for people eating meat from implanted cattle; however, abuse of anabolic steroids by direct consumption or injection is a concern and a reason to closely regulate the use of growth-promoting hormones in animals.

12.5.4 Estrogens

These female hormones are produced by the ovary and to a lesser extent by the adrenals. In pregnant females estrogens are produced by placental tissues. Estrogens are known to aid in regression and closure of the epiphysial cartilage plate of long bones, thus slowing growth. Increased estrogen secretion occurs with the onset of puberty. This fact partially explains why girls stop growing appreciably in height soon after the onset of menstruation (coinciding with the secretion of estrogen).

12.5.5 Insulin

Insulin is a protein hormone secreted by beta cells of the islets of Langerhans in the pancreas (Chapter 11). This hormone increases the uptake of amino acids and glucose into muscle. Insulin also stimulates growth by increasing the synthesis of RNA and protein in a number of tissues including muscle and bone.

12.5.6 Glucocorticoids

Glucocorticoids, produced by the adrenal glands, are inhibitors of growth. One of these hormones, cortisol, decreases synthesis of DNA and protein in a number of tissues.

12.6 NUTRITION AND GROWTH

Although higher animals including farm mammals have very complex physiological systems, such as the endocrine glands and a central nervous system, they are unable to manufacture nutrients needed for their everyday life. The nutrients must be obtained from a source outside the body for proper growth to occur (Chapter 18).

The mother carefully protects the nutritional needs of the young as it develops in her uterus. She often draws on body reserves to supply the needs of the developing fetus. If nutrients supplied to the mother are severely deficient during pregnancy, birth weights of the young will be subnormal and their vigor may be reduced. The lack of **vitamins** and minerals in the mother's diet during pregnancy may have a marked effect on the vigor of the young without causing a great effect on birth weights. A lack of vigor is usually followed by significant death losses of the young at birth or shortly thereafter. Frequently poor prenatal growth rate (light birth weight) has little effect on mature size in animals if postnatal (after birth) nutrient supplies are adequate.

The effect of undernutrition after birth on mature size depends on (1) age at which underfeeding occurs, (2) length of the underfeeding period, and (3) kind of underfeeding (energy, vitamin, or other deficiency). A diet may provide adequate energy but be inadequate in vitamins and minerals, resulting in undernutrition. Early Missouri studies (1919) showed the effect of underfeeding in cattle. **Yearling steers** were fed a diet to maintain, but not increase, their weight. Skeletal growth continued in the steers and they increased in height and length. Their fat reserves were depleted and they became very thin.

Research reports differ as to whether underfeeding can permanently stunt an animal, thus preventing it from attaining its potential mature size. Severe underfeeding beginning at birth and extending for several weeks may prevent the individual from attaining its normal mature size. If an underfed animal is placed on a full **ration** again it often grows more rapidly than is normal for that particular species. (This type of growth is called *compensatory gains.*) Underfeeding when an animal is young results in a longer time required to attain normal mature size when it is again placed on **full-feed.** This result suggests that limited feeding delays maturity in some way. The life span of rats may be extended by limiting caloric intake but not by protein, vitamin, or mineral undernutrition.

The plane of nutrition can be used to adjust the growth rate in different stages of the growing–fattening period, especially in swine. It also can be used to limit the amount of carcass fat present at slaughter (Table 12.2). Because fattening in swine usually occurs rapidly between 100 and 200 lb, limiting the ration energy content of the diet at this time will also limit the amount of carcass fat.

A balanced diet is essential to attain optimum inherent growth. We often think of certain races and nationalities as tall or short. However, children of Japanese immigrants born in California tend to be taller than their parents. Improved nutrition is credited for the increase in the height of these American-born Japanese children over their parents (Chapter 3).

The energy consumed per kilogram (kg) in doubling birth weight is approximately the same for most mammals (except humans), namely about 4000 calories (cal) (Table 12.3).

12.7 HEREDITARY MECHANISMS IN GROWTH

Variations in growth within a species reflect both hereditary and environmental influences because both have large effects on growth in farm animals and humans. Environmental factors affecting growth include nutrition, exposure to disease, **parasites,** and injuries. Hereditary factors include those in which a single gene (or group of genes) may have a large effect on the size and growth rate of an individual. A good example is dwarfism, which occurs in several animal species. Hereditary factors also include those in which each of many genes has a small individual effect while the overall effect of these may be quite large. Examples of this type of heredity are the *large* and *small* strains of swine, sheep, cattle, horses, and **poultry.**

TABLE 12.2	**The Effect of Limited Feeding on Gains, Amount of Fat, and Skeletal Growth in the Carcass of Swine at Slaughter**	
	Full-fed	**Limited-fed**[*]
Initial weight, lb	38.7	39.1
Final weight, lb	205.1	216.7
Average daily gain, lb	1.5	1.3
Body length, mm	717	722
Leg length, mm	520	533
Backfat thickness, mm	46.2	42.9
Depth of fat on ham, mm	40.0	35.5
Weight of fat cuts, %	29.7	27.5
Age at slaughter, days	167	193

*Included 12 pigs per lot. Those limited-fed were given 85 percent of a full-feed.
Source: Missouri Agr. Expt. Station.

TABLE 12.3	**Energy Consumption per Kilogram Required to Double Birth Weight**

Species	**Energy Required, Cal**
Swine	3754
Sheep	3926
Cow	4243
Dog	4304
Horse	4512
Cat	4554
Rabbit	5066
Human	28,864

Source: M. Rubner, "Das Problem der Lebensdauer und seine Beziehungen zum Wachstum und Ernährung," *Arch. Hyg.* **Vol. 66** p. 45 Berlin, 1908.

12.7.1 Prenatal Growth

Prenatal growth may be inhibited by external factors even though the developing fetus has a genetic potential for rapid growth and a larger size. Prenatal growth in chickens is limited by egg size or, more specifically, by the amount of food available to the developing chick. In litter-bearing animals, extremely large litters may result in some developing fetuses being undersized at birth because of a lack of uterine space and/or available nutrients. The transplantation of fertilized rabbit ova from a small to a large strain increased the birth weight of the developing fetuses about 40 percent. However, young from the transplanted small strain did not weigh as much at birth as the young of the large rabbit breed (large mothers) that were allowed to develop in the uterus of large mothers. This is because the size potential of the small strain was limited by heredity. Similar results have been obtained when fertilized eggs have been exchanged between small and large strains of female mice.

12.7.2 Growth from Birth to Weaning

Growth during this period may be affected by the maternal environment to which the young are exposed before and after birth. In chickens effects of the nutrients supplied by the egg may be continued beyond **hatching,** but these effects appear to be largely **transitory** and disappear later in life if the individual has the inherent capability for fast growth. In lactating mammals the inherited growth impulse possessed by the young may be temporarily overshadowed by the amount of milk supplied by the mother. Many studies with swine indicate that up to 20 percent of the variation in weaning weight is due to the animal's own genes and 35 to 50 percent to the environment to which the litter is exposed. This environment includes such factors as available milk, litter size, and other factors peculiar to each litter. The growth of cattle and sheep between birth and weaning appears to be more closely related to the genetic makeup of the individual calf and lamb (20 to 30 percent) than is the growth of litter-bearing animals. Research data suggest that birth weights in cattle are more highly heritable than weaning weights and that many of the genes responsible for fast growth before birth are also responsible for fast growth between birth and weaning.

12.7.3 Postweaning Growth

After weaning, maternal influences become relatively less important than the individual's own genetic makeup and they may disappear completely at maturity. Although the maximum genetic size of the individual is fixed at conception, many environmental factors such as inadequate nutrition, disease, and accidents may prevent the individual from achieving its genetic potential.

The genetic influence on the mature size of an individual within a species has been demonstrated many times by selection experiments. Large and small strains of chickens, mice, rats, rabbits, dogs, swine, sheep, and cattle have been developed by selection for size differences over a period of years.

The mature size of farm mammals is related to the rate and efficiency of gain from birth to market weight, although other factors are also involved. In general, late-maturing cattle are still growing at the typical market weight of 1000 to 1200 lb and do not possess as high a proportion of fat in their bodies as do animals of smaller earlier-maturing strains and breeds at the same weight. It is interesting that some animals such as beavers never stop growing. Their life span in the wild is about 12 years. Captives have lived 19 years.

12.7.4 Genetic Control of Growth Mechanisms

Although it is known that genes affect growth in almost all animals including humans, only fragmentary evidence is available to illustrate the physiological pathways of gene action. Experiments have shown that breeds of chickens, rats, and inbred strains of swine vary in their needs for certain nutrients, especially the B-complex vitamins. Some strains show deficiency symptoms for certain nutrients whereas others fed the same diet show no adverse effects.

Hereditary dwarfism in humans and mice has been shown to result from a lack of growth hormone production indicating the relationship between genes and the synthesis of protein hormones. Chickens have shown wide genetic differences by the response of the **comb,** thyroid, and gonads to administration of anterior pituitary hormones. This suggests there are genetic differences in the sensitivity of certain target organs to hormones. In one strain of yellow mice there is a greatly increased rate of fat deposition, which has been attributed to a hormonal alteration of carbohydrate metabolism. Specific genes are responsible for this hormonal abnormality. Some research suggests that increased rate and efficiency of gain in swine due to **hybrid vigor** result from a more efficient metabolic system. Hybrid vigor is due to a nonadditive type of gene action (Chapter 9).

12.7.5 Association between Growth and Other Traits

In some species growth appears to be genetically associated with other vital body processes. An increase in degree of inbreeding is accompanied by a decline in growth and an increase in mortality rate. **Crossbreeding** usually increases growth rate and reduces mortality. Both inbreeding and crossbreeding provide examples of a genetic effect. Many of the genes responsible for rapid growth are also responsible for efficient gains. Faster growth in several species appears to be related to a higher **ovulation** rate, although in some experiments in which selection was for either heavier or lighter mature weights, lowered fertility was observed in both. Among beef cattle on a New Mexico range large-type Hereford cows produced significantly more and heavier calves in their lifetime than smaller **blockier-type** Herefords.

It is interesting that the metabolic rate of animals does not change proportionately with body weight. Brody,[4] in his

---

[4]S. Brody, *Bioenergetics and Growth*, Reinhold Publishing, New York, 1945.

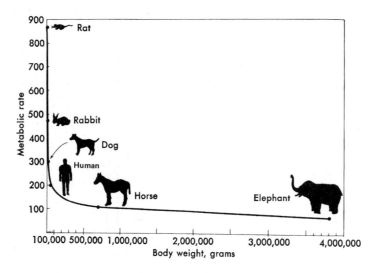

Figure 12.8 Graph depicting the inverse relationship of relative metabolic rate and body size in selected mammals. Metabolic rate is given in cubic millimeters of oxygen per gram of body weight per hour.
W. T. Keeton, Biological Science, 3d ed., Norton, New York, 1980.

monumental text, concluded that a doubling of body weight was associated with a metabolic increase of approximately 73 percent. This means that basal metabolism varies to the 0.73 power of body weight ($W^{0.73}$). The relation of relative metabolic rate and body size is shown in Figure 12.8.

12.8 SENESCENCE (AGING)

And so from hour to hour we ripe and ripe.

William Shakespeare (1564–1616)
English dramatist and poet

Life is most delightful when on its downward slope.

Lucius A. Seneca (4 B.C.–A.D. 65)
Roman statesman, dramatist, and philosopher

Senescence is the process of growing old. People refer to *growing old*, suggesting that it is recognized as a gradual and continuous process rather than one that occurs suddenly. The term *aging* encompasses the complex of changes that eventually lead to deterioration of the animal and ultimately to its death.

Problems of aging are especially important in human medicine and much research is currently being conducted in this field. Such problems are of less importance in farm animals because breeding animals are often **culled** well before they reach old age. This is particularly true of swine and poultry. Cows, however, are kept to an older age if they are not removed from the herd for reasons unrelated to aging. Many beef cows are kept until they are 9 to 12 or more years of age. Nevertheless, there is an age of each farm animal at which productive performance peaks. This is shown in Table 12.4 for selected farm animals. These data show that egg-laying rate is highest during the hen's first year and declines gradually thereafter (Chapter 16). Maximum litter size of **sows** occurs between 3

TABLE 12.4 | **Age and Performance of Selected Female Farm Animals**

Age of Females, Years	Weaning Wt of Calves, Lb (Hereford)	Average Litter Size Farrowed (Duroc and Poland Sows)	Egg Production per Year (Leghorn Hens)*
1	360	7.8	**169**
2	389	9.4	146
3	411	**9.8**	124
4	427	9.5	109
5	438	9.2	95
6	**443**	9.2	86
7	**442**	7.3	66
8	434		67
9	421		51
10	402		41

*Present annual egg production is higher; however, the same trend prevails.
Note: Boldface numbers represent peak productive performance.
Source: Missouri Agr. Expt. Station bulletins.

and 4 years of age, although a limited number of sows are kept in production that long. This is because sows become extremely large and are more expensive to feed and manage. Moreover, large sows present more problems at **farrowing** than smaller and lighter ones because they are more likely to accidentally overlay and smother their piglets.

. . . it is appointed unto men once to die.

Hebrews 9:27

. . . the days of our years are threescore years and ten.

Psalms 90:10

It has been said that an animal begins to die as soon as it is born. This is true in the literal sense because each passing day brings it one day closer to the actual day of its death. It is true also in the physiological sense because shortly after fertilization of the ovum, cells of certain tissues cease their division. Cell division ceases progressively among other tissues as the young develop until division is finally limited to cells in tissues such as the skin surface (epidermis) and blood in which periodic cell replacement is essential to protect the health and well-being of the individual.

Why does the mayfly live but a few hours while the pike (fish) may live 2 centuries? The rate of decline in the velocity of growth with increasing age is generally inversely proportional to the length of life (Figure 12.9). In a species that reaches maturity in a few years (cattle) the life span is shorter than in a species such as the human in which many years are required to attain maturity. Certainly species vary in their life span. The average life span of a rat is 3 years, the cow 20 to 25, the horse 30 to 35, humans 80 to 90 years, and the Galápagos tortoise 500 years. Humans have always searched for the fountain of youth. Most humans are not content with living just threescore and ten (70) years but rather would like to live forever, leading to a vast amount of research in this area.

> The slowest beasts are always strongest and manage, too, to live the longest.

> **Benjamin Franklin (1706–1790)**
> **American statesman and philosopher**

Of academic interest is the belief of some scientists that the life span of individuals is more or less fixed at conception by the metabolic potential the individuals receive from their parents. Furthermore, the rate at which individuals utilize their metabolic potential determines to a certain extent their length of life. This suggests that the "candle should not be burned at both ends" or a shorter life will result. As noted by Brody,[5] Rubner[6] calculated the quantity of energy metabolized per kilogram of body weight from maturity to death and found it is nearly the same in **homeothermic** animals (except humans). This is illustrated by the following estimates:

Species	Body Wt, Kilograms	Maximum Life Span, Years	Calories Expended During Lifetime per Kilogram Adult Body Wt
Horse	450	38	170,000
Cow	450	28	141,000
Dog	22	20	164,000
Cat	3	30	224,000
Guinea pig	0.6	6	266,000

[5] Ibid.
[6] M. Rubner, "Das Problem der Lebensdauer und seine Beziehungen zum Wachstum und Ernährung," *Arch. Hyg.* **Vol. 66** p. 45 Berlin, 1908.

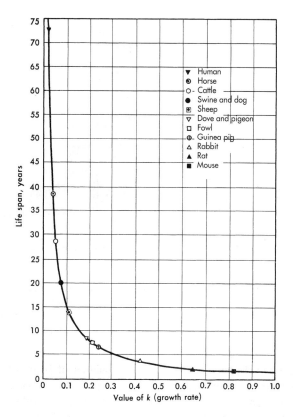

Figure 12.9 Estimated life span of selected species as a function of the growth rate k of approach to mature weight. Usually, the longer the growth period (the smaller the k value), the longer the life span. *Data from* Missouri Agr. Res. Bull. *97, 98, 101, and 102. Redrawn from Bulletin 102.*

Unlike the above species, which expend about 200,000 cal/kilogram of body weight during the life cycle, humans were estimated to expend about 800,000 cal/kilogram. However, if humans extend their days and shorten their periods of rest they must **metabolize** more energy per day. Some scientists theorize that this results in a decreased life of one or more body parts and/or systems.

Many physiological functions deteriorate with advancing age of humans and animals (Figure 12.10). The gonads secrete smaller amount of hormones, muscular strength and speed of motion decline, reaction time is increased, and there appears to be a longer time required for recovery when body substances become unbalanced. For example, when **intravenous** injections of glucose are administered to older persons a longer period of time is required to restore the normal blood glucose concentration. Collagen becomes less elastic with increasing age as shown by less elasticity of the skin and blood vessels compared with those of younger individuals. Some people believe that with advancing age there is a gradual loss of functional units in the organs or a gradual impairment of those remaining; both processes could be involved. A gradual breakdown of neural and endocrine control in the aging animal may be responsible in large part for changes associated with senescence.

Many factors are known to affect the life span of individuals. Longevity is related to heredity. Benjamin Franklin, a great

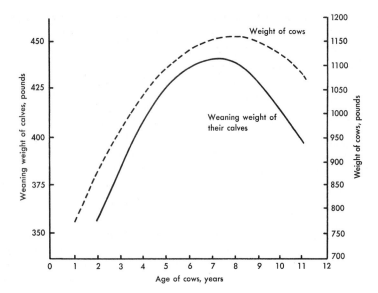

Figure 12.10 Graph showing the peak of physical maturity (maximum body weight of cows) and of physiological maturity (maximum weaning weight of calves) of Hereford cattle. Note the decline in both after 7 to 8 years of age.
Courtesy of Missouri Agr. Expt. Station.

American statesman, lived to age 84 closely approximating his mother's and father's ages at death, 85 and 89 years, respectively. Females in most species live longer than males. Restricted-calorie diets increase the life spans of mice and dogs. In studies at Cornell University rats fed high-protein diets had decreased life spans (14 percent dietary protein appeared to be the most favorable to longevity). In **poikilothermal** animals (animals in which the body temperature varies with environmental temperature) life span is shortened by increasing the temperature and lengthened by lowering it. In humans the vascular and nervous systems are usually the first to suffer from senescence.

12.9 SOME HYPOTHESES OF AGING

> Consequently to preserve life is to use meats and drinks according to the age of the person. For the dyet of youth is not convenient for old age nor contrariwise.
> **Thomas Cogan, *The Haven of Health*, London: Imprinted by Richard Field for Bonham Norton, 1596 p. 273**

Several hypotheses of why aging occurs have been proposed although no single one appears to explain it adequately. Aging is a complex process with many factors and physiological functions being involved.

12.9.1 Genetic Hypothesis

The genetic hypothesis states that in aging there is an accumulation of mutations within body cells that interfere with their normal function. Some people believe that rapidly dividing tissue cells are able to discard new mutations, whereas slowly dividing cells have an accumulation of mutations that causes

organ degeneration. Evidence for this theory is that nonlethal doses of radiation shorten the life span in some animals, presumably because they increase the number of somatic mutations within cells. Opposing this theory is evidence that aging changes occur in fixed postmitotic cells, whereas radiation primarily affects dividing cells. Also, radiation does not appear to affect the physical properties of collagen.

Another closely related hypothesis involves *telomeres.* Telomeres are the ends of eukaryotic chromosomes. The telomeres are thought to have important functions in the protection, replication, and stabilization of the ends of chromosomes. Some recent research indicates that telomeres may also function as a biological "clock" or "counter." When cells divide, DNA replicates (makes a new copy). The very ends of the DNA strands in most cells are not replicated; therefore, each time cells divide and the DNA is replicated the telomeres become shorter. The dividing cells in tissues from young animals (which have divided relatively fewer times) may have longer telomeres than the dividing cells in tissues from elderly animals (which have divided more times). Once the telomeres reach a certain length the cells can no longer divide, the cells age and eventually die leaving no daughter cells. It has thus been hypothesized that if telomere length can be maintained, a population of cells could be able to sustain division for a longer time and possibly slow aging of that tissue.

12.9.2 Immunological Hypothesis

According to the **immunological** hypothesis, organisms have the ability to produce **antibodies** against foreign cells and substances to which they are exposed if exposure occurs before a certain critical period in life. After this time the individual gradually loses its ability to resist invasion of foreign substances. For example, with advancing age humans become more susceptible to certain infectious diseases to which they are immune at a younger age. This suggests that with advancing age they lose their ability to cope with an adverse environment.

12.9.3 Developmental Hypothesis

The developmental hypothesis states that aging results from what is often called *overdifferentiation* (extreme cellular chemical specialization). For example, nerve cells of an adult animal cannot divide and there are no other cells that can divide and then differentiate into nerve cells. Thus when nerve cells die or become nonfunctional, some aspect of control over the body is lost. This loss of ability to coordinate the regulation of the body may, in due time, result in its destruction.

12.9.4 Biochemical Hypotheses

The biochemical hypotheses of aging propose that rare and irreparable nongenetic metabolic "accidents" occur. For example, the accumulation of oxidative products in tissues of older animals has been shown to occur. The products of such accidental chemical reactions gradually accumulate in cells and may be insoluble and have no function although they may interfere with normal cellular metabolism. Many of these products are referred to as "aging pigments" and include lipofuscin and ceroid. The association of these aging pigments with deterioration of cellular function appears to support the pigment theory of biochemical aging.

A second biochemical hypothesis of aging is referred to as the free radical theory. Lipids in cell membranes are continuously exposed to free radicals from superoxide anions and radiation. Free radicals oxidize carboxyl groups in the cell membranes forming lipid peroxides. Lipid peroxides are unstable and cause the cell membranes to rupture leading to cell death. The loss of functional cells leads to deterioration in vital organs and ultimately to death of the animal.

A third biochemical hypothesis of aging is the glycosylation theory. Glucose can react with proteins and nucleic acids forming glycosylated cross-linkages. These cross-linkages result in changes in protein solubility and function. Over time a buildup of glycosalation end products results in a deterioration of organ function. This mechanism is thought to play an important role in the pathogenesis of Alzheimer's disease and atherosclerosis.

12.10 SUMMARY

In livestock production *growth* is often defined as *the normal increase in weight and body size over a period of time.* This is not an accurate definition, however. True growth from a chemical standpoint is *an increase in the amount of body protein and mineral matter.* Deposition of fat and accumulation of body water are not true means of growth.

The cell is the basic unit of growth. In growth, cells increase in numbers (hyperplasia) and size (hypertrophy). Early embryonic growth is due to an increase in cell numbers, whereas most growth near birth or shortly thereafter can be due to increases in cell size and number. Cell division ceases in some organs much before it does in others.

The postnatal growth curve of all animals is sigmoid (S-shaped). The growth rate of animals decreases shortly after puberty. Many factors are known to affect growth. These include (1) the genetic constitution of the individual, (2) nutrients and environmental conditions encountered during its lifetime, and (3) hormones secreted by the individual's endocrine glands. The life span in multicellular animals is generally related to their age at maturity. A species that reaches maturity more quickly has a shorter life span than one that matures at an older age. As the animal grows older it loses some of its ability to cope with the environment. Although there is not complete agreement on the causes of aging or senescence, most authorities agree that cells either die or lose their ability to function properly. Both processes may be important. In any event, death occurs when one or more vital organs cease to perform their required functions.

Humans and the animals that serve them are by nature predestined, after growth and maturation, to decline and finally die. Their structure and functions are more complex than those of any other living creature. Life is like the flame of a candle whose form is constant while each particle changes every moment. Death takes small bites during the aging process and eventually destroys enough cells that the body tissues fail to function as an integrated entity, resulting in death of the entire organism.

STUDY QUESTIONS

1. Define growth, differentiation, morphogenesis, regeneration, hypertrophy, hyperplasia, intussusception, permanent cells, stable cells, and labile cells.

2. Growth is not just an increase in body weight. Explain.

3. What factors may influence the weight of the young at birth?

4. How do muscle cells increase in size?

5. How is the protein:DNA ratio related to the growth of tissue?

6. Explain how bones grow in length and width.

7. How do cancer cells differ from normal cells in their division process?

8. What are some body cells that divide in a way similar to that of cancer cells?

9. What are the two main periods of growth in mammals?

10. Describe the postnatal growth curve in populations and in individual animals.

11. What forces are responsible for the typical kind of growth curve?

12. What factors are responsible for cessation of growth in populations of single-celled individuals?

13. List factors responsible for the maximum size of an individual.

14. What hormones are related to growth?

15. Describe the source, function, and chemical composition of growth hormone (GH).

16. Is GH secreted in the normal mature individual?

17. What is the relationship of nutrition to growth and life span of animals?

18. What evidence is there that hereditary mechanisms affect the rate and amount of growth in farm animals?

19. Is growth related to other traits in mammals? Explain.

20. What is senescence?

21. Describe some changes within the body that occur with advancing age.

22. List some factors that are known to affect the life span of animals.

23. Discuss hypotheses of aging. Does any one hypothesis adequately explain all the processes that occur in aging? Why or why not?

ANATOMY AND PHYSIOLOGY OF REPRODUCTION AND RELATED TECHNOLOGIES IN FARM MAMMALS[1]

Let the earth bring forth the living creature after his kind, cattle, and creeping thing, and beast of the earth after his kind, and it was so.

Genesis 1:24

13.1 INTRODUCTION

Reproduction is a complicated process in all **species** of animals. It is complicated because it is dependent on the proper functioning of biochemical processes of many organs. The anatomy of the male and female reproductive systems must be compatible within each species to make efficient reproduction possible.

Physiological compatibility on the part of the male and female is also necessary for reproduction. The female must be willing to accept the male in the act of mating **(copulation)** when the egg is released from the **ovary (ovulation).** Similarly the male must be willing and able to deliver **spermatozoa** to the proper site in the female reproductive tract at the proper time for **conception** to occur. Synchronization of **ovum** release from the ovary and introduction of spermatozoa into the female reproductive tract is necessary because the lives of the released ovum and the ejaculated spermatozoa are limited within the reproductive tract of the female (Table 14.5).

Maturation of the ovum, ovulation, estrus, **estrous cycle,** pregnancy, **parturition,** and **lactation** are all dependent on the proper function of various **hormones** and organs (Chapter 11). Any abnormality in the anatomy or physiology of the reproductive system results in lowered **fertility,** infertility, or **sterility. It** is amazing, indeed, that the fertility rate in farm animals is as high as it is in view of the many complicated biological and physiological processes involved and necessary for conception and birth of the new individual(s).

In this chapter the anatomy and physiology of reproduction in male and female farm **mammals** are discussed. The physiology

of lactation and of egg laying are discussed in Chapters 15 and 16, respectively. The anatomy and reproductive physiology of the male and female chicken are discussed in Chapter 16.

13.2 ANATOMY OF THE MAMMALIAN MALE REPRODUCTIVE TRACT

The role of the male is to produce large numbers of viable male sex cells (spermatozoa) and to deliver them to the proper place in the reproductive tract of the female at the proper time. Inability of the male to perform either of these functions results in infertility.

The male farm mammal contributes approximately one-half the inheritance to each of his **offspring.** The possible exception to this is that he supplies something less than one-half the inheritance to his male offspring because he transmits to them only the **Y chromosome.** The Y chromosome is considerably smaller than the **X** chromosome and therefore carries fewer **genes** (Figure 8.2). After the male mammal delivers viable spermatozoa to the reproductive tract of the female and the ovum is fertilized, his role in reproduction is completed.

The male reproductive organs in various species of farm mammals are similar in the parts they contain, but there are species differences in their size and degree of development (Figures 13.1 and 13.2).

The testicles (testes) in the male are the *primary sex organs.* They produce spermatozoa and the male hormone *testosterone.* Each male normally possesses two testicles carried in the scrotum. The scrotum in the **ram, bull,** and **stallion** is located in the **ventral** portion of the body **anterior** to the hind

[1]The authors acknowledge with appreciation the contributions to this chapter of Dr. D. J. Kesler, Department of Animal Sciences, University of Illinois at Urbana–Champaign.

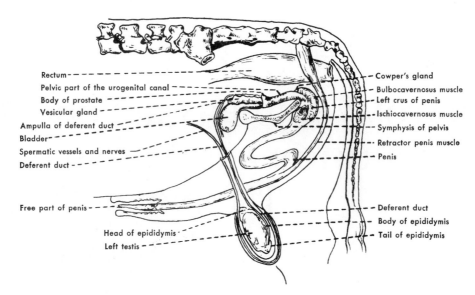

Figure 13.1 Diagram of the reproductive tract of the bull showing major parts.
Courtesy of University of Missouri Extension Service.

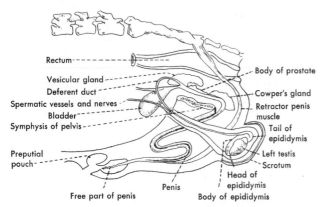

Figure 13.2 Diagram of the reproductive tract of the boar showing major parts.
Courtesy of University of Missouri Extension Service.

legs. In the boar the scrotum is located ventral to the anus (Figure 13.2). The scrotum functions as a heat-regulating mechanism in all male farm mammals, especially the ram and bull. The testicles are drawn close to the body when the outside temperature is cold; they are relaxed when the outside temperature is hot. When environmental temperatures permit, the scrotum keeps the testicles 39 to 41°F (4 to 5°C) below normal body temperature. This lowered temperature is essential for optimal spermatogenesis.

In some male farm mammals one testicle may fail to descend into the scrotum being retained in the body cavity at normal body temperature. Such a male is a unilaterally **cryptorchid** (*crypto* meaning "hidden" [from Greek *kryptos,* hidden], *orchid* meaning "testes"); however, because he has one testicle in the scrotum he is usually fertile. The testicle in the body cavity does not produce spermatozoa but does produce testosterone. Some males have both testicles in the body cavity

and are known as bilateral cryptorchids. Bilaterally cryptorchid males do not produce spermatozoa and are sterile, although they possess normal sexual activity and the outward appearances of the male because they secrete testosterone.

Castration in the male consists of removing both testicles, usually within the first few days of life. Such males do not develop male characteristics and lack sex drive **(libido).** A castrated male lamb is called a **wether;** a castrated male calf, a **steer;** a castrated male pig, a **barrow;** and a castrated male horse, a **gelding.** A male castrated after sexual maturity retains some of its secondary male characteristics and libido and is often known as a **stag.** When unilaterally cryptorchid males are castrated usually the testicle in the scrotum is removed but not the one in the body cavity. When such a castrated male reaches the age of sexual maturity the testicle in the body cavity produces the male hormone testosterone, and the individual possesses normal sexual activity and sex characteristics even though he is infertile.

A diagram of a cross section of a testicle is shown in Figure 13.3. Each testicle contains long tubules known as *seminiferous tubules.* The seminiferous tubules join in the *rete testis,* which in turn connects with the head of the epididymis. Between the seminiferous tubules are found connective tissue, blood capillaries, and interstitial cells (cells of Leydig).

A portion of the testis of a ram showing cross sections of seminiferous tubules is seen in Figure 13.4. The seminiferous tubules in the sexually mature male possess spermatozoa in all stages of development (Figure 8.6) and *Sertoli,* or nurse, cells. The **lumen** of seminiferous tubules possesses spermatozoa, which have the general shape of those in semen. Spermatozoa in the testes do not possess motility and are incapable of fertilization. The **morphology** of mature spermatozoa of different species of farm mammals is similar, and mature spermatozoa of different farm mammals possess the same parts as ram sperm shown in Figure 13.5. Semen characteristics for the various farm mammals together with other pertinent information are given in Table 14.5.

The *epididymis* is a long, continuous, tortuous tube often measuring 400 to 500 ft in length. It consists of a head, body, and tail. The head portion is connected with the testis (Figure 13.6). The tail appears as an enlargement on the ventral portion of the testis in the scrotum of the ram and bull. It is inverted, or on the **dorsal** portion of the testis, in the scrotum of the boar (Figure 13.2). The tail of the epididymis is a storage depot for spermatozoa and as many as 200 billion sperm have been collected from the epididymides of a sexually mature boar at cas-

tration. Spermatozoa mature physiologically as they progress through the epididymis.

The tail of the epididymis widens into a larger tube leading to the *urethra*. This long tube is known as the *vas deferens* (Figure 13.6). A male may be rendered infertile if a portion of each vas deferens is surgically removed (**vasectomy**). This prevents passage of spermatozoa into the urethra and therefore they do

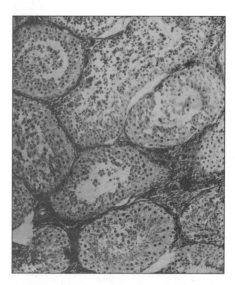

Figure 13.4 Cross section of seminiferous tubules from the testis of a mature ram. Within the seminiferous tubules are male gametes in different stages of development. Between the seminiferous tubules are the cells of Leydig, which produce testosterone, the male hormone. *Courtesy of Missouri Agr. Expt. Sta. Bull. 265.*

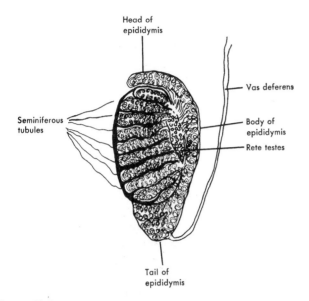

Figure 13.3 Cross section of a testis showing different parts.

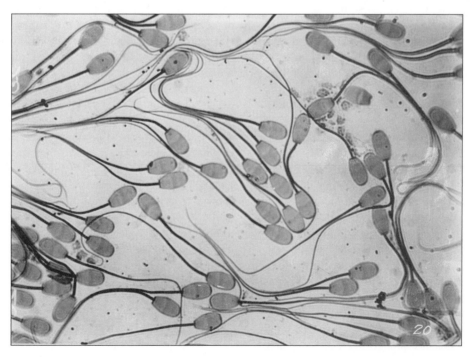

Figure 13.5 Microphotograph of spermatozoa of the ram. Spermatozoa from the semen of other farm mammals are similar in structure and appearance. *Courtesy of Missouri Agr. Expt. Station.*

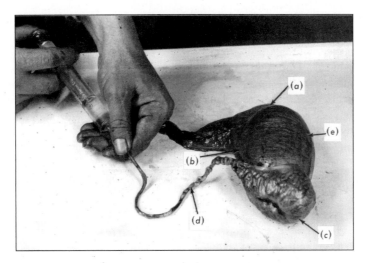

Figure 13.6 The testicle of a boar. (a) Head of the epididymis; (b) body of the epididymis; (c) tail of the epididymis; (d) vas deferens; (e) testicle.
Courtesy of Missouri Agr. Expt. Station.

not get into the seminal fluids at **ejaculation.** If a vasectomy is performed correctly the male is infertile but retains normal sex drive and male characteristics. Vasectomies are often performed in males for the purpose of using them as estrous detector animals. A vasectomy may be performed in the human male as a means of birth control where a pregnancy in his spouse might be dangerous and/or unwanted.

The vas deferens widens and becomes enlarged in the ram and bull at the point where it joins the urethra. This enlargement, known as the *ampulla,* may serve as a temporary storage depot for spermatozoa in species such as the bull and ram in which semen ejaculation is rapid.

The *urethra* begins at the opening of the bladder and is continuous with the *penis.* It acts as a passageway for both urine and spermatozoa. In the sexually mature male, especially the bull, the **posterior** portion of the urethra is S-shaped and is known as the *sigmoid flexure.* At the time the male mates with the female, the penis becomes erect and engorged with blood. The sigmoid flexure straightens, extending the penis so semen can be deposited into the female reproductive tract. The retractor muscle relaxes at this time allowing the sigmoid flexure to straighten at the time of copulation. After copulation this muscle contracts (retracts), drawing the penis back into the sheath where it is protected from injury. Some bulls cannot perform natural service because the retractor muscle will not relax and allow extension of the penis. Even though such bulls possess **viable** spermatozoa in the testes and epididymides they are infertile in natural service because they cannot deliver spermatozoa to the proper site in the female reproductive tract.

The penis of male farm mammals varies in shape among species so as to be compatible with the reproductive tract of the female in order to aid efficient reproduction. The boar penis possesses a corkscrew shape at its end so it fits into the interlocking cervix of the sow during copulation. This results in semen being forced into the sow's uterus.

The male reproductive tract includes several glands known as the *accessory sex glands.* These include the prostate, vesicular glands, and Cowper's glands (bulbourethral glands).

The **prostate gland** is located near and around the opening of the bladder. Its secretions nourish and stimulate sperm activity. In the human male and in males of some other species, the prostate gland often becomes enlarged and inflamed causing pain and difficult urination. The prostate gland is commonly surgically removed when so affected.

Two *vesicular glands* enter the urethra near the vas deferens and near the neck of the bladder. The vesicular glands are large in the mature boar, measuring 2 to 4 inches in diameter. They are smaller in males of other farm mammals. Functions of the vesicular glands are to neutralize urine residues, add volume to semen, and possibly to stimulate the activity of spermatozoa.

The two *Cowper's glands* are located posterior to the bladder, prostate, and vesicular glands. They resemble small walnuts in shape in the bull but are about 4 to 5 in long in the sexually mature boar. Also known as the bulbourethral glands, Cowper's glands secrete a fluid that becomes a gel in boar semen when it contacts secretions of other accessory sex glands. This gelatinous material is the last to be ejaculated and forms a plug in the cervix of the sow, which helps prevent the backflow of semen from the vagina at copulation and aids in more efficient reproduction. Secretions of the Cowper's glands also add volume to the ejaculate and neutralize the detrimental effects of urine in the urethra.

13.3 ANATOMY OF THE MAMMALIAN FEMALE REPRODUCTIVE TRACT

The female farm mammal plays an especially important role in reproduction. Through the ovum she supplies one-half the inheritance of each of her offspring, she nourishes the young

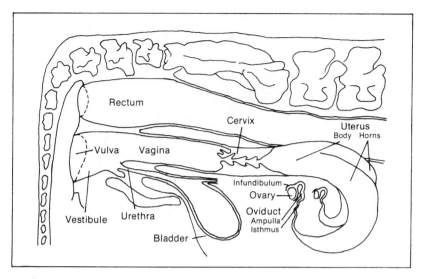

Figure 13.7 Drawing of the major reproductive organs of the cow.
From T. R. Troxel and D. J. Kesler, "The Bovine Estrous Cycle: Dynamics and Control," University of Illinois Coop. Ext. Cir. *1205.*

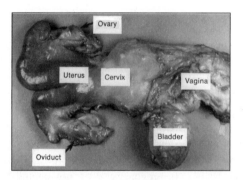

Figure 13.8 Photograph of the major reproductive organs of the cow.
From T. R. Troxel and D. J Kesler, "The Bovine Estrous Cycle: Dynamics and Control," Univ. of Illinois Coop. Ext. Cir. *1205.*

within her body (in the uterus) until birth, then nourishes (through her milk) and cares for the young until they are weaned.

The anatomy of the reproductive tracts of female farm mammals is similar, but there are differences in size and shape among species, especially in the size and shape of the uterine horns.

The female reproductive tract (Figures 13.7 and 13.8) consists of the vulva, vestibule, vagina, cervix, uterus, uterine horns, oviducts, and ovaries (Figure 13.9a and b). The *vulva* is the exterior portion of the female reproductive tract. It serves as the entrance to the internal organs and it also accommodates the passage of urine during urination.

The *vestibule* is the general passageway to the urinary and reproductive tracts. It extends inward from the vulva for about 4 in (10 centimeters, cm) to where the urethra opens into its ventral surface from the bladder. The clitoris, which has the same embryonic origin as the penis, lies on the base of the vestibule.

The *vagina* is that portion of the reproductive tract between the vestibule and the cervix. It is about 12 in (30 cm) long in most mature female farm mammals. In the cow and ewe, semen

is deposited by the male during copulation in the anterior portion of the vagina. In the mare and sow a portion of the semen is deposited in the uterus.

The *cervix* is the opening into the uterus through which sperm must pass to meet the egg in the process of fertilization. It is also the opening through which the young must pass from the uterus at birth. The cervix is about 2 to 3 in (4 to 7 cm) long in the cow and ewe but longer in the sow and mare. The cervices of the cow and ewe possess circular folds that make it difficult to pass a pipette (an instrument for passing fluids) through them into the uterus. It is nearly impossible to do this in the ewe but is possible in the cow. Artificial insemination in the cow is accomplished by placing semen in the cervix or the uterus. The inseminator places one hand into the rectum and feels the cervix via the rectal wall. The insemination tube is then inserted into the vagina to the cervix. A gentle pressure on the pipette and a rolling movement of the cervix through the rectal wall cause the pipette to pass through the cervix in most cows (Chapter 14). The sow has an interlocking type of cervix that also makes it difficult to pass a pipette through it into the uterus. In the mare the opening through the cervix is not difficult to pass and even capsules an inch or more in diameter can be deposited in the uterus during estrus.

The reproductive tracts of the cow and sow are shown in Figure 13.9a and b. A female mammal normally possesses two *uterine horns* (except the human female), although one or both may be absent in abnormally developed individuals. The uterine horns of the sow are much longer than those of other female farm mammals. The sow's uterine horns may measure 5 to 7 ft (2 meters, m) in length when the broad ligament is removed and the horns are stretched to their full length. The sow possesses longer uterine horns because she normally gives birth to 10 or more young, whereas the cow and mare possess shorter uterine horns and usually give birth to only one young. The ewe often

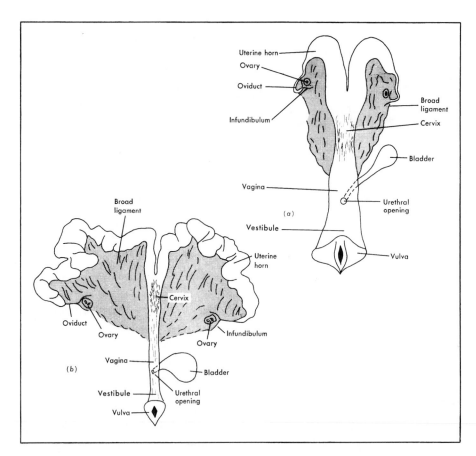

Figure 13.9 (a) Drawing of the reproductive organs of the cow showing major parts; (b) drawing of the reproductive organs of the sow showing major parts.

gives birth to one young but it is not unusual for her to have twins, triplets, or quadruplets.

The anterior end of each uterine horn becomes much smaller and blends into a small tube known as the *oviduct*. The oviduct is lined with microscopic cilia (hairlike projections) that assist in movement of the ovum toward the uterine horns. The oviducts are 4 to 5 in (13 to 11 cm) long and end in a funnel-shaped membrane known as the *infundibulum*. The infundibulum receives the ovum when it is released from the ovary at ovulation and guides it into the oviduct.

The normal female farm mammal possesses two ovaries, one located in proximity to each oviduct. The *ovary* possesses thousands of female sex cells in different stages of development. Female farm mammals as well as human females, possess all the potential ova they will ever have at birth. Only a relatively few of these ova are ever ovulated.

A diagram of a maturing follicle and ovum is shown in Figure 13.10 and an actual photograph of a sow ovary is shown in Figure 13.11. Ovaries of mature female farm mammals possess *follicles* in different stages of development. During estrus the ovary possesses mature follicles within which the ovum matures. When fully mature the follicle ruptures and releases the ovum (ovulation). The ruptured follicle then becomes filled with blood and is known as a *corpus hemorrhagicum* (blood-filled body). Cells then gradually replace the blood clot in the

ruptured follicle forming a *mature corpus luteum* (Figures 13.11 and 13.12). If pregnancy does not occur the corpus luteum regresses, becoming small and white, and is known as a *corpus albicans* (white body). The corpus albicans gradually disappears and can no longer be seen on the ovary. If pregnancy occurs the corpus luteum remains functional throughout the pregnancy.

13.4 PHYSIOLOGY OF REPRODUCTION IN FARM MAMMALS

The physiology of reproduction in farm mammals involves different parts of the reproductive tract mentioned previously as well as endocrine glands and their secretions discussed in Chapter 11.

13.4.1 Reproduction in Males

Gonadotropic hormones secreted by the anterior pituitary gland are involved in reproduction in the male. They are the same hormones from the anterior pituitary gland that affect reproduction in the female. The follicle-stimulating hormone (FSH) from the anterior pituitary gland stimulates sperm production by the seminiferous tubules. The luteinizing hormone (LH) stimulates

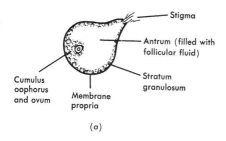

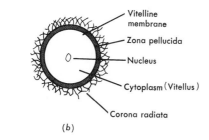

Figure 13.10 (a) Diagram of a maturing follicle; (b) diagram of a mature ovum.

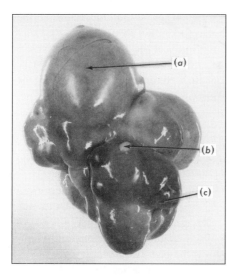

Figure 13.11 A sow ovary. (a) Large follicle; (b) corpus albicans; (c) corpus luteum.

Courtesy of Missouri Agr. Expt. Station.

the interstitial cells (cells of Leydig) to produce the male hormone testosterone.

Testosterone causes growth, development, and secretory activity of the accessory sex glands, which include the prostate, vesicular glands, and Cowper's glands. Testosterone also stimulates secretory activity of various other cells in the male reproductive tract including those in the epididymis. The hormone also causes growth and development of other parts of the reproductive tract including the penis, and is necessary for maturation and survival of spermatozoa stored in the epididymis. Secondary sex characteristics such as the male voice, the crest in the bull and boar, thick horns in the bull, whiskers in the human male, as well as many other masculine characteristics are dependent on secretion of testosterone. This is evident when males are castrated at a young age because the secondary sex characteristics never fully develop in such individuals. Sex drive (libido) in the male is also dependent to a great extent on testosterone production, although certain psychological factors are involved as well. The role of psychological factors is demonstrated in males castrated after sexual maturity because they often retain varying levels of libido following castration. Testosterone secretion may also play a significant role in social dominance (peck order) among animals (Chapter 24).

The amount of semen produced and the number of spermatozoa contained in each semen sample vary among species. Semen consists of spermatozoa mixed with secretions of various accessory sex glands and those of the epididymides. The average volume of the normal ejaculate (semen) of the ram is about 1 cc; of the bull, 5 cc; of the stallion, 75 to 100 cc; and of the boar, 200 to 300 cc. The average number of spermatozoa per ejaculate varies from about 2.5 billion in the ram to 20 billion in the boar (Table 14.5). Semen contains some spermatozoa that are dead, some that are alive but nonmotile, others that are weakly motile, and some that show strong, forward progressive motility. Proportions of different kinds of motile and nonmotile spermatozoa from many samples of normal bull semen before and after storage are shown in Figure 13.13. Spermatozoa of abnormal shapes often appear in the ejaculate; a high percentage of these is indicative of an upset in spermatogenesis in the testes and is often related to low fertility (Figure 14.3).

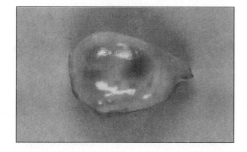

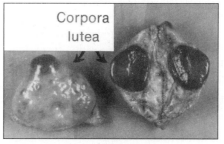

Figure 13.12 Ovary of the cow with developing follicles (left) and corpora lutea (right). The ovary on the right was cut to indicate how deep the corpus luteum grows.

From T. R. Troxel and D. J. Kesler, "The Bovine Estrous Cycle: Dynamics and Control," Univ. of Illinois Coop. Ext. Cir. *1205.*

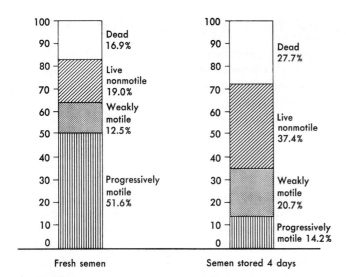

Figure 13.13 The percentage of dead and live spermatozoa in fresh and stored semen 32°F (0°C) of the bull as determined by means of the live–dead stain and actual counts of the percentage of motile sperm made with a hemocytometer.
Courtesy of Missouri Agr. Expt. Station.

Spermatozoa may be stored and retain their fertilizing capacity for several days in the epididymis within the body of a male. Their fertile life is limited, however, in some species when stored outside the body at temperatures slightly above freezing. Bull spermatozoa have been collected and stored outside the body with considerable success for use in artificial insemination (AI) (Chapter 14). Bull semen can also be frozen and stored indefinitely at very low temperatures. The fertile life of boar, stallion, and ram spermatozoa stored at temperatures slightly above freezing is usually limited to 48 h. Spermatozoa from these species have not been frozen and stored as successfully as those from bulls. The life of the spermatozoa of farm mammals in the female reproductive tract is generally limited to 24 to 40 h in cattle, sheep, and swine and 96 to 144 h in horses (Table 14.5).

Many experiments have shown definite adverse effects of thermal stress on reproduction in male farm animals. Adverse effects appear to be more common in rams than in males of other species. Some rams are low in fertility or temporarily infertile during the summer months. Poor quality of semen during periods of thermal stress is reflected in the production of smaller numbers and in larger percentages of abnormal spermatozoa. This indicates an adverse effect on spermatogenesis. Bulls also may show a decline in sperm production during the hot summer months but infertility seldom results. Increasing the temperature of the testes by almost any means (e.g., cryptorchidism and insulation) causes decreased sperm production and sometimes infertility in males of several species.

13.4.2 Reproduction in Females

The ovary of the sexually immature female farm mammal is quiescent for a period of several months and then suddenly follicles mature, the female comes into estrus (heat), an egg is ovu-

TABLE 13.1	Age of Puberty in Farm Mammals	
	Approximate age of puberty, months	
Species	**Males**	**Females**
Cattle	6–9	5–14
Sheep	6–8	7–10
Horses	18–24	15–24
Swine	5–8	5–8

lated, and the female has reached the age of sexual maturity (**puberty**). The approximate age of puberty for females of different species of farm mammals is given in Table 13.1. The cause of sudden onset of puberty in the female is unclear. Ovulation can be induced in sexually immature female farm mammals by injecting gonadotropins. This shows that the ovaries of these immature females are capable of responding to gonadotropins but the hormones probably are not present in the anterior pituitary gland or are not released in sufficient amounts to cause the onset of puberty. Most females within a given species reach puberty near a certain general age and weight although breed, family, and individual differences exist. Breed and family differences indicate a genetic effect on this trait.

Immature **gilts** near the age of puberty have been brought into estrus when administered a commercial product containing both pregnant mare serum gonadotropin (PMSG) and human chorionic gonadotropin (hCG). Primarily the PMSG has FSH activity and hCG primarily has LH activity. Combined they stimulate follicular development capable of resulting in a normal ovulation.

Endocrine Control of the Ovary

Follicle-stimulating hormone from the anterior pituitary gland, when released into the bloodstream, initiates development of ovarian follicles. Leutinizing hormone acts in synergism with FSH to stimulate maturation of the follicles. As a follicle develops and matures the egg also matures, and the cells lining the follicle multiply and secrete fluid until the follicle becomes a fluid-filled cavity resembling a blister (Figure 13.12). At this point the cells lining the follicle wall begin to produce estrogens.

Estrogens have several important functions. These include (1) the induction of estrus (heat) in the female; (2) an increase of motility of the uterus, oviducts, and infundibulum during estrus, which aids in bringing the sperm and egg together in the process of fertilization; (3) dilation of the cervix; (4) synthesis and secretion of cervical mucus; (5) initiation of duct growth in mammary glands; and (6) development of secondary sex characteristics.

Additionally, estrogen acts in a unique manner in the hypothalamus and the anterior pituitary gland to stimulate a massive release of LH called the LH surge. This stimulates follicle(s) to rupture and release the egg (ovulation). The interval from the LH surge to ovulation has been demonstrated to be about 24 to 32 h in the cow. The infundibulum then picks up the egg and rapidly transports it to the ampullary–isthmic junction for potential fertilization.

TABLE 13.2	Length of Estrus and Estrous Cycle, Time of Ovulation, and Length of Gestation in Selected Female Farm Mammals			
Mammal	**Length of estrus**	**Length of estrous cycle**	**Approximate time of ovulation**	**Length of gestation**
Cow	12–24 h	19–21 days	10–14 h after end of estrus	272–292 days
Ewe	24–36 h	15–17 days	24–30 h from beginning of estrus	140–150 days
Mare	4–9 days	19–21 days	24–48 h before end of estrus	335–345 days
Sow	24–72 h	19–21 days	35–45 h from beginning of estrus	110–115 days

After ovulation, estrogen levels drop rapidly and cells in the ruptured follicle begin to grow and divide to form a new structure called the corpus luteum (Figure 13.12), which begins producing progesterone. If conception occurs the corpus luteum will be maintained and will continue producing progesterone because progesterone is required to maintain pregnancy. If conception does not occur progesterone levels drop and the corpus luteum will regress. In the cow this regression is due to an increase in $PGF_{2\alpha}$ that destroys the corpus luteum. The $PGF_{2\alpha}$ is synthesized by the uterine endometrium and released in response to oxytocin that is synthesized by the corpus luteum. The manner in which the embryo signals establishment of pregnancy to the dam varies among species. About 12 to 17 days after conception the embryos in the cow synthesize an antiluteolytic protein known as bovine interferon-tau (bINF-τ). The bINF-τ binds to uterine interferon receptors and inhibits uterine secretion of $PGF_{2\alpha}$. This results in persistence of the corpus luteum, which then continues secretion of progesterone required for maintenance of pregnancy. In the human the embryo synthesizes hCG, which is luteotropic and stimulates maintenance of the corpus luteum and continual synthesis of progesterone. In both species the end result is the continued synthesis of progesterone.

Progesterone has several functions: (1) it maintains pregnancy by blocking uterine contractions, (2) it prevents ovulation by blocking the LH surge, and (3) it combines with estrogens to stimulate growth of the lobule–alveolar system of mammary glands.

Follicles grow in waves as has been demonstrated in the cow. Most cows have two follicular waves whereas most **heifers** have three. Follicles must reach 8 to 10 mm in size before they are capable of ovulating; however, without exogenous therapy most follicles ovulate at a size of about 14 to 16 mm. Only follicles that are developing during the follicular wave (when progesterone concentrations fall) are capable of ovulating. Follicles that reach maturity in the presence of elevated progesterone concentrations undergo atresia (degeneration).

Estrus and the Estrous Cycle

Estrus is the time a female will receive the male in the act of mating. The *estrous cycle* begins at puberty and is the interval between two estrous periods when the female is not pregnant. The length of estrus and the estrous cycle for selected farm mammals are presented in Table 13.2. Estrus and ovulation are synchronized in most species so that introduction of spermatozoa into the reproductive tract of the female and release of the ovum from the ovary

occur at about the same time. This increases the probability of successful fertilization because the life of the ovum and sperm in the female reproductive tract is limited to a few hours.

Some female farm mammals show obvious behavioral changes associated with estrus. The cow will often have a clear mucus on the vulva and exhibit certain behaviors: she may become unusually nervous; have reduced appetite and milk production; and butt, bellow, and walk about more than usual. She may also attempt to mount other cows; however, because this behavior is also seen in cows that are not in estrus it is not a reliable indicator of estrus. The most reliable sign of estrus is that the cow will stand to be mounted by other cows or a bull. Mares in estrus urinate frequently and exhibit rhythmic vulvar contractions called winking. Sows show an immobility response: standing still if pressure is applied to their backs. Mares and sows demonstrate these behaviors especially in the presence of stallions or boars. Estrus is less obvious in ewes but experienced producers can detect ewes in estrus. (The ewe in estrus will stand when mounted by a ram. A vasectomized ram wearing a marking harness may be used.)

The estrous cycle in farm mammals has four phases: proestrus, estrus, metestrus, and diestrus. *Proestrus* is that phase of the cycle that occurs just before estrus when the reproductive system is preparing for release of the ovum. There is a thickening of the vaginal wall, increased vascularity of the uterine mucosa, and maximum follicular growth. *Estrus* is the period of the estrous cycle when the female will accept the male in the act of mating and when the follicle matures and releases the ovum. *Metestrus* is the phase immediately following estrus, when the uterus makes preparations for pregnancy and the corpus luteum forms and begins secreting progesterone. *Diestrus* is usually the longest phase of the estrous cycle (occurs between metestrus and proestrus). During the first part of diestrus the corpus luteum becomes fully developed. In the latter part there is regression of the corpus luteum and uterine mucosa.

Nonpregnant female farm mammals usually express both inward and outward signs of the estrous cycle throughout the year, or during the breeding season if they are seasonal breeders. In some species of animals such as most primates (including humans), inward signs of the estrous cycle occur but no specific outward signs are readily evident. Growth of follicles on the ovary and **proliferation** of the endometrium of the uterus to prepare for pregnancy occur periodically even though no estrus occurs. Menstruation (a period of bleeding) is characteristic of primates and is due to a cyclic sloughing off of the portion of the endometrium of the uterus that was built up to receive and nourish the fertilized ovum in anticipation of pregnancy. Such cycles are called menstrual cycles. Commonly, when the human female reaches the age of 45 to 50 years the menstrual cycle gradually ceases. This is known as *menopause* (a pause in the occurrence of menstruation), or the *change of life,* after which the normal events of the menstrual cycle no longer occur and the female is infertile. Menopause does not occur in female farm mammals.

Continuous estrus occurs in the nonpregnant, sexually mature female rabbit, at least during the breeding season. Copulation in the rabbit induces ovulation by releasing LH from the anterior pituitary gland. When ovulation occurs corpora lutea are formed on the ovary and the reproductive system is brought under the influence of progesterone as in other species.

Monoestrous and *polyestrous* are terms sometimes used to describe the yearly sexual activity of females. Certain wild animals such as the fox have only one estrous period each year and thus are referred to as *monoestrous* (*mono* meaning "one") animals. Other animals have many estrous periods annually (if they do not become pregnant) and are called *polyestrous* (*poly* meaning "many") animals. Still other species such as sheep are definite seasonal breeders because they show sexual activity during only a particular season. Many breeds of sheep do not have estrous cycles during the late spring and summer months but sexual activity begins abruptly in late summer and early fall, during a period when daylight is decreasing. Some breeds of sheep mate during most seasons of the year. Other species of animals such as birds show sexual activity in early spring, when daylight is increasing, but are sexually inactive at other times of the year (Chapter 16). Seasonal breeding activity appears to be related to the effect of day length on melatonin secretion by the pineal gland. Melatonin has a role in the regulation of secretion and release of gonadotropins by the anterior pituitary gland.

Wild animals mate at a time when their young will be born in the most favorable environmental circumstances for survival; that is, in the spring when there is an ample supply of food and the weather is warm. The **domestication** of animals has tended to reduce seasonal breeding in some species such as swine and cattle so that young can be produced at any time of the year. Even in these species, however, there appears to be a seasonal pattern of greater sexual activity and more efficient reproduction at the period of the year corresponding to the natural mating season in wild animals of that particular species.

Estrus usually does not occur during pregnancy, although exceptions have been observed in cows. After pregnancy is terminated by birth of the young, the normal estrous cycle is initiated again in most female farm mammals. However, suckling of the young may delay the interval from parturition to resumption of ovarian cycles especially in beef cattle and swine. Estrus occurs in sows 3 to 5 days after the offspring are weaned.

Mares usually come into estrus 5 to 10 days after **foaling.** This estrous period is called *foal heat* and is accompanied by ovulation. Foal heat usually lasts from 1 to 10 days and mares are sometimes bred on the ninth day after foaling.

Cows nursing calves will usually show their first estrus 40 to 80 days following parturition. When feed is limited the period between calving and the first estrus may be prolonged to 80 to 100 days, or even longer. Milked cows will generally ovulate 20 to 30 days after parturition; however, these are often not in association with outward signs of estrus and are called "silent estrus." Silent estrus is most frequently observed in dairy cows; however, this phenomenon has been reported in other species as well. Although cattle may ovulate without displaying estrus, many "silent" heats are a result of inadequate observation by the herdsperson.

Most breeds of sheep are seasonal breeders and cannot be bred immediately following lambing (giving birth to young) because lambing generally occurs in the presence of increasing day length.

Abnormalities of estrus and the estrous cycle are sometimes observed. Some females show almost continuous estrus for many days rather than the normal rest between estrous periods. Such females are called *nymphomaniacs,* a term that indicates abnormal sexual desire. The condition is caused by cystic follicles that fail to ovulate.

Ovarian cysts are structures that develop from follicles that fail to ovulate. Cystic follicles appear to result from the lack of certain hormones or from an imbalance of two or more hormones. Photographs of ovarian cysts in cattle and pigs are shown in Figure 13.14. They occur in about 5 to 15 percent of dairy cows but seldom in heifers and beef cows. Single or multiple ovarian cysts have also been detected in sows and gilts. In cattle the cysts are usually 2.5 to 4.0 cm in diameter, but may be larger. Although some cows with ovarian cysts are nymphomaniacs most are anestrus. Estrous cycles can be reestablished for about 80 percent of cows with this condition by treating them

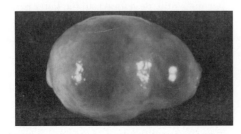 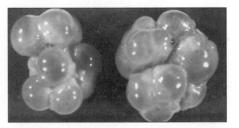

Figure 13.14 Ovarian cysts on an ovary from a dairy cow (left) and multiple ovarian cysts on ovaries from a gilt (right).

From D. J. Kesler and H. A. Garverick, "Ovarian Cysts in Dairy Cattle: A Review," J. Animal Sci., 55 (1982):1147.
Photographs courtesy of University of Missouri, left, and University of Illinois, right.

with 100 µg of GnRH (gonadotropin-releasing hormone). The tendency to develop ovarian cysts is probably inherited.

In some females (especially cows) there is an interval of only 8 to 10 days between estrous periods rather than the usual 21 days. This is due to the abnormally early degeneration of the corpus luteum with an insufficient amount of progesterone being produced to inhibit production of LH by the anterior pituitary gland. An early degeneration of the corpus luteum allows growth of a new group of follicles sooner than normal and estrus occurs within a shorter period of time. Other females may not show signs of estrus at the first expected estrous period following breeding, but will show estrus at the second. Thus one estrous period appears to be skipped. One explanation for this situation is that pregnancy actually might have been initiated but the developing embryo died and was resorbed after a few days. Another possible explanation is that the female came into estrus but it was not observed because estrus was limited to a very short period of time or was mildly expressed.

Exposure to high **ambient temperatures** often lowers reproductive efficiency of females. This is especially true if their body temperature is increased above normal. Adverse effects from such exposure are observed in tropical areas, especially in regions where summer temperatures normally exceed 100°F, and in other regions in which abnormally high temperatures, often associated with periods of prolonged drought, may occur during the summer months.

Adverse effects of high ambient temperatures on reproduction have been noted in several species, especially the ewe. Extreme thermal stress may cause the ewe to fail to exhibit estrus. Thermal stress at the time of mating or shortly thereafter can cause a failure in normal development of the fertilized ovum. Later in gestation thermal stress may cause fetal deaths and resorptions, abortions, and birth of abnormally formed young. Under laboratory conditions extremely high temperatures have caused abortions in cows. Abortions were noted in some sows at the Missouri Agricultural Experiment Station in the summer of 1954 when ambient temperatures during a heat wave reached 112 to 114°F. Smaller litters were **farrowed** that fall, more mummified fetuses were observed at birth, and variation in litter size at birth was greater than when the same sows farrowed the previous spring. Some experiment stations have reported that litter size at farrowing was larger among sows that had been cooled by sprinkling during the hot summer months than among controls.

Ovulation

Ovulation is the process by which the ovum (egg) is released from the ovary. The usual time of ovulation in four species of farm mammals is given in Table 13.2. Among farm mammals the cow and mare usually ovulate just one ovum per estrous period, although multiple births sometimes occur because of the ovulation of two or more ova. It is also possible for two or more offspring to develop from a single ovum. Such individuals are genetically alike and thus are identical.

Multiple births are common among sheep and are the usual type of birth in swine. Ewes may ovulate from one to five ova during a single estrous period. In swine, gilts may ovulate 10 to

12 and sows 15 to 20, or more. The number of ova developing into young at birth depends in part on the amount of space and nutrients available to the ova in the uterus. Production of more than the usual number of ova may be induced in females by administration of gonadotropins containing FSH during proestrus (the time follicles normally develop). Production of more than the usual number of ova is known as **superovulation.**

Ovulation among different species is often referred to as *spontaneous* or *induced*. The cow, ewe, mare, and sow ovulate spontaneously, that is, when follicles develop and the ova mature. Ovulation occurs naturally without the influence of an external force. In the female rabbit and in cats, ovulation does not occur without copulation. This type of ovulation is referred to as induced ovulation. Copulation in the rabbit sends a neural impulse to the hypothalamus that causes the anterior pituitary gland to release LH. This hormone causes ovulation. If copulation does not occur mature follicles eventually regress and another crop of follicles develops in the ovary. Thus, at least in the active mating season in the mature female rabbit, there are overlapping crops of follicles and ripened ova present in the ovary if copulation and pregnancy do not occur.

Fertilization

*Fertilization is the union of male and female **gametes** (sex cells) to form a new individual.* The unfertilized egg has the **haploid** number of chromosomes, as do spermatozoa, and their union in fertilization restores the **diploid** number of chromosomes in the fertilized egg.

Fertilization in farm mammals usually occurs in the upper one-half of the oviduct. This means that spermatozoa have a considerable distance to travel from their point of deposition in the female reproductive tract at the time of mating to the site of fertilization. In some species of mammals spermatozoa arrive at the site of fertilization in the oviducts 15 to 20 min after mating. Because the rate of travel of sperm from their own motion appears to be only 3 to 4 mm/min, another force must be responsible for their rapid transport within the female reproductive tract. Uterine contractions are probably the most important force. Freshly ejaculated spermatozoa from some species must undergo changes within the female reproductive tract before they are capable of fertilizing the ovum. These changes are referred to as *sperm capacitation.*

Although many millions of spermatozoa usually are introduced into the reproductive tract of the female at mating, only a few (1000 or less) actually reach the site of fertilization. This number is sufficient, however, because only one viable sperm is required to fertilize each ovum.

Microscopic studies of the union of the sperm and egg in prepared solutions outside the body (**in vitro** technique) have led to the conclusion that the sperm and egg meet by chance alone. However, inseminations of the female with a mixture of semen from two or more males indicate that sperm from some males are more likely to fertilize the ovum than sperm from others. Some research suggests that there may be selective fertilization in certain instances because litter size in swine, or conception rate in cattle, may be higher when a male is mated to females of his own breed than when mated to females of another breed.

In some species the ovum possesses a cumulus mass of cells (cumulus oophorus) when it reaches the site of fertilization. The spermatozoon penetrates this mass of cells by means of its own motility and possibly by means of the enzyme (hyaluronidase) it contains, which dissolves this mass of cells. After penetrating the cumulus mass the sperm also must penetrate the *zona pellucida*. Many sperm may enter the zona pellucida (Figure 13.15), but normally only one penetrates the *vitelline membrane* to unite with the **nucleus** of the ovum. Once the sperm penetrates the vitelline membrane this membrane sets up a "fertilization block" that prevents the entrance of other sperm. The fertilization block appears to diminish in its effectiveness under certain conditions. One of these conditions is delayed copulation leading to fertilization of aged ova. In such instances, more than one sperm may penetrate the vitelline membrane leading to the condition known as **polyspermy.** This would lead to the condition known as **polyploidy** and the developing embryo would possess more than the 2*n* number of chromosomes. *Polyploidy* among larger mammals may be a cause of some embryonic death loss. Possibly, but rarely in farm mammals, some triploid individuals (possessing the 3*n* number of chromosomes) survive to birth.

Development of young from unfertilized ova (parthenogenesis) has been reported among turkeys. **Parthenogenesis is** a common, and even the usual, means of reproduction in some species of insects. It is extremely rare or nonexistent in farm mammals. Parthenogenetic turkeys are always males. Some are fertile and capable of producing viable offspring. (See Section 8.2 for discussion of sex chromosomes of birds.)

The transport of fertilized ova (zygotes) through the oviduct requires about 2 to 3 days in swine. Fertilized ova in farm mammals can be recovered and transplanted from one female to another (Figure 13.16).

Pregnancy

The period of *pregnancy* (**gestation**) is *the time in which the female carries her developing young within the uterus.* The ges-

tation period ranges from 110 to 115 days in the sow, 140 to 150 days in the ewe, 272 to 292 days in the cow, and 335 to 345 days in the mare (Table 13.2).

While the fertilized ovum is moving through the oviduct to the uterus, it is nourished by yolk and uterine secretions (uterine milk) until contact is made between the maternal and fetal membranes (*implantation* or nidation). After implantation the developing young obtain nourishment from the mother.

Implantation is a gradual process in farm mammals. Research indicates that implantation occurs 10 to 18 days after fertilization in the ewe, 12 to 24 days after in the sow, 20 to 32 days after in the cow, and 35 to 60 days after in the mare. In certain wild species such as the weasel, mating occurs in the summer but implantation does not occur until late winter. This situation is known as *delayed implantation* and is another timing mechanism in the reproductive process of animals to ensure the birth of young when the environment is favorable for survival.

Several types of abnormal implantations or pregnancies have been observed in humans and may occur in other animals. These include ovarian, tubal, and abdominal pregnancies. The young are seldom, if ever, carried to full term in such pregnancies.

Early in pregnancy the young develop certain membranes to provide for their protection and nourishment. These are known as *fetal (extraembryonic) membranes;* those of the pig are shown in Figure 13.17. The **amnion** is the innermost membrane surrounding the fetus. It is filled with a fluid known as *amniotic fluid,* within which the developing fetus is suspended. The outer layer of fetal membranes is called the **chorion.** It makes contact with the maternal uterine tissues. **In ruminants** including the cow and ewe, the chorion attachment to the uterus is of a **cotyledonary type** (Figure 13.18) in which contact is made only at certain points on the uterus rather than over its entire surface area. The placental attachment in the mare and sow is of a diffuse type (Figure 13.19) in which contact with the uterus is made over most of the surface area of the chorion. The outer portion of the allan-

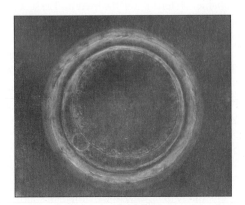

Figure 13.15 A sow ovum, showing many sperm in the zona pellucida.

Courtesy of Dr. B. N. Day and Missouri Agr. Expt. Station.

Figure 13.16 Inbred Yorkshire pigs from purebred Hampshire sows. They resulted from the transplantation of fertilized ova from Yorkshire to Hampshire sows.

Courtesy of Missouri Agr. Expt. Station.

tois is fused with the chorion and the inner layer with the amnion, forming a sac or space filled with allantoic fluid.

The placenta is formed by fusion of the chorion and the uterine mucosa. It has several important functions: (1) transmission of **nutrients** from mother to young, (2) transmission of wastes from young to mother, (3) protection of young from shock and adhesions by means of amniotic fluid, (4) prevention of the transmission of bacteria and other large molecular substances from mother to young, and (5) secretion of hormones. The placental barrier in domestic mammals prevents large **molecules** such as **antibodies** and some of the fat-soluble **vitamins** (vitamin A) from passing in large amounts from mother to young. Some **viruses** are small enough to penetrate the placental barrier and may cause defects in the young if they reach it in the stage of pregnancy when certain body parts are being developed. The developing young are particularly susceptible to viral infections because they have not produced antibodies of their own and have received few, if any, from their mother. Certain other chemical substances in the mother's diet penetrate the placental barrier in some farm mammals and may cause defects during fetal development.

Each developing embryo usually has its own set of membranes, although those of two individuals may fuse resulting in their having a common blood supply. Fetal membranes anastomose in twin bovine pregnancies in which one fetus is a female and one a male. Because of this anastomosis, there is an exchange of cells and blood between the two fetuses. Compounds from the male fetus entering into the female inhibit normal development of the female reproductive anatomy. This causes the female offspring to be a **freemartin.** A heifer born as a twin with a bull is a freemartin and infertile more than 90 percent of the time.

In the sow, where many young are nourished within the uterus during the same pregnancy, the young are spaced at definite intervals within the uterine horns (Figure 13.20). If the young are too close together or if the uterus is crowded, some of the developing young fail to survive at birth. Another phenomenon also occurs in the sow to help ensure an approximately equal number of young in both uterine horns: if one ovary produces only 1 ovum and the other ovary 12, some fertilized ova from the side where the 12 were produced will move across to the uterine horn on the opposite side where only 1 ovum was produced. This mechanism tends to distribute the young evenly within the two uterine horns and is known as *intrauterine migration.*

Even though the gestation period varies greatly in length among species, all young in female farm mammals are born at about the same degree of development or maturity. This is not true of all species of animals, however. For example, baby opossums are born about 12 days after the ova are fertilized. They have no fetal membranes at birth. They find their way into the pouch of the mother after birth, each fastens its mouth to one of the nipples, and remains attached to that nipple until it is fully developed. The female opossum normally ovulates 45 to 50 ova and gives birth to 25 or more young, but there are only 12 to 13 nipples available in the pouch, so no more than this number of young survive. The baby mouse, rat, and rabbit are more mature than the opossum at birth; nevertheless, they are quite helpless. Their eyes are closed, they have no hair, and they cannot walk. They grow and develop quickly, however, if properly nourished

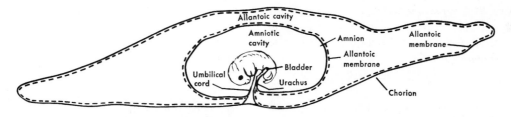

Figure 13.17 Drawing of the fetal membranes of the pig. Placental membranes are of the diffuse type, making contact with the uterus over the entire surface of the chorion.

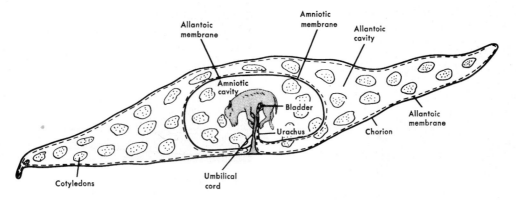

Figure 13.18 Drawing of the fetal membranes of the calf. Placental membranes of the ruminants (cattle, sheep, and goats) are of a cotyledonary type, making contact with the uterus only at certain points (the cotyledons).

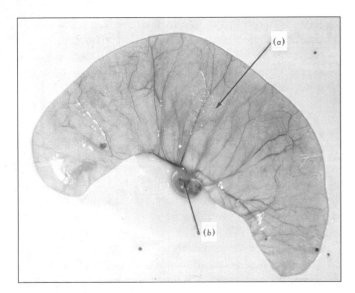

Figure 13.19 Fetal membranes of the pig at 25 to 30 days of pregnancy. (a) The allantoic sac filled with fluid; (b) the fetus encased in the amnion. Note the blood supply to different parts of the chorion leading to the umbilical cord of the fetus.
Courtesy of Missouri Agr. Expt. Station.

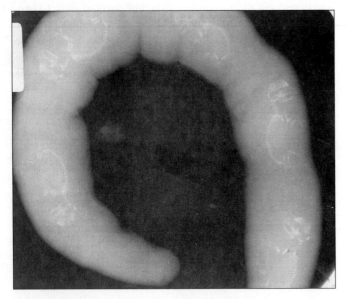

Figure 13.20 Radiograph (X ray) of the uterine horn of the sow at midpregnancy showing spacing of the fetuses. Only bony parts of the body of the fetuses show in this radiograph.
Courtesy of Missouri Agr. Expt. Station.

by the mother. They may be weaned at about 21 days of age. The young of the guinea pig are much more mature at birth and can care for themselves, if necessary, when they are only a few hours old (Figure 12.4).

Parturition

Parturition is *the act of giving birth to young by the mother.* It is preceded by specific activities in females. For example, females of most species lose their appetite and usually try to find a

secluded area where they will have privacy during the birth process. Sows will often build a nest before giving birth to young (Chapter 24). The **ligaments** and muscles around the pelvic region of most females (especially the cow and mare) relax, and usually milk is present 24 to 48 h before the young are born.

Parturition in farm mammals is initiated by the fetal anterior pituitary gland. The role of the fetal anterior pituitary gland was first demonstrated to be involved in parturition in cattle with prolonged gestation. Pregnancy in these cattle was generally prolonged from 3 to 4 weeks, resulting in birth weights of calves varying between 110 and 168 lb (the usual birth weight is 65 to 135 lb). In such instances the calves were too large for normal birth and were dismembered or delivered by cesarean operation. It was demonstrated that this prolonged gestation was caused by the absence of the fetal anterior pituitary gland. It was later shown that experimental destruction of the pituitary gland of the pregnant female also caused prolonged gestation. Animals without a pituitary gland, however, would deliver their young if they were administered ACTH (adrenocorticotropic hormone). Therefore ACTH produced by the fetal anterior pituitary gland appears to be the hormone that initiates the process of parturition. The same phenomenon has also been demonstrated in sheep. Further research is underway because there appears to be certain events that precede the release of ACTH.

Other hormones involved in parturition are estrogen, $PGF_{2\alpha}$, oxytocin, relaxin, progesterone, and glucocorticoids (Chapter 11). $PGF_{2\alpha}$ is released to destroy the corpus luteum. This is necessary because high levels of progesterone inhibit uterine contractions. After progesterone levels are reduced, oxytocin and $PGF_{2\alpha}$ stimulate uterine contractions. Estrogen sensitizes the uterus to oxytocin, and relaxin causes relaxation of the birth canal. These events are altered in species having no corpus luteum present at parturition. In species such as the horse and human, progesterone necessary to maintain pregnancy is synthesized by the placenta. At the time of parturition placental progesterone synthesis is shifted to estradiol synthesis by events initiated by ACTH release in the fetus.

Females should be observed carefully at parturition to determine if the birth process is progressing properly. This is especially important in young females producing their first offspring. The normal presentation of the newborn in cattle, horses, and sheep is front feet first with the outstretched head resting on the feet. In swine both head- and hind-feet-first presentations are normal. Various abnormal presentations are encountered among animals. These include one or both legs folded back, the neck bent backward, or a posterior presentation (except in swine where the young are usually born hind feet first). Posterior presentations (breech) in cattle are of particular concern because when the connection of the umbilical cord with the uterus is severed, fetal blood flow from the mother ceases and the young has no oxygen supply. Unless delivered promptly it may perish from lack of oxygen.

Abnormal presentations may be implicated when labor lasts several hours without progress being made in delivering the young. One may assist the female as needed by pulling gently, but firmly, on the young at the time *labor pains* (uterine contractions) occur in the mother. If this assistance is ineffective an experienced person should be called. It is sometimes necessary to dismember the young or deliver them by a cesarean operation (named after Julius Caesar who supposedly was delivered this way).

Following parturition it is important that the newborn receive colostrum rich in antibodies and vitamin A (Chapter 15). The young particularly need antibodies from their mother's milk to combat infections because they possess few, if any, antibodies at birth and it is several days before they can build their own (Chapter 22). It is also important to observe mothers (especially cattle and sheep) closely following parturition to make certain the afterbirth is completely expelled from the uterus. The **afterbirth** is *the placenta and the extraembryonic membranes with which the fetus is connected.* In some instances retained afterbirths can cause infections and subsequent reproductive problems.

13.5 APPLICATION OF RECENT RESEARCH AND RELATED TECHNOLOGIES IN THE PHYSIOLOGY OF REPRODUCTION

As discussed in Chapter 11, hormones and other biological compounds have been used to control various aspects of reproduction in farm mammals. Certain hormones and chemicals used experimentally have not been approved by the FDA for commercial use on the farm or ranch. In Sections 13.5.1 through 13.5.8 several applications of recent research discoveries and technologies in the field of reproductive physiology will be discussed.

13.5.1 Synchronization of Estrus and Ovulation

Estrous synchronization involves manipulating the reproductive processes so females can be bred during a short, predefined interval with normal fertility. This application of technology facilitates breeding in two important ways: (1) it reduces, and in some cases eliminates, the labor of estrous detection; and (2) it enables the producer to schedule the time of breeding. For example, if all animals in a herd can be induced to exhibit estrus concurrently, the producer can arrange for a few days of intensive artificial insemination. Although the total amount of labor involved with insemination may not be reduced significantly, it is concentrated into a much shorter period of time. Other advantages of estrous synchronization include the following: (1) it creates a more uniform group of offspring, (2) it enables the producer to breed more females to a select male, (3) it reduces the time span of breeding and parturition seasons, and (4) with certain compounds heifers can be bred without detecting estrus.

One compound that has been used to synchronize estrus and ovulation in cattle, horses, and sheep with consistent success is $PGF_{2\alpha}$, which is effective in synchronizing estrus by lysing the corpus luteum. If $PGF_{2\alpha}$ is injected 5 or more days after ovulation it lyses the corpus luteum and causes estrus to occur 2 to 3 days after treatment. Because $PGF_{2\alpha}$ is effective only between 5 and 17 days postestrus, one injection will synchronize only about 65 to 70 percent of estrus-cycling bovine females. It has minimal effect on anestrous heifers.

One of the more popular methods of synchronizing heifers involves the concurrent administration of melengestrol acetate (MGA) and $PGF_{2\alpha}$. The MGA is an orally active compound fed to heifers daily for 14 days. The heifers are then maintained for an additional 19 days without MGA feeding and on day 19 they are administered an injection of $PGF_{2\alpha}$. Following the injection of $PGF_{2\alpha}$ most heifers will exhibit estrus within 72 h. One effective method of ensuring a high pregnancy rate involves breeding heifers observed in estrus during the first 72 h after $PGF_{2\alpha}$ injection. Heifers not observed in estrus during the 72 h interval are then mass inseminated. The basis behind this logic is that although the heifers were not observed in estrus, it is likely that they are approaching ovulation and estrus was either expressed during darkness or expressed with minimal overt signs. Results using this strategy have been excellent and 60 to 80 percent of the treated heifers become pregnant. A major advantage of this strategy is that all heifers will be inseminated within a short period of time.

A procedure that has been used effectively to synchronize estrus in cows (both beef and dairy) is the Ovsynch procedure. This procedure involves the administration of GnRH followed in 7 days by an injection of $PGF_{2\alpha}$. The GnRH is administered to synchronize the follicular wave. The GnRH either ovulates or causes atresia of antral follicles and a new wave is initiated about 2 days after administration. As mentioned above, follicles grow in waves. Therefore it is not only necessary to control the timing of luteolysis but also necessary to control the timing of follicular development so a follicle of ovulation size is available at the appropriate time. The $PGF_{2\alpha}$ is administered to control the timing of luteolysis. A second injection of GnRH is administered 36 to 48 h after the injection of $PGF_{2\alpha}$. This injection of GnRH provokes a preovulatory LH surge so ovulation occurs at a synchronized time. In dairy cattle where animal handling is less objectionable, cows are inseminated about 16 h later. In beef operations, cows are inseminated at the time the second GnRH injection is administered. Although this may compromise the pregnancy rate somewhat, the small decrease is outweighed by the lower animal-handling requirement.

Another procedure being considered by the FDA for approval in beef and dairy heifers and in beef cows is the concurrent administration of an intravaginal progesterone insert (also called CIDR) and an injection of $PGF_{2\alpha}$ (at the time of insert removal). This procedure effectively synchronizes estrus and requires minimal lead time. Although the procedure is just now being considered for approval in the United States, it has been used in New Zealand for several years where the procedure has been used successfully in dairy cows, as well.

13.5.2 Superovulation

The ovaries of female farm mammals possess many thousands of potential ova at birth and produce no new ones thereafter. Because female livestock produce and utilize a limited number of ova to produce young during their lifetime most of the original ova in the ovary do not produce young and are in effect wasted.

Superovulation (actually superstimulated follicular development) consists of injecting the female with fertility drugs causing a larger-than-normal number of follicles (each of which contain one oocyte, or egg) in the ovaries to mature and ovulate (rupture and release the egg). Follicle-stimulating hormone is injected twice daily. Additionally, on day 3 or 4 of FSH injection $PGF_{2\alpha}$ is injected. This results in lysis of the corpus luteum on the ovary thereby triggering the next estrus, commonly 2 days after administration of $PGF_{2\alpha}$.

When a bovine female is treated with a fertility drug she will usually produce and ovulate more eggs than normal. The eggs can be fertilized inside the female or removed and fertilized *in vitro* where "matings" can be planned for each egg.

If eggs are fertilized inside the female the resulting embryos can be left to develop into twins or triplets, or they can be collected and transferred to other females called *surrogate mothers* or *recipients*. When surrogate mothers are not readily available the embryos can be frozen to await transfer later.

Even a single embryo can be made to produce additional offspring. A single-cell fertilized egg divides into two cells, which divide into four cells, then eight cells, and so on. At this early stage of development all cells in the embryo are exactly alike genetically. This means that splitting a four-cell embryo into four separate cells by microsurgery or by chemical dispersion and then transferring these cells to surrogate mothers results in identical quadruplets.

The potential economic benefits of this technology are tremendous. For example, a bovine female on fertility drugs can produce eight or more embryos. If these embryos are separated at the four-cell stage, each cell is grown in culture to the four-cell stage and the process repeated 6 times, theoretically more than 4000 potential offspring could be produced, all genetically identical. But the potential for manipulating embryos extends even further, as will be discussed later.

13.5.3 *In Vitro* Fertilization (Test-Tube Babies)

The first reported *in vitro* fertilization (IVF) "test-tube" human baby was born in July 1978 in Great Britain. The procedure involves, to sum: (1) recovering a mature egg from the ovary (in live cattle this can be done surgically or via transvaginal ultrasound-guided needle aspiration of a follicle), (2) placing the egg in a petri dish enriched with a culture medium similar to that of the oviducts, (3) collecting and placing viable spermatozoa with the egg in the culture medium and incubating at body temperature, and (4) placing the **morula** intravaginally into a recipient uterus following proper hormonal preparation of the recipient female to accept the new embryo. This procedure is complicated; however, the technology is commercially available throughout much of the world.

Approximately 15 percent of married couples in the United States are infertile generally for reasons such as blockage of the woman's oviduct or a condition known medically as *oligospermia* (insufficient sperm count to effect conception). For infertile couples the above procedure of embryo culture and transfer offers promise as a means of bearing children that are genetically their own. The procedure is not only used in humans but in farm animals as well. In humans the procedure has had unique and novel success. In one case a 63-year-old woman gave birth to a child. In another the mother of a woman who was having difficulty sustaining a pregnancy to term served as the surrogate mother and gave birth to her grandchild. Numerous cases of multiple offspring after IVF have been reported. One woman gave birth to eight children with one pregnancy. The limits of IVF are minimal and, together with modifications of the technology, it enables many infertile couples to bear children.

13.5.4 Embryo Transfer (ET)

Embryo transfer (ET), a recent application of science, is the placing of an embryo into the lumen of the oviduct or uterus. More broadly ET includes the sequence of steps for transferring embryos from one female to another (superovulation, embryo recovery, and storage of embryos *in vitro*). The donor is the genetic mother from which embryos are recovered and the host or surrogate mother is called a recipient (Figure 13.21).

The first step in embryo transfer involves estrous synchronization and superovulation using the procedures discussed previously. After estrus and insemination or *in vitro* fertilization, the developing embryos are ready to be transferred. The embryos of pigs and sheep are generally transferred by surgical methods whereas those of cattle are usually transferred utilizing nonsurgical techniques.

Although the first ET was performed successfully in rabbits nearly a century ago at Cambridge University, the first successful bovine ET was reported in 1951 at the University of Wisconsin. More recently ET has been successful in humans, sheep, goats, and horses.

Normally cows produce only one calf per year; however, with embryo transfer it is common for a cow to be the genetic mother of 12 to 15 or more calves born within a few days of one another (Figure 13.21). Thus embryo transfer enables the animal breeder to proliferate offspring from cows identified as genetically superior and to selectively mate each cow with several superior sires in a given year. Another use of embryo transfer is to obtain offspring from otherwise infertile donors. Such infertility often results from a senescent uterus, which is circumvented by transferring the embryos to a younger, more active and environmentally sound uterus. Further application of this technology is the export of embryos, which is less costly and involves less risk of transmitting disease than exporting live animals and allows the offspring to be born in their final environment, reducing the stress associated with acclimation.

Somewhat analogous to the placing of whooping crane eggs into nests of the sandhill crane, embryo transfer could be used to spare rare species or breeds from extinction. An example is the current effort to save the Angora breed of sheep from extinction in Australia. The breed is being reestablished from only a few animals by transferring Angora embryos into other breeds of sheep. The same technique has been used with other endangered species.

Figure 13.21 The Holstein cow (upper right) is the genetic mother of the 10 calves shown above. She was superovulated and the embryos were recovered from her uterus 7 days after conception. After 3 to 10 h of culture in vitro, the embryos were transferred to the uteri of the 10 recipient cows (left) for gestation to term. The surrogate mothers may be of any bovine breed; however, before the embryos can be transferred successfully, the estrous cycles of the surrogate mothers must be in synchrony with that of the genetic (donor) mother.

Photograph was kindly provided by Dr. G. E. Seidel, Jr., Embryo Transfer Laboratory, Colorado State University, Fort Collins.

Other merits of ET include (1) the use of valuable or "elite" genotypes to produce a larger proportion of desirable phenotypes such as cows producing milk of a more favorable protein-to-fat ratio, and twinning in selected species of livestock could be accomplished more rapidly. Similarly the impact of sex-linked genes could be increased through ET. (2) ET techniques can improve the intensity of selection of cows to be the dams of replacement heifers and of bulls used in artificial insemination. (3) ET techniques can improve the accuracy of selection through progeny testing of females. (4) The application of ET could extend the reproductive life of females via frozen embryo processes. Such a procedure would facilitate increased selection intensity because elite dams would be available for producing sons beyond their lifetime. (5) ET procedures may also offer the opportunity to decrease generation interval if such a change would improve genetic change per unit of time. (6) One additional benefit of ET application to the production of young sires is increased concomitant production of females with superior pedigrees (see also Chapters 8 and 9).

Testing for Mendelian Recessive Alleles

If a homozygous recessive individual is mated to a suspected carrier of a Mendelian recessive allele, and eight normal offspring result, the probability that the individual is not a carrier is 99.6 percent. Without superovulation and embryo transfer, testing for a particular undesirable trait would require essentially the full reproductive life of a cow. This compressing of an 8- or more-year test period into 1 year is especially important when an outstanding cow is a daughter of a carrier.

Using ET in Genetic Testing

Embryo transfer has been used to diagnose and study the inheritance of both syndactyly and polydactyly in cattle. Syndactyly, or mule-foot, is the most common genetic defect of cattle in the United States. It has been observed in Angus, Chianina, Hereford, Holstein, Simmental, and crossbred cattle. Because it is a single-gene recessive, congenital defect, it can be carried and transmitted by adults normal in appearance. Thus all suspect carrier AI sires should undergo test matings before being used extensively. Using embryo transfer from syndactylous donors bred to suspect sires provides a relatively economic and fast way of progeny-testing bulls. Suspect females can be tested by inseminating them with semen from a mule-foot bull before embryo recovery. Embryos recovered from syndactylous donors or suspect cows are transferred into recipients for fetal development. Two embryos can be transferred into each recipient to reduce the cost of the test. Cesarean removal of the fetuses from the recipients at 60 days following conception reduces the time required to complete the test.

Bovine polydactyly (three or more toes per limb) is much less common than syndactyly. It has, however, been observed in Holstein and Simmental cattle. The mode of inheritance of polydactyly is not well understood but it appears to involve at least three genes, one dominant and two recessive.

Collecting Ova and Embryos from Donor Cows

Nonsurgical methods are utilized to recover ova on days 6, 7, or 8 after estrus. The donor cow is restrained and given an epidural block in the area of the tail head. The local anesthetic injected into the spinal column prevents the animal from straining when the arm is in the rectum. A flexible rubber tube

(Foley catheter) with three passageways is then passed through the cervix and into the uterus (Figures 13.22 and 13.23). It is temporarily stiffened with a metal or plastic rod to enable it to pass through the cervix and into the body or horn of the uterus. A rubber balloon, which is built into the anterior end of the tube, is then inflated via passageway B with approximately 20 to 25 milliliters (ml) of air (to about one-half the size of a golf ball) so that it extends to fill the uterine lumen and prevents fluid from escaping around the edges. There are two holes in the tube anterior to the balloon, which lead into separate passageways: one (A) for fluid entering the uterus, the other (C) for fluid draining from the uterus. Embryos (about 1/200 inch in diameter) are collected in a balanced salt solution to which **antibiotics** and heat-treated serum are added. A phosphate-buffered saline is the most commonly used solution. The solution is held in a container about 3 ft above the cow where it is connected to the Foley catheter by an inflow tube (Figure 13.23). A second tube is connected to passageway C of the catheter to collect the medium that has flushed (washed) the ova and embryos out of the uterus. This solution is then collected in cylinders holding approximately 2 pints (908 ml) of fluid. Unfertilized ova and embryos can be allowed to settle to

the bottom and the bulk of the medium is siphoned off or collected via a filtration device. The last cupful of fluid is searched under the microscope for the treasured embryos. Because bovine embryos form no intimate attachment to the uterus before day 18 they can be recovered nonsurgically for up to about 14 days with no apparent damage, although research indicates that a larger number of normal embryos can be obtained 6 to 8 days after estrus than at other times.

The embryos are identified, classified, washed through a sterile medium and then stored until transfer. They commonly live in this solution at 99°F (37°C) for about 24 h without loss in viability, or they can be frozen to –320°F (–196°C) in liquid nitrogen. The current state of the art has not perfected the freezing and thawing process, so up to half of the embryos may be killed. Those retaining viability can be stored indefinitely.

For high pregnancy rates the recipient, or surrogate mother, must be at the same stage of the reproductive cycle as the donor. Embryos are usually transferred nonsurgically using artificial insemination equipment. Pregnancy rates are increased slightly if the embryos are transferred directly into the uterus surgically through a small incision in the flank of the recipient under local anesthesia. With the surgical method pregnancy rates of 60 to

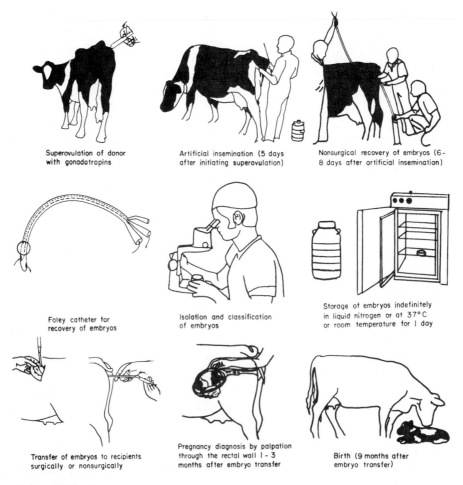

Figure 13.22 Schematic presentation of bovine embryo transfer procedures.
From G. E. Seidel, Jr., Science, 211 (January 23, 1981):351–358. Copyright 1981 by the American Association for the Advancement of Science. Used by permission.

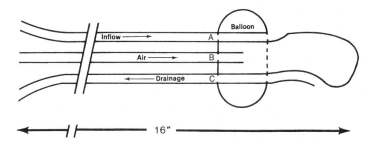

Figure 13.23 Diagrammatic depiction of a three-way Foley catheter.

70 percent can be achieved, approximating those obtained through normal breeding.

Embryo transfer techniques have been used in many species including humans. The most important commercial species after cattle in the United States are horses and swine, although a limited amount of commercial work is also being done with sheep, goats, rabbits, and other species .

Embryo transfer in cattle currently costs from $500 to $1500 per pregnancy. Firms providing such services have grown rapidly in recent years. An estimated 50,000 calves were born as a result of this technology in North America in 1982. By 1992 there were 58,740 Holstein males and 103,212 Holstein females registered that resulted from ET. Add the number of ET calves from other dairy breeds and those from all of the beef breeds as well as ET offspring of sheep, goats, swine, horses, rabbits, and other animals and the important technologies associated with ET in animal production become readily apparent. Indeed, data provided by the International Embryo Transfer Society (IETS) indicates that in 1999 more than 520,000 cattle, 60,000 sheep, 20,000 goats, 10,000 swine, 3000 rabbits, 2000 deer, and 1000 horses were born worldwide as a result of ET. Clearly global interest in ET is growing rapidly and together with newer technologies including that of sexed semen, for example, that generate predetermined gender embryos, will heighten the demand from animal producers to take advantage of the economic, genetic, health security, and other benefits of using ET.

Freezing Embryos

The freezing of bovine semen has been perfected. However, the freezing and subsequent thawing of viable embryos has been more difficult because embryos are composed of masses of cells rather than being a single cell, as is the case with spermatozoa. Fortunately researchers have been successful in discoveries that led to practical and reliable procedures being available today.

13.5.5 Preselecting Offspring Sex

Methods to control the sex of offspring have been in folklore since the time of ancient Greeks. This technology has long been awaited in the livestock industry and is now nearly perfected. Of the numerous procedures that have been attempted, flow cytometry is the procedure that has been adopted. This procedure was

first demonstrated to successfully preselect the sex of rabbit offspring and subsequently has been demonstrated to effectively preselect the offspring sex in humans, swine, sheep, and cattle.

Field studies have been conducted in cattle; however, the most efficient use of sorted spermatozoa is in the *in vitro* production of embryos where a limited number of sperm can be used to fertilize a large number of oocytes. This is because currently only about 6 million X- and Y-bearing spermatozoa can be sorted per h with a purity of about 90 percent. Faster sorting rates of 20 million spermatozoa per h can be achieved in situations where lower purity (75 to 80 percent) is acceptable.

Flow cytometry sorts spermatozoa on the basis of the difference in DNA content of X and Y chromosome–bearing spermatozoa. This difference in DNA content is 3 to 4 percent in most mammals. In this procedure spermatozoa are incubated with a florescent dye that binds to DNA. The spermatozoa are then individually oriented to pass by fluorescence detectors where intensity of florescence is measured to determine whether the spermatozoa is sorted as X or Y chromosome bearing: the X chromosome will have 3 to 4 percent more florescence than the Y chromosome.

Eleven field trials have been conducted in which 1000 heifers were inseminated with sexed semen. In most trials the pregnancy rates resulting from insemination with sexed, frozen sperm were within 90 percent of the unsexed, frozen controls. In these trials accuracy approaching 90 percent males or females was achieved. Embryonic death losses between 1 to 2 months of gestation were not increased with pregnancies from sexed sperm.

Therefore technology is being perfected to achieve the long-awaited goal of offspring sex selection. This technology is particularly attractive to dairy producers where most bull calves have limited value, and the production of increased numbers of heifer calves would provide increased opportunity for selection pressure among females retained for breeding purposes. Human IVF clinics are increasingly adopting the technology (some IVF clinics have already made the technology available to clients).

13.5.6 Cloning and Transgenics

In February 1997 the media presented a story that fascinated the world: the first mammal cloned from an adult cell. However, **cloning** was nothing new. Mammals had been cloned from embryonic cells for years. Indeed, natural cloning occurred since identical twins are actually clones of each other. The revolutionary scientific breakthrough with regard to the cloning of the sheep named Dolly was the fact that she was cloned using an adult cell.

Since the elucidation of DNA by Watson and Crick, researchers had hypothesized that adult cells contained all the genetic information necessary to duplicate an individual; however, after years of intense research successful cloning had not been accomplished. In fact, many scientists had essentially given up on being capable of cloning an individual from an adult cell. But Ian Wilmut, a determined animal scientist from the Roslin Institute (formerly the Animal Breeding Research Station) in Roslin, Scotland, persisted in studying the subject further.

The procedure used to create Dolly was: (1) oocytes were recovered from ewes following GnRH-induced ovulations; (2) quiescent, diploid donor cells were produced in cell culture by reducing the concentration of serum causing the cells to exit the growth cycle; (3) fusion of the donor cell to the enucleated oocyte and activation of the oocyte were induced by electrical pulses; and (4) the embryo was cultured in ligated oviducts of sheep and then transferred to a recipient for development to term.

Many researchers have developed creative ideas as to why Dolly was created and what the future of cloning technology will be. However, Dolly was created to answer the scientific question, "Can an adult cell be used to develop an identical offspring?" Scientific query was not the only driving force behind research in cloning; funding for Dr. Wilmut's research came from industry sponsors who envisioned clones being used in human medicine.

Following the success of cloning using genetic material from adult cells, biotechnology companies have further developed this technology to make it commercially viable. The 2000 World Dairy Expo featured a clone of "Lauduc Broker Mandy," one of the most admired cows in the Holstein breed (a two-time All-American and three-time All-Canadian winner with an outstanding pedigree, body type, and production record). "Mandy II" was sold for $82,000 at the World Classic, making history as the first cloned heifer calf to be sold in a public auction.

Aside from cloning, another relatively new technology is that of developing **transgenic animals.** Transgenic animals have a genetic makeup of more than two parents. This technology is actually the extension of the recombinant DNA technology used to develop bovine GH (growth hormone) to enhance milk production of dairy cattle. The production of a transgenic cell involves the inclusion of foreign DNA into the DNA of another cell. In recombinant DNA technology this usually involves the splicing of a human gene for the synthesis of a specific protein such as GH or insulin into the DNA of the bacteria *E. coli*. To produce a transgenic animal the foreign DNA, such as the human gene for the synthesis of GH or insulin, is spliced into the DNA of an animal (e.g., a pig). In fact the first transgenic pig contained a gene for human GH, and following birth these transgenic piglets grew more rapidly than contemporary pigs. However, the amount of expression of the GH gene was unpredictable and the use of this technology for animal production has taken a backseat to the development of transgenic animals to produce substances needed for human medicine.

A possible application of transgenic technology would be as follows: Some humans have a form of hemophilia called hemophilia B because of a deficiency of the blood-clotting factor IX. Treatment with blood-clotting factor IX is needed to prevent death from bleeding. Human blood-clotting factor IX is in very limited supply. It may be possible to create a transgenic cow by inserting the human gene for the production of human blood-clotting factor IX. When this cow lactates her milk may contain large quantities of blood-clotting factor IX that could then be separated from the milk by chromatography procedures and used in human medicine. Furthermore, once the ideal transgenic for a specific substance is found, that animal can be cloned for production on a larger, infinite basis.

The next step not only envisioned but also in progress is the production of human organs using transgenic technology. If one were to transplant a pig heart into a human it would be immediately rejected, a condition called *acute rejection syndrome*. This is due to surface cell proteins that are recognized by the human's immune system. However, using a spin-off of transgenic technology one can sequence the DNA and remove a segment or a specific gene that encodes for the production of this surface cell protein. An offspring created by this procedure is called a knockout individual because one of its genes has been "knocked-out." This yields a pig with a heart that contains neither the porcine surface cell protein nor the potential for acute rejection syndrome. Great strides in this area are expected during the next decade.

In effect, researchers are creating genetically engineered animals for human medicine. Since the creation of Dolly, advances have been occurring at an incredibly fast rate. Although 277 attempts were required to produce Dolly, the success rate of cloning is improving. Even at that rate the production of one individual with the ability to produce cures for human disease is justified. Today scientists are cloning not only sheep but cattle, mice, and pigs as well. Discussions related to the cloning of humans continues. Would one's clone be one's identical? One's clone would likely have a different uterus for gestation, a different set of parents or at least parents that have aged—those born during a different time and in a different environment. Therefore the couple that has paid $2.3 million to clone their dog will likely be disappointed if its clone is produced. Certainly from the scientific perspective this is not the primary reason for cloning.

Cattle are being cloned for farm animal production; however, here again the technology has limits. In one case an aged bull that produced a large number of valuable offspring has been cloned with the thought that his clone will have essentially identical sperm as the original bull. Some researchers envision using clones to reduce the effects of individual variation on outcomes of research. Numerous other applications will be created now that we are able to clone an animal using an adult cell.

13.5.7 Ultrasound and Pregnancy Diagnosis

Ultrasound technology has aided greatly in the scientific progress made in reproductive biology during the past decade. Information provided on follicular waves in Section 13.4.2 was discovered with the aid of ultrasonography. However, one of the most valuable commercial applications of ultrasonography is for determining pregnancy in cattle, swine, and horses. Ultrasound technology utilizes the basis that tissues have different abilities to propagate or reflect sound waves. Liquids do not reflect sound waves whereas dense tissues reflect much of the sound beam. These waves are then made visible on an ultrasound screen. This permits viewing an embryo during very early gestation. In cattle the embryo can be viewed as early as 20 days. A heartbeat can also be detected at this time. On a practical basis the use of ultrasonography is generally not recommended until about day 28 of gestation.

A note of caution is appropriate. The early monitoring of pregnancy enables one to observe a greater incidence of embryonic loss. Although ultrasonography does not contribute to embryonic loss, it demonstrates that even though pregnancy may have occurred not all pregnancies will be maintained to term. The incidence of embryonic loss is greater in dairy cows than in heifers and beef cows.

Ultrasonography is also used in pigs and horses. Because ultrasonography is done transrectally using a gloved hand and arm to control the transducer, the procedure must be modified for swine in which a holding device has been created that not only permits early pregnancy diagnosis (about 20 days into gestation) but also permits users to determine return estrus by monitoring follicle size.

13.5.8 Induced Parturition

Glucocorticoids, a family of hormones secreted by the adrenal glands, will induce parturition in the ewe, mare, sow, and cow. High dosages are required in the mare and sow.

Dexamethasone and flumethasone, two synthetic glucocorticoids now available commercially, induce calving in the cow at the rate of 85 percent or greater. Induced parturition should not be attempted until the last 2 to 3 weeks of pregnancy because calves are more likely to live at that stage of gestation. Cows may be expected to calve 34 to 60 h after treatment.

Induced parturition in cows is used to control the time of calving. This makes it possible to observe the cow or heifer closely so that assistance can be rendered if needed. It is especially useful in heifers that are calving for the first time and in those that were bred to large bulls of another breed. Pregnancy may also be terminated when abnormalities have been detected. Induced parturition can be used to lower birth weights and shorten gestation periods when these traits appear to be greater than normal, thus helping reduce death losses at calving. Properly induced parturition has no significant detrimental effects on milk production, survival of calves, or their subsequent growth rate.

One disadvantage of induced parturition is that retained placentas are more common than with normal gestation periods and parturitions. This is one reason the procedure has not been widely accepted by cattle producers.

Prostaglandins have also been used to induce parturition in female livestock. Intravenous infusions or intramuscular injections of $PGF_{2\alpha}$ have been used to induce parturition in sows. Treatment is followed by birth of the first piglet in 24 to 28 h. The gestation period is shortened but survival rate of the piglets appears normal. Injections of $PGF_{2\alpha}$ in heifers also commonly shortens the gestation period and induces parturition. Prostaglandins seem to increase the incidence of retained placentas in cows and heifers.

13.6 SUMMARY

Farm mammals must be born before they can grow and develop to the age at which they can provide food, service, and pleasure for humans. For that reason efficient reproduction is one of the most important aspects of animal production. Profitable livestock production requires that a maximal number of young be marketed per breeding female. Therefore efficient livestock production requires knowledge of the production and management of farm mammals as well as of their anatomy, physiology, genetics, and nutrition and of disease and parasite control among them.

STUDY QUESTIONS

1. Define the following: FSH, LH, seminiferous tubule, cells of Leydig, estrus, estrous cycle, ovulation, fertilization, freemartin, gestation, parturition, follicle, progesterone, estrogens, cryptorchid, and vasectomy.

2. Draw a diagram of the male reproductive system and label the parts.

3. What are the primary sex organs in the male? In the female?

4. What are the secondary sex organs in the male? The accessory sex organs?

5. What are the main functions of the testes? The scrotum?

6. What does production of a large proportion of abnormally formed spermatozoa indicate?

7. Can a male have all the normal sex characteristics and still be sterile? Explain.

8. What hormone stimulates sperm production by the testes?

9. What hormone stimulates testosterone production in the male?

10. Diagram the reproductive tract of the female and name the parts.

11. Describe differences between the reproductive tracts of the cow and the sow.

12. Outline the endocrine control of the ovary.

13. What hormone does the follicle secrete?

14. What hormone is secreted by the corpus luteum?

15. Outline the functions of LH and FSH in the female.

16. List functions of estrogens, progesterone, and testosterone.

17. What is meant by sperm capacitation?

18. What are some irregularities of estrus and the estrous cycle in farm mammals, and what causes each?

19. If a sexually mature female remains in almost constant estrus, what is a probable cause and a suggested treatment?

20. If a sexually mature female does not exhibit estrus, what are the probable causes and suggested treatments?

21. What is the normal length of the estrous cycle in the cow, sow, ewe, mare, and human female?

22. Describe changes that occur within the female reproductive tract between estrous periods.

23. Which species may be classed as a monoestrous animal? As a polyestrous animal?

24. What is the menopause? Does it occur in female farm mammals?

25. Name a spontaneously and a nonspontaneously ovulating species.

26. What is meant by superovulation? By superfetation? (cf. Glossary also).

27. How do the sperm and egg meet in the process of fertilization?

28. Where is the usual site of fertilization in farm mammals?

29. Name a species of animal that has follicles on the ovary but no corpus luteum.

30. Describe some abnormal pregnancies that may occur in animals.

31. What is polyspermy? Why may it occur in farm animals? What is the usual result when it occurs?

32. What is parthenogenesis?

33. What is meant by intrauterine migration?

34. What are some functions of the placenta?

35. What are the fetal membranes in the young of farm animals?

36. Describe differences between the placental membranes of the calf and the pig.

37. Is it possible to determine the number of ova produced in female farm animals at a single heat period? Explain.

38. What is a cesarean operation? Why is it so named?

39. Why is it necessary for newborn farm mammals to get colostrum? Is this necessary in humans? Why?

40. Why is a posterior presentation abnormal in cattle?

41. What is meant by the synchronization of estrus and ovulation? Discuss the fundamental physiology involved.

42. What is superovulation? Discuss its merits and limitations.

43. What are the basic procedures used in the technology of *in vitro* fertilization, or so-called test-tube babies?

44. Discuss the procedures and applications of embryo transfer.

45. What is the current state of the art of freezing embryos?

46. Of what potential significance is microsurgery with embryos?

47. What is the current state of the art of "sexed semen"? Of sexing embryos?

48. What is meant by induced parturition? Discuss its merits and limitations.

49. What was unique in the cloning of Dolly?

50. Outline the procedure used in producing insulin using recombinant DNA technology.

51. Outline a possible method of producing human blood-clotting factor IX using a transgenic animal.

52. Outline a method of increasing growth rates of swine using transgenic technology.

ARTIFICIAL INSEMINATION[1]

The environment fosters and selects; the seed must contain the potentiality and direction of the life to be selected.

George Santayana (1863–1952)
American poet and philosopher

14.1 NOMENCLATURE AND DEFINITION

Artificial insemination (AI) is the introduction of male reproductive cells into the female reproductive tract by an artificial means. It is commonly abbreviated AI when associated with domestic animals and wild birds. In humans artificial insemination is abbreviated AIH when the husband's semen is used and AID when the semen is that of a donor.

The foremost value of artificial insemination in farm animals lies in its use as a tool for the genetic improvement of **livestock** (especially traits having economic importance) on a mass basis. In 2000 approximately 65 percent of the dairy cattle and 8 percent of the beef cattle in the United States were inseminated artificially.

14.2 HISTORY AND DEVELOPMENT OF ARTIFICIAL INSEMINATION

Humans may not have originated the practice of artificial insemination. Males of certain **species** of spiders possess no **copulatory** organs. They deposit semen on a small mat spun by themselves, dip their feelers into the semen, and plunge the **sperm**-laden feelers into the abdomen of the female.

An old Arabian document dated 700 of the Hegira, which corresponds to the year 1322 of our era, records that an Arab chief of Darfur, who owned a prized **mare,** introduced a wad of wool into the animal's **genitals** and left it there for 24 hours. He

knew that a neighboring hostile tribe (specifically an enemy chieftain) had an excellent **stallion** in its possession. At night he crept into their camp and held the odorous wad under the stallion's nostrils, whereupon the horse became sexually excited and subsequently ejaculated on a piece of cotton held in readiness by the owner of the mare. He then hurried home and introduced the cotton into the mare's vagina. The mare then came in **foal.**

A second incident recorded in the fourteenth century further substantiates the early use of AI. While at war with a neighboring tribe an Arabian sheik, knowing that his enemies had exceptionally good mares that gave them a military advantage, sent some of his men into the enemy's camp by night with orders to fertilize the mares with semen of an old, lame, stabled stallion of inferior genetic merit.

In 1677 Anthony van Leeuwenhoek, a Dutch lens maker in Delft, Holland, first reported the discovery of human spermatozoa through a microscopic lens. He described what he saw as "man swimming in his own pool." In reporting his findings related to sperm cells and semen research to the Honorable Viscount Brouncker, president of the Royal Society, he said, "If your Lordship should consider that these observations may disgust or scandalize the learned, I earnestly beg your Lordship to regard them as private and publish or destroy them, as your Lordship thinks fit."

In 1777 Lazzaro Spallanzani, a priest of Modena, Italy, and a **physiology** professor at the University of Pavia, began a series of successful experiments using AI on reptiles. In 1780 he used AI successfully to inseminate a Spanish **bitch** that had produced one previous **litter.** He later reported:

> When she showed signs of coming in **heat,** I shut her up, fed her myself and kept the key on me. When she had been thus isolated for 13 days, I saw clear signs of heat, the external sex organs

[1] The authors acknowledge with appreciation the contributions to this chapter of Dr. J. D. Sikes, Department of Animal Sciences, University of Missouri–Columbia, and Dr. G. A. Doak, President, National Association of Animal Breeders, Columbia, Missouri.

241

becoming moist and excreting engorged fluid. On the 23rd day she seemed to me to be ready for an artificial **fecundation.** By **spontaneous ejaculation,** a young dog of the same **breed** provided me with 19 grains (about 1.2 cc) of seminal fluid which I immediately injected into the uterus by means of a syringe. As fecundation depends on the natural warmth of the semen, I took care to have the syringe at the same temperature as that of the animal, namely 30°C. After two days, I saw that she was no longer in heat and after 20 days that her belly was swollen; so on the 26th day I set her free. It was 62 days after my injection that she **whelped** three living pups, two males and one bitch, resembling the parents in color and shape. I found that six grains of the seminal fluid had remained behind in the syringe and concluded from this that a very small amount of semen is required in nature. This discovery has persuaded me that we shall be able to do the same thing with larger animals.

He further concluded:

> The day will come when this discovery will acquire immense importance for the human society.

In subsequent studies Spallanzani found that the fertilizing power of semen resided in sperm carried by the spermatic (seminal) fluid. When semen was filtered he observed that the liquid that passed through was incapable of causing fertilization, but the residue on the filter was high in fertilizing capacity. This discovery gave rise to intensive investigations of sex cells.

An American dog breeder, Everett Millais, artificially inseminated 19 bitches and successfully impregnated 15 during the period from 1884 to 1896. Concurrently Professor Hoffman of Stuttgart, Germany, recommended AI following natural matings in horses. He recovered semen deposited by the stallion into the vagina of the mare by using the speculum-and-spoon technique. The semen was then diluted with cow's milk and injected into the uterus using a syringe. He concluded, however, that this was impractical and discontinued his studies.

In 1899 Elias I. Ivanoff, a Russian researcher, began a series of studies using AI. He was successful in pioneering the artificial insemination of birds, horses, cattle, and sheep and apparently was the first to artificially inseminate females of the latter two species successfully. Mass breeding of cows through artificial insemination was first accomplished in Russia, where 19,800 cows (an average of 100 cows per bull) were bred in 1931. The Russian scientist Dr. V. K. Milovanov stated that the year 1931 marked the end of the experimental stage in artificial insemination of cows and its beginning as a "powerful zootechnical tool." Thenceforth he concluded the objective of AI was utilization of spermatozoa from the best sires on the largest possible number of cows. Its other uses, **disease prophylaxis** and the therapy of sterility, became of secondary importance.

In Denmark in 1936 the first cooperative artificial-breeding association in the world was organized.

In 1935 Dr. C. L. Cole and Professor L. M. Winters artificially inseminated sheep at the University of Minnesota. In May 1936 Dr. Cole discussed the use of artificial insemination with Dr. L. O. Gilmore (Ohio Experiment Station) who had a breeding problem—his albino **bull** would not serve an albino **heifer.** Cole offered to collect semen from the bull and insemi-

nate the heifer and Gilmore accepted. On February 25, 1937, the first calf to result from artificial insemination in the United States was born.

Professor E. J. Perry visited one of the cooperatives in Denmark and saw the possibilities of artificial insemination. In 1938 he established the first cooperative artificial-breeding association in the United States. It was organized by the Cooperative Extension Service of the New Jersey State College of Agriculture with the aid of the New Jersey Holstein breeders. However, artificial insemination of cows was already being practiced in a few herds in the United States.

In 1939 a Jersey **cow** on exhibition at the World's Fair in New York was artificially inseminated and conceived to a bull that was on exhibition at the San Francisco Fair. This event generated much interest and, as shown in Tables 14.1 and 14.2, the use of artificial insemination in cattle in the United States has since been widely accepted. In 1999 the United States custom froze and sold 22,680,000 units of dairy and 3,974,000 units of beef semen.

TABLE 14.1	Early Sire Use through Artificial Insemination in the United States		
Year	Number of sires	Cows per sire	Sires per stud
1938[*]			
1939	33	227	4.7
1943	574	318	9.7
1948	1745	982	19.2
1953	2598	1865	27.1
1958	2676	2483	37.7
1963	2559 (incl. 401 beef)	3250	50.2
1968	2380 (incl. 352 beef)	3303	72.1
1971[†]	2514 (incl. 347 beef)	3402	96.8

[*]Initiated in May.
[†]The above data have not been compiled since 1971.

TABLE 14.2	Units of Beef and Dairy Semen Sold and Custom Frozen in the United States (Thousands)				
Year		Domestic use	Export sales	Custom frozen	Total
1979	Beef	1086	240	1125	2451
	Dairy	12,467	1836	682	14,985
	Total	13,553	2076	1807	17,436
1989	Beef	822	243	1069	2134
	Dairy	12,770	4307	761	17,837
	Total	13,592	4550	1830	19,971
1999	Beef	898	771	2304	3974
	Dairy	13,621	8421	638	22,680
	Total	14,519	9192	2942	26,654

Source: Dr. G. A. Doak, President, National Association of Animal Breeders, Columbia, Missouri, personal communications.

Artificial insemination is currently used in dairy and beef cattle, goats, sheep, swine, horses, turkeys, bees, dogs, red fox, fish, mink, humans, and many other species. A summary of its worldwide use in cattle is given in Table 14.3. Research leading to the development of artificial insemination on a commercial basis in the United States (especially in dairy cattle) is almost exclusively a contribution of the colleges of agriculture of land-grant universities. Pioneering research was done at the Illinois, New Jersey, Missouri, Michigan, Minnesota, New York, Pennsylvania, and Wisconsin experiment stations.

14.3 IMPORTANCE AND IMPLICATIONS OF ARTIFICIAL INSEMINATION

It is now well known that artificial insemination provides an important tool in livestock breeding. It was initially organized in the United States as a means of making available the service of superior **purebred** dairy **sires** to all dairy cattle breeders, especially those with **grade** cattle. However, it is currently commonplace in purebred **herds.** Also, more recently its use in breeding beef cattle, swine, goats, and horses throughout the United States has increased. Since 1956 millions of units of frozen semen have been exported annually to other countries (Table 14.4). The utilization of frozen semen for improving cattle, including water buffalo, in many lands offers a great challenge and opportunity.

14.3.1 Advantages and Benefits

Artificial insemination increases the usefulness of superior sires. The average bull in commercial AI service is typically mated to thousands of cows annually, in contrast with only 30 to 50 cows annually bred to bulls under natural mating conditions. A few Holstein bulls have produced over 1,000,000 straws of semen during their years of AI service. Artificial insemination provides a fast increase in production potential. A recent USDA study indicates that when superior sires are used 3 to 4 times more rapid genetic improvement can be made through AI than through natural service.

It is estimated that the genetic superiority of dairy cows sired by bulls in AI service, compared with the same number of cows sired by natural service, results in approximately 8 billion lb of additional milk being produced annually in the United States. Furthermore, the genetic superiority of slaughter steers currently sired by AI results in an estimated production of over 40 million additional lb of edible meat annually. This extra milk and meat increase the efficiency of production and results in lower per serving cost to consumers.

The time required to establish a reliable sire proof on a young bull is greatly reduced through AI. Therefore, the transmitting ability is determined more quickly and as a result of bulls being mated to cows in many herds the sire proof is more meaningful.

Artificial insemination prevents the exposure of a healthy sire to diseases such as vibriosis, trichomoniasis, leptospirosis, brucellosis, BVD virus, and tuberculosis. It makes possible the use of sires unable to serve a cow naturally and the mating of large sires to small females without injury or mating problems. The danger to the caretaker in handling bulls is greatly reduced, often eliminated.

Artificial insemination permits small purebred breeders and all grade and/or commercial breeders to have the services of superior sires. Some bulls have been brought into artificial

TABLE 14.3	World Use of Artificial Insemination
Region	**Estimated number of cows (1979)**
Europe	63,952,150
Former U.S.S.R. (Europe and Asia)	25,679,000
North America	10,876,000
Central and South America	4,533,617
Asia	4,758,396
Africa	365,551
Oceania (Australia and New Zeland)	2,034,000
World totals	112,198,714
Selected country	**Estimated percent cattle bred with AI (1994)**
Finland	99
Israel	99
Hungary	98
Japan	96
Czechoslovakia	94
Bulgaria	92
Denmark	90
Norway	90
New Zealand	90
Poland	87
Sweden	80
France	75
Australia	70
Germany	70
Netherlands	70
Former U.S.S.R.	70
Great Britain	65
United State: dairy	65
beef	6
Canada: dairy	60
beef	11
Ireland	55
Selected species	**Estimated world use of AI (1979)**
Cattle	112,198,714
Sheep	67,421,811
Goats	68,322
Swine	1,714,414
Mares	926,117
Total	182,329,378

Sources: H. A. Herman, *Improving Cattle by the Millions—NAAB and the Development and Worldwide Application of Artificial Insemination,* University of Missouri Press, Columbia and London, 1981; various publications of Foreign Agricultural Service, USDA; and Dr. G. A. Doak, President, National Association of Animal Breeders, Columbia, MO, personal communications.

TABLE 14.4	United States Export of Dairy and Beef Semen by Region, 1999 (Thousands)

Region	Dairy Units	Dairy $ value	Beef Units	Beef $ value	Total units	Total $ value
North America	1421	8212	56	344	1477	8556
Caribbean	27	151	2	8	29	159
Central America	152	782	25	85	177	867
South America	2904	8192	609	1178	3513	9370
European Community	1920	19,998	10	45	1930	20,043
Other western Europe	143	1286	0.3	3	144	1289
East Europe and former U.S.S.R.	289	2068	8	21	297	2089
Middle East	425	2440	23	42	448	2482
North Africa	77	464	—	—	77	464
Other Africa	232	1269	2	19	234	1288
South Asia	4	12	—	—	4	12
Other Asia	352	3703	0.5	3	353	3706
Oceania	400	2258	36	319	436	2577
World total	8346	50,835	772	2067	9118	52,902

Source: Dr. G. A. Doak, President, National Association of Animal Breeders, Columbia, MO, personal communications.

insemination service at a cost in excess of $250,000. Yet their semen is commonly available for only $5 to $25 per mating. The average artificial insemination cost per mating (semen plus inseminator fee) in the United States is about $8 to $15. Furthermore, AI permits mating of outstanding individuals though great distances apart. Moreover, the mating to an outstanding sire after his death is made possible through the freezing of semen and use of artificial insemination. One example of this is cited in which an AI **stud**[2] achieved a successful mating more than 25 years after the sire's death.

Frozen semen of production-tested bulls can be used as a control in studies of genetic progress. Progeny resulting from periodic matings to the bull initially used are compared with progeny of production-tested sires selected later. For example, assume a selection experiment was initiated in 1980 using semen from production-tested bull A. Progeny resulting from matings in 1990 and 2000 using semen from bull A can be compared with progeny of other bulls in service in 1990 and 2000 to determine the extent of progress made through selection.

Because semen is examined regularly infertile bulls are likely to be detected earlier than with natural mating. This procedure encourages the keeping of better breeding, calving, and fertility records. Artificial insemination increases fertility in some species, for example, domestic turkeys, which are often clumsy and lack the ability to mate naturally. It serves as a research tool for studies in reproductive physiology, fertility, sterility, and cellular physiology.

Artificial insemination offers a wider choice of sire selection and individual matings. Many problems and decisions

associated with buying a bull are eliminated. Furthermore, it provides a means of attaining more uniformity of market cattle and permits better control of seasonal reproduction. In beef AI has provided the only means of accessing certain recently introduced beef breeds.

14.3.2 Limitations

Artificial insemination can quickly disseminate undesirable genetic traits. As a result of mating more cows per sire, the frequency of a specific detrimental **recessive gene** may be increased. However, AI can identify undesirable recessive traits earlier, which can help minimize detrimental effects. Mass mating with a given bull tends to narrow the genetic base and thereby restrict variation, which is a tool important to animal breeders. Although these are potential limitations, the authors wish to point out that most males used in AI programs are carefully selected (often through inbred progeny testing) to minimize the spread of undesirable traits.

Special training in semen collection, processing, storage, and use is essential with AI. Additionally, AI requires special equipment and facilities (especially in working with beef cattle). Artificial insemination requires experienced and well-trained inseminators to achieve **conception** rates equal to those with natural service. Caution is essential in processing semen to prevent errors in labeling, handling, shipping, and using frozen semen.

Although New Year's Day, Easter, the Fourth of July, Labor Day, Thanksgiving, and Christmas are the only "no service" days for most commercial AI service providers, the demands during periods of heavy breeding may cause undesirable delays in obtaining service. Furthermore, it is often difficult to detect cows in heat, especially among beef cattle, whereas the bull maintains a 24-h vigil for animals in heat and is naturally much

[2] A bull stud, or semen-producing business, is any individual or business entity owning or leasing one or more bulls from which the individual or business entity collects, processes, and distributes semen for use in the insemination of animals owned by others.

better adapted to recognize symptoms of **estrus** than are humans. It should be noted, however, that two bulls may chase the same cow while other cows in estrus are nearby.

Artificial insemination requires more labor than does natural service. It requires a high cow population per square mile to be profitable for the inseminator and economical for the livestock breeder. It reduces the demand for bulls for breeding purposes. However, many bulls are not of sufficient genetic merit to be used for breeding purposes.

14.4 SEMEN COLLECTION[3]

The primary objective in semen collection is to obtain maximum output of high-quality spermatozoa per ejaculate. As the frequency of ejaculation increases the volume, sperm concentration, and total number of spermatozoa per ejaculate generally decrease; however, total weekly sperm harvest increases. Proper stimulation and preparation of the bull before semen is collected increases the number of spermatozoa obtained per ejaculate.

Several studies indicate that oxytocin and certain other hormones influence sperm output. Sperm output in **rams,** rabbits, and many other species increases when males are injected with oxytocin. Research at Michigan State University showed that anterior pituitary hormones are released into the blood at ejaculation; luteinizing hormone increased slightly and growth hormone concentration doubled within 5 min after ejaculation. Curiously, growth hormone was higher in mature than in young bulls and highest when ejaculation followed sexual preparation. Prolactin increases sixfold and testosterone twofold within 5 min after ejaculation.

Cornell researchers found that weekly sperm output of Holstein bulls may average between 30 and 35 billion cells. Sperm output was positively correlated with maximum testes–scrotal circumference.

14.4.1 Semen Output and Frequency of Ejaculation

Factors known to influence semen output are age of the bull (maximum output commonly occurs between 4 and 8 years of age), season of the year (December is the best month for total sperm production in the United States), and frequency of ejaculation (increased frequency of ejaculation results in more total sperm being obtained, but with fewer sperm per collection). Bulls collected 6 times per week ejaculated 3.3 times more sperm than those collected only 1 time weekly. In Cornell University studies when bulls were properly stimulated, first ejaculates produced 78 percent more sperm than second ejaculates.

Researchers at The Pennsylvania State University found when semen was collected from bulls frequently they could be ejaculated up to 77 times in 5 h with no apparent ill effects. Their studies found a continuous high frequency of ejaculation (6 times weekly) from bulls 1 to 7 years of age greatly increased

the harvest of semen without impairing growth, reproductive capacity, or fertility.

14.4.2 Testicular Development

Testicular weight is an important trait that provides an accurate estimate of the amount of sperm-producing parenchyma in the testis. Because testicular weight cannot be measured directly in breeding bulls, a scrotal circumference (SC) measurement is used as an accurate and repeatable estimate. The relation between SC and paired testes weight is $r = 0.95$. Canadian scientists found that the heritability of testicular size in yearling beef bulls completing 140-day growth-performance tests is 0.69.

Factors affecting the rate and extent of testicular development include age, breed, body weight, and nutritional regime. Research at North Carolina State University suggests that selection for increase in scrotal size at 365 days should increase testes size and weight and sperm number in bulls. Because sperm production by a male is, in large part, a function of testis size, the number of units of semen that can be obtained is closely related to testicular size. Research is under way at Louisiana State and Michigan State Universities to develop ways of increasing testicular size in bulls using techniques involving immunization and hormones.

Research at Colorado State University also showed that scrotal circumference or testicular size is correlated positively with fertility ($r = 0.58$). Because the correlation of SC with actual testicular weight is $r = 0.95$, this means that SC is a reliable predictor of the amount of sperm-producing tissue within the testes. Therefore, the larger the testes the greater the sperm production potential. This is of economic significance whether a bull is used on a conventional or on an artificial insemination basis.

Another interesting finding of the Colorado studies was a negative genetic correlation of -0.71 between age at first estrus in heifers and SC of half sibling males. This suggests that the female progeny from bulls with above-average testicular size for their age would begin to cycle and exhibit estrus earlier than average, which would be beneficial in breeding programs where heifers are bred to calve at 2 years of age. Additionally, it suggests that selection for early age at puberty in heifers should result in a more fertile cow herd with higher lifetime productivity.

14.4.3 Preejaculation Sexual Preparation

Sexual preparation of bulls may be defined as *prolonging the period of stimulation beyond that adequate for mounting and ejaculation.* Studies at Michigan State University showed that motile sperm output per ejaculation in bulls can be increased as much as twofold when two or three false **mounts** (and active restraint) are given as rapidly as possible before ejaculation.

Stimuli other than false mounting may also be used to increase sperm output in bulls. These stimuli include moving the stimulus animal, exchanging stimulus animals, changing locations of preparation, changing the personnel handling the bulls, and combinations of these. Additionally, undefined stimuli from a teaser animal near a bull cause the bull to ejaculate more

---

[3] For additional information, the reader is referred to H. A. Herman, J. R. Mitchell and G. A. Doak, *The Artificial Insemination and Embryo Transfer of Dairy and Beef Cattle,* 8th ed., Interstate Publishers, Inc., Danville, IL, 1994, pp. 45–55.

TABLE 14.5	Semen Production in Selected Mature Males					
Class of animal	Semen volume/ ejaculate, ml	Sperm concentration, $\times 10^6$/ml	No. of sperm/ ejaculate, $\times 10^9$	Life of sperm in female tract, h	Survival time with fertilizing capacity (fresh)	No. of females per ejaculate
Boar	200–300	25–1000	20	24–40	1–2 days	15–25
Bull	5–6	800–1200	4–6	28–30	4–6 days	300–500
Cock	0.2–2.0	0.5–60	0.03	168–504	1–2 days	8–12
Human	2–6	50–150	3.5	24–96	1–3 days	
Ram	0.7–2.0	800–4000	2–3	34–40	5–7 days	40–100
Stallion	50–150	30–800	6	96–144	1–2 days	8–12

Figure 14.1   An artificial vagina used to collect bovine semen.
Courtesy of Missouri Agr. Expt. Station.

sperm. The undefined stimuli are not visual because blinded bulls also respond to sexual preparation with increased sperm output. Part of the stimuli must involve olfactory (sense of smell) mechanisms because bulls mount **estrous** cows twice as fast as nonestrous cows and yield larger volumes of semen and greater sperm numbers.

Apparently sexual preparation is less effective, as measured by time to first mount, in beef than in dairy bulls. Pennsylvania studies showed that sexual preparation significantly increases both sperm concentration and semen volume for dairy and beef bulls. However, it took nearly 10 times longer to stimulate beef (10.9 min) than dairy (1.1 min) bulls based on time to first mount. Semen harvest and related data are presented in Table 14.5.

14.4.4 The Artificial Vagina (AV)

In 1930 Russian scientists developed the artificial vagina (Figure 14.1). The AV has proved to be the most practical and satisfactory means of collecting semen in most domestic animals. Its use aids in obtaining a normal ejaculate in which the semen is clean and free of extraneous secretions. Additionally, a quick measure of quantity is obtained in the graduated semen collection vial.

The artificial vagina should have an internal temperature of about 109 to 122°F (43 to 50°C) and should be the appropriate length for each bull. It is important that the bull ejaculate near the end of the AV into a directacone to reduce sperm loss on the surfaces of the AV and to prevent semen from being in contact with the warmer temperature of the AV.

During collection the artificial vagina should be positioned parallel to the mount (teaser) animal in a slanting position near the anticipated path of the bull's penis. The operator, wearing plastic gloves, should guide the penis into the AV by grasping the sheath well behind the orifice. To avoid possible retraction, pre-

mature ejaculation, and spread of disease the operator should not physically touch the protruded penis. As the bull thrusts forward for ejaculation the operator should allow the AV to move forward and tip it slightly to allow the semen to flow into the collecting vial. Special care must be taken to avoid bending the penis. This may injure or cause discomfort to the bull. Semen collection from the ram and stallion via the artificial vagina is similar to that in the bull; however, in the **boar** the emission of semen continues for 5 to 20 min and thus prolongs the collection period.

The bull, ram, and stallion are sensitive to the temperature of the AV but not particularly pressure sensitive. The boar is much more pressure sensitive in collecting normal ejaculates.

Freshly harvested semen must be protected against heat or cold shock to ensure motility and **viability,** which are essential to fertility. Sanitation is important with respect to equipment, personnel, handling, and processing of semen to protect against bacterial contamination and reduced semen quality.

14.4.5 Mechanical Manipulation

This method of collecting semen has had only limited success in bulls but is of importance in birds. Because bird testicles are retained within the body, gentle pressure is applied in the abdominal–cloacal area to stimulate semen release, which is collected in a small vial or syringe. This technique is utilized in domestic fowl as well as in wild birds. Artificial insemination is used as a method of increasing fertility in zoo-raised and endangered species of wild birds.

Boars are more sensitive to pressure than bulls and often do not respond satisfactorily to AV utilization for semen collection. Therefore, collecting boar semen commonly involves the technician applying manual pressure with a gloved hand directly on the penis.

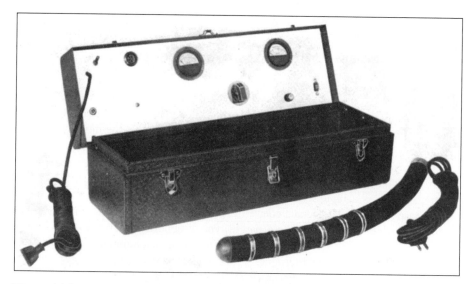

Figure 14.2 An electroejaculator used to collect semen from the bull.

H. A. Herman and F. W. Madden, The Artificial Insemination of Dairy and Beef Cattle, *6th ed., Lucas Brothers Publishers, Columbia, MO, 1980.*

In bulls gentle pressure is applied through the rectum on the ampullae and vas deferens. The primary disadvantages are that semen is often contaminated with bacteria and extraneous matter, small ejaculates are common, and many animals do not respond well. However, this method may be used to obtain semen from bulls unable to mount and serve a cow or artificial vagina. Samples obtained by this method have only about one-half the number of sperm as semen collected using the artificial vagina.

14.4.6 Electrical Stimulation

Through the use of a voltmeter, transformer, milliammeter, and one or more electrodes, an alternating current of 3 to 30 volts may be produced that will cause an electrical excitation of the ejaculatory nerve centers resulting in partial to complete ejaculation (Figure 14.2). Repeated rhythmic stimulation periods are alternated with short rest periods. This method of semen collection has been used routinely in rams and to a lesser extent in other farm animals. It results in the collection of clean semen and may be used on males that have been injured or are lacking in **libido.** It may also be used on animals who were "psychologically conditioned" by natural matings and refuse to serve the artificial vagina.

In the bull it is possible to obtain accessory sex **gland** secretions at lower voltage than is required for ejaculation. As a result large semen samples are often obtained from a bull with an electroejaculator, but these samples have lower sperm concentrations. The ram requires a low-peak voltage (2 to 8 volts) whereas the boar requires a higher (25 to 30 volts) voltage than the bull for ejaculation by this method.

Cornell studies showed that semen obtained by electroejaculation was lower in solids than that obtained with an artificial vagina. (The total weight of solids of **bovine** semen, as measured by refractometry, averages about 10 g of solids per 100 g of seminal plasma.)

Tranquilization appears to aid in the collection of higher-quality semen in the bovine when electroejaculation is employed. Moreover, less electrical stimulus is needed to obtain an ejaculate in the tranquilized bull.

14.4.7 Vaginal Collection

Perhaps the oldest, yet least desirable, method of harvesting semen is to recover it from the anterior end of the vagina following copulation. This may be accomplished by using a syringe, sponge, or vaginal spoon. It is not recommended, however, because the semen is contaminated with mucus and urine, there is increased danger of spreading genital diseases, and it is impossible to recover an appreciable amount of the ejaculate.

14.5 EVALUATION OF SEMEN[4]

The best indication of fertility in a bull is a live calf born to his mating. This evidence is never available in young bulls and breeding records are often incomplete in older ones. Therefore, certain physical and chemical tests have been developed to assist in determining semen quality.

Appearance

This can be noted at the time of collection. The color of semen is usually milky white and the fluid is quite viscous. Usually the greater the viscosity the higher the sperm concentration.

Enumeration of Spermatozoa[5]

The concentration, or number of spermatozoa per unit volume of semen, is an important consideration in determining the optimum

---

[4] Ibid., pp. 63–71.
[5] Ibid., pp. 73–79.

dilution ratio of semen to be processed. It is primarily the number of spermatozoa, the percent of mobile spermatozoa, and the degree of progressive motility that are used to appraise the estimated semen fertility.

There are a number of methods for estimating spermatozoa number. In one, photoelectric colorimetry, the amount of light passing through a standard dilution of semen is used to estimate sperm number. This method is rapid, highly repeatable, and is used extensively. Hemocytometry, which is also used to count red and white blood cells on a calibrated slide, is a second method of counting spermatozoa. A third method involves visually making an opacity rating, which is a rapid estimate of sperm concentration. In the process, semen is diluted and compared with previously prepared opacity standards of known concentration.

Photometric techniques are quite variable in the determination of sperm concentration in the boar and stallion because their ejaculates vary substantially in the nonsperm cellular semen component and therefore decrease the accuracy of readings. Researchers at Colorado State University found great monthly variation of seminal characteristics among stallions.

Motility[6]

This is rate of movement of spermatozoa and can be observed microscopically. A drop of semen is placed on a clean, clear glass slide that has been warmed to about 100°F (38°C). Using a low-power objective (100–125X) of the microscope, motility can be observed and a motility value may be assigned on the basis of ratings of 0 to 5.

Because it is simple, quick, and inexpensive percent motile spermatozoa is probably the most widely used assay for evaluating semen. However, evaluations of motility are subjective and are influenced by the thickness and temperature of the sample on the slide. Research indicates that the accuracy of estimating motility of spermatozoa is increased by (1) using a semen sample of 8 to 10 μm in thickness, (2) viewing it near an air bubble and/or away from the edge of the slide, and (3) maintaining clean slides and coverslips. Most research indicates that motility estimates alone are not reliable predictors of semen fertility.

Commonly used motility ratings with their respective descriptions are:

5 = *excellent motility*. 80 percent or more of the spermatozoa are in highly vigorous motion. Swirls and eddies are rapid and changing.

4 = *very good motility*. Approximately 70 to 80 percent of the spermatozoa are in vigorous rapid motion. Waves and eddies form and drop rapidly.

3 = *good motility*. 50 to 70 percent of the spermatozoa are in motion. Motion is vigorous, but waves and eddies formed move slowly across the microscopic field.

2 = *fair motility*. 30 to 50 percent of the spermatozoa are in motion. No waves and eddies can be observed.

1 = *poor motility*. Less than 30 percent of the spermatozoa are in motion, resulting in sluggish motility.

0 = *no progressive motility*.

It is interesting that the motility of boar semen decreases rapidly when the semen is exposed to light but increases in the presence of oviductal fluids. Researchers at the University of Georgia found that the motility of boar semen increased from 70 percent at entry to 90 percent after being held 1 h in the oviduct.

Morphology[7]

A microscopic determination can be made of the percent and type of abnormal spermatozoa present in semen. High-quality semen contains a minimal number (5 to 15 percent) of abnormal spermatozoa, whereas low-quality semen frequently contains a larger number (up to 30 or more percent) of **morphologically** abnormal spermatozoa. Deformed spermatozoa may result from cold or heat shock, x-rays, and nutritional or endocrine imbalances that disrupt normal **spermatogenesis.** First ejaculates of bulls following sexual rest are characterized by low motility and many abnormal sperm cells. Normal and abnormal spermatozoa are shown in Figure 14.3.

Dead–Alive Stain[8]

This test was initially developed at the Missouri Agricultural Experiment Station using ram semen and is based on differences among living and dead spermatozoa in permeability to a dye. Live spermatozoa do not absorb dye and remain light in color. Dead sperm cells accept dye and appear dark under the microscope (Figure 14.4). The dead–alive stain is most meaningful when used with the microscopic motility examination. A fairly high percentage of dead sperm may not be readily apparent in the motility check because many inactive spermatozoa are swept about by movements of the live and active sperm cells.

Dead sperm dilute the concentration of fertile sperm of semen; they are also toxic to live sperm. However, addition of the **enzyme** catalase apparently eliminates the toxic effect of killed sperm on **livability** of bovine spermatozoa. Scientists at Virginia Polytechnic Institute and State University showed a high correlation ($r = 0.60$) between fertility and average intact acrosomes, the part of the spermatozoan that releases egg-penetrating enzymes. The correlation between fertility and abnormal sperm morphology ranged from $r = 0.27$ to $r = 0.37$.

Cold Shock

This is a simple and quick check of spermatozoan vigor. The percentage of resistant spermatozoa may be determined by comparison of the number of sperm that are alive before and after cold shock. This test aids in estimating storage life and fertilizing capacity of semen. Cold shock of semen occurs when sperm are rapidly cooled to a temperature near their freezing point.

[6] Ibid., pp. 63–71.

[7] Ibid., pp. 85–91.
[8] Ibid., pp. 81–83.

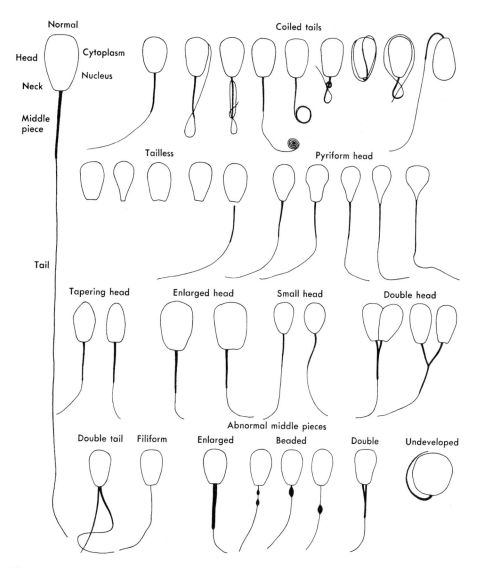

Figure 14.3   Normal and abnormal types of bovine spermatozoa.

H. A. Herman and F. W. Madden, The Artificial Insemination of Dairy and Beef Cattle, *6th ed., Lucas Brothers Publishers, Columbia, MO, 1980.*

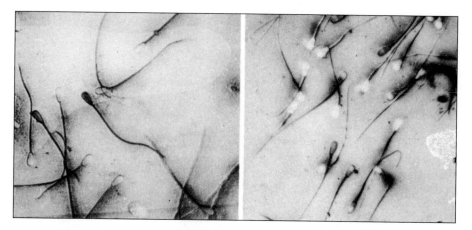

Figure 14.4   Stain showing dead (dark color) and alive (light color) sperm cells.

H. A. Herman and F. W. Madden, The Artificial Insemination of Dairy and Beef Cattle, *6th ed., Lucas Brothers Publishers, Columbia, MO, 1980.*

Hydrogen-Ion Concentration

Bovine semen has an average **pH** of 6.7. Semen obtained by the massage method often has above-normal pH (8.0), whereas semen collected in the artificial vagina after intense precollection teasing or long intervals between collections is lower than normal in pH (6.3). Changes in the pH of semen after collection may be regarded as a rough measure of the **metabolic** activity of spermatozoa, producing lactic acid from **glycolysis.** The resulting lowered pH causes a decrease in semen motility.

14.6 EXTENSION OF SEMEN[9]

The main objective in extending semen is to increase the volume of the ejaculate so that a large number of females may be mated to a given male. In natural mating one ejaculate, and often more, is used to inseminate one female, whereas through artificial insemination and the extension of semen one ejaculate may be used for several hundred females. In the bovine each ejaculate averages 5 ml of semen, containing between 0.8 and 1.2 billion spermatozoa per milliliter. Research has established that 12 million or less normal live spermatozoa in a volume of 0.5 ml will result in satisfactory conception rates among females inseminated artificially. It is apparent, therefore, that the average ejaculate from a normal bull can be used to inseminate 300 to 500 cows if the volume is increased by appropriate extension.

Semen of the ram is highly concentrated (Table 14.5), and even though ejaculate size is small, a large number of **ewes** (40 to 100) can be inseminated from one ejaculate. The volume of the boar ejaculate exceeds that of all other farm animals (Table 14.5); however, because sperm concentration is lower and spermatozoa must travel a long distance through the uterine horns to fertilize the ovum, one boar ejaculate can be used for only about 20 **sows** through artificial insemination. Semen from the stallion and **cock** also appear to have limited extension qualities, each ejaculate being useful in the breeding of about 8 to 12 females through artificial insemination.

14.6.1 Characteristics of a Good Extender

The ideal medium for semen extension will not only increase ejaculate volume but will also be favorable for both survival and longevity of spermatozoa and for maintenance of their viability and fertility over extended periods of time. A number of media have been developed that provide adequate nutritional components, pH (6.5 to 6.7), **buffering** capacity, and protection against bacterial contaminants and temperature shock. In addition, the effects of electrolytes, nonelectrolytes, and viscosity should be considered. An extender must be free of toxic substances (e.g., bacterial products or infectious organisms) harmful to the spermatozoa, the female reproductive tract, the fertilization process, and the implantation and development of the fertilized ovum. An ideal extender should be simple to prepare, readily reproducible, reasonably inexpensive, and readily available.

The semen and extender are at the same temperature when mixed. Semen is extended to provide a desired number of services, each containing an adequate number of live sperm cells to ensure conception. For example, plastic straws are usually processed to provide 0.25 to 0.5 ml, and glass vials (ampules) 0.7 to 1.0 ml, of extended semen per service. Both contain approximately the same total number of spermatozoa.

Egg-Yolk Extenders

In 1934 Milovanov observed the beneficial effects of adding egg yolk to extending fluids. Today egg yolk citrate is one of the commonly used semen extenders. Only the yolk portion is used because egg white contains a substance (lysozyme) toxic to spermatozoa. Variations from 5 to 50 percent in the amount of egg yolk used appear to give good results; however, lower concentrations (20 percent or less) are commonly used today.

Gelatin-Containing Extenders

These were developed by Danish workers. They employ a method of inseminating with semen gelled in cellophane straws; however, commercial artificial insemination in the United States does not employ the use of diluents containing gelatin. Gelatin capsules can be used as semen containers in artificial insemination of mares.

Milk Extenders

In recent years heated (197.6°F [92°C] for 10 min) whole milk or skim milk diluents have been employed in commercial artificial insemination. These diluents are economical, easily prepared, provide good protection for spermatozoa, and give satisfactory fertilizing capacity to semen. The foremost objection to whole milk as an extender is that the fat globules make microscopic examination difficult. This problem is partially solved when skim or homogenized milk is used. Some processors use the milk of goats, which is "naturally" homogenized. Cow's milk has been successfully used as an extender for bull, ram, boar, and stallion semen. Glucose is often added to supplement the low sugar content of stallion semen. Studies at Texas A&M University indicate that evaporated milk can be used with glycerol as an extender in freezing stallion semen.

Fruit and Vegetable Juices

Juices have been used in some countries for extending bovine semen with apparently acceptable results. Such juices as tomato broth and carrot juice, which are often abundant and economical, have been used. Coconut milk is being used to extend fresh semen in the tropics where refrigeration is at a premium and coconuts are plentiful. Bovine semen may be held at room temperature for a few days in coconut milk.

Extenders for Frozen Semen

The use of glycerol and selected sugars is important in the extension of semen to be frozen (Section 14.7.2).

[9] Ibid., pp. 101–116.

Others

Blood **plasma** and **serum,** Tyrode's solution, starch solution, seminal fluid, various solutions of alcohol glycerols, and sugars have been used. However, most of these are presently used only in research.

14.6.2 Coloring of Semen[10]

To assist the AI technician in identifying semen by breeds and thereby safeguard against error, some AI organizations use certified food and vegetable dyes. The addition of one drop of coloring per 50 ml of extender does not affect livability and fertility of the sperm. Colors used by some AI studs to help identify selected dairy and beef breeds are given below.

Ayrshire—purple	Angus—orange
Brown Swiss—brown	Beef Shorthorn—magenta (purplish red)
Guernsey—yellow	Charolais—pink
Holstein—green	Hereford—tan
Jersey—red	Santa Gertrudis—blue
Milking Shorthorn—lime	Simmental—yellow

Some AI studs use colors to label semen vials (ampules) and straws. (More recently colored straws are used to distinguish breeds.)

14.6.3 Antibiotics[11]

Antibiotics commonly used in frozen semen are Gentamicin, Lincomycin, Spectmomycin, and Tylosin as recommended by CSS (Certified Semen Services, Inc.). Most extenders include **antibiotics** (Section 14.15.1).

It is interesting that the microfloral content of stallion semen is greater than that of bull semen. This is because the prepuce of stallions is more highly contaminated with bacterial flora. Studies at Cornell University showed that the antibiotic Amikacin inhibited bacterial growth in stallion and bull semen and had no significant effect on the motility of stallion and bull spermatozoa, even at relatively high concentrations.

Trichomonas fetus can survive semen freezing and processing (with antibiotics) procedures presently used in the AI industry.

14.7 SEMEN STORAGE

The cardinal concern in extending the service of superior males to large numbers of females is to preserve fertile spermatozoa successfully.

14.7.1 Fresh Liquid Semen[12]

Semen of several species may be extended and successfully stored at 41°F (5°C) for 1 to 4 days. During the infancy of artifi-

cial insemination semen was shipped in therm with cracked ice. Often it was flown by li dropped by parachute in the vicinity of the lo Currently all AI studs (semen-producing bus United States are freezing semen and only a few continue to use fresh liquid semen in the insemination of cattle. Conversely, fresh semen is extensively used in the artificial insemination of sheep, swine, and turkeys. Many American boar studs are now shipping fresh cooled boar semen throughout the United States. Specialized containers are now available for shipping cooled stallion semen worldwide (Chapter 5).

14.7.2 Frozen Semen[13]

In 1776 Spallanzani observed that freezing stallion semen in either snow or the winter cold did not kill the "spermatic vermiculi" but rather held them in a motionless state until exposed to heat, after which they were motile for several hours. Davenport reported in 1897 that human spermatozoa survived freezing at 1°F (–17°C).

In 1949 British scientists reported the discovery that extenders containing glycerol could be used to freeze fowl semen. This discovery resulted when a laboratory technician mistook a bottle containing glycerol for a diluting-medium bottle. He subsequently observed that the addition of glycerol enhanced the resistance of spermatozoa to freezing. This has proved to be the most revolutionary procedure relative to artificial insemination in cattle that has been developed to date. It reduced the barriers of time, distance, selective matings, and semen losses (resulting from unused fresh semen) in artificial insemination.

Frozen semen is currently being used successfully in cattle, dogs, fish, goats, horses, sheep, and humans. Further research is needed before it can have practical significance in poultry and swine.

The principal method of storing frozen semen is in liquid nitrogen (N_2) at –320°F (–196°C). Liquid nitrogen is supercold. It boils at –320°F and turns into gas when released from pressure. It is the fourth-coldest known substance. Studies at Cornell University indicate that there is little or no decline in fertility of frozen bovine semen stored in liquid nitrogen for 2 years. Apparently after 2 to 3 years there is a slight loss of fertility with time.

One concern in freezing spermatozoa is preventing crystallization of cellular water. Therefore, an extender containing glycerol[14] is added to dehydrate the sperm cells prior to freezing and alter the formation of ice crystals, which would kill the cells. Bulls vary considerably in the ability of their sperm to withstand freezing and storage. A substantial amount of research effort is being expended to elucidate the reasons.

Current methods of freezing and thawing bovine semen recommend rapid rewarming, which results in better spermatozoa survival than does slow rewarming. Thawing frozen semen

---

[10] Ibid., pp. 113–114.
[11] Ibid., pp. 112–113.
[12] Ibid., pp. 114, 131.

[13] Ibid., pp. 117–136.
[14] Minnesota studies showed that more of the enzymes important to the sperm's gaining entrance into the sow's ova for conception were retained when glycerol was not used in freezing boar semen.

in ice water for 8 to 10 min before use is no longer recommended, especially for semen frozen in straws.

The quality of frozen–thawed semen, particularly viability, is closely related to the biological performance of that semen as measured by heterospermic insemination. *Heterospermic insemination* refers to the use of semen from more than one male in the insemination dose. Here semen is commonly mixed so that sperm numbers from each male are estimated to be equal. *Homospermic insemination* refers to the conventional use of only one male in each insemination dosage. Performance of the male is expressed as conception rate. Studies at the Virginia Polytechnic Institute and State University Agricultural Experiment Station and at several other stations show a clear relation between the heterospermic performance of males (i.e., the ability to sire offspring competitively) and their homospermic performance (i.e., the ability to sire offspring when used singly). Thus males that are superior homospermically are also superior heterospermically. In cattle the reported correlation between the logarithm of the heterospermic index and the homospermic index of stud bulls is positive and significant, $r = 0.69$.

Studies at the University of Minnesota have involved the collecting and freezing of spermatozoa from fish. Dr. E. F. Graham and his research associates achieved an 80 to 90 percent fertility rate with frozen and thawed fish spermatozoa. Subsequent research has involved the freezing of fertilized fish eggs. Perfection of this technique would free fish culturists from the necessity of scheduling their seasonal work around the natural spawning seasons. It could also facilitate the use of larger fish to stock lakes thereby giving a size advantage over natural predators and increase the rate of fish survival.

The world trend is toward freezing semen in the concentrated form. Several concentrations and techniques are being studied (Figure 14.5).

Ampule

This is a container with enough semen for only one service. For cattle each ampule commonly contains 0.5 to 1 ml of diluted semen. Glass ampules have been used extensively in artificial insemination. However, the trend is toward greater use of straws.

Pellet

The pelleting of semen was first reported by Japanese researchers in 1962. It is a concentrated method of storing semen but it presents problems in identification, automation, and sanitation. Pellets are thawed in physiological saline (0.9% NaCl).

Straws[15]

Maximum use of spermatozoa can be obtained through inseminating from the straw in which the semen was frozen. Plastic straws are being used extensively in the United States (more than 95 percent of all units of frozen semen are now marketed in straws). They are about 2.5 to 5 inches in length and commonly hold from 0.25 to 0.5 ml of semen. In comparison with glass ampules, plastic straws require less storage space, yield a higher recovery rate of motile sperm when thawed, and fewer sperm are lost during the insemination process (e.g., fewer sperm adhere to the container).

Rapid rates of thawing are beneficial to postthaw motility, acrosomal retention, and fertility of bovine spermatozoa packaged in plastic straws. Straws are commonly thawed in clean water at 90 to 95°F for 40 s.

Shell-Freezing

This method was developed in 1966 by Graham at the University of Minnesota. It results in a higher percentage of live sperm than was possible using previous processes. Graham also found a means for removing dead and abnormal sperm from semen. This is accomplished by passing semen through a glass-fiber filter. Eggs fertilized by abnormal cells usually abort prematurely. Therefore it is important that those sperm be removed. This technique increases the viability of spermatozoa and ensures a higher birthrate.

Lyophilization of Semen

This technique has become a challenge to scientists. Using a high vacuum, semen moisture is changed from a solid to a vapor, bypassing the liquid form (**sublimation**). Semen so

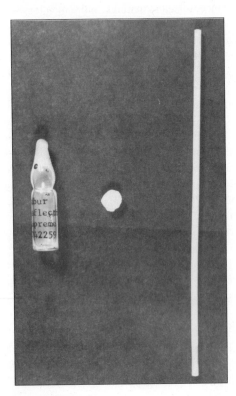

Figure 14.5 Methods of storing frozen semen (left to right): ampule, pellet, and straw.
Courtesy of Missouri Agr. Expt. Station.

[15] Herman, Mitchell, and Doak, p. 118.

TABLE 14.6	Summary of AI Regulations for Registry in Selected Beef-Breed Associations

Association	Blood-typing of bull required	Must be owner of bull	Nonowner certificate required	Use of semen after death of bull
American Angus Association	Yes	No	Yes	Yes
American Brahman Breeders Association	Yes	No	Yes	Yes
American Chianina Association	Yes	No	No	Yes
American Gelbvieh Association	Yes	No	Voluntary	Yes
American Hereford Association	Yes	No	Yes	Yes
American International Charolais Association	DNA	No	Yes	Yes
American Maine–Anjou Association	Yes	No	No	Yes
American Polled Hereford Association	Yes	No	Yes	Yes
American Salers Association	Yes	No	Yes	Yes
American Shorthorn Association	Yes	No	Yes	Yes
American Simmental Association	Yes	No	No	Yes
Beefmaster Breeders United	Yes	No	No	Yes
International Brangus Breeders Association	Yes	No	Yes	Yes
North American Limousin Foundation	Yes	No	No	Yes
Red Angus Association of America	Yes or DNA	No	Yes	Yes
Santa Gertrudis Breeders International	Yes	No	Yes	Yes

Source: "Artificial Insemination Regulations of Beef Breed Associations," Certified Semen Services, subsidiary of National Association of Animal Breeders, Columbia, Missouri, August 1997.

treated for 24 h may lose 90 to 95 percent of its moisture. However, only 5 to 15 percent motile sperm are recovered when the semen is placed in an extender for use. To date only one successful pregnancy has been reported from the use of **lyophilized** semen in the United States and other scientists have been unable to duplicate successfully the procedure reported.

On the basis of recent research studies conducted by scientists in the Department of Anatomy and Reproductive Biology at the University of Hawaii School of Medicine, Honolulu, and the Department of Veterinary Anatomy at the University of Tokyo, Japan, the following conclusions were reported in *Nature Biotechnology,* vol. 16 (July 1998), pages 639–641:

Freeze-dried mouse spermatozoa are all motionless and dead in the conventional sense. When injected into oocytes, however, their nuclei can support embryonic development even after three months' preservation in a dried state. Although the freeze-drying protocol reported here will need further improvement, the results suggest that it may be possible to store the male genomes at room temperature. (e-mail yana@hawaii.edu)

14.7.3 Custom Freezing[16]

For a fee many specialized bull studs collect, freeze, process, store, and dispense semen from bulls owned by private breeders. Such semen is generally used to service cattle belonging to the owner of the sire. This practice will likely expand in the future.

14.8 REGULATIONS GOVERNING ARTIFICIAL INSEMINATION IN CATTLE

The artificial insemination of purebred dairy cattle must be done in compliance with regulations developed by the Purebred Dairy Cattle Association (PDCA) (a federation of the principal dairy breed registry organizations in the United States) and Certified Semen Services (CSS) a wholly owned subsidiary of the National Association of Animal Breeders (NAAB). A summary of AI requirements of members of PDCA is available from CSS. All bulls from which semen is frozen, along with their living untyped parents, must be blood-typed before offspring from the use of such semen are eligible for registration. Blood-typing, as well as DNA typing, is a means of verifying bull identity.

Semen must be labeled to include the complete registration name and number of the sire from which it was collected. Other records relating to female ownership and identity may be required. Health standards for bulls producing semen for AI have been set forth by CSS.

Registration requirements of selected beef-breed associations relating to artificial insemination are summarized in Table 14.6.

14.9 ARTIFICIALLY INSEMINATING THE COW[17]

A cow usually remains in heat for 12 to 24 h and should be inseminated within 6 to 10 h following the first visible signs of standing estrus. Insemination equipment commonly includes

[16] Ibid., pp. 137–139.

[17] Ibid., pp. 143–160.

straw insemination guns, disposable plastic sheaths, and plastic gloves. Semen may be deposited in the vagina, cervix, uterine body, or uterine horns. Semen deposition deep in the uterine body is preferred on first service. Cervical semen deposition is preferred on repeat services because pregnancy is less likely to be interrupted in cows showing signs of estrus during pregnancy when semen is deposited in the cervix rather than in the uterus. Semen deposition in the uterine horn is undesirable because the nonovulating uterine horn might be selected or injury may occur. (Pregnancies occur more frequently in the right uterine horn of cows.) Deposition of semen in the vagina results in dilution, contamination, and lowered conception rate. Of course, when a bull inseminates a cow naturally, approximately 5 billion spermatozoa are deposited in the vagina. However, when semen is deposited artificially into the cervix considerably fewer sperm are required to achieve conception.

14.10 ARTIFICIAL INSEMINATION IN SWINE

The recent expansion of AI in swine has been closely related to changes in the structure of the U.S. swine industry. A litter farrowed in Minnesota in 1970 was apparently the first successful AI of a sow using boar semen preserved by freezing. In the mid-1980s it was estimated that in Iowa (state having the largest swine population) approximately 8 percent of sows and gilts were bred by AI. Meanwhile, a major production restructuring has occurred with large units becoming a dominant force. At the end of the twentieth century Purdue University scientists estimated that approximately 60 percent of swine were bred by AI. Many boar studs have been established in recent years to ship fresh cooled semen throughout the United States.

Although there are numerous AI studs collecting boar semen, many of the larger swine operations collect boars on-site. Compared with bovine AI it is much more common for boar semen to be used fresh and unfrozen. Lower conception rates and smaller litter sizes are economic limitations to the use of frozen boar semen.

14.11 ARTIFICIAL INSEMINATION IN HORSES

Use of AI in the horse industry has been impacted as much by industry regulation as by physiological limitations. Although the horse was reported to be the first recorded species to successfully utilize AI, and notwithstanding the fact that freezing technology for stallion semen was developed many years ago, widespread use was limited by breed association rules and restricted development of technology for extending, freezing, and storing stallion semen.

Until relatively recent years most equine breed associations would not permit the registry of a foal born from semen coming from an off-site stallion. But, as numbers of certain breeds of horses declined and concerns for the need to expand some breed registries became apparent, breed associations began accepting new rules related to use of AI.

Currently most breeds of light horses permit registry of foals born from AI, although there are differences in the autho-

rized use of frozen versus fresh semen. By contrast, registered Thoroughbreds must result from the natural mating (a "cover" mating) between stallion and mare, although fresh semen may then be used via AI to reinforce the mating.

14.12 ARTIFICIAL INSEMINATION IN POULTRY[18]

The primary objective of artificial insemination in poultry is to achieve better fertility, in contrast with the objective of disseminating better genetic characteristics in other species. It is especially beneficial to breeders of broad-breasted turkeys, which have low fertility in natural matings.[19] More than 90 percent of the turkey breeders in the United States use artificial insemination in conjunction with natural mating to increase fertility substantially. Semen is "milked" from trained **toms** into thermos bottles. About 0.3 cc of semen can be obtained twice weekly. The semen is pooled and commonly used within 4 h following collection because most results with stored turkey semen have been disappointing. The use of artificial insemination in the chicken hen, using semen from the cock, is increasing.

14.13 ARTIFICIAL INSEMINATION IN BEES

Bee breeders are using artificial insemination to develop and maintain superior strains of bees. Double-**hybrid** bees resulting from artificial insemination may produce up to 50 percent more honey than comparable commercial lines. Bees possessing greater longevity and activity and less sting have been developed through artificial insemination. Semen is taken from the everted penis after the **drone** (12 or more days of age) has been placed in chloroform fumes so that partial eversion is induced. Semen is then deposited in the oviducts via a special syringe. An experienced technician can inseminate 100 queen bees hourly.

Anesthesia of queens with CO_2 hastens the onset of egg laying and keeps queens quiet during instrumental insemination. If CO_2 were administered to an unmated queen she would begin laying eggs. Because such eggs would be unfertilized the resulting offspring would be drones. Then, by collecting semen from these drones and inseminating the virgin queen artificially, an inbred **generation** of bees could be obtained. It is possible for four sperm cells to penetrate an egg laid by the queen bee. Therefore bee geneticists have been able to transmit eye characteristics of four different males in new bees.

When antibiotics are added to bee semen it remains capable of fertilization for 16 to 18 weeks and has been shipped from the United States to Europe and Latin America.

---

[18] Ibid., pp. 315–316.

[19] Research at the University of Illinois showed that roosters fed caffeine became sterile in about 3 weeks. Apparently, libido was unaffected by ingesting caffeine, which is believed to inhibit spermatogenesis. The Illinois studies showed that when roosters were fed caffeine for 63 days (three cycles) and then had the caffeine withdrawn from their diets, sperm production was resumed in about 3 weeks.

14.14 ARTIFICIAL INSEMINATION IN HUMANS

It is every woman's heritage to bear children.

Jane Seymour (1509–1537)
Third wife of Henry VIII of England
and mother of Edward VI

In the Bible infertility is looked upon as a divine punishment (Genesis 30:1–2). The barren marriage is a problem nearly as old as humanity. In some countries a barren woman met the fate of losing her life; in others she was returned to her parents. Through the years the onus of **barrenness** commonly fell on the female. More recently people have learned to appreciate that childless marriages can be attributed to the male, female, or both. About 10 to 15 percent of all marriages are childless for reasons of infertility.

According to a 1977 publication of the American College of Obstetricians and Gynecologists entitled "Infertility Causes and Treatments," both the man and the woman have problems contributing to infertility in 15 to 20 percent of all infertile couples, whereas 35 to 40 percent of the cases of infertility are due to problems in the man and another 35 to 40 percent result from problems in the woman. In the balance of cases no known cause can be determined.

Pregnancy by Proxy

The first recorded case of pregnancy by proxy was biblical. Abraham's barren wife, Sarah, sent Abraham to mate with Hagar, her handmaiden, who bore him Ishmael. And notwithstanding all the emotional stresses and strains such arrangements may cause, childless couples have been quietly making them ever since. In recent years, however, AID has removed at least some of the sting from planned surrogate motherhood.

In 1785 J. Hunter recommended that the technique AIH be employed in a case where the husband (an English merchant) had a physical deformity preventing normal **coitus.** Results were successful. *Medical World* published its first discussion of AID in April 1909.

An estimated 50,000 children are born annually in the United States through artificial insemination (AIH and AID). There are legal, social, psychological, ethical, moral, theological, and cultural problems to be considered before AID is chosen.

14.14.1 Situations Encouraging AIH (AID)

Several factors might prompt a couple to consider artificial insemination: (1) The husband is infertile, and the couple prefers AID (or pooled semen) to an adoption. The woman is fertile by all standards of investigation. (2) The doctor may recommend AIH as a means of increasing conception. Several ejaculates of the husband's semen are centrifuged to concentrate the spermatozoa. The couple should have attempted spontaneous conception for at least 18 months prior to AIH. Multiple ejaculates from oligospermic (low sperm count) husbands can be collected, frozen, stored over a period of several months and then thawed, pooled, concentrated by centrifugation, and inseminated (AIH) at spermatozoa concentrations many times greater than that of the original single ejaculates. This technique has proved successful when a reduced number of mobile spermatozoa was the primary cause of male infertility. Spermatozoa normally constitute about 5 percent of the volume of the human ejaculate. The normal concentration of spermatozoa in human semen varies between 50 and 150 million per ml (Table 14.5). (3) There may be an Rh incompatibility, the husband being **homozygous** positive and a series of pregnancies having been lost. (4) A somewhat rare situation of inborn errors of metabolism (transferred by recessive genes), which results in a condition such as phenylketonuria, or faulty metabolism of phenylalanine, may encourage the use of AID. (5) Social pressure may be involved and the couple may feel inadequate in relation to their neighbors, family, and friends unless they bear children. (6) Stored semen collected prior to male sterilization (**vasectomy**) and/or chemotherapy may be used through AIH to allow family planning to be continued.

The disclosure that a California businessperson set up an exclusive semen bank named Repository for Germinal Choice, which offers the sperm of Nobel Prize winners to carefully selected women, rekindled the scientific and moral controversy sparked by advocates of human genetic engineering movements even before World War II. One sperm contributor, Dr. William B. Shockley, a 1956 physics laureate, said, "I welcome this opportunity—and endorse the concept of increasing the people at the top of the population."

14.14.2 Donor Characteristics

It is essential that the donor remain anonymous. He must be fertile, free of disease, of aptitude equal to or above that of the couple in consideration, of similar physical proportions to the husband, and compatible with the Rh genotype of the wife. It is a common practice to mix donor semen with that of the husband.

14.14.3 Legal Agreements

For the protection of the doctor, husband, and wife certain statements must be signed and witnessed. These statements commonly include an expressed desire for AIH (AID) by both husband and wife, an acceptance of the resulting child by the husband as a type of semiadoption, and the exclusion of adultery, rape, or any other involvement consideration that might be considered harmful to marital happiness. The husband is asked to be present when his wife receives an artificial insemination.

14.14.4 Freezing and Thawing of Human Semen

Both foreign and domestic researchers have successfully frozen human semen with subsequent **impregnations.** Glycerol is commonly used to protect the spermatozoa during freezing after which they are stored in liquid nitrogen. Agents used to increase sperm recovery after thawing include human follicular fluid, human milk, human blood serum, and egg yolk.

There were about 2600 births reported worldwide from the use of frozen human semen in 1998. The longest reported period that frozen human sperm have been preserved is 10.5 years.

14.14.5 Timing of AIH (AID)

It is impossible to perceive precisely the time of **ovulation** in women, whereas signs accompanying estrus and ovulation are visible in most domestic animals. Therefore certain measures are often used to assist the detection of ovulation, which in turn increase the likelihood of conception. The **Papanicolaou stain,** basal body temperature charts, and formation of a **spinnbarkeit,** when interpreted by a competent **cytologist,** provide the most practical and accurate indications of ovulation. Most ovulations occur from day 9 to day 21 of the cycle with about one-third occurring on day 14.

It was observed in 1867 by W. Squire in Great Britain that basal body temperatures (BBT) could be correlated with changes in the menstrual cycle of women. The concept that the biphasic response was related to ovulation was introduced by T. H. Van de Velde of The Netherlands in 1904. He later suggested that the temperature elevation observed in the second half of the cycle was caused by progesterone. In 1974 F. P. Zuspan and P. Rao correlated urinary catecholamine excretion with BBT patterns suggesting that hypothalamic thermoregulatory control is mediated by norepinephrine. Several researchers have proposed that BBT can be used either as an indicator of ovulation or to provide a means of fertility control. The consensus is that the procedure is useful clinically but that the current state of the art does not render it a reliable method of fertility regulation.

At the time of ovulation in women the cervical mucus becomes clearer, more watery, and less tenacious. Its "stringiness" and acidity can be measured.

Blood plasma progesterone—the hormone produced by the ovary (corpus luteum) following ovulation—can also be measured to determine if ovulation has occurred in women.

14.15 THE NATIONAL ASSOCIATION OF ANIMAL BREEDERS, INC.

This nonprofit organization was founded in 1946 and has its national offices in Columbia, Missouri. The basic objective is to effect mass livestock improvement through artificial insemination. It works closely with breed associations, AI studs, and the USDA–APHIS Veterinary Services to establish and publish codes for animal identification,[20] health, and other rules governing the use of artificial insemination. Nearly all organizations engaged in the commercial operation of AI businesses in the United States are members of NAAB.

14.15.1 Certified Semen Services, Inc. (CSS)[21]

This organization was formed in 1976 as a subsidiary of the NAAB. Initially CSS provided a semen identification auditing program involving an annual on-site visit by a CSS representative. During this visit procedures and records relating to the iden-

tification of semen—from the time a bull is bought or custom collected through processing and distribution—are reviewed.

Subsequently the CSS program was expanded to include sire health. The CSS Minimum Health Requirements outline specific testing procedures for bulls and mount animals before they enter isolation, during isolation, and for bulls housed in a central location following isolation. Additionally they outline a protocol for treating neat (raw) semen and tested/approved extenders with Gentamicin, Tylosin, Linco-Spectin (GTLS), the combination of antibiotics used to control specific microorganisms. Specific diseases tested for in the CSS Sire Health Program are brucellosis, leptospirosis (five serotypes), BVD virus, trichomoniasis, tuberculosis, and campylobacteriosis. Health tests related to the above diseases are conducted in accordance with procedures described in the "Recommended Uniform Diagnostic Procedures for Qualifying Bulls for the Production of Semen" published by the American Association of Veterinary Laboratory Diagnosticians. In the CSS program semen packages containing semen from bulls meeting the CSS Minimum Health Requirements are imprinted with the CSS logo (**CSS**®), which signifies CSS Health Certified Semen.

14.16 THE FUTURE OF ARTIFICIAL INSEMINATION

Few advances in science have had the impact that artificial insemination has had on animal breeding. During the past seven decades artificial insemination has advanced from an unfamiliar term to a well-accepted practice in the mating of domestic animals throughout the world. Yet the potential that artificial insemination presents as a means of improving animals on a mass basis remains virtually untapped in many species. Lack of communications, roads, transportation, and other factors impede the increased use of AI in developing countries.

Estrous synchronization among cattle, sheep, and swine is now practical on a large scale (Section 13.5.1). With the synchronization of estrous cycles in many females it will be feasible to artificially inseminate large numbers of females with fresh, as well as stored, semen from outstanding males in a relatively short period of time. This would be impossible through natural mating.

Estrous Synchronization and AI[22,23]

Artificial insemination has received wide acceptance and use in the dairy industry and its use is expanding in the beef industry, largely because of management constraints. The availability of prostaglandin $F_{2\alpha}$ and other synchronizing compounds for regulating ovulation in cattle has resulted in a slight increase in the number of beef cows inseminated artificially.

Unlike dairy producers, beef producers derive the major portion of their income from the calf crop, thereby making fertility a

---

[20] As part of positive identification, blood-typing of a bull to be used in AI and of his living parents that have not been blood-typed is required by breed associations.
[21] Herman, Mitchell, and Doak, pp. 247–258.

[22] Ibid., pp. 168–173.
[23] "Improving Reproductive Performance," Symposium in Roanoke, Virginia, sponsored by the National Association of Animal Breeders, June 1999; and D. Patterson, J. Whittier, and A. Williams, "Getting in Synch," *ABS Breeders Journal,* Spring 2000.

highly important consideration. Results of a recent study placed the economic importance of (1) fertility, (2) growth rate, and (3) carcass quality in the ratio of 10:2:1. This indicates that to beef producers fertility may be 5 times more important economically than growth and 10 times more important than carcass quality.

Progeny testing is well established in dairy cattle breeding programs. Its use is expanding rapidly in the breeding of beef cattle, sheep, and swine. Swine testing stations, for example, have identified boars capable of siring lean, meaty, well-muscled pigs that gain rapidly and use feed efficiently. Performance-tested boars can be mated with 2000 or more sows annually through AI, whereas by natural service a given boar may sire only about 80 litters per year.[24]

Progeny testing will become even more important as a means of reinforcing the merits of AI in **domestic** animals. Breeders using progeny-performance data to select and breed improved livestock find their males in demand by AI organizations. Computer matings have become a reality. Frozen semen will make germ plasm from outstanding males available to purebred and commercial breeders throughout the world.

Through fundamental research additional advances in the physiology of reproduction will be forthcoming. Among these advances will be measures and techniques giving an increased impetus to AI.

Genetic Engineering and AI

With genetic engineering techniques for transferring specific genes within and among species, it may be possible to transfer the gene responsible for twinning in sheep to cattle and thereby greatly increase the reproductive performance of cattle. Equally as exciting is the possibility of transferring genes in the armadillo that control cleavage of the ovum to produce identical quadruplets. It may also be possible to use genetic engineering techniques to transfer the genes that regulate the synthesis of the hormones that control growth and metabolic processes. Once these genes are transferred and become permanently fixed in the recipient animal, they can be spread quickly and widely throughout the population through the use of AI.

Sperm-Typing for Sex Preselection[25]

Research indicates that it is possible to separate X and Y sperm cells using a laminal flow method of sperm fractionation.

The X sperm, which carry the larger of the two sex chromosomes, swim in a more inclined path away from the flow axis and fractionate at the bottom of the flow cell, whereas the lighter Y sperm are enriched at the top fraction.

The potential merits of sex preselection techniques in livestock and poultry are indeed immense and could have a tremendously favorable impact on animal food production and efficiency in both developed and developing countries. It is

entirely possible that in the near future scientists will use genetic engineering techniques to isolate or synthesize genes controlling desirable production characteristics of farm animals in the laboratory. These genes could then be transplanted into Y sperm and transmitted via AI, along with the usual set of sperm chromosomes, to impregnate vast numbers of females. Once the desirable genes became established in the germ line of a breeding animal, they would have tremendous potential for propagation.

Milk Progesterone and Pregnancy of Cattle[26]

It is important to determine pregnancy early. One means of doing this is assay for milk progesterone. Concentrations of estrogen and progesterone are inversely related in the peripheral circulation of cattle. Estrogen levels are highest during the follicular phase, whereas progesterone is elevated during the luteal phase. Measurement of either hormone will provide a reflection of ovarian status. Morphological changes in the corpus luteum (CL) are accompanied by fluctuations in progestin concentrations of peripheral blood. If pregnancy occurs the CL persists, thus preventing cyclicity. By day 19 after breeding, plasma progesterone concentrations are significantly higher in pregnant animals. Cyclic changes in progesterone that occur in blood plasma also occur in milk; thus the testing of milk for progesterone 19 to 25 days after insemination is 80 to 90 percent accurate in diagnosing pregnancy of lactating cows, and 95 to 100 percent accurate in detecting animals not pregnant. Milk progesterone as a pregnancy test was offered commercially in Great Britain in 1975 and it is now available through several laboratories in the United States. Radioimmunoassay is the procedure commonly employed to measure milk progesterone.

Detecting Estrus in Cattle[27]

The detection of estrus is extremely important in conjunction with the use of AI. The following two approaches recently reported in cattle are of interest.

In USDA studies utilizing dogs previously trained to detect a characteristic odor in the cow's reproductive tract at estrus, it was observed that the estrous odor emerges slowly during the 3 days before estrus, reaches a definite peak in intensity on the day of estrus, and disappears within 1 day thereafter. The dogs were more than 80 percent accurate in detecting estrus in cows.

Canadian scientists at the University of Guelph have developed a technique for detecting estrus in cows with 95 percent accuracy. It is based on the use of a mercury gauge to record the frequency of movements of the cow's neck.

Reproductive Physiology Research and AI

Additional research is needed to improve present and develop new techniques of heat (estrous) detection, ovulation detection, breeding or insemination per se (e.g., Where is the best place and when is the best time to deposit semen in the female reproductive tract?), early pregnancy detection, evaluating and predicting the

---

[24] Studies at The Ohio State University indicate that litter size may be increased in gilts (10.4 as opposed to 8.4 pigs per litter in the Ohio studies) by intrauterine treatment with boar semen 3 weeks before insemination.
[25] G. E. Seidel and L. A. Johnson, "Sexing Mammalian Sperm—Overview," *Theriogenology* 52 (1999):1267–1272.

[26] Herman, Mitchell, and Doak, pp. 185–195.
[27] Ibid., pp. 144–149.

fertility of males, perfecting fertility matching in animals (e.g., Are certain bulls more effective in achieving conception in certain cows?), contact capsule insemination (e.g., Could a female be inseminated at any time during the estrous cycle with a fertile time-released dose of semen?), fertility prediction (e.g., Is it possible to identify, and thereby cull, animals with important reproductive problems, both males and females, at an early age?), sexing semen, improving techniques for freezing embryos, and subsequent embryo transfer (see also Chapter 13).

14.17 SUMMARY

Artificial insemination was used in mating horses and dogs at least two centuries ago. However, its commercial application for mating farm animals was not made practical until improved communications, refrigeration, transportation, and especially advances in reproductive physiology research paved the way.

Artificial insemination has helped facilitate tremendous improvements in the genetic merit of farm animals during the past seven decades. The use of AI enables the animal breeder to assist the best in multiplying the fastest.

Largely because of the sizable number of females that can be mated to outstanding males, many livestock breeders use AI as a tool in improving animals. They thereby increase profits and provide consumers with more acceptable animal products and at reduced costs. In poultry the use of AI increases producer profits through improved fertility; in horses it offers a means of mating females to outstanding stallions that are often great distances away.

In the effort to provide more meat and milk for humans, the use of AI to spread superior genetic germ plasm holds great opportunities, especially in light of the fact that only about 5 percent of the cattle of developing countries (in comparison with more than 40 percent of the cattle of developed nations) are artificially inseminated.

STUDY QUESTIONS

1. Distinguish between AID and AIH.

2. Who first observed human spermatozoa? When?

3. What contributions to the progress of artificial insemination did Spallanzani make?

4. What approximate percentage of dairy and beef cows in the United States are inseminated artificially?

5. Name 10 species in which artificial insemination is now used.

6. Which state agricultural experiment stations pioneered research relating to artificial insemination?

7. What are some merits of artificial insemination of farm animals?

8. What are some limitations of artificial insemination of farm animals?

9. Name four methods of collecting semen, and briefly list the advantages and disadvantages of each.

10. What is the best indication of fertility in a male?

11. Why is the enumeration of spermatozoa important?

12. Of what use is a morphological study of semen?

13. What is the scientific basis of the dead–alive stain of spermatozoa?

14. How manyfold can the average ejaculate of bovine semen be extended? Boar semen? Stallion and cock semens?

15. Describe a good semen extender (diluter).

16. Name several semen extenders. Which ones are preferred?

17. Why is semen often color-coded for use in artificial insemination?

18. Can fresh stallion semen be stored for longer or shorter periods than bull semen?

19. What is the principal method of storing frozen semen?

20. Name three ways of storing frozen semen in the concentrated form.

21. What is lyophilized semen? Is the technique practical?

22. Do all beef breeds allow semen to be used indefinitely after a bull is dead?

23. Discuss the purposes of the organization Certified Semen Services, Inc.

24. What is the primary objective of artificial insemination in poultry? Is it the same in cattle?

25. Describe situations that might prompt a couple to consider AID.

26. Why is a legal agreement important in AID?

27. Can human semen be frozen successfully?

28. What is the relation between estrous synchronization and artificial insemination?

29. Discuss the relation between artificial insemination and genetic engineering.

30. What are the implications of sperm-typing for sex preselection?

31. Why is the detection of estrus in farm mammals important to the increased use and success of artificial insemination? Discuss two or more recently reported techniques of detecting estrus in cows.

32. What are some areas of reproductive physiology in which additional research is needed to more fully realize the potential of artificial insemination in food-producing animals?

33. What, in the opinion of the authors, does the future hold for artificial insemination of farm animals?

PHYSIOLOGY OF LACTATION[1]

Thy breasts shall be as clusters of the vine.

The Song of Solomon 7:8

15.1 INTRODUCTION

During the past century science has exposed many secrets related to the **physiology** of **lactation.** There are about 5000 species of animals with true mammary **glands** that perform activities vital to the preservation of **mammalian** life. Humans and elephants have only two mammary glands, which are located in the **pectoral** (chest) region. There are four mammary glands in the **inguinal** (groin) area of the mare; however, they are served by only two teats, each containing two **ducts** (one per gland). Sheep and goats possess two inguinal mammary glands, each with one teat. The cow has four glands, which terminate in four teats. **Litter**-bearing species (**sow, bitch,** and other **multiparous** animals) often have 10 to 18 mammary glands along the abdominal wall from the pectoral to the inguinal region.

Monotremes (duck-billed platypus, echidna) are egg-laying mammals that have mammary glands contained in an abdominal pouch and which consist of a large number of small cutaneous glands without a nipple. Newly hatched monotremes obtain nourishment by licking the region of the maternal abdomen in which these glands are situated. Mammary glands of women and bitches have several ducts opening on the teat surface; those of cows, goats, and sheep have a single opening. Teats of the sow and mare are traversed by two streak canals. A woman's breast consists of some 20 distinct lobes. Each lobe is a distinct gland with a **lactiferous,** or milk, sinus opening by a separate **orifice** onto the summit of the **mammilla.** There are no cisterns in the woman, sow, or bitch; therefore, milk ducts or sinuses open directly to the surface of the nipple. The capacity of their mammary glands is increased by a dilation of the milk ducts.

The highly developed mammary gland of the cow is the sole basis for the existence of the dairy industry and contributes immeasurably to the health and **nutrition** of humans. It is no wonder the **bovine udder** has appropriately been called the world's greatest food factory. Because lactating cows produce more than 90 percent of the milk supply of the United States, discussion in this chapter will be primarily related to the physiology of lactation in the bovine species.

> The cow is the foster mother of the human race. From the day of the ancient Hindoo to this time have the thoughts of men turned to this kindly and beneficient creature as one of the chief sustaining forces of human life.
>
> **William Dempster Hoard (1836–1918)**

15.2 MAMMARY GLAND DEFINED

Mammary glands are one of the major features that distinguish mammals from other kinds of animals. Mammals derived their name from the Latin word *mamma,* which means "breast." Because the mammary gland secretes outward from the body it is an *exocrine gland.* The mammary gland also generally is called a *merocrine gland,* even though parts of the cell may be secreted with the milk components.

[1]The authors acknowledge with appreciation the contributions to this chapter of Dr. Walter L. Hurley, Department of Animal Sciences, University of Illinois at Urbana–Champaign.

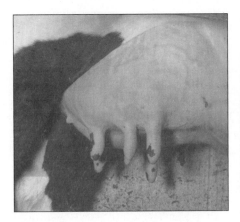
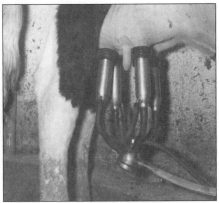

Figure 15.1 Although almost all cows have only four functional mammary glands, the above photographs depict a cow with six functional glands.
Photographs by John R. Campbell.

15.3 ANATOMY AND ARCHITECTURE OF MAMMARY GLANDS

The four mammary glands of the cow[2] are collectively called the *udder* which, exclusive of milk, commonly weighs about 30 to 40 lb (varying from 10 to 120 lb). A high-producing cow may accumulate 60 to 90 lb of milk in her udder prior to milking. (Only about 80 percent is normally removed at milking.) A desirable udder should be capacious, possess a relatively level floor, and be strongly attached. A high-quality udder should contain a maximum amount of secretory (glandular), a minimum amount of connective, and essentially no fatty tissue. The connective tissue might be likened to the walls, floor, and ceiling of a large house. The secreting cells fill the rooms. After milking, the normal high-quality udder feels soft and pliable with no apparent lumps or knots, which would be indicative of connective tissue resulting from injury or **disease.** However, it is impossible to predict accurately how productive an udder is by visual inspection or by its size.

The principal supporting structures of the udder are the **median** and **lateral** suspensory **ligaments** (Figure 15.2). The median suspensory ligaments contain elastic fibers that are stretched laterally and vertically as the udder fills with milk; this stretching allows the squarely placed, perpendicularly hanging teats to protrude obliquely outward and downward. Probably less than 50 percent of the milk secreted can be stored in the natural storage areas of the udder. The balance is accommodated only through stretching of the udder. High milk production may cause these ligaments to become permanently lengthened and may result in a "breaking away" of the udder from the body causing a pendulous udder. This condition may be partially alleviated by affixing udder supports (Figure 15.3). The udder is not directly connected with the abdominal cavity except through the inguinal canal.

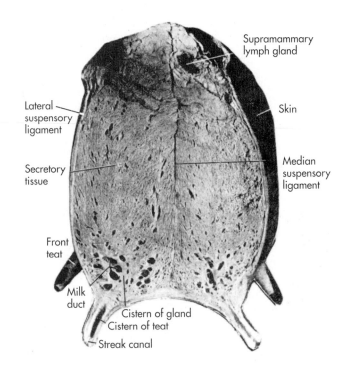

Figure 15.2 Cross section of the rear quarters of a cow's udder.
Courtesy of Missouri Agr. Expt. Sta. Bull, 344 (see Fig. 15.5 legend).

15.3.1 Internal Structure

The udder is divided into right and left halves by a heavy membrane. Front and rear quarters are separated and normally represent about 40 and 60 percent, respectively, of the secreting capacity. The milk from each quarter can be removed only from the teat of that quarter. Infusion of different-colored stains into the teats of a cow will show a clear division between fore and rear quarters when the udder is surgically sectioned.

At the end of each teat is the opening (streak canal, or meatus) through which milk is removed (Figure 15.2). It is largely the strength (tone) of the **sphincter** muscles surrounding the streak canal that prevents milk from flowing out and determines

---

[2]As with many biological phenomena in nature, there are exceptions in the number of mammary glands of cows (Figure 15.1).

Figure 15.3 Dairy cows on a farm near Copenhagen. Note the harnesslike device (cow brassiere) attached to minimize the stress given to the udder attachments of the high-producing cow in the foreground.
Photograph by John R. Campbell.

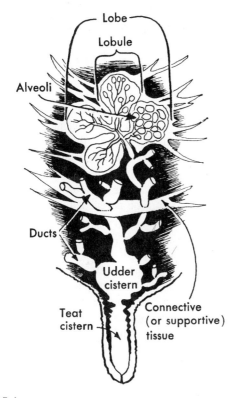

Figure 15.4 Each mammary gland in the cow consists of many lobes, each of which contains smaller lobules, which in turn contain many alveoli.
C. W. Turner, Harvesting Your Milk Crop, 3d ed., Babson Bros., Chicago, 1973.

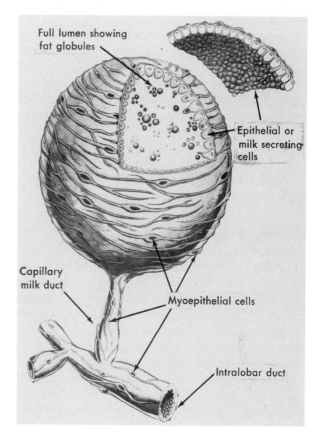

Figure 15.5 An alveolus where milk is secreted.
Courtesy of Missouri Agr. Expt. Sta. Bull, 793.

the ease or difficulty with which milk may be withdrawn. Weak and incompetent sphincters that fail to exercise control of the closing of the streak canal contribute to a condition called **patency** and increase susceptibility to **mastitis.**

The teat widens into its cistern and in turn into the gland cistern (Figures 15.2 and 15.4). There are 10 to 20 large ducts (**galactophores**) branching into all parts of the gland to collect and convey milk downward. Each gland is divided into lobes separated by connective tissue membranes. The lobes are each

drained by a single duct and further branch into lobules (sacs), which are drained by ducts (tubes). The lobules resemble bunches of grapes and are composed of a large number of hollow spherical structures called *alveoli.* Milk constituents are formed in the epithelial cells lining each alveolus and secreted into its **lumen** (Figures 15.4 and 15.5).

There are millions of alveoli in high-producing, efficient mammary glands. An alveolus is of microscopic size (approximately 60,000 per cubic centimeter) and varies from 0.1 to 0.4 millimeter (mm) in diameter. Its single layer of epithelial cells varies from 0.007 to 0.010 mm in height depending on its milk-secreting status. Surrounding the epithelium of an alveolus are groups of myoepithelial cells. These cells stretch out as milk is secreted, and when a cow is properly stimulated to let down her milk they contract, forcing out the milk of the alveolar lumen (milk letdown, Section 15.8).

15.3.2 The Circulatory System

The blood supply to the udder during lactation is profuse. There are 300 to 500 volumes of blood passing through the udder for each volume of milk secreted. Therefore the udder must possess an extensive **vascular** bed. Arterial blood enters the top of each half of the udder through an external **pudic** artery (via the inguinal canal). These branch into the **cranial** (supplies forequarters) and the **caudal** (supplies rear quarters) arteries. There are three primary routes for venous blood to return to the heart for oxygenation. The first route is the external pudic vein (parallels the external pudic artery). The second route is through the **subcutaneous** abdominal vein (milk vein). The third possible route is the perineal vein through the pelvic arch. Some have attempted to use the size and prominence of the milk veins (superficial veining) as an index of milking ability; however, the milk veins have been **ligated** experimentally without appreciably affecting milk production.

Surrounding each alveolus is a network of tiny capillaries carrying blood to the base of the epithelial cells that line the alveolus. The materials in blood from which milk is made pass through the capillaries and are taken up by the milk-making cells.

15.3.3 The Lymphatic System

The lymphatic system is composed of lymphatic vessels that carry **lymph** from mammary tissues to the heart via the venous blood system. In the cow the flow is upward and toward the rear of each half of the udder where it passes through the **supramammary lymph nodes**. The nodes act as filters that remove or destroy foreign substances (bacteria) that may have gained entrance into the tissues. It is through the effective action of lymph nodes pouring leukocytes (white blood cells) into the blood vascular system that infections are kept localized and invading microorganisms are destroyed. These lymph nodes and vessels are especially important in controlling **inflammation** at **parturition** and in removing sloughed or injured tissue. Congestion (swelling) in the udder at the time of parturition results from the accumulation of large quantities of lymph and is called **edema** (Figure 15.6 a and b).

The lymph capillaries surround the alveoli and absorb part of the **plasma** and products returning from the **intercellular** spaces into which they have flowed from the secreting cells.

Edema in the cow's udder tends to inhibit milk ejection. This may be a built-in safeguard of nature that minimizes the intensity of early postpartum milk secretion until the parathyroid gland can become fully functional in its regulation of blood calcium and phosphorus. This may help to prevent **parturient paresis** (Section 15.5.4) in high-producing animals.

15.4 GROWTH AND DEVELOPMENT OF MAMMARY GLANDS

Mammary glands of the bovine female begin developing in early **fetal** life. The teats are apparent at birth. The teat cisterns and gland cisterns are also present. As the young immature female grows, udder size increases in proportion to the increase in body size. Prior to **puberty** there is little duct development or growth of glandular (secretory) tissue. When the **heifer** reaches sexual maturity, estrogen (produced by the follicles on the **ovaries**) stimulates development of the large duct system. With each recurring estrous cycle the glandular tissue is stimulated to grow rapidly. After heifers have passed through a number of estrous cycles the ducts show substantial branching into the fat pad of the udder (Figure 15.7).

Normally there is little or no lobule–alveolar growth prior to pregnancy. High blood concentrations of both estrogen and progesterone are needed to stimulate the lobule–alveolar development characteristic of mammary growth and development during pregnancy. However, research indicates that some lobule–alveolar development can be induced in nonpregnant females. When **ovulation** occurs the follicle quickly develops into a corpus luteum and secretes progesterone. In this case virgin heifers can be stimulated to secrete milk by the use of only estradiol injections during the luteal phase of the estrous cycle. Mammary growth also can be induced in nonlactating, nonpregnant animals by injections of a combination of estradiol and progesterone. Hormonal stimulation of mammary growth and induction of lactation can be successful in virgin heifers or in nonlactating mature cows.

The **morphogenesis** of the mammary gland does not terminate when the adult stage is reached because pregnancy is essential for the development of its full functional potential.

During the first pregnancy growth of the udder becomes continuous rather than cyclic as observed during recurrent estrous cycles. Progesterone causes rapid development of the fine ducts and, in conjunction with the high blood concentrations of estrogen during pregnancy, causes development of the lobule–alveolar system. During the first and second trimesters of pregnancy the fine ducts and alveoli grow (**proliferate**) and replace fatty (**adipose**) tissue in the udder, giving little external

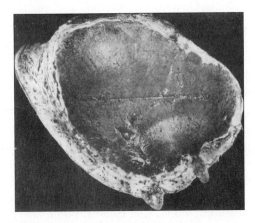

Figure 15.6a Cross section of an edematous cow's udder at parturition. Note the thickness of the subcutaneous area.
C. W. Turner, The Mammary Gland, *Lucas Brothers Publishers, Columbia, MO, 1952.*

Figure 15.6b An udder showing edema at parturition. Note the accumulation of lymph fluids (apparent swelling) forward of the udder.
Courtesy of Missouri Agr. Expt. Station.

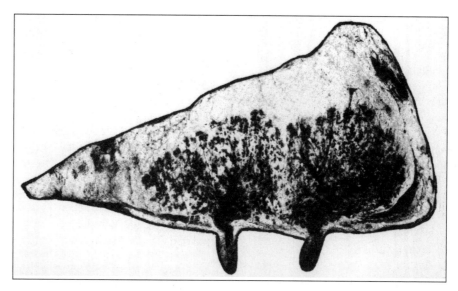

Figure 15.7 A section of the bovine udder of a sexually mature heifer. Note the growth of the duct system and early lobule–alveolar development. Considerable adipose tissue remains into which the duct and lobule–alveolar system have not yet penetrated. *Courtesy of* Missouri Agr. Expt. Sta. Bull. *793.*

visual evidence of growth. At this time milk is not produced and those alveoli which have been formed remain collapsed. During late pregnancy (last **trimester**) the epithelial cells lining the alveoli and fine-duct system begin to function and secrete **colostrum,** particularly during the final weeks before parturition. The colostrum is secreted into the lumen of the alveolus and into the duct system. This accumulation of colostrum in the alveoli in late pregnancy accounts in part for the apparent rapid enlargement of the udder prior to parturition. Mammary gland growth in the cow continues through the third trimester of pregnancy and during early lactation.

In some species such as rodents and swine, the total amount of mammary gland **DNA** can double from parturition to the time of peak lactation. The amount of DNA in a gland is an indicator of the number of cells in that gland. Measuring the amount of DNA in a tissue after application of experimental treatments to the animal can provide a comparison of how much a treatment enhances cell division and tissue growth. It is less clear how much the mammary gland of the dairy cow grows during early lactation but the extent of growth would be substantially less than in a rat or sow. Mammary size in dairy cattle is stimulated further with each subsequent pregnancy (usually through about the fifth pregnancy).

In humans the breasts of the newborn (male and female) may be observed to express a small amount of milklike fluid commonly called "witch's milk" (described by Aristotle). This phenomenon results from the influence of maternal **hormones** on fetal breasts. These hormones circulate in the blood and apparently transfer freely across the placenta to enter the fetal circulation.

At puberty the human female breast begins to grow rapidly.[3] Within about 2 years after the onset of menstruation the breasts have about reached their full growth. (The weight of each gland varies from 140 to 300 g or more.) At this time there is growth and branching of the duct system but little **maturation** of glandular tissue. The breasts undergo cyclic changes associated with menstruation. Just prior to menstruation the breasts become firmer and somewhat enlarged and may become tender. The soreness women often experience in their mammary glands at each menstrual period is due perhaps to an increased blood supply.

The most pronounced alterations in the breasts occur during pregnancy. Duct development occurs rapidly during the first few months of pregnancy. Later the glandular tissues develop rapidly and begin to swell and secrete milk. The nipple and an area of skin surrounding it (areola) are darkly pigmented. The pigmentation deepens in the early months of pregnancy and only partially returns to its virgin color after **gestation.**

There is a steady **retrogression** of the mammary glands following menopause, which marks the end of reproductive life and diminution of the ovarian secretion.

Hope springs eternal in the human breast.
Alexander Pope (1688–1744)
English poet

The number of women breast-feeding their babies in the United States decreased from approximately one-half in 1936 to about one-fourth in 1974. However, the trend has substantially reversed. By 1998 it was estimated that nearly 60 percent of all American women breast-fed their children.[4] It is conservatively

---

[3]Puberty in humans occurs late in the growth period, when approximately two-thirds of mature weight is achieved, whereas in most other species of mammals it occurs when about one-third of mature weight is achieved. The mechanism, probably endocrine in nature, involved in this unique delay of puberty in humans is of considerable interest.
[4]Interestingly, if the mammary glands of women become functionally active in secreting milk before 28 years of age, they have, on the average, 90 percent less breast cancer than those who have not lactated.

estimated that 85 percent of all women have the capacity to breast-feed their infants 6 months or longer. In general, most mammals are breast-fed until they have tripled their birth weight, which in human infants occurs at about 1 year of age.

Six decades ago Chinese and Japanese mothers nursed their children up to 5 to 6 years of age, Caroline Islanders up to 10 years, and Eskimos up to 15 years. One reason for the longer nursing period was to extend the natural spacing of children. But today women have a number of means of conception control, which brings us to the question, "What effect do birth control pills have on lactation?" Reports indicate that when the high-estrogen pill is taken soon after the baby's birth, suppression of lactation may be 50 percent or more. However, more recent low-estrogen contraceptive pills have little or no effect on the ability to breast-feed.

15.4.1 Milk without Motherhood

Stilbestrol (a synthetic estrogen) has been used to initiate lactation. In one Missouri study, an ointment containing 1 percent stilbestrol was applied three times weekly to the mammary glands of a virgin goat. After 30 days (and a rapid growth of the mammary duct system comparable to that caused by natural estrogen) the secretion of milk increased to 1500 ml daily.

More recently use has been made of the hormones estrogen and progesterone to grow and develop mammary glands of virgin bovine females to such an extent that they have been made to lactate. Thus it is possible to artificially stimulate mammary gland growth and to initiate lactation in nonpregnant females. By stimulating lactation in infertile heifers and cows, dairy farmers may be able to obtain high milk production in the future from animals that might otherwise be slaughtered. Maximizing milk yield by hormonal induction of lactation is a current area of research.

15.5 Hormonal Regulation of Lactation

Mammary gland physiology is closely related to hormonal and neurohormonal mechanisms. The mammary gland is a secondary sex characteristic of which the development, initiation, and maintenance of activity, and finally its **involution,** are dependent on hormonal balances. A number of hormones influence the growth of the gland, initiation of lactation, and maintenance of milk production (Figures 15.8 and 15.9).

15.5.1 Prolactin (Lactogen)

Prolactin is a protein hormone secreted by the **anterior** pituitary gland that stimulates the initiation of lactation (**lactogenesis**) in mammary glands and proliferation of the epithelium lining the crop glands of male and female doves and pigeons. It increases the production of a caseous material consisting of **desquamated** epithelial cells known as "crop milk" with which the young are fed. Prolactin also has been called *lactogen, luteotropin, galactin,* and *mammotropin.* This hormone induces **broodiness** (maternal drive, or maternal instinct) in hens, arouses the maternal instinct

in virgin rats, and is essential to the maintenance of lactation (**galactopoiesis**). There are **enzymes** within the epithelial cells that are essential to the activation of cells to convert blood constituents into milk. Prolactin functions to stimulate increased enzyme activity, which in turn stimulates milk secretion. Mammary gland cells fail to secrete milk in the absence of prolactin. This hormone has a major role in maintenance of lactation in many nonruminant species and in some ruminants including sheep. There is a gradual rise in the secretion of prolactin during late pregnancy that is stimulated by estrogen.[5] Following parturition the release of oxytocin at the time of each milking is believed to stimulate the secretion of prolactin, which in turn is discharged quickly into the blood following the stimulus of milking. It then passes via the blood to the udder to stimulate the epithelial cells to secrete milk during the interval between milkings.

15.5.2 Growth Hormone (Somatotropin)

The anterior pituitary gland secretes growth hormone. It largely regulates the growth rate in young animals and also influences milk secretion. This hormone appears to influence the substances from which milk is made by increasing the availability of blood amino acids, fats, and sugars for use by the mammary gland cells in milk synthesis. Growth hormone is a major factor required for maintenance of lactation in many ruminants such as cattle and goats, but is also necessary for full maintenance of lactation in nonruminants.

The advent of genetic engineering has allowed for production of large quantities of this hormone for use in cattle. Injections of bovine somatotropin (bST) can increase milk production of dairy cows by 15 percent around peak lactation and even greater in late lactation when compared with uninjected cows. Injection of bST every 2 weeks beginning at 9 weeks of lactation has become a common procedure for enhancing milk production in dairy cattle. To be effective bST administration must be accompanied by an increase in feed intake. Administration of bST does not affect reproductive performance of cows or their susceptibility to mastitis. Increases in reproductive problems or mastitis may, however, occur in association with high milk yield in cows. The increased usage of bST for enhancing milk production has been an important contribution to the productivity of the U.S. dairy industry in recent years.

15.5.3 Thyroid Hormones

Lactation is influenced greatly by the **thyroid** hormones, thyroxine and triiodothyronine, which are secreted by the thyroid gland. Thyroid hormones regulate the **metabolic** processes of the body and influence the general well-being of animals (Figure 15.10). They increase appetite, heart rate, flow of blood to the udder, and rate of milk secretion. If thyroid hormone secretion is low, a cow cannot secrete a large quantity of milk regardless of

---

[5]The quantity of estrogen secreted by beef cows in late pregnancy is considerably lower than its average secretion by dairy cows. The low-estrogen secretion by beef cows may explain, in part, one of the causes of low secretion of prolactin and lower milk production of beef cows.

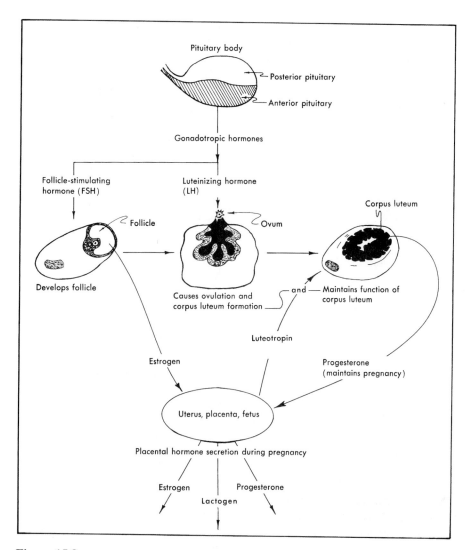

Figure 15.8 Endocrine control of udder growth. The pituitary secretes gonadotropic hormones, which act on the ovary to stimulate the estrous cycle. First, follicle-stimulating hormone (FSH) causes the ovarian follicle to develop. At this time estrogen is secreted, which acts on the duct system of the udder. In addition, the egg or ovum is caused to mature. Then luteinizing hormone (LH) is discharged from the pituitary, causing ovulation (discharge of the ovum) and the formation of the corpus luteum. If the animal becomes pregnant, the functional activity of the corpus luteum is maintained (Section 13.4.2) and the secretion of progesterone continues. Progesterone prepares the uterus for the reception of the fertilized egg and maintains the growing embryo and fetus in the uterus. In some species, the placenta secretes a luteotropic hormone, which helps maintain the corpus luteum. The placenta also produces estrogen, progesterone, and placental lactogen, which prepare the mammary glands for lactation (Section 15.5) through stimulation of the lobule–alveolar system of the udder.
Adapted from Missouri Agr. Expt. Sta. Bull. *793.*

the size of udder she possesses or what the other hormone secretion rates may be. Thyroidectomy of a cow causes a marked decrease (up to 75 percent) in milk secretion, which can be restored by administering **desiccated** thyroid tissue or thyroxine. Research has revealed that thyroxine secretion rates show seasonal trends with increases in the fall and winter and declines in the spring and summer. In Missouri dairy cows secrete only one-third as much of this iodine-containing hormone in the summer as in the winter. This may contribute to the reduced milk secretion observed in hot weather.

15.5.4 Parathyroid Hormone

The hormone that regulates blood levels of calcium and phosphorus, which are abundant in milk, is parathyroid hormone. It is also related to parturient paresis (milk fever). On initiation of lactation there is a rapid withdrawal of calcium and phosphorus from blood by the mammary glands. Animals with adequate parathyroid hormone secretion and good nutrition do not reveal symptoms of parturient paresis. Administration of vitamin D (D_2 or D_3) will maintain normal lactation in parathyroidectomized animals. This result further documents the observation

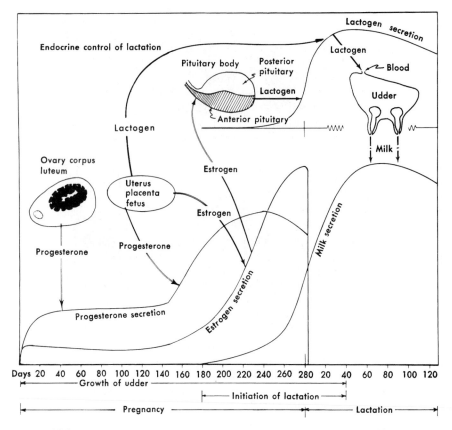

Figure 15.9 Endocrine control of the growth of the udder, which is stimulated by progesterone and estrogen during pregnancy. Later the increasing secretion of estrogen by the placenta stimulates the secretion of prolactin (lactogen), which stimulates the initiation of lactation by the epithelial cells of the udder. The rapid rise of estrogen secretion in late pregnancy also overrides the physiological function of progesterone (which normally maintains pregnancy) and aids in the initiation of parturition. In experimental animals, the withdrawal of progesterone and the injection of estrogen similarly stimulate the secretion of prolactin and initiate lactation. Following parturition and expulsion of the placenta, secretion of estrogen quickly drops, and maintenance of the secretion of prolactin is caused by the regular discharge of the hormones stimulated by milking. Placental tissue of several species secretes a placental lactogen, which may be involved in the growth of mammary glands and in the initiation of lactation.

Adapted from Missouri Agr. Expt. Sta. Bull. *836.*

that feeding high levels of vitamin D (20 million IU 3 to 7 days prepartum) will significantly reduce the incidence of parturient paresis in lactating cows. Stimulation of parathyroid hormone secretion by feeding a calcium-deficient diet during the 2 weeks prior to calving also aids in preventing parturient paresis in cows.

15.5.5 Adrenals

Whereas small amounts of adrenal hormones are essential to milk production, the secretion or administration of high quantities may depress lactation. The administration of cortisone and deoxycorticosterone in adrenalectomized animals will maintain lactation. These hormones are secreted more profusely when animals are subjected to stress of various kinds. Above-normal adrenal hormone secretion rate is beneficial in overcoming **stress** conditions at the expense of milk secretion.

15.5.6 Oxytocin

This hormone is synthesized in the hypothalamus (in the paraventricular and supraoptic nuclei). Oxytocin is transported through the neuronal axons from the hypothalamus to the **posterior** portion of the pituitary gland where it is stored. Oxytocin is released into the blood in response to stimulation of the teats. It is essential for ejection of milk (Section 15.8). It also stimulates the uterine **musculature** to contract, induces expulsion of the egg in the hen, and is used to induce active labor in women or to cause contraction of the uterus after delivery of the placenta.

15.5.7 Placental Lactogen

Placental lactogens are synthesized and secreted from the placenta. Generally they have both prolactin-like and growth hormone–like activities. However, there is great variation among species in the activities of the hormone and even in its

Figure 15.10 Comparison of a Jersey heifer (left) thyroidectomized at 54 days of age with a control (right) when both were 40 months of age. Note retardation of growth and development. Thyroidectomy resulted in dry, scaly skin, short hair, little horn growth, and absence of estrus. *Courtesy of Missouri Agr. Expt. Station.*

synthesis. For example, pigs and rabbits apparently do not have a placental lactogen. In rats and goats secretion of placental lactogen into the maternal blood during pregnancy may contribute to growth of mammary glands. Because the source of placental lactogen is lost when the placenta is expelled at parturition its effects must occur during pregnancy. Placental lactogen probably synergizes with estrogen, progesterone, prolactin, and growth hormone in mammary development during pregnancy in certain species. However, in the cow the level of placental lactogen in maternal blood is low and does not seem to contribute to mammary growth.

Administration of bovine placental lactogen (bPL) to dairy cattle during lactation can result in increased milk production. The mechanism for this effect is unknown but it seems to be separate from the effects of administering bST. Injecting both bST and bPL results in an additive effect on milk production.

15.5.8 Insulin

Secreted by the pancreas, insulin is involved in the movement of glucose into body cells. Insulin regulates glucose uptake in the mammary gland. Although an absolute requirement of insulin for lactation has not been demonstrated *in vivo,* the hormone is required to maintain mammary cells in tissue culture. In a culture of mammary tissue collected during gestation or lactation, insulin promotes mammary cell growth and cell division. Synthesis of milk components by mammary cells in a tissue culture requires a minimum of insulin, hydrocortisone, and prolactin included in the culture medium. These hormones are

often referred to as the *lactogenic complex of hormones.* Alone any one of these three hormones usually does not induce lactation, but rather the presence of all three hormones is required.

15.6 HOW MILK IS MADE

One of the oldest theories advanced relating to the formation of milk was that the blood of menstruations suspended during pregnancy was transported to the mammary gland for the formation of milk. One of the best-known drawings of Leonardo da Vinci shows ducts connecting the mammary gland directly to the uterus of a woman.

Later milk secretion was found to result from the combination of (1) filtration or transport of certain constituents from the bloodstream, (2) synthesis of other constituents by cellular metabolism, and (3) cell degeneration. Present concepts of lactation primarily involve a combination of the first two processes with little cell degradation.

A variety of techniques have been used to demonstrate the utilization of selected nutrients for milk synthesis and to characterize the processes of milk synthesis and secretion. By collecting arterial and venous blood samples, it is possible to **assay** constituents of arterial blood supplying milk **precursors** to the mammary glands and constituents of venous blood as it leaves the udder. Constituents taken up by the mammary gland will be in lower concentrations in the venous blood compared with arterial blood. This technique determines the arteriovenous (AV) difference of the precursor.

Another technique is the *in vitro* culture of mammary tissue explants (slices or pieces of tissue). This often is combined with the use of radioisotopes to trace or follow the precursor molecule. A substance believed to be a milk precursor or **substrate** is labeled with a **radioactive isotope** and added to the culture medium. Tissue explants are incubated at body temperature in a medium rich in oxygen and potential milk precursors. After periods of incubation the tissue is analyzed to determine the fate of the precursor by identifying the intermediate or end product molecules which contain the radioisotope.

Another approach to studying substrate utilization by mammary glands is the perfusion technique. It is possible to remove (**excise**) the udder surgically and suspend it in a normal position in the laboratory. Blood can then be circulated through the glands (**perfusion**) by an artificial heart and purified by an artificial lung. Although this technique is difficult to maintain and seldom used in current studies, it can enable researchers to introduce substances believed to be essential to lactation into the system and to follow their uptake into milk. Electron microscopy can be used to follow milk constituents during their formation in the cell or their secretion into the lumen of the alveolus. Cell images are magnified many thousandfold and some constituents can be identified in cells by labeling the constituent with enzymes or antibodies.

15.6.1 Milk Precursors and Composition

Water

The most prevalent milk component is water (87 percent by weight for cow's milk). It is filtered from blood. Milk and blood are **isotonic** [both have 6.6 atmospheres (atm) of **osmotic pressure**]. This means there is no osmotic pressure developed on either side of the semipermeable membranes of the milk-secreting cells. Yet milk contains about 90, 13, 10, 9, and 5 times as much sugar, calcium, phosphorus, lipids, and potassium but only one-half and one-seventh as much protein and sodium, respectively, as does blood plasma. This apparent paradox is possible because when many of these constituents are synthesized into milk components they are in a form that does not influence the osmotic pressure to any appreciable extent. For example, calcium is complexed with casein to form calcium caseinate in milk. Nearly 75 percent of the osmotic pressure on the milk side results from lactose (which is in true solution), whereas the minerals are in a nonionic state. On the blood side the chlorides account for approximately 75 percent of the osmotic pressure.

It should be noted that when mammary tissue is injured or diseased, as with mastitis, the permeability of the epithelial barrier between blood and milk in the alveolar lumen is altered and certain blood constituents filter between the cells into milk in greater amounts, for example, the chlorides and blood serum proteins. In addition, synthesis of certain constituents (e.g., lactose) within the gland is slowed.

Lactose

The principal carbohydrate of milk is lactose, a disaccharide which consists of one molecule of glucose and one of galactose.

Glucose is a normal blood component whereas lactose is not. Determinations of arteriovenous differences have shown glucose to be taken up by lactating mammary tissues. Arterial blood loses about 25 percent of its glucose content while passing through the lactating gland. Incubated lactating mammary tissue slices have been shown to synthesize lactose from glucose. Glucose is the precursor for lactose in experiments using **labeled** carbon **atoms.** Lactose is synthesized from glucose in the epithelial cells of the mammary gland. The milk whey protein, α-lactalbumin, is important for the regulation of lactose synthesis. Because lactose is the major *osmole* in milk its synthesis is closely tied to the fluid volume of the milk produced. Synthesis of this nondiffusible sugar in the epithelial cell causes water to be osmotically drawn into the cell.

Protein

Proteins found in milk result from synthesis in epithelial cells or from passage of preformed blood serum proteins. Because the major milk proteins **casein,** α-lactalbumin, and β-lactoglobulin are not present in blood they must be synthesized from blood precursors (amino acids). These proteins constitute about 94 percent of the protein nitrogen in cow's milk. Caseins are proteins which can form complexes with themselves and with calcium and phosphorous (the major milk minerals). These complexes (chemically defined as calcium caseinate) are secreted as insoluble aggregates called casein micelles. To the consumer of milk (young or adult) the casein micelle is a readily available source of amino acids from protein digestion and calcium and phosphorous. These nutrients are particularly important for growth of the newborn. Disruption of the casein micelle structure by acid precipitation or enzyme hydrolysis results in formation of a curd. This is what happens to milk when it is ingested and arrives in the acidic stomach of the animal. It also is the basis for making cheese and many cultured milk products. Curd formation from whole milk results in trapping fat droplets in the curd.

Once curd is formed the remaining water-soluble components are collectively called *whey.* Whey is composed of water, lactose, noncasein proteins (the whey proteins), water-soluble vitamins, and some minerals. α-Lactalbumin and β-lactoglobulin are the major whey proteins in cow's milk. α-Lactalbumin is important for lactose synthesis. The function of β-lactoglobulin remains one of the mysteries of milk proteins. Milk also contains many minor whey proteins such as immunoglobulins and **serum** albumin. Unlike the major milk proteins, which are synthesized in the mammary epithelial cell from amino acids, the immunoglobulins and serum albumin are found preformed in blood and are transported across the epithelial barrier into colostrum or milk. Immunoglobulins are bound to receptors at the serum (basal) side of the alveolar cell and transported through the cell. This selective process is particularly important during colostrum formation when large amounts of immunoglobulins are transported through epithelial cells. For example, normal milk contains only 0.05 to 0.11 percent immunoglobulins whereas colostrum may contain more than 15 percent. Serum albumin is not selectively transported through epithelial cells. Much of the serum albumin found in colostrum or milk diffuses between the cells or is transported through the cell in a nonselective manner.

Fat

Most milk fat is chemically in the form of *triglycerides.* Each triglyceride molecule is composed of three fatty acid molecules bound to one glycerol molecule. Approximately 75 percent of milk fat is synthesized in the mammary gland of the cow. In the **ruminant,** acetate is the principal precursor of milk fatty acids of chain lengths up to and including the 16-carbon palmitic acid. This explains why cows on high-grain and low-**forage** rations often secrete milk having a low fat content (this diet results in the reduced production of acetate in the rumen). Oleic, stearic, and fatty acids of longer chain lengths originate primarily from absorption of blood fatty acids and glycerides into the alveolar cells. Glucose is important in synthesis of the glycerol portion of the fat molecule. In nonruminants, glucose also acts directly as a precursor for synthesis of fatty acids.

Human milk fat usually contains a higher concentration of unsaturated fatty acids than does cow's milk. Moreover, the fatty acid composition of human or cow's milk can be significantly altered without affecting either milk yield or total fat content. Fatty acid composition of milk closely resembles dietary fatty acid composition during energy equilibrium, whereas on an energy-deficient diet milk fat is nearer the composition of human or cow depot fat.

Minerals

The minerals of milk are derived from the blood and reach milk through filtration. Colostrum milk has a much higher mineral content than normal milk (Table 15.4, Section 15.11). The major milk minerals are calcium and phosphorous but milk also contains magnesium, potassium, sodium, and numerous other minerals.

Vitamins

Vitamins enter milk unchanged from their form in blood. Those vitamins synthesized by the **rumen flora** through fermentation processes (B complex) are in rather constant amounts in milk, whereas the concentrations of fat-soluble vitamins, especially vitamins A and D, in milk are dependent on quantities in the ration and the body stores of the lactating animal. Any variation in the concentration of B-complex vitamins in milk is determined largely by factors such as **breed** and stage of lactation rather than by diet.

15.7 HOW MILK IS DISCHARGED (SECRETED)

Essentially all the milk that can be obtained at the time of milking is present in the udder when milk evacuation begins. Immediately after a milking the alveolar epithelial cells are shallow and **cuboidal** in form. The secreting cells lengthen as they synthesize more milk constituents, and fat globules begin to collect in the half of the cell closest to the alveolar lumen.

The rate of milk secretion is most rapid immediately following milking. As milk synthesis in the cell progresses and milk is discharged into the alveolar lumen, two things cause a reduction in further milk secretion: (1) the accumulation of a chemical inhibitory factor in the alveolar lumen and (2) a gradual rise in the **intramammary** pressure. A feedback inhibitor of lactation (FIL) has been identified as a low-molecular-weight peptide secreted from mammary epithelial cells as part of the normal process of milk secretion. Once secreted from the cell into the alveolar lumen, this chemical factor feeds back on the cell to gradually suppress further milk synthesis and secretion.

Between milkings the mammary glands become engorged with milk as the alveolar lumina fill. This results in an increase in fluid pressure in the gland. Increased intramammary pressure slows the rate of further milk secretion by decreasing mammary blood flow, which in turn causes decreased availability of nutrients and stimulatory hormones. Milk secretion each hour following milking is in the magnitude of 90 to 95 percent of that of the preceding hour. About 35 h after milking a cow milk secretion is appreciably reduced, if not stopped entirely, and intramammary pressure is nearly maximal. The concept of cyclic changes in intramammary pressure between milkings is diagrammatically illustrated in Figure 15.11.

Specific areas of the cell synthesize casein and lactose and form vesicles containing these constituents surrounded by a membrane (Figure 15.12). Secretory vesicles (elaborated by the Golgi apparatus) are the major means by which most nonfat milk constituents (water, lactose, proteins, minerals, and water-soluble vitamins) are discharged from the cell. After the secretory vesicles are formed and filled they migrate to the apical region (closest to the lumen) of the cell, their membranes fuse

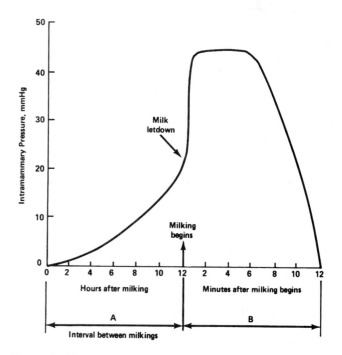

Figure 15.11 Changes in intramammary pressure in relation to milking times.

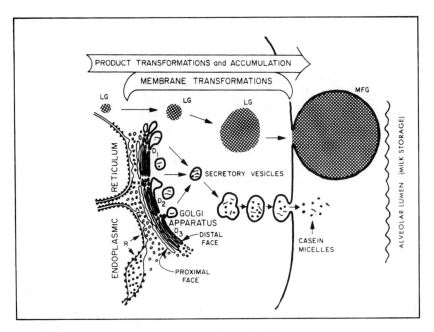

Figure 15.12 Diagrammatic summary of the structural and functional relations among the endoplasmic reticulum (ER), Golgi apparatus, secretory vesicles, and plasma membrane in membrane transformations and the secretion of lipid (fat) and protein (largely casein) of milk in the lactating secretory (epithelial) cells of the alveolus. In the representation of Golgi apparatus function diagrammed above, new membrane material is received from the ER at the forming, or proximal, face and is transformed and utilized in the elaboration of secretory vesicles at the distal, or maturing, face. Secretory proteins are contained in these vesicles, and the caseins appear to undergo micelle formation or aggregate within the secretory vesicle. The vesicles are also believed to contain lactose. Fusion of secretory vesicles with the plasma membrane provides new plasma membrane to replenish that lost from the envelopment of fat droplets during milk fat globule formation. Lipid globules (LG) emerge covered by a layer of apical plasma membrane to form the milk fat globules (MFG). Export milk proteins synthesized in the ER, shown above with bound ribosomes (R), move to the proximal face of the Golgi apparatus. Secretory vesicles bleb off the distal face of the Golgi cisternae and move to the apical surface, where the vesicular membrane fuses with the apical plasma membrane. Contents of the vesicles include casein micelles, other proteins, and most of the constituents of milk, which are discharged into the alveolar lumen. D_1, D_2, and D_3 represent dictyosomes, or Golgi bodies.
Modified from T. W. Keenan et al., in B. L. Larson and V. R. Smith (eds.), Lactation: A Comprehensive Treatise, Vol. II, Academic, New York and London, 1974.

with the plasma membrane, and their contents are discharged into the alveolar lumen.

Lipid droplets (fat globules) form near the endoplasmic reticulum (Figure 15.12). As they increase in size they migrate to the apical region of the secretory cell where they are progressively enveloped, or surrounded, by the plasma membrane. Eventually the membrane-bound globule is "pinched off" into the alveolar lumen. Formation of a fat globule requires approximately 5 h.

Secretory vesicle membranes replenish that portion of the plasma membrane given up to the fat globule, which indicates that the secretion of fat and that of protein and lactose are closely linked. Vesicles are abundant in mammary secretory (epithelial) cells, and the amount of membrane they can contribute to the cell surface appears to be more than adequate to replenish plasma membrane lost to milk fat during droplet secretion.

Secretory vesicles are formed by the Golgi apparatus (Figure 15.12). The vesicles arise along the distal (maturing or exit) face of the Golgi bodies and are formed at the expense of cisternal membranes. Cisternae are the flattened vesicles or tubules that make up the endoplasmic reticulum (ER). Milk proteins are synthesized in the ER and transported to the proximal face of the Golgi bodies by small vesicles. Newly formed Golgi cisternae are gradually displaced toward the distal face through loss of cisternae resulting from secretory vesicle formation. As the cisternae migrate across the stack their membranes are gradually transformed from being ER-like to being vesicle- and plasma-membrane-like. The Golgi apparatus serves as the intermediate link between the ER and the plasma membrane. Secretion of milk involves not only discharge of lactose, protein, and fat globules from the epithelial cell but also the discharge of large quantities of cellular membranes.

Secretion of the three major components of milk (protein, fat, and lactose), are interrelated through membrane flow and differentiation. Thus there is concomitant synthesis of both secretory and membrane proteins in mammary cells and milk fat secretion cannot proceed without concomitant synthesis of intracellular and membrane proteins. Furthermore, lactose synthesis cannot occur without the synthesis of a specific milk protein (α-lactalbumin).

15.8 THE PHENOMENON OF MILK LETDOWN

The evacuation of milk from the udder involves more than pressure (as in hand milking) or vacuum (as in machine milking) to open the teat canal. Vacuum per se will cause the galactophores and larger ducts to collapse and prevent emptying of the fine ducts and alveoli.

Unless proper stimulation is provided only a limited amount of milk may be removed at the onset of milking. This is because milk is held back in the gland by the small size of the capillary duct leading from each alveolus, the constrictions of each branch of the duct system, and the sag in the ducts at their points of suspension by connective tissue (due largely to the weight of milk). When a cow is properly stimulated there is a sudden expulsion of milk from the alveoli, which causes a filling and distension of the large ducts and udder cisterns. This is commonly referred to as *letdown,* or milk expulsion. What phenomenon is involved in milk letdown?

Erection Theory

An early theory of milk letdown was that erection of the teats, comparable to that occurring in the penis, was involved. Massaging the teats, which are richly **innervated** and thus easily excited (much like the erectile tissue of the penis), was believed to be the stimulus that resulted in contraction of the smooth muscle cells. The veins would thereby be occluded resulting in the udder becoming engorged with blood. The pressure created would force milk from the alveoli and smaller ducts into the larger ducts and udder cisterns. *Subsequent studies have rendered this theory untenable.*

The nipple of the breast in a woman is one of the most richly innervated structures in the body. Nerve fibers, both somatic and sympathetic, are found in profusion in and beneath the skin. The nipple contains smooth muscle that is arranged both circularly and vertically. The stiffening and erection of the nipple that results from mild stimulation is caused by the contraction of these strands of muscle. Nature likely provided nipple erection to facilitate grasping of the nipple by the baby during **suckling.** Milk removal is accomplished by the stripping motion of the infant's tongue against the hard palate, evacuating the milk from the lactiferous sinuses through openings in the nipple.

Neurohormonal Theory

Ott and Scott[6] injected extracts from the posterior pituitary gland into the ear veins of lactating goats and observed an increased milk flow through **cannulated** teats. Gaines[7] associated the nervous system with the milk letdown when he **anesthetized** a bitch and found that milk expulsion was inhibited. He then injected extracts from the posterior pituitary **intravenously** (IV) and observed milk letdown. These extracts were later purified and the hormone involved was identified as oxytocin (pitocin).

The natural stimulus for milk letdown by a cow is the nursing, or suckling, act by the calf. However, massaging the udder and teats will cause the same reflex. Also **auditory, olfactory,** and visual stimuli are often effective. Massaging the uterus and ovaries of the cow often causes the release of oxytocin. The word *oxytocia* means *rapid birth*. The hormone oxytocin was so named because it provokes vigorous uterine contractions and thus aids in expelling the fetus and placenta from the uterus. Oxytocin is used to induce and assist labor. Uterine contractions often occur in women when their babies nurse. This effect has prompted some doctors to recommend that mothers nurse their babies because it may speed the return of the uterus to normal size.

Milk may frequently be observed to flow freely from the teats of a lactating cow or mare or the nipples of a lactating woman during **coitus.** This flow of milk results from the stimulus given the uterus that provokes the release of oxytocin. Pregnant women have an enzyme called oxytocinase (pitocinase) in their blood plasma that inactivates oxytocin. This enzyme hydrolyzes oxytocin released during pregnancy and prevents premature uterine contractions. It first appears in blood about 3 weeks after fertilization of the ovum. From weeks 4 to 38 of pregnancy there is more than a thousandfold increase in plasma oxytocinase. Thus its concentration parallels placental growth and it has been used to estimate the stage of pregnancy. Its activity disappears entirely very shortly after parturition.

Contractile Tissue as the Basis for Milk Letdown

On proper stimulus, oxytocin is discharged into the bloodstream and reaches the udder 30 to 40 s later. There it causes the myoepithelial cells surrounding the alveoli and ducts to contract (Figure 15.13). These contractile cells in turn squeeze the alveoli, much like squeezing the bulb of a syringe, and expel milk from the lumina down the ducts and into the cisterns of the glands and teats. Cistern milk pressure is almost doubled. (The pressure becomes the equivalent of 50 to 60 mmHg.)

Oxytocin is usually effective in maintaining myoepithelial cell contraction for 8 to 12 min. The effect is thus **transitory.**[8] After this time there is a decrease in the amount of milk obtained, a decrease in the rate of milk flow, and an increase in the percentage of fat in the residual milk. The *residual,* or complementary, *milk* is that milk remaining in the mammary glands following the completion of milking and normally averages about 15 to 20

---

[6] I. Ott and J. C. Scott, *Proc. Soc. Exp. Biol. Med.,* 8 (1910):48.

[7] W. L. Gaines, *Amer. J. Physiol.,* 38 (1915):285.

[8] There is sufficient oxytocin stored in the posterior pituitary to give 10 or more letdowns in succession. However, most cows will respond with only one letdown at a given milking. It is said that many cows in India will let down only a portion of their milk for humans at milking and thereby save milk for a second letdown for the suckling calf.

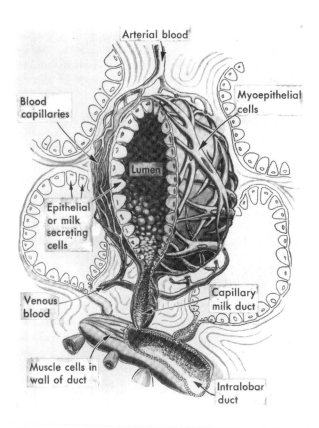

Figure 15.13 The blood vessels and myoepithelial cells surrounding an alveolus.

C. W. Turner, The Mammary Gland, *Lucas Brothers Publishers, Columbia, MO, 1952.*

percent of the total yield and about one-fourth of the milk fat. It is possible to obtain residual milk by injecting 10 IU of oxytocin intravenously or by injecting 20 IU of oxytocin **subcutaneously.** The latter method of injecting the hormone is much less disturbing to the cow than the former method. The practice of removing residual milk is often followed by veterinarians prior to antibiotic therapy for mastitis. This practice permits maximal effectiveness of antibiotics and minimal dilution of them with milk. Residual milk increases in quantity with higher milk production and with advancing age but decreases with advancing lactation.

In 1889 Babcock[9] observed that the last quarter milked yielded the least milk. This led to the recommendation of fast milking to obtain the maximum amount of milk. Does the transitory effect of oxytocin result from contractile fatigue of the myoepithelial cells or from **dissipation** or inactivation of oxytocin? Because administering additional oxytocin will result in the release of most residual milk it does not follow that contractile fatigue can account for the transitory effectiveness of oxytocin. A portion of the residual milk may be accounted for by a gradual collapse in the duct system as milk is removed from the udder. A calf reduces the amount of residual milk in the collapsed duct system through a vigorous massaging of the udder during the suckling act.

Cows vary considerably in their rate of milk ejection. Peak milk flow (occurring between the first and second minutes of milking) may reach 12 to 15 lb/min in fast-milking cows and be as low as 2 to 6 lb/min in slow-milking ones. Milk flow rate is highly **heritable** (between 0.54 and 0.8). Milking rate is largely controlled by the anatomy of the teat. Furthermore, milking rate for individual cows remains rather constant throughout life.

Oxytocin Effects Neutralized

A cow does not voluntarily withhold her milk from humans. In 1660 Markham offered the following advice: *The milkmaid whilst she is milking, shall do nothing rashly or suddenly about the cow which may afright or amaze her; but as she came gently, so with all gentleness she shall depart.*[10] This statement acknowledged that fright, anger, noise, pain, irritation, or embarrassment (in women) may block milk letdown. This inhibition results from the release of **epinephrine** (adrenalin) from the adrenal medulla. This hormone increases blood pressure, heart rate, and cardiac output. In the mammary gland it causes the tiny arteries and capillaries to constrict and thus prevents oxytocin from reaching the myoepithelial cells in concentrations sufficient to cause contraction. Apparently it is more

[9]S. M. Babcock, *Sixth Ann. Rept. Wisconsin Agr. Expt. Sta.,* 1889, pp. 42–62.

[10]G. A. Markham, *A Way to Get Wealth,* Book II, *The English Housewife,* E. Brewster and G. Sawbridge, London, 1660, pp. 143–144.

potent on a per unit basis than oxytocin and thereby quickly and effectively overrides the effects of the latter. Epinephrine may remain in the blood for a longer period of time than oxytocin. Hence a frightened cow should be allowed to stand for 20 to 30 min before one attempts to gain milk letdown.

It is possible for a cow to eject milk normally from three quarters and not from the fourth. One may observe this by placing a teat cannula in an injured teat. If one then stimulates milk letdown the cow can be milked from the normal quarters and concurrently milk will flow out through the cannula. Then when the injured teat is touched, an abrupt cessation of milk flow results for about 30 s, while the other three teats continue to give milk normally. Milk flow was interrupted by a nervous reflex contraction and not by the release of epinephrine.

The sow usually yields less milk in the rear teats; hence, a possible basis for the commonly held view that the **runt** of the litter nurses the rear teat. But this may reflect the rigor of piglets nursing the more easily accessible teats and their doing a more complete job of extracting milk and, thereby, stimulating those mammary glands to secrete more milk, which in turn supports faster growth (Figure 15.14).

15.9 REGRESSION (INVOLUTION) OF THE MAMMARY GLAND

Failure to evacuate milk from the mammary gland causes involution, even when adjacent glands are evacuated or suckled. An example is the **atrophy** of unnursed glands of sows having small litters of pigs. Moreover, following peak production after parturition there is a gradual reduction in mammary size and in the quantity of milk secreted until the animal goes dry (nonlactating). Peak production in cows normally occurs about 6 to 12 weeks into the lactation and may reach 120 to 200 lb daily in very high-producing cows, whereas in women peak flow is from 1000 to 3000 cc daily at about week 25 of lactation. Some women lactate for 15 or more years. Sows nursing 14 piglets can yield over 2 lb of milk per piglet daily with peak production occurring about 3 to 4 weeks postpartum. Piglets normally nurse about 30 ml from their mothers at an average interval of about 50 min. The frequency of nursing decreases as piglets become older.

The gradual reduction in milk secretion after peak production may result in part from a slow decline in the secretion rates of various hormones or a reduction in hormone receptors, but there is also a gradual loss of secretory epithelial cells. It is possible that one or more hormones causes a reduction in the rate of cellular division following maximum milk yield resulting in reduced multiplication of epithelial cells. Thus after several months of lactation the total number of secreting cells would be appreciably reduced. The slow loss of cells continues until the end of the lactation period.

Following the cessation of milking there initially is a swelling of the udder from accumulation of milk, but this is followed by a rapid shrinking of the mammary glands as water is resorbed and some milk components are hydrolyzed. Loss of epithelial cells and regression of alveoli continue for several weeks after drying off the cow. This process is called *involution*. Following completion of mammary gland involution in the

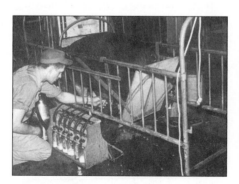 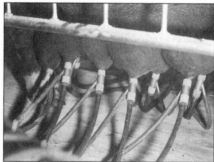

Figure 15.14 Milking machine for sows. This 12-teat-cup milking machine (left and right) was designed by Dr. D. A. Hartman during research at Cornell University on milk yields of sows, nursing frequencies and growth rates of piglets, and other pertinent parameters. Utilizing a movie camera–time clock apparatus, Hartman observed that piglets nursed on average at intervals of 43.5 min the first week, and that the interval gradually lengthened until the sixth week, when they nursed every 58.2 min. Piglets select a certain teat within a few days after birth and nurse it throughout the lactation. The larger, more vigorous pigs did a more complete job of extracting milk and thereby stimulated the mammary glands to higher milk production. Smaller, weaker piglets did not empty the glands as completely and consequently those glands did not produce as much milk, resulting in small pigs at weaning. Hartman noted that the runts (smallest pigs) do not always nurse the hind teats, a commonly held view. There is considerable variation in milk production among mammary glands, but low-producing glands are as likely to be near the front as the rear of the sow. Milk production per sow ranged from 398 to 723 lb of milk during the 6-week experimental periods. This variation explains in large part why certain litters are healthier and grow much more than others.
Courtesy of Dr. D. A. Hartman and Cornell University.

cow, ducts and some collapsed alveolar structures are maintained. In other species such as the mouse and rat, weaning of the young results in rapid and extensive loss of the alveolar structures (within 4 or 5 days) because of the death of the epithelial cells. In the nonlactating gland of these species few alveoli are found because the gland consists largely of ducts, and connective and fatty tissues. Mammary gland involution in the sow, as initiated by weaning the piglets, also results in extensive loss of alveolar structures over a 5- to 7-day period. Women have nearly full maintenance of the lobule–alveolar system during the nonlactating state, whereas the cow has intermediate maintenance and the rat minimal maintenance of the lobule–alveolar system during the dry period. In women degeneration, or involution, of mammary glands occurs following menopause.

When milk withdrawal is incomplete the drying-off process is hastened. When alveoli are not emptied milk secretion is further reduced, thus reducing the total amount of milk secreted between milkings. Three common methods employed to dry off a high-producing dairy cow are intermittent milking (e.g., once daily or on alternate days), incomplete milking, or abrupt cessation of milking. The swelling of the udder, which occurs after abrupt cessation of milking at dry off, is not harmful to the gland directly. However, leakage from teats caused by elevated intramammary pressure for the first few days may increase susceptibility to mastitis. The dairy cow's udder is highly susceptible to contracting new intramammary infections during the first 2 weeks following drying off. Infusion of antibiotics into the teats at dry off (dry cow antibiotic therapy) is an important method of reducing the incidence of mastitis in dairy cattle.

15.10 Factors Affecting Lactation

Several factors influence the intensity of lactation. The role of hormones was discussed previously (Section 15.5).

Inheritance

The ability to secrete milk is dependent on the genetic constitution of the animal. Control of lactation requires complex regulation of numerous genes including those which code (genetically) for hormones that stimulate milk synthesis, enzymes involved in milk synthesis, milk proteins, proteins involved in cell structure, and many others.

Secreting Tissue

The most fundamental factor limiting lactation is the amount of glandular tissue. The milk-producing potential is proportional to the amount of milk-secreting tissue. Small mammary glands are likely to be a disadvantage in lactation because of their inability to secrete and to store enough milk. British researchers found a significant correlation between mammary gland weight and milk yield at peak lactation in goats.

Stage and Persistency of Lactation

There is considerable variation in the *persistency* of milk secretion following peak production in early lactation. Some cows are very persistent and their rate of milk-secretion decline is slow (2 to 4 percent of their previous month's production). The production of other cows declines more rapidly (6 to 8 percent of their previous month's production) so that they show poor persistency. Figure 15.15 illustrates a second method of expressing persistency. The average percentage decrease in milk production per month is used to indicate lactation persistency. Cows of high persistency produce more milk in a lactation than do cows of low persistency when their maximum production is equal. Cows that milk out rapidly are usually more persistent. One cow at the Missouri Agricultural Experiment Station milked for about 8 years without calving, and a New Zealand cow milked continuously and without a pregnancy for 12 years.

Frequency of Milking

As milk accumulates in the lumen of the alveoli and storage areas of the udder it tends gradually to inhibit milk secretion (Section 15.7). More frequent removal of milk permits maximum intensity of the milk-manufacturing process. The completeness of milk removal at milking is also important. Milk left in the udder at milking time tends to check the secretion of milk during the interval between milkings. This explains the recommendation of incomplete milking for a few days postpartum; it lessens the requirement for calcium and allows the parathyroid gland to become active in producing the parathyroid hormone, which is closely related to prevention of parturient paresis (Section 15.5.4).

Pregnancy

During the first 5 months of pregnancy the decline in milk yield in pregnant cows is similar to the equivalent lactation period in nonpregnant ones. However, following the fifth month of pregnancy cows begin to decline more rapidly in milk yield (decreased persistency). Missouri studies showed a 450-lb difference in milk yield allegedly due to the increasing divergence of nutrients from the mammary glands to the uterus for fetal growth and maintenance.

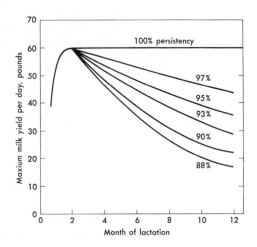

Figure 15.15 The persistency of milk secretion, defined as the degree with which the rate of secretion is maintained as lactation advances.

Courtesy of Missouri Agr. Expt. Sta. Bull. *836.*

Age

It is believed that there is a slight additional growth of secreting cells during each pregnancy until cows reach about 7 years of age or the fifth pregnancy. This is manifested by the increase in yearly milk production to that age. After about 7 years there usually is a gradual decline in total milk secreted during each subsequent lactation. Some cows remain productive to 20 or more years of age (the record is believed to be 27 years); however, most are retired from active production by the age of about 10 to 12. In women milk output also decreases with increasing age.

Animal Size

Larger cows normally secrete more milk but they may not be any more efficient in converting food nutrients into milk than smaller ones. Cows normally will not secrete more milk daily than the equivalent of 8 to 12 percent of their body weight, whereas goats may secrete enough milk daily to equal 20 or more percent of their body weight. Missouri studies found milk production to vary with approximately the 0.7 power of body weight. Thus a 1 percent increase in body weight among animals tends to be associated with an increase of 0.7 percent in milk production.

Estrus

Milk secretion in cattle generally is reduced when the cow is in **heat.** This occurs primarily from the physical activity associated with estrous behavior rather than from a direct effect of reproductive hormones on lactation. The reduced milk production is temporary. To minimize milk loss cows should be confined during **estrus.**

Disease

A number of diseases may significantly reduce the amount of milk secreted. Disease may affect heart rate and, therefore, the rate of blood circulation through the mammary gland, which influences milk secretion. It has been estimated that 400 volumes of blood pass through the mammary gland for each volume of milk secreted. The major disease in dairy cattle is mastitis, defined as *an inflammation of the mammary gland.* Mastitis is caused most often by bacterial infection. Milk production is reduced when a cow has clinical mastitis, which is recognized when there are clots or flakes in the milk and often accompanied by increased body temperature and reduced appetite. Subclinical mastitis is more difficult to recognize because changes in the milk are not obvious and systemic symptoms are minimal. When cows have subclinical mastitis significant loss in milk yield may occur without producer awareness.

Dry Period

Cows are normally bred 60 to 90 days after parturition. It is expected that they will lactate for about 305 days and then be given a 60-day dry period before the next calving. The optimum length of dry period occurs when an extra day of dry period causes the loss in milk secreted in the current lactation period to be balanced by a gain in milk secreted in the following lactation period. Of interest is one milk record in which 42,805 lb of milk were produced in 365 days without the benefit of a dry period. There is no known physiological or endocrinological reason for believing that milk-secreting cells are benefited by a lengthy rest period. Breeding and managing cows for high persistency raises the prospect of extended lactations (greater than 18 months). Successful extension of lactations in dairy cattle could result in reevaluating current management practices which are based upon a 12- or 13-month lactation. Current practices for the time of rebreeding following parturition, nutrition before and after peak lactation, and the need for a dry period may all be changed if the average lactation length of a high-producing herd is 18 months or greater.

Body Condition (Fatness) at Parturition

The storage of body fat during late lactation and the dry period, and its subsequent use in the following lactation, is an inefficient use of feed energy. It is perhaps only two-thirds as efficient as the direct use of dietary energy for milk production. Additional body weight (body **condition**) at parturition generally results in additional milk secretion during the following lactation. Cornell researchers found that 100 lb of body weight can yield about 400 lb of milk and 100 lb of body fat can yield about 880 lb of milk.

United States Department of Agriculture (USDA) scientists found that it is possible for a high-producing cow to secrete more calories into her milk than she consumes in her feed. This is explained by a loss in body weight. For example, in one experiment a Holstein cow (housed in an energy–metabolism laboratory) produced 19,331 lb of **FCM** (4 percent fat-corrected milk) in 305 days. During the first 89 days of lactation she secreted 9433 lb of milk, which required 15,000 more **calories** (kilocalories) of estimated net energy per day than she received from her feed. Her weight loss during this period was 330 lb. This was calculated to account for nearly 3000 lb of milk that could not have been secreted from energy realized from her feed. During days 90 to 175 her body weight remained steady, and from day 176 to 305 she consumed substantially more feed than was needed for body maintenance and milk production. Her **ADG** (average daily gain) was 4.1 lb during that period. Her ration was 80 percent **concentrates.**

Subsequent USDA studies in an energy–metabolism laboratory demonstrated that high-producing dairy cows can utilize body fat for up to half their energy needs during early lactation. However, the amount of protein reserves that can be utilized is not proportional to the amount of body fat. Therefore the dietary energy:protein ratio should be considered when cows are withdrawing body reserves during intense lactation.

Environmental Temperatures

Temperatures above 80°F for Holsteins and 85°F for Jerseys reduce milk production. This may result partially from a decreased appetite and/or reduced thyroxine secretion. Optimum temperature is about 50 to 60°F, the **comfort zone** being 50 to 80°F for most lactating cows (Chapter 17).

Feed

The speed of synthesis and diffusion of various milk constituents is dependent on the concentration of milk precursors in blood, which reflects the quality and quantity of the food supply. Nature provides for maintenance, growth, and reproductive needs before energy is made available for lactation. Inadequate feed **nutrients** probably limit the secretion of milk more than any other single factor in the dairy cow.[11] Although good nutrition alone cannot guarantee high milk production, poor nutrition can prevent attainment of a cow's full potential just as surely as poor management, low genetic potential, or an unfavorable environment. The maintenance of lactation (galactopoiesis) is closely related to an adequate feed intake by lactating animals. Inadequate nutrition limits milk secretion in cows to a greater degree than in women. For example, inadequately nourished, low-income mothers of India often yield as much milk daily as the average cow of the area (about 2 qt daily).

Preparation for Milking

Of foremost significance in obtaining complete letdown of milk is proper preparation of the cow for milking. Improper preparation will result in incomplete milk evacuation, which increases residual milk and decreases the intensity and amount of subsequent milk secretion. Premilking stimulation causes an earlier release of oxytocin from the neurohypophysis, which results in shorter time required for milk out.

Stress

Secretion of milk can be decreased by stress. High secretion of stress hormones can affect milk synthesis and secretion. Additionally stress can inhibit milk letdown (Section 15.8). As animals are selected to secrete higher levels of milk, stress may play an increasingly important role in achieving maximal milk production.

Suckling Stimulus

Many cases have been reported in which nonlactating women who have previously borne children have initiated lactation by allowing a baby to suckle their breasts. Such lactation occurs more commonly in young women but has been reported in postmenopausal women as well. The same phenomenon has been observed in virgin heifers suckled by other heifers. Suckling can induce secretion of fluid from the udder and may result in mastitis. Suckling-induced lactation has been observed in a number of species including the dwarf mongoose, prairie dog, lion, wolf, and dolphin. Many of these species live in communal systems in which subordinate females are observed to suckle offspring from the dominant females of the group. Typically this suckling-induced stimulation of lactation in a nonpregnant female requires several weeks of repeated suckling. In the case of a newborn animal suckling a nonlactating female, the newborn

must receive nourishment from another source until sufficient milk production is stimulated in the suckled female.

15.11 FACTORS AFFECTING THE COMPOSITION OF MILK

Milk may be defined as *the whole lacteal secretion obtained by the complete milking of one or more healthy lactating females.* According to G. Seiffert, "Mother's milk is a treasure impossible to replace." The term *milk,* unqualified, has come to be accepted in the United States as meaning cow's milk. The same constituents are present in the milks of all species of mammals; however, as shown in Table 15.1 there is considerable variation in the average milk composition of various mammals. Variations in the composition of milk may result from one or more of a number of causes.

Breed

As shown in Table 15.2 milk composition varies among breeds of dairy cows. The within-breed variation in milk composition is proportional to the percent of a given milk component for a given breed. Jerseys therefore, because of their high fat test, have a greater variation in percent fat than Holsteins. As fat content changes by 1 percent, average changes in nonfat solids are Holsteins, 0.55 percent; Jerseys, 0.25 percent; and Guernseys, 0.36 percent. Breed or strain differences in milk composition have also been noted for rodents, swine, and other species.

Inherited Factors

Milk of some cow **families** contains greater quantities of certain milk components than others. Heritability estimates for milk yield and selected milk constituents are given in Table 15.3. Since the protein, fat, and nonfat solids (NFS) components are highly heritable (about 0.5), fast genetic progress in selective matings for these milk constituents can be expected. About one-half the superiority of selected parents should be reflected in records of their **progeny.** A negative genetic correlation is given for the pounds of protein, fat, and NFS, which

[11]This is also true in sows. For example, research at the University of Illinois showed that milk production was significantly reduced in sows as dietary protein was reduced from 18 to 10 percent. There was, however, little change in the composition of milks among sows fed 10, 14, and 18 percent protein diets.

TABLE 15.1	**Average Composition of Milk from Various Mammals, Percent**				
Mammal	**Milk fat**	**Protein**	**Lactose**	**Minerals**	**Total solids**
Woman	3.7	1.6	7.0	0.2	12.5
Cow	4.0	3.3	5.0	0.7	13.0
Mare	1.3	2.2	5.9	0.4	9.8
Sow	5.3	4.9	5.3	0.9	16.4
Cat	3.3	9.1	4.9	0.6	17.9
Dog	8.3	7.1	3.7	1.3	20.4
Sheep	5.4	4.8	4.6	0.9	15.7
Goat	4.1	3.7	4.2	0.8	12.8
Elephant	15.2	4.9	3.4	0.8	24.3
Reindeer	18.7	11.1	2.7	1.2	33.7
Whale	22.2	12.0	1.8	1.7	38.1

TABLE 15.2	Average Composition of Milk among Major Dairy Breeds, Percent					
Breed	Fat	Protein	Lactose	Ash	Nonfat solids	Total solids
Ayrshire	4.0	3.5	4.7	0.68	8.9	12.9
Brown Swiss	4.0	3.6	5.0	0.73	9.3	13.3
Guernsey	5.0	3.9	4.9	0.74	9.5	14.5
Holstein	3.4	3.3	4.9	0.68	8.9	12.3
Jersey	5.4	3.9	4.9	0.71	9.5	14.9

Source: Missouri Agr. Expt. Sta. Bull. 365.

TABLE 15.3	Heritability Estimates and Genetic Correlations of Milk Components			
		Genetic correlations		
Milk yield and component	Heritability	Milk	Fat	Nonfat solids
Milk yield (ME)	0.32			
Fat, %	0.58	− 0.20		
NFS, %	0.59	− 0.20	+ 0.50	
Protein, %	0.50	− 0.20	+ 0.48	0.94
Lactose, %	0.53			

Source: Adapted from J. R. Campbell and R. T. Marshall, *The Science of Providing Milk for Man,* McGraw-Hill, New York, 1975, p. 127; and R. C. Laben, *J. Dairy Sci.,* 46 (1963):1293–1301.

means that as the total pounds of milk increases per lactation, the percentage of these components in milk decreases but the total pounds of these milk components increases. A high genetic correlation usually exists, therefore, between total milk yield and yield of any given component. It is thus evident that milk composition can be changed by selection.

Stage of Lactation

As shown in Table 15.4 colostrum milk of cows is especially high in **NFS** because of the high protein content, which is primarily contributed by immunoglobulins.[12] Lactose concentration is lower in colostrum than in normal milk. Following peak milk production the protein and NFS content remains relatively stable until late lactation when a gradual rise begins (see pregnancy effects). Most cows' milk contains 0.5 to 1.5 percent less fat during the first 2 months than during the last 2 months of lactation. The lactose content of milk is highest early in lactation and declines linearly during the balance of lactation.

[12]Researchers at Rutgers University found that the composition of colostrum of the sow can be affected by the prepartum protein intake. However, they concluded that sow's milk is not affected materially in composition by dietary factors, especially protein. They reported that weight gains of piglets nursed by sows fed a 5 percent protein diet were significantly less than those of piglets nursing sows fed 10 and 15 percent diets for the first 14 days. Weight gains among piglets nursing sows fed 10 or 15 percent protein rations were not significantly different.

TABLE 15.4	Average Composition of Colostrum and Normal Milk of the Cow, Percent	
Component	Colostrum	Normal
Water	71.7	87.0
Solids (total)	28.3	13.0
Milk fat	3.4	4.0
Casein	4.8	2.5
Globulin and albumin	15.8	0.8
Lactose	2.5	5.0
Minerals	1.8	0.7

Condition of the Cow at Parturition

Cows carrying considerable body fat at the time of parturition will secrete more fat and NFS into their milk than when in marginal body condition.

Season of the Year

The fat percentage of milk usually increases in the fall and decreases in spring. Protein and NFS show less seasonal fluctuation than fat; however, both normally reach a high for the year in May and June and decrease in late summer. There appears to be an inverse relation between the protein and lactose contents of milk. The apparent seasonal fluctuations may reflect changes in feed, environmental temperature, and/or other contributing factors. The unsaturated fatty acids are higher in fat produced in summer than in winter.

Completeness of Milkings

The fat content of milk varies appreciably between the first and last milk removed from the udder, whereas other constituents of milk are fairly constant at different stages of milking. The first streams of milk may contain only 1.0 to 1.5 percent fat whereas the strippings may contain 6.0 to 8.5 percent. Where milking is incomplete or when cows are improperly stimulated prior to milking the quantity of residual milk increases and is richer in fat. The percentage of NFS is not appreciably affected.

There is also a difference in fat content in the successive portions of milk evacuated from women and other species whose milk is capable of *creaming* (fat globules can be aggregated into more compact clusters). This property is not found in milk of the sow, goat, and water buffalo in which the fat is more evenly distributed throughout (much as in homogenized milk).

Interval between Milkings

Cows are normally milked at 12-h intervals. Uneven milking intervals affect the fat more than the protein or NFS percentages. With a longer interval the milk will be lower in fat and lactose, it will be slightly higher in protein, and, as a result of the counterbalancing (fluctuations of protein, lactose, and minerals), the NFS content will remain stable.

Age of Cow

Protein, fat, and NFS decline with age. The NFS declines more than fat. In one study decreases from the first to the ninth lactation were 0.08, 0.19, 0.25, and 0.34 percent for protein, fat, lactose, and NFS, respectively.

Feed

Currently there are no feeds, additives, or feeding practices that will profitably change the composition of milk. Most changes in milk composition resulting from feeding are temporary and limited in nature. Vegetable oils have been shown to cause a temporary increase in milk fat. Conversely fish oils usually depress the fat content of milk from 0.1 to 0.5 percent for as long as they are fed. Underfeeding reduces the protein and NFS content somewhat, but especially lowers milk yield. Milk protein is higher among cows fed high-energy rations than among those fed low-energy diets. High-grain feeding coupled with very low **forage** feeding results in a significant reduction of milk fat in most cows. The basis for this was explained previously (Section 15.6.1). The fat-depressing effects of high-grain restricted-roughage diets can be alleviated by including sodium bicarbonate or other buffering compounds in the diet. Completely pelleted **rations** or grinding the forage too fine also reduces fat content. Nonfat solids and protein are not appreciably affected.

One of the wonders of the cow is that her milk is nearly the same day after day whether her feed is good or poor. It is chiefly the *quantity* of her milk that varies with her feed, not the *quality*. A few exceptions have been noted but most of these are minor. For example, the vitamin A potency of milk reflects the body stores of a cow and the amount of carotene or vitamin A in the ration. While pasturing on fresh, green, lush grasses the cow may secrete milk containing a lower fat percentage but as much as 2500 IU of vitamin A per quart, whereas when dry feeds predominate the vitamin A may fall to 400 to 800 IU/qt. Vitamin D may vary in milk from 5 IU/qt in winter day-length to 50 IU in summer sunshine and 160 IU/qt when cows are fed 200 g of irradiated yeast daily. **Fluid milks** commonly have vitamins A and D added (Chapter 3).

Excitement or Stress

This may result in incomplete milk removal and thus a lower fat test. The NFS content is unchanged.

Disease

Little is known about the influence of certain diseases on protein and NFS. An increase in body temperature frequently is accompanied by an increase in fat percentage and a decrease in milk yield and NFS content. Udder infections such as mastitis cause a decrease in milk fat, NFS, protein, and lactose and a notable increase in the mineral and chloride contents of milk. Milk from cows with mastitis frequently has a salty taste, which is attributed to a higher concentration of chlorides. During mastitis the change in concentration of ions, particularly sodium, results in a change in the electrical conductivity of the milk.

Advantage has been made of this property by including electrodes in milking machines to measure electrical conductivity as milk leaves the udder. Combining this information with a computer-assisted data analysis system allows for automated identification of cows in the herd that have mastitis. Scientists at Texas A&M University and others found that milk from cows with mastitis also contains more sodium, copper, iron, zinc, and magnesium but less calcium, phosphorus, molybdenum, and potassium than normal milk. Mastitis also results in greater hydrolysis of milk casein. Yield of cheese is dependent on content of intact casein molecules (Section 15.6.1, Protein). Therefore breakdown of casein resulting from mastitis reduces cheese yield.

Pregnancy

The increase in NFS and protein late in lactation must be associated with pregnancy since **open** (nonpregnant) cows do not show increases in these milk components with advancing lactation. Pregnancy has no apparent effect on the fat content of milk.

Environmental Temperature

Temperatures below 30°F and above 70°F cause an increase in the fat content of milk, whereas protein and NFS quantities decline at higher temperatures and increase at lower ones.

Exercise

Slight exercise apparently increases the fat content from 0.2 to 0.3 percent without reducing the quantity of milk secreted. Moderate to heavy exercise in high-producing cows, however, will result in reduced milk secretion and a corresponding increase in percent fat. Exercise causes no apparent change in the NFS content of milk.

Drugs

Most drugs have little or no effect on milk composition, although several will cause a decrease in the amount secreted. Many drugs and other compounds (e.g., **pesticides**), when consumed in the feed, may be secreted into the milk. Also, cows treated with **antibiotics** may shed them into milk. Generally, when antibiotic **therapy** is used in the treatment of mastitis, the milk should be discarded from cows so treated for at least 72 or 84 h following intramammary or intramuscular treatments, respectively. It is important to read the label on the antibiotic container.

15.12 IMMUNOLOGICAL ASPECTS OF COLOSTRUM

The manner in which developing animals achieve **immunologic** competence and the timing of this development have attracted much interest for theoretical as well as practical reasons. The transfer of immunity from mother to offspring differs among animal species. Young mammals depend on a **passive immunity** (acquired immunity produced by administration of preformed **antibodies**) for their resistance to infectious diseases (Chapter 22). Mumps, measles, diphtheria, scarlet fever, and certain other infections of humans and animals rarely attack the very young. This suggests that the mother has an active **immunity** against the specific infective agent and that this immunity is transmitted to the offspring. **Neonatal** mammals are unable to produce antibodies within their own bodies for some time following birth. They must acquire these antibodies from their mother either while in the uterus prepartum or through colostrum postpartum.

The route of transfer varies among species (Table 15.5). Farm animals should receive colostrum during the first 12 to 24 h after birth if they are to acquire passive immunity. As shown in Table 15.4 colostrum is especially high in immunoglobulins (antibodies). Immunoglobulins are optimally absorbed by the calf's intestine during the initial hours following birth. The ability to absorb large protein molecules decreases during the first day after birth and **gut** closure is complete at about 24 h postpartum. After that time the **neonate** digests the antibodies, thus destroying their immunization properties. This apparently results from the neonate being unable to absorb the large protein molecule (molecular weight about 900,000 compared with 6000 for the protein hormone insulin). Humans acquire passive immunity while **in utero** by transport of antibodies from the maternal blood across the placenta and into the fetal blood. The immunoglobulins in human colostrum may benefit the infant by protecting the intestinal lining from external pathogenic bacteria.

In women colostrum accumulates in the mammary glands in late pregnancy; normal milk does not appear until 2 to 3 days following parturition. Milk secretion is then usually profuse and dramatic. Relief of tension by suckling is imperative, but if not allowed lactation will cease in a few days. The act of suckling (or otherwise emptying the breasts of milk) is a specific stimulus to further lactation, acting indirectly through hormones of the anterior pituitary gland (Section 15.5.1).

Because newborn calves, foals, kids, lambs, and piglets have low innate resistance to disease, and because these animals do not acquire passive immunity while in utero, transfer of immunoglobulins via colostrum is of special significance. Moreover, because intestinal permeability to immunoglobulins persists for only 12 to 24 h after birth it is imperative that the newborn of these farm animals receive colostrum early if they are to acquire disease resistance. Should the mother die at parturition the calf should receive colostrum from another cow or from a supply that has been frozen for such emergencies. Because the calf can absorb antibodies prepared in other animals such as sheep or horses, colostrum from those sources may also be fed and vice versa. However, certain diseases are species-specific and colostrum from another species might not afford the desired protection. Research conducted at the Max Planck Research Institute in Germany and at North Carolina State University showed that piglets artificially raised and fed cow's milk or cow's colostrum outgained control animals that suckled sows naturally (Figure 15.16). This research could have

TABLE 15.5	Time of Transfer of Passive Immunity from Mother to Offspring	
Animal species	**Prenatal**	**Postnatal**
Horse, pig, ox, sheep, and goat	–	+ to 36 h
Dog and cat	+	+ to 10 days
Human	+	–
Rat and mouse	+	+
Rabbit and guinea pig	+	–

Figure 15.16 An automated device used to raise piglets from birth to 2 weeks of age at the Max Planck Research Institute, Mariensee, West Germany. Milk is warmed and fed mechanically at preset time intervals.
Photograph by John R. Campbell.

practical significance because cows secrete milk with greater **feed efficiency** than do sows. Additionally, such a management practice could free sows to farrow three rather than the more common two litters annually.

A substance foreign to the tissues of an animal can be injected into that animal to stimulate the formation of a specific antibody in the blood. The foreign substance is called an *antigen*. Animals can be immunized against pathogenic bacteria, viruses, toxic substances, and even hormones or other physiologically active substances. Binding of the antibody to its specific antigen can result in neutralizing or clearing the antigen from the animal. This field of study presents significant opportunities in preventive medicine (Sections 22.2.4 through 22.2.7). For example, antigens specific for certain types of mastitis-causing bacteria can be used to successfully immunize dairy cattle. Such immunizations do not necessarily prevent mastitis but rather speed recovery from the infection when it occurs. Mammary gland immunity of the cow or other species can be manipulated to provide a source of antibodies specific against intestinal diseases of newborn human infants or adults.

15.13 TRANSGENIC ANIMALS AND LACTATION

Transgenic animal technology takes advantage of the ability to transfer specific genes into the genome of an animal. This powerful technique provides significant opportunities for biomedical research, pharmaceutical production in animal systems, and enhancement of food animal production. Tissue-specific expression of *transgenes* in the mammary gland of an animal is of particular interest. Secretion of pharmaceutically valuable proteins in the milk of various species is being investigated. Large quantities of the valuable protein may be produced from

a limited number of lactating animals. Other research is aimed at using transgenic technology to enhance livestock production characteristics. Recent research at the University of Illinois has demonstrated that transgenic technology can be used to enhance production characteristics of swine. Milk production is a limiting factor to growth of the newborn piglet. α-Lactalbumin is a regulator of milk production because of its role in lactose synthesis (Section 15.6.1). Transgenic sows that express the bovine α-lactalbumin transgene in their mammary glands during lactation have greater milk production and enhanced growth rate of their piglets. Other studies at Illinois are aimed at improving newborn piglet growth and health by increasing the concentration of an intestinal growth factor in sow's milk. Transgenic technology aimed at improving lactation characteristics of livestock can result in enhanced growth and health of offspring in economically valuable food animals.

15.14 SUMMARY

The presence of mammary glands and their ability to secrete milk distinguishes mammals from all other forms of animal life. It is fortunate, indeed, that nature provided humans with many alternative means of obtaining the product of lactation. But especially important are the mobile, miniature milk-making factories that constitute the udders of cows and provide over 90 percent of the milk supply of humans in the United States.

Although percentages of milk constituents vary among species, the milk of all mammals contains similar components. Milk of each species was designed by nature to be especially beneficial in rearing the young of that species (Section 3.2.4). Many factors influence the composition of milk and the intensity of its secretion. However, milk composition of a given animal is nearly the same day after day and similar throughout the world.

STUDY QUESTIONS

1. How many mammary glands does the mare have?

2. Give two anatomical characteristics in which mammary glands of cows and women differ.

3. Which species provides the most milk for humans?

4. Define a mammary gland.

5. What proportion of milk is commonly secreted in the fore and rear quarters of the cow?

6. Distinguish between secretory and connective tissues of mammary glands.

7. Is there a possible relationship between high milk production and a pendulous udder? Explain.

8. Can milk from the right forequarter be removed by milking the left forequarter? Explain.

9. What determines how easy it is to remove milk from a cow?

10. What are galactophores?

11. Distinguish between epithelial and myoepithelial cells.

12. Where is milk secreted? What is an alveolus?

13. Approximately how much blood is circulated through the mammary gland for each volume of milk secreted?

14. Is the size of milk veins a good indicator of a cow's milk-secreting ability?

15. In what direction does lymph flow in the cow's udder?

16. What is edema? Is it totally undesirable at parturition? Explain.

17. What are the roles of estrogen and progesterone in the growth and development of mammary glands?

18. How may one experimentally quantify the degree of mammary gland growth?

19. What is "witch's milk"?

20. How is it possible to obtain milk without motherhood?

21. What is the role of prolactin in lactogenesis?

22. Do dairy cows ordinarily secrete more or less thyroxine in winter than in summer? Relate this to the level of milk production in those two seasons.

23. How is growth hormone related to milk secretion?

24. Which vitamin is related to the parathyroid hormone?

25. Name three species in which oxytocin is important. What are its effects?

26. How is milk made? How is it discharged?

27. Give several techniques for determining which substances are milk precursors.

28. What milk protein is important for lactose synthesis? What is the significance of lactose synthesis in the mammary epithelial cell for milk secretion?

29. From where or from what are the following milk constituents derived: (a) protein, (b) lactose, (c) fat, (d) minerals, (e) vitamins, and (f) water?

30. In what form is casein secreted? What other milk components is casein associated with when secreted?

31. How can milk and blood be isotonic when milk contains *more* sugar, calcium, phosphorus, lipids, and potassium and *less* protein and sodium than does blood plasma?

32. When is milk secretion the most rapid? Explain.

33. Why are the strippings richer in milk fat than the first few streams of milk?

34. Can vacuum per se be used to fully evacuate milk from mammary glands? Explain.

35. What contributions to the understanding of milk letdown were made by Ott and Scott?

36. What is oxytocinase? Is it present in all species?

37. What is meant by the transitory effect of oxytocin?

38. What is complementary (residual) milk?

39. To obtain the most milk, should cows be milked quickly or slowly? Explain.

40. What is meant by peak milk flow? Is this trait heritable?

41. What is the relationship between epinephrine and oxytocin in milk letdown?

42. What is mammary gland involution? How may it be hastened?

43. Briefly discuss factors affecting lactation.

44. Is it possible for a cow to secrete more calories into her milk than she consumes in her ration? Explain.

45. What factors affect the composition of milk?

46. Is the percentage of milk protein, lactose, and fat of low or high heritability? Of what practical significance is this?

47. Why is it important for the newborn calf to receive colostrum milk?

48. Does the neonatal human benefit immunologically from mother's colostrum milk? Explain.

49. Distinguish between *in vivo* and *in vitro*. (See Glossary.)

50. How may transgenic animal technology be used to benefit humanity?

16

PHYSIOLOGY OF EGG LAYING[1]

Whereas the nourishment milk is produced for mammals in the breast, nature does this for birds in the egg.

Aristotle (384–322 B.C.)
Greek philosopher

16.1 INTRODUCTION

Most students think of an egg only as a kind of food produced by chickens. Actually, all birds lay eggs but few other than chicken eggs are used for human consumption in the United States. Egg laying by birds results from a complex natural endowment whose prime aim is procreation. Many other kinds of living organisms including insects, worms, fishes, reptiles, and **mammals** produce eggs. In all levels of animal life (above the very lowest) it is impossible for young to be produced except from eggs.

> It has, I believe, been often remarked that a hen is only an egg's way of making another egg.
>
> **Samuel Butler (1612–1680)**
> **English satirical poet**

The foremost purpose of eggs in all **species** is to perpetuate life. Eggs, therefore, are an essential link in the reproductive cycle of animal life. An egg, like milk, is a secretory product of the reproductive system and the **endocrine, metabolic,** and physiochemical mechanisms of egg laying are similar to those of lactation (Chapter 15). The **bird's egg** is much larger than the mammal's egg because the bird's egg must contain food for the embryonic development of the young while it is growing outside its mother's body. Conversely, most young mammals develop embryonically within the mother's body and obtain food from their mother (after implantation) until they are ready to be born. The **cleidoic** (closed environment) arrangement of

the egg demands that egg **nutrients** be concentrated. Because fat is more concentrated calorically (per unit bulk) than sugar the egg is rich in fat rather than in carbohydrates.

16.2 EGG COLORS AND SHAPES

Eggs vary greatly in color, shape, and size. However, nature provided these several kinds of eggs with one common objective: to provide for the future form and existence of the species.

16.2.1 Egg Color

The eggs of wild birds have a wide range of colors from the white of the flicker's egg to an almost solid-black egg laid by some ducks. Leghorn **hens,** the predominant type of chicken used for table egg production, lay white-shelled eggs although most **breeds** of chickens lay brown-shelled eggs. The Araucana hens of South America lay blue-shelled eggs. Contrary to the belief of some there is no relation between eggshell color and the nutritional content of eggs.

16.2.2 Egg Shape

Because albumen is added to the yolk as it spirals through the oviduct, why are eggs not all spherical? In general, the smaller the egg laid by a given hen the more nearly round it is. This shape likely results from the ease with which a small egg passes through the oviduct. Large eggs are subjected to greater pressure exerted by oviductal muscles so that they frequently have conical, oval, elliptical, or biconical shapes. Normal egg shape

---

[1]The authors acknowledge with appreciation the contributions to this chapter of Dr. J. D. Firman, Department of Animal Sciences, University of Missouri–Columbia.

is determined in the magnum (Figure 16.3, Section 16.4.2); however, the shape may be modified by abnormal or unusual conditions in the isthmus or uterus. The vast majority of eggs laid are of normal shape. Eggs which are misshapen will generally have weak areas and may not remain intact during the typical cleaning and handling that eggs go through in a commercial situation. This is why one rarely sees an abnormally shaped egg at the supermarket.

16.2.3 Bird Eggs

Usually the larger the bird the larger the egg. For example, the African ostrich, the male of which may be 8 ft tall and weigh 300 lb, lays an egg much larger than the adult female hummingbird whose egg weighs about as much as a copper penny (Table 16.1).

The number of eggs that the different kinds of wild ducks lay varies greatly. The hornbill lays one egg a year. Pigeons usually lay 2 to 4 eggs a year and gulls lay 4. The graylag goose lays 5 to 6 eggs, the mallard duck 9 to 11, the ostrich 12 to 15, and the partridge 12 to 20 eggs annually. **Domesticated** hens and ducks may lay 350 or more eggs in 1 year.

16.3 THE STRUCTURE OF AN EGG

When the contents of a fresh egg are exposed by careful removal of a sizable portion of the shell and shell membrane from the upper half, an **opaque,** circular white spot is commonly visible on the yolk's surface. In the unfertilized egg this spot is called the *blastodisk* and its homologous structure in the fertilized egg is called the *blastoderm.* The blastodisk is only 3 to 4 mm in diameter and within it are the chromosomes. The most important part of an egg is the **nucleus** (germ). This part develops into the new animal. The other egg components provide food and protection for the young animal.

A bird's egg has five principal parts: (1) shell, (2) shell membranes, (3) albumen (from Latin *albus* meaning "white") called *egg white* because of its color after coagulation, (4) yolk, and (5) germinal disk (Figure 16.1).

The shell is composed of two main layers, which contain pores so water and gases can pass through. For many years scientists thought the contents of chicken eggs were sterile, and in most cases this is true. However, we now know it is possible for bacteria to pass through the shell and contaminate the egg. As many as 8000 pores have been observed in an egg. The shell of a hen's egg is **translucent** as laid but becomes opaque as soon as it dries. A thin film called the cuticle covers the outside of a fresh egg. The cuticle soon dries and tends to seal the pores of the egg thereby reducing the loss of water and gases and invasion by bacteria. During the shell-formation process pores are formed in places where the eggshell is in contact with the uterine epithelium. The shell is more porous on the large end of the egg. During the 3-week period of incubation about 15 percent of the egg's moisture evaporates. There is a direct linear relation between egg volume and shell thickness. The shell contains

TABLE 16.1	Proportional Parts of Selected Bird's Eggs				
			Proportional parts, %		
	Weight of egg, g	Incubation period, days	Albumen	Yolk	Shell
Precocial birds*					
Ostrich	1400	42	53.4	32.5	14.1
Goose	200	28	52.5	35.1	12.4
Turkey	85	28	55.9	32.3	11.8
Duck	80	30	52.6	35.4	12.0
Chicken	58	21	55.8	31.9	12.3
Pheasant	32	24	53.1	36.3	10.6
Partridge	18	24	50.8	37.0	12.2
Average	268	—	53.4	34.4	12.2
Altricial birds[†]					
Golden eagle	140.0	28–36	78.6	12.0	9.4
Dove	22.0	13–14	72.4	18.1	9.5
Pigeon	17.0	14–18	74.0	17.9	8.1
Starling	7.0	11–14	78.6	14.3	7.1
Robin	2.5	14	70.3	24.2	5.5
Wren	1.0	10	71.0	24.1	4.9
Hummingbird	0.5	16	69.7	25.3	5.0
Average	27.1	—	73.5	19.4	7.1

*Quite mature at birth. Note the high level of yolk in their eggs.

[†]Very undeveloped and immature at birth. Note the relatively low percentage of yolk in their eggs.

Sources: Compiled from several sources, including J. R. Tarchanoff, *Arch. Ges. Physiol.* **33**(1884):303–378; F. Groebbels and F. Möbert, *J. Ornithol.* **75**(1927):376–384; and A. L. Romanoff and A. J. Romanoff, *The Avian Egg,* Wiley, New York, 1949.

similar levels of calcium, the primary shell constituent, whether the hen's egg is small or large. Thus larger eggs will generally have thinner shells. The eggshell must be thick enough to support its contents and yet fragile enough to crack easily when the young bird hatches.

Hens often lay proportionately more soft-shelled eggs during periods of elevated environmental temperatures. Breakage is a multimillion-dollar problem of the poultry industry and hot weather, particularly above 86°F (30°C), is a major cause. Chickens cool themselves by panting and this in turn changes the chemical composition of their blood. Carbonate is lost during panting and as a result less calcium is available for deposition in eggshells. Research at the University of Illinois showed that more calcium can be made available for eggshell production by providing carbonated water to laying hens in hot weather.

Just inside the shell are two thin *shell membranes,* which surround the white (albumen) portion of an egg. The outer shell and inner shell membranes are bonded together, except at the large end of the egg where they separate to form the air cell. When laid the egg contains no air cell until it cools and contracts forming a space between the two membranes. The air cell

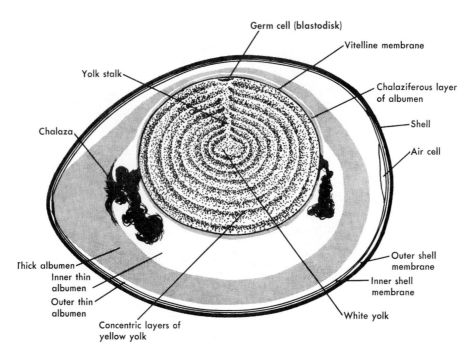

Figure 16.1 The major component parts of a bird's egg.
After F. B. Adamstone.

increases in size with time (especially in a dry, warm place). Enlargement of the air cell results when water and gas escape from the egg.

16.3.1 The White (Albumen)

The albumen is an elastic, shock-absorbing, insulating, semi-solid protein mass having a high water content. It has four principal layers: (1) outer thin white, (2) thick white, (3) inner thin white, and (4) thick white that surrounds the yolk. This innermost layer of white is twisted in a ropelike structure at each end of the yolk and forms the *chalaza*. The chalazas help anchor and stabilize the yolk near the geometric center of the egg but allow it to turn and twist (Figure 16. 1).

16.3.2 The Yolk

The yolk is composed of a series of concentric, alternating dark and light layers and is contained in a thin yolk sac called a *vitelline membrane*. The specific gravity of the yolk is less than that of the white, therefore it stays slightly above the center in a freshly laid egg. The *germ cell* (germ spot) is a tiny area on the upper surface of the yolk. It is lighter in color than the rest of the yolk and thus easily identified. The germ cell (blastoderm) in a **fertile** egg develops into the embryo under proper environmental conditions. A two-yolk (double-yolked) egg may be fertile but seldom hatches. Double-yolked eggs probably result from the almost simultaneous **ovulation** of two yolks. Infertile eggs are preferred for table consumption.

Nature provided the egg to maintain continuity of life. Therefore its contents give a well-balanced diet to the developing embryo and it provides an excellent nutrient source for humans, as well. The shell is largely (93 to 98 percent) calcium carbonate ($CaCO_3$) and provides the embryo with calcium through diffusion for bone formation and other body-building purposes. The shell also contains a small amount of protein and other minerals.

When eggs of birds are grouped according to the relative amounts of yolk and albumen there are two broad classes. Those in which the yolk constitutes about 35 percent of total weight belong to *precocial* birds; eggs in which the yolk constitutes a lower percentage (about 20 percent) of total weight belong to the *altricial* group (Table 16.1). Because the yolk has the greatest food value, a relatively large yolk ensures a more advanced stage of development in the young at birth (hatching), which characterizes precocial birds. In the small-yolked eggs of altricial birds, the young are helpless nestlings. Moreover, most altricial birds lay eggs having relatively thin shells. In proportion to total egg weight, eggshells of precocial birds are considerably heavier.

16.4 REPRODUCTION AND EGG FORMATION

The formation of eggs and their subsequent utility is an essential link in the reproductive cycle. In many species **fertilization** is not an essential preliminary to egg laying. The domestic hen can lay eggs continuously without being mated or without being stimulated to lay by the presence of a male. This biological phenomenon has been utilized advantageously by humans in producing infertile eggs for human consumption. However, in nature eggs must be fertilized to perpetuate life within each species. Hence the need for the **rooster** (male **fowl**) remains.

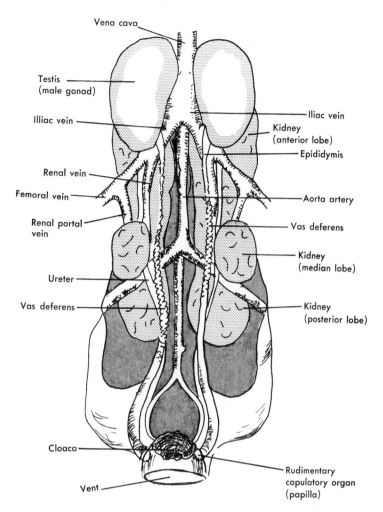

Figure 16.2   The avian male reproductive system.
After L. V. Doom, University of Chicago.

16.4.1 Anatomy and Architecture of the Avian Male Reproductive System

The **avian** male possesses two testes (**gonads**). They are normally yellow and situated high in the abdominal cavity near the **anterior** ends of the kidneys (Figure 16.2). Unlike those of human males and other higher animals, gonads of male birds do not descend into an external scrotum. The testis consists of a large number of slender convoluted **ducts** from which the lining gives off male reproductive cells (**spermatozoa**). These narrow ducts (seminiferous tubules) lead to paired vas deferens, tubes that convey spermatozoa and seminal fluid outside the body.

Each vas deferens terminates near the small papilla, which together serve as an intromittent organ. This rudimentary **copulatory** organ is used to classify baby chicks according to sex on the basis of cloacal examination. Additionally, geneticists have introduced the fast-feathering sex-linked characteristic in many heavy meat strains. In chickens with this characteristic the primary wing feathers grow faster than the covert feathers in the female. In the male the covert feathers are the same length or longer than the primary feathers.

During the process of mating the spermatozoa are introduced by the papillae into the oviduct opening in the cloacal wall of the female.

16.4.2 Development of the Bird's Egg

The reproductive organs of the hen normally consist of a *left* **ovary** and a *left* oviduct. The right ovary and oviduct are present early in life but later regress (**atrophy**).

Egg development begins with the formation of the germ and yolk in the female ovary (from the Latin *ovarium* meaning "egg holder"). The fowl ovary contains a large number (3000 or more) of spherical bodies (Graafian follicles) and resembles a cluster of grapes.

Each reproductive cell is called an *ovum* (Latin meaning "egg"). It is contained in a thin envelope of the ovary called the *follicle.* Concurrent with egg development is the deposition of yellow and white yolk granules. When the yolk is fully formed the follicle that houses it ruptures along a streak (a nonvascular suture line) called the *stigma.* The yolk is freed from the ovary and, with its germ cell, escapes into the infundibulum (funnel) of the oviduct.

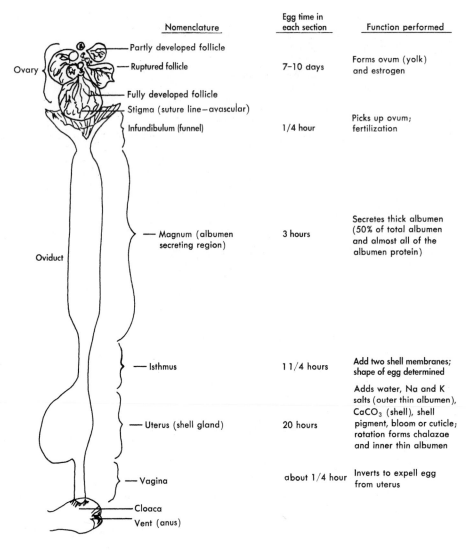

Nomenclature	Egg time in each section	Function performed
Partly developed follicle		
Ruptured follicle	7–10 days	Forms ovum (yolk) and estrogen
Fully developed follicle		
Stigma (suture line—avascular)		
Infundibulum (funnel)	1/4 hour	Picks up ovum; fertilization
Magnum (albumen secreting region)	3 hours	Secretes thick albumen (50% of total albumen and almost all of the albumen protein)
Isthmus	1 1/4 hours	Add two shell membranes; shape of egg determined
Uterus (shell gland)	20 hours	Adds water, Na and K salts (outer thin albumen), CaCO₃ (shell), shell pigment, bloom or cuticle; rotation forms chalazae and inner thin albumen
Vagina	about 1/4 hour	Inverts to expell egg from uterus
Cloaca		
Vent (anus)		

Ovary

Oviduct

The time from oviposition to ovulation = about 1/2 hour
The time from ovulation to oviposition = about 25 hours

Figure 16.3 Reproductive system and egg formation in the hen.

Semen can be stored for many days in the oviduct where fertilization occurs. Within half an hour of the egg's arrival it departs for a journey of about 25 h in assembly-line fashion. The yolk is grasped by the mouth of the oviduct (infundibulum) and travels slowly through the magnum (about 3 h), isthmus (about 1.25 h), uterus (about 20 h), and vagina (about .25 to .5 h).

As the yolk traverses the magnum most of the albumen is secreted and deposited around it. It is interesting that most of the proteinaceous albumen is deposited in only 3 h. The albumen is filtered from the bloodstream into the glands of the magnum. The magnum is incapable of distinguishing between an egg yolk and a foreign body. Thus it will deposit albumen around a marble, a cork, or even a round ball of paper.

The shell is formed in the lower region of the oviduct (uterus).[2] In the domestic hen about 24 to 26 h elapse from the time the yolk enters the oviduct until the completed egg is laid (Figure 16.3). It should be noted that there is *no cervix* in the

reproductive system of birds. The lower end of the oviduct opens into the *cloaca* (Latin for common sewer), which contains openings for the reproductive, digestive, and urinary tracts.

The ancestors of domestic hens normally laid eggs during the spring months only. Through scientific selection and mating systems (Chapter 9) and modern, improved poultry management domestic hens now lay eggs throughout the year. Under natural conditions egg laying is most intense during the spring months. Nearly all wild birds lay in the spring. Birds respond to the gradual increase in the amount of daylight during the spring months and thus exhibit **photoperiodism.**

The bird's eye is sensitive to light intensity. As the light increases it causes an increase in the activity of the pituitary gland and its hormonal secretion. The higher **hormone** levels

[2]Research at The Pennsylvania State University indicates that calcite crystal (calcium carbonate) seeding begins in the isthmus. Eggshell ash is 98 percent calcite.

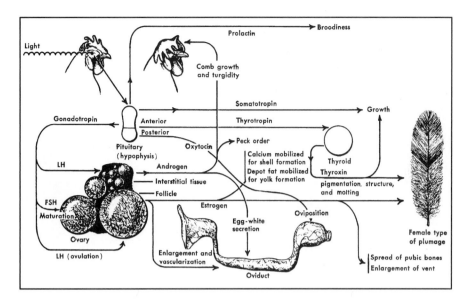

Figure 16.4 Diagram showing the principal effects of endocrine secretions and their interrelations in the female fowl.

Adapted from L. E Card and M. C. Nesheim, Poultry Production, *11th ed., Lea & Febiger, Philadelphia, 1972.*

speed ova development and subsequent egg laying. The spectral quality of light is important. Ultraviolet light apparently has no effect on sexual maturity. Some species of birds (e.g., certain owls) are more responsive to red than to green light. The great bald eagle has unusually large and powerful eyes in comparison with its body weight (approximately the size of an orange when compared with the size of human eyes as part of body weight), which aid greatly in its preeminent position of power as monarch of the sky. Indeed, the exceptionally keen eyesight of the bald eagle spawned the term "eagle eye."

16.4.3 How Avian Embryos Develop

Chickens and turkeys and **drakes** usually mate with 10 to 20 or more females. Chickens mate at any hour of the daylight period but most frequently immediately following laying (when the uterus is empty).

The fertilization of bird eggs occurs in the upper portion of the oviduct. Normally only one male reproductive cell fertilizes each female **gamete** to form the basis for the development of the embryo. For good fertility in poultry the ratio of males to females (natural matings) should be about 1:15 and 1:10 in the light and heavy breeds of chickens, respectively. Turkeys have been selected for broad breasts and large size. Natural mating is ineffective in today's commercial turkeys, which necessitates an artificial insemination (AI) program. Fertility is excellent for 7 to 10 days after AI and may continue for 50 or more days.

Eggs of the domestic hen must be kept at the proper constant temperature of 98 to 100°F (37 to 38°C) for 21 days before chicks will hatch. Turkey, guinea, and most duck eggs incubate for 28 to 30 days. (Muscovy duck eggs require 35 days.) Goose eggs incubate about 28 to 32 days, quail eggs 22 to 24 days, and pheasant eggs 23 to 28 days depending on the variety. Eggs

may be incubated under hens (natural incubation) or in incubators (artificial incubation).

16.5 HORMONAL REGULATION OF EGG LAYING

Both the physical appearance and the functioning of birds are profoundly affected by hormones. (The various endocrine glands and their secretions were discussed in Chapter 11.) Few endocrine effects result from the direct action of a single hormone. Instead the physiological activity of the chicken, particularly the female, is dependent on a complex interrelation of glandular effects, as exemplified in the complex hormonal control of ovulation and egg formation.

The interrelations of these various hormones are shown diagrammatically in Figures 16.4 and 16.5. The entire process of egg laying is dependent on hormone synchronization and balance. If an individual endocrine gland releases a hormone without awaiting the proper signal, abnormalities such as yolkless eggs, soft-shell eggs, and eggs within eggs are likely to result.

16.5.1 FSH and LH

Follicle-stimulating hormone (FSH), from the anterior pituitary gland, stimulates growth of ovarian follicles with their contained ova. When a follicle becomes mature another anterior pituitary hormone (luteinizing hormone, or LH) is released and causes ovulation. Functional ovarian activity in the mature hen can be inhibited by starvation. This effect is attributed to a failure of the pituitary gland either to produce or to release gonadotropins because the administration of these substances will cause follicular growth and ovulation.

The oviduct is also under hormonal control and is stimulated at the most appropriate time to receive the released ovum. Ovarian

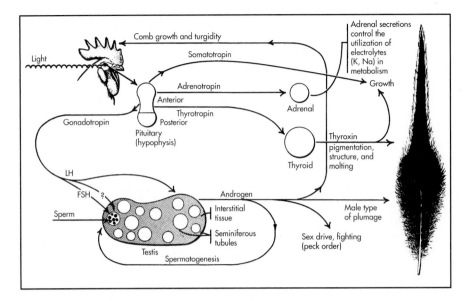

Figure 16.5 Diagram showing the principal effects of endocrine secretions and their interrelations in the male fowl.

Adapted from L. E. Card and M. C. Nesheim, Poultry Production, *11th ed., Lea & Febiger, Philadelphia, 1972.*

follicle secretions are responsible for (1) enlargement of the oviduct to functioning size, (2) spread of the **pubic** bones and enlargement of the vent, (3) mobilization of depot fat for yolk formation and calcium for shell formation, and (4) female plumage. The secretion of albumen is apparently under the control of a hormone (androgen) secreted by the ovarian interstitial tissue.

16.5.2 Parathyroid Hormone

Formation of the eggshell is partially controlled by hormones secreted by the parathyroid glands. As discussed in Chapter 11 these hormones maintain normal calcium and phosphorus levels in the blood.

16.5.3 Thyroid Hormone

In addition to influencing body growth and feather color and formation, the **thyroid** gland is partially responsible for seasonal changes in egg laying, body weight, and egg weight. The latter effect presumably results from the varying amount of light to which hens are subjected with the changing seasons. The thyroid gland increases in size and hormone secretion during that portion of the year when **molting** normally occurs. The feeding of large doses of either fresh or **desiccated** thyroid results in rapid molting. Thyroxine also regulates body metabolism (Chapter 11).

16.5.4 Oxytocin

This hormone is secreted in the hypothalamus but is stored in and released from the posterior pituitary gland. It causes **oviposition,** or the contraction of muscles of the shell gland (uterus), and hence expulsion of the fully formed egg.

16.5.5 Prolactin

This hormone is responsible for the manifestation of the maternal instinct. Prolactin is believed to give a hen the urge to seek a nest and lay eggs in it. If prolactin is administered to a rooster in sufficiently high doses he will become broody and set on eggs as a normal broody hen does. If discontinued the effect soon disappears and such roosters are likely to kill the **chicks** that they previously cared for and defended.

16.5.6 Sex Hormones

The sex hormones affect the social status of birds. The male sex hormone (androgen, or testosterone) is secreted by the testis and is responsible for the red, waxy **comb** and wattles. In the normal laying hen the female hormone (estrogen) controls the secondary sex characteristics such as normal female plumage, absence of spurs on the legs, and female sexual behavior.

Chickens (unlike farm mammals) *never exhibit estrus* and are *never pregnant.* Because birds produce no corpora lutea following ovulation they can ovulate daily. In farm mammals corpora lutea are present and secrete progesterone, which prepares the uterus for the reception and development of the fertilized **ovum** by a glandular **proliferation** of the endometrium. The secretion of the corpus luteum (progesterone) also prevents further ovulations when pregnancy occurs (Chapter 13).

16.5.7 Adrenal Hormones

Secretions of the adrenal glands are involved in the metabolism of **carbohydrates,** sodium, and potassium and in the regulation of blood pressure.

16.6 How an Egg is Laid (Oviposition)

Eggs are usually formed small end **caudal,** that is, with the small (sharp) end first, as they move in assembly-line fashion down the oviduct. This is true of wild birds and of domestic fowls. However, if the hen remains quiet during the act of laying most eggs are presented large (blunt) end first. What explains this apparent paradox?

Just prior to being laid the fully formed egg is turned horizontally (not end over end) 180°. For this to be accomplished the egg must drop from its normal position high between the ischia (analogous to the hipbones) to a point opposite the tips of the pubic bones. This preliminary voyage is necessary because the normal egg is too long to turn in a horizontal plane within the pelvic arch. After the egg has been thus turned it is in position ready to be laid. The hen accomplishes this egg-turning feat in only 1 to 2 min. Should the hen be physically disturbed as she raises her body slightly when the egg is about to turn, she is likely to expel it prematurely and in that event it will be presented small end first. Occasionally the egg is retained and goes back up the oviduct (reverse peristalsis) in which case the hen puts on a second shell. If it reverses to the magnum the hen will add a second white and shell (an egg within an egg), as shown in Figure 16.6.

The act of laying an egg is made possible through the influence of the release of the hormone oxytocin from the **posterior** pituitary gland. Oxytocin causes the uterine **musculature** to contract vigorously and expel the egg. It seems reasonable to conclude that the uterine muscle pressure required for expelling an egg is more effectively applied to the small end; hence, the basis for the reversal of the egg just prior to laying. The uterus expands to about 3 times its normal size during passage of an egg. The uterine wall everts (**prolapses**) through the vagina and cloaca allowing the egg to drop from the cloacal **orifice.** The egg does not physically contact the wall of either the vagina or the cloaca. The vaginal musculature contracts voluntarily about the prolapsed uterus thereby assisting in oviposition.

Both egg laying and ovulation are subject to the external influence of light and darkness. Normally about 30 min after a fully developed egg is laid, a new egg is ovulated from the ovary. It is at this time that a hen cackles, not when she lays the *hard-shelled* egg. This delay from the time of oviposition to ovulation allows the bird time (in its natural environment) to get away from the nest before she calls for a mate.

16.7 Factors Affecting Egg Laying

The most basic factor limiting the number of eggs a hen lays is her genetic makeup, which greatly influences her physiological efficiency and metabolic activity.

The average annual number of eggs laid per hen in the United States doubled from about 140 in 1935 to 280 in 2000, which is an average increase of about 2.1 eggs per hen each year from 1935 to 2000.

16.7.1 Removal of Eggs from the Nest

The regular removal of eggs increases the rate of egg laying. Even wild birds such as the sparrow have the inherent ability to lay several times their usual seasonal egg output if the eggs are removed from the nest. In the wild a hen laid about a baker's dozen (13 eggs) in a nest, and the pressure of the eggs against her breast caused her to realize that that was as many eggs as her body could cover and hatch effectively. This recognition caused her endocrine system to switch from secreting the gonad-stimulating hormones to secreting prolactin, which is the hormone responsible for the manifestation of the maternal instinct. When the chicks are **hatched** and the effect of the eggs' pressure on the breast is removed the bird's pituitary resumes the secretion of FSH and LH and egg laying continues. This phenomenon explains why the removal of eggs from the nests of laying hens aids in maintaining egg laying.

16.7.2 Age at Sexual Maturity

Sexual maturity is reached when a **pullet** lays her first egg (commonly at about 5 months of age). The onset of lay is closely controlled in commercial situations by close control of day length (increasing day length above 12 h will stimulate a hen to lay), body weight, and body fat. Increasing day length hastens sexual maturity whereas decreasing day length delays sexual maturity (Figure 16.7). Modern methods of light control in poultry rearing and laying can largely eliminate seasonal effects on age to sexual maturity and minimize variations in egg-laying patterns. Achieving the correct timing for onset of laying is critical to achieving maximum egg output from commercial hens. Those that lay too early may produce a number of very small, unsaleable eggs and begin to lay prior to being moved to the cage house thereby disrupting the laying cycle.

Figure 16.6 An egg within an egg.
Courtesy of Columbia Missourian.

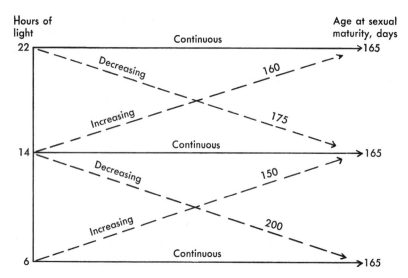

Figure 16.7 Diagram illustrating the control of sexual maturity in chickens by day length (hours of light). Absolute light level is not as important as the change in the amount of light. Note (a) continuous light at the same level results in about the same age at sexual maturity, (b) increasing light hastens sexual maturity, (c) decreasing light delays sexual maturity, and (d) increasing or decreasing light below 14 h has a greater effect than when light exposure is more than 14 h daily.
Courtesy of Professor Q. B. Kinder, University of Missouri.

Pullets that begin laying late may miss a number of eggs early and have their cycle shortened when the entire flock's lay is ceased.

16.7.3 Season and Light Patterns

Because hens are very sensitive to light, maintenance of laying is at least partially done through increasing and maintaining hours of day length. Most hens will be kept with 14 to 18 h of day length. As the ovulatory cycle is generally slightly longer than 24 h, most egg-producing hens at peak production would lay a **clutch** of 10 to 14 eggs. The first egg might be laid very early (4 A.M.) with each subsequent egg laid somewhat later in the day until she reaches early afternoon. She will then skip a day and resume laying early in the morning again.

16.7.4 Broodiness

Brooding is the act of hens **setting** on and attempting to incubate eggs. This trait is undesirable in egg producers because the hen neglects egg laying and favors setting on a nest when she goes broody. Modern breeds of chickens are less broody than those of years past. Broodiness is not a significant problem in the chicken industry.

Even though the natural tendency to incubate eggs (broodiness) has been decreased through selective matings of chickens, it still recurs cyclically in turkeys. If allowed to run its full course each broody period lasts about 4 weeks, which is the length of time corresponding to the turkey's natural incubation period. Because broody hens seldom lay eggs this nonproductive period is undesirable in commercial breeders. Broodiness may be induced but generally occurs in a small percentage of turkey hens.

Broody hens should be removed immediately from the flock because it appears they may induce other hens to go broody. Broodiness reduction strategies include prompt removal from nests, careful collection of eggs not laid in nests, and placing removed hens under 24-h light stimulus.

16.7.5 Clutch Length and Intensity of Egg Laying

The interval between the ovulation of two successive eggs is usually 24 to 26 h. A hen normally lays one egg a day for several successive days before missing a day. This uninterrupted series of successive eggs laid is referred to as a *clutch*. This trait is heritable and hens laying more eggs per clutch (6 or more) produce more eggs than hens laying fewer eggs (1 to 3) because they have fewer nonproductive days. Hens having a short interval (less than 2 h) between oviposition and ovulation lay more eggs per clutch and have higher annual egg production.

Most wild birds lay only one clutch of eggs during the breeding season. However, should the clutch be destroyed the bird may lay another clutch. Some wild birds such as the robin and bluebird lay, incubate, and brood two and sometimes three clutches in one season. Some wild birds may abandon the nest if any eggs are removed from it, whereas others may continue laying in an attempt to establish a clutch for hatching. In one study, when the eggs of a flicker were removed from the nest as soon as they were laid, the bird produced 71 eggs in 72 days. Data on normal clutch size and annual egg laying are presented in Table 16.2.

16.7.6 Egg Laying Cycles

A young hen or pullet will begin laying eggs at 18 to 20 weeks of age. The egg-production cycle starts slowly and then rapidly

TABLE 16.2	Egg Laying by Selected Birds	
Species	**Average clutch size (no. of eggs)**	**Maximum no. laid annually when eggs are removed from nests**
Chickens		
Egg-producing	10–14	300–365
Meat-producing	10–14	190–230
Game or fancy	10–14	60
Ducks		
Egg-producing	14–20	250–310
Meat-producing	14–20	120
Turkeys	15–20	220
Geese	12–15	100
Ostriches	12–15	100
Pheasants (in confinement)	10–12	104
Quails (in confinement)	12–20	130
Pigeons	2	50
Canaries	4–6	60

increases in a few months to peak production. Peak production in a Leghorn-type hen will reach 90 to 95 percent of hens laying on a given day (hen-day production). After peak production, egg production will gradually decrease over many months. Generally by 12 to 14 months of lay, egg production decreases to the extent that the birds are no longer economical to maintain and will be replaced. In some cases hens will be molted and put through another egg-laying cycle. Broilers and turkeys used for breeding purposes will follow a similar egg-laying cycle although it will start later (24 to 30 weeks), be shorter, and result in fewer total eggs produced (100 to 200). As the hen ages she lays larger and larger eggs as well as increases her body size.

16.7.7 Molt

The shedding and replacing of feathers is a natural physiological phenomenon in birds associated with a decline in the functional activity of the reproductive organs. Egg laying ceases as molting begins, and there is a marked regression in the secondary sex characteristics. Therefore molting corresponds to a period of physiological rest. Persistent layers molt last and require the shortest time to shed their old feathers and grow new ones. Moreover, birds that do not exhibit a complete molt may lay without lengthy interruptions for 2 or more years.

Molting can be induced by one or more of the following conditions: (1) withholding feed and/or water, (2) an abrupt decrease in the daily amount of artificial light, and (3) administration of derivatives of thyroxine (these also cause depigmentation of feathers).

In the commercial poultry industry molting is generally induced (referred to as forced molting) to provide a rest period before a new egg-laying cycle occurs. In years past molting was not nearly as common as it is today. This is due to the decreased use of laying hens for human food. In the past hens were processed for meat after the first laying cycle. The meat was generally used in soups and the old or spent hens were referred to as soupers. Today spent hens are no longer used for this and thus have lost the value inherent with their use for meat. Currently most hens are rendered into an animal feedstuff.

Hens that are molted are based on an economic decision analysis. If a hen is to be molted the first egg cycle is generally shortened. The hens are molted (6 to 8 weeks) and would then have another cycle similar to the first with the exception that it will be shorter, reach lower peak production, and yield fewer eggs overall. Although hens lose weight during the molt they will generally be larger and lay larger eggs during a second cycle.

16.7.8 Nutrition

The egg laying of the domestic hen's ancestor, which laid only 25 to 30 eggs annually, was probably not particularly affected by **nutrition.** However, the hen of today is expected to lay 300 or more eggs annually and must therefore receive special nutritional considerations. To lay an egg a day a hen needs approximately twice as much protein,[3] carbohydrate, and fat as is required for body maintenance and she needs a substantial increase in dietary mineral intake. During the year a 300-egg hen secretes approximately 4 times as much dry matter as is contained in her body. The 38 to 40 lb of eggs that she lays are composed of about 4.6 lb of protein, 4.1 lb of fat, 4.2 lb of minerals, and 25.1 lb of water. Thus adequate nutrition and a balanced diet are imperative for the laying hen. Of the 103 known chemical elements, 36 have been found in bird eggs.

The blood proteins, calcium, phosphorus, and **lipids** in the **laying** hen are about two-, two-, three-, and fourfold, respectively, higher than in the nonlaying hen. The increase in these blood constituents is explained by the need for great quantities of proteins for yolk and albumen formation, for an abundance of calcium and phosphorus in shell formation, and for lipids in the formation of egg yolk. When the hen ceases to lay these blood constituents are soon reduced to the level prior to laying.

16.7.9 Environmental Changes

Extreme heat or cold, rain, wind, or sudden changes in temperature and humidity may have a temporary adverse effect on the reproductive activity of the hen by (1) causing physical discomfort, (2) lowering vitality, (3) reducing food intake, or (4) restricting exercise and thereby resulting in reduced egg laying.

---

[3]Studies at the University of Missouri–Columbia demonstrated that the protein level in diets of laying hens could be reduced to 15 percent when proper amounts and balance of amino acids were provided.

16.8 Factors Affecting the Composition and Characteristics of Eggs

The average gross composition (by weight) of hen eggs is about 10 percent shell, 30 percent yolk, and 60 percent albumen (white). Egg composition was further defined in Table 3.1. Large eggs contain proportionally more albumen and less yolk than small ones. The percentage of shell remains nearly constant, except that the first pullet eggs contain relatively more shell and less yolk than is characteristic of hen eggs in general. The last egg of a clutch is usually smaller and has a thicker shell than the egg that preceded it. The quantity of shell secreted per egg remains relatively constant; consequently, the larger the egg the thinner the shell.

Interestingly, hen eggs are of fairly constant composition from day to day. Should the actively laying hen be deprived of important dietary essentials, nature allows her to withdraw from body reserves to produce an egg of full nutritional value. When calcium is not provided in the diet, body stores are soon depleted to such an extent that the hen stops laying.

A medium-sized hen producing 300 eggs annually deposits about 3.75 lb of calcium carbonate (1.5 lb of calcium) in the eggshells. This represents about 30 times the amount of calcium in her entire body and points to the need for an abundance of calcium in her feed. Present evidence indicates that only about 60 percent of the calcium in a particular egg comes directly from the feed; the balance comes from deposits in the long bones. In wild birds, which lay only a few eggs, all the calcium may come from bones. However, when laying hens are suddenly deprived of all feed calcium their eggs promptly show thin shells and the hens stop laying in 10 to 14 days. Of all egg components the shell is the most variable. In one study shell weight decreased about 25 percent when the hen's environmental temperature was increased from 66 to 102°F (19 to 39°C). The albumen also decreased significantly whereas the yolk weight remained about constant.

16.8.1 Yolk Color

Yolk color is influenced mostly by the hen's diet. Highly pigmented feed ingredients, especially alfalfa, can produce dark orange yolks whereas less pigmented feeds result in a relatively light yellow yolk, which is preferred by most egg consumers.

Gossypol (a toxic dye) is present in raw cottonseed meal and may cause olive-colored egg yolks, especially in stored eggs. This color is attributed to the reaction between the ferric iron in the yolk and the yellow pigment gossypol. Apparently the ferric **ions** in fresh eggs are insufficiently dissociated to enter into a visible reaction.

Blood spots, or meat spots, probably occur in about 3 percent of all chicken eggs. They are more frequent in eggs of the heavier breeds (e.g., Rhode Island Reds) than in those of the lighter breeds (e.g., White Leghorns). There is evidence indicating that blood spots result from intrafollicular hemorrhage several days prior to ovulation. Meat spots result from the sloughage of small bits of flesh from the reproductive tract before the shell is laid down.

16.8.2 Feed

Recent interest in the possible association of **cholesterol** and **atherosclerosis** (Chapter 3) has prompted those involved in egg production to seek ways to lower the cholesterol content and/or increase the degree of unsaturation of the fatty acids of egg (Section 3.2.2). The latter change has been accomplished by feeding diets containing higher levels of unsaturated **fats** to hens.

The cholesterol content of eggs is not easily changed by alterations in the hen's diet. Feeding a high-cholesterol diet to a hen significantly increases the amount of cholesterol in the egg (by 20 to 30 percent); however, feeding a low-cholesterol diet will not significantly reduce the level of cholesterol in eggs. Working with normal human subjects a scientist at the University of Missouri–Columbia demonstrated that the body reduces cholesterol synthesis to compensate for increased dietary intake thereby maintaining a normal level of blood cholesterol (Section 3.5.1).

Birds eat more in relation to their body size than do most humans. The smaller the bird the more it eats in proportion to its weight. A 14-oz pigeon eats food equal to one-twentieth of its weight daily. To equal this a 180-lb man who "eats like a bird" would have to eat 9 lb of food daily. (The average is 3 lb daily for a man.) The greater food intake per unit of body weight required by small animals results from a higher metabolic rate among small than among large animals (see Chapter 12 and Figure 12.8).

16.9 Factors Affecting Egg Size

The size of eggs is commonly expressed in terms of weight, which provides a more convenient basis of comparison than dimensions or volume. Egg weight varies immensely among species (Table 16.1). Hen eggs normally weigh from 40 to 80 g each.[4] The highest recorded chicken-egg weight was 320 g, the smallest was 1.3 g.

16.9.1 Genetic Influences

The average egg weight per hen is highly heritable (about 40 percent) and the trait of laying large eggs has become rather permanently fixed in domestic birds. Obviously the commercial hen has been bred in part to produce not only a large number of eggs but large-sized eggs, as well.

16.9.2 Age of Bird

Pullet eggs normally are about three-fourths the maximum size reached at maturity. The first egg laid by a pullet is almost always her smallest, and the size of the first egg is a good indication of the relative size of eggs that a bird will lay in the future. If the first egg is large the bird usually continues laying large eggs. Data related to hatching date, age at sexual maturity, and egg size are presented in Table 16.3.

---

[4]Minimum egg weights per dozen, determined by the USDA, are Jumbo, 30 oz; Extra Large, 27 oz; Large, 24 oz; Medium, 21 oz; Small, 18 oz; and Pee Wee, 15 oz.

TABLE 16.3	Relation of Hatching Date to Days to Sexual Maturity and Egg Size in Chickens

Month hatched	Light per day in month of hatch, h	Days to sexual maturity (natural lighting)	Large eggs per year, %
January	9.5	164	87
March	11.2	184	89
May	13.8	189	94
July	14.9	200	94
September	13.0	190	93
October	11.8	179	72

Source: W. C. Skoglund, New Hampshire Expt. Station.

16.9.3 Environmental Temperature

Eggs laid by mature hens are larger in the colder months (December to February) and smaller in hotter ones (June to August).

16.9.4 Size of Bird

Larger hens usually lay larger eggs. Of interest is the kiwi bird of New Zealand (weighs about 2000 g), which lays an egg weighing nearly one-fourth its body weight. Because adult bird body weight is highly heritable (about 50 percent) the influence of bird size on egg size is closely related to genetic makeup. Much of the early improvement in domestic-hen egg size resulted from selection for larger body size; however, today's commercial egg strains have been successfully selected for small body and large egg size concurrently. The greatest increase in egg size has been in fowls that have been under domestication the longest (some more than 5000 years); the least is observed in turkeys, which were domesticated more recently (Chapter 2). An increase in egg size throughout the first year is associated with a gain in body weight.

16.9.5 Ovum Size

Egg size is largely determined by the size of the ovum passing through the oviduct. Relatively less albumen is deposited around a small yolk (as in pullets) and the finished egg is smaller than an egg laid by a mature hen, which normally ovulates a larger yolk.

The most common abnormality in eggs is the phenomenon of two yolks. In a double-yolked egg two ova are ovulated simultaneously; however, double-yolked eggs may also result from the premature ovulation of one of the two ova or from the retention of one yolk in the body cavity for a short period of time.

16.9.6 Intensity of Egg Laying

Although many contradictory reports regarding the relation between egg weight and the intensity (rate) of egg laying have been published there is sufficient evidence to warrant the conclusion that if intensity only is selected for, egg size tends to decline, and if large egg size only is selected for, the intensity, or laying rate, declines.

16.9.7 Nutrition

A deficiency of feed nutrients tends to reduce egg weight and curtail the number of eggs laid. A deficiency in vitamin D tends to reduce egg weight. This may be related to the role of vitamin D in calcium metabolism, so important for eggshell formation.

16.10 IMMUNOLOGICAL AND MEDICAL ASPECTS OF EGGS

The egg possesses antigenic properties. Egg **antigen–antibody** reactions are manifested by such commonly observed phenomena as allergic symptoms, **anaphylactic shock** (caused by an exaggerated reaction of the organism to a foreign protein), **complement fixation,** and precipitation (the formation of a visible precipitate caused by an antibody to **soluble** antigen that specifically aggregates the **macro**molecular antigen). Thus egg proteins are used to test whether a person is hypersensitive to certain substances.

Egg albumen may cause **allergy.** The term *allergy* refers to *clinical symptoms resulting from specific antigen–antibody reactions within the body.* Skin tests may be employed to diagnose possible adult and child allergies to egg albumen.

The hen transmits some antibodies to the chick through the egg. If diphtheria **antitoxin** is present in egg protein before incubation it may be detected in the blood of newly hatched chicks. Tetanus antitoxin is similarly transmitted to the chick as are antibodies against fowl pox and Newcastle disease. This **transitory immunity** (also known as passive or transferred immunity) disappears within the first month of life.

Egg albumen is often used as an antidote when a person orally consumes a poison. The albumen apparently prevents absorption of the poison by coating the mucous membrane of the stomach and inactivates some poisons by combining with them chemically. Egg albumen is also used to increase the reliability of positive reactions in the Wassermann test (a test for syphilis based on the fixation of complement). Eggs and chick embryos are also used as **culture** media for the biological manufacture of over 30 **viral vaccines.**

When heated (above 135°F, 57°C) with a liquid, egg albumen either settles to the bottom or forms a scum at the top. Egg albumen may be added to coffee to settle precipitates. Egg albumen is also sometimes used in fruit and alcoholic beverages as a clarifying agent. A 2 percent solution of dried albumen is added to the beverage and the mixture is then heated. The albumen coagulates and settles out carrying with it the fine particles that otherwise would cause cloudiness of the beverage.

16.11 SUMMARY

There are many colors, shapes, and kinds of eggs. Nature designed the eggs of each species especially for the perpetuation of that species. Eggs of precocial birds contain more yolk (it contains the greatest food value) than those of altricial birds. This correlates with the greater maturity of precocial birds at birth.

Humans have found the egg both to possess fine flavor and to be a rich source of important nutrients. By means of domestication and continued genetic improvement through **selection,** humans have developed highly productive birds. Unlike their ancestors they lay generous quantities of eggs throughout the year and thereby enhance the nutritional status of humanity. The egg-laying patterns of birds are closely related to changes in light. Through research humans have discovered ways of applying artificial light and thereby minimizing seasonal fluctuations in egg production.

Although percentages of egg constituents vary among species the eggs of all species contain the same components. Many factors influence the composition, size, and the intensity with which eggs are laid. However, egg composition of a given bird species is nearly the same day after day and throughout the world.

STUDY QUESTIONS

1. Why are bird eggs larger than those of mammals?

2. Why do you think nature provided for variation in eggshell color?

3. Differentiate between precocial and altricial birds; also the relative levels of albumen and yolk in their eggs. Review both Table 16.1 and Section 16.3.2.

4. What is the blastodisk? Blastoderm?

5. Name the five principal parts of a bird's egg.

6. What is the function of the chalazas?

7. Why does the yolk remain above center in a freshly laid egg?

8. Can double-yolked eggs be fertile? Will two chicks hatch from them?

9. Is mating essential to egg laying? To egg fertility?

10. How is it possible to classify baby chicks according to sex?

11. Are both the right and left ovaries and oviducts functional in most domestic birds?

12. Briefly relate how an egg is developed as it traverses the oviduct. How much time is required for this process?

13. Do domestic birds have a cervix?

14. What is photoperiodism? Is the spectral quality of light related to sexual maturity and egg laying in domestic birds?

15. When do chickens most frequently mate?

16. What is the length of the incubation period in chickens? In turkeys?

17. How are the hormones FSH and LH related to egg laying?

18. Which hormone is closely related to eggshell formation?

19. Briefly discuss the functions of thyroxine in poultry.

20. Which hormone is involved in the expulsion of an egg?

21. What is prolactin? What is its relation to hatching of eggs naturally?

22. Are hens ever pregnant? Why?

23. How is it possible to have an egg within an egg?

24. Does the regular removal of eggs from a nest influence egg laying?

25. What is the effect of increasing light on sexual maturity in chickens? Of decreasing light?

26. Under normal conditions are eggs laid during the day or night? How can this laying pattern be changed?

27. What is broodiness? How may it be induced in hens?

28. What is a clutch? Is clutch size heritable?

29. Is there more variation in clutch size among wild or among domestic birds?

30. At what age is egg laying most intense in chickens? Is this true of geese?

31. What is the relation of clutch size to life span of birds?

32. Is molting natural? How can molting be induced? Can it be prevented?

33. Approximately how much of the dietary protein, carbohydrate, and fat intake of a laying hen is used for body maintenance? For egg laying?

34. What is the average gross composition of hen eggs? Does the composition vary much from day to day? Which egg component is the most variable?

35. Does a hen's feed affect egg yolk color?

36. Does dietary fat affect the degree of unsaturation of fatty acids of eggs? Explain.

37. Do birds eat more or less in relation to their body size than humans? Why?

38. Briefly discuss factors affecting egg size.

39. How can the formation of double-yolked eggs be explained?

40. Can the hen transmit antibodies to the chick through her egg?

ECOLOGY AND ENVIRONMENTAL PHYSIOLOGY[1]

The biotic power of an animal is limited only by the repressive forces in the environment.

Samuel Brody (1890–1956)

17.1 INTRODUCTION

More than 2000 years ago Aristotle said that certain conditions of **climate** and weather produce the most energetic people and the finest **livestock;** that the rise or fall of mighty empires is due to changes in either climate or knowledge of protection from climate and weather; and that the nation that will lead the world is the one that will best mitigate negative impingements of its weather. The pertinence of this ancient prediction is evidenced today by the urgency of such problems as a tendency toward higher summer temperatures and the phenomenal growth of populations throughout the world. In tropical and arctic climates **adaptation** is difficult for the most productive livestock and poultry. In these areas, also, the population/food ratio is unfavorable to the nutritional well-being of people. Moreover, the increase in population is often greater than the rate of increase in food production. The wisdom of Aristotle's foresight can be further illustrated by the development of only a few mighty empires because approximately three-fourths of the world's population lives outside the temperate zones, in areas where air temperature often rises above 100°F. This temperature is far above the maximum of the **thermoneutral zone** for humans and most animals. Today war and peace efforts are global in scope, embracing weather extremes that stress humans and machines to the limits of their endurance.

A **symbiotic** relationship exists between humans and animals. Although each may live independently of the other, reciprocal benefits result from their interactions. The plight of

humanity is better because of animals and vice versa. Animals tend to serve as a buffer between humans and their **environment.** Because of this there is a need for additional knowledge of the symbiosis among human beings, animals, and the environment.

> Human ecology and animal ecology have developed in curious contrast to one another. Human ecology has been concerned almost entirely with the effects of man upon man, disregarding often enough the other animals amongst which we live.
>
> **Charles Elton**

Progress during the twentieth century in acquiring a better understanding of the relations among climate, humans, and animals was very significant. One of the first practical applications of managing the environment was the use of artificial light in egg-laying houses to increase the hours of light and thereby stimulate fall and winter egg laying (Figure 17.1). Artificial light schedules also may be used to hasten or delay sexual maturity in poultry (Chapter 16). Conversely, decreasing light stimulates sheep to mate in the fall, allowing the **lambs** to be born in the spring when the plant **nutrient** supply and weather are more favorable for survival, growth, and development of the young. When sheep are transported across the equator from one hemisphere to another they change (by 6 months) the time of the calendar year at which they mate to conform with the seasons in their new environment. Thus light is known to influence reproduction in poultry and sheep and in goats, horses, cattle, and swine, as well. The effects of increasing natural day length may be reflected in part by hormonal changes that promote the increase in **lactation** milk yield when cows are put on new pasture in the spring.

[1] The authors acknowledge with appreciation the contributions to this chapter of Dr. S. E. Curtis, Department of Animal Sciences, University of Illinois at Urbana–Champaign.

Figure 17.1 This hen is believed to hold the world record for egg laying in 1 year. In research at the Missouri Agricultural Experiment Station, Dr. Harold Biellier selected birds that had the biological capability of responding, productionwise, to light–dark cycles of less than 24 h and, in addition, had the genetic capability of laying eggs at intervals of less than 24 h over an extended period of time. Dr. Biellier demonstrated experimentally that the length of environmental light to which birds are exposed has a profound effect on the level and persistency of egg production. When exposed to 22- and 23-h light–dark cycles, rather than the usual 24-h day length for part of the year, this hen laid 371 eggs in 365 days and went on to lay 448 eggs in 448 days. Such research has many important applications, as well as economic and production implications, in providing nutritious eggs at economical prices for the consuming public.
Courtesy of Dr. H. V. Biellier, University of Missouri.

Samuel Brody, a keen-minded pioneer in the field of environmental physiology and animal energetics at the Missouri Agricultural Experiment Station from 1921 to 1956, gave both leadership and scholarship to the science of comparative animal physiology. The methods he and his research colleagues developed to measure physiological responses of horses and mules to various workloads under outdoor summer conditions are clas-

sic. Figure 17.2 illustrates the use of an ergometer to measure work rate and a respiration apparatus to measure energy expended for the work.

Data obtained demonstrated there were significant individual differences among animals in pulse rate, work efficiency, respiration rate, and other cardiorespiratory measures, and further that these differences were correlated with the oxygen debt observed following work. Such individual differences could be used in subsequent animal selection and breeding programs. These experimental techniques, and others that followed, gave impetus to additional research studies related to the physiology of growth, reproduction, lactation, nutrition, and senescence as well as to endocrine aspects of farm animal production. Much of the material presented in this chapter had its genesis in Brody's early research endeavors.

The need for more food has resulted in the migration of both humans and animals from one region to another as discussed in Chapter 18. This migration in many cases exposes humans and animals to a new environment, which may require adaptation. There are many challenging aspects of **ecology,** the branch of biological science that deals with the relations of living things to their environments and of *environmental physiology,* which deals with the surrounding conditions that affect structures and organ functions of humans and animals. An attempt has been made to include many of the interesting aspects of ecology and environmental physiology in this chapter. Behavioral adaptations of animals to their environments are explored in Chapter 24.

17.2 HEREDITY AND ENVIRONMENT

When the early American settlers first brought animals from Europe many of the animals lacked resistance to the environmental extremes they encountered here and died prematurely. Yet how many plants and animals would there be in America today if the pioneers had concluded that they could not exist and grow here because none were here at the time they came? The common farm animals are among the most adaptable birds and mammals on earth; that is why they became domesticated in the first place.

An animal's nature is a product of its **heredity** and environment. A given hereditary pattern is itself, however, the product of environment, because **mutation** rates and selective survival are conditioned by the environment. In this way the different climatic regions cause animals to adapt to life in them. This adaptation may be illustrated by the fact that animals evolved in cool regions are adapted to cold weather, often having an abundance of wool or hair and **subcutaneous** fat, whereas animals evolved in hot regions are adapted to hot weather by the sparsity of wool or hair and subcutaneous fat. The fat resources of animals in hot climates may be stored in humps (cattle) and tails (fat-tailed sheep). Such fat stores do not interfere significantly with heat **dissipation** from the body.

As will be discussed in more detail later, dark surfaces tend to absorb more solar radiation than light-colored ones. If dark

Figure 17.2 Field apparatus used in early research studies to test cardiorespiratory and metabolic functions in horses and mules during rest, work, and recovery from work. Data collected or generated included oxygen consumption, carbon dioxide production, pulmonary ventilation rate (volume of air exhaled and inhaled per unit of time), pulse rate, rectal temperature, energy expended, and work efficiency.
Courtesy of Missouri Agr. Expt. Station.

skin absorbs more solar-radiant heat than light skin and if it is true that humans and animals best suited to the environment of a particular region have survived, how can the apparent paradox and anomalous adaptation be explained that darker-skinned people commonly populate the hottest regions? Could this be due to a more uniform distribution of pigment throughout the **epidermis** or to a greater thickness of the corneum? Or, are there other roles for pigment in the epidermis in addition to heat absorption?

Skin pigments include **melanin** (black to yellow), which is produced in the body. Light-skinned persons develop temporary melanin pigmentation (tanning) when exposed to sunlight whereas dark-skinned persons possess melanin or melanoid pigments as a genetic characteristic. Ultraviolet (UV) radiation is **bactericidal** and toxic to skin and nerve tissues. Melanin protects nerves and skin from the injurious UV radiation. In this sense pigmentation of tropical races has adaptive value. Animals indigenous to the tropics often have light-colored hair and dark skin to afford protection both from sunshine heat (light hair) and its UV component (melanin in the skin).

Environment includes the physical, chemical, and biological elements that surround animals. With the exception of feedstuffs, which in the case of domesticated animals are commonly studied by themselves, all components of the environment are included in the realm of animal ecology.

Various environmental factors may either promote or impair animal performance by facilitating or inhibiting productive and reproductive processes. Scientific investigations of environmental effects on animal production are of recent vintage, probably because most people believed that breeding, feeding, and management research had greater economic impact than improvements in environmental factors. However, as animal production becomes more intensive in space and time, environmental aspects of food–animal management become more and more important. The work of Dr. Stanley E. Curtis and his co-workers at the University of Illinois in this rapidly growing field of science is both timely and noteworthy,

and his monumental textbook provides an excellent overview and summary of the present state of the art.[2]

17.2.1 Regional Adaptation of Cattle

Animals are now transported long distances and in large numbers from temperate to tropical zones and from semitropical zones to areas where cold conditions prevail. Moreover, frozen semen is now available for international shipment (Chapter 14) making it easier to transport new genetic materials from one place to another. However, transporting animals or germ plasm does not change the inherent genetic characteristics that make it possible for the animal to adapt to its environment.

Most breeds of livestock were developed in temperate zones and were naturally selected to fit the environment. Several features distinguish Indian from European cattle: (1) Indian cattle have excellent radiators—enormous **dewlaps** (loose, pendulous skin under the throat extending back between the legs and along the belly), sheaths (navel flap in females), and long ears.[3] Their bodies tend to be small resulting in a relatively large surface area per unit weight and their skin has little hair. Conversely, European cattle have fat, hairy, tight hides, undeveloped dewlaps, and relatively small ears. (2) Indian cattle have a more extensive system for evaporative cooling, which allows them to adapt better to hotter climates. Missouri studies showed that in the zone of **thermoneutrality** (Section 17.9) Shorthorns had higher skin, respiratory, and total vaporization rates than Brahmans. However, when exposed to temperatures high enough to cause a rise in rectal temperature, the opposite

[2] S. E. Curtis et al., *Environmental Aspects of Animal Care,* Iowa State University Press, Ames, IA, 2002.
[3] McDowell (R. E. McDowell, *Improvement of Livestock Production in Warm Climates,* Freeman, San Francisco, 1972) reported that removal or reduction of the appendages did not bring about a significant change in the response of animals to thermal stress. The hump of Zebu cattle is high in fat and not well supplied with blood. Moreover, the Zebu has less **subcutaneous** fat than European breeds. This suggests that one reason the Zebu is more heat-tolerant than European cattle may be that it stores much of its depot fat in the hump, rather than as subcutaneous fat, which tends to impede the flow of warm blood to the surface for cooling.

was true. (3) The hair color of Indian cattle tends to be lighter; thus it reflects more sunlight than does that of European cattle. (4) Indian cattle are more resistant to **ticks,** flies, mosquitoes, and other pests than are European cattle.

17.2.2 Genetics and Heat Tolerance

Is it possible to combine the *high productivity* of the *heat-intolerant* European breeds with the *high* **heat tolerance** and *low productivity* of the Indian-evolved cattle to obtain a highly productive, heat-tolerant animal? Wide genetic differences are determined by a series of **genes** of multiple types, some genes having an *additive* and others a *nonadditive* effect (Chapter 8). When breeds are crossed the genes recombine in new patterns. By combining genes from two or more breeds backlogs of the evolutionary resources of both may be utilized in developing new breeds. An example is the Santa Gertrudis cattle developed on the Kleberg family's King Ranch in Texas. These cattle possess the heat tolerance of their Indian ancestors (Brahmans) and the high-meat-production characteristics of their European ancestors (Shorthorns).

Genetic combinations having favorable hereditary characteristics also may be obtained within a breed, but progress is slower. As discussed in Chapter 9, the more homozygous the breed, the more limited the progress.

17.3 ADAPTATION TO ENVIRONMENT

> Complete adaptation to environment means death. The essential point in all response is the desire to control environment.
>
> **John Dewey (1859–1952)**
> **American philosopher and educator**

Climate (from the Greek *klima* meaning "slope" or "tilt") depends on topography, on distribution of vegetation (especially forests), and on large bodies of water. (Seventy-one percent of the earth's surface is covered with water.) The environment is largely reflected in the climate of a particular area.

Animals themselves are entities through which materials and energy flow before eventually returning to the environment. Without the ability to adapt animals are at the mercy of the environment. When animals are repeatedly or continuously exposed to major environmental challenges they may develop functional and structural changes that result in an increase in their ability to live without **stress** in the new environment. These changes are collectively referred to as the adaptive process of **acclimatization.**

Animals best adapted to a particular environment have survived and, unless "transplanted" by human beings, populate areas best suited for their survival. However, modern transportation has increased the movement of animals and placed selection of breeding stock on a global basis, forcing consideration of the question, "How do the changed conditions affect the productivity, health, and longevity of animals?"

Environmental temperature affects the secretion of certain hormones. For example, laying hens secrete over twice as much thyroxine in winter as in summer and the same trend has been observed in lactating dairy cows. There is a parallelism among egg laying, lactation, growth, and indeed all productive processes.

The depressing effects of hot weather on body activities of animals may be termed "hot-weather laziness." But when viewed **homeostatically** the decreased activity is a biological mechanism for preventing overheating of the body. This is an adaptive mechanism enabling survival in hot weather. Survival responses by an animal to environmental stress are more immediately crucial to life than the productive and reproductive processes exploited by humans in animal production.

17.3.1 Physical Environmental Factors in Adaptation

Physical environmental effects on animals may be classified as *natural* and *artificial.* The *natural factors* include (1) temperature (hot and cold), (2) air **humidity,** (3) air movement (wind and draft), (4) barometric pressure, (5) rainfall and water, (6) altitude, (7) dust, (8) radiation (visible, ultraviolet, and infrared), (9) **cosmic radiation,** and (10) atmospheric electricity.

The *artificial factors* include (1) atmospheric pollution from industry and smog, (2) toxic compounds in water, (3) sonic factors (noise, ultrasonics), (4) **ionizing radiation (isotopes, x-ray),** and (5) artificial **ionization** of air.

Animals have certain inherent mechanisms that enable them to adapt to changes in various natural and to a lesser degree artificial physical environmental forces.

Altitude

Did you ever wonder why trees and animal life thin out as you drive up the mountains in the western United States? Did you ever climb a high mountain? If so you noted a decrease in temperature, vegetation, and animal life, but more noticeable was the increased respiration rate necessary to acquire sufficient oxygen to continue the climb. Climatic effects of altitude changes are important. Atmospheric pressure is reduced by about one-half at 18,000 ft compared with that at sea level. The llama is well adapted to altitudes up to 17,500 ft. The lower affinity of their hemoglobin for oxygen increases the delivery of oxygen from their red blood cells to other tissues.

17.3.2 Natural Adaptations of Animals to Their Environment

The razorback conformation of the wild pig was a natural adaptation enabling it to penetrate dense vegetation in search of food. Desert animals commonly must traverse large territories in search of food because of the effects of seasonal rainfall on vegetative growth, so they have relatively long legs. The large, flat foot of the camel enables it to move readily over shifting sand. Similarly the surefootedness of the mountain goat, which enables it to graze on steep and rocky hillsides, is not an accident of nature.

Animals having a relatively small surface area with respect to body mass are best adapted to cold climates. Coat thickness and hair characteristics in mammals and plumage in birds are greatly influenced by climate.

17.3.3 Natural Variation in the Heat and Cold Tolerance of Animals

The long-haired yak cattle of Tibet and western China and the woolly cattle of the Scottish Highlands are almost as cold-tolerant as the arctic-dwelling caribou and reindeer although they employ different strategies of adaptation. Even European-evolved cattle are cold-tolerant. Conversely, the Indian-evolved cattle, called **Zebu** or Brahman, are heat-tolerant but cold-intolerant.

Missouri studies found the thermoneutral zone (temperature interval in which no demands are made on the body's temperature-regulating mechanisms) of European cattle is between 30 and 60°F whereas for Indian cattle it is between 50 and 80°F. Therefore, beginning at about 60°F in European and about 80°F in Indian cattle, the thermoregulative mechanisms become active as demonstrated by an increase in respiration and vaporization rates. Beginning at about 80°F in European and 95°F in Indian cattle these mechanisms become incapable of meeting the demands for heat dissipation resulting in (1) a rise in body temperature, (2) decreases in feed intake and metabolic rate, and (3) decreases in milk secretion and other productive processes. Often the animal loses weight as a consequence.

17.3.4 Comparison of the Climatic Adaptive Physiology of Humans and Cattle

The close symbiotic association between humans and cattle causes one to wonder what climatic adaptations they have in common. Because of modern science and technology human beings are climatically the most adaptable species. In the nude, however, humans are uncomfortable at temperatures around 60°F or below. (This lower critical temperature depends on air temperature, humidity, air velocity, and other factors.) Humans are much more heat-tolerant than cattle. This is illustrated by their near-normal and the cow's 5 to 7°F above-normal rectal and skin temperatures when exposed to an environmental temperature of 105°F. In a 105°F environment the cow's skin and rectal temperatures were 105 and 107°F, respectively, whereas those of humans were 95 and 99°F, respectively.

Human sweating increases at the rate of 9 to 13 percent per 1°F temperature rise between 85 and 105°F. This accounts for their ability to withstand high temperatures. Conversely, the cow has very limited means of evaporative cooling in the same temperature range, which results in a rapid rise in rectal temperature. Swine—the sweat glands of which do not respond to heat stress—are even less able to cool themselves evaporatively in a hot environment.

17.4 STRESS

A large number of environmental factors may cause an animal to respond homeostatically. Exposure to stressful conditions sometimes results in nonspecific increases in secretion of glucocorticoid hormones by the adrenal glands and frequently in other more specific changes as well.

In farm animals one of the most important stress factors from an economic standpoint is *heat stress*. When exposed to high temperatures animals exhibit **polypnea** (panting). This condition of increased respiration rate results in the increased dissipation of heat in two ways: (1) by warming the inspired air and especially (2) by increasing **evaporation of water** from the respiratory passages and lungs.

At high temperatures livestock and poultry lower feed consumption and thereby reduce heat production. As a result of this, productivity (egg, meat, milk, and wool) is also reduced.

Measuring Thermal Stress

The stressfulness of an environment can be measured by the animal's response to it. There are several means of detecting whether an animal is under thermal stress. The most obvious stress index is body temperature response. An elevated body temperature in turn reflects certain other detectable indices such as heart rate, respiration rate, and **heat production (HP).**

17.5 HOMEOSTASIS AND HOMEOTHERMY

Homeothermic animals of large body size have no need of annual migration. **Homeostasis** enables them to adjust to the winter cold. But small homeotherms (e.g., birds) find seasonal migration necessary. This is because small animals have relatively more surface area per unit of body weight than do large ones and heat dissipation is proportional to body surface area.

Disappearance of large reptiles in the Cretaceous and Eocene Periods may have resulted from unusual temperature changes that homeotherms but not **poikilotherms** survived.

Homeotherms vary in normal body temperature from 96°F (36°C) in elephants to about 109°F (43°C) in small **avian** species. The respective normal rectal temperatures of selected species may be grouped as follows:

96 to 101°F: elephant (96.0), mouse (97.3), human (98.6), rat (99.0), and horse (100.0)

101 to 104°F: cattle (101.0), monkey (101.1), cat (101.5), dog (102.0), sheep (102.3), pig (102.5), rabbit (103.1), and goat (103.8)

104 to 106°F: turkey (104.9), goose (105.0), owl (105.3), and duck (106.0)

107 to 109°F: chicken (107.0), English sparrow (107.0), hummingbird (108.0), and robin (109.4)

17.5.1 Homeotherms (Warm-Blooded Species)

Homeothermic animals tend to maintain constant *internal* temperature even when *external* temperature varies. For example, the annual atmospheric temperature range in Montana may be more than 150°F (– 40 to +110°F) yet the body temperatures of cattle, horses, and sheep wintering outdoors are much the same as those during the summer. How is it possible for animals to adapt to the vicissitudes of weather?

Homeotherms are much better able to protect themselves against cold than against heat. As *cold weather* approaches the cold-weather homeothermic mechanisms include (1) growth of insulating hair and subcutaneous fat; (2) increase in thyroid activity; (3) consumption of great quantities of food, which, because of its **heat increment (HI),** warms animals; (4) seeking protective shelter and warming solar **radiations;** (5) grouping together (huddling); and (6) increases in activity,[4] both voluntary and involuntary (shivering). All these mechanisms increase heat production. A means of *heat conservation* is the constriction (narrowing) of superficial blood vessels (**vasoconstriction**) reducing blood flow to the skin from which heat is lost via **conduction,** convection, evaporation, and radiation.

As *hot weather* approaches cooling mechanisms become important. These include (1) moisture **vaporization,** (2) avoidance of solar radiation, (3) depression of thyroid activity, and (4) refraining from work and reducing productive functions (includes the agriculturally important egg laying, lactating, and meat production because these processes increase heat production).

Moisture vaporization from skin and respiratory passages is the most important cooling mechanism and is the only one when environmental temperature equals or exceeds body surface temperature. Profusely sweating species can therefore withstand high environmental temperatures. Most nonsweating species attempt to compensate for their inability to sweat by panting, often protruding their tongues and blowing air rapidly over its moist surface thereby accelerating evaporation rate. Swine will die when exposed to an atmospheric temperature of 100°F in a dry, sunny lot whereas they can withstand that temperature indefinitely when given access to a mud wallow; water evaporates from mud on a pig's skin at a very high rate cooling the pig in the process.

17.5.2 Partly Homeothermic, Hibernating Species

Animals in this category include the bat, woodchuck, dormouse, ground squirrel, hedgehog, groundhog (marmoset), opossum, and prairie dog. These animals maintain a fairly constant body temperature during the summer, but as winter approaches and their temperature-regulating mechanisms begin to fail they retire into burrows or migrate into caves below the frost line and enter their winter sleep (**hibernation**) until the coming of spring. Body temperature falls during hibernation to a few degrees above freezing and metabolic rate and pulmonary ventilation (respiration) may decline to about 1 percent of normal summer levels.

In times of famine human beings have been known to show **dormancy** during most of the winter months. Unlike that of the marmoset, however, the body temperature of humans does not fall appreciably below normal during their "winter sleep." The dormancy (not true hibernation) of the bear is probably of the same category—not a loss in ability to maintain normal body temperature but rather a way of saving energy by sleeping deeply. Hibernation probably is an evolutionary adaptation to prevent starvation during periods of food scarcity.

Cold Habituation

When suddenly subjected to a relatively severe cold stress, animals that have been experiencing a milder cold stress for an extended period sometimes delay for several hours their thermogenic reaction to the acute stress. In fact, body temperature may actually decrease for a short period. This *cold habituation* is useful because it enables animals to withstand acute cold stresses of short duration, which frequently occur in nature (e.g., at night), without raising metabolic rate unnecessarily. This enables animals to conserve feed energy. If the stress persists for longer than a few hours, heat production increases to effect heat balance at normal body temperature.

17.5.3 Poikilothermic, Estivating Species

Estivating animals, which sleep during a hot, dry summer, are exemplified by frogs, crocodiles, and alligators. Estivation probably is an evolutionary adaptation to periods of water scarcity. Hibernation and estivation of **insects** and soil inhabitants are of considerable agricultural interest.

Poultry embryos might be classed as cold-blooded until they reach a certain age. Even after hatching they must be kept under **brooding** heater units until they perfect their homeothermic mechanisms. Similarly, newborn rats, puppies, kittens, and **piglets** are very susceptible to cool environments.

It is fortunate indeed that humans and the animals that serve them can maintain a near-constant body temperature. At extremely low temperatures reactions slow down or cease and at too high a temperature total destruction of organic complexes occurs and death ensues.

17.5.4 Age and Homeothermy

Homeothermic mechanisms are not required by the developing mammal before birth. Depending on **ambient temperature,** children become fully homeothermic between 12 and 24 months of age, rat pups in about 3 weeks, chicks in 3 to 4 weeks, calves and lambs within a day after birth, and piglets 2 to 3 days **postnatum.**

17.5.5 Experimental Hypothermy

Until recently it was assumed that nonhibernating mammals could not tolerate significantly reduced internal temperature. However, survival of "supercooled" bats, cats, and rats prompted surgeons to attempt this procedure (**hypothermy**) in brain and **cardiac** surgical operations. An example is cited of a woman who was cooled to a "core" temperature of 48°F and recovered. Her heart was at a standstill for 45 min. Several investigators have observed **ventricular fibrillation** in humans and dogs during cooling, usually at a body temperature of about 25°F below normal.

---

[4] It should be noted that animals commonly decrease their activity in the huddle; so much in fact, that even though they need to increase feed intake, sometimes they actually decrease it—preferring to huddle rather than to eat.

17.6 TEMPERATURE REGULATION

The body is able to employ physiological, behavioral, and anatomical mechanisms to accomplish homeothermy. For example, the **endocrine** *system* aids the animal in adjusting to seasonal and short-term temperature changes by its secretion of thyroxine and other hormones. Thyroxine is associated with seasonal temperature changes, epinephrine and growth hormones enable animals to cope with rapid temperature changes.

The temperature-regulating center is in the brain's hypothalamus. It is an astonishingly accurate and quick thermostat. The hypothalamic thermoregulatory center is itself sensitive to the temperature of the blood flowing through it. The hypothalamus integrates this information with other inputs on the entire body's thermal status as it decides whether thermoregulatory responses are needed and, if so, which ones.

Humans can maintain a normal body temperature when exposed for short periods to an atmosphere of 300°F or higher (a temperature that will grill a beefsteak) provided the air is dry, thus permitting evaporative heat loss at a very high rate. Conversely, a damp (high relative humidity) atmosphere with a temperature of 120°F causes body temperature to rise rapidly and cannot be endured for more than a few minutes. This is because the difference between the vapor pressure at skin and respiratory passages and the air is too narrow to permit a high evaporation rate.

The nervous system also controls the "ruffling" of feathers or "raising" of hairs to increase insulation and thereby decrease heat loss (**thermolysis**). Contraction of smooth muscles of the skin gives rise to "goose flesh." Furthermore, shivering (tensing of muscles), chattering of teeth, and other muscular activities involved in acute heat-production responses are under nervous control.

Animals try to reduce heat loss in cold weather by devices such as (1) reduced vaporization (no sweating); (2) lowered respiration rate; (3) shunting of blood from surface to deep body; (4) hunching and huddling; (5) production of warm coats of feathers, fur, hair, or wool; (6) deposition of subcutaneous fat; and (7) sheltering from the wind and the cold, clear sky.

Chemical temperature regulation adds to heat-preservation processes through (1) increased **thermogenesis** (by increasing metabolic rate or exercising), (2) acute shivering and chronic muscle tension, (3) increased food intake (increasing the **heat increment of feeding**), and (4) increased adrenal and thyroid activity.

The temperature-regulating mechanisms are depressed by **anesthetics,** hypnosis, and general body fatigue.

17.6.1 Body Temperature–Regulating Mechanisms

Brody[5] summarized the rules governing climatic adaptations and named them after four pioneer investigators:

Bergmann Rule (1847)

This rule relates body size to climate. It is a geometric fact that small, *light* animals have larger surfaces per unit weight and

therefore lose heat more rapidly than larger, *heavy* animals, which have less surface area per unit weight. Body conformation influences this relationship.

Breeds of larger size are usually found in colder regions and smaller breeds in warmer climates. This is illustrated by the fact that a larger breed of dairy cattle (Holstein) is better adapted to the cooler northern areas of the United States whereas a smaller breed (Jersey) is better adapted to the warmer southern areas. Further examples of the natural adaptation of small animals to warm climates are the pygmy races of elephants, hippopotamuses, and buffalo found in warmer regions of the world. Dairy producers, however, have successfully adapted the large Holstein cow to the south by providing shade and various other means of decreasing the problems of high heat gain and compromised heat dissipation.

A seemingly contradictory example of the above rule can be found in the large Brahman cattle, which inhabit hot areas, and the smaller (with respect to overall body conformation) Shorthorn cattle, which evolved in a cooler climate. However, careful examination reveals that the Brahman, while possessing a large periphery and thereby a large surface area, is not an especially heavy animal, has a low metabolic rate, and possesses an extensive heat-dissipation system per unit weight (Figure 17.3). Conversely, the Shorthorn is short and "blocky" resulting in a lower body surface area:weight ratio.

The Bergmann rule is modeled after Newton's law of cooling, which indicates that the larger the surface area of a given body the greater its rate of heat loss in a given cooler environment via conduction, **convection,** radiation, and vaporization. Newton's law of cooling may be written as $q = kA (t_1 - t_2)$, where q is the rate of heat flow between the body surface and the environment; A is the surface area of the body; t_1 and t_2 are, respectively, the temperatures of body surface and environment; and k is the coefficient of heat transfer determined empirically.

Although Brody did not mention the *Allen rule* (1877) it is closely related to the Bergmann rule. Whereas Bergmann deals with body bulk, Allen states that in a given species of warm-blooded animals those living in cold areas tend to have shorter extremities than those in warm climates. This applies only to the living appendage and not to the fur or feathers covering the animal. This rule deals with the physics of heat loss from a warm body to a cooler surrounding. The Allen rule also can be stated as: *The nearer the body is to being a perfect sphere, the smaller is its surface area per unit volume.* Each species has its own mechanisms for heat regulation so that **interspecies** comparisons should not be made.

Wilson Rule (1854)

This rule relates insulating cover to climate. Breeds that evolved in cold climates have a dense, heavy, thick external coat (300 g/m² in Shorthorns) whereas those from warm climates have less hair that is short, glossy, stiff, and thin (10 g/m² in Zebus).

There is probably little difference in skin thickness between European and Indian cattle, but there is a difference in the thickness of the subcutaneous fat (thick in European cattle, thin in Indian).

[5] *J. Dairy Sci.,* **39** (1956):715–725.

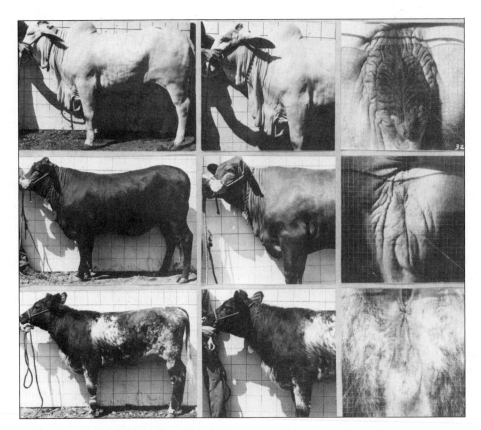

Figure 17.3 The differences in general conformation, and especially in relative heat-dissipating surface areas, in the Brahman (top), Santa Gertrudis (center), and Shorthorn (bottom) breeds of cattle at age 1 year. Note the larger and more corrugated and vascular vulva in the heat-tolerant Brahman.
Courtesy of Missouri Agr. Expt. Station.

Gloger Rule (1833)

This rule relates color to climate. Skin pigments protect against ultraviolet radiation. Conversely, the hair of animals in the tropics tends to be light-colored, which reflects solar radiation and serves to protect against overheating. Skin pigmentation of cattle gradually darkens with increasing temperature and humidity. Summer lightening and winter darkening of hair supports this rule (see Figure 17.9). Also there is increased **sebum** secretion with increasing temperature, giving the hair a reflective sheen protective against solar radiation.

Claude Bernard Rule (1876)

This rule relates climatic changes to internal body changes. Bernard observed that blood flow increased in rabbit ears during hot weather for cooling purposes and increased again during cold weather to warm the ears. The mechanism involved is analogous in principle to the valves in thermostatic regulators that control the steam supply to radiators. It is the vasomotor (controlling the size of blood vessels) dilating and constricting device and certain anatomical features that govern the rate of blood flow through the blood capillaries in the tissue layers near the skin in order to maintain skin temperature. By using these mechanisms animals can tolerate very low temperatures at their extremities. Humans, however, do not have an effective vascu-

lar thermoregulator and would find it fatal to imitate the arctic gull, which parades its slim naked legs at −40°F, or the Eskimo dog or arctic bear, which walk on their naked footpads in regions where the temperature is −60°F. Livestock and poultry also possess some of these same mechanisms. Shunting of blood in arteriovenous anastomoses is particularly important in thermal regulation in swine. Arteriovenous anastomoses are connections between arteries and veins that allow arterial blood to enter the venous circulation without passing through capillary beds.

17.7 Nutritional Aspects of Environmental Conditions

Brody told his students that scientists often attempt to prove scientifically what farmers and others have known for a long time.[6] This idea is illustrated by the conventional wisdom that when exposed to cold conditions farm animals will increase their feed intake and therefore their heat production. Cattle will increase **forage** consumption (even if it is of inferior quality) during extremely cold weather. This allows them to take advantage of

---

[6] Personal experience of the senior author.

its high heat increment (Chapter 18). They also tend to increase their tolerance to cold by increasing the deposition of insulating subcutaneous fat. Furthermore, farm animals may increase heat production by shivering and conserve heat by huddling. Changes also occur in the body's nutrient requirements at extremely low and high temperatures.

17.7.1 Changing Diet Composition to Fit the Weather

Lowered feed consumption is one of the principal causes of decreased productive synthesis during hot weather. Dairy cows yield less milk and consume less **TDN** for each degree of rise in body temperature.

Diets fed to lactating cows during periods of high environmental temperatures should be relatively low in fiber, high in energy, and moderately low in protein. Such diets have a lower heat increment so less metabolizable energy is lost as heat. Consequently, more nutrients are available to support lactation. Possible dietary effects on milk production during the summer are depicted in Figure 17.4. The chief components of high-energy diets are cereal grains. However, grains are low in potas-

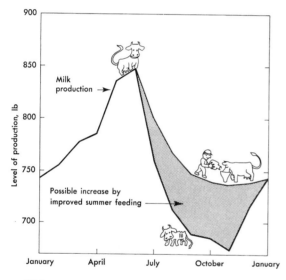

Figure 17.4 Milk yield declines during late summer, mostly as a result of high temperatures, but also in part because of the deterioration of the food supply, including the lignification of forages. *USDA.*

sium content (Chapter 21) and this, coupled with increased urinary potassium loss because of increased water consumption in hot weather, makes it necessary to provide supplemental potassium in the cow's diet.

The greatest heat increment of feeding is caused by **protein-rich foods**. Actual heat production from nutrient metabolism of food to yield 100 kcal of net energy from protein, carbohydrate, and fat fed separately to a mature animal whose standard metabolism is 100 kcal daily will be approximately 130, 106, and 104 **kilocalories,** respectively. So a prizefighter should eat beefsteak after the fight and not before because he will be sweating profusely during the increased activity within the ring and have no need for the added heat of protein digestion. High-protein foods are preferred in cold weather because of their high heat increment of feeding whereas low-protein diets (but ones that meet daily protein needs) are more suitable from this standpoint in hot weather.

Notwithstanding the heat increment gain favoring the increased use of protein in cold weather, protein is an expensive nutrient, thus the current trend in animal production is to increase the energy/protein ratio in cold weather and to decrease it in warm weather, thereby sparing the burning of expensive protein as simple fuel. The actual crude protein **concentration** in this case is higher in summer and lower in winter.

17.7.2 Comparative Feed Efficiency of Homeotherms and Polkilotherms

Why do humans produce **bird eggs** rather than reptile eggs for food? Although it has a long **generation interval** *Chelonia* (green turtle) produces eggs and meat that have been eaten by humans for many years. A green turtle can lay as many as 200 eggs in 1 day. Many tortoises that **graze** and do not compete with humans for cereals have been used for meat.

The **fowl** was **domesticated** largely for **cock**fighting and religious reasons (Chapter 2), not for egg laying. Development for efficient egg laying occurred later. Because poultry compete with humans for cereal grains why not substitute a poikilotherm, which does not burn large quantities of food to maintain its body temperature? Poikilotherms have low metabolic rates but because they grow so slowly their net **feed-conversion** efficiency is inferior to that of homeotherms (Table 17.1).

TABLE 17.1	Time and Maintenance Energy Expended for Doubling Body Weight in Homeotherms and Poikilotherms								
	Body wt, g			**Time required to double body wt, days**			**Maintenance energy expended (metabolism, cal)**		
	Cat	Dog	Fish (pike)	Cat	Dog	Fish (pike)	Cat	Dog	Fish (pike)
Birth weight	87	225	70						
Doubling once	174	450	140	8	10	270	3200	3000	3300
Doubling twice	348	900	280	12	13	300	4000	3070	2350

Source: Modified from M. Rubner, *Biochem. Z.,* **148** (1924):222, 268.

Both growth rate and maintenance cost are higher in the homeothermic animal (cat and dog) than in the poikilotherm (pike) but the overall efficiency is apparently about the same in both (Table 17.1). This might be expected because temperature affects **anabolic** (productive) and **catabolic** (destructive and maintenance) processes to a similar degree. In poikilotherms the rates of life processes such as aging, eating, growth, and metabolism increase and decrease with temperature.

17.8 FEVER

When body temperature is elevated it is said a **fever (pyrexia)** is present. This causes dilation of the skin capillaries so the skin appears flushed. With recovery, profuse sweating ensues to enable the body to dissipate heat rapidly and body temperature falls.

Fever has been recognized as a cardinal sign of infection since the time of Hippocrates. Most fevers reflect a disturbance of the normal equilibrium between thermogenesis and thermolysis because of deranged signals from the hypothalamic thermoregulatory center. This derangement can be caused by pyrogens and other toxins of microbial origin.

Fever increases an animal's metabolic rate by around 7.2 percent per 1°F rise (13 percent for each 1°C); hence the basis for the old adage, "feed a fever." During severe infections in humans as much as 400 g of body protein may be destroyed daily, which is roughly 4 times normal dietary intake (Chapter 3). **Spontaneous** temperature variations of 2°F or more are normal within a day in all animals and the peaks should not be considered "fever." Animals can tolerate a body temperature only around 10°F above normal.

Most diseases have characteristic temperature curves. Fever may be caused by (1) disease organisms (dead or alive), (2) chemical pyrogens (fever-producing substances) such as dinitrophenols, (3) thyroxine, (4) surgery, and (5) other foreign bodies.

An elevation in body temperature is often associated with the terms *heat exhaustion,* **dehydration** *exhaustion,* heat stroke, and *sunstroke.* These conditions may be precipitated by a hot environment, excessive heat production consequent to strenuous physical exercise, exposure to sunshine, or any combination of these. Collapse may result from the high temperature per se but more likely from circulatory failure. The latter results largely from peripheral vasodilation when much of the blood is at the body surface for cooling and therefore not available for oxygen transport throughout the body, and from the blood becoming "sludgy" and therefore difficult to move through the blood–vascular system.

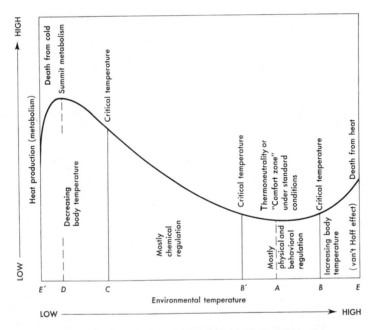

Figure 17.5 Generalized diagram of the environmental–temperature influence on heat production in homeotherms (warm-blooded animals). Note the broad range of accommodation to low temperatures in contrast with the restricted range of accommodation to high temperatures. The increased heat production with increasing cold (from *B′* to *D*) is a biological adaptation that provides protection to the animal. The increasing heat production corresponding to increasing heat (from *B* to *E*) results from a physiochemical necessity expressed by the van't Hoff–Arrhenius generalization (Section 10.5) and is destructive to the animals, ending in death (*E*).

Courtesy of Missouri Agr. Expt. Sta. Bull. *423, 1948.*

17.9 THE THERMONEUTRAL ZONE

The thermoneutral zone of environmental temperature for different animals varies with many factors including age, body size, natures of protective coverings, and adaptation status. It also depends on feed-intake rate and activity level.

The thermoneutral zone as defined by segment $B' B$ of Figure 17.5 may not be optimal for highest animal productivity because productive processes involve a heat increment not easily dissipated in a warm or hot environment. A cool or cold environment stimulates feed intake and product synthesis.

In Figure 17.5 heat production is plotted against environmental temperature to depict the relations between chemical and physical heat regulation. The physical temperature-regulation segment includes the zone of thermoneutrality (BB') where the animal does not employ extraordinary chemical thermoregulatory devices.

Thermogenesis starts increasing in direct response to cold as ambient temperature decreases below the "lower critical temperature" B' (Figure 17.5) in an attempt to offset the increasing thermolysis. When environmental temperature reaches point C the metabolic response has reached a plateau and is unable to cope further with the cold, whereupon body temperature starts to decrease despite the increased thermogenesis.

The French physiologist Jean Giaja called environmental temperature D (Figure 17.5) the point of *summit metabolism* (maximum sustained heat production) below which point a decrease in temperature results in breakdown of the homeothermic mechanisms, resulting in declines in both heat production and body temperature and eventually in death of the animal.

17.10 HEAT PRODUCTION

In cold conditions (defined as environmental temperatures below the lower critical temperature) animals need to both conserve and produce heat. They absorb heat from surrounding surfaces with higher temperatures and from direct or indirect sunshine. As shown in Figure 17.6, balance exists between those factors increasing heat production and those enhancing heat loss. Animals have a huge "maintenance cost" for circulation, respiration, ingestion, excretion, muscular activities, and many other bodily functions. The maintenance–energy expense is eventually given off as heat.

As discussed in Chapter 18 water and certain vitamins and minerals are dietary-essential nutrients for animals but they provide no metabolizable energy. Standard *heat production* is derived from *carbohydrates, fats,* and *proteins.* Selected factors increasing heat production over the standard metabolic rate (SMR) are depicted in Figure 17.6. Factors include (1) physical activity, (2) shivering, (3) erection of hairs or feathers (primarily to conserve heat), (4) subconscious tensing of muscles (muscle tone), (5) vasoconstriction, (6) fever, (7) disease resistance, (8) specific dynamic activity of food (heat increment of feeding), (9) results of secretion of epinephrine and thyroxine, and (10) increased metabolic rate. Heat gains may also be derived from radiation, conduction, and even convection.

17.10.1 Shivering

Shivering is an early, temporary response to acute cold by an unadapted organism and may increase heat production in animals by threefold or more.

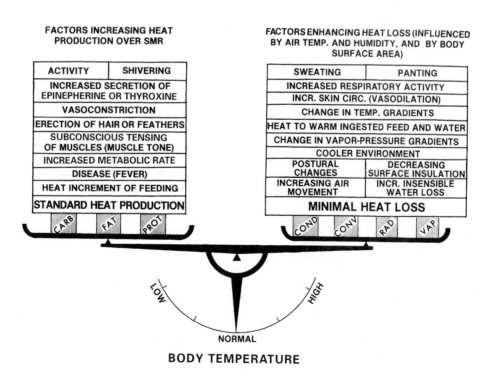

Figure 17.6 Balance between the factors increasing heat production and those enhancing heat loss in humans.

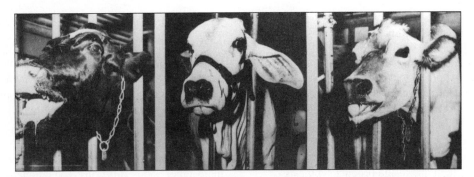

Figure 17.7 The heat tolerance of the Brahman (center). Note the panting and salivation in the Holstein (left) and to a lesser degree in the Jersey (right) at 105°F.
Courtesy of Missouri Agr. Expt. Station.

17.10.2 Variations in Heat Production among Animals

In one study heat-production rate per unit weight of Zebu **heifers** was found to be about 20 percent lower than that of Brown Swiss heifers of similar weight and age. Lesser heat production by Indian cattle partially explains their stronger heat tolerance and weaker cold tolerance (Figure 17.7).

Males normally have a higher metabolic rate than females or **castrated** males. Furthermore, animals heavily engaged in product synthesis have higher heat increments of feeding and of syntheses. A lactating dairy cow, for example, may produce twice as much heat per unit of time as a nonlactating cow of the same size.

17.11 HEAT DISSIPATION

Heat produced in domestic animals must be passed on to the environment and is dissipated primarily in four ways: (1) conduction, (2) convection, (3) radiation, and (4) evaporation (Figure 17.6). In cool or moderate environments small amounts of heat are lost by warming and humidifying inspired air and with feces and urine. Each of these avenues of heat exchange is important to an animal's heat regulation and is closely related to environmental temperature and humidity levels (Figure 17.8).

17.11.1 Conduction

Conduction of heat is based on the principle that heat flows from warm to cool objects or places. Conduction therefore depends on (1) physical contact of the animal with surrounding surfaces, (2) temperatures of the body and environmental surfaces (temperature gradient), and (3) thermal conductivity and contact area of the surfaces.

Cold water is an effective means of cooling by conduction. Metals also *conduct* heat readily whereas air, fat, feathers, hair, nylon, silk, wood, and wool do so less readily. Heat loss from animal to environment via conduction is minimized by insulation of the coat and of the environment. Animals can increase

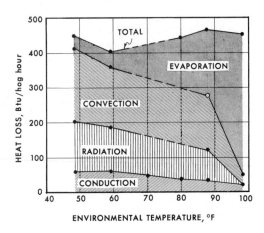

Figure 17.8 The effect of environmental temperature on heat loss in swine (75 to 125 lb) by the four major methods of heat dissipation.
Modified from Agr. Engr., *33 (1952):150.*

heat dissipation by lying on cool floors and heat gain by lying on warm ones.

17.11.2 Convection

Factors influencing the effectiveness of convection in heat dissipation include (1) body surface area, (2) velocity of air movement (wind or draft), (3) temperature of the animal's surface, and (4) air temperature. The air immediately adjacent to an animal tends to warm up. Replacement of this layer of warm air surrounding the animal by cooler air away from the animal carries heat from the animal by convection.

In still air a relatively warm animal causes the nearby air to be warmed. Because this warmer air is less dense it will rise, carrying heat away from the animal. This is called "free convection." Convection is enhanced by air movement, which may either cool or warm the animal depending on whether the air is cooler or warmer than the animal's surface. This is called "forced convection."

Convection, then, is the mechanism by which heat is transferred by being carried away. Although not shown in Figure 17.8, a higher air velocity would have increased the rate of heat loss by convection giving a lower hog surface temperature. (This increases the temperature gradient for further convective and radiant heat exchange.)

17.11.3 Radiation

Radiation is an important means of heat loss from animals to cool surfaces and heat gain from warm ones. Like a fire the sun radiates its energy in straight lines as **electromagnetic** waves of various wavelengths at the speed of light (186,000 mi/s). **Radiant energy** does not heat the air directly but indirectly by heating solid receiving surfaces such as soil, water, buildings, trees, clouds, dust, and animals. In this way radiant energy is changed into thermal energy, which in turn warms the surrounding air by conduction and convection. Solid objects are also heated by reflected radiation.

The loss of heat by infrared waves (wavelength above 700 nanometers) depends on radiator temperature. When people sit near a large window with a glass temperature of 0°F they may feel cold. Having no appreciation for this principle they frequently check the thermostat only to find that it indicates a room temperature of 72°F. However, if they understand the radiation principle they realize that heat rays are leaving their body and striking the cold glass window resulting in a heat drain. Animals also receive infrared thermal radiation from their surroundings. All surfaces at a temperature above absolute zero (−460°F or −273°C) radiate energy.

Factors influencing heat loss by means of radiation include (1) animal surface area, (2) animal skin temperature, (3) temperature of air surrounding the animal, and (4) **emissivity** of the animal's cover and skin (the ability to absorb and emit heat).

Most organisms have an emissivity for thermal radiation between 0.95 and 1 and are called *blackbodies.* The ground surface also radiates very nearly as a blackbody. The atmosphere radiates infrared radiation toward the ground as well as toward outer space. Water vapor, carbon dioxide, and ozone have strong absorption bands at specific infrared wavelengths and therefore provide an efficient means of radiating streams of energy groundward and spaceward. This explains in part why clear, dry nights tend to be chilly whereas humid, overcast nights tend to be warmer. Clouds also shield animals from excessive heat loss to outer space.

Highly polished metals such as aluminum foil or copper have a low thermal-radiant emissivity of about 5 percent as contrasted with 100 percent emissivity of solid-black material or 98 percent emissivity of skin (dark or light). This means that if the temperature of the surface radiating in the far-infrared range is higher than that of skin, the skin absorbs 98 percent and reflects only 2 percent of the radiated heat. Conversely, if skin temperature is higher than that of the environment it approximates a heat-radiating blackbody and emits heat.

Aluminum paint absorbs only 40 percent of solar radiations compared with 70 percent absorption by asbestos shingles, unpainted wood, stone, brick, and red tile and 90 percent by black surfaces such as slate or tar. A white surface may absorb only 20 percent of the visible radiation (only around half the energy in the solar spectrum is in the visible portion) falling on it, hence the basis for changing *coat color* with changing environmental temperature (Figure 17.9). A white or yellow/tan coat with a smooth, glossy texture is best to minimize adverse effects of solar radiation.

Cosmic Radiation

When considering the possibility of animal life on other planets one must make reasonable estimates of the radiation spectrum and heat load relationships at the surface to be reasonably assured that *constructive* radiations are present and *destructive* radiations absent. In 1927 Herman J. Muller, a geneticist, reported that the frequency of gene mutations was affected by **irradiation.** For this observation he was awarded a Nobel prize. The earth's surface is constantly being bombarded with high-energy cosmic rays similar to those used by Muller. Could cosmic radiation have played a role in the natural selection of animals through its influence on gene mutation?

Animal life has been able to evolve over the earth's surface because of protection from the cosmic cold of outer space and from the ionizing radiations of the sun by an *envelope of atmosphere.* Furthermore, animals have been shielded from cosmic particles and solar storms by magnetic fields.

Sunburn

Although the ultraviolet portion (short wavelengths) of solar radiation does not greatly increase the heat load on humans it is important because when it is absorbed by tissues it causes damage to their cells (sunburn of the skin). These short wavelengths also are reflected by snow, light clouds, and still water.

Several pathological disturbances are associated with exposure of animals to intense sunlight. For example, when lightly pigmented cattle are exposed to intense sunlight containing a relatively large ultraviolet component they may develop ocular squamous cell carcinoma (cancer eye). Hereford cattle—the breed most susceptible to cancer eye—vary in pigmentation of the eyelid and eyeball and there is strong correlation between low pigmentation and high incidence of cancer eye. The **heritability** of eyeball pigmentation approximates 0.6 and that of eyelid pigmentation 0.4.

Photosensitization can occur in any animal. It results from hypersensitivity of the skin to sunlight caused by a blood component that is derived directly or indirectly from certain feeds. When this compound is acted on by ultraviolet radiation it is transformed into a toxic substance that harms and can even kill an animal. Heavily pigmented skin protects the blood from ultraviolet rays. One remedy for mild photosensitization is to shade the affected animal from direct sunlight.

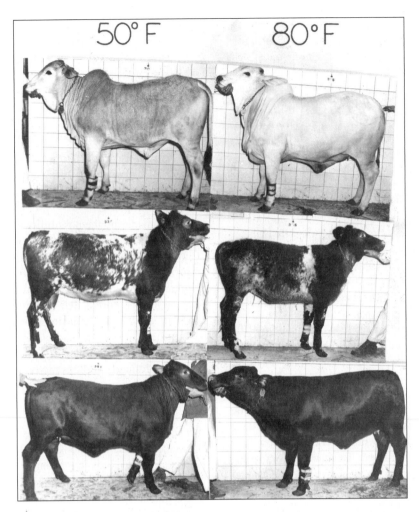

Figure 17.9 The effect of environmental temperature on growth and hair color in the Brahman (top), Shorthorn (center), and Santa Gertrudis (bottom) breeds. Brahmans had darker hair at 50°F than at 80°F. The Shorthorns at 80°F were stunted. Other breeds did equally well at 50 and 80°F.
Courtesy of Missouri Agr. Expt. Station.

A strain of Southdown sheep has, for example, been found to be congenitally photosensitive. These sheep have an impaired ability to clear phylloerythrin, a photosensitive end product of chlorophyll metabolism, from their blood. Phylloerythrin is formed from chlorophyll consumed in green plants. Newly sheared sheep are especially vulnerable to photosensitization and sunburn. Photosensitizing agents may also be derived directly from feedstuffs. For example, the buckwheat plant is known to contain three such compounds.

Seasonal variation of sunlight is enormous in the polar regions, slight in equatorial regions. Changes in sunlight over a day are dramatic in the tropics whereas they are slight in the polar regions.

17.11.4 Evaporation

Evaporation is the most important means of heat dissipation in hot environments because the temperature gradient between an animal's surface and the environment becomes increasingly narrow as environmental temperature rises. Evaporation of water from the skin depends on (1) temperature of and moisture on the skin; (2) cover (hair, wool, feathers); (3) humidity, velocity, and temperature of the surrounding air; (4) respiratory rate and volume; (5) water available for evaporation; and (6) animal surface area.

A thick cover may reduce air velocity on the skin and entrap a layer of water vapor over the surface of the animal thereby reducing the efficiency of evaporative heat dissipation.

Three types of sweating are common to sweating animals: (1) **insensible** perspiration (diffusion water), (2) thermal sweat (from sweat glands), and (3) nonthermal sweat *(emotional sweat)*. *Insensible perspiration* is continuous (unless ambient and skin vapor pressures are the same). Water either diffuses through the skin or is exhaled from the lungs. *Thermal sweat* is expressed onto the skin by the skin's sweat glands. Vaporization of water is the principal mechanism available for cooling the body when environmental temperature is above that of the body, preventing sensible heat flows (conductions, convections, and radiations, which are driven by temperature gradients). Actually heat may be gained by conduction and radiation under these conditions. *Nonthermal sweat* or "cold sweat" is of little importance to maintaining heat balance.

Sweat gland population density varies with species and body regions. Horses are profuse sweaters. Cattle sweat to some extent, sheep less. Pigs' sweat glands do not respond to heat stress. Bird skin has no sweat glands.

Animals that sweat little or not at all are called "panters." These include all poultry species, pigs, sheep, goats, dogs, and cats.

The proportionate amount of heat lost by vaporization can be altered in animals by **shearing** or otherwise reducing the cover. For example, in a moderately warm environment a normally feathered chicken loses about half its heat by vaporization, whereas a frizzle-feathered fowl may lose only 15 to 20 percent of its heat by evaporative cooling because the frizzle-feathered fowl loses more heat by sensible means.

Evaporative heat loss increases with an increase in environmental temperature (Figure 17.8), especially at temperatures higher than around 65°F.

Tree Shade

Shade from deciduous trees is beneficial to animals during hot periods. Moisture evaporates from the undersides of leaves cooling the shaded area without interfering with air circulation.

17.11.5 Cooling in Panting Species

There is little moisture loss from the skin of panting animals. Only that which reaches the surface by passive diffusion is evaporated.

Birds have no sweat glands and therefore accomplish evaporative cooling mostly by panting. Similarly there are few functional sweat glands in the dog (except on the pads of the feet). Therefore the dog makes good use of evaporative cooling by panting, moving large volumes of air over its moist tongue and air passages. The blood is cooled as it flows through these areas.

Because overheating in panting species may result from lack of skin moisture for evaporation, the application of water to the body surface is beneficial during hot periods. Under natural conditions of hot weather swine resort to the moisture and coolness of mud, cattle stand in cool water and wallow in mud, and sheep migrate to cooler places.

Fanning cools sweating animals because it accelerates the rate of evaporation. However, except when environmental temperature is lower than skin temperature, fanning is of little benefit to cattle because they sweat very little. In dry regions evaporative coolers can reduce peak temperature 10 to 12°F below surrounding temperature (Figure 17.10). Such a system moves the air, which is cooled to a temperature below that of the animal's surface providing forced convection. Reproductive efficiency increased from 30 percent in controls to 60 percent in the "cooled" group during summer months. Moreover, daily milk yield averaged about 4 lb higher in cows with access to evaporative coolers.

In all animals blood is shunted to the surface where it can be cooled by vaporization of water. The blood deficiency in the core organs resulting from this may be partly compensated for by increased heart rate. This redistribution of the blood is a cause of fetal stunting in cows and ewes under heat stress during pregnancy.

The normal body temperature of a dairy cow is 101 to 102°F. She has a tremendous capability to produce heat but limited ability to lose it. When exposed to an environmental temperature of 80 to 85°F the lactating cow quickly responds with panting and experiences elevated body temperature.

When Zebu and European cattle were exposed to environmental temperatures of up to 105°F moisture vaporized at a lower rate (per unit surface area or per unit weight) from the Zebus. At temperatures above 105°F the reverse was true, demonstrating the Brahman's reserve capacity to tolerate high temperature.

When the ambient temperature exceeds body surface temperature an animal is unable to dissipate heat to the surrounding air through its skin by sensible means (conduction, convection, and radiation). In this condition vaporization becomes all-important. Respiratory vaporization is increased by panting; the animal's tongue protrudes, saliva flows, and the animal may become heat-exhausted (Figure 17.7). An ambient temperature of 107 to 108°F for extended periods of time can be fatal to cattle and sheep.

The use of air-conditioning by refrigeration in the commercial production of meat, milk, and eggs is still too expensive. Cooling only the heads of lactating dairy cows has been found beneficial. Animals were exposed to an 85°F ambient temperature except for the heads of the test group, which were cooled to 65°F (Figure 17.11). Milk yield of control cows decreased 25 percent from normal production at 65°F whereas that of the experimental group decreased by only 10 percent. Zone cooling is used commonly at the sow's head in farrowing houses and cooled roosts have been used in chicken houses.

Evaporative cooling also has proved successful in environmentally controlled laying-cage houses. When outside temperature is 100°F or above use of large exhaust fans to pull air through moistened pads (water is circulated through large areas of wood-wool or excelsior in the side walls) can reduce house temperature by 10 to 15°F depending on outside vapor pressure.

Figure 17.10 Cattle shades equipped with evaporative coolers.
Courtesy of Dr. G. H. Stott, University of Arizona.

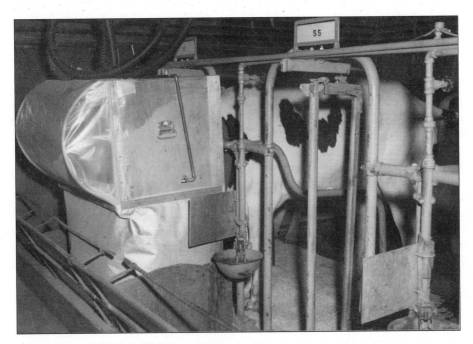

Figure 17.11 Cooling the head of a lactating dairy cow in an air-conditioned box. The balance of the cow is exposed to ambient temperature.
Courtesy of Missouri Agr. Expt. Station.

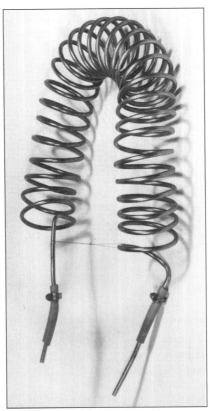
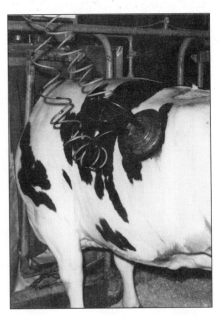

Figure 17.12 A technique for circulating water (warm or cold) through the rumen to study the effect of changing rumen temperature on VFA production and feed digestibility. Water-storage tank with heater and temperature controls (far left) is connected with a 24-ft (1/4-in) section of stainless-steel tubing (middle) by high-pressure plastic tubing (far right).
Courtesy of Dr. F. A. Martz and Missouri Agr. Expt. Station.

This more favorable environment is reflected in an increase in egg production.

17.12 EFFECTS OF CLIMATE ON PRODUCTION

High ambient temperatures depress growth, reduce production, and lower reproductive efficiency. During periods of hot weather hens eat less and lay fewer and smaller eggs with thinner shells (Chapter 16).

A comfortable cow converts feed into milk more efficiently than one exposed to high temperatures (Chapter 15). In one study when the environmental temperature was increased from 60 to 95°F, rectal temperature of cows increased from 101 to 104°F and daily milk yield declined by nearly 40 percent. Moreover, growth rate is greatly reduced in most European breeds of cattle during hot weather (Figure 17.9). There is a change in the proportions of **VFA** (volatile fatty acids) (acetate, propionate, and butyrate) produced by cows when the **rumen** temperature is elevated (Figure 17.12). This has practical significance because the VFA provide the **ruminant** animal 60 to 80 percent of its energy needs (Chapter 19) and cows often drink warm water in summer and cold water in winter.

High ambient temperatures have an adverse effect on **spermatogenesis** and are also detrimental to **ova** and **sperm** in the female reproductive tract. The results are lower fertility and higher embryonic mortality. In one study semen quality of European dairy bulls was reduced more than that of the Brahman when the animals were exposed to high temperatures.

How single climatic factors such as temperature, relative humidity, air movement, or radiation affect animals can be studied under controlled environmental conditions in **psychrometric** and other kinds of environmental *chambers* or *climatic laboratories* (Figure 17.13).

17.13 SUMMARY

Humans learned early on to wear clothes and modify their environment by using fire for warmth when they became chilled. But only recently have humans been able to cool themselves using mechanical refrigeration. Animals can employ many mechanisms to conserve heat and increase heat production but they possess limited means for increasing heat dissipation. Whether we can genetically select animals to fit the regions where we want and need them or develop an economical means of cooling them depends on future research. However, the future success of the animal industry is dependent on effects of the environment on producing animals. Environment includes not only the climatic complex but also nutrition, disease, management, and other related factors. The genetic makeup of the animal determines the limits of its ability to produce whereas

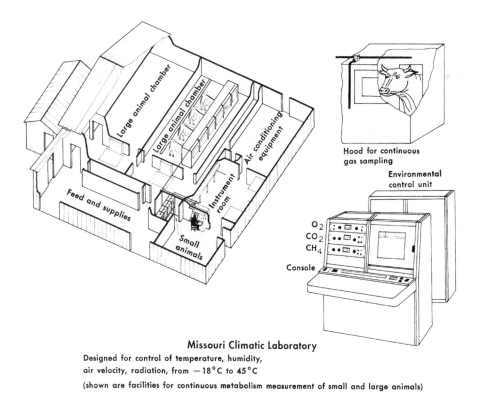

Missouri Climatic Laboratory
Designed for control of temperature, humidity, air velocity, radiation, from −18°C to 45°C

(shown are facilities for continuous metabolism measurement of small and large animals)

Figure 17.13 Animal psychroenergetic laboratory for conducting studies under controlled environmental conditions of the effects of temperature (0 to 110°F), humidity (0 to 100 percent RH), air movement, light (fluorescent and incandescent), ventilation rate, and other environmental factors on the health and efficiency of growth, production, and reproduction of animals.

Courtesy of Missouri Agr. Expt. Station and USDA.

the environment determines how much of its inherited capabilities the animal will actually achieve. From 40 to 90 percent of **phenotypic** variability typically results from environmental factors.

The "climatic complex" includes many factors in addition to temperature, humidity, barometric pressure, wind velocity, and solar radiation. *Ecology,* the scientific study of interaction between animals and their environments, must be appreciated from the standpoint of energy flow, animal temperature, diffusion theory, chemical processes, molecular biology, biological rhythms, air pollutants, social factors, pathogenic microbes, insects, and other parasites to name a few.

The complexity of biological problems associated with growth, reproduction, productivity (meat, milk, eggs, wool), longevity, genetic mutations, genetic differences among individuals and species, physical activity, adaptation, and many others must come to be understood with all the sophistication of modern science if humankind is going to continue to "be fruitful, and multiply, and replenish the earth, and subdue it; and have dominion over the fish of the sea, and over the fowl of the air, . . . and over the cattle, and over all the earth, . . . and over every living thing that moveth upon the earth," as instructed by its Creator (Genesis 1:26, 28).

STUDY QUESTIONS

1. What were Aristotle's thoughts pertaining to the effects of climate on people and their livestock?

2. Is there a symbiotic relation between humans and animals? Explain your answer.

3. What is the effect of light on reproduction in poultry? In sheep?

4. Define ecology. What does environmental physiology deal with?

5. Cite examples of animals that are better adapted to some climatic regions than others.

6. Why are Indian cattle better adapted to warm climates than European cattle?

7. What is acclimatization?

8. Distinguish between natural and artificial physical environmental factors affecting humans and the animals that serve them.

9. Cite several examples of the natural adaptation of animals to their environment.

10. What is polypnea? What causes it?

11. What is meant by homeothermy? Homeostasis?

12. Differentiate between estivation and hibernation.

13. What is meant by poikilothermy? Name some poikilotherms.

14. Why are homeothermic mechanisms not required by developing mammals before birth?

15. Where is (are) the heat-regulating center(s) of the body?

16. By what means can heat loss be reduced in cold weather?

17. Discuss briefly the following rules governing climatic adaptations: (a) the Bergmann Rule, (b) the Wilson Rule, (c) the Gloger Rule, (d) the Claude Bernard Rule.

18. What is meant by heat increment of feeding? By standard metabolic rate?

19. Would a high-protein diet be beneficial to humans in cold weather? Why?

20. Is there much difference between homeothermic and poikilothermic animals in their overall efficiency (maintenance energy expended)? Why?

21. What is thermogenesis?

22. What is meant by the thermoneutral zone? Cite some factors that influence it.

23. What are the principal factors that cause an increase in heat production over standard metabolic rate?

24. Discuss briefly the four primary means of heat dissipation in domestic animals.

25. What is insensible perspiration?

26. When exposed to a high ambient temperature, which animal would benefit most from an electric fan: (a) a chicken, (b) a sow, or (c) a cow? Why?

27. In general, is the production of farm animals in the United States depressed more by cold or hot weather?

28. For what purposes are psychrometric chambers used?

PRINCIPLES OF NUTRITION
Plant and Animal Composition[1]

So basically influenced are we by the matter of food and drink that revolutions, peace, war, patriotism, international understanding, our daily life and the whole fabric of human social life are profoundly influenced by it—and what is the use of saying "peace, peace" when there is no peace below the diaphragm. This applies to nations as well as individuals—men refuse to work, soldiers refuse to fight, prima donnas refuse to sing, senators refuse to debate, and even presidents refuse to rule the country when they are hungry.

Lin Yutang (1895–1976)

18.1 INTRODUCTION

Nutrition is the science that deals with food and the **nutrients** it contains. It includes the process of providing body cells with the proper amount and balance of nutrients to enable the cells to function in the many metabolic and endocrine processes involved in growth, body maintenance, work, production (meat, milk, eggs, wool), reproduction, and optimal formation of the body's defense mechanisms against disease. Nutrition, therefore, has an inveterate and profound effect on the health of humans and animals.

Rapid strides have been made in recent years in establishing nutrient requirements of humans and animals. Basically the science of nutrition tries to answer several questions about nutrients: What are they? What do they contribute? Why do humans and animals need them? How do they function within the body? How does the body utilize food substances for the various body processes?

A fundamental factor underlying the behavior of humans and animals has been their desire and need for food. Availability of food has caused the migration of both humans and animals from one region to another and it has been largely responsible for the settling and agricultural development of new lands. A scarcity of food has been the major cause of war among races of people.

Most things except agriculture can wait.

Jawaharlal Nehru (1889–1964)
First Prime Minister of India (1947–1964)

Even though primitive humans knew that food was essential for survival, they knew nothing about the specific nutrients required by the body. They probably knew only that air, food, and water were essentials of life. It has been said that the history of humanity perhaps could be written in terms of success in fulfilling nutritional needs.

The modern era of the science of nutrition was pioneered by the French chemist A. L. Lavoisier in the 1770s. Lavoisier was the first to recognize that animal heat is derived from the **oxidation** of body substance. He compared animal heat with that produced by a candle. The general form of the apparatus Lavoisier used in his experiments was illustrated in two drawings made by Madame Lavoisier. The methods of study he used, however, are unknown because Lavoisier was executed on May 8, 1794, at the age of 51 by the Paris Commune. (He was found guilty of allowing the collection of taxes on water contained in tobacco.)

The animal body is a remarkable chemical laboratory that operates at relatively low temperatures and in which **genes** guide growth and development. Genes cause specific cells to produce **enzymes** that help digest food, enable muscles to contract, and assist **cells** in carrying out their varied and complex processes. Food is basic to life and is perpetually related to body chemistry and health. **Diseases** such as arthritis, **atherosclerosis,** and cancer have their genesis in the biochemistry of the body. Eventually better understanding of how all these body processes work will be achieved. Only then will health and life itself be better understood.

The period of the Industrial Revolution has been paralleled by a biological revolution of comparable significance. During the past century approximately 50 nutrients have been identified as essential for farm animals, and the integration of nutritional knowledge with genetic advances and modern

[1] The authors acknowledge with appreciation the contributions to this chapter of Dr. G. C. Fahey, Jr., Department of Animal Sciences, University of Illinois at Urbana–Champaign.

production practices has produced remarkable achievements indeed. A notable example is the production of more than 70,600 lb (more than 32,000 kg) of milk in a single lactation by a Holstein cow.

Because the dietary needs of high-producing animals differ greatly from those of low-producing ones, future increases in the productivity of food-producing farm animals will depend in large part on the progress that can be made in the food and nutritional sciences. The challenge is great but so is the opportunity to improve the nutritional status of animals that can in turn provide a reliable supply of high-quality foods beneficial nutritionally to humanity.

18.2 COMPOSITION OF PLANTS AND ANIMALS

> All flesh is grass.
>
> **Isaiah 40:6**

Plant and animal tissues are composed of *water,* **carbohydrates,** *proteins, lipids* (including **fat** and related substances), **vitamins,** and mineral matter. Plants contain the same substances as animals but in different amounts. The main difference between plants and animals is that plants usually contain large amounts of carbohydrates whereas animals contain only traces. The cell walls of plants are composed largely of fibrous carbohydrates; cell membranes of animals consist mostly of proteins. Plants generally store their reserve food as starch; animals store theirs as fat. Animals depend on plants for energy; plants derive energy from the sun and manufacture certain nutrients that cannot be manufactured by animals.

18.2.1 Water

Chemically water consists of two **atoms** of hydrogen and one of oxygen (H_2O). It is the most abundant and most important constituent of plant and animal tissues. Green plants contain 70 to 80 percent water whereas animals contain from 40 to 75 percent. The amount of water an animal's body contains varies with its age and degree of fatness. For example, an embryonic **calf** is 90 percent or more water, a newborn calf is slightly more than 70 percent water, and a steer ready for market contains only 40 to 45 percent water.

Animals have three sources of water. The first is that which they drink. The second is that **ingested** as a component of food. The third is known as *metabolic water,* which is derived from the breakdown of carbohydrates, fats, and proteins. Metabolic water is the chief source of water for animals during **hibernation** (sleep or rest during winter).

Water has many functions in plants and animals. It transports nutrients from the soil to plant leaves where various organic substances are formed. It also transports nutrients from one part of plants and animals to another. Water distends the cells in plants and animals thus helping them hold their shape. It is used in many biochemical reactions in the body. It helps regulate the temperature of plants and animals through **evaporation** and other processes. For example, when exposed to high environmental temperatures horses and humans sweat (perspire) to maintain body temperature, whereas swine cannot sweat appreciably and so wallow in water to cool themselves. Water also is the principal constituent of body substances (synovial fluid) that lubricate the joints. Moreover, an aqueous medium is necessary for most metabolic reactions of the body.

It is obvious that large amounts of water are needed by plants and animals in the performance of their various functions. In fact, an animal will die more quickly from lack of water than from lack of any other dietary substance.

18.2.2 Carbohydrates

Carbohydrates are organic compounds that contain carbon (C), hydrogen (H), and oxygen (O). They are the main organic compounds found in plants and commonly represent 50 to 75 percent of the total dry matter in **livestock** feeds. Carbohydrates are abundant in most seeds, fruits, roots, and tubers of plants. They are formed by the process of photosynthesis, which involves the action of sunlight on chlorophyll. Chlorophyll is the active photosynthetic material in plants. It is a complex molecule with a structure similar to that of **hemoglobin,** which is found in the blood of animals. Chlorophyll contains magnesium; hemoglobin contains iron. More specifically, carbohydrates are formed of water (H_2O) from the soil, carbon dioxide (CO_2) from the air, and energy from the sun. A simple chemical reaction in which a carbohydrate (glucose) is synthesized by photosynthesis in plants is:

$$6CO_2 \ + \ 6H_2O \ + \ 673\,cal \ \rightarrow \ C_6H_{12}O_6 \ + \ 6O_2$$

| Carbon dioxide | Water | Energy from the sun | Glucose | Oxygen |

Photosynthesis is an extremely important process because through it comes all food for humans and animals and fossil fuels providing heat for houses and fuel for engines. Energy contained in each of these substances came originally from the sun and was made available by the process of photosynthesis.

Carbohydrates are classified as monosaccharides (simple sugars), disaccharides (two **molecules** of simple sugars), trisaccharides (three molecules of simple sugars), and polysaccharides (many molecules of simple sugars). Selected carbohydrates of food are grouped in Table 18.1.

Monosaccharides are simple sugars containing five or six carbon atoms in the molecule. They are water-soluble. The monosaccharides that contain six carbon atoms have the molecular formula $C_6H_{12}O_6$. They include glucose (also known as dextrose) found in plants, ripe fruit, honey, sweet corn, and other foods. In animals glucose is found mainly in the blood where a certain concentration is vital for life as well as being an immediate source of energy. An ill person can be fed by having glucose infused directly into the bloodstream.

Disaccharides (*di* meaning "two") are carbohydrates that contain two molecules of simple sugars. They have the general

Classification	Name	Monosaccharide unit	Proximate analysis portion
	Arabinose	$C_5H_{10}O_5$	Nitrogen-free
	Deoxyribose*		extract
	Ribose		
	Xylose		
Monosaccharides	Fructose (levulose)	$C_6H_{12}O_6$	
	Galactose		
	Glucose (dextrose)		
	Mannose		
Disaccharides	Lactose	Galactose, glucose	
	Maltose	Glucose, glucose	
	Sucrose	Glucose, fructose	
Trisaccharide	Raffinose	Galactose, glucose, fructose	
Polysaccharides (usually contain more than 10 monosaccharide units)	Dextrins Glycogen Gums Starch	$(C_6H_{10}O_5)n$	
	Cellulose	$(C_6H_{10}O_5)n$	Crude fiber
	Hemicellulose		

TABLE 18.1 Classification of Selected Carbohydrates of Food

*Deoxyribose has the formula $C_5H_{10}O_4$. The prefix deoxy- means minus one oxygen atom.

formula $C_{12}H_{22}O_{11}$. Therefore they represent two molecules of simple sugar minus one molecule of water (two hydrogen atoms and one oxygen atom). The most important disaccharides are sucrose, maltose, and lactose. Sucrose is found in beet or cane sugar and contains one molecule of glucose (dextrose) and one of fructose (levulose). Sucrose is very sweet and is commonly used to sweeten foods. Maltose is found in germinating seeds and contains two molecules of glucose. It is only about one-third as sweet as sucrose. Lactose (milk sugar) is found only in milk (or milk products). It contains one molecule of glucose and one of galactose. It is even less sweet than maltose.

Polysaccharides have the general chemical formula $(C_6H_{10}O_5)n$, which means that they contain many molecules of simple sugars. The two principal groups of polysaccharides are starch and cellulose, although there are other important ones including the hemicelluloses (fibrous carbohydrates) and pectin.

Starch is the most prevalent polysaccharide in plant seeds, fruits, and tubers and is easily digested by animals. When starch is broken down it yields many molecules of glucose. Glycogen, "animal starch," is found in small amounts in liver, muscles, and other tissues of an animal's body. It also contains many molecules of glucose.

Cellulose is a polysaccharide having the same general formula as starch $[(C_6H_{10}O_5)n]$. It is found mostly in the cell walls and woody portions of plants. Cotton is almost pure cellulose. Cellulose and hemicelluloses are not digested because animals lack the required enzymes. Animals can utilize these carbohydrates only if they are first degraded by microorganisms that produce cellulose- and hemicellulose-degrading enzymes. All animals have microorganisms in their intestinal tract. In most animals there is only slight utilization of fibrous carbohydrates. However, **ruminants** (cattle, sheep, and goats) have evolved with a rumen (paunch) that harbors millions of microorganisms that degrade fibrous carbohydrates. Horses have a large cecum, which likewise harbors millions of microorganisms. The microbes digest the fibrous carbohydrates and make their bodies and their end products of digestion (largely **volatile fatty acids**) available for their host (Chapter 19).

The foremost functions of carbohydrates in animal diets are to supply energy and heat for body processes. (Special carbohydrates such as ribose are also needed for production of substances in the animal's body such as **RNA**.) Carbohydrates also aid in the utilization of body fat and exert a sparing effect on protein. Surplus energy is converted into fat, which is stored as a potential energy source for the animal. Crude fiber (including cellulose) is a source of heat and energy when digested. It also prevents **compaction** by providing a laxative effect and by maintaining the proper muscular tone in the alimentary tract. This helps prevent **bloat** in ruminants and colic in horses.

18.2.3 Fats and Oils (Lipids)

Fats and oils contain carbon, hydrogen, and oxygen, as do carbohydrates, but they contain more carbon and hydrogen in proportion to oxygen than do carbohydrates. For example, sucrose has the general formula $C_{12}H_{22}O_{11}$ whereas a fat, tristearin, containing stearic acid has the composition $C_{57}H_{110}O_6$ (see example on page 316).

A fat molecule is formed by the combination of three fatty acid molecules with one molecule of glycerol, as follows:

$$CH_2O[H\ HO]OC\text{-}R \quad CH_2OOC\text{-}R$$
$$CHO[H\ HO]OC\text{-}R \longrightarrow CHOOC\text{-}R + 3H_2O$$
$$CH_2O[H\ HO]OC\text{-}R \quad CH_2OOC\text{-}R$$

GLYCEROL FATTY   FAT   WATER
          ACIDS

The letter R represents the fatty acid radical, which may be composed of several combinations of carbon, hydrogen, and oxygen. If R represents the hydrocarbon portion of stearic acid $(C_{18}H_{36}O_2)$ minus the carboxyl group (COOH), the general chemical composition of the fat is $C_{57}H_{110}O_6$. Thus the fat tristearin is a triglyceride with three molecules of stearic acid and one molecule of glycerol.

Fats furnish about 2.25 times as much energy as carbohydrates when they are **metabolized** because they have a higher proportion of hydrogen to oxygen. For this reason they have a greater caloric food value per pound. Fats are insoluble in water but soluble in ether. When plant and animal tissues are extracted with ether the substances that are soluble in this compound are called *lipids* (ether extract).

Nearly all fat found in food is digestible. However, it takes considerable time for digestive juices to break down fats. Therefore fried foods (coated with fats) will be digested more slowly than broiled foods. Fatty foods thereby keep a person from feeling hungry for a longer period of time than do nonfatty foods.

Fats have different melting points. Some are solids at room temperatures; others, the oils, are liquids. Fats that are solids at ordinary room temperatures are composed largely of **saturated fatty acids.** Most oils are composed principally of *unsaturated fatty acids.* Unsaturated fats consist of fatty acids that have fewer hydrogen atoms attached to the carbon atoms with double bonds between adjacent carbon molecules. For this reason they can readily accept hydrogen or other elements (such as iodine). The **iodine value** is a measure of the degree of unsaturation. Measurement of the iodine value is possible because an unsaturated fat can easily unite with iodine. (Two atoms of iodine are added to each double bond in the fat.) Examples of a saturated and an unsaturated fatty acid are:

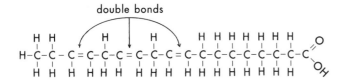

Stearic acid, saturated

double bonds

Linolenic acid, unsaturated

Beef fat, or tallow, is composed primarily of saturated fatty acids and is a solid at ordinary room temperature. Linseed oil contains a high proportion of unsaturated fatty acids (polyunsaturated) and is a liquid at room temperature.

It is generally true that unsaturated fatty acids (such as linoleic, linolenic, and arachidonic) are necessary for proper nutrition. They must be supplied in the diet because the animal cannot synthesize them. For this reason they are called *essential fatty acids.* Essential fatty acids are required for the construction of cell membranes and for the production of **prostaglandins,** which are hormonelike substances that have many effects on body cells and are involved in the immune response. A deficiency of linoleic acid leads to dermatitis, poor

wound healing, impaired growth, reproductive failure, and water loss from the skin.

Human infants on a fat-free diet have developed an **eczema** that can be cured by adding fat to the diet. A similar condition has been produced by feeding rats a fat-free diet. It is believed that certain essential fatty acids are necessary for the normal maintenance of skin. Other experiments show that **calves, lambs, pigs,** and chickens need a minimum amount of fat for normal growth (Figure 18.1), and dairy cows secrete less milk when fed a diet too low in fat. Fat is also required for **absorption** of fat-soluble vitamins. Adequate amounts of the essential fatty acids are supplied in most practical farm diets.

Nearly all fats in grains and seeds extractable with ether are true fats. In grass, hay, and other **forages** ether-soluble substances that are not true fats are found in small amounts. These include **sterols,** carotenes, chlorophyll, **phospholipids,** waxes, and essential oils.

Sterols

These are complex alcohols found in both plants and animals. **Ergosterol** is found in plants and is especially important in livestock nutrition because when activated by sunlight it forms vitamin D_2. **Cholesterol** is widely distributed in animal tissues such as the brain, nerves, and blood. It seems to aid in the transport of fat from one part of the body to another. The action of sunlight on 7-dehydrocholesterol results in the formation of vitamin D_3 (Chapter 20).

Carotenes

These are yellow substances found in carrots, sweet potatoes, and milk fat. They are also present in green plants although the yellow color is masked by the green color of chlorophyll. Carotenes serve as a precursor of vitamin A inasmuch as animals have the ability to convert them into this essential vitamin.

Chlorophyll

This is the green substance in plants. It has minimal food value for animals because it is destroyed by digestive juices in the gastrointestinal tract. It has received considerable attention and publicity because of its deodorizing properties imparted to products designed for human use.

Phospholipids (Phosphatides)

These are present in small amounts in the bodies of animals. They contain various combinations of fatty acids, glycerol, phosphoric acid, and nitrogenous groups. Phospholipids are vital in living protoplasm and cell membranes. One of the most important of these compounds is lecithin found in egg yolk, blood, and liver.

Waxes

These occur in small amounts in plants, forming tiny films on the surface of plant stems and fruits. Their main function appears to be in protecting the plant surface from the weather. They are relatively unimportant as a source of nutrients for livestock.

Figure 18.1 The pig on the bottom was fed a diet adequate in fat (5 percent ether extract). The one on the top was fed a diet deficient in fat (0.06 percent ether extract). Note the loss of hair and scaly, dandruff-like dermatitis, especially on the feet and tail.
Courtesy of Dr. W. M. Beeson, Purdue University.

Essential Oils

These give plants their characteristic odor and taste. They probably function to protect the plant from insect and microbial predators. It is interesting that sagebrush is used to deodorize feces. (The compound responsible for the odor of sage is an essential oil.)

The essential oil of onion or garlic contains volatile flavor compounds high in sulfur. They include hydrogen sulfide, thiols, disulfides, trisulfides, and thiosulfinates. These compounds pass from the ingested plant via the animal's bloodstream into milk resulting in an undesirable "flavor" to milk if intended for human consumption.

18.2.4 Proteins

Proteins are organic compounds that contain carbon, hydrogen, nitrogen, oxygen, sulfur, and phosphorus. They are the principal organic food compounds that contain nitrogen. Proteins are essential for life because they constitute the active protoplasm in all living cells. They are found in the reproductive and actively growing parts (such as the leaves) of plants. In animals the muscles, internal organs, skin, hair, wool, horns, hooves, and bone marrow are composed, in large part, of proteins, as are other body parts.

The functions of proteins in the body include (1) repair of tissue, (2) growth of new tissue, (3) metabolism (**deamination**) for energy, (4) metabolism into substances vital in body functions (these vital substances include blood **antibodies** that fight **infection**), (5) enzymes that are essential for normal body function, and (6) many **hormones.**

Proteins consist of long chains of amino acids, and can include hundreds and even thousands of the 20 or more naturally occurring amino acids. Proteins are formed by the ribosomes in the cell **cytoplasm** according to the code transmitted from the gene by means of mRNA (Chapter 8). The chemical reaction in which amino acids are linked together in a chain (dipeptide) is:

$$R-\overset{\overset{H}{|}}{\underset{\underset{NH_2}{|}}{C}}-\overset{\overset{O}{\|}}{C}-OH + R-\overset{\overset{H}{|}}{\underset{\underset{N}{|}}{C}}-\overset{\overset{O}{\|}}{C}-OH \rightarrow R-\overset{\overset{H}{|}}{\underset{\underset{N}{|}}{C}}-\overset{\overset{O}{\|}}{C}-\overset{\overset{H}{|}}{N}-\overset{\overset{R}{|}}{\underset{\underset{H}{|}}{C}}-\overset{\overset{O}{\|}}{C}-OH + H_2O$$

Amino acid   Amino acid   Dipeptide   Water

The $\begin{bmatrix} \overset{O}{\overset{\|}{}} & \overset{H}{\overset{|}{}} \\ -C & -N- \end{bmatrix}$ linkage is known as the *peptide linkage.* The different proteins depend on the kind, number, and arrangement of the 20 or more amino acids linked together in the molecule, much as the meaning of different words in the English language depends on the kind, number, and arrangement of the 26 letters of the alphabet they contain.

Plants have the ability to form amino acids (and proteins) from nitrogen, sulfur, phosphorus, and water from the soil and carbon dioxide (CO_2) from the air through photosynthesis. Animals cannot synthesize amino acids. Therefore they must obtain them directly from the feed they consume or digestion of bacteria that contain them and are present within the animal's own digestive tract (in ruminants). Bacteria within the digestive tract synthesize proteins from ammonia. The ammonia may come from degradation of dietary protein or from nonprotein nitrogen compounds such as urea or ammonium salts.

Animals have a limited capacity to change surplus amino acids into those required by the body. They can convert only a few of the simpler amino acids into others needed by the body. Those they cannot synthesize and are essential for growth must be supplied in feed. Amino acids have been classified as *essential,*[2] *semiessential,* and *nonessential* according to their effect on growth in rats. An **essential amino acid** is one that is not synthesized in the body at a rate necessary to meet requirements for normal growth and well being. Nutrition literature indicates that the following 10 amino acids are essential (Figure 18.3, p. 320): (1) lysine, (2) tryptophan, (3) phenylalanine, (4) leucine, (5) isoleucine, (6) threonine, (7) methionine,[3] (8) valine, (9) arginine, and (10) histidine. Whether an amino acid is essential will vary depending on **species,** animal's age, and level of performance. For this reason there may be differences within and among species (Figure 18.2 a and b).

Ruminants can obtain amino acids synthesized by microorganisms in the rumen from simpler nitrogen-containing compounds (Chapter 19). Microorganisms build these nitrogen

---

[2] Note that *essential* means the amino acid is *essential in the diet.* However, all amino acids are needed for protein synthesis. Thus it is important that the animal either consumes or produces the nonessential amino acids also.
[3] Methionine is the least expensive and most commercially available amino acid.

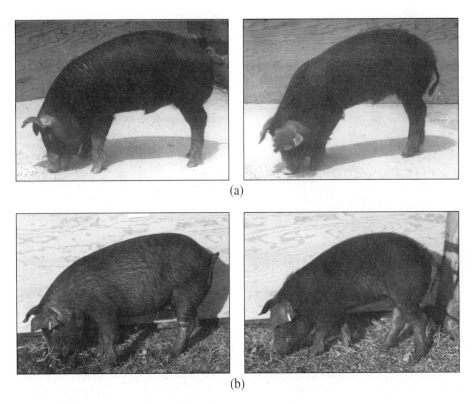

(a)

(b)

Figure 18.2 (a) The pig on the left was fed a diet containing adequate amounts of lysine (2 percent DL-lysine). It gained 25 lb in 28 days. The pig on the right was fed a diet deficient in lysine. It lost 2 lb in 28 days. (b) The pig on the left was fed a diet containing adequate amounts of tryptophan (0.4 percent DL-tryptophan). It gained 25.5 lb in 21 days. The pig on the right was fed a ration deficient in tryptophan. It lost 8 lb in 21 days.

Courtesy of Dr. W. M. Beeson, Purdue University.

compounds into proteins in their own cells and then they are digested by the animals. By this procedure ruminants convert inferior proteins and even nonprotein nitrogenous compounds (such as urea) into the superior proteins present in milk and meat. Thus ruminants occupy a unique position in providing the best-quality proteins (milk and meat) to humans. Horses do not possess a rumen but rather have an enlarged cecum where some protein synthesis is accomplished by microorganisms. However, the feeding of urea to horses is *not* recommended. The synthesis of proteins from simple nitrogenous compounds by microorganisms is limited in nonruminant animals such as swine, poultry, and humans.

Ruminants are less likely to develop certain nutritional deficiencies than are other farm animals because they can obtain essential amino acids, the B-complex vitamins, and vitamin K from microbial synthesis. They are usually followed in this respect by the **herbivores** (e.g., horses and rabbits), then by **omnivores** (e.g., swine and poultry), and finally by such species as dogs and rats.

The quality of proteins in a feed is usually referred to as being high or low depending on whether the feed contains the essential amino acids in the proper proportion or balance. Proteins from animal sources (meat, milk, eggs) are of high quality whereas those from plant sources such as cottonseed meal, lin-

seed meal, and wheat shorts are of low quality. An exception among plant proteins is soybean meal, which is of high quality (Chapter 3).

The nutritive value of a protein may be determined by evaluating responses obtained using experimental animals in the laboratory. Nutritive value is an expression of the percentage of protein in a feed digested by an experimental animal and the amount used for metabolic purposes. Nutritive value is also called the **biological value (BV)** of a protein. A feed having a biological value of 100 percent would be one in which all the digested protein contained in it was utilized for metabolic purposes. Feeds seldom, if ever, have a biological value of 100 percent. Some have a value of 80 to 90 percent and they are classed as sources of high-quality proteins. A low BV indicates low protein quality in feed or small quantities of protein usable for metabolic purposes (largely growth and maintenance).

Nonprotein nitrogen (**NPN**) substances are those found in plants and seeds containing nitrogen but not in the form of protein. Urea, an end product of metabolism, is a compound containing nonprotein nitrogen. When fed to ruminants it supplies part of the animal's protein needs because it is converted (an energy source such as corn or molasses is required) into protein by microorganisms in the rumen (Chapter 19). Urea is now produced on a commercial scale by processes that utilize nitrogen from the air.[4] Its structural formula is:

$$\begin{array}{c} H \qquad\quad O \qquad\quad H \\ \diagdown \qquad\quad \| \qquad\quad \diagup \\ N - C - N \\ \diagup \qquad\qquad\qquad \diagdown \\ H \qquad UREA \qquad H \end{array}$$

Protected Proteins

Much of the dietary protein consumed by ruminants is destroyed through fermentative digestion in the rumen. On average only about 30 to 40 percent of the dietary protein escapes destruction and is available for absorption from the gut. But the major source of protein for the ruminant animal consists of microbial cells produced in the rumen. The microbial cells contain about 50 percent protein and actually synthesize most of this protein from degraded dietary protein. The microbes eventually pass from the rumen to the intestine where they are digested and absorbed. Approximately two-thirds of the protein of meat and milk resided at one time in microbial cells within the rumen.

Ruminant digestion facilitates the upgrading of low-quality dietary proteins to high-quality microbial protein. Additionally, bacteria can utilize nonprotein sources of nitrogen such as urea or ammonia to make microbial protein. It is possible, however, for rumen microbes to destroy more protein than they make resulting in a net loss of protein. It has been estimated, for example, that as much as 25 percent of protein fed to high-producing dairy cows may be lost because of excessive destruction of dietary protein by rumen microbes.

One possible solution to this problem is to feed proteins less susceptible to breakdown by rumen microbes. Certain feedstuffs such as brewers' grains, distillers' grains, and corn gluten meal contain proteins resistant to microbial degradation. Approximately one-half of the protein in these feedstuffs will escape breakdown in fermentative digestion. Unfortunately some of the most important and more commonly used protein sources such as soybean meal, alfalfa, and cottonseed meal are more extensively degraded (less than one-third of the proteins in these feedstuffs escapes destruction in the rumen).

Several approaches have been researched and used to decrease protein destruction in the rumen. Heat treatment of feeds will lower protein losses and there is commercial interest in marketing heat-treated proteins designed specifically for ruminants. Chemicals such as formaldehyde and tannins have been used successfully to protect protein. It is easy, however, to overprotect protein thus making it unavailable for absorption from the intestinal tract. This results in rich manure and poor producers.

Feeding protected proteins, or "escape" proteins as they are called, can reduce the amount of dietary protein that must be fed

to obtain the same animal performance. Another potential benefit derived from feeding relatively resistant proteins is that it affords greater opportunity for efficient use of NPN in the diet. When there is extensive degradation of protein in the rumen, the rumen microorganisms are content to use the resulting raw materials to make their cellular protein and they do not make good use of the NPN that may have been added to the diet. If, however, less dietary protein is degraded the rumen bacteria need, and do make use of, the nonprotein nitrogen added to their diet. The goal of an ideal protein-feeding strategy is to have most of the dietary protein escape destruction in the rumen and be used directly by the ruminant. Nonprotein nitrogen would be included in the diet for bacterial use and conversion into protein in the rumen.

18.2.5 Minerals

It has long been known that inorganic minerals play an important role in animal nutrition. Minerals constitute about 3 to 5 percent of the animal body. Animals cannot synthesize minerals and therefore must obtain them from food. Research has shown that these minerals must be provided in proper proportions as well as in adequate amounts (Chapter 21). An excess of certain minerals can be detrimental to the individual resulting in poor performance. Fortunately there is wide latitude in the amounts of most minerals that can be supplied in the diet without harming the individual. A mineral deficiency seldom causes death, but performance is often reduced to such a level that economic losses result. These losses are often **insidious,** which means that their effect is more serious than is readily apparent.

At least 15 mineral elements are recognized as performing essential functions in the body (Chapter 21). Proof of their necessity has been derived through careful experimentation. New experimental evidence occasionally adds other minerals to this list. Some essential minerals are sodium, potassium, phosphorus, calcium, chlorine, magnesium, iron, sulfur, iodine, manganese, copper, cobalt, molybdenum, selenium, and zinc. Possible essential minerals include fluorine, chromium, barium, bromine, strontium, vanadium, and silicon (Figure 18.3). Many other mineral elements are present in the body because they are present in feed when it is ingested and are not completely excreted in the urine and **feces.** In general, good practical diets contain most minerals at the needed level. But it is often necessary to supply some minerals (mineral supplements) in amounts above those present in the feed. This is especially true of calcium, sodium, chlorine, and phosphorus. Additional mineral elements (**supplements**) must be included in diets in areas where they are known to be deficient in the soil and plants (Chapter 21).

Minerals perform many functions in the body. They (1) constitute a portion of the structural materials (skeleton), teeth, and hemoglobin; (2) function in maintaining the optimum acid–base balance in body fluids and therefore are essential for life; (3) maintain the proper cellular **osmotic pressure** necessary for the transfer of nutrients across the cell membrane; (4) maintain the optimum acidity of digestive juices so that digestive enzymes can function; (5) maintain the proper contractility of muscles, especially those of the heart, and play an important role in the

---

[4] In 1828 a German chemist, Friederich Wöhler, discovered that he had made urea, a substance that a French chemist had found in urine 50 years earlier. Wöhler was trying to make another compound but instead synthesized urea.

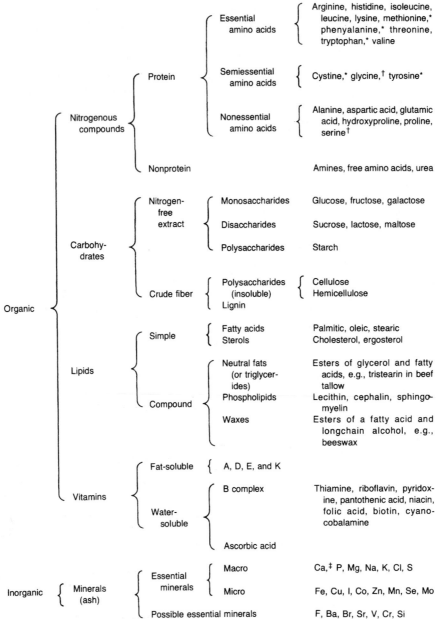

*Cystine will spare, but not completely replace, methionine. Tryptophan will spare, but not completely replace, nicotinic acid. Similarly, tyrosine will spare, but not completely replace, phenylalanine.
†Glycine is essential for young poultry (the chick can synthesize glycine, but not at a rate sufficient for maximum growth). Serine will spare glycine.
‡See Appendix E for elements represented by these symbols.

Figure 18.3 Classification of food nutrients.

normal functioning (**irritability**) of nerves; (6) prevent convulsions; and (7) are related to the function of certain vitamins in bone building. Various minerals also have other specific and essential functions within the body. They are discussed in Chapter 21.

Symptoms of mineral deficiencies include a general lack of thrift, slow and inefficient weight gains, impaired reproduction, and decreased production of milk, meat, eggs, wool, and work. These symptoms are usually not recognized until the deficiencies are extreme. Marginal deficiencies are often insidious and not recognized because no extreme symptoms are observed, yet they deprive farmers and livestock producers of millions of dollars annually. Hence the old English proverb, "An ounce of prevention is worth a pound of cure," is truly applicable. Mineral sup-

plements are usually relatively inexpensive, so deficiencies can be avoided by making the proper amounts available to animals. Mineral balances and interrelations are discussed in Chapter 21.

18.2.6 Vitamins

Vitamins are organic substances required in very small amounts containing various combinations of carbon, hydrogen, oxygen, and nitrogen. Vitamin B_{12}, or cyanocobalamin, also contains the mineral cobalt. The substances called *vitamins* are not closely related chemically. They are divided into two groups: (1) fat-soluble vitamins (A, D, E, and K) and (2) water-soluble vitamins (the B complex and C). Vitamins differ in their physiological functions and distribution in feedstuffs.

Some vitamins are required by all species of animals whereas others are required by only a few. It is not necessary to include certain vitamins in the food of some species because they are synthesized either by bacteria within the digestive tract or by the animal itself. Some foods contain **precursors** from which the animal derives the needed vitamin. For example, carotene in food is converted into vitamin A within the animal's body. Also, certain sterols are changed into vitamin D within the body through the action of sunlight. Synthesis of some B-complex vitamins such as folic acid and inositol is known to occur in the intestines of many animal species.

Vitamins function in many ways but probably most important are their actions as organic **catalysts** or as the essential components of catalysts that play important roles in many biological oxidations. Vitamins function in normal digestion, absorption, and utilization of food. They function in growth and reproduction and are important for maintaining the general health of humans and animals. Certain vitamins are also essential for regeneration of visual purple in the eye and are necessary for the formation of prothrombin, which is essential for normal blood clotting. Some vitamins are coenzymes (work with enzymes), which are necessary for the utilization of energy and the synthesis of proteins from amino acids. More specific functions of each vitamin are discussed in Chapter 20 and are summarized in Tables 20.1 and 20.2. A classification of food nutrients was presented in Figure 18.3.

18.3 ANALYSIS OF FOODSTUFFS

Farm **mammals** depend on plants for the nutrients necessary for maintenance, growth, reproduction, lactation, and other body functions. Some foods from plant sources possess more nutrients than others. For this reason it is important to know their chemical composition. Such information is useful in formulating diets for livestock. The composition of feeds is determined and grouped according to the percentage of dry matter, crude protein, fat, crude fiber, **nitrogen-free extract (NFE),** and mineral matter (ash). Please refer to the Glossary for additional information pertaining to these compositional values.

18.3.1 Dry Matter

The percentage of dry matter is determined by finding the percentage of water in a food and subtracting that value from 100 percent. The percentage of water is determined by drying a finely ground sample of food in an oven until a constant weight is attained.

The calculations involved are:

$$\% water = \frac{\left(\begin{array}{c}\text{wt of food sample}\\\text{before drying}\end{array}\right) - \left(\begin{array}{c}\text{wt of food sample}\\\text{after drying}\end{array}\right)}{\text{wt of food sample before drying}} \times 100$$

$$\% \text{ dry matter} = 100 - \% \text{ water}$$

All foods contain some moisture but the amount they contain depends on the length of time and way they are stored and the amount of moisture in the air. Most dry forages and grains contain 8 to 12 percent moisture.

18.3.2 Crude Protein

The **crude protein (CP)** in a food includes all the nitrogenous compounds. To estimate the percentage of crude protein, the nitrogen content of a food is chemically determined and this figure is multiplied by the factor 6.25. This factor is used because nitrogen represents approximately 16 percent of a protein ($100 \div 16 = 6.25$). The approximate protein content is important nutritionally and because foods rich in protein are the most expensive.

A formula for calculating the percent of crude protein in a food is:

$$\% \text{ crude protein} = \frac{\text{wt of nitrogen} \times 6.25}{\text{wt of original sample}} \times 100$$

18.3.3 Crude Fat (Ether Extract)

The percentage of fat in a food is determined by extracting a finely ground sample continuously with ether for several hours in a suitable apparatus. The ether and the ether-soluble components are separated from the food. The ether then is evaporated and the fatty residue is weighed and compared with the weight of the sample before extraction began. The fatty residue remaining after the ether is evaporated contains many substances other than true fat. This fatty residue is referred to as *crude fat*. Seeds of soybeans, flax, and cotton contain large amounts of crude fat (20 to 35 percent); those of corn and wheat contain only 4 to 5 percent.

18.3.4 Crude Fiber and Nitrogen-Free Extract (NFE)

Crude fiber contains the more insoluble and nondigestible carbohydrates such as cellulose. The **crude fiber (CF)** in a food is determined by boiling a sample first in a weak acid, then in a weak alkali, and removing and washing out the dissolved material. The weight of the original sample and its dry weight after the dissolved material is removed give the basis for calculating the percentage of crude fiber. **Concentrates** such as corn and wheat contain only 2 to 3 percent crude fiber; roughages may contain more than 20 percent.

Feeds having more *nitrogen-free extract* contain more digestible carbohydrates. The most important of these is starch. The NFE is calculated by subtracting the percentages of water, ash (see Section 18.3.5), crude protein, fiber, and fat from 100 percent.

For example, the percent NFE of yellow corn would be calculated as follows:

	Percent
Water	15.0
Ash (minerals)	1.2
Crude protein	8.6
Crude fiber	2.0
Fat	3.9
Total	30.7

The percent NFE of this corn would be 100 – 30.7, or 69.3 percent.

18.3.5 Mineral Matter (Ash)

Mineral matter is determined by weighing a food sample and then burning (ashing) it until all the carbon has been removed. The weight of the residue compared with the original sample weight gives the mineral content. The mineral content of foods such as corn and wheat is low (between 1 and 2 percent); alfalfa hay contains about 8 percent mineral matter.

A chemical analysis is important because it gives useful information about the composition of a food. It is made even more valuable, however, when linked with a digestion trial (Section 18.4). A **proximate analysis** of food reveals nothing about its vitamin content, the *quality* of its protein, digestibility, palatability, the presence of injurious (toxic) substances, or whether the feed may have been spoiled.

18.4 DETERMINATION OF THE DIGESTIBILITY OF FEEDS

Although a chemical analysis of a food may relate something about its nutritive value for livestock, it does not actually indicate its degree of **digestibility.** This must be determined for a particular feed by conducting digestion trials (Figure 18.4 a and b). Many digestion trials have been conducted and the percent digestibility of certain foodstuffs has been determined. These figures are average values, however, and may not apply exactly to an individual animal or to different species of animals. The digestibility will also vary with different varieties of plants and differing levels of maturity at the time of harvest. Additionally, it is important that the dietary intake be adequate when the nutritive value of feeds is being determined while avoiding excessive intake.

In a digestion experiment the percentage of each class of nutrient present in a food is determined by chemical analysis. For a few days an animal is fed weighed quantities of the food to be tested so that residues of previously fed foods are removed from the digestive tract. The experimental animal is then fed the same amount of weighed food each day. The feces are collected, weighed, and analyzed to determine their nutrient content. The difference between the nutrients in the food consumed and the nutrients in the excreted feces gives the amount retained in the body of the animal, or the amount of digested nutrients.

The fraction of a nutrient that is digested is called the **digestion coefficient** and may be calculated for a feed (hay) as follows:

Figure 18.4 (a) Cow in digestion stall. Feed (amount offered and refused) is weighed to determine intake. Note the fecal collection pan at the rear of the belt. The urine flows into a collection pan at the front of the belt-motor apparatus (not visible). (b) An alternative method of urine and feces collection.
Courtesy of Missouri Agr. Expt. Station.

	Pounds of nutrients					
Amounts	Dry matter	Protein	Fiber	Fat	NFE	Ash
15 lb of hay	13.50	1.50	4.50	0.60	6.00	0.90
20 lb of feces	5.43	0.60	2.25	0.24	1.80	0.54
Pounds digested	8.07	0.90	2.25	0.36	4.20	0.36
Digestion coefficient	0.60	0.60	0.50	0.60	0.70	0.40

Thus

$$\text{Digestion coefficient for protein} = \frac{0.90 \text{ lb digested}}{1.50 \text{ lb consumed}} = 0.60$$

The digestion coefficient multiplied by 100 gives the percent digestibility, or 60 percent in the example above. Obviously a determination of the percentage of each nutrient in a feed that is digested is a more accurate estimate of the value of a feed than is a chemical analysis. Many experiments have been conducted using different feeds and the digestion coefficient has been determined and average values recorded.[5]

[5] See F. B. Morrison, *Feeds and Feeding,* 22nd ed., Morrison Publishing, Claremont, Ontario, Canada, 1959.

18.4.1 Nutritive Ratio (NR)

The **nutritive ratio** is the ratio of the sum of digestible carbohydrates and fat to **digestible protein (DP).** It may be calculated as follows:

$$NR = \frac{\text{digestible NFE} + \text{digestible fiber} + (\text{digestible fat} \times 2.25)}{\text{digestible protein}}$$

A diet providing a large amount of protein in relation to other nutrients would possess a *narrow nutritive ratio* whereas one containing a relatively small amount of protein and a large amount of fat and carbohydrates would be described as having a *wide nutritive ratio.* The correct NR is necessary for maximum and efficient growth of animals.

18.5 THE ENERGY CONTENT OF FOODS

Energy is necessary for an animal to perform work and other productive processes. All forms of energy are converted into heat. Thus energy as related to body processes is expressed in heat units (**calories**). The *small* calorie is the amount of heat required to raise the temperature of 1 g of water 1 degree centigrade (specifically from 14.5 to 15.5°C). The **kilocalorie (kcal)** is the amount of heat required to raise the temperature of 1 kg of water 1 degree centigrade and is equal to 1000 small cal. A **megacalorie** (or **therm**) is 1000 kcal. One kilocalorie is approximately 4 British thermal units (Btu); 1 Btu is the amount of heat required to raise the temperature of 1 lb of water 1 degree Fahrenheit.

Several systems have been developed for expressing the energy content of foods and the energy requirements of animals. They are discussed here briefly.

18.5.1 Total Digestible Nutrients (TDN)

The **TDN** content of a food is expressed as a percentage and can be determined only by a digestion trial, as described previously. A formula for determining the percentage of TDN in a feed is:

$$\% \text{ TDN} = \frac{\begin{array}{c}\text{dig. protein} + \text{dig. crude fiber} + \text{dig. NFE} + \\ (\text{dig. EE} \times 2.25)\end{array}}{\text{unit of feed}} \times 100$$

In this formula lipids (ether extract, EE) are multiplied by 2.25 because they possess about 2.25 times more energy per unit weight than do carbohydrates. This is because carbohydrates have sufficient **molecular** oxygen present to balance the hydrogen so that heat is released only from the oxidation (combination with oxygen) of carbon. In lipids there is much less oxygen present in the molecule and oxygen outside the lipid molecule unites with both the carbon and hydrogen in the lipid yielding more heat. The combustion of 1 g of hydrogen yields about 4 times more heat than does the combustion of a similar weight of carbon. Oxidations of carbohydrates, fats, and proteins yield about 4100, 9450, and 5650 cal/g, respectively.

A limitation of TDN as a measurement of food energy is that it does not account for losses such as those of combustible gases and the **heat increment** that occur when a feed is consumed by an animal. These losses are much larger for forages than for concentrates. This means that a pound of TDN from a forage has considerably less value for productive purposes in an animal than does a pound of TDN in a concentrate (grain). For example, 1 lb of TDN in corn yields about 1 megacalorie of net energy. In the better hays (forages) 1 lb of TDN yields about 0.75 megacalorie and in low-quality roughages about 0.5 megacalorie. The TDN content of feeds is commonly related inversely to their fiber content. Concentrates are low in fiber and high in TDN.

18.5.2 Gross Energy (GE)

The **gross energy** (heat of combustion) of a food can be determined by completely burning a sample of food until the oxidation products (carbon dioxide, water, and other gases) are formed. A *bomb calorimeter* is used to measure the heat liberated from such a combustion. A bomb calorimeter consists of a closed vessel (bomb) in which the food is burned. The bomb is enclosed in an insulated jacket containing water that absorbs the heat (calories) produced (Figure 18.5). The procedure is as follows: (1) dried feed to be tested is weighed and placed in the bomb, (2) cover is screwed down tight, (3) bomb is charged with 25 to 30 atm of oxygen, (4) bomb is then placed in calorimeter jacket where it is surrounded by a known volume of water at a known temperature, (5) water is stirred until a constant temperature is attained, (6) charge then is ignited by an electric current, and (7) readings are taken on the thermometer to determine maximum temperature rise of the water. The GE determination reveals the total caloric energy in the food being tested.

Figure 18.5 A bomb calorimeter. Note bomb being gassed (lower center).

Courtesy of Missouri Agr. Expt. Station.

18.5.3 Digestible Energy (DE)

The digestible energy of a food is represented by that portion of the food consumed which is not excreted in the feces. The food is fed to an animal and the energy present in the feces is determined by means of bomb calorimetry. The difference between gross energy (Section 18.5.2) in the food and that in the feces is called *digestible energy*. The DE also can be calculated from TDN values by using the factor of 2000 kcal/lb of TDN or from digested nutrients by using the gross caloric factors for proteins, carbohydrates, and fats of 5.65, 4.10, and 9.45 kcal/g, respectively.

18.5.4 Metabolizable Energy (ME)

The **metabolizable energy** of a food is determined by subtracting the energy losses in the feces, urine, and combustible gases (primarily methane in ruminants) from the total energy in the food. A desirable characteristic of ME is that more of the gross energy not available to the animal for productive purposes is accounted for than by the measurement of TDN or digestible energy.

18.5.5 Net Energy (NE)

Determination of the **net energy** content of a food goes one step beyond determining the ME because the heat increment (heat resulting from the digestion and **assimilation** of feeds) is determined and subtracted from the total energy in the feed. Heat increment is measured by the amount of heat given off by an animal's body. Dr. Samuel Brody, world-renowned scientist in the area of bioenergetics and growth and formerly of the Dairy Science Department of the University of Missouri, called heat increment the "food utilization tax." Heat increment is approximately twice as high in lactating as in nonlactating cows but this difference varies with the lactation level and the amount of food consumed by the **lactating** cow. During cold weather a portion of the heat increment is used to keep the animal warm and this is advantageous. Heat increment is a disadvantage to animals in hot weather (at least in daylight hours) when they have the problem of keeping cool (Chapter 17).

The NE of a food can be determined in practical feeding experiments. A food such as corn in which the net energy is known is compared with a food of which the NE value is not known. Net energy also can be determined by the combined use of a bomb calorimeter and a respiration calorimeter. The former is used to determine the gross energy in a food and in the feces after the feed is fed; the latter is employed to determine the heat increment and losses of combustible gases.

Net energy, then, is that portion of the total (or gross) energy that is retained in the body for useful purposes such as maintenance, growth, milk production, egg laying, and muscular work.

18.5.6 Estimated Net Energy (ENE)

The estimated net energy value of a particular food may be calculated by comparing the production from it with that of food whose NE value is known (such as corn). Estimated net energy values are used in evaluating feeding programs for **DHIA** records processed by an electronic system. A factor favoring the ENE system is that it is more accurate when the feeding values of concentrates and forages are compared in attempts to develop the most economical diet.

18.5.7 Uses of Energy by Animals

> If you do not supply nourishment equal to the nourishment departed, life will fail in vigor; and if you take away this nourishment, life is utterly destroyed.
>
> **Leonardo da Vinci (1452–1519)**
> **Italian painter, sculptor, architect, and engineer**

What happens to dietary food energy (gross energy) in the dairy cow is shown in Figure 18.6. This figure also summarizes the different kinds of energy discussed previously.

Maintenance energy is that which is required to keep an animal in energy equilibrium and prevent the loss of body tissues or weight. The amount of energy consumed by an animal above that required for maintenance is used for *growth, reproduction, body fat, lactation,* and/or *egg laying* in the various farm animals. Young, growing animals have much higher dietary requirements in relation to body weight than mature ones for protein, energy, vitamins, and minerals. The amount of energy required for these various body processes has been determined and average values have been summarized. Such information, along with data pertaining to the composition of foods, is helpful in formulating diets for different groups of animals.[6]

18.6 FEED ADDITIVES

Various compounds have been added to livestock and poultry diets to increase the efficiency of food utilization. Most of these additives do not supply nutrients although they affect food utilization in some species.

Protamone

Also called thyroprotein this **iodinated casein** product has thyroxine[7] activity and has been fed to animals to increase the level and efficiency of production. Small amounts of this substance fed to young animals have produced, in some instances, more rapid growth. When fed to dairy cows at a stage of lactation when milk production normally declines, lactation increases of 20 percent have been reported. Some reports indicate that the feeding of thyroxine compounds may increase egg laying in hens. These drugs are effective in small amounts but can be detrimental if overdoses are given. They are not used on a practical scale in the feeding of livestock and poultry. Moreover, their use is prohibited in dairy cows enrolled in **official production**-testing programs.

---

[6] The nutrient requirements of domestic animals are given in publications available through the National Academy of Science–**National Research Council,** 2101 Constitution Ave. N W, Washington, DC 20418.
[7] Thyroxine is a thyroid hormone involved in the regulation of metabolism and body functions (Chapter 11).

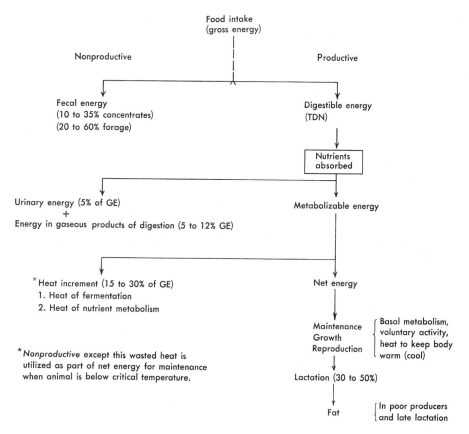

Figure 18.6 Utilization of food energy by the dairy cow.

Diethylstilbestrol

This additive, a synthetic estrogen, was used earlier to increase rate and efficiency of weight gains in steers by 10 to 15 percent but its use in meat-producing animals has been banned by the FDA.

Zeranol (trade name Ralgro)

This is an anabolic agent that promotes growth and improves feed efficiency in cattle. It is a chemical derivative of resorcylic acid lactone, a product of fermentation. Research indicates that one of its modes of action is to stimulate the pituitary gland to secrete increased amounts of somatotropin (growth hormone). Zeranol is implanted subcutaneously on the back side of the ear. Implanted animals commonly increase body weight about 10 percent faster and have a feed efficiency of about 8 to 10 percent greater than that of nonimplanted animals.

Monensin (trade name Rumensin)

Monensin is a polyether antibiotic useful as an anticoccidial agent for broilers and lambs. It is also used widely in the diets of beef cattle in which it increases the efficiency of feed utilization by modifying rumen fermentation. The molar percentage of propionate is commonly increased about 25 to 40 percent whereas that of acetate is usually decreased about 15 to 30 percent. Body weight gain is usually not affected significantly but the amount of feed required per unit of weight gain is commonly decreased by about 10 percent.

Another feed additive, *lasalocid* (trade name Bovitec), apparently acts in a similar manner to Rumensin in improving feed efficiency; it is also used widely in feedlot cattle.

Antibiotics

These compounds are widely used in swine and poultry diets. Antibiotics have been effective in increasing the rate and efficiency of gain in swine and poultry especially when the incidence of disease is high. They have also been used in feeding baby calves before they have a fully developed rumen and have been fed in some instances to growing, fattening cattle. Their major use, however, has been at relatively low concentrations in diets of nonruminant animals such as swine and poultry. Medicative levels in swine diets have been used effectively under some conditions.

Thiouracil

This additive is a drug that has been experimentally used in efforts to increase the rate and efficiency of gain in animals. Favorable results have been reported, mostly for swine. Thiouracil depresses the production of thyroxine (the **thyroid** hormone) and lowers the metabolic rate of animals. This conserves energy. Feeding thiouracil to swine for about 30 days after they reach a weight of 165 lb has significantly increased the rate and efficiency of gain. It has not been used on a practical scale on the farm.

Others

Dozens of feed additives have been studied in various experiments. More will be tested in the future but probably few will have an impact on livestock production equal to that of antibiotics and the synthetic estrogens.

18.7 SUMMARY

An understanding of the deep-seated principles of nutrition is important because humans and animals are dependent on food nutrients for the processes of life. We have attempted to present the basic fundamentals of animal nutrition, giving special emphasis to the composition of plants and animals. A classification of food nutrients was presented in tabular form. Methods of analyzing foods, means of expressing their energy, and uses of food nutrients were discussed.

Much remains to be learned about the nutrient requirements of farm mammals and poultry. Many feeding standards were established early in the twentieth century before recent selection techniques provided humans with a means of developing animals that possess greater inherent producing ability. Significant advances in animal and human nutrition will be forthcoming as humans learn how to better utilize the animals that serve them.

STUDY QUESTIONS

1. Is level of nutrition related to the behavior of humans and animals? Explain.

2. Who was Lavoisier? What did he contribute to the science of nutrition?

3. List the major substances found in the bodies of animals.

4. How do plants and animals differ in their chemical composition?

5. What are carbohydrates? What chemical elements do they contain?

6. List the main classifications of carbohydrates.

7. Which carbohydrates are easily digested by animals? Which ones are not?

8. What is the major nutritional contribution of carbohydrates?

9. How do fats and oils differ from carbohydrates in their chemical composition?

10. Why do fats supply more energy than carbohydrates when they are metabolized?

11. Describe the differences between saturated and unsaturated fatty acids. Which fatty acids are essential?

12. What is the major function of fatty acids in the body?

13. What are proteins? How do they differ chemically from carbohydrates and fats?

14. List the major functions of proteins in the body.

15. What determines if an amino acid is essential or nonessential?

16. Why do ruminants and nonruminants differ in their dietary protein requirements?

17. What is meant by the biological value of a protein? What is meant by the term "protected protein"?

18. List the major functions of minerals in the body. What are some symptoms of mineral deficiencies?

19. What are vitamins? What are their two major classifications?

20. What are some major functions of vitamins in the body?

21. Describe the major classifications of feedstuffs.

22. What major chemical substances are found in ether extract? In nitrogen-free extract?

23. A finely ground sample of food weighed 100 g. After being dried in an oven until it reached a constant weight, it weighed 88 g. What was the percent of water in the original sample? (*Answer:* 12 percent.) What was the percent dry matter (DM) in the feed? (*Answer:* 88 percent.)

24. The nitrogen content of a 200-g sample of food was determined and found to be 3.6 g. What is the crude protein content of this food? (*Answer:* 11.25 percent.)

25. A chemical analysis of a food sample gave the following results: ash, 2.6 percent; crude fiber, 10.6 percent; crude protein, 12.0 percent; fat, 4.1 percent; and water, 8.9 percent. What was the percent NFE in this sample? (*Answer:* 61.8 percent.)

26. A 100-g sample of food was burned until all the carbon was removed. The weight of the burned residue was 6.8 g. What was the percent of mineral matter (ash) in this sample? (*Answer:* 6.8 percent.)

27. A ration was found to contain 66.83 g of digestible NFE, 1.31 g of digestible fiber, 3.64 g of digestible fat, and 7.37 g of digestible protein. What is the nutritive ratio of this ration? (*Answer:* 1:10.35.) What does the nutritive ratio mean?

28. Another ration was found to contain 26.69 g of digestible NFE, 3.81 g of digestible fiber, 4.86 g of digestible fat, and 37.66 g of digestible protein. What is the nutritive ratio of this ration? (*Answer:* 1:1.1.)

29. A diet has the following composition:

	Percent
Digestible protein	7.6
Digestible crude fiber	1.7
Digestible NFE	66.0
Digestible fat (EE)	2.7

What is the percent TDN in this diet? (*Answer:* 81.4.)

30. Define or describe the following terms: kilocalorie, megacalorie, gross energy, digestible energy, metabolizable energy, net energy, and digestion coefficient.

31. What is meant by the term feed additive?

32. What are some feed additives that have been used on a practical scale?

THE PHYSIOLOGY OF DIGESTION IN NUTRITION[1]

Before hazarding a theory we propose to multiply our observations, to investigate the phenomena of digestion, and to analyze the blood both in health and disease.

A. L. Lavoisier (1743–1794)
French chemist

19.1 INTRODUCTION

Some species of animals depend entirely on plants for food. They are called **herbivores.** Other species feed almost entirely on flesh of other animals. They are called **carnivores.** Still other species consume both plants and flesh. They are called **omnivores.** Regardless of their eating habits all animals depend on plants (directly or indirectly) for food. Moreover, it can be said that all animal life depends indirectly on the sun for food because it is through the action of sunlight on the chlorophyll of plants and the process of photosynthesis that plants convert elements from the air and soil into the **nutrients** that supply food for animals. Thus without energy from the sun there would be no food for plants, animals, and humans.

Animals do not use all plant nutrients for the various body processes exactly as they come from the plant. Most complex nutrients must be broken down (digested) to simpler compounds before they can be absorbed and utilized. The different species of animals have digestive tracts adapted to break down the type of food they consume. Thus herbivores differ from carnivores and omnivores in the anatomy and physiology of their digestive system.

It is the purpose of this chapter to discuss the anatomy and physiology of the digestive systems of animals, especially of farm mammals and poultry.

19.2 TYPES AND CAPACITIES OF DIGESTIVE SYSTEMS

Anatomical and capacity differences in the digestive systems among species are more significant physically than nutritionally because foods in the digestive tract are still, in a sense, outside the body. In the process of digestion, nutrients enter the body by **absorption** through the gut wall. The metabolic processes that then utilize the absorbed nutrients are essentially the same for all species.

19.2.1 Anatomy and Types of Digestive Systems

The alimentary tract extends from the lips to the anus. The principal parts include the mouth, pharynx, esophagus, stomach, and the small and large intestines.

The length and complexity of the tract vary greatly among species. The tract is relatively short and simple in carnivores but is much longer and more complex in herbivores. In some herbivores (e.g., horse and rabbit) the stomach is relatively simple and comparable with that of carnivores whereas the large intestine and cecum are much more capacious and complex than those of carnivores. Conversely, in the group of herbivores known as ruminants (cow, goat, sheep) the four-chambered stomach is large and complex whereas the large intestine is long but less functional.

The three general types of digestive systems and the foremost anatomical differences are depicted in Figure 19.1.

The anatomy of the **avian** digestive system is shown in Figure 19.2. The digestive system of birds differs from that of mammals in that birds have no teeth to physically break

---

[1] The authors acknowledge with appreciation the contributions to this chapter of Dr. Neal R. Merchen, Department of Animal Sciences, University of Illinois at Urbana–Champaign.

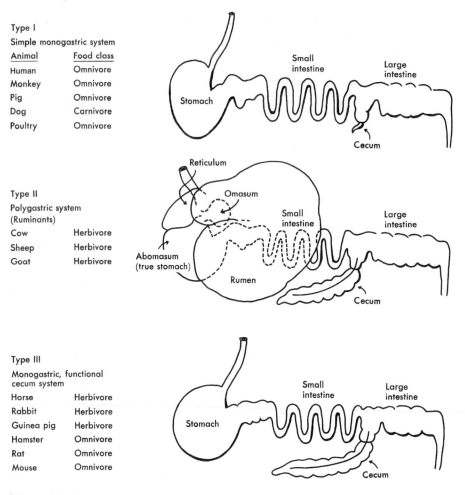

Figure 19.1 Classification of animals according to type of digestive system.
Modified from E. W. Crampton and L. E. Lloyd, Fundamentals of Nutrition, *W. H. Freeman, San Francisco, 1959.*

down their food. The glandular stomach in the bird is known as the *proventriculus.* Between the proventriculus and the mouth is an enlargement of the esophagus called the *crop.* The **ingested** food is stored temporarily in the crop where it is softened before moving to the proventriculus. Food passes quickly through the proventriculus to the *ventriculus,* or *gizzard.* The main function of the gizzard is to crush and grind coarse food simulating the role of the teeth in mammals. This is aided by grit and gravel, which the bird ingests and stores in its gizzard.

19.2.2 Capacity of Digestive Systems

The average capacities of the different parts of the **gastrointestinal** (GI) tract in selected animals and humans are given in Table 19.1.

The total capacity of the entire digestive system in the human, who is omnivorous, is only about 6.34 qt (6 liters). The adult human stomach holds about 5 cups, or about 1.2 liters. Even allowing for the prompt passage of many foods into the intestines it is obvious that there is a physical limit to the quantity of food humans can consume and process in a given time. The dog, a carnivore, also has a small digestive system

although it is somewhat larger per unit of body weight than that of the human.

The pig, an omnivore, possesses a larger digestive capacity per unit of body weight than either the dog or human. Even so, the digestive tract in this species is limited in capacity. The digestive system of the pig is better adapted to the use of **concentrated** feeds such as grains although limited amounts of **forages** can be included in the diet. The horse, a herbivore, possesses a much larger digestive system than the pig and is able to utilize large amounts of **roughages** in its diet because of a greatly enlarged cecum. The horse, however, is considered a nonruminant.

Cattle and sheep are herbivores and have greatly enlarged digestive tracts designed to process bulky feeds. Both of these species are called **ruminants** because they possess three stomach compartments (**rumen, reticulum,** and **omasum**) not found in nonruminants. Such a digestive system provides the space for processing large quantities of bulky forages to provide energy and nutrients to sustain maintenance and high levels of production (milk, meat, and wool). For example, the cow has about 9 times more digestive tract capacity than humans in proportion to body weight. Not only is every digestive tract component that has a counterpart in humans 3 or 4 times more capacious, but the ruminant stomach of the cow has a total capacity per unit of body weight exceeding that of the simple human stomach by more than 30-fold. A cow weighing 1200 lb may have a stomach capacity of 300 lb.

The relative sizes of the four stomach compartments in the ruminant vary with age. The first three compartments are small in the newborn and very young ruminant representing less than 30 percent of the total stomach capacity. As the young animal is switched from a milk diet to one containing cereal grains and forage the rumen develops, and by the time the animal reaches

maturity the rumen accounts for about 80 percent of the stomach capacity (Figure 19.3). In mature animals food normally passes first into the rumen. Under normal conditions the **calf** rumen becomes functional in about 6 to 8 weeks. Evidence of a functional rumen is based on (1) rumen odor, which is indicative of

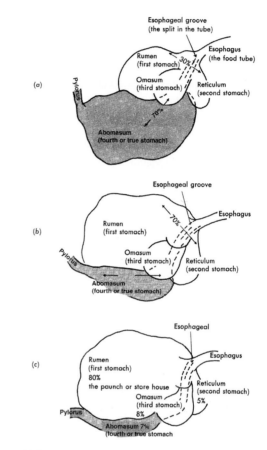

Figure 19.3 The development and relative capacities of the various stomach compartments of (a) the calf at birth, (b) the calf at 2 months, and (c) the mature cow.

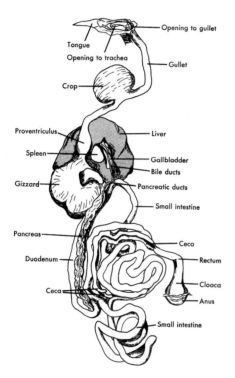

Figure 19.2 Major components of the avian digestive system. *After F. B. Adamstone.*

			Animal			
Part	Ox (cattle)	Sheep and goat	Horse	Pig	Dog	Human
Gastric compartment						
Rumen	202	23				
Reticulum	8	2				
Omasum	19	1				
Abomasum (simple stomach)	23	3	18	8	4	1
Subtotal	252	29	18	8	4	1
Small intestine	66	9	64	9	2	4
Cecum	10	1	33	1		
Large intestine	28	5	96	9	1	1
Total	356	44	211	27	7	6

TABLE 19.1 Capacity, in Liters, of the Digestive System of Selected Animals

Source: Adapted from H. H. Dukes, *The Physiology of Domestic Animals,* Comstock Publishing Associates, Ithaca, NY, 1970.

fermentation; (2) a decline in blood glucose; and (3) the production of **volatile fatty acids.**

19.3 THE PROCESS OF DIGESTION

Digestion in higher animals includes all the activities of the alimentary tract and its associated glands and organs in the conversion of food into compounds available for absorption and **assimilation.** It also includes the rejection of unabsorbed food residues (**fecal** materials). Most foods when consumed are too complex to be absorbed into the blood and **lymph** without preliminary digestive changes. Glucose, soluble salts, water, and a few other nutrients are exceptions.

The processes of digestion are chemical, mechanical, microbiological, and secretory in nature. Chemical factors include **enzymes** and certain nonenzymatic chemical substances (such as hydrochloric acid) produced by the digestive glands. The main mechanical processes are **mastication, deglutition, regurgitation,** gastric and intestinal motility, and **defecation.** Microbiological processes reflect the activities of bacteria and, in some species of animals, **protozoa** within the digestive tract. The secretory contributions to digestion include the enzymes and chemicals produced (secreted) by the digestive glands.

19.3.1 Digestion in Monogastric (Nonruminant) Animals

Chemical and secretory factors of digestion, especially enzyme action, play an important role in the digestion of foods in humans, poultry, swine, and other nonruminant animals. Digestion of food (especially fiber) by means of microorganisms is of lesser importance in these species. However, in the horse digestion of fiber by microorganisms is of great importance because a great deal of digestive activity occurs in the cecum.

19.3.2 Digestion in Ruminant Animals

Within the digestive compartments of the ruminant are large populations of microorganisms (approximately 1 billion bacteria and 1 million protozoa per milliliter) that break down **cellulose** to form the short-chain volatile fatty acids (acetate, propionate, and butyrate commonly called **VFA**). These volatile fatty acids are absorbed largely through the rumen wall and provide the ruminant with 60 to 80 percent of its energy needs.

Although cellulose is digested in the rumen we should hasten to add that starches, sugars, and proteins are also largely broken down here. In fact, an average of 60 to 90 percent of the total digestion within ruminants occurs in the rumen. It is therefore a remarkable fermentation vat.

In payment for their rumen-housing privileges the microorganisms perform some special favors for their host. These include synthesis of all the B-complex vitamins (Chapter 20) and all the **essential amino acids** (Chapter 18). The latter can be synthesized from nonprotein nitrogen (NPN) compounds such as **urea** or diammonium phosphate or from proteins that are deficient in one or more essential amino acids. Finally the microor-

Figure 19.4 The reticulorumen from a lamb fed a completely pelleted diet (left) and one fed shelled corn and chopped hay (right). Note that the papillae from the rumen epithelium of the lamb receiving the pelleted diet are longer and wider.
Courtesy of Dr. G. B. Thompson and Dr. W. H. Pfander, University of Missouri.

ganisms give their life in payment for food and shelter, being digested by their host farther down the gastrointestinal tract.

19.3.3 Special Features of Ruminant Digestion

The nonglandular epithelial membrane of the developed rumen is lined with papillae that aid in absorption from the rumen (Figure 19.4). Although Figure 19.4 shows the increase in length and width of papillae in lambs fed completely pelleted diets, it does not clearly show that there is frequently a **keratinization** of the rumen epithelium on such diets. This latter effect may result in reduced absorption of nutrients from the rumen.

A substantial amount of gases (methane and CO_2) is produced in the rumen and if allowed to accumulate causes **bloat,** or an excessive distension of the rumen. However, these gases are normally expelled by **eructation** (belching) and to a lesser extent by absorption into the blood, followed by elimination through exhaled air from the lungs.

19.4 APPETITE

A fundamental factor involved in the voluntary intake of food is appetite. The human center within the hypothalamus is the most important area of the brain in regulating food intake. Other factors include (1) blood glucose level, (2) amount of fill (quantity of **ingesta** in the stomach), and (3) environmental temperature (decreased appetite in hot weather).

When the appetite is given full control of what shall be eaten, it is surprising to note how pigs naturally select the specific feeds which swine herdsmen have long since approved of as the best, and, what is equally surprising, the pigs show a marked avoidance of those feeds usually considered as ill adapted to swine.

J. M. Evvard (stated in 1915)

19.5 THE PREHENSION OF FOOD

Animals differ greatly in the manner by which they harvest and convey food to their mouths **(prehension)**. Domestic animals use their lips, teeth, and tongue for the prehension of food. The upper lip is the chief prehensile structure in the horse whereas the tongue performs that function in cattle (cattle have an especially strong tongue). The sheep has a **cleft** upper lip (the goat does not) and uses its incisor teeth and tongue to procure food. A pointed lower lip, teeth, and tongue serve as prehensile structures in swine whereas the beak is used for this purpose in poultry.

19.6 THE MASTICATION OF FOOD

Mechanical reduction of particle size of food is accomplished in the mouth by chewing, or *mastication.* This act aids the digestive process by (1) reducing food-particle size resulting in a greater surface area for the digestive juices to act on, and (2) mixing the food with saliva to facilitate easier swallowing.

Remastication of food is important in ruminants in which large quantities of food may be consumed in a relatively short period of time. The animal can then relax under a shady oak, regurgitate food (in the form of a **bolus**), and **ruminate.**

While most dentists agree that mastication is essential to digestion, Spallanzani proved nearly two centuries ago that certain foods such as meat could be digested without having been chewed. He observed meat to disappear from tiny linen bags that he swallowed and later recovered following their tour of the digestive tract.

In birds the food enters the mouth and passes through the esophagus to the crop where it is temporarily stored and softened. It then passes rather quickly through the proventriculus (true stomach) to the ventriculus, or gizzard. Here the food is ground and crushed into smaller particles (inasmuch as birds have no teeth).

19.7 ENZYMES OF THE DIGESTIVE TRACT

An enzyme is an organic compound produced by specific cells within the body that speeds biochemical reactions at ordinary body temperatures without being used up in the process. Enzymatic activity is responsible for most of the chemical changes occurring in foods on their movement through the digestive tract. A summary of the enzymes contributing to the digestive process is presented in Table 19.2.

19.7.1 Digestion of Carbohydrates

In some species the digestion of **carbohydrates** begins in the mouth where the foods come in contact with ptyalin, an enzyme secreted by the salivary glands. The saliva of humans, swine, and dogs contains small amounts of amylase (or ptyalin), that of the horse contains very small amounts, and that of ruminants (cow, sheep, and goat) contains none. Amylase digests starch to maltose and **dextrins.** Mucin in the saliva does not digest starch but it lubricates the food so that it is easily swallowed.

As was mentioned previously, microorganisms in the rumen break down cellulose to form volatile fatty acids. They also digest starches, sugars, fats, proteins, and nonprotein nitrogen to synthesize microbial protein, B vitamins, and vitamin K. No enzymes from the ruminant gastric secretions are involved in microbial synthesis.

Amylase from the pancreas is secreted into the first portion of the small intestine (duodenum) where it digests starches and dextrins to simpler dextrins and maltose. Farther along in the small intestine other enzymes from the intestinal juices also digest carbohydrates. These enzymes include *sucrase* (invertase), which breaks down sucrose into glucose and fructose; *maltase,* which breaks down maltose into glucose; and *lactase,* which breaks down lactose into glucose and galactose.

Microorganisms in the cecum and colon also digest cellulose forming VFA. As is true in the rumen, enzymes secreted by the digestive tract of the animal are not involved in the digestion of cellulose by the microorganisms of the cecum and colon.

19.7.2 Digestion of Proteins

The true stomach is the site of the beginning of protein digestion in several species. The abomasum of ruminants is comparable with the true stomach in other mammals and the proventriculus in birds. Hydrochloric acid is produced by cells in the true stomach thus supplying an acid medium that activates pepsin and rennin, aiding in protein digestion. The initial step in protein digestion occurs when foods come in contact with the enzyme *pepsin* from the gastric juice. Pepsin splits proteins into less complex compounds known as peptides. In young nursing animals the enzyme *rennin* causes milk to coagulate forming *paracaseinate,* which stays in the stomach longer than if the milk had remained a liquid. Therefore more complete digestion results.

Pancreatic juice is secreted into the duodenum and contains the enzymes *trypsin, chymotrypsin,* and *carboxypeptidase.* These enzymes continue protein digestion initiated in the stomach by breaking down the more complex substances into peptides and finally into amino acids, which are absorbed from the GI tract.

19.7.3 Digestion of Fats

The enzymatic digestion of fats begins in the stomach where *lipase* from the gastric juice converts fats into fatty acids and glycerol. (This process is limited in many species.) In the duodenum, bile from the liver emulsifies fats breaking them down

TABLE 19.2	Digestive Processes in Farm Mammals and Poultry			
Region	**Secretion**	**Enzyme**	**Substrate or function**	**End products**
Mouth	Saliva	Amylase (ptyalin)	Starch and carbohydrates	Maltose and dextrins
			Urea (recycles)	
	Mucin	—	Lubrication of food	
Crop (birds)	Mucus	—	Lubricates food	
Rumen	—	Enzymes from microorganisms	Cellulose	VFA
			Polysaccharides and starches	Microbial protein,
			Sugars, fats, and proteins (urea)	B vitamins, vitamin K
Stomach (abomasum	Gastric juices and acids	Pepsin	Proteins	Proteoses and peptones
in mammals and		Lipase	Fats	Fatty acids and glycerol
proventriculus in birds)	Mucus	—	Coating of stomach lining and lubrication of food	
Gizzard (birds)	—	—	—	Ground foods
Nursing animals	Gastric juice	Rennin	Milk protein (casein)	Paracasein
Duodenum	Pancreatic juice	{ Trypsin		
		{ Chymotrypsin	Proteins, proteoses, peptones, and peptides	Peptones, peptides, and amino acids
		Amylase	Starch, dextrins	Dextrins and maltose
		Lipase	Fats	Higher fatty acids and glycerol
		Carboxypeptidase	Peptides	Amino acids and peptides
	Bile (liver)	—	Fats	Emulsion of fats
Small intestine	Intestinal juice	Peptidase (erepsin)	Peptides	Amino acids and dipeptides
		Sucrase (invertase)	Sucrose	Glucose and fructose
		Maltase	Maltose	Glucose
		Lactase	Lactose	Glucose and galactose
		Polynucleotidase	Nucleic acid	Mononucleotides
{ Large intestine { Cecum and colon	—	Cellulase from microorganisms	Cellulose	VFA

into smaller globules and increasing the total surface area on which lipase acts. Pancreatic lipase continues the digestion of fats where it is not complete, breaking them down into fatty acids and glycerol.

19.7.4 Digestion of Other Nutrients

Minerals are dissolved from foods in the hydrochloric acid secretions of the stomach. They are also released from the organic compounds that are digested by the various enzymes. Water requires no digestion before being utilized by the animal. Vitamins can be used as such within the body without conversion into simpler compounds.

19.8 AVIAN DIGESTION

No digestive enzymes are secreted in the gizzard of fowls. The main function of this organ is to reduce the size of food particles. The digestibility of coarse foods is greatly reduced in gizzardectomized (gizzard-removed) birds but in spite of this they remain healthy for many years.

From the gizzard food passes through the intestinal loop, called the *duodenum,* which parallels the *pancreas* anatomically.

The pancreas has an important role in avian digestion as it has in other species. It secretes an abundant supply of pancreatic juice containing amylolytic, lipolytic, and proteolytic enzymes, which **hydrolyze** starches, fats, and peptides, respectively. Liver bile also enters the duodenum and aids in the digestion of lipids.

Food materials move through the small intestine whose walls secrete intestinal juices containing peptidases and some sugar-splitting enzymes. Peptidases complete protein digestion, yielding amino acids, and the sugar-splitting enzymes convert disaccharides into simple sugars (monosaccharides) that can be assimilated by the body. Absorption is accomplished through the *villi* of the small intestine.

No liquid urine is voided by birds. The urine is discharged into the cloaca and excreted with the feces. The white material in bird droppings is largely uric acid whereas the urinary nitrogen of mammals is mostly urea. The relative shortness of the avian digestive tract is reflected in a rapid digestive process (about 4 h).

19.9 ABSORPTION OF FOOD NUTRIENTS

The process whereby food nutrients, following digestion, are transferred from the **lumen** of the gastrointestinal canal to the blood or lymph is called *absorption.*

Most absorption of nutrients in carnivores and omnivores, and a significant amount in herbivores, occurs in the small intestine. Probably no food nutrients are absorbed from the mouth and esophagus and very few from the stomach. (An exception is the absorption of VFA across the rumen wall in ruminants.) Except for water absorption from the colon very little absorption occurs from the large intestine of carnivores and humans. Conversely, the large intestine is the site of substantial absorption of volatile fatty acids in many herbivores.

The mucous membrane of the small intestine is richly supplied with minute, finger-shaped projections, called *villi,* which absorb food nutrients. Within the villi are tiny capillary blood vessels and lymph **ducts,** which collect the absorbed nutrients. Fats are absorbed as fatty acids and glycerol, principally by the lymph, whereas water, inorganic salts, and the end products of carbohydrate digestion (monosaccharides and VFA) and of protein digestion (amino acids and peptides) are absorbed largely by the blood.

Absorption completes the digestive process and makes nutrients available to support the process of life.

19.10 FACTORS AFFECTING THE DIGESTIBILITY OF FEEDS

Knowledge of factors affecting the **digestibility** of feeds is important because it can be effective in increasing the efficiency of **feed conversion.**

19.10.1 Temperature

The environmental temperature may have a pronounced effect on the appetite of an animal and the amount of feed it consumes. This could have an indirect effect on the digestibility of a food. Experiments with dairy cattle indicate that increases in temperatures up to 90°F are accompanied by corresponding increases in the apparent digestibility of feeds. This is largely because feed intake is reduced and food remains in the gastrointestinal tract for a longer period of time.

19.10.2 Rate of Passage through the Digestive Tract

If for various reasons the food consumed should pass through the digestive tract too quickly there may not be enough time for complete digestion of the nutrients by the digestive enzymes. It is also possible, however, that if the **rate of passage** of the food were too slow fermentation losses could be greater than desired. In general, however, experimental data indicate that a more rapid passage of food is related to a lower digestibility of the feed consumed.

19.10.3 Level of Feeding

The digestibility of feeds in ruminants is higher at a level of feeding near maintenance than it is when animals are on **full-feed.** Many experiments with swine suggest that the efficiency of food conversion is greater on **limited-feeding** than on full-

feed. In spite of these observations experiments have failed to show any difference in the digestibility of feeds fed swine on either full or limited diets. Greater feed efficiency in animals fed limited amounts of feed could be the result of less wastage of feed or a more efficient utilization of the feed consumed rather than of improved digestibility.

19.10.4 Physical Form of a Feed

Whether grinding feed affects its digestibility depends on how well animals chew their grain before it passes through the digestive tract. Very young and very old animals, which do not have good teeth, do not masticate (chew) their food as well as mature animals with sound teeth.

Grinding Grain

Grinding grain for animals exposes a larger surface to the digestive juices and could therefore increase its digestibility. Cattle do not chew their grain as thoroughly as sheep and swine and much of the whole grain passes through the digestive tract into the feces without being digested. Therefore grinding grain for cattle usually increases its digestibility.

Grinding Hay

Grinding hay increases its rate of passage through the digestive tract and thereby reduces its digestibility somewhat because there is less time for microbial fermentation of cellulose and hemicellulose. Most forage-consuming animals chew hay well enough that there is optimum exposure to the digestive process.

Pelleted Diets

Pelleted diets have gained in popularity for many species of livestock in recent years. Pelleted or cubed diets have increased the rate and efficiency of gain in lambs and cattle (and possibly swine), but there is not complete agreement among research workers as to whether feed digestibility is improved.

Heating

Heating per se does not greatly affect digestibility of feeds. However, heat may be used to destroy a growth inhibitor. For example, raw soybeans possess a trypsin-inhibiting factor that prevents the proper utilization of the protein by nonruminants such as pigs and poultry. Proper heating (but not overheating) of raw soybeans increases the availability of the proteins for these species by inactivating the inhibitor. Cooking most other feeds has little or no effect on their digestibility.

19.10.5 Composition of the Diet

In vitro experiments have indicated that the extent of digestion of cellulose in high-quality hay is higher than in low-quality roughages such as **corn stover,** wheat straw, corncobs, and mature timothy–bluegrass hay. The addition of minerals and nitrogen greatly increases cellulose digestion in low-quality roughages but has little or no effect on the digestion of cellulose in high-quality clover, rye, and alfalfa hays.

Considerable attention has been given to the needs of ruminal microorganisms for maximal digestion of feeds. Urea has been added to ruminant diets in varying amounts for synthesis into proteins by rumen microorganisms. A certain amount of protein and urea in the diet appears to increase the digestibility of cellulose in roughages. Small amounts of minerals such as phosphorus, iron, sodium, potassium, calcium, magnesium, sulfur, and chlorides are essential for the growth of rumen microorganisms and, indirectly, are necessary for greater cellulose digestion. The addition of B vitamins (especially biotin) has increased the rate of cellulose digestion in the artificial rumen but has not shown a beneficial effect in *in vivo* studies.

Small amounts of readily available energy such as are found in glucose and starch appear to be required for the normal growth of rumen microorganisms and thus aid in the digestion of cellulose. Large amounts of glucose and starch may decrease cellulose digestion, however.

The addition of corn oil or lard to diets of ruminants that contain large amounts of coarse roughages (such as cottonseed hulls) appears to decrease the digestibility of the roughage. A limited amount of fat can be included in diets of swine and poultry without affecting the digestibility of other nutrients.

Feeding **antibiotics** to swine and poultry has been shown in many experiments to increase rate and efficiency of gain. Such increases are due to a reduced incidence of **subclinical** infections rather than increased digestibility of feeds. Feeding antibiotics to young calves before their rumen has fully developed has also increased the rate and efficiency of gain. This is probably due to better utilization of feed nutrients rather than to increased digestibility of nutrients in the diet. Feeding certain antibiotics to mature ruminants may be detrimental to the growth of rumen microorganisms resulting in poor cellulose digestion. The level of antibiotics fed is, of course, important.

19.11 Efficiency of Food Conversion

Animals convert only a portion of the feed nutrients they consume into food for humans. For this reason a larger population could be fed if humans ate plant products directly rather than consuming animals and their products (Chapter 1). However, humans cannot utilize some plants, especially grass and other forages. Therefore it is essential to use certain animals as "middlemen" for food production from forages. Animal products supply large quantities of good-quality proteins and other important nutrients vital to the health and well-being of humans (Chapter 3). In addition, humans have had an appetite for animal products through the years. It is doubtful they will lose this appetite in the years ahead, especially as standards of living improve.

The ability of animals to convert nutrients in the feed they eat into animal products is referred to as *efficiency of food conversion* and depends on (1) their ability to digest nutrients in feed; (2) their requirements for energy and protein for growth, maintenance, and other body functions; (3) the amount of these nutrients lost in metabolic end products and nonproductive work; (4) the type of feed consumed; and (5) the composition of the animal products produced.

The digestibility of air-dried feed varies with the species of animals that consume them. Nonruminant animals such as swine and poultry digest approximately 75 to 85 percent of the nutrients in concentrates but very little of the nutrients in forages such as hay and silage. Ruminants, however, digest about 50 to 55 percent of the total nutrients in alfalfa hay and 75 to 80 percent of the nutrients in concentrates such as ground corn. Furthermore, ruminants digest 75 to 80 percent of the dry matter in *immature* grasses.

Not all the digestible nutrients in a feed consumed by animals are used for productive purposes. Some of the digestible protein in a food is not incorporated into body tissues because there is not a proper balance of amino acids in the feed for building body proteins. Where there is a lack of balance the extra amino acids are deaminated and used as a source of energy. The end products of protein metabolism such as urea and uric acid are eliminated in the urine and feces and are lost to the animal for productive purposes. In general, the higher the protein content of a feed the larger the amount of energy lost through these channels.

As is true of proteins, not all the digestible energy in other nutrients such as carbohydrates and fats is used for productive purposes. Some energy is lost in the feces, urine, combustible gases, and the heat of digestion (Figure 18.6). The proportion of digestible energy taken into the body used for productive purposes varies with the species, individual, and kind of feed consumed.

19.12 Factors Affecting the Efficiency of Food Conversion

The efficiency of food conversion in meat-producing and **lactating** animals is usually expressed as the units of body weight gained or pounds of milk secreted by an animal per unit of feed consumed. A more accurate method would be to express the efficiency of food conversion in terms of the amount of feed required to produce a unit of milk or meat, or better yet the amount of feed required to produce a unit gain of **carcass** or a pound of 4 percent **FCM**. Expressing efficiency on the basis of carcass weight, however, requires slaughtering the animal and recording accurate weights of carcass parts. This is time-consuming and expensive and has the disadvantage that breeding stock cannot be selected on such a basis. Recently attempts have been made to develop measurements of muscle mass in the live animal by various means including low-level radiation counters. If these attempts prove successful it will be practical to express the efficiency of feed conversion on a carcass basis.

The efficiency of feed conversion into milk and eggs is rather easily measured. Perhaps this is one reason that the production efficiency of these two animal products has greatly increased. The greatest improvement in feed conversion in recent years among domestic animals has been in **broiler** and egg production. This improvement has been brought about by improved breeding, feeding, and management practices.

Many factors affect the efficiency of feed conversion. Some of them will be discussed in the following sections.

19.12.1 Inheritance

Data from many experiments show that feed conversion among farm mammals is from 30 to 40 percent heritable. This means that genes with an additive type of expression affect this trait and selecting the best and mating the best to the best should improve it. However, **genes** with a nonadditive effect have little influence on the efficiency of feed conversion. This conclusion is based on the fact that feed efficiency declines very little when **inbreeding** is practiced and improves little when animals are **crossbred** (Chapter 9).

19.12.2 Age and Weight

Younger animals require less feed per unit of gain than do older, more mature ones. The main contributing factor to more efficient feed conversion in young animals is the composition of body mass, which constitutes the weight gain. As shown in Table 19.3 the percentage and pounds of fat increase as pigs grow older and become heavier. Because more energy is required to deposit a pound of fat than a pound of protein, less gain is made per unit of feed as a pig becomes older and heavier. Although not shown in Table 19.3 the protein mass in the pig also increases with advancing age and increasing body weight; however, it decreases as a percentage of total body composition.

Another limiting factor is that the maintenance requirement increases with the 0.75 power of body weight ($W^{0.75}$), which means that there is less feed in addition to that required for maintenance to use for growth as body weight increases. The amount of food consumed per unit of body weight also decreases as the animal becomes older and heavier.

19.12.3 Level of Feeding

Limited feeding, if not too drastic, often results in increased efficiency of gain. This is true in different species of animals and may result from an increased digestibility of the diet or from less feed wastage. Moreover, animals provided with limited amounts of feed usually have lower percentages of fat in the weight they gain.

19.12.4 Average Daily Gain

Faster-gaining animals on full-feed usually make more efficient gains than slower-gaining ones on full-feed. Many of the same genes that cause fast gains also cause more efficient gains so that the improvement of one trait through selection will cause an improvement in the other. The **correlation** between the two traits is high but not perfect.

When all are full-fed, faster-gaining animals make more efficient gains than slower-gaining ones because they (1) use a smaller percentage of their total feed intake for maintenance, (2) are usually healthier, (3) are growing rather than depositing fat, and (4) have a more efficient metabolic system, which allows better utilization of the food they consume for body weight increases.

19.12.5 Other Factors

Many other factors may affect the efficiency of feed conversion in livestock. These include the diet quality and its **palatability** to the animal. Feeding certain **hormones** and drugs may also increase the economy of gains under some conditions (Section 18.6).

19.13 SUMMARY

All animals depend directly or indirectly on plants for their supply of nutrients. The digestive system of each species is adapted to the kind of feeds normally consumed. Ruminants possess a complex digestive system consisting of four stomach compartments in which microorganisms help in digesting cellulose and other fibrous plant compounds that cannot be broken down by the animal's own digestive enzymes. This enables them to utilize large quantities of forages. Other species such as swine and poultry consume larger proportions of grains and other concentrates and depend almost entirely on digestive enzymes to chemically break down the nutrients in these feeds.

TABLE 19.3	Effect of Body Weight on Efficiency of Feed Conversion, Daily Feed Consumption, and Pounds of Fat in the Carcass of Swine			
Weight of pig, lb	**Feed per 100-lb gain**	**Pounds of feed consumed daily**	**Percent of fat in body**	**Pounds of fat in body**
Birth (2.5 lb)	—	—	2.4	0.06
50	300	3.00	9.7	4.85
100	385	4.00	16.2	16.20
150	430	5.63	29.1	43.65
200	455	7.00	28.5	57.00
250	505	8.15	32.1	80.25
300	570	8.40	42.6	127.80

Source: Chemical composition data from *Missouri Agr. Res. Bull.* 73. Percentage of carcass composition may vary somewhat from these figures in modern meat-type swine.

Foods must be broken down into simpler compounds before they can be absorbed into the bloodstream where they are used for maintenance energy; body repair; growth of new tissues; reproduction; work; and the production of meat, milk, eggs, and wool. In the processes of digestion, absorption, and utilization many plant nutrients are lost to the animal for productive pur-poses. Although animals are somewhat inefficient converters of plant nutrients into animal products they possess digestive sys-tems designed to enable them to serve as "middlemen" in con-verting many plants (especially those containing large amounts of cellulose, which is not utilized by humans) into animal prod-ucts useful to the health and nutrition of humans.

STUDY QUESTIONS

1. Define the following: herbivores, carnivores, omnivores, proventriculus, ventriculus, monogastric, and ruminant.

2. Compare the capacities of the digestive tracts of several farm animals.

3. Name the major parts of the digestive systems of farm animals.

4. How do monogastric (nonruminant) and ruminant animals differ in their ability to digest forages and foods high in fiber? Why do they differ?

5. Why do monogastric (nonruminant) animals not bloat as readily as ruminant animals? What causes bloat?

6. What affects and/or controls appetite in farm animals?

7. Distinguish between prehension and mastication of food.

8. Outline the processes and enzymes that digest carbohydrates in farm animals.

9. Discuss the process of protein digestion in farm animals.

10. What are the major enzymes involved in the digestion of fats?

11. What is an enzyme and what produces it?

12. Where are most food nutrients absorbed into the body in farm animals?

13. How does the digestive system of birds differ from that of mammals?

14. Discuss some factors that may affect the digestibility of feeds.

15. List and discuss some factors that affect the efficiency of feed conversion into animal products.

THE NUTRITIONAL APPLICATION OF VITAMINS TO HUMAN AND ANIMAL HEALTH[1]

There is still an unknown substance in milk, which, even in very small quantities, is of paramount importance to nutrition. If this substance is absent the appetite is lost and with apparent abundance the animals die of want.

C. A. Pekelharing (1905)

20.1 INTRODUCTION

Prior to the twentieth century **carbohydrates, fats, proteins,** and a few minerals were generally considered to be the only nutrients required for normal body functions. Centuries before, however, certain observations suggested that other organic compounds are essential to the maintenance of good health. For example, it has been known for some 300 years that eating fresh fruits and vegetables is effective both **prophylactically** and **therapeutically** against scurvy. It has also been appreciated for a long time that rickets can be cured by the oral consumption of cod liver oil. Also, in 1897 the Dutch physician Eijkman found that the disease beriberi, which afflicted humans consuming diets of polished rice, could be cured by eating rice polishings (Section 20.4.2).

These observations suggested that there were other naturally occurring food components that were indispensable for sound health and well-being but that were not carbohydrate, fat, or protein in nature. In 1912 Funk, a Polish biochemist working in London, first introduced the term *vitamine* (a vital amine), which later became **vitamin** (from the Latin *vita* meaning "life"), to denote this group of organic compounds.

20.2 VITAMINS DEFINED

Vitamins are organic compounds essential for normal growth and maintenance of animal life. They are required in only minute amounts. Some vitamins are required for metabolic reactions essential for the transformation of energy but they do not themselves contribute energy. Some are metabolic essentials but not dietary essentials because they are synthesized within the bodies of many species.

20.2.1 Nomenclature

Vitamins were originally categorized as (1) *fat soluble* and (2) *water soluble* because fat solvents could extract the former from foods and water could be used to extract the latter. The fat-soluble vitamins include A, D, E, and K and contain only carbon, hydrogen, and oxygen. The water-soluble ones include ascorbic acid (C) and the B complex (so called because they were originally described as unknown factors B_1 to B_{12} before their chemical nature was understood); they contain carbon, hydrogen, and oxygen and may also contain nitrogen, sulfur, or cobalt. Pertinent information about the fat-soluble vitamins is summarized in Table 20.1 and water-soluble vitamins in Table 20.2.

20.3 THE FAT-SOLUBLE VITAMINS

Prior to the discovery of the fat-soluble vitamins it was assumed that fats served only as a source of energy for animals. The fat-soluble vitamins are usually stored in the body and therefore need not be consumed on a daily basis.

20.3.1 Vitamin A

The earliest descriptions of symptoms now attributable to vitamin deficiency relate to vitamin A. Night blindness and its cure, feeding liver, is mentioned in the Ebers papyrus (ca. 1600 B.C.), in Chinese writings of the same period, and later by Hippocrates and Roman writers. Celsus (25 B.C.–A.D. 50) is believed to have

---

[1] The authors acknowledge with appreciation the contributions to this chapter of Dr. Neal R. Merchen, Department of Animal Sciences, University of Illinois at Urbana–Champaign.

TABLE 20.1	The Fat-Soluble Vitamins		
Nomenclature	**Function**	**Clinical deficiency symptoms**	**Common sources**
Vitamin A (carotene)	Important in cellular metabolism. Regeneration of visual purple in eyes. Essential for normal epithelial tissue lining the digestive, respiratory, and reproductive tracts. Required for proper function of the immune system.	Night blindness, **keratinization** of epithelium, retarded growth and appetite, xerophthalmia, muscle incoordination, rough hair coat or plumage, reduced fertility.	Yellow corn (beta-carotene), alfalfa, and grasses; egg yolk, liver, fish liver oils, milk fat, green vegetables; commercial preparations. Preformed vitamin A is found only in animal products or in vitamin supplements.
Vitamin D	Role in calcium and phosphorus absorption and metabolism. Normal **calcification** of bones.	Rickets, **osteomalacia,** decreased egg laying and hatchability in poultry.	Fish liver oils, irradiated yeast, egg yolks, milk fat, field-cured hays, commercial preparations. Can be formed in skin by exposure to ultraviolet light.
Vitamin E (tocopherols)	Normal reproduction in some species, antioxidant, hatchability of eggs, activities of cell nucleus.	Infertility in rats, mice, guinea pigs, and possibly swine. Skeletal muscular dystrophy, **exudative diathesis, encephalomalacia,** liver **necrosis,** cardiac muscle abnormalities, dental depigmentation.	Cereal grains (mostly in germ); egg yolk; oils of soybean, peanuts, and cottonseed; alfalfa; beef liver; wheat germ oil; commercial preparations.
Vitamin K	Required for prothrombin formation, which is essential for normal blood clotting.	**Subcutaneous** and **intramuscular** hemorrhages, especially in poultry.	Green, leafy plants; liver, egg yolk, fish meal; synthetic form (menadione).

been the first to use the term *xerophthalmia,* which literally means "dry eye," a condition resulting from a vitamin A deficiency.

While feeding a purified diet (containing all the known essential nutrients) to white rats at the University of Wisconsin in 1913, Dr. E. V. McCollum observed that the rats became sick and ceased to grow, their eyeballs became dry, their tear glands did not act, their eyelids became swollen and inflamed, and finally blindness occurred. When Dr. McCollum added milk fat to the diet the rats soon became healthier and their eye condition improved. The milk fat was a good source of fat-soluble vitamin A.

Vitamin A is essential to growth and efficient food utilization. It is important in the resistance of the body to **infection,** keeps epithelial tissues normal, and prevents night blindness. Vitamin A occurs in plants as carotene (**precursor** of vitamin A).

The value of grass was recognized in the time of Jeremiah who recorded

> Yea, the hind [a female deer] also calved in the field, and forsook it, because there was no grass. And the wild asses [four-footed, hoofed mammals related to the horse] did stand in the high places, they snuffed up the wind like dragons; their eyes did fail, because there was no grass. . . .
>
> **Jeremiah 14:5–6**

Although Jeremiah associated good eyesight with the availability of grass, it was several thousand years later before scientists found the scientific explanation that grass contains carotene. Vitamin A is essential in the eye's regeneration of visual purple (rhodopsin), which is instrumental in preventing night blindness. Vitamin A and carotene are absorbed from the small intestine and are stored primarily (approximately 80 to 90 percent) in the liver.

Medical research demonstrated that high levels of vitamin A aid in the healing of wounds. Physicians noted slow healing of skin grafts among patients severely burned. This was especially marked among those receiving steroids. When high lev-

els of vitamin A were administered, however, the speed of healing among the burn victims was near normal.

Clinical Deficiency Symptoms

1. Impaired growth results because vitamin A is essential in the building of new cells.
2. Keratinization of epithelial tissues occurs. Vitamin A is essential to the health and integrity of epithelial tissues. A deficiency results in its keratinization. (Tissue changes into **keratin,** which is an insoluble protein and the chief structural constituent of horns, nails, hair, and feathers.) Keratinized tissues in the various tracts of the body (digestive, genitourinary, reproductive, and respiratory) lower the resistance of the affected epithelial tissues to infective **organisms.** Pneumonia, diarrhea, kidney stones, and bladder stones are more prevalent and reproductive efficiency is greatly reduced.
3. Night blindness occurs. In prolonged vitamin A deficiency a more severe visual disorder (xerophthalmia), which can result in blindness, may occur.
4. Rough hair coat and skin diseases (e.g., acne vulgaris, which is a chronic **inflammatory** disease of the sebaceous glands) are often closely associated with vitamin A deficiency. Studies at the University of Missouri indicate that vitamin A may be beneficial in the prophylaxis and therapy of ringworm in cattle (Figure 20.1).

Sources

Sources of vitamin A include fish liver oils, butter (milk fat), egg yolk, cheese, liver, green vegetables (in most of which it exists as carotene), and synthetic preparations. The green color of plants and dry forage is a good indicator of the carotene content. (Dark-green color indicates high carotene content.) However,

TABLE 20.2	The Water-Soluble Vitamins and Related Compounds		

Nomenclature	Function	Clinical deficiency symptoms	Common sources
Vitamin C (ascorbic acid)	Formation and maintenance of intercellular material in bones and in soft tissues. Acts as a tissue **catalyst** (aids in healing). Antioxidant.	Scurvy, hemorrhages throughout body, swollen and bleeding gums, **anemia.** Loss of teeth.	Citrus fruits; tomatoes; green, leafy vegetables; potatoes. Synthetic preparations.
Thiamine (B$_1$)	Component of two **coenzymes,** essential in CHO metabolism and energy transfer. Promotes normal appetite and digestion. Helps keep the nervous system healthy and prevents irritability.	Beriberi in humans, polyneuritis in birds, lack of appetite, hyperirritability, incoordination. Chastek paralysis in foxes. Reproductive failure in horses.	Milk and milk products, brewers' yeast, wheat germ, unmilled cereals, grain by-products, lean pork, liver, kidney, egg yolk. Good-quality hay. Synthetic preparations.
Riboflavin (B$_2$ or G)	Forms a part of two flavoprotein coenzymes, role in energy transfer (helps cells use oxygen), function in protein metabolism, component of xanthine oxidase. Helps keep skin healthy.	Curled-toe paralysis in the chick; **lesions** of skin, eye, and nervous system; depressed appetite; retarded growth.	Milk, cheese, liver, kidney, eggs, fish, green forage, oil meals, fermentation products. Commercial preparations.
Pantothenic acid	Component of coenzyme A, which is involved in many metabolic reactions.	Retarded growth, skin lesions, **gastrointestinal** troubles, lesions of nervous system and adrenal gland, skin and hair depigmentation. Goose-stepping in pigs.	Liver, egg yolk, milk, alfalfa hay, peanut meal, cane molasses, yeast, rice, wheat brans. Cereal grain and by-products. Royal jelly of bees.
Nicotinic acid (nicotinamide or niacin)	Component of two coenzymes; energy transfer. Can be spared by the amino acid tryptophan. Health of digestive and nervous systems.	Pellagra or black tongue, **dermatitis,** diarrhea, **dementia,** loss of appetite and weight, vomiting, anemia. Enlarged hocks in turkey poults.	Milk, meat, eggs, green vegetables, peanut butter, animal and fish by-products, distillers' grains and yeast, fermentation solubles, oil meals. Hays and grains are fair sources.
Pyridoxine (B$_6$)	Part of enzyme concerned with protein metabolism. Essential for normal metabolism of tryptophan.	Dermatitis and convulsions in rats, pigs, and poultry. Microcytic hypochromic anemia in puppies, pigs, and rats.	Yeast, liver, muscle, meat, egg yolk, milk, cereal grains, vegetables.
Biotin	Functions in enzyme systems. Fat synthesis, **deamination** of certain amino acids.	**Perosis** in chicks, **dermatitis,** loss of hair, disturbances of nervous system.	Yeast and organ meats, whole grains, molasses, milk.
Folic acid	Role in transfer of single-carbon units. Synthesis of purines and certain methyl groups. **Erythropoiesis.**	Retarded growth and **macrocytic,** anemia. Poor feathering and pigmentation of feathers.	Green, leafy materials; organ meats; cereals; soybeans; animal by-products.
Cyanocobalamin (B$_{12}$)	Methyl group synthesis, purine synthesis, CHO and fat metabolism. Synthesis of nucleic acids. Known as animal protein factor (**APF**).	Pernicious anemia in humans. Retarded growth, low hatchability of eggs, posterior incoordination, unsteadiness of gait.	Milk, meat, fish meal, animal by-products. Commercial preparations.
Related compounds			
Choline	Component of **phospholipids,** essential in building and maintenance of cell structure. Transmission of nerve impulses. Fat metabolism in the liver.	Fatty livers (**cirrhosis**), renal tubular degeneration, enlarged spleen and hemorrhagic condition of kidneys. Perosis in chicks.	Milk, meat, eggs, fish, all naturally occurring fats.
Inositol	Lipotropic action in certain rat diets, where other vitamins are deficient.	**Alopecia** (loss of hair).	Occurs in plant products in the organic phosphorus substance phytin.
Para-aminobenzoic acid (PABA)	Anti-gray-hair factor in mice and rats. A growth stimulant in chicks.	Graying of hair (**achromotrichia**) in animals other than humans.	Synthetic preparations.

this visual evaluation of the carotene-containing qualities of forages is limited because the green pigment (chlorophyll) is not destroyed at the same rate as carotene.

20.3.2 Vitamin D

Prior to the twentieth century humans resembled the living plant in being dependent on sunshine for health and well-being. Why the Egyptians worshiped sun gods is unknown but their religion may have been a result of their observing the favorable effects of sunshine on the rachitic children of the royal fami-

lies.[2] Rickets is a condition (caused by vitamin D deficiency) that results in a disturbance of bone development. The disease is characterized by insufficient deposition of mineral in bones, consequent bending and distortion of bones under muscular action, and formation of nodular enlargements on the ends and

[2] The role of vitamin D in the proper development of bones has a long history. The Greek historian Herodotus visited a battlefield where the Persians had defeated the Egyptians in 526 B.C. He observed that the skulls of the slain Persians were fragile whereas those of the Egyptians were strong. The Egyptians attributed the difference to the effect of sunlight; they were bareheaded from childhood, whereas the Persians wore turbans to protect themselves from the sun.

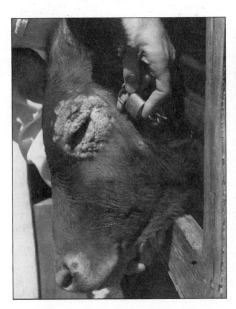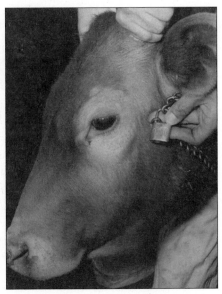

Figure 20.1 Effect of vitamin A on the health and integrity of the skin and on the fineness of the hair coat. Note the favorable effect of vitamin A injections on ringworm infection in this 6-month-old Guernsey bull. After four weekly injections (300,000 IU each), note the nearly complete recovery (right) and especially the change in skin and hair coat appearance. *Courtesy of Missouri Agr. Expt. Station.*

sides of bones.[3] An early concept of the cause of rickets was a lack of exercise. One of the early (about 1650) remedies for rickets was to "take roots of Smallage, Parsley, Fennel, and Angelica [all herbs], slice and boil them in distilled water of Angelica, unset Hyssop, and Collsfoot, of one part each, until they are tender; then strain and boil it to a syrup with white honey."

Later cod liver oil was observed to be beneficial in the therapy of rickets. In 1922 Dr. McCollum and his associates at Johns Hopkins University in Baltimore, Maryland, identified the antirachitic element in cod liver oil and named it vitamin D. In 1924 Dr. A. F. Hess (physician) and Dr. H. Steenbock (biochemist at the University of Wisconsin) discovered that rickets could also be prevented by **irradiating** certain foods for animal consumption with ultraviolet light.

Although nature provided little vitamin D in common foods it gave humans a way of forming it in their bodies. The ultraviolet radiation in sunlight (wavelength 280 to 320 nanometers) serves as a source of the radiant energy necessary to convert 7-dehydrocholesterol (an animal sterol stored beneath the skin surface) into vitamin D_3 (cholecalciferol). *Ergosterol* (a plant sterol), on irradiation, yields ergocalciferol (vitamin D_2). There are some 10 *sterol* derivatives having vitamin D activity; however, ergosterol and 7-dehydrocholesterol are the chief provitamins D, which yield D_2 and D_3, respectively. Vitamins D_2 and D_3 are further converted into even more biologically active forms of vitamin D in the liver and kidneys.

Apparently vitamin D_2 (the plant form of the vitamin) and vitamin D_3 (the animal form) have the same antirachitic values

for the dog, pig, rat, **ruminant,** and human, whereas D_3 is more beneficial than D_2 for poultry. Recent evidence indicates that D_3 is more active than D_2 in promoting calcium absorption in Cebus monkeys.

Rickets seldom occurs in the tropics where exposure to sunshine is maximal; it is principally confined to the temperate zone where **radiation** may be markedly reduced, particularly in winter. (Arctic residents consume substantial amounts of vitamin D–rich fish.) Darkly pigmented skin filters out much (as much as 65 percent) of irradiation; therefore, blacks are especially susceptible to rickets. Could the need to regulate absorption of ultraviolet radiation partially explain why Caucasians are white in the winter (white skins allow maximum photoactivation of 7-dehydrocholesterol into vitamin D at low intensities of ultraviolet radiation) but pigmented in the summer when more sunlight is available? In almost all races the skin is lighter in the **neonate** and gradually darkens as the individual matures, a change that parallels the declining need for vitamin D. Is this phenomenon an inherent safety feature provided by nature to ensure ample vitamin D for the growth, development, and well-being of the young? Examples of rickets are shown in Figure 20.2a and b.

The Role of Vitamin D

This vitamin is required for normal **calcification** of growing bones. It serves an important role in the metabolism of calcium and phosphorus in the body throughout life but especially in growth, reproduction, and **lactation.** Vitamin D enhances the absorption of calcium and phosphorus from the small intestine and helps maintain normal blood levels of these minerals. Being fat-soluble, some vitamin D is stored in the liver; however, this

---

[3] The discovery of vitamin D was one of the greatest scientific and medical discoveries of all time. An estimated 90 percent of the young children of Europe had rickets as late as 1900.

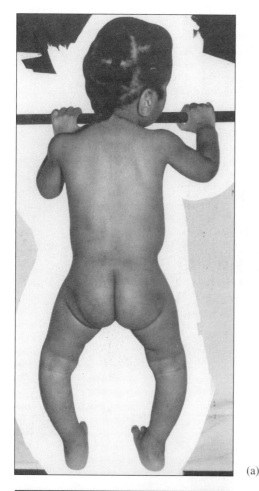

(a)

(b)

Figure 20.2 (a) Rickets in an infant. Note the bowlegs that curve laterally, indicating that the weakened bones have bent as a result of standing.
Courtesy of Dr. R. L. Nemir, New York University, and the Upjohn Company, Kalamazoo, MI.
(b) Advanced stage of rickets caused by a lack of dietary vitamin D when a pig was fed indoors. Note the leg abnormalities.
Courtesy of Dr. J. M. Bell, University of Saskatchewan, Canada.

amount is limited and the body needs a more regular and consistent **exogenous** source of vitamin D than of vitamin A.

Excessive vitamin D may be harmful in that it mobilizes calcium and phosphorus from the bony tissues. These minerals are redeposited in soft tissues, principally in the walls of blood vessels, but also in kidney tubules, bronchi, and the heart.

Clinical Deficiency Symptoms

1. Rickets, a bone deformity that results from the lack of calcium and/or phosphorus deposition. It occurs primarily in young animals or infants (Figure 20.2a). Only mammals and birds are subject to rickets.
2. **Osteomalacia,** a condition characterized by a partial decalcification of normal bones that leads to a softening and brittleness of bones. It is more common in older persons (senile osteomalacia) and animals whose bones are fully grown but whose diets are deficient in vitamin D or in calcium and phosphorus.
3. Low concentration of blood **plasma** phosphorus.
4. Thickening and swelling of joints.

Sources

Sources of vitamin D include liver oils of various fish, fortified milks, butter, egg yolk, irradiated animal and plant sterols, and commercial preparations. The supplementation of diets with vitamin D is recommended for animals spending the winter months in northern latitudes when there is less sunlight and its efficiency in converting vitamin D precursors into vitamin D is greatly reduced and for animals confined indoors during any season.

20.3.3 Vitamin E

This vitamin was discovered in 1924 and was called *tocopherol* (from the Greek *tocos* meaning "childbirth," and *phero* meaning "to bring forth").

A deficiency of vitamin E leads to sterility in the rat and to muscular dystrophy in the dog, guinea pig, rabbit, and certain other **species.** An early sign of deficiency in male rats is the loss of **spermatozoan** motility. Pregnancy may occur in the female but embryonic development is retarded and often results in fetal resorption.

Tocopherol isomers are strong *antioxidants.* (They hinder **oxidation,** which chemically consists of an increase in positive charges on an **atom** or the loss of negative charges.) Vitamin A and carotene are susceptible to oxidative destruction in the presence of unsaturated fats (Chapter 18), both in the cell and in the digestive tract, and are protected against this change by vitamin E.

The vitamin is distributed throughout the body, stored principally in fat, and therefore not a daily dietary requirement. Vitamin E appears to be relatively nontoxic although some evidence indicates that large dosages may mobilize phosphorus and cause bone decalcification.

Vitamin E passes through the placental membranes and also into the mammary gland. Therefore the mother's dietary level of vitamin E is reflected in body stores of the young at birth and similarly by the amount it obtains from its mother's milk.

It is now known that the trace element selenium and vitamin E have interrelated functions. An example is the protection afforded rats by selenium against **necrotic** liver degeneration, which normally results from a vitamin E–deficient diet.

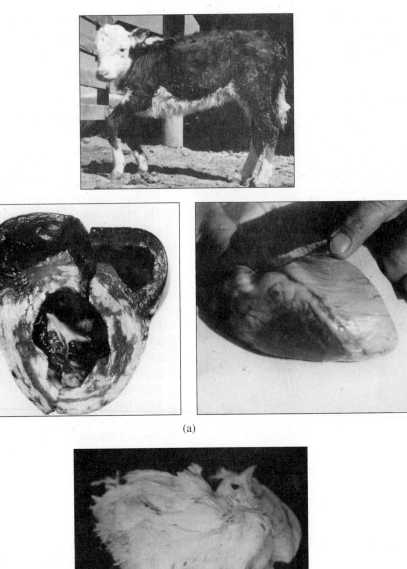

(a)

(b)

Figure 20.3 (a) White muscle disease of a 12-week-old calf caused by a vitamin E deficiency. Weak muscles, lameness, and abnormal locomotion characterize the calf. The other photographs show abnormal white areas in the cardiac muscles of calves 5 to 6 weeks old.
Courtesy of Dr. O. H. Muth, Oregon State University.
(b) α-tocopherol deficiency in the chick. Note the loss of control of legs and the head retraction.
Courtesy of Cornell University, Poultry Science Department.

Clinical Deficiency Symptoms

The most dramatic characteristics of a vitamin E deficiency are the changes in various body tissues. Those related to this discussion are as follows:

1. Infertility in guinea pigs, hamsters, mice, rats, and possibly swine.
2. A brown discoloration of the rat's uterus and **adipose** tissue.
3. Skeletal muscular dystrophy in guinea pigs, hamsters, **lambs,** rabbits, rats, and swine. It is also called "stiff lamb disease," a form of muscular dystrophy in sheep, which is similar in appearance to "white muscle disease" caused by a deficiency of vitamin E and/or selenium in cattle (Figure 20.3a).
4. Cardiac muscle abnormalities in cattle, lambs, monkeys, **poultry,** rabbits, and rats (often characterized by "brown fat" in the heart of pigs and sometimes referred to as "mulberry heart").
5. Nutritional **encephalomalacia** (edema of the brain) in **chicks,** which is also called "crazy chick" disease. Symptoms include lack of coordination, head retraction, convulsions, and twitching of the limbs (Figure 20.3b).
6. Liver **necrosis** (death of liver cells) in rats and liver and muscular degeneration in pigs.

Sources

Sources of vitamin E include wheat germ oil, cereals, egg yolk, beef liver, and high-quality **forage.** It is also prepared synthetically.

20.3.4 Vitamin K

Vitamin K is essential to the formation of prothrombin by the liver. Blood coagulation consists of two major steps: (1) prothrombin (in the presence of thromboplastin, calcium, and other factors) is converted into thrombin, and (2) fibrinogen (acted on by thrombin) is converted into the fibrin clot (Figure 20.4).

Vitamin K was discovered in Denmark (1934) when chicks fed an ether-extracted (fat-free) diet developed a hemorrhagic

condition. A deficiency may occur in humans, mice, rabbits, and rats in the presence of abnormal fat **absorption.**

Vitamin K is stored in appreciable amounts in the liver. Massive doses of vitamin K are apparently nontoxic. The bacteria of the intestinal tract of humans and most **domesticated** animals are capable of synthesizing vitamin K; hence, a hemorrhagic tendency induced solely by the dietary lack of vitamin K rarely occurs. (Poultry have a short intestinal tract and host so few microorganisms that they must be provided with a dietary vitamin K source.) Infants, however, are an exception because the (1) prenatal stores of the vitamin are small, (2) intestinal bacterial **flora** necessary to synthesize vitamin K is not established at birth, and (3) the usual diet of the neonate does not supply it. The result is that a mild vitamin K deficiency is common in the human neonate. Occasionally the blood prothrombin concentration reaches such a low level that hemorrhages occur, usually within the first week of life, commonly starting with a hemorrhage from the **gastrointestinal** (GI) tract. Hemorrhagic disease of the neonate can usually be prevented by the administration of vitamin K to the expectant mother **prepartum** or to the infant at birth. Oral administration of sulfonamides and other **antibiotics** may destroy vitamin K–synthesizing organisms in the GI tract. Therefore the prophylactic administration of vitamin K is recommended to avoid the prothrombin-reducing effect of certain drugs.

Missouri studies indicate that female rats are less susceptible than males to a vitamin K deficiency. Apparently this is related to the sex **hormones** because when the male sex hormone testosterone is administered to females they soon develop a vitamin K deficiency, whereas the males are less susceptible to vitamin K deficiency when they receive the female sex hormone estrogen.

Dicoumarol

This compound was isolated originally from spoiled sweet clover which causes *hemorrhagic sweet clover disease* and later made synthetically. It is used clinically as an anticoagulant in thrombotic states and acts by depressing the factors concerned with the formation of thrombin. Because vitamin K will overcome the effect of **dicoumarol,** the action of the latter is similar to that of any **antimetabolite.**

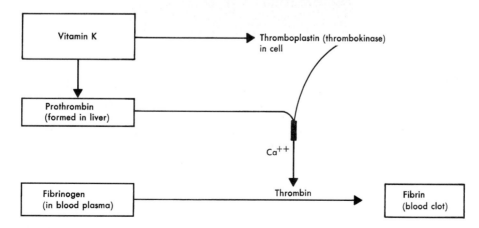

Figure 20.4 The role of vitamin K in blood clotting.

Figure 20.5 Vitamin K deficiency in chicks. Note the spontaneous hemorrhages under the skin of the chick at left, which was fed a vitamin K–deficient diet for the first 15 days. A normal chick the same age but fed an adequate diet is shown at right.
Upjohn Company Vitamin Manual, *1965.*

Clinical Deficiency Symptoms

1. Hypoprothrombinemia, a deficiency of prothrombin in the blood (Figure 20.4).
2. Subcutaneous and intramuscular hemorrhages, especially in poultry (Figure 20.5).

Sources

Sources of vitamin K include alfalfa, spinach, cabbage, egg yolk, fish meal, and pork liver. Its synthetic **analogues** are prepared commercially. Naturally occurring vitamin K is fat-soluble but some of the synthetic preparations are water-soluble.

20.4 THE WATER-SOLUBLE VITAMINS AND RELATED COMPOUNDS

The vitamins that are soluble in water include ascorbic acid (vitamin C), which is apparently required only in the diets of primates and guinea pigs, and the B complex, which is required only in the diets of **monogastric** animals. The ability of the ruminant to use microbially synthesized B-complex vitamins was discussed in Chapter 18.

The B vitamins may be subdivided into two groups: (1) those involved in the release of energy from food (thiamine, B_1; riboflavin, B_2; nicotinamide; pantothenic acid; and biotin) and (2) the **hematopoietic** vitamins, or those related to the formation of red blood cells (folic acid and B_{12}, sometimes called *cobalamine*). Pyridoxine, B_6, functions so that it fits both the energy-releasing and hematopoietic categories of vitamins.

B vitamins are involved in the absorption of food from the intestine and are therefore concerned in the first stage of the nutritional process. Oxidation of absorbed nutrients is accomplished enzymatically and certain B vitamins are involved in the process as essential components of coenzymes. Cellular energy is supplied by this oxidation of food within the cell.

20.4.1 Vitamin C (Ascorbic Acid)

This is a wonderful secret of power and wisdom of God, that He has hidden so great an unknown virtue in this fruit—sour oranges and lemons—to be a certain remedy for this infirmity [scurvy].

Anonymous (sixteenth century)

A tour of the *Mayflower* in Plymouth or of *Old Ironsides* in Boston reminds one of the tales recorded about the perils of mutiny and famine. Another peril, less recorded but equally dangerous on early ocean voyages, was disease among sailors. Of the 160 men who sailed with Vasco da Gama around the Cape of Good Hope in 1497, 100 died of scurvy.

Records of early voyages reveal that transportation was slow and there was no way to keep fresh fruits and vegetables for extended periods of time. Early symptoms of scurvy were loss of appetite and fatigue. Later the victim's mouth became sore, teeth became loose, bones broke easily, and small blood vessels burst **subcutaneously;** the majority of those afflicted eventually died of the disease.

After a voyage from France to Newfoundland in 1536, Jacques Cartier's men became sick with scurvy and 26 men died of the disease. One day Cartier noticed that a native Indian who had had scurvy earlier appeared cured. He asked about the Indian's therapy and discovered that the Indian had made a drink by boiling the bark and leaves of a selected tree (probably an American spruce). When Cartier's men drank this "tea" they too were cured of scurvy.

In 1735 four ships (*Dragon, Hector, Susan,* and *Ascension*) sailed for the West Indies. On all ships except the *Dragon* men developed scurvy and died. There were no casualties on the *Dragon* whose men had fruit juice in their meals. Why were these observations of the benefits of fruit juices in the prevention of scurvy not quickly applied in human nutrition? Science advances slowly with each experience and discovery contributing to previous knowledge. Furthermore, people are often slow to accept and apply new findings. Resistance to change often impedes progress.

About two centuries after Cartier's curious observation of the benefits of "Indian tea" on scurvy victims, Dr. J. Lind, a British navy surgeon and perhaps the first true "experimental nutritionist," applied Cartier's findings experimentally in 1747. On one ship (the warship *Salisbury*) he used six pairs of sailors who were ill with scurvy. He gave cider, cream of tartar, oranges and lemons, and other special doses to pairs one, two, three, and four to six, respectively. In only 6 days experimental pair three (who had received oranges and lemons) were greatly improved whereas the others remained ill. The results of this simple experiment prompted Dr. Lind to recommend that all British sailors receive orange and lime juice daily. By 1795 this recommendation was adopted by the British navy and from that day to this sailors have often been called "limeys" because limes were chiefly used at that time. Thus Dr. Lind laid the foundation for the concept that deficiency diseases result from the lack of essential food constituents.

The famous voyages and adventures of the hardy mariners played an important role in establishing the basis of vitamin C. Perhaps history will record equally important contributions of our space travelers to the science of **nutrition** and **physiology** as humans continue to triumph over distance and disease.

Vitamin C is involved in the absorption of iron from food. Experiments with radioiron incorporated into food show that **ingestion** of ascorbic acid and food containing ascorbic acid enhanced the absorption of iron (Figure 20.6). Thus ascorbic acid is indirectly effective in preventing iron-deficiency anemia. Fever increases the need for vitamin C. Vitamin C seems to help maintain the metabolic and functional properties of phagocytes, because consumption of too little of the vitamin or too much in supplemental form may impair phagocytic function.

Clinical Deficiency Symptoms

Scurvy occurs only in the human, other primates, and guinea pig;[4] other mammals are apparently capable of synthesizing ascorbic acid in the body. Scurvy is characterized by hemorrhages throughout the body (result of increased fragility of blood capillaries), swollen and bleeding gums, and **anemia.**

Sources

Sources of vitamin C include lemons, limes, oranges, grapefruit, and tomatoes. Other fruits and vegetables also furnish appreciable amounts of vitamin C. In addition, it is manufactured synthetically (the first vitamin to be synthesized in the laboratory). It is apparently not stored in appreciable amounts in the body inasmuch as blood plasma concentrations of vitamin C reflect dietary intake very closely. Ascorbic acid is absorbed from the small intestine and excreted via the urine.

20.4.2 Thiamine (Vitamin B$_1$)

For many centuries people believed that persons who ate meat of certain animals would act like those animals. The person

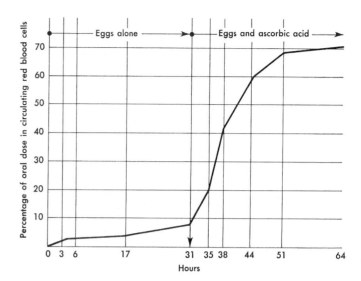

Figure 20.6 Effect of ascorbic acid on absorption of radioactive iron (^{59}Fe) from food in humans. This man, whose blood values were normal, was first given eggs containing 3.5 mg ^{59}Fe. Less than 10 percent of this radioactive iron could be subsequently measured in the blood. When 1 g of ascorbic acid was added to a comparable amount of radioactive food iron, absorption was increased to more than 70 percent. *Courtesy of Dr. C. V. Moore and Dr. R. Dubach, Washington University, St. Louis, MO.*

who ate snake's meat would be as sly and cunning as a snake. Because the heart was regarded as the center of courage, those wishing to be courageous considered the hearts of animals as choice meat. Similarly those who wanted to run fast consumed meat from the legs of deer.

Today advertising, promotion, and habit largely govern what people eat. Humans have eaten bread for centuries. During the Middle Ages white bread was rare and expensive and, probably for those reasons, was preferred. A person's wealth often determined the kind of bread he or she ate and black bread was generally considered a badge of poverty. This prompted people to find faster and easier ways to remove the **germ** and **bran** from grain to produce a white flour. However, they were unknowingly throwing away valuable vitamin B$_1$ with the germ and bran. (Some 80 percent of this vitamin is lost from the flour in the milling process.) Today most white flour is *enriched*[5] with B$_1$ and other nutrients that are lost in milling.

People of those nations whose diets consisted largely of rice developed a serious nerve disease called *beriberi* when they switched to white (polished) rice. (Vitamin B$_1$ was discarded when the germ and skin were thrown away.) Beriberi in Singhalese means "I cannot" signifying that the person is too ill

---

[4] Certain other animals apparently need dietary vitamin C. For example, catfish develop broken-back syndrome when kept for extended periods in water devoid of vitamin C. In streams, they apparently obtain vitamin C from algae.

[5] *Enriched* refers to the practice of adding certain dietary nutrients to food products. For example, most white flour is *enriched* with the B vitamins thiamine, niacin, and riboflavin, and with iron. Research has shown, however, that "enrichment" of white flour with these vitamins and iron does not compensate for all losses in milling. The "enriched" bread is deficient in the amino acids lysine and valine, as well as in certain vitamins and minerals present in whole wheat flour. Incidentally, oatmeal (rolled oats) is a whole grain (rather than a milled one like decorticated wheat) and, nutritionally considered, it is richer in vitamins, minerals, and protein than even whole wheat. Nutritionally fortunate, then, are children whose parents prepare warm oatmeal and milk for breakfast, rather than serving the more convenient (and more costly) sweet rolls made with white flour, or the highly advertised sugar-coated, dry cereals.

to do anything. The disease existed as early as 2600 B.C. in the Orient where it was called the "scourge of the Orient." It is characterized by spasmodic rigidity of the lower limbs with muscular **atrophy,** paralysis, and anemia. There is a degeneration of the nerves that control the body (especially the lower limbs). Frequently the heart becomes enlarged.

Early experiments that eventually led to the association of thiamine (B_1) with beriberi are of interest. From 1878 to 1882 about one-third of the enlisted Japanese naval force of 5000 men were victims of beriberi. In 1882 Dr. Takaki, a naval medical officer, initiated an experiment using two crews of 270 men each going on a 9-month voyage over the same route. One crew (control group) was given the regular diet consisting largely of polished (**decorticated**) rice and the diet of the other crew consisted of less rice and more whole-grain barley. In addition, the second crew had meat, milk, and vegetables added to their diet. Among the 270 control men receiving the regular navy rations there were 169 cases of beriberi, resulting in 25 deaths. On the second vessel there were only 14 cases of beriberi and no deaths. In each case of beriberi on the second ship the sailor had not eaten his allowance of new foods.

The next important observation about beriberi was made in 1897 by Dr. Christian Eijkman, a Dutch physician and medical officer for a prison in the Dutch East Indies, who later received the Nobel prize in medicine. While walking through the prison yard one day Dr. Eijkman noticed chickens that seemed paralyzed, exactly as did the patients he was attending. Could these **fowls** have beriberi as his patients in the prison had? The chickens were fed leftovers from the table and so they were being fed the same food as the men. Prisoners had beriberi and so did the prison chickens, he concluded. He could not recall having observed these symptoms of beriberi in chickens outside the prison yard, which prompted him to postulate that the *food* the prisoners were eating was causing the disease.

Dr. Eijkman began an experiment in his laboratory by feeding different foods to chickens and then keeping careful records of their growth and body condition. He made two important observations: (1) chickens fed diets composed entirely of polished rice developed a disease very similar to beriberi, and (2) diseased chickens fed rice polishings (the outer coating of the rice, which was normally discarded) recovered. Moreover, he proved that chickens fed unpolished rice never developed the disease. From these observations Dr. Eijkman concluded that beriberi must be caused by some poisonous substance in rice that rice polishings could destroy. Conversely, Dr. Funk, who originally named vitamins, postulated that rice polishings contained a substance—namely vitamins—essential to the good health of chickens. Somewhat later Dr. R. R. Williams, an American chemist working in the Philippines, isolated vitamin B_1 (thiamine).

Thiamine is essential to carbohydrate **metabolism.** The principal role of thiamine is as a part of a coenzyme in the oxidative **decarboxylation** of alpha-keto acids. The deficiency state, as it occurs **spontaneously** in humans, is called *beriberi.* It is characterized by the accumulation of pyruvic and lactic acids, particularly in the blood and brain, and an impairment of the cardiovascular, nervous, and GI systems.

Raw fish contain the **enzyme** thiaminase, which inactivates thiamine. A thiamine deficiency may exist in foxes that have eaten raw fish causing a condition called *Chastek paralysis* (a thiamine-deficiency **syndrome** resembling Wernicke's disease in humans that is due to alcoholism).

Thiamine is synthesized by bacteria in the intestinal tract of ruminants but biosynthesis is not a dependable source for humans and other monogastric animals. The liver stores a limited amount of thiamine; however, it is generally considered to be a daily dietary essential for humans. Considerable thiamine is stored in the bodies of pigs. Excess dietary thiamine is excreted via the urine.

Bacteria synthesize some thiamine in the horse cecum but apparently not enough to meet daily requirements.

Clinical Deficiency Symptoms

1. Classical beriberi.
2. **Edema,** especially in the legs (also called *wet beriberi*).
3. **Polyneuritis** in rats and birds (Figure 20.7a and b).
4. Loss of appetite, decreased growth (Figure 20.7c), muscle weakness, incoordination.

Sources

Sources of thiamine include brewers' yeast, wheat germ, and milk; egg yolk; meat (especially pork and organ meats such as liver and kidney); whole-grain, enriched cereals and breads; nuts; and dried legumes such as peas and beans.

In Europe and the United States where diets usually provide an abundance of thiamine, the occurrence of beriberi is rare except in alcoholism (Wernicke's disease[6]).

20.4.3 Riboflavin (Vitamin B_2)

This vitamin was originally labeled vitamin G. It is a component of the greenish-yellow fluorescent pigments of milk (lactoflavin), eggs (ovoflavin), liver (hepatoflavin), and grass (verdoflavin). It was first reported in milk (1879) by A. W. Blyth who called it *lactochrome.*

Riboflavin functions as a coenzyme and is essential in normal energy transfer in the body. It is also important in protein metabolism.

Clinical Deficiency Symptoms

1. Leg paralysis in poultry, "curled-toe paralysis" (Figure 20.8b).
2. Stiff and crooked legs in swine.
3. Impaired growth.
4. In mammals there are eye, lip, skin, and tongue **lesions;** seborrhea (functional disturbance of the sebaceous glands); **dermatitis** (Figure 20.8a); and scrotal skin lesions.

Sources

Sources of riboflavin include milk and milk products; eggs; lean meats; legumes; distillers' solubles; and green, leafy vegetables.

---

[6] Wernicke's disease is a neurologic disorder caused by a thiamine deficiency. It is secondary to alcoholism. Alcohol interferes with GI transport of thiamine and also blocks the activation of thiamine-containing enzymes.

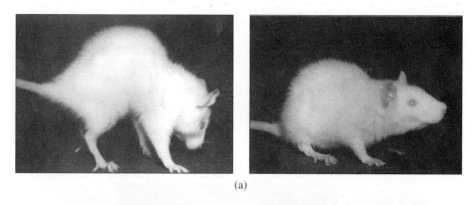

(a)

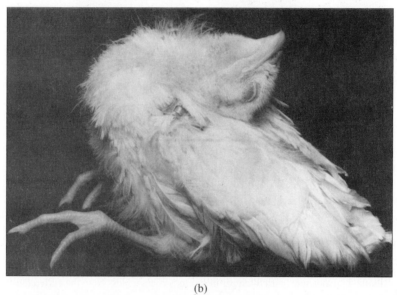

(b)

(c)

Figure 20.7 (a) Polyneuritis in the rat. This thiamine-deficient rat shows the typical arched back and hyperextended hind legs (left). Such rats have a spastic gait, turn awkwardly, and lose balance. At right, the same rat (8 h after receiving thiamine hydrochloride) has normal use of its hind legs.

Upjohn Company Vitamin Manual, *1965.*

(b) Thiamine deficiency (polyneuritis) in the chick. Note the retraction of the head.

Courtesy of Dr. H. R. Bird, University of Wisconsin.

(c) Thiamine deficiency. Contrast the growth of these littermates. The pig on the left received the equivalent of 2 mg of thiamine per 100-lb live weight, while the one on the right received none.

Courtesy of Dr. N. R. Ellis and the USDA.

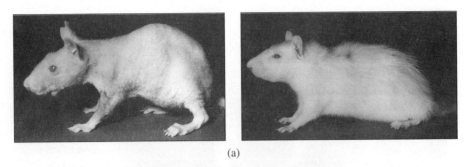

(a)

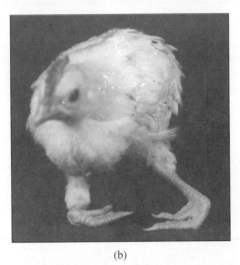

(b)

Figure 20.8 (a) Generalized dermatitis and growth failure in a riboflavin–deficient rat. Note the marked keratitis of the cornea (left). Same animal after 2 months of treatment (right).
Upjohn Company Vitamin Manual, *1965.*
(b) Riboflavin deficiency (curled-toe paralysis) in a young chick. Note the curled toe and tendency to squat on hocks.
Courtesy of Cornell University, Poultry Science Department.

It was first synthesized in 1935. Being light-sensitive, riboflavin is destroyed by blue or violet rays. It is stored in limited amounts in the liver and kidneys; however, it is needed in the diet on a daily basis.

Information pertinent to the water-soluble vitamins and related compounds is presented in Table 20.2.

20.4.4 Pantothenic Acid

The root word in the derivation of this term is Greek and means "from everywhere" because it is found so widely in living organisms. The universal distribution of pantothenic acid in all living material suggests that it performs a vital function in cellular metabolism. Like many other B vitamins, pantothenic acid is a component of an enzyme, coenzyme A, an important factor in the regulation of carbohydrate and fat metabolism. A dietary deficiency of pantothenic acid is rare in humans and animals. It is a component of almost all foods but in varying concentrations.

Pantothenic acid is required in the chick, dog, pig, and rat. It is synthesized by **rumen** microorganisms in cattle, goats, and sheep and to an appreciable extent in the intestines of humans, horses, and rabbits.

Clinical Deficiency Symptoms

1. A pantothenic acid–deficient rat shows lesions of the adrenal cortex (**hypertrophy**), depletion of lipid materials, and ultimately hemorrhage and necrosis. Other symptoms in deficient rats include impaired growth, graying of the hair, testicular degeneration, impaired antibody formation, duodenal ulcers, and fetal abnormalities.

   Adrenalectomy restores the hair and skin pigment in black rats made gray by pantothenic acid deficiency. This phenomenon apparently relates to the involvement of adrenal hormones and pantothenic acid in the production of **melanin.**

2. In chicks the principal deficiency symptom is dermatitis (especially of the eyelids, vent, corners of the mouth, and feet); however, spinal cord lesions, **involution** of the thymus, and fatty degeneration of the liver may also occur. Feathering is retarded and rough in appearance (Figure 20.9a).

3. Pantothenic acid–deficient swine develop an abnormal gait called *goose-stepping* (Figure 20.9b). They also may develop gastrointestinal ulcers.

Sources

Sources of pantothenic acid include liver, egg yolk, milk, potatoes, cabbage, and peas. The royal jelly of bees and the ovaries of codfish are exceptionally rich sources of pantothenic acid.

20.4.5 Niacin (Nicotinamide)

This vitamin earlier was called vitamin B$_5$. It is associated with *pellagra,* which the Italian scientist Frapoli appropriately named (from the Italian *pella agra* meaning "rough skin") in 1771.

The story of the search for the cause and cure of pellagra numbers among the great stories of modern medicine and is as exciting as reports of FBI person hunts. Because the common stable fly (*Stomoxys calcitrans*) displayed certain salient characteristics that seemed to qualify it for the role of a transmitter of pellagra it was implicated as a **vector** of the disease. It was later cleared of this indictment. In 1735 Don Gasper Casal, the brilliant young physician of King Philip V of Spain, wrote about pellagra. In 1740 Jujati, an Italian scientist, confirmed the symptoms described by Casal: brown or red scaly patches appeared on the skin (especially on those parts of the body exposed to the sun), the tongue became sore and irritated, diarrhea and indigestion developed, a loss in body weight occurred, and finally the mind became affected.

Pellagra was first reported in Europe just before the American Revolution and was common in the United States a century later. After 1907 pellagra spread with speed throughout the southern United States. It became common among people who had no milk or fresh meat but consumed corn bread and fat salt pork. Pellagra was commonly referred to as the disease of the "three D's"—*dermatitis, diarrhea,* and **dementia**

(a)

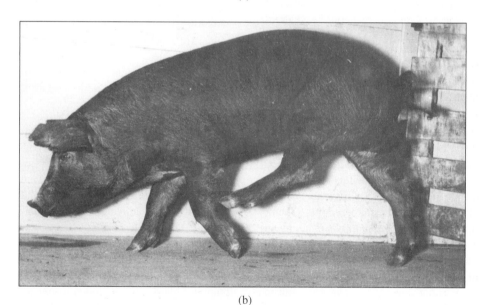

(b)

Figure 20.9 (a) Pantothenic acid deficiency in the chick. The eyelids, corners of the mouth, and adjacent skin areas are involved. Feathering is retarded and rough (left). The same bird (right) after receiving calcium pantothenate for 3 weeks.
Upjohn Company Vitamin Manual, *1965.*
(b) Pantothenic acid deficiency in the pig. The abnormal gait (goose-stepping) develops on a low pantothenic acid diet.
Courtesy of Dr. R. W. Luecke, Michigan State University.

(mental deterioration)—and was associated with a diet of "three M's"—*maize meal* (cornmeal), *molasses,* and *meat* (salt pork).

Changes in societies often result in nutritional problems. As populations move, economic, environmental, and social conditions may evolve that do not always allow people to consume diets that their ancestors found to support good health.

Pellagra was first believed to be an infectious disease rather than a nutrient deficiency. For 2 years Dr. Joseph Goldberger of the U.S. Public Health Service attempted to isolate the **"germ"** that caused pellagra. Then he began to study the reports of southern hospitals for clues that might suggest that pellagra resulted from something other than a germ. He noted that although a high percentage of the patients in three southern hospitals had pellagra the doctors, nurses, and orderlies who handled and even slept in the same ward with those ill with the disease were not afflicted. From this fact Dr. Goldberger concluded that pellagra was not **contagious.** He searched for further clues to explain why patients had pellagra and hospital personnel did not. He found that the hospital staff did not eat the same food as the patients. Could diet be related to this disease, which was causing thousands of deaths annually? That answer appeared too simple; however, he initiated an experiment to see.

Dr. Goldberger selected an orphanage in Missouri where the diets of children consisted of corn pone (corn bread made without milk or eggs), salt pork, hominy, and molasses. To these foods he added milk and eggs. The results were indeed pleasing, for in less than a year pellagra was eliminated among children in the orphanage. In other orphanages in which the dietary improvement had not been made the occurrence of pellagra continued to be high.

In the true scientific spirit, Dr. Goldberger conducted a second experiment. This time his objective was to see if a poor diet per se would *cause* pellagra in healthy men. But where does a scientist get healthy men to volunteer for such an experiment? On promise of pardon 12 healthy convicts on the Rankin Prison Farm in Georgia volunteered to serve as experimental subjects (11 were selected for the experiment). They were fed maize (corn) meal, white bread, flour, potatoes, salt pork, and syrup. After 6 months seven had developed pellagra.

After experimenting in state hospitals, orphanages, and prisons, Dr. Goldberger then established the fact that pellagra resulted from an inadequate diet. He and 15 of his associates (in 1916) tried in a number of ways to transmit pellagra to themselves by a series of inoculations with blood, nasopharyngeal secretion, **feces,** urine, and **desquamating** epithelium. The results of this heroic experiment were negative. In 1926 Dr. Goldberger and his associates demonstrated similarities between pellagra in humans and black tongue in dogs and in 1937 Dr. C. A. Elvehjem and his co-workers at the University of Wisconsin found niacin to be the active anti-black-tongue factor. It was soon found to be the specific cure and preventive for pellagra.

By 1947 it had become well established experimentally that the amino acid tryptophan was a precursor for niacin synthesis in the body (mostly in the intestine). Corn protein is low in tryptophan; thus individuals consuming a predominantly corn diet are predisposed to pellagra. Such a dietary regime increases the nicotinamide requirement. Thus pellagra might be considered a dual deficiency of both nicotinamide and tryptophan. Nicotinamide is a component of two coenzyme systems and is important in cell respiration and energy transfer.

The horse, rat, and neonatal calf apparently have little dietary need for niacin provided adequate amounts of tryptophan are available in their diets.

Research at Purdue University demonstrated that the feeding of 100 ppm of niacin significantly increases nitrogen retention in growing lambs.

Clinical Deficiency Symptoms

1. Dermatitis, dementia, diarrhea, loss of appetite and weight, vomiting, and anemia are commonly associated with a niacin deficiency. Usually the first lesions are those of the mucous membranes of the mouth, tongue, and vagina. The dermatitis is bilaterally symmetrical and is especially prominent on the exposed areas of the face, neck, hands, forearms, and feet. It also occurs on the elbows and knees and under the breasts. Precipitating factors are sunlight, heat, and friction. For these reasons most pellagra occurs in the warm months (Figure 20.10a).
2. As is true of other B vitamin deficiencies, pellagra often accompanies poverty, chronic alcoholism, fever, hyperthyroidism, pregnancy, and the **stress** of injury or surgical procedures.
3. Reduced growth occurs (Figure 20.10b and c).

Sources

Sources of niacin include milk (see page 55), lean meat, eggs, green vegetables, cereals (except corn), and peanut oil meals. It is not stored in appreciable amounts in animal tissues.

20.4.6 Pyridoxine (Vitamin B$_6$)

There are three related compounds with vitamin B$_6$ activity, namely pyridoxine, pyridoxal, and pyridoxamine. Pyridoxal is a component of several enzymes. Pyridoxine functions in the conversion of tryptophan into niacin and in the metabolism of amino acids and thus affects protein metabolism.

In laboratory animals fed diets deficient in pyridoxine (vitamin B$_6$), the lymphoid organs show signs of marked immune incompetence. For example, the thymus, spleen, and lymph nodes are smaller and not as well developed as organs of animals fed adequate levels of pyridoxine. Because of underdeveloped lymphoid tissue, pyridoxine-deficient animals are less able to mount an immune response to infecting stimuli. Pyridoxine-deficient animals have decreased cell-mediated immunity and are slow in producing antibodies.

Severe deficiency is uncommon in humans but pyridoxine status may be less than optimal in certain groups. For example, pregnant women and women using oral contraceptives need additional pyridoxine. Recent research showed that

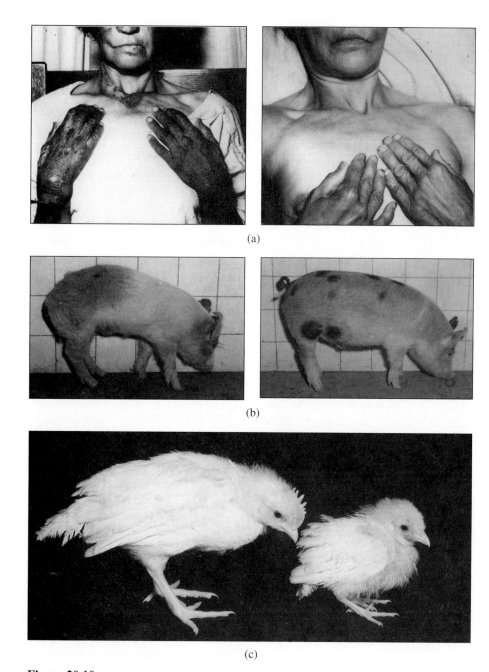

(a)

(b)

(c)

Figure 20.10 (a) Characteristic lesion of pellagra (left) on the back of the hands and dermatitis outlining the exposed area of the neck. The same patient (right) after nicotinamide therapy.
Upjohn Company Vitamin Manual, *1965.*
(b) Niacin deficiency in a pig (left) that received a diet containing 80 percent yellow corn. The growthier and normal-appearing pig (right) had niacin added to the diet.
Courtesy of Dr. D. E. Becker, University of Illinois.
(c) Effect of niacin deficiency on chick growth.
Courtesy of Dr. H. R. Bird, University of Wisconsin.

women who use oral contraceptives over extended periods of time secrete significantly lower levels of vitamin B_6 in their milk. Alcoholics risk a deficiency of the vitamin because of poor diets.

The diet of people residing in the United States commonly provides ample amounts of vitamin B_6 and deficiencies are seldom seen.

Clinical Deficiency Symptoms

1. Convulsive seizures (rats, poultry, dogs, infants, and swine).
2. Arterial lesions (monkeys).
3. Anemia (dogs and swine).
4. Dermatitis and edema of paws and nose (rats) as shown in Figure 20.11.
5. Impaired growth (all young animals).

Figure 20.11 Pyridoxine deficiency in the rat. The condition is characterized by edema; acanthosis; and denuding of the ears, paws, and snout (left). The same rat (right) after 3 weeks of treatment with pyridoxine hydrochloride.
Upjohn Company Vitamin Manual, *1965.*

Sources

Sources of pyridoxine include milk, muscle meats, liver, green vegetables, whole-grain cereals, and yeast.

20.4.7 Biotin

Originally called vitamin H this is a component of a coenzyme system and is concerned with CO_2 fixation, decarboxylation, and **deamination.**

A biotin deficiency can be induced in rats by feeding them raw egg white, which contains a biotin antagonist, **avidin,** which inhibits biotin absorption from the GI tract.[7] The deficiency state in the rat is characterized by dermatitis around the eye. This "egg white injury" has been produced in dogs, humans, and monkeys. Temperatures sufficient to coagulate egg albumen inactivate avidin.

Biotin is widely distributed in foodstuffs and therefore a deficiency of it is rare. Biotin is synthesized in the intestines of humans, chickens, dogs, rats, and ruminants (rumen). It has been shown that humans will often excrete (via feces and urine) more biotin than was ingested.

Clinical Deficiency Symptoms

1. Symptoms of an induced biotin deficiency include growth retardation, dermatitis (Figure 20.12a and b), loss of hair, and disturbances of the nervous system. In rats and dogs an ascending paralysis may be observed with accompanying growth cessation and a spectacled condition (depigmentation around the eyes).
2. In chicks biotin deficiency has been associated with **perosis,** as are manganese, choline, and folic acid.

Sources

Sources of biotin include milk, meats, yeast, and whole grains.

20.4.8 Folic Acid

This is needed to build red blood cells (**erythropoiesis**). The primary role of folic acid is apparently in the synthesis of nucleoprotein. It is especially important to mammalian cells during

mitosis (required to carry **metaphase** to **anaphase**). It appears to be protective against certain spinal diseases of babies when given as a supplement in the diet of pregnant women.

Folic acid–deficient chicks grow slowly, are incompletely feathered, have a reduced number of red blood cells and **leukocytes,** and have lower **hemoglobin** and hematocrit values.

Folic acid is interrelated with ascorbic acid and vitamin B_{12}. Apparently the chick, because of its short digestive tract, is the only domestic animal that requires a dietary source of folic acid. Ruminants synthesize the vitamin in their rumen and sufficient intestinal synthesis apparently occurs in other species to meet requirements.

Clinical Deficiency Symptoms

1. Growth retardation (Figure 20.13).
2. Abnormal blood cells (red and white).
3. Poor feathering in chicks.
4. Poor pigmentation of colored feathers.

Sources

Sources of folic acid include liver, dark green leafy vegetables, whole grains, and cereals.

20.4.9 Cyanocobalamin (Vitamin B_{12})

This vitamin is also called *cobalamin* and, like folic acid, participates in nucleic acid synthesis. Vitamin B_{12} is closely related in function to folic acid. Apparently three vitamins in particular (B_{12}, folic acid, and ascorbic acid) are involved in blood cell formation. The element cobalt is a component of vitamin B_{12}.

Pernicious anemia, resulting from a vitamin B_{12} deficiency, has not been observed in species other than humans. Rumen bacteria synthesize the vitamin if sufficient cobalt is present in the diet. Intestinal synthesis also occurs in humans, rats, and pigs but the synthesized vitamin is not absorbed and is excreted in the feces.

Clinical Deficiency Symptoms

1. Pernicious anemia is the most severe form of vitamin B_{12} deficiency. *Pernicious* stems from the Latin *perniciousus* meaning "fatal," and *anemia* means *a reduction below normal in the number of erythrocytes.* Thus the name *pernicious anemia* was originally selected because the

---

[7] A dramatic illustration of balance between vitamins is the interrelation between biotin and avidin. Many types of necessary balance between foods or nutrients are not entirely understood. For example, dogs digest raw starch and raw egg when they are fed together but cannot digest raw starch or raw egg when each is fed alone.

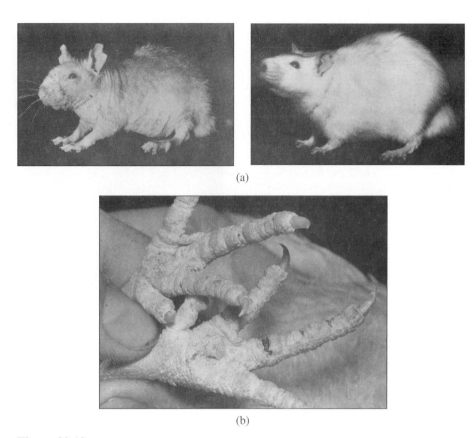

(a)

(b)

Figure 20.12   (a) Biotin deficiency in a rat. Note the dermatitis, which begins around the eye (left). The same rat (right) 3 weeks after adequate biotin was added to the diet.
Upjohn Company Vitamin Manual, *1965.*
(b) Biotin deficiency in the chick. Note the severe lesions on the bottom of the feet.
Courtesy of Dr. H. R. Bird, University of Wisconsin.

Figure 20.13   Folic acid–deficient chick (left) is stunted, poorly feathered, and severely anemic. The healthy chick (right) received the same ration supplemented with 100 μg of folic acid per 100 g of ration. Both are 4 weeks old.
Upjohn Company Vitamin Manual, *1965.*

disease was inevitably fatal before the discovery that liver consumption would cure the disorder. It is sometimes called *Addisonian pernicious anemia,* after Thomas Addison who first described its clinical characteristics in 1855. It usually occurs after the age of 30 years. Addison's description included the following: "The countenance gets rather pale, the whites of the eyes become pearly, the general frame flabby rather than wasted, the pulse perhaps large but remarkably soft and compressible, and occasionally with a slight jerk, especially under the slightest excitement—the lips, gums, and tongue seem bloodless."

Failure of gastric mucosal secretion of "intrinsic factor" leads to reduced ability to absorb vitamin B$_{12}$ from the GI tract. This precipitates pernicious anemia. **Desiccated** pig stomach is a good source of intrinsic factor and is **orally** effective in the **therapy** of pernicious anemia. Also effective therapeutically are liver extracts and the **parenteral** injection of crystalline vitamin B$_{12}$.

2. Vitamin B$_{12}$ deficiency causes low **hatchability** of eggs and low vitality and retarded growth of chicks.

Sources

Sources of vitamin B$_{12}$ include milk, meat (especially organ meats), and fish products.

20.4.10 Choline

This organic compound is synthesized in the body from methionine and is a structural component of fat and nerve tissue. It is now known to participate as a **catalyst** in metabolic functions. It occurs along with the B-complex vitamins in milk, meat, eggs, fish, and most cereals. Choline is present in all fat-containing foods. As a component of **phospholipids,** choline is essential to the building and maintenance of cell membranes.

No disease attributable to a choline deficiency has ever been demonstrated in humans. Choline prevents fat deposition in the liver. It has been called the "lipotropic factor," which means that it enhances the deposition of body fat (but not in the liver).

Choline is closely related to the amino acid methionine and may yield methyl groups (CH$_3$) to combine with homocysteine to form methionine. Choline is also closely associated with biotin and folic acid inasmuch as a deficiency of any one of the three will cause perosis in chicks.

Clinical Deficiency Symptoms

1. Impaired growth, fatty livers, enlarged spleen, and kidney hemorrhage.
2. Perosis in chicks (Figure 20.14) and reduced egg laying in hens.

Sources

Sources of choline include milk, meat, eggs, fish, and naturally occurring fats.

20.4.11 Inositol

This compound is closely related to the B-complex vitamins. It is a cell component in animal tissues and is found in especially high concentrations in many organ tissues (heart, kidneys, spleen, **thyroid,** and testicles). It is apparently not a dietary requirement for humans and most domestic animals.

Clinical Deficiency Symptoms

Retarded growth and a loss of hair in mice and rats.

Sources

Sources of inositol include meats, nuts, fruit, and whole grain.

Figure 20.14 Perosis (slipped tendon), which may result from a deficiency of choline, biotin, folic acid, or manganese.
Courtesy of Cornell University, Poultry Science Department.

20.4.12 Para-Aminobenzoic Acid (PABA)

When rescued at Cape Sabine my hair was entirely white, probably due to semistarvation, and darkened again within a year.
From Adolphus Greely's account
of his 1881 Arctic expedition

PABA holds tentative vitamin status because it acts as an anti-gray-hair factor in mice and rats and as a growth-stimulating factor in chicks. It is a molecular component of folic acid.

Para-aminobenzoic acid counteracts the **bacteriostatic** effect of sulfonamides. Present evidence indicates that PABA increases the physiological potency of insulin and penicillin and that it may inhibit the production of thyroid hormones. There are no established dietary requirements for PABA in humans or domestic animals.

20.5 VITAMIN ASSAYS

Biological Assay

The oldest and still most commonly used technique to determine the vitamin potency of a food is the **bioassay.** Usually animals are depleted of the vitamin by withholding its intake. Known amounts of the purified vitamin are then fed or administered (in a series of levels) to groups of the depleted animals. From the growth rates a standard response curve is prepared. The vitamin potency of the food in question is then determined in a similar way, comparing the animal response with the standard response curve. This technique is somewhat limited by the expense of labor, equipment, animals, and feed.

Microbiological Assay

The second assay technique for vitamin potency, **microbiological assay,** involves the use of microorganisms as test subjects.

This method is faster and less costly; however, it has the disadvantage that the vitamin must first be extracted from the foodstuff before it is added to the assay medium.

Chemical Assay

A third test method for the concentration of a vitamin is based on the chemical characteristics of the particular vitamin. This technique is fast but should be confirmed with bioassays to make certain the substance has vitamin activity in the body.

20.6 EXPRESSING VITAMINS A AND D QUANTITATIVELY

The vitamin A potency of food is expressed in International Units (IU) or U.S. Pharmocopeia units (USP). One IU is defined as 0.344 µg of crystalline all *trans*-vitamin A acetate, equivalent to 0.3 µg of crystalline vitamin A alcohol. Beta-carotene is used as the standard for provitamin A, 0.6 µg being equivalent in activity to 0.3 µg of vitamin A, using the rat as the assay animal. Beta-carotene is split enzymatically to yield vitamin A. In humans 1 IU of carotene is only about one-half as valuable as 1 IU of vitamin A, depending on its food source.

An IU (USP) of vitamin D is defined as the antirachitic activity, measured in a bioassay with rats, of 0.025 µg of crystalline vitamin D_3. Vitamin D potency is expressed in rat units per gram, except in poultry where the International Chick Unit (ICU) is used to express the activity produced in chicks by 0.025 µg of crystalline vitamin D_3. Vitamins D_2 and D_3 have equal value for mammals, but D_2 has only one-thirtieth the activity of D_3 for poultry.

20.7 SUPPLYING VITAMINS TO FARM MAMMALS AND POULTRY

Dietary vitamin requirements of farm animals[8] can be met largely through the consumption of the common feedstuffs grown under normal conditions. Vitamins are needed in very small amounts (e.g., the addition of only 4 to 8 mg per ton of **ration** provides sufficient vitamin B_{12} for poultry). Most vitamins are now available commercially in a stabilized form and at a nominal cost.

20.8 SUMMARY

A brief review of the principal roles, sources, and clinical deficiency symptoms of the fat- and water-soluble vitamins has been presented. Research leading to the identification and importance of vitamins is primarily a contribution of the twentieth century.

Some essential vitamins are stored in considerable quantities within the body during periods of adequate intake. These stores are drawn on later to meet requirements during periods of shortage. Others must be supplied regularly because storage in the body is limited. Moreover, there is a considerable variation of vitamin needs among species. Some species can synthesize adequate quantities of certain vitamins within their bodies. Additional research is needed to determine the exact requirements for various vitamins and vitamin-related compounds in many species.

STUDY QUESTIONS

1. How can certain vitamins be metabolic essentials and yet not be required in the diet?

2. Name the fat-soluble vitamins. Name the water-soluble ones. Which group contains the most elements?

3. Why are fat-soluble vitamins not required in the diet on a daily basis?

4. Give at least one important function, one or more **clinical** deficiency symptoms, and two or more dietary sources of vitamins A, D, E, and K. Construct a table, and refer to Table 20.1 only as needed.

5. Differentiate between 7-dehydrocholesterol and ergosterol.

6. What mineral has recently been shown to perform some functions of vitamin E?

7. Which species have a dietary need for ascorbic acid? Does this mean that animals of other species have no need for vitamin C? Explain.

8. What significant contribution did Dr. Christian Eijkman make to the field of vitamins?

9. Give at least one important function, one or more clinical deficiency symptoms, and two or more dietary sources of ascorbic acid, the B-complex vitamins, and the related compounds. Construct a table, and refer to Table 20.2 only as needed.

10. Why do healthy ruminants have no apparent dietary requirement for the B-complex vitamins?

11. Write a paragraph on the story of the search for the cause and cure of pellagra.

12. Which amino acid serves as a precursor for the synthesis of niacin in the body? Relate this to the antipellagric properties of milk. See Section 3.2.3.

13. What methods are used to determine the vitamin potency of foods for humans and animals?

14. How is the vitamin potency of food expressed?

15. What is a good reference for obtaining information pertaining to vitamin needs of farm animals? (Hint: See **NCR** in the Glossary.)

---

[8] Information pertaining to the specific vitamin requirements of the various species of farm mammals and poultry is available in *Nutrient Requirements of Domestic Animals,* National Academy of Sciences–**National Research Council,** Washington, DC 20418.

THE NUTRITIONAL CONTRIBUTIONS OF MINERALS TO HUMANS AND ANIMALS[1]

In all science, error precedes the truth, and it is better it should go first than last.

Horace Walpole (1717–1797)
English author

21.1 INTRODUCTION

Advances in the science of mineral **nutrition** during the twentieth century were greater than in all the previous time of humans on earth. Nutritional studies clearly showed that complex interrelations exist among mineral **elements.** Thus the mineral requirements for a given feeding regimen vary depending on the presence of certain **inorganic** and **organic** compounds. A classic example is that high levels of calcium and/or of phytic acid increase the zinc requirements of swine and dogs. Therefore the requirement for a given mineral element must be considered in relation to other components of the diet. Feeding an excess of iron ties up phosphorus and results in iron rickets. Feeding a high level of copper results in a significant decrease of liver zinc stores. These are only a few illustrations of the delicate relationships among minerals. And it should be readily appreciated that fortifying animal diets with excessive amounts of either **micro-** or **macro**mineral elements may prove to be more detrimental than beneficial.

There seems to be little **species** difference with respect to the essential minerals. An apparent exception is cobalt, an essential dietary requirement for **herbivorous** animals only.

To understand better the nutritional contributions of minerals to humans and animals it is necessary to consider them individually, as will be done in this chapter.

21.2 THE MACROELEMENTS

The macrominerals (from the Greek *macro* meaning "major") include calcium, magnesium, sodium, and potassium as the principal **cations** and phosphorus, chlorine, and sulfur as the principal **anions.** Pertinent information related to these elements is summarized in Table 21.1.

21.2.1 Calcium and Phosphorus

More than 70 percent of the total body ash is calcium and phosphorus. About 99 percent of the calcium and 80 percent of the body phosphorus are located in bones and teeth. Thus it is readily apparent that calcium and phosphorus are important in the formation and maintenance of the skeleton of humans and animals. The Ca:P ratio by weight in bone is about 2:1.

Radioisotope studies have shown that there is a continuous exchange of calcium and phosphorus between the bones and soft tissues (about 1 percent daily). If the dietary intake of calcium is inadequate the animal can draw on its soft bone (vertebrae, skull, mandibles, and ribs) reserves. This is especially important during pregnancy, **lactation,** and egg laying. Parathyroid hormone (Chapter 11) and vitamin D (Chapter 20) are closely related to this calcium-mobilization mechanism. However, if an individual continues indefinitely to consume a diet with a negative mineral balance, **osteomalacia** may develop.

Subnormal **calcification** of bones in growing animals may cause *rickets* (Figure 21.la and b). This condition may result from an inadequate dietary level of calcium and/or phosphorus or from decreased **absorption** of these inorganic elements, as in vitamin D deficiency.

---

[1]The authors acknowledge with appreciation the contributions to this chapter of Dr. G. C. Fahey, Jr., Department of Animal Sciences, University of Illinois at Urbana–Champaign.

TABLE 21.1	The Macrominerals		
Nomenclature	**Function**	**Clinical deficiency symptoms**	**Major sources**
Calcium (Ca)	Bone and tooth formation, blood clotting, enzyme activation, muscle contraction.	Rickets, slow growth and bone development, **osteomalacia, tetany,** thin-shelled eggs.	Milk, legumes, steamed bone meal, calcium phosphates, ground limestone, ground oyster shells.
Phosphorus (P)	Bone and tooth formation, component of many enzyme systems, release of body energy, part of **DNA** and **RNA.**	Rough hair coat, **pica,** lowered appetite, slow growth and low utilization of feed, lowered blood **plasma** phosphorus.	Milk and eggs, oilseeds and hulls of cereals, steamed bone meal, dicalcium phosphate, tripolyphosphate, defluorinated phosphate.
Magnesium (Mg)	Enzyme activator, constituent of skeletal tissue.	**Anorexia** (lowered appetite), hyperirritability, muscular twitching, **tetany** (convulsions), profuse salivation, **opisthotonos** (muscle spasms).	All feeds, particularly plant products (especially leafy vegetables and cereal grains).
Sodium (Na)	Muscle contraction; maintenance of **osmotic pressure** of body fluids; component of bile, which aids in fat digestion.	Loss of weight, craving for salt, eating of soil, reduced appetite.	Common salt; cured meats, cheese, many canned vegetables, soups.
Potassium (K)	Maintenance of **electrolyte** balance, **enzyme** activator, muscle function.	Heart **lesions,** loss of weight, reduced appetite, muscle weakness, poor wool growth.	Normal rations, widely distributed.
Chlorine (Cl)	Acid–base relations, maintenance of osmotic pressure of body fluids, used to make hydrochloric acid, necessary for digestion.	Craving for salt, reduced appetite, decreased blood chloride level.	Common salt.
Sulfur (S)	Synthesis of amino acids in ruminants (component of sulfur-containing amino acids).	Slow growth, low **feed efficiency,** slow wool growth in sheep.	Protein supplements, forages, cereals.

Increasing the proportion of cereal (especially oatmeal) in the human diet has been observed to increase the tendency to develop rickets. Much of the phosphorus in cereals is present as *phytin,* which is poorly absorbed by humans and poultry. Phytin combines with dietary calcium in the intestine and lowers its availability. The customary use of milk with porridge provides extra calcium needed under these conditions. Excessive intakes of aluminum, iron, and magnesium interfere with phosphorus absorption by forming insoluble phosphates.

Calcium is a normal and essential constituent of all living body cells but it is more concentrated in blood. Parathyroid hormone regulates blood calcium and phosphorus levels. Calcium is essential to normal muscle contractions; therefore, when the blood calcium concentration drops markedly, **tetany** will result (as in **parturient paresis**). As is noted in Figure 20.4, calcium is also essential for blood clotting.

Phosphorus is a component of the energy transfer system of the body and calcium activates the **enzyme** ATPase. Also, phosphorus is a part of the genetic materials **DNA** and **RNA.** Hence it is concerned with the **metabolism** of almost all **nutrients** through its vital role in both vitamin and enzyme activity.

Calcium:Phosphorus Ratio

Because calcium and phosphorus are present in the body on a 2:1 basis, it would seem reasonable to assume that their intakes should be in about the same ratio. (Ca:P ratio range of 1:1 to 3:1 in diets is satisfactory.) If much more calcium than phosphorus is consumed the excess calcium is not absorbed in the proximal part of the small intestine but accompanies the phosphorus to the lower portion of the small intestine (the point where most phosphorus is absorbed). This excess calcium combines with

phosphorus to form insoluble tricalcium phosphate thus interfering with phosphorus absorption. Conversely, an excess of dietary phosphorus over calcium will in the same way decrease the absorption of both calcium and phosphorus.

In general, **forages** (especially **legumes**) are high in calcium and low in phosphorus, whereas the cereal grains are high in phosphorus and low in calcium. Therefore the **livestock** feeder must consider the Ca:P ratio when making major shifts in the feeding regimen.

Milk is an excellent source of both calcium and phosphorus (Chapter 3). Eggs are rich in phosphorus but low in calcium. Meats are low in calcium but provide significant amounts of phosphorus. **Bone meal** and dicalcium phosphate are good **supplemental** sources of calcium and phosphorus for farm mammals and ground limestone and oyster shells are good supplemental calcium sources for poultry.

In countries where milk is not available bones are often consumed as such and in bone soups (cooked with vinegar, which disintegrates the hard bone and renders its calcium available).

Clinical Deficiency Symptoms

1. Rickets.
2. Osteomalacia.
3. **Pica** (depraved appetite), especially noted in phosphorus deficiency. The animals may chew wood, bones (Figure 21.2), and rocks.
4. Impaired appetite (**anorexia**), slow growth, rough hair coat.
5. Easily fractured bones.
6. Lowered blood **plasma** phosphorus.
7. Thin-shelled eggs; with severe deficiency, hens stop laying eggs.

(a)

(b)

Figure 21.1 (a) Rickets in a young Jersey bull (left) and a Missouri mule (right).
Courtesy of Dr. W. A. Albrecht, University of Missouri.
(b) Calcium deficiency in a pig. Note the abnormal bone development and rachitic condition. A lack of calcium retards normal skeletal development but does not usually depress total weight gain.
Courtesy of Dr. N. R. Ellis and the USDA.

Figure 21.2 Phosphorus-deficient cow (left) showing unhealthy appearance and bone-chewing tendency. Same cow (right) after phosphorus was added to the diet.
Courtesy of Dr. W. E. Petersen, University of Minnesota.

21.2.2 Magnesium

About three-fourths of the body's magnesium is found in the skeleton. The balance is distributed throughout the body fluids. A close biological relation between calcium and magnesium has recently been demonstrated experimentally by the substitution of magnesium for calcium in bone formation without sig-nificantly changing the **x-ray** differentiation pattern. However, the body apparently exhibits "selective absorption" for calcium over magnesium when both are present in the diet. An increase of calcium or phosphorus (or both) in the ration apparently increases the minimum magnesium requirement and may result in symptoms of magnesium deficiency.

Magnesium is essential as an activator of many enzyme systems, especially those involved in carbohydrate metabolism. It is an essential constituent of bones and teeth. Magnesium is a component of chlorophyll (as iron is of **hemoglobin**). It is also vital in the proper functioning of the nervous system.

In general, leafy vegetables and cereal grains are rich sources of magnesium whereas most animal products contain lesser amounts. Magnesium tetany has been observed in calves and children fed milk alone for extended periods without magnesium supplementation. Most of the commonly fed rations of farm animals supply ample magnesium without supplementation.

Clinical Deficiency Symptoms

1. Tetany, a disease of cattle called "grass tetany" or "grass staggers" (frequently observed in the Netherlands and New Zealand), associated with a magnesium deficiency. It is most commonly observed among dairy cattle subsisting on lush spring pastures that have had high nitrogen applications. Figure 21.3a depicts a lamb in tetany resulting from a magnesium deficiency.
2. Hyperirritability; magnesium, like calcium, depresses **irritability** of nerves.
3. Muscular twitching, convulsions, weak pasterns, excessive salivation.
4. **Opisthotonos** (a form of spasm in which the head and the heels are bent backward and the body bowed forward).
5. Retarded growth and reduced feed efficiency.
6. Calcification of soft tissue in certain species as depicted in Figure 21.3b.

21.2.3 Sodium, Potassium, and Chlorine

These three essential dietary minerals are present largely in the body fluids and soft tissues. They function to (1) maintain **osmotic pressure** and acid–base equilibrium, (2) control the movement of food nutrients into cells, (3) control nerve and muscle function, and (4) regulate water metabolism.

Because these minerals are not stored to any appreciable extent in the body there is a regular dietary need. A deficiency in any of these minerals results in reduced appetite, decline of growth, weight loss, decreased production (meat, milk, eggs, and wool), and decreased blood plasma concentrations. A rough hair coat often characterizes a salt (NaCl) deficiency. If allowed free access to salt, farm animals will not suffer a salt deficiency. However, **ruminants** fed little or no **roughage** may exhibit a potassium deficiency (Figure 21.4).

Sodium

This element is essential for normal muscle contraction. It is also essential for maximal utilization of dietary energy and protein and for efficient reproduction. Deficiency of it in laying hens results in a loss of weight, decreased egg laying, and **cannibalism.**

Potassium

This mineral plays a vital role in muscle function. A potassium deficiency causes heart **lesions** and degeneration of the kidney tubules. Low levels of blood potassium also result in muscle weakness and depression. Excessive levels of potassium can result in muscle tremors, collapse, and death from heart failure. High blood levels of potassium are generally due to kidney failure or endocrine disease rather than from excessive dietary intake. In horses a genetic defect in potassium metabolism results in hyperkalemic periodic paralysis (HYPP). Horses with HYPP should be fed diets low in potassium (e.g., oat or timothy hay rather than alfalfa).

There is an interrelation between sodium and potassium in metabolism. At inadequate levels of either the deficiency symptoms are aggravated by a large excess of the other. Research indicates that potassium may be required by one or more enzyme systems.

Chlorine

The primary role of chlorine is in control of acid–base equilibrium and osmotic-pressure regulation. However, it is also an

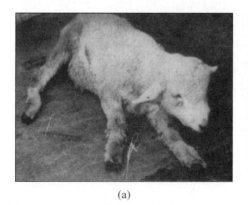

(a)

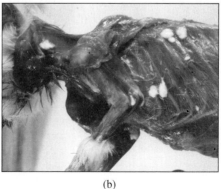

(b)

Figure 21.3 (a) Magnesium deficiency in a lamb. Note stiff legs.
Courtesy of Dr. U. S. Garrigus, University of Illinois.
(b) Large calcium phosphate deposits on the rib cage of a guinea pig fed a diet deficient in magnesium. Frequently, such deposits form in and around joints, giving rise to an arthritis-like condition. The condition is entirely prevented by feeding adequate levels of magnesium.
Courtesy of Dr. B. L. O'Dell, University of Missouri.

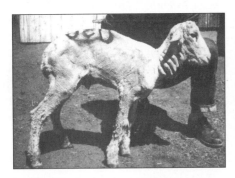

Figure 21.4 The lamb at left received a potassium-deficient ration (0.1 percent K), whereas the lamb at right received sufficient dietary potassium (0.6 percent K).
Courtesy of Dr. R. L. Preston and the University of Missouri.

important component of gastric secretions. A limited amount of chlorine is stored in the skin and **subcutaneous** tissues.

Salt (NaCl)

The inclusion of "common salt" in the diets of humans and animals has been practiced for a long time. Early humans used salt as a condiment (primarily as a seasoning) and probably were unconcerned about its nutritional and physiological support. Salt stimulates salivary secretion and promotes the action of diastatic enzymes. It is the mineral compound most often needed as a dietary supplement for most farm animals.

When the dietary intake of salt is low, the human kidney may excrete as little as 1 g of sodium chloride per day. Conversely, it may excrete a large quantity (as much as 40 g daily) when salt intake is high. The latter situation requires an increased water intake.

When the rigors of exercise and/or heat result in profuse perspiration animals may consume large quantities of water in relatively short periods of time. In such situations cramps may result unless salt is taken concurrently with the water.

An excessive salt intake may result in water retention, which causes **edema.** A salt deficiency is more prevalent in herbivores than in other animals because forages and grains contain negligible amounts of salt.

Salt-deficiency symptoms include (1) lack of appetite, (2) rough hair coat, (3) eating of soil, (4) unhealthy appearance, (5) decrease in production, and (6) loss of weight (Figure 21.5).

21.2.4 Sulfur

Most body sulfur is found in proteins and more specifically in the amino acids *cystine* and *methionine.* Sulfur is present in insulin and glutathione as well as in wool. The majority of dietary sulfur is derived from an organic source (protein). Only a small fraction is **ingested** in the inorganic form, which in most animals is ineffective in satisfying body requirements for sulfur (although research indicates that inorganic sulfur may be useful in poultry diets).

Sulfur functions in the synthesis of sulfur-containing amino acids in the **rumen** and certain other sulfur compounds

Figure 21.5 Extreme salt deficiency in a cow.
Courtesy of Dr. S. E. Smith, Cornell University.

of the body. Small amounts of inorganic sulfur supplements have been observed to increase the utilization of **urea** as a source of nitrogen for ruminants. A sulfur deficiency may result in reduced growth, lower **feed efficiency,** and slow wool growth in sheep (Figure 21.6).

21.3 THE MICROELEMENTS (TRACE ELEMENTS)

The microminerals (from the Greek *micro* meaning "minor") that have been shown to be essential for cattle, sheep, and swine include cobalt, copper, iodine, iron, manganese, molybdenum, selenium, and zinc. In experiments in which animals were fed highly **purified diets** one or more of the following elements have been classified as essential trace minerals for certain species: arsenic, chromium, nickel, tin, and vanadium. Pertinent information related to the microelements is summarized in Table 21.2.

21.3.1 Iron

Although needed in small quantities, iron plays an essential role in the life of humans and animals. Iron is present as an

TABLE 21.2	The Microminerals		
Nomenclature	**Function**	**Clinical deficiency (or excess) symptoms**	**Major sources**
Iron (Fe)	Component of **hemoglobin,** component of many enzyme systems.	Nutritional anemia, **thumps** in pigs, diarrhea, loss of appetite.	Eggs, soil, forages and grains, iron injections; liver, pork; ferrous sulfate.
Copper (Cu)	**Erythropoiesis, coenzyme** system, hair pigmentation, reproduction, collagen and elastin synthesis, iron utilization.	Depraved appetite, stunted growth, diarrhea, **osteomalacia** (in mature cattle), bleached hair and wool, **ataxic** gait, anemia, loss of **condition,** aortic rupture in swine and poultry.	Feedstuffs and $CuSO_4$ (0.25 to 0.5% $CuSO_4$ added to salt fed free-choice).
Iodine (I)	Synthesis of thyroxine.	Enlarged necks in calves and lambs, goiter, hairless pigs and woolless newborn lambs, dead or nonviable calves.	Iodized salt (KI in salt), cod-liver oil.
Cobalt (Co)	Component of vitamin B_{12}, RBC formation, proper function of rumen microorganisms.	Loss of appetite, weakness, **emaciation,** rough hair coat, anemia, reproductive failure.	Cobalt pellets (for ruminants); 0.5 ppm of cobalt salt added to ration (vitamin B_{12} injection to relieve cobalt deficiency).
Zinc (Zn)	Carbonic anhydrase, enzyme activator.	Retarded growth, **anorexia, parakeratosis** in swine, hyperkeratosis in chicks, poor feathering, poor **hatchability.**	ZnO, $ZnCO_3$, or $ZnSO_4$ added to ration; forages.
Manganese (Mn)	Growth and bone formation, enzyme activator.	**Perosis** (slipped tendons) in poultry, lowered hatchability and eggshell strength, lameness, stiffness.	$MnSO_4$ at 100 g/ton of feed; widely distributed in feeds, nuts and seeds, milk, legumes, and cereals.
Selenium (Se)	Destroys peroxides; related to vitamin E, which prevents peroxide formation.	**Necrosis** of liver, white muscle disease in sheep (deficiency); "alkali disease" or "blind staggers" (excess, above 5 ppm).	Oil meals and grains.
Molybdenum (Mo)	Enzyme systems. Affects copper absorption and availability to tissues.	Excess: **teart,** diarrhea, loss of weight, emaciation.	Widely distributed; rarely a problem.
Fluorine (F)	1 to 2 ppm in water added to aid in preventing tooth decay.	Excess: chalky and mottled teeth (fluorosis), decreased appetite, and slow growth.	Water.

Figure 21.6 Lambs fed a low-sulfur diet. Lamb 6 received 3 g of sulfur per pound of diet, whereas lamb 5 received none. Note the excessive salivation, lacrimation, and shedding of wool by lamb 5. *Courtesy of Dr. U. S. Garrigus, University of Illinois.*

iron–porphyrin nucleus (heme) in the hemoglobin molecule, in the protein fractions of cytochrome C, and in other important enzymes. Hence iron is a component of oxygen carriers (hemoglobin in red blood cells) and of oxidizing **catalysts** (enzymes), which are essential for cellular **oxidation.** Therefore iron is essential for life.

More than 60 percent of the body's iron is in the form of hemoglobin. Iron is stored in the liver, spleen, kidneys, and, to some extent, the bone marrow. However, because the destruction and formation of red blood cells (RBC) is continuous (the average life of an RBC is about 3 months), there is a constant turnover of iron in the body.

Iron-Deficiency Anemia

The **anemia** resulting from an iron deficiency may occur anytime the iron intake becomes deficient relative to the needs for hemoglobin formation. It is most likely to occur during the **suckling** period of mammals because milk is deficient in iron. In Missouri studies calves receiving iron dextran injections gained 22 percent faster to 12 weeks of age than noninjected controls. In the seventeenth century, physician Thomas Sydenham made a tonic of iron and wine and used it in the treatment of anemia.

In baby **pigs** a condition commonly called **thumps** soon develops (within 3 to 4 weeks) if the **piglets** are confined without access to soil and do not receive supplemental iron during the suckling period. (Because many pigs are now raised in confinement iron injections are routine.) Human babies also soon deplete their body stores (within 6 months) unless supplemental iron is provided to the normal high-milk diet. **Calves, foals,** and **lambs** seldom demonstrate iron deficiency because they usually have access to forage early in life.

Both pregnancy and egg laying increase the need for additional dietary iron. Studies utilizing **radioactive** iron (^{59}Fe) have shown that there is an increased absorption of iron from the small intestine during pregnancy to meet the needs for fetal growth. Feeding supplemental iron to a **lactating** female does not increase the iron content of the milk.

It should be noted that there are many kinds of anemia not caused by iron deficiency. Hereditary anemia (sickle-cell anemia) was mentioned in Chapter 8. Deficiencies of copper, protein, and certain vitamins can cause anemia. Certain pathological conditions may result in anemia. It may result from an interference with or cessation of hemoglobin production or from an excessive loss or destruction of blood.

Relative to the amount needed by the body, most animal foods (except milk) contain liberal amounts of iron. Eggs, meat, and many leafy vegetables provide good sources of iron for humans. However, the iron of many leafy plants is utilized poorly whereas iron from inorganic sources is utilized more completely. Hemoglobin iron is the best-utilized source.

Excessive dietary iron interferes with phosphorus absorption by forming an insoluble phosphate and this may cause rickets. Iron toxicosis is uncommon in farm animals but has been reported in children as a result of accidental excessive intake of iron pills; in Bantus of South Africa because of cooking in iron pots and drinking Kaffir beer, which contains a high level of iron; and in certain genetic conditions.

An iron deficiency in cattle and sheep may cause pica, similar to that caused by a phosphorus deficiency. Iron deficiency is characterized by diarrhea, loss of appetite for the normal foods, and anemia.

21.3.2 Copper

Like iron, copper is important in hemoglobin formation. It too is stored in the liver and, to a lesser extent, in the spleen, kidneys, heart, lungs, and bone marrow. Because milk is low in both copper and iron, nature provides liver stores at birth to supply the mammal with copper throughout the suckling period. Copper is not an essential component of the hemoglobin molecule but it is needed as a catalyst in its formation. Therefore anemia may result from a copper or an iron deficiency or both. Animals suffering from inadequate copper intake are unable to utilize iron at a normal rate and a deficiency in hemoglobin synthesis exists.

Figure 21.7 A typical copper-deficient poult at left. Note the depigmentation of the feathers. Age is 4 weeks.
Courtesy Dr. J. E. Savage and Dr. B. L. O'Dell, University of Missouri.

Copper, like iron, is vital to certain cellular enzyme systems involved with oxidation-**reduction** reactions.

A copper deficiency in sheep is reflected in changes in the **fleece.** The fibers become progressively less crimped (the wool becomes steely or stringy), wool growth is slowed, and black wool turns white. Depigmented feathers or a bleached hair coat is a common symptom of copper deficiency in poultry and other animals (Figure 21.7). Copper deficiency interferes with synthesis of **keratin,** the principal constituent of hair and wool. These changes in hair and wool reflect a failure of enzyme activities for which copper is essential. Diarrhea, loss of appetite, and swelling about the pasterns may also occur in copper-deficient animals. Furthermore, in an extreme copper deficiency the bones become fragile and the animal exhibits an **ataxic** (uncoordinated) gait. A disease of lambs, called "swayback" characterized by nervous symptoms, is caused by a copper deficiency and results from damage to nerve tissue during embryonic development.

Although essential in small amounts, excessive copper is toxic. Copper toxicity is accentuated by low molybdenum diets. The addition of molybdenum to diets (currently not approved by the **FDA**) containing toxic levels of copper will counteract possible copper poisoning. Thus it is apparent that copper and molybdenum are biologically antagonistic. Research indicates that high levels of dietary copper tend to deplete the liver stores of zinc.

Missouri studies have shown that when chickens are fed a purified diet deficient in copper during their early growth and development, they may die of aortic rupture. This results from defective elastin (scleroprotein) in the walls of the aorta. Copper is essential to the enzyme system that aids in the use of the amino acids needed to synthesize the elastin fibers (Figure 21.8).

21.3.3 Iodine

The animal body contains a very minute amount of iodine, approximately 60 percent of which is found in the **thyroid** gland. The foremost need for dietary iodine is in the synthesis of thyroxine. The discovery of iodine in the thyroid gland was made by Baumann in 1896.

Activity of the thyroid gland is regulated by thyrotropin (TSH, or thyroid-stimulating hormone) from the **anterior** pituitary (Chapter 11). When the blood concentration of thyroxine declines the anterior pituitary increases its output of TSH, which in turn causes the thyroid to increase its thyroxine output. The thyroxine then enters the bloodstream and performs its physiological function of regulating the metabolic rate of the body. Thus iodine is indirectly involved in control of the rate at which the body uses energy.

When dietary iodine is insufficient the thyroid cannot synthesize sufficient thyroxine and the blood concentration declines causing the release of thyrotropin. This hormone of the anterior pituitary gland in turn increases the activity of the thyroid gland causing it to enlarge (Figure 21.9). This condition is called *goiter* (compensatory **hypertrophy,** or an enlargement involving the formation of more tissue in an effort to secrete more thyroxine). Cleopatra and the model for the Mona Lisa reportedly each had a small goiter.

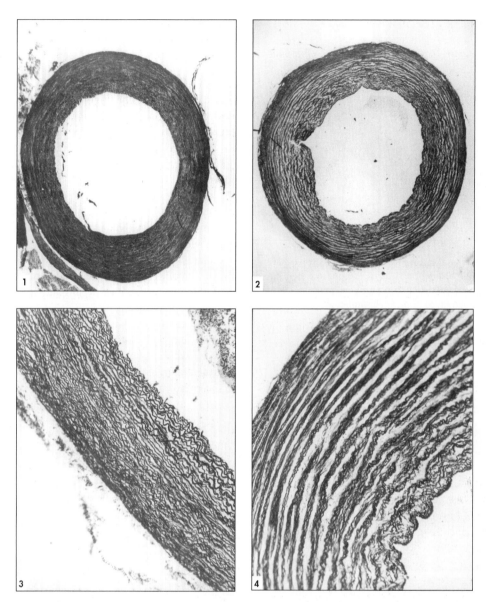

Figure 21.8 Cross section through arch of the aorta from a control turkey poult fed a milk diet with 50 ppm of copper (upper left, × 24). Cross section through arch of the aorta from copper-deficient poult (upper right, × 24). Lower left and right show, respectively, control and copper-deficient aorta magnified 120 times. Note thicker aortal wall and the accumulation of nonelastin material and focal breaks in the copper-deficient poult.

Courtesy of Dr. J. E Savage and Dr. B. L. O'Dell, University of Missouri.

Goiters are more likely to occur during times of increased metabolic rate such as **puberty** and pregnancy. In farm animals the newborn may exhibit a goiter resulting from an iodine deficiency in the ration of its mother during pregnancy. Typical iodine-deficiency symptoms are hairlessness and abnormal growth and development in young pigs (Figure 21.10); enlarged necks (goiters) in calves, lambs, and **kids;** and weak foals. Sheep showing a typical goiter are depicted in Figure 21.11. Iodine deficiency is a geographic problem[2] and people living in goitrous areas observed the benefits of seaweed in preventing goiter hundreds of years before the discovery of iodine.[3] Crops reflect the level of iodine in the soil on which they are grown. When farm animals drink only rainwater they are likely to be deficient in iodine unless they are

---

[2]For example, the Great Lakes region of the United States and Canada, the Pacific northwestern states, and Switzerland are areas in which an iodine deficiency is likely to occur unless supplemental dietary iodine is provided.

How to provide iodine for the general population was at first a problem. Three proposals were advanced: (1) Because goiter is prevalent among iodine-deficient children, why not add it to candy? This method was tried in some schools and failed because not all children liked the same kind of candy. (2) Add iodine to drinking water. But then what should be done about some rural areas? (3) Add iodine to salt. This method was and continues to be successful. More than a dozen countries now iodize their table salt. The amount added varies from 1 part in 10,000 in the United States and Canada to 1 part in 200,000 in Poland. In at least two countries, Canada and Switzerland, all salt for home use must be iodized. This requirement reduced the incidence of goiter by more than 85 percent in Switzerland.

[3]Seaweed provides a reliable source of iodine and has been used in the treatment of goiter for more than 3000 years.

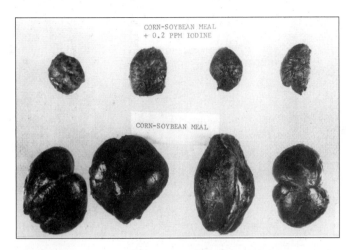

Figure 21.9 Thyroid glands of pigs fed a corn–soybean meal basal diet with or without 0.2 ppm iodine for 51 days. The goitrous thyroid glands of pigs fed the basal diet (bottom row) weighed approximately 6 times more on average than those of pigs fed supplemental iodine (top row).

Courtesy of Dr. Gary L. Cromwell, University of Kentucky, Lexington.

Figure 21.10 The pig in front was fed a basal diet plus 0.5 percent potassium thiocyanate for 51 days. The littermate in back was fed the basal diet plus 0.2 ppm iodine for the same period. The pig fed thiocyanate showed symptoms of hypothyroidism, including shortened legs and extreme lethargy.

Courtesy of Dr. Gary L. Cromwell, University of Kentucky, Lexington.

fed iodized salt or an inorganic form such as either potassium or sodium iodide.

21.3.4 Cobalt

This element was not recognized to be essential for growth and health until 1935. As with iodine, cobalt deficiency is a regional problem.

Cobalt is an integral component of the vitamin B_{12} **molecule.** It is essential in the synthesis of this vitamin by rumen **microflora.** Thus it is actually a deficiency of vitamin B_{12}, not of cobalt per se, that is responsible for the metabolic failure observed in cobalt-deficient ruminants. This explains why no essential role for cobalt has been demonstrated in swine and poultry. (They require dietary B_{12}.) It is believed that cobalt stimulates the appetite of ruminants through its action on the rumen **flora.** Research suggests that cobalt may also be associated with the synthesis of pyridoxine, niacin, and riboflavin by rumen microorganisms.

Clinical symptoms of a cobalt deficiency include loss of appetite and body weight; **emaciation;** long, rough hair coat; retarded wool growth (and weak fibers); scaliness of the skin; abortion; reduced milk secretion; and anemia. Cobalt deficiencies in sheep and cattle are shown in Figure 21.12a and b, respectively.

21.3.5 Zinc

Most of the body's zinc is found in the liver, bones, and **epidermal** tissues (skin, hair, and wool). Because zinc absorption from grains and legume seeds by swine and poultry is very poor, diets based on these feedstuffs frequently must be supplemented with zinc salts.

The primary physiological role of zinc is related to enzymatic activity. The most fundamental physiological action of zinc is in its role in protein synthesis. Zinc is required for the formation of DNA and RNA, the compounds that control cellular multiplication and growth through their influence on protein synthesis (see Chapter 8).

Clinical symptoms of a zinc deficiency include (1) **parakeratosis (dermatitis)** in swine (Figure 21.13a) and calves, (2) retarded growth, (3) decreased feed efficiency, and (4) increased susceptibility to infections. Parakeratosis is

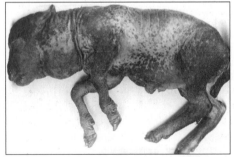

Figure 21.11 Ewe showing a typical goiter that is due to an iodine deficiency (left) and woolless neonatal lamb (right) born of an iodine-deficient ewe.

Courtesy of Dr. J. E. Catlin, Montana State University.

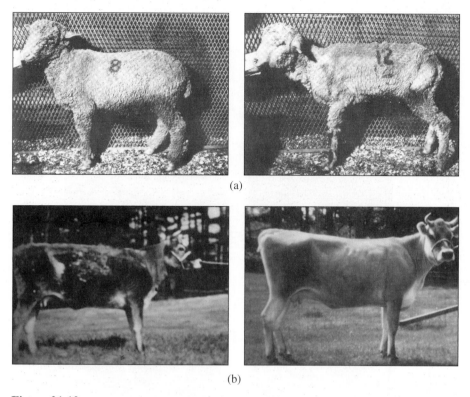

(a)

(b)

Figure 21.12 (a) Comparison of a sheep fed a cobalt-adequate diet (left) with one fed a cobalt-deficient diet (right).
Courtesy of Dr. S. E. Smith, Cornell University.
(b) The calf at left shows effects of a cobalt-deficient diet. The same calf is pictured at right a few weeks after cobalt was added to the diet.
Courtesy of Dr. R. C. Carter, Virginia Polytechnic Institute and State University.

accentuated by rations high in calcium (especially noted when animals receive rations containing primarily plant proteins, because both excessive calcium and phytates in plants interfere with zinc absorption).

Calves fed a zinc-deficient diet also develop parakeratosis. They are characterized by an unhealthy appearance; rough hair coat; stiffness of the joints; dry, scaly skin on the ears; and a thickening and cracking of skin around the nostrils (Figure 21.13b).

Zinc-deficiency symptoms in the chick include slow growth, shortened and thickened long bones, poor feathering, reduced **hatchability,** and embryonic **anomalies.** In severe deficiency keratosis occurs. Missouri studies, using purified diets, have shown zinc to be especially important to proper structural bone development in chickens (Figure 21.14). Zinc is closely related to the **assimilation** and proper use of other minerals.

21.3.6 Manganese

Manganese is found in the liver, bone, muscle, and skin. Probably the most important function of manganese in the body is to activate several enzymes concerned with carbohydrate, fat, and protein metabolism. Manganese has been shown experimentally to be essential for normal reproduction in several laboratory animals. Manganese-deficient diets delay sexual maturity in females, cause irregular **ovulations** and weak young at birth, and may cause sterility in males.

In poultry a manganese-deficient diet causes **perosis** (slipped tendons), a malformation of the leg bones of growing chicks (Figure 21.15), and lowered hatchability. In rabbits bone malformations may occur. Perosis is not caused solely by a manganese deficiency because a vitamin deficiency of biotin, folic acid, or choline will also result in perosis (Chapter 20). Diets high in calcium and phosphorus predispose poultry to perosis, probably by interfering with manganese absorption.

21.3.7 Selenium

Only recently has selenium been considered an essential micronutrient for animals. It is now well established that selenium can perform some of the functions of vitamin E. In vitamin E deficiency, hydroperoxides are formed during metabolism of unsaturated fatty acids. A selenium-containing enzyme, glutathione peroxidase, can destroy these hydroperoxides thereby preventing them from damaging tissues. A low dietary selenium level (< 0.05 ppm) may cause white muscle disease in sheep and cattle.

Basic research studies using the bull have shown that selenium is associated with the reproductive system. However, its physiological role in reproduction has not been clearly defined. Research at the University of Illinois indicates that low levels of selenium are anticarcinogenic in mice. The mode of action has not been elucidated.

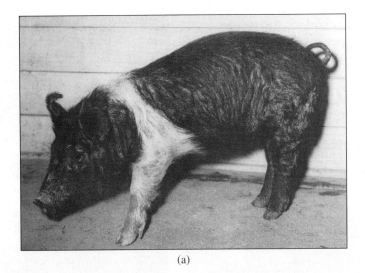

Figure 21.13 (a) Parakeratosis in swine.
Courtesy of Dr. R. W. Luecke, Michigan State University.
(b) Calf showing loss of hair on legs and severe scaliness, cracking, and thickening of the skin as a result of zinc deficiency at age 15 weeks (left). Same calf (right) 5 weeks after zinc was added to the diet.
Courtesy of Dr. W. J. Miller, University of Georgia.

Studies have shown that crops grown on soils high in selenium (> 5.0 ppm) may contain toxic levels of that element. In some regions (South Dakota, for example) livestock consuming such feeds may develop "alkali disease" or "blind staggers" because of interference with oxidation catalysts. There is commonly a loss of hair in cattle, horses, and swine; the hooves slough off, lameness occurs, appetite diminishes, and growth (in young animals) is retarded (Figure 21.16). Studies at The Ohio Agricultural Research and Development Center in Wooster indicate that dietary selenium levels are most critical during early stages of growth and when cattle are fed diets marginal or deficient in protein.

Selenium is a cumulative poison so that toxic symptoms may be observed only after an extended period of its consumption in low quantities. Small amounts of arsenilic acid or arsenic compounds are effective in reducing the toxicity of selenium.

A selenium deficiency may be reflected in a condition known as *nutritional myopathy,* or white muscle disease, in lambs and cattle and as heart and skeletal muscular dystrophy in mink. Large doses of vitamin E will normally correct these abnormal conditions.

Approval was gained from the FDA in 1974 to add selenium to chicken, swine, and turkey feeds in a premix as either sodium selenite or sodium selenate. Similar approval for the addition of selenium to ruminant feeds was obtained in 1979.

21.3.8 Molybdenum

This mineral is known to be a component of one or more enzymes. When cattle graze on vegetation grown on soils high in molybdenum (especially noted in Canada and England) they may develop a condition called **teart,** which is caused by molybdenum poisoning. It is also recognized as "peat scours" in New Zealand. The chief symptoms of molybdenum toxicity are diarrhea, loss of weight, emaciation, anemia, and stiffness. These symptoms may be cured by the administration of copper sulfate. This fact points again to the interrelation between molybdenum and copper discussed previously (Section 21.3.2). Toxic levels of molybdenum

Figure 21.14 A malformed chick embryo (note missing appendages) caused by a zinc-deficient diet (left). Both control (right) and zinc-deficient embryos were incubated 18 days. *Courtesy of Dr. J. E. Savage and Dr. B. L. O'Dell, University of Missouri.*

Figure 21.15 Manganese deficiency caused the "slipped tendons" in the chick at right. A normal chick is pictured at left. *Courtesy of Dr. J. E. Savage and Dr. B. L. O'Dell, University of Missouri.*

interfere with copper metabolism, thus increasing the copper requirement. (1 g $CuSO_4$ per animal daily will prevent and cure symptoms of molybdenum toxicity.) Tolerance to molybdenum is believed to be affected by the intake of methionine and inorganic sulfate as well as by the copper content of the diet.

21.3.9 Fluorine

The essentiality of this mineral has not been clearly established. It is found in very minute amounts throughout the body but notably in hair, bones, and teeth. It is known that fluorine, at appropriate intake levels, aids in preventing dental caries.[4] Conversely, higher levels will cause *fluorosis* (chalky and mottled teeth) as shown in Figure 21.17. Mottled teeth are structurally weak.

The effects of fluorine are cumulative so that the intake of small quantities over an extended period of time may produce toxic effects. Rock phosphates generally contain 3 to 4 percent fluorine (toxic amounts). Proper defluorination procedures are necessary to render this mineral safe for supplemental mineral purposes. Fluorine inhibits several enzyme systems, which may explain in part its toxic effects in animals.

21.3.10 Vanadium

Vanadium has recently been described by USDA scientists as an essential element for the health of chicks and rats. Their research showed that certain physiological functions of test animals become impaired when they consume a diet low in vanadium. Effects include reduced feather and body growth, impaired reproduction and survival of the young, altered RBC numbers and iron metabolism, impaired tissue metabolism, decreased insulin activity, and altered blood lipid levels.

Researchers at Colorado State University reported that the required vanadium level may be between 50 and 500 ppb when the element is consumed in a purified diet.

21.3.11 Arsenic, Nickel, and Tin

When extreme care in ration preparation is taken and mineral contamination of caging, bedding, and air supply are avoided, it can be shown that arsenic, nickel, and tin are essential for normal

---

[4]The rate of dental caries in certain areas of the United States and Canada was reduced by as much as 65 percent following water fluoridation.

Figure 21.16 Ewe (left) suffering from selenium toxicity. Loose wool is typical of sheep afflicted with chronic selenium poisoning. The lambs (right) show congenital deformity traceable to selenium injury during fetal development.
Courtesy of Dr. P. O. Stratton, University of Wyoming.

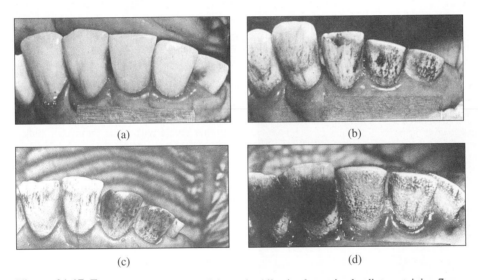

(a) (b)

(c) (d)

Figure 21.17 Effects of fluorine on teeth in cattle. All animals received a diet containing 7 ppm fluorine. In addition, animals b through d had, respectively, 30, 50, and 100 ppm fluorine added to their diets. Note the increasing discoloration of the teeth with increasing dietary fluorine levels in comparison with the teeth of the control cow (a).
Courtesy of C. S. Hobbs and G. M. Merriman, University of Tennessee, Agr. Expt. Sta. Bull. *351.*

animal development although natural deficiencies have not been reported. In addition to reduced growth, lowered reproductive performance and changes in hemoglobin concentration and liver function have been observed in experimental studies. University of Illinois researchers reported that a specific effect of nickel deficiency on urease activity in the rumen may explain the improved growth response of sheep and steers on low-protein diets that are given nickel supplements.

21.4 SUMMARY

A review of the principal roles, sources, and clinical deficiency symptoms of the macro- and microminerals has been presented. Research is rapidly accumulating regarding the significance of

minerals in animal nutrition. The need of an adequate and balanced supply to ensure good health and performance of farm animals is of special interest.[5] Certain elements are needed in small amounts but are toxic to animals in larger quantities.

Some essential mineral elements are stored in considerable quantities within the body during periods of adequate intake. These stores are drawn on later to meet requirements during periods of shortage. Other minerals must be supplied regularly because storage in the body is limited. Life cannot exist without mineral matter because many of the important body functions depend on mineral compounds.

[5]Information pertaining to the specific mineral needs of the various species of farm mammals and poultry is available in *Nutrient Requirements of Domestic Animals,* National Academy of Sciences–**National Research Council,** Washington, DC 20418.

STUDY QUESTIONS

1. Cite one or more examples of the interrelation among minerals.

2. What are the macrominerals? The microminerals?

3. Name a hormone and a vitamin that are closely related to calcium mobilization.

4. Why is the dietary Ca:P ratio important?

5. Are appreciable amounts of sodium, potassium, and chlorine stored in the body? Of what practical significance is this?

6. Which mineral compound is most often needed as a dietary supplement for most farm animals?

7. Give at least one major function, one or more clinical deficiency symptoms, and one or more dietary sources of each macromineral. Construct a table, and refer to Table 21.1 only as necessary.

8. Which two minerals are important in hemoglobin formation?

9. What is the relation of iodine and thyroxine?

10. Give at least one major function, one or more clinical deficiency symptom, and one or more dietary sources of each micromineral. Construct a table, and refer to Table 21.2 only as necessary.

11. Cobalt is an integral component of vitamin B_{12}. Cobalt is now believed to be associated with the synthesis of what other B vitamins by the rumen microorganisms?

12. Is a zinc deficiency likely to occur among farm animals under normal feeding conditions? Why?

13. Selenium can perform certain functions of which vitamin?

14. Cite two examples of minerals whose toxic effects may not be readily apparent at first, but rather only after extended periods of their consumption in low quantities.

15. What are the effects of feeding high levels of fluorine to cattle?

16. Where can one obtain more information pertaining to the mineral needs of farm animals? (See **NRC** in the Glossary.)

22

ANIMAL DISEASE AND THE HEALTH OF HUMANS[1]

He who has health, has hope; and he who has hope, has everything.

Arabian proverb

22.1 INTRODUCTION

The health, fitness, and ingenuity of people depend on an adequate diet of wholesome, nutritious foods. In the development of modern civilizations, humans have come to rely heavily on meat, milk, and eggs as major sources of dietary nutrients. In the United States animal products are the foundation of a diet that has helped build a strong nation. Animal health is perhaps the most significant factor in the production of wholesome meat, milk, and eggs. Moreover, for aesthetic reasons, pleasure, sports, and other purposes people frequently associate closely with animals (Chapters 4 and 5). Because many **diseases** of animals may be transmitted to humans (called **zoonoses**) the relationship between animal health and **public health** becomes readily apparent. The prevention, control, and elimination of animal diseases contribute important safeguards to the health of humans. Crowding increases the prevalence of animal disease by increasing the exposure to infectious agents (Figure 22.1). The same principle applies to humans. Thus as animal and human populations of the world increase it becomes imperative that students and the public become more informed about the relationship of animal diseases and human health.

Many **insects** and other **arthropods** are **parasitic** on humans and (farm) animals. Some **ectoparasites** visit the **host** only for a blood meal. Several **species** of mites invade the skin and cause **dermatitis** in humans, dogs, and other animals. Some mites are **vectors** of transmissible diseases. A large and important group of diseases are transmitted to humans by vectors (invertebrates) such as mosquitoes; these diseases include malaria and forms of **encephalitis.** Other rickettsial, bacterial, and viral diseases may be transmitted to humans by arthropod parasites, for example, rodent fleas transmit plague, and ticks transmit Rocky Mountain Spotted Fever, Lyme disease, and tularemia.

It is often said that the United States is the healthiest place in the world in which to raise **livestock.** However, according to United States Department of Agriculture (**USDA**) surveys there are still many animal losses. For example, probably 10 percent of all young **pigs** die between **farrowing** and weaning time, an estimated 10 to 15 percent of all calves and **lambs** die before marketing age, and the annual losses of chickens and turkeys due to diseases average 10 percent or more. Moreover, death losses are only part of the picture. **Morbidity** losses from diseases are often greater than those from **mortality.** For example, **bovine mastitis** seldom leads to death yet it is the most costly disease of dairy cattle in the United States.[2] **Intestinal** worms cause the death of relatively few animals but render thousands unthrifty and unprofitable. A recent governmental report estimated the total annual losses of livestock and their products to be $6.9 billion (about 10 percent of the total annual income from livestock). There is no way to accurately determine the losses caused by parasites or digestive disorders.[3]

Provisions for supplying people with safe animal products were discussed in Chapter 3. In this chapter animals (**domestic**

[1]The authors acknowledge with sincere appreciation the contributions to this chapter of Dr. K. L. Campbell, Professor of Veterinary Clinical Medicine, College of Veterinary Medicine, University of Illinois at Urbana–Champaign; and Dr. P. L. Nicoletti, Professor, Department of Pathobiology, College of Veterinary Medicine, University of Florida, Gainesville.

[2]J. R. Campbell and R. T. Marshall, *The Science of Providing Milk for Man*, McGraw-Hill, New York, 1975.
[3]An estimated 1.6 billion pounds of animal protein, which is needed immensely to improve human nutrition, are lost annually worldwide as a result of animal disease.

Figure 22.1 The spread of infectious diseases is enhanced by certain management practices, such as close confinement and crowding of animals (left). The photograph at right depicts an aborted fetus in a brucellosis-infected Florida dairy herd. The natural curious tendency of cows to lick a *Brucella*-laden fetus contributes to the spread of brucellosis.

Courtesy of Dr. P. L. Nicoletti, College of Veterinary Medicine, University of Florida, Gainesville.

and wild) that are sources of **infections** for humans and those that harbor (serve as **reservoirs**) **organisms pathogenic** to humans are discussed. Certain insect-borne diseases are included although most **life cycles** and other pertinent information related to insects and other arthropod vectors are presented in Chapter 23.

22.1.1 Food-Related Illnesses

More than 200 diseases are transmitted through food. Many of these are of animal origin. The pathogens include viruses, bacteria, parasites, and **toxins.** More than 90 percent of those caused by a microscopic germ are bacterial in origin. Symptoms of foodborne illness range from mild gastroenteritis to life-threatening neurologic, hepatic (liver), and renal (kidney) disorders or dysfunction.

According to the Centers for Disease Control (CDC) foodborne illnesses in the United States have been estimated to cause 6 to 81 million illnesses and up to 9000 deaths annually. These data may underestimate the actual incidence because many cases of human illness are not properly diagnosed or reported.

Some of the most important bacteria of animal origin that contaminate food are *Campylobacter* spp, *Salmonella* spp, *Escherichia coli* O157:H7, *Listeria monocytogenes,* and *Yersinia enterocolitica.*

Modern methods of food production and processing permit increased opportunities for food contamination. For example, chilling water vats in poultry carcass processing may become contaminated from a single source but can subsequently contaminate thousands of carcasses. Improper food preparation in the home, undercooking, or equipment contamination may lead to illness from a nonanimal source such as salads.

There are many possible ways to control and reduce foodborne illnesses. These include reduction of infections in animals, processing modifications, improved storage refrigeration, and education of those who prepare food. Techniques such as carcass decontamination and irradiation are being tested and approved (Chapter 3).

22.2 DISEASE AND HEALTH

However secure and well-regulated civilized life may become, bacteria, **protozoa, viruses,** and infected fleas, lice, ticks, mosquitoes, and bedbugs will always lurk in the shadows ready to pounce when neglect, poverty, famine, or war lets down the defenses. About the only genuine sporting proposition that remains is the war against these ferocious fellow creatures, which stalk us in the bodies of rats, mice, and all kinds of domestic animals; which waylay us in our food and drink and even in our love.

Hans Zinsser (1878–1940)
American bacteriologist

The World Health Organization (WHO) of the United Nations defines *health as a state of complete physical, mental, and social well-being, and not merely the absence of disease and infirmity.* In this chapter *health* refers to a state in which all parts of the body are functioning normally whereas *disease* refers to a disturbance in function or structure of any organ or body part. *Public health* refers to the health of human populations, especially on a community basis.

22.2.1 Types of Disease

There are two general types of disease. *Infectious* diseases are caused by pathogens or **germs** (i.e., disease-producing microorganisms: viruses, bacteria, rickettsia, and **fungi**). *Noninfectious* diseases may result from one or more of the following conditions: mechanical injuries such as flesh wounds; digestive disturbances resulting from **bloat, ingestion** of hardware, or dental failure; poisoning from chemicals (lead, arsenic, nitrates, **insecticides**) or intoxications of plant origin (black nightshade, hemlock, double-leaf stage of cocklebur, larkspur, lily of the valley, and others, see Section 22.6); nutritional deficiencies (lack of fat, proteins, vitamins, minerals, see Chapters 18, 20, and 21) or excesses; abnormal cell growth, whether malignant or nonmalignant; genetic disorders (e.g., **hip dysplasia**); or metabolic disorders such as diabetes mellitus or **acetonemia.**

22.2.2 Modes of Spreading Disease

There are a number of important means by which infectious diseases are transmitted. These include contact with diseased animals, polluted water, contaminated vehicles used to transport animals, **carrier** animals (e.g., asymptomatic or nonclinical cases), **carrion** feeders (e.g., dogs, foxes, or birds may carry bits of infected **carcasses** to clean farms), insects (especially flies and mosquitoes), ticks, air pathways, and contaminated facilities and handling equipment (**cattle** chutes, surgical equipment, **poultry** crates).

22.2.3 Modes of Pathogen Entry

Certain pathogens affecting people and animals are present throughout the environment. Some important means by which they gain entrance into the body are through the respiratory tract; the digestive tract; wound contamination; the mucous membranes of the eye, for example, pinkeye and leptospirosis (the latter may be acquired when the urine of an infected animal is introduced into the eye); the **genital** tract (especially during mating or **parturition**); the teat canal (especially during **lactation**); the navel cord (in the **neonate**); contaminated instruments (syringes and/or surgical); and insect bites.

22.2.4 Body Defenses against Disease

Fortunately nature provided mechanisms whereby animal and human life are sustained in the presence of pathogens. These defensive mechanisms include the skin and mucous membranes as the first line of defense (certain body secretions—e.g., tears—contain **lysozyme**,[4] which has **antiseptic** properties); and the digestive tract (gastric juices and stomach acids depress bacterial growth), tissue fluids (**lymph** contains **leukocytes** such as **macrophages,** which destroy many pathogens by engulfing them), and lymph nodes and liver, which mechanically trap pathogens until macrophages can destroy them. The defensive mechanisms also include the reactive defenses. The latter include *inflammatory reactions* characterized by the following five cardinal signs: (1) increased blood supply (redness, Latin *rubor*), (2) increased temperature of the part (heat, Latin *calor*), (3) swelling of the part (**edema,** Latin *tumor*), (4) increased sensitivity (tenderness or pain, Latin *dolor*), and (5) loss of function (Latin *functico laesa*). The reactive defenses also include the **febrile** *reaction* characterized by an increase in body temperature (caused by effects of toxins of microorganisms on the heat-regulating mechanism of the hypothalamus) and an increase in **metabolic activity,** and *immune reaction* (such as development of **antibodies**).

Inflammation

A type of body reaction to injury (mechanical, chemical, or infectious) is called **inflammation.** It is an attempt by body cells to destroy injurious agents. Inflammation begins with an accumulation of fluid and white blood cells around the area of injury or infection. There is often an accumulation of **exudate (pus),** which consists of fluid containing dead cells, fibrin, leukocytes, and the causative organism.

22.2.5 The Resistance of Animals and People to Pathogens

> Complete and lasting freedom from disease is but a dream remembered from imaginings of a Garden of Eden.
> **René J. Dubos (1901–1982)**

Immunity

Immunity is the ability to resist and/or overcome an infection. Following exposure to an infectious disease or artificial immunization an animal develops an increased resistance to the disease. This results largely from the production of antibodies that aid in the defense of the host by reacting or uniting with the causative microorganism or its toxin. Production of antibodies is stimulated by a variety of substances known collectively as *antigens.* An **antigen** (from the Greek *anti* meaning "against," and *geneo* meaning "produce") is a substance that stimulates the formation of specific antibodies when it is introduced into an animal. Antigens are usually proteins, foreign to the animal body, soluble in body fluids (either *in vivo* or *in vitro*), and react or unite with a specific antibody. Each pure antigen is specific and stimulates the production of antibodies against itself but usually not against other antigens.

Introduction of an antigen into an animal stimulates the production of immune substances called *antibodies,* which may be demonstrated in both the tissues and blood of the recipient 7 to 14 days postinoculation. Antibodies may be obtained from the blood **serum** of hyperimmunized animals or humans, known as *immune serum* or **antiserum.** Antibodies are quite antigen-specific; that is, they often react with only one antigen. Serum antibodies are proteins known as serum globulins. It is chiefly the *gamma* globulin that is increased in immune animals and humans.

Natural (Inherent) Resistance

Natural (inherent) resistance is the normal or innate resistance of animals to infection. This includes both *mechanical* (such as the skin) and *physiological* barriers to aid the host in resisting microorganisms. The latter include the unfavorable acidity of certain body secretions and unfavorable body temperature. Moreover, blood is **bactericidal** and may contain specific antibodies and other substances. A few natural antibodies are inherited; for example, those responsible for the human blood groups. *Species immunity* is the natural resistance of a given animal species to microorganisms and/or parasites that may infect another species. An example is the malarial parasite, which produces disease primarily in humans (also in certain birds and primates). People are naturally immune to **canine** distemper and certain other animal diseases whereas animals such

---

[4]Lysozyme has an interesting history. In 1922 Alexander Fleming discovered a substance in his own nasal mucus capable of dissolving, or *lysing,* certain bacteria. The substance was identified as an enzyme and named *lysozyme.* Fleming believed that some organisms produced antibacterial substances and he went on to discover penicillin, the first true antibiotic.

as cats, cattle, and horses possess natural resistance or immunity to measles and other diseases of humans.

Acquired Resistance or Immunity

Acquired resistance or immunity is that gained by having had the disease (actively acquired immunity, e.g., swinepox) or through artificial immunization (actively or passively acquired immunity). Active immunity can be induced by receiving a **vaccine** (e.g., tetanus toxoid) that causes the production of antibodies (actively acquired artificial immunity). Passive immunity can be acquired by being injected with immune serum (e.g., tetanus **antitoxin**) or serum containing antibodies produced in another host (passively acquired artificial immunity). Passive immunity may also be acquired by ingestion of colostrum.

Active acquired immunity is produced in response to the entrance of a specific antigen into the host. Following recovery from an infectious disease the individual is commonly resistant to that disease for varying lengths of time. For example, a dog that recovers from clinical distemper is immune to subsequent infection. Recovery from swinepox virus infections is followed by lifelong immunity whereas recovery from contagious viral pustular dermatitis results in an immunity lasting only 1 to 3 years.

Active acquired artificial immunity results from the injection of an immunologically active form of the infectious, or specifically related, agent (vaccine). There are several types of vaccines: suspensions of killed bacteria (**bacterin**) such as blackleg and typhoid vaccines; microorganisms that are **attenuated** or of reduced **virulence** (modified live vaccines) such as the virus of canine distemper; the products of bacterial growth such as the toxins of tetanus; and polyvalent vaccines such as the triple vaccine of typhoid, paratyphoid A, and paratyphoid B bacteria. Use of the latter results in the concurrent production of antibodies against antigens of all three organisms. Live bacteria and viruses usually give stronger and more lasting immunity

than killed **cultures.** The efficacy of vaccines varies from rather poor to near complete protection.

Passive acquired immunity occurs in the neonate as a result of antibody transfer from the mother to the offspring through either the placenta, as in humans, or the **colostral** milk, as in cattle, horses, sheep, and swine (Section 15.12). Calves are born without immunity from the dam and it is important that they receive antibodies through intestinal absorption of colostrum.

Passive acquired artificial immunity is possible through the injection of *immune serums* (antibacterial or antiviral serums or antitoxins) but the resistance lasts only a short while. In passive immunity, antibodies produced in one animal are transferred in serum to a recipient animal or person that does not participate in production of the antibodies. Antitoxins against tetanus result in passive acquired artificial immunity. Injection of antiserum affords immediate increased resistance (passive immunity). The major differences between active and passive immunity are presented in Table 22.1.

22.2.6 Recent Application of Technology to Protection from Disease

Viral Vaccine Controls Cancer

The concept of protecting one animal with a viral vaccine from another originated with Edward Jenner in 1798 when he observed that milkmaids survived **epidemics** of smallpox. He demonstrated that the cowpox virus produced little or no disease in humans yet protected humans from smallpox. This concept served as the basis for development of a commercially produced, federally licensed (fowlpox) vaccine for Marek's disease, a form of cancer in chickens, the first vaccine to prevent cancer of any farm animal.

Before the Marek's vaccine was developed almost all chicken flocks were affected and in some instances over half the flock died. The cancer can now be prevented by a highly effective vaccine, which is administered to most chickens in developed countries.

Monoclonal Antibodies

Poultry producers lose millions annually to a disease called coccidiosis because affected birds fail to grow as they should. Additionally they spend millions on medicine to control the disease. An effective vaccine against coccidiosis would reduce costs of poultry production resulting in more profits to producers and lower costs to consumers.

Vaccines in general could greatly reduce the billions of dollars spent annually for drugs and medication to control a number of human and animal diseases. It is important that the vaccines contain specific antigens, which will cause production of antibodies against these antigens.

New technologies have permitted the identification of specific antigens which may be inserted in other organisms such as poxvirus and which will result in protection against the original pathogenic organism. An example is the oral vaccine used for control of rabies in wildlife (see Section 22.3.1). Many future vaccines will be developed using genetic engineering techniques.

TABLE 22.1	Major Differences between Active and Passive Immunity	
	Active immunity	**Passive immunity**
Body participation	Forms own antibodies	Recipient of antibodies produced in another animal
Material introduced into body	Antigens	Antibodies
How immunity is acquired	Naturally, in response to disease or **subclinical** infection	Naturally, by placental transfer of antibodies and/or from colostral milk
	Artificial exposure by inoculation with vaccines (immunity is produced naturally)	Artificially, by injection of antiserum (antitoxin; antibacterial serum)
Duration of immunity	Months and years	Relatively short (often only a few weeks)
Usefulness	Primarily **prophylactic** (preventive)	Primarily **therapeutic**; temporarily prophylactic

Another use of modern technology to produce antibodies is in hybridoma cells that produce specific antibodies. Hybridomas are made in the laboratory by fusing an antibody-producing cell from the spleen of a mouse with a cell from a mouse tumor (myeloma), which will multiply fast and indefinitely outside the mouse's body in tissue culture but will not grow in any other species of animal. Although the spleen manufactures a wide range of antibodies individual spleen cells produce only one specific antibody. Thus in a hybridoma the spleen cell provides the design of the product and the tumor cell provides for mass production of the product, which is called a **monoclonal antibody.** Because of their purity these antibodies are especially useful in the diagnosis of diseases.

Genetic Resistance to Disease

Many noninfectious diseases result from hereditary weaknesses. For example, it has been shown that in poultry and laboratory animals some inbred lines are more susceptible than others to certain nutritional diseases. Differences among inbred lines are largely genetic. Additionally many hereditary metabolic and nervous disorders have been reported in animals and humans. Genetic resistance to disease, however, usually refers to resistance to infectious diseases. **Zebu** cattle are more resistant to certain infectious diseases and parasites than are British breeds.

Genetic resistance to disease may result from the ability of an individual to prevent the entrance of disease organisms or from the ability to destroy these organisms effectively if they gain entrance into the body. Failure to combat organisms properly after they have entered the body is illustrated in humans by the genetic defect known as agammaglobulinemia. This defect results from failure of the body to produce gamma globulins (antibodies) in normal quantities. In addition some animals are more susceptible to diseases because of reduced resistance as a result of exposure to certain immunosuppressive organisms such as BVD (bovine virus diarrhea). A synergism develops among pathogens resulting in a more severe disease than occurs with one pathogen. Another example occurs when humans are infected with HIV (human immunodeficiency virus), which reduces resistance to many other diseases.

22.2.7 Antigen–Antibody Reactions

The most common antigen–antibody reaction is a highly specific chemical union in which **molecules** are held together by strong attracting forces. The antibody molecules are large, complex proteins with receptor sites on their surfaces that unite chemically with reciprocal reactive sites on the antigen. The building of a large mass, or latticework, of these **inter**locking antigen and antibody units results in **agglutination** or clumping (together) of cells containing the antigen. Agglutination tests are used in blood-typing and in diagnosing such diseases as typhoid fever, tularemia, and brucellosis.

Precipitation Reaction

A precipitin reaction results from the precipitation of a **soluble** antigen with its specific antibody (antiserum). The precipitin reaction is used in the laboratory identification of different species of meats or in the identification of bloodstains in medicolegal situations. The *precipitation reactions* are similar to *agglutination reactions* except they involve precipitation of molecules from a solution rather than the clumping together of cells and are observed in a gel. Examples are tests for equine infectious anemia and bovine leukosis.

Complement Fixation Reaction

A complement fixation reaction is based on the principle that complement, which is bound in one reaction, cannot react in a second (Figure 22.2). In the complement fixation test, complement is allowed to react with one antigen–antibody system and then a second indicator system (sensitized red blood cells) is added. If **hemolysis** of the red blood cells occurs complement has not been fixed by the first antigen–antibody mixture and the test is negative. Conversely, in a positive test complement is fixed in the first reaction and therefore is unavailable to produce hemolysis of the red blood cells. The Wassermann test for syphilis is a widely known complement fixation reaction.

Toxin–Antitoxin Reaction

A toxin–antitoxin reaction helps explain the protective action of immune serum against the toxins of diphtheria and tetanus. Antitoxins (from the Greek *anti* meaning "against," and *toxin* meaning "poison") like other antibodies are found in the blood serum of immune animals and react specifically with antigen (toxin). The protective effect of antitoxin results from the direct neutralization of toxin.

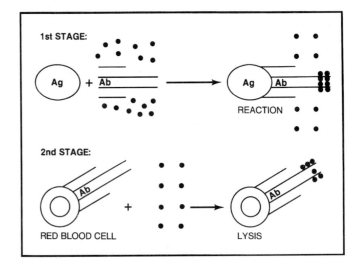

Figure 22.2 Principles of complement fixation. In the first stage, antigen (Ag) and antibody (Ab) react in the presence of complement (•). The interaction of antigen and antibody fixes some, but not all, of the complement available. In the second stage, the residual or unfixed complement is measured by adding antibody-sensitized red blood cells, which are lysed by residual complement. Thus, a reciprocal relation exists between the amounts of lysis in the second stage and the amount of antigen present in the first stage.

22.2.8 Testing for Disease

The veterinarian has several means of diagnosing diseases. These include physical examination, antigen–antibody reactions of the blood (Section 22.2.7), skin testing with antigen (e.g., tuberculosis of cattle), microscopic examination of blood (used to identify the type of infectious organism and also to make red and white blood cell counts), skin scrapings (especially useful for identifying fungus or parasitic mite infections), bacterial and/or fungal cultures, microscopic and/or visual **fecal** examination (primarily for parasites), chemical tests of blood and urine, body temperature records, **biopsy,** and/or **necropsy.**

22.3 SELECTED ANIMAL DISEASES TRANSMISSIBLE TO HUMANS

> Unhealthy farmers are poor producers. In many parts of the world ill health due to infection and ill health due to bad diet are inextricably interconnected. Acute protein **malnutrition** is so often precipitated by infectious diseases that there is a case for saying that an attack on these diseases should be the first step toward preventing it. Certainly public health measures against infectious diseases and measures to produce more protein should be undertaken simultaneously.
>
> **W. R. Aykroyd (1899–1979)**
> ***The Conquest of Famine,* 1975**

22.3.1 Viral Infections

These infections are caused by a group of minute agents called *viruses.* A virus is characterized by a lack of independent metabolism and can therefore replicate only within living cells. The individual viral particles consist in part of either **DNA** or **RNA** (but not both).

Cells can be grown or cultivated under artificial conditions, which is called *tissue culture.* Such cultures can be used to support viral growth in the production of vaccines. These cultures are also used in diagnostic tests in which viruses produce cellular changes *in vitro.*

Rabies (hydrophobia)

This is one of the oldest and most feared diseases known to humans. It was described in the fourth century B.C. by Aristotle who wrote, "Dogs suffer from a madness which puts them in a state of fury, and all of the animals that they bite when in this condition become also attacked by rabies." *Rabies* is derived from the Latin word *rabere* meaning "rave" or "fury." It probably received its name because infected animals often become excited and attack nearby objects or animals.

All **mammals** are believed to be susceptible to rabies, which may be transmitted to humans through the saliva of a rabid animal (via a bite or skin laceration). The most common vectors of rabies are the **carnivorous** or biting animals such as cats, coyotes, dogs, foxes, raccoons, skunks, and wolves. Rabies in the United States has become largely a disease of wildlife including bats (Figures 22.3 and 22.4). Wild animals accounted for over 91 percent of the 7067 reported cases in 1999, only 8.5 percent (603) were domestic animals. There were five human cases in 2000. Bats are important vectors in the United States and certain other countries and the source of most human infections in the United States. The incidence of rabies in dogs in the United States has decreased from 5688 of the total 8837 cases (64 percent) in 1953 to 111 of 7067 cases (1.6 percent) in 1999. In 1999 the leading animal species was the raccoon (2872), followed by skunks (2072), bats (989), foxes (384), and cats (278). Since 1981 rabid cats have outnumbered rabid dogs in the United States because of the large number of unvaccinated cats. Outdoor free-roaming cats are at highest

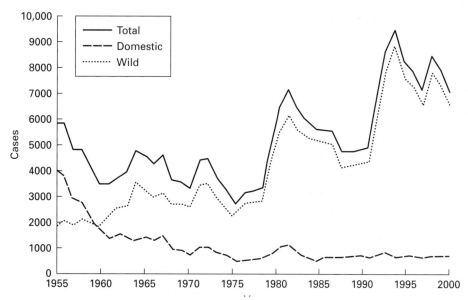

Figure 22.3 Cases of animal rabies in the United States, by year, 1955 through 2000.
Courtesy of Centers for Disease Control, Atlanta, GA.

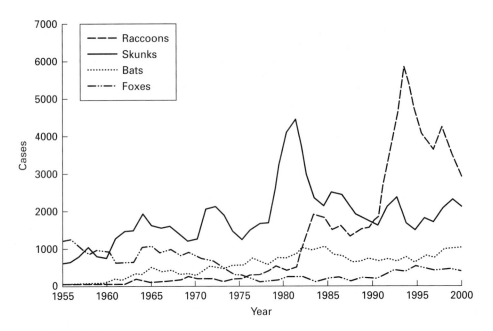

Figure 22.4 Cases of rabies in wild animals in the United States, by year and species, 1955 to 2000.
Courtesy of Centers for Disease Control, Atlanta, GA.

risk; however, indoor cats may acquire rabies from contact with rabid bats getting into houses. Rabies has been rather effectively controlled in dogs through animal control and very effective vaccines, which are required in many areas.

The causative virus multiplies and causes degenerative changes in brain tissue, which results in convulsions, excessive salivation, madness, paralysis, and finally death of the (infected) animal.

The most widely used vaccine for humans in the United States is developed from growth of virus in human diploid cells (HDCV). In 1997 the Food and Drug Administration (FDA) approved another vaccine for humans, which is grown in chicken embryo cells. Both vaccines are inactivated and used for both pre- and postexposure vaccination.

Severed chicken heads laced with rabies vaccine, developed by zoologist Alexander Wandeler and his co-workers at the University of Bern in Switzerland, have been scattered as bait for foxes throughout valleys in Switzerland and other European countries in a highly successful campaign to impede the spread of rabies. The effort marks one of the first times that an attempt has been made to immunize wild animals. An estimated 60 percent of the foxes in the test areas ate the bait. Similar wildlife vaccination projects are under way in Texas (targeting coyotes) and selected areas in many east and northeastern states (targeting raccoons). More recently special baits such as fish meal have been used instead of chicken heads. The vaccine is a live product and genetically engineered using vaccinia (pox) virus as a vector.

The average incubation period of rabies in humans is from 30 to 60 days. (The range is 10 days to more than 8 months.) Humans infected with rabies virus commonly develop a hydrophobia, or the "fear of water," phenomenon. This results from painful spasms of the **pharyngeal** muscles caused by the act of drinking. The condition causes the victim to avoid swallowing saliva, which then drools from the mouth. Rabies almost always terminates in the death of infected animals. Identification of the disease may be made by examination of histologic sections of brain (cerebellum) tissue of the suspected animal for the presence of the rabies virus on the basis of a specific antigen–antibody reaction using direct fluorescent microscopy (a fluorescent compound is linked to the rabies antibody which can then be seen with a special microscope).

Rabies is an infection which requires broken skin for entrance of the virus. If a person who has been bitten has the wound properly treated and quickly accepts treatment he or she rarely succumbs to rabies. The first rabies treatment, a brain-derived killed virus developed by Louis Pasteur in 1883 requiring a large number of injections into abdominal muscles, is no longer used in the United States and rarely in other countries. Treatment today is three or five intramuscular injections of a very effective, inactivated vaccine. People working in a high-risk occupation should be **immunized** against rabies before exposure. Estimates are that not more than 15 to 20 percent of the persons bitten by rabid animals develop the disease even when no antirabies treatment is given. However, because of the high mortality associated with rabies a person exposed to the disease should be immediately placed under the care of a physician.

A current common control in dogs and cats is the injection of inactivated rabies vaccine at 3 and 12 months of age followed by booster vaccinations every 1 to 3 years depending on the vaccine type and local rabies control regulations. Vaccines to protect horses and cattle are not routinely used; however, they are available for selected situations.

TABLE 22.2	Selected Arthropod-Borne Diseases Transmitted from Animals to Humans*		
Name of disease	Vector or means of transmission	Animal affected	Animal hosts (reservoirs)
Western encephalitis	Mosquitoes to birds to mosquitoes	Mosquitoes to humans or horses (terminal hosts)	Birds
Eastern encephalitis	Mosquitoes to birds to mosquitoes	Mosquitoes to humans or horses (terminal hosts)	Birds
St. Louis encephalitis	Mosquitoes	Chiefly humans	Fowl, wild birds
Venezuelan encephalitis	Mosquitoes	Horses and other equines, humans, fowl	Unknown
Western Nile virus	Mosquitoes	Humans, horses, birds	Birds

*More than 200 arthropod-borne viruses (**arboviruses**) have been identified and associated with an animal host. More than one-third of these cause a febrile or more severe disease in humans.

Newcastle Disease

This is a **contagious** and often fatal viral infection of poultry and many wild birds, which is characterized by neurologic and respiratory disturbances. Newcastle disease represents a mild potential hazard to persons exposed to infected carcasses and to those involved in the production of vaccines against it. The disease is characterized in humans by a severe conjunctivitis (inflammation of the membranes lining the eyelids) and a syndrome of fever, chills, headache, and general malaise (weakness and/or discomfort). **Killed virus** vaccines and those of a modified live or attenuated virus of avian embryo origin are capable of stimulating immunity in poultry and wild birds. One form of the disease (velogenic) is considered an exotic disease in the United States and all exposed birds are slaughtered by order of the USDA.

Encephalitis

This viral disease found in humans and horses appears primarily in two forms, a Western form (WE) and a more severe Eastern form (EE). Its principal reservoir is wild birds. The disease is communicable to humans (only from a bite by the vector mosquitoes) in whom an inflammation of the brain and spinal cord develops. It is characterized by fever, incoordination, and in advanced stages convulsions and coma. The St. Louis encephalitis strain is found chiefly in humans and has no apparent important animal reservoir except mosquitoes and wild birds. The virus is transmitted by the *Culex pipiens* and *C. tarsalis* mosquitoes. Selected arthropod-borne diseases transmissible from animals to humans are presented in Table 22.2.

West Nile Virus

An outbreak of arboviral encephalitis was recognized in human patients in New York City in late August 1999. It was initially attributed to St. Louis encephalitis on the basis of positive serologic findings in cerebrospinal fluid. Using a virus-specific IgM ELISA, the cause was confirmed as a West Nile–like virus. The patients had muscle weakness and encephalitis. Aerial and ground applications of mosquito adulticides were instituted.

Before and concurrent with the outbreak in humans local health officials and Bronx Zoo workers noted increased fatalities among several avian species, especially crows. When tests were negative for Eastern and Western encephalitis viruses further tests at the CDC on viral isolates from birds and humans indicated they were closely related to West Nile virus, which had never previously been reported in the Western Hemisphere.

In 2000 there were 52 confirmed cases in humans including 9 deaths. There were 60 equine cases in 2000 of which 23 (38 percent) died or were euthanized. The majority of cases were in the Eastern United States; however, the disease is spreading westward. Horse-to-horse transmission has not been established.

Surveillance in birds and pools of mosquitoes found many species of birds and *Aedes* and *Culex* mosquitoes positive. The virus source is suspected to have originated from migratory birds, possibly from Israel.

Bluetongue

Although not transmissible to humans this viral disease is transmitted by insects (species of *Culicoides,* biting gnats or midges) and affects sheep, cattle, goats, and wild ruminants. The clinical signs are most severe in sheep. Symptoms may include a swollen and cyanotic ("blue") tongue, oral ulcers, lameness, wool loss, pneumonia, and deformed offspring. Cattle and wild ruminants are reservoirs and infected animals do not always show symptoms. Bluetongue disease can be transmitted through fresh or frozen bovine semen. Only 5 of the 24 known **serotypes** of the virus have been isolated in the United States. There is no effective treatment for this difficult-to-control disease. A vaccine is available in the United States but protects only against one of the strains. The major impact of bluetongue on the U.S. cattle industry has been its constraint on exports of breeding animals. The ban on bovine semen exports from the United States to the United Kingdom, Australia, and New Zealand has resulted in an estimated annual loss of approximately $24 million. Ruminants imported from countries with bluetongue must be tested and found negative prior to entry into breeding units. Hopefully these measures will prevent the dissemination of the bluetongue virus.

The Pox Diseases of Humans and Animals

Human **ecology** and animal ecology have developed in a curious contrast to one another. Human ecology has been concerned almost entirely with . . . the effects of man upon man, disregarding often enough the other animals amongst which we live.

Charles Elton

Smallpox (Variola)   Smallpox (variola) is caused by a virus and is usually transmitted from person to person. It may be transmitted by inoculation into several animal species and thereby lose much of its virulence for humans. When carried back to humans a mild disease known as *vaccinia* confers immunity to smallpox. Until a satisfactory prophylactic vaccination for smallpox was developed the disease was feared by all.

A disease known as *cowpox* is characterized by lesions on the teats and skin of the udders of cows. The viruses of smallpox and cowpox are immunologically similar. In 1798 Jenner reported that persons who milked cows infected with cowpox (vaccinia) had resistance to smallpox. Smallpox vaccine was commonly produced by propagating the virus on the scarified skins of **calves.** The resulting vaccine virus was proven to be effective in the prophylaxis of smallpox in people for more than a century. The word *vaccination* is derived from the Latin *vacca* meaning "cow." (The term originally referred only to the use of cowpox virus to prevent smallpox.)

On December 9, 1979, the WHO's Global Commission for the Certification of Smallpox Eradication declared that smallpox eradication had been achieved throughout the world. This has resulted in discontinuation of the practice of routine smallpox vaccination by all countries.

Foot-and-Mouth Disease (FMD)   Loffler and Frosch found in 1899 that this highly infectious disease of cattle and other **cloven-footed** animals is caused by a filterable virus. It is only one-millionth (0.000001) of an inch in diameter and is believed to be the smallest of viruses which affect humans or animals. Cattle and swine are more susceptible than sheep and goats. Wild ruminants (antelope, buffalo, camels, deer, llamas) are also susceptible. Rarely humans may become infected with a mild form of the disease (mouth lesions only). The disease is transmissible from infected animals to people, presumably by contact. The virus can be carried in meat or milk products, by birds or other animals, on vehicles, on the clothing or shoes of humans, and through the air. The virus is concentrated in lymph nodes, blood, and bone marrow. Studies have shown that the FMD virus can survive for about 3 months on boots that become contaminated on infected premises, on hay for about 4 months, on hair for a month, and on refrigerated carcasses for up to 4 months. The cool, damp climate of Britain provides conditions in which it flourishes. Spreading like a plague, foot-and-mouth disease caused 2,471,000 animals (1,927,000 sheep, 431,000 cattle, 112,000 hogs, and 1000 goats) to be killed (then burned or buried) in England and Wales during the first 5 months of 2001. Direct costs attributable to eradication were estimated at $820 million. Additionally an estimated $6 billion were lost to tourism by the end of summer 2001. Recent outbreaks of FMD caused substantial losses in the Netherlands, South America, and other parts of the world.

In countries where FMD is endemic it is controlled by various methods including slaughter of all susceptible animals on affected farms, control of animal movements, and by vaccination using killed vaccines, which have a short duration of effectiveness. There are 7 major viral types and over 50 subtypes of the virus and vaccines must contain the specific viral type the animal is exposed to for effectiveness.

Many countries including the United States are free of FMD. Stringent measures are taken to prevent its entry such as forbidding importation of fresh or frozen meat that has not been heated. Animals may be imported but conditions are very strict. Because of the economic impact of these barriers and the losses due to the disease many countries have very active programs to eliminate the virus. FMD is probably the most important economic disease of animals in the world because of the direct loss of animals and the establishment of commerce barriers.

In order to be considered free of foot-and-mouth disease, vaccination must cease for several years because some vaccinated animals may show no signs of infection but shed virus to other susceptible animals.

Vesicular Stomatitis (VS)   This disease is caused by a virus. In 1997 the USDA reported VS occurred on 380 premises in the southwestern United States; 273 of these in Colorado and the remainder in Arizona, New Mexico, and Utah. Approximately 90 percent of the cases were in horses. It causes blisterlike lesions (vesicles) in cattle, horses, sheep, swine, and humans. The clinical symptoms of VS closely resemble those of foot-and-mouth disease but VS is usually short-lived and mortality rates are low. VS generally occurs at 10- to 15-year intervals and is most commonly diagnosed in animals that have been near low-lying marshes, swamps, and similar areas following periods of heavy rainfall and high humidity. These conditions favor increased populations of mosquitoes and gnats that may spread the disease.

Humans affected by the virus commonly have blisters on the lips, tongue, and foot and have flulike symptoms in the respiratory tract. The disease is rare in humans. Vaccines may be used to control VS. These are used primarily during outbreaks.

Contagious Ecthyma of Goats and Sheep   More commonly known as *sore mouth, scabby mouth,* or *orf* this is a highly communicable disease of goats and sheep caused by a filterable virus. In humans it causes vesicles on the hands and face where the skin is scratched but the symptoms are usually mild and often last about 3 weeks. Lambs and **kids** may be protected by **vaccination.**

22.3.2 Rickettsial and Chlamydial Infections

Soldiers have rarely won wars. They more often mop up after the barrage of epidemics. Typhus with its brothers and sisters—plague, cholera, typhoid, dysentery—has decided more campaigns than Caesar, Hannibal, Napoleon and all the inspector generals of history. The epidemics get the blame for defeat, the generals get the credit for victory.

Hans Zinsser (1878–1940)
American bacteriologist

First described in 1909 by Ricketts, the organisms that cause rickettsial diseases are neither true bacteria nor filterable viruses. **Rickettsiae** are generally considered to cause animal diseases that are transmitted from animal to animal (including humans) by arthropods, especially ticks. Rickettsiae are parasites of arthropods (commonly **intracellular**), and humans and animals become infected when infected arthropods feed on them or when a wound is contaminated with the **feces** of an infected arthropod.

Rocky Mountain Spotted Fever (RMSF)

In 1909 Ricketts discovered that this rickettsial disease is transmitted to humans by two or more tick species, which included the Rocky Mountain wood tick (*Dermacentor andersoni*) and the American dog tick (*D. variabilis*). The disease is most common among persons whose activities expose them to tick bites. When in areas infested with ticks one should check frequently for tick presence (the tick must be attached for 5 to 20 hours to transmit the organism), even if insect repellents are used.

The two ticks indicated above are the most common vectors (the brown dog tick *Rhipicephalus sanguineus* can also transmit the disease) of this disease. They are "three-host" ticks and spend the periods between feedings away from the hosts. The larval and nymphal stages are found on smaller animals (especially **rodents**) whereas the adults prefer larger animals (both wild and domestic). It is the adult tick that transmits the infection to humans (see also Section 23.5.3 and Section 22.3.3, Lyme disease). (The two immature stages are rarely found on humans.) Dogs may carry ticks into the house and should therefore be inspected frequently. The CDC reported 579 cases of RMSF in 1999.

Q Fever

First recognized in Australia in 1935 this disease is classified as a rickettsial disease although the mode of infection to humans differs from that of other infections in this group because arthropod transmission may not be involved. The causative organism of Q fever is *Coxiella burnetii* (originally named *Rickettsia burnetii*). Ticks are the most important vector. People may acquire Q fever through the inhalation of contaminated dust (containing tick feces). However, most people become infected through exposure to livestock, especially sheep, or rarely through the ingestion of raw milk. The disease has been identified in cattle in some 35 states within the United States and therefore is recognized as **endemic**.

Preferred prophylactic measures are avoidance of ticks, care in aiding animals during parturition, and proper pasteurization of milk. The armed forces have studied this disease from the standpoint of biological warfare.

Psittacosis (Ornithosis)

This chlamydial[5] infection, found in many species of birds, may be transmitted to humans. Psittacosis is most common in parakeets, pigeons, parrots, and other related pet birds. However, other species of domestic poultry (turkeys) and nonpsittacine wild birds may become infected. In the latter species the infection is generally termed *ornithosis*.

The disease may be transmitted to humans by inhalation of the organisms from airborne droplets of nasal secretions or contaminated dust, feathers, and urine. In people the disease is manifested by influenza-like symptoms 7 to 15 days after contact with infected birds. There were 16 human cases reported in 1999 but it is believed that the incidence is higher because many cases are not diagnosed.

22.3.3 Bacterial Infections

The discussion of infectious diseases of bacterial origin will be confined to the more common ones affecting the health of humans and animals.

Tuberculosis

The ease of transmissibility of pathogenic species of the genus *Mycobacterium* between animals and humans makes it significant to public health. There are three common species of tubercle bacilli responsible for tuberculosis in **homeotherms.** *Mycobacterium bovis* (cause of bovine tuberculosis) is capable of infecting several species including dogs and humans. Similarly, *M. tuberculosis* (human types of tubercle bacilli) may be transmitted from humans to caged (or those in zoos) primates and dogs. *Mycobacterium avium* (usual avian tuberculosis organism) may also infect swine.

Tuberculosis has been observed widely in most homeotherms and in such **poikilotherms** as alligators, fish, frogs, snakes, and turtles. Hence it is doubtful that any species of animal has an absolute resistance to tuberculosis. Fortunately the occurrence of bovine tuberculosis in the United States has been greatly reduced through the joint cooperative efforts of farmers, veterinarians, and government agencies to eliminate tuberculin-reacting cows and, in some cases, the entire herd. The tuberculosis eradication program in cattle was initiated in the United States in 1917 when about 5 percent of the cattle were infected. By 1936 over 3.5 million infected cattle had been slaughtered. Today less than 0.01 percent of cattle in the United States are tuberculin-positive.

Methods of Controlling Infection from Cattle to People
These methods include the maintenance of healthy cows through testing programs and sanitation and the pasteurization of milk (Chapter 3). The USDA meat-inspection service is the primary surveillance method of locating herds that may be infected. Federal legislation requires inspection of meat sold both intrastate and interstate. In countries having a high occurrence of tuberculosis the use of a vaccine to immunize humans against the disease is sometimes practiced. However, in the United States where human infection of bovine tubercle bacillus is rare (most human cases are from other humans, usually direct transmission by inhalation) and the disease has been virtually eliminated in cattle,

---

[5]Members of the genus *Chlamydia* are agents having a cell structure intermediate between that of viruses and bacteria.

and because vaccination renders the diagnostic test with tuberculin invalid, vaccination in humans is not advisable.

Tuberculosis in Mammals other than Cattle Dogs are resistant to *M. avium* but susceptible to *M. bovis* and *M. tuberculosis*. Most tuberculosis of swine is of the avian type.

The federal meat inspection records indicate that about 15 percent of the swine slaughtered 60 years ago had tuberculosis lesions (primarily found in lymph nodes of the head, neck, and **mesentery**) whereas today less than 1 percent have the characteristic lesions. This striking reduction is attributed largely to increased confinement of poultry and restriction of their freedom to roam farms and hog lots where they deposited large amounts of fecal material that possibly contained tubercle bacilli. Requirements that garbage be cooked have aided in reducing the prevalence of swine tuberculosis.

Monkeys are popular mammals at the zoo, both for children and the organisms causing tuberculosis. Monkeys rarely contract tuberculosis in the wild; however, when they are exposed to infected humans or other **primates** in captivity their susceptibility to the disease is greatly enhanced and presents a hazard to public health.

Chickens are very susceptible to *M. avium* but through confined housing and sanitary precautions its prevalence has been greatly reduced in recent years. The relatively short time that poultry are now kept on farms has also helped to reduce the occurrence of avian tuberculosis in the United States. However, tubercle bacilli may remain **viable** and pathogenic in chicken litter and/or infected soil for several years. Humans are relatively resistant to this pathogen but *M. avium* may be transmitted to them. Badgers are known carriers of tuberculosis in England and therefore many are killed. More recently feral deer have been found to be infected in Michigan. Several cattle herds have been infected. The presence of the bovine type of tuberculosis in wild animals presents a difficult challenge in efforts to eradicate this disease.

Brucellosis

This bacterial infection derives its name from a British Army surgeon, Sir David Bruce. In 1887 he discovered the bacteria that later was named *Brucella melitensis*. It is also called "Bang's disease" in cattle (after a Danish veterinarian, Bernard Bang, who isolated *B. abortus* in 1897). Bang established that the organism commonly causes abortion in cows.

Brucellosis is found throughout many countries of the world. There are three species of the genus *Brucella* of major public health significance: *B. abortus* (of bovine, or cattle, origin), *B. melitensis* (of **caprine** [goat] or sheep origin), and *B. suis* (of **porcine,** or swine, origin). The latter was isolated in 1914. The correlation of Malta fever in humans with infected goats and of undulant fever in humans with infected cattle was not proven until early in the twentieth century.[6] The *Brucellae* have a variety of hosts including cattle, goats, horses, humans, sheep, and swine. They have also been isolated from reindeer,

caribou, bison, elk, camels, and yak. Hence there is a large potential animal reservoir of infection to people. The disease has been experimentally transmitted to chickens, guinea pigs, hamsters, mice, monkeys, rabbits, and rats. Infected caribou or reindeer sometimes transmit *Brucella* infection to humans.

Occurrence of Brucellosis Data obtained by the Animal and Plant Health Inspection Service (APHIS) of the USDA indicate that the incidence of brucellosis in cattle (based on blood agglutination tests) in the United States decreased from 11.5 percent in 1935 to less than 0.01 percent in 1999. Much of this decrease is attributable to surveillance programs of the USDA (such as the brucellosis ring test and market cattle testing program), which involve testing for *Brucella* agglutinins (types of antibody). The market cattle test program is a system of applying back tags to market beef cattle and testing the blood at slaughter or at auction markets. Cases of infection are traced to the farm of origin where further testing is done. Those animals found to be **reactors** are sold for slaughter.

Brucellosis Card Test (BCT) This test is based on the use of a disposable card on which blood serum or **plasma** is mixed with a buffered whole-cell suspension of *B. abortus* (antigen), which reacts (agglutinates) with antibodies in the blood serum of animals infected with brucellosis. The test may be conducted in the laboratory or on the farm or ranch and is read as either negative or positive (there is no suspect classification). The BCT is especially useful as a diagnostic test in cattle populations in which the prevalence of brucellosis is high because of its simplicity. However, it is too sensitive to use exclusively in vaccinated cattle or in herds or areas where the prevalence of brucellosis is low. It is also useful in identifying infected swine, sheep, and goats.

Brucellosis Milk Ring Test A modification of the blood serum agglutination test, called the brucellosis ring test (BRT), is used to detect *Brucella* agglutinins in milk. It serves as a screening test for brucellosis in cattle by detecting antibodies in bulk milk (a pooled sample of herd milk). The BRT test is performed by mixing nonhomogenized milk in a test tube with stained (by means of a purple dye) *Brucella* organisms. If the milk contains antibodies agglutination occurs and as the cream layer rises the agglutinated *Brucella* cells are carried with the milk fat globules. A purple cream layer with white milk below indicates a positive test. Conversely, in a negative test the cream layer is white and the milk is diffusely purple because the stained bacterial cells remain in suspension. In all 50 states milk samples are tested by the BRT at least 4 times annually. Reduction in the number of **herds** positive to the BRT reflects the efforts of the cooperative state–federal brucellosis-eradication program (Figure 22.5).

Brucellosis in Cattle, Swine, and Goats The *Brucella* organism is an extracellular and intracellular parasite. In pregnant animals the organism often localizes in the placenta. The **chorionic** tissue of the placenta reacts to the infection with inflammation, becomes edematous, and finally decreased circulation to the **fetus** results in an abortion. At the time of abortion

---

[6] The first well-described cases of brucellosis in humans were in goats-herdsmen on the island of Malta. The disease has also been given the name of *undulant fever* because it causes a sluggish, undulating, low fever in people.

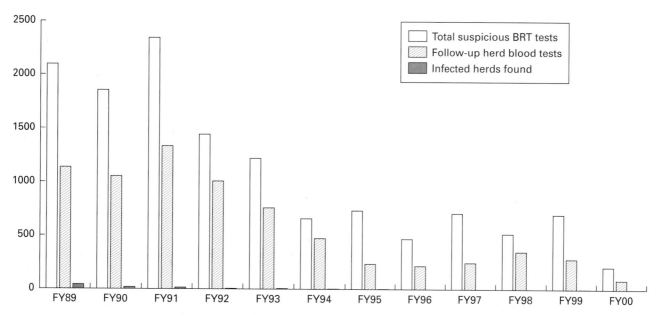

Figure 22.5 Brucellosis milk surveillance test results, FY 1989 to FY 2000.
USDA.

the uterine fluids and tissues are heavily contaminated with organisms, which may contaminate food, bedding, and the surroundings. Animals ingesting contaminated materials readily acquire brucellosis. Water contaminated with *Brucella* is also a potential source of infection. The fact that the *B. abortus, B. suis,* and *B. melitensis* organisms usually become localized in the mammary **glands** and may be secreted into milk is an important potential health hazard to humans. (Once infected the mammary glands of most cows remain infected.) The three species of *Brucella* are relatively species-specific but there can be cross species infections. Routes of natural *Brucella* infection to humans are depicted in Figure 22.6.

The revised USDA classifications for brucellosis control became effective in 1982 and by the end of 2000 all states were classified as Free or Class A (Figure 22.7). Class Free states must have remained without infection for at least 12 months. Class A status means that the infection rate for cattle may not exceed 0.25 percent (25 cases per 10,000 cattle tested) during the previous 12 months.

Brucellosis in Sheep and Goats   Some breeds of sheep are susceptible to infection by *B. melitensis* and occasionally by *B. abortus*. The milking breeds in other countries are especially susceptible. Several countries in the Mediterranean and Middle East have serious infections in sheep, goats, and humans. Infections in people have been traced to sheep or goat milk and to cheese manufactured from unheated milk in countries other than the United States.

Brucellosis in Humans   The bacteria may be transmitted to humans in whom it causes brucellosis or undulant fever (also called *Malta fever* and *Mediterranean fever*). Undulant fever is so named because the body temperature of infected persons varies or *undulates*. Acute brucellosis in people is characterized

by fever, chills, headache, night sweats, and weakness. These symptoms may extend from a few days in mild cases to several months in acute cases depending upon treatment used. The fever normally ranges from 100 to 103°F but may reach 104 to 105°F. Recurring fevers are common.

The occurrence of brucellosis in humans continues to be high in many countries except in the Scandinavian countries, the Netherlands, Switzerland, Canada, Australia, New Zealand, West Germany, and the United States. The reported incidence of human brucellosis in the United States decreased sharply from 1947 (6400 cases) to 1980 (183 cases) as a result of the widespread adoption of the pasteurization of dairy products and the favorable results achieved through the cooperative efforts of the bovine brucellosis eradication program. The downward trend has continued but at a slower rate. There were 185 cases of human brucellosis reported to the CDC in 1981, only 82 in 1999 (Figure 22.8). Brucellosis is most prevalent among adult males (especially veterinarians and others handling infected animals). Other potential sources of infection include drinking unpasteurized milk from infected animals, inhaling contaminated air (especially with *B. suis*), handling infected animals (especially swine) in an **abattoir** or meat-packing plant, and being accidentally exposed to *Brucella* as may occur when laboratory personnel work with *Brucella* cultures.

Prophylaxis and Control   The two basic control measures responsible for the marked reduction of brucellosis among farm animals during the past 35 years have been the (1) slaughter of infected animals (as detected by the blood agglutination test) and (2) immunization with living cultures of *B. abortus* strains 19 and more recently RB51 (RB stands for "rough *Brucella*"). The latter vaccine does not cause production of agglutinins (a type of antibody), and thus does not interfere with diagnostic serologic tests.

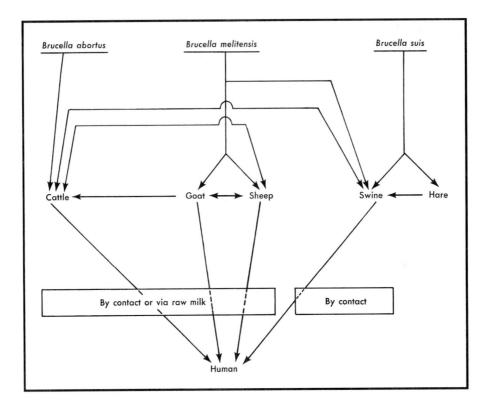

Figure 22.6 Routes of natural *Brucella* infection to humans.

Figure 22.7 State classifications of brucellosis, December 2000.
USDA.

Therapy of Brucellosis No single dependable and effective therapeutic agent has been developed for animals because long-term **therapy** is economically nonfeasible. Also, therapy is relatively ineffective because the bacteria are intracellular. In humans, however, several drugs and combinations of antibiotics have good therapeutic value in the treatment of brucellosis.

Leptospirosis

This infectious disease is produced by *Leptospira* (from the Latin *lepto* meaning "thin," and *spira* meaning "spiral") bacteria. Domestic animals susceptible to leptospirosis include cattle, dogs, goats, horses, sheep, and swine. Many wild animals may also be infected. The chief ports of entry for leptospirae are

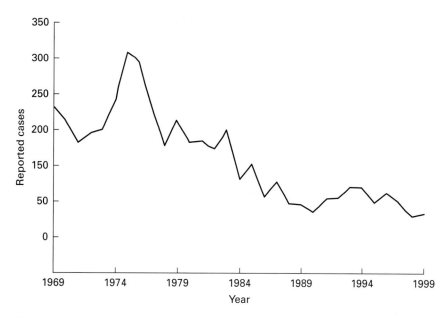

Figure 22.8 Reported cases of human brucellosis in the United States from 1969 to 1999.
Courtesy of Centers for Disease Control, Atlanta, GA.

through mucosal surfaces (conjunctival, nasal, and oral) and broken skin. Modes of infection of humans include occupational assignments like abattoir worker, fish cleaner, or sewer worker; exposure to infected cattle, goats, horses, rodents, sheep, and swine; and swimming in ponds or slow-moving streams contaminated by diseased domestic or wild animals. Rats are the most important reservoir of the disease in humans.

Leptospirae thrive in a moist environment but do not survive in dry environments or under acid or highly alkaline conditions. *L. hardjo* is the chief causative organism in cattle, *L. pomona* in swine. These organisms may be shed in the urine, which may be a "port wine" color (because of hemoglobinuria from destruction of red blood cells). Renal **dysfunction** is common in infected animals and in humans. Abortions (most common in late pregnancy) may occur in cows and **sows.** Leptospirosis in humans often resembles influenza. Leptospirosis cases in humans are no longer reported to the CDC. At the time of the last report in 1994 there were approximately 100 cases per year. Vaccination in domestic animals with multivalent bacterins at 6- to 8-month intervals is a good prophylactic measure. No vaccines are used in humans.

Anthrax

Some historians believe that anthrax was the fifth plague of Egypt mentioned in the Bible. Several authors have noted that Moses threatened Pharaoh with a disease that would infect his livestock:

> The hand of the Lord will fall with a very severe plague upon your cattle which are in the field, upon the horses, the donkeys, the camels, the oxen, and upon the sheep: there *shall* be a very grievous murrain [a pestilence, or plague].
>
> **Exodus 9:3**

Although anthrax is chiefly a disease of animals it was known at the end of the sixteenth century that it could be trans-

mitted to humans. It was the first animal and human disease to be traced to a microbial source. German bacteriologist, Robert Koch, identified the anthrax bacillus in 1876. Moreover, it was the first disease against which a bacterial vaccine was found to be effective. Pasteur produced a preventive vaccine in 1880. Anthrax is caused by the bacterium *Bacillus anthracis.* Its **spores** are very resistant to cold, heat, and drying and may even be viable and virulent for 50 years or longer.

Humans and swine possess considerable natural resistance to anthrax whereas cattle, goats, sheep, and wild **herbivores** (e.g., antelopes and camels) are quite susceptible. Dogs and wild animals may also become infected with the disease (especially from eating infected animals). Insects may carry and mechanically transmit the disease. Farm animals may become infected with anthrax by grazing on contaminated pasture, ingesting contaminated feed (especially **bone meal,** fish meal, and **tankage** that have not been subjected to high temperatures sufficient to kill the spores) or water, or being bitten by infected flies. The organisms may also be spread by birds and small animals that have fed on infected **cadavers.**

Anthrax infection in humans is rare in the United States. A person may become infected through a skin lesion or a small wound on an exposed surface or in some countries by eating improperly cooked meat (especially of sick animals or those that have died of the disease). Most cases of human anthrax may be acquired from direct contact with animal excretions, secretions, or infected tissues and by contact with animal skin, hair, or hide. In the United States it is primarily a rare occupational hazard. In one Australian factory goat hair is now **irradiated** to destroy the harmful anthrax spores before the hair is sold for use in making cloth and rugs. Certain areas of the Americas, Africa, and Asia still have dreaded animal losses resulting from anthrax. Infected animals should be isolated and if they die the carcasses should be cremated or properly buried.

Today an attenuated vaccine (called Sterne's) is used to control anthrax in farm animals (in areas where climatic and soil conditions are favorable for the spores to survive).

Many aspects of anthrax have led to fears that the causal organisms could be used in bioterrorism. During the Gulf War it was discovered that huge quantities of bacteria had been cultivated in Iraq. This led to a decision by the U.S. armed forces to vaccinate troops in a controversial program with an inactivated product.

The use of anthrax in bioterrorism became a prominent public health concern in the fall of 2001 following the World Trade Center calamity in New York City. Anthrax spores in the form of white powder were found in letters sent to a publishing company in Florida, selected Congressional and other governmental offices, offices of national news media personnel, hospitals, and numerous others. Several humans, including postal workers, became infected with the cutaneous (skin) or inhalation (pulmonary) forms of anthrax. Fatalities occurred among the latter. Testing of persons possibly exposed to anthrax spores included bacteriologic cultures of nasal cavities. Persons suspected of having been exposed to the pathogen were advised to start taking "cipro" (ciprofloxacin) or doxycycline, antibiotics generally effective in prophylaxis against anthrax.

Salmonellosis

There are more than 2300 serotypes of *Salmonella* organisms and many may cause gastroenteritis in humans. Five serotypes cause about 60 percent of the reported isolations from humans. Of these *S. typhimurium* is responsible for approximately 30 percent. In the United States the reported cases of salmonellosis increased from 723 in 1942 to 40,596 in 1999. Salmonellosis is now estimated to affect more than 2 million persons (nearly 1 percent of the United States population) annually. The high occurrence of the disease is attributed to poultry and egg products and the centralization of food-processing operations. *S. enteritidis,* an infection of poultry that can be transmitted through eggs, is an important cause of disease in humans. Swine may also host *Salmonella* organisms, which may infect people when they eat improperly cooked meat.

Septicemia in calves may be caused by foodborne salmonellae or contaminated water. *Salmonella dublin* is an increasingly common cause of diarrhea and death in young calves. To prevent the possible cycling of *Salmonella* through swine to people, raw garbage intended for use as food by swine should be cooked. As a further precaution, eating improperly cooked meat and eggs and drinking unpasteurized, **raw milk** should be avoided. Rats, other rodents, cockroaches, houseflies, and many other animals may be mechanical carriers of salmonellae.

Tularemia

This disease of mammals was first described in 1911 in Tulare County, California (from which the name is derived), as a "plague-like disease of rodents." In 1919 it was described in Utah as "deerfly fever." Today it is known that tularemia is caused by *Francisella tularensis.* The chief vectors of tularemia

are ticks (especially the dog tick and wood tick) and bloodsucking flies (especially the deerfly and horsefly). The bacteria have been isolated from more than 100 species of mammals, birds, and reptiles. Primary **modes** of transmission to humans include bites from infected ticks or other arthropods, handling of infected animals, and inhalation of airborne *F. tularensis.* In people the disease is characterized by irregular fever, which lasts for several weeks if improperly treated. Cooking readily renders the infected tissues safe for human consumption. There were 132 cases of tularemia reported in the United States in 1993; however, the CDC no longer requires doctors or hospitals to report all cases of the disease.

Plague

This is an **acute** infectious disease divided primarily into the two types of *bubonic plague,* in which the causative organism *Yersinia pestis* is transmitted chiefly from the rat reservoir of infection to humans by the rat flea, and *pneumonic plague* in which transmission is direct through airborne infection from person to person. Since the dawn of history bubonic plague has been regarded in some regions of the world as one of the major pestilences of humanity. The ill effects of the disease were first recorded in the Bible (I Samuel 6:4) and have been a serious threat to humans ever since. In the 1300s a form of bubonic plague called the *black death* (so called because it caused formation of spots of blood or hemorrhages that turned black under the skin) killed a fourth of the population of Europe. From 1347 to 1350 about 20 million people in Asia, Africa, and Europe died from plague.

Plague is primarily a disease of rodents. Bubonic plague occurs more frequently in tropical or semitropical climates whereas pneumonic plague in epidemic form is a disease of cold and temperate regions. The control of fleas and rats (which are responsible for more than 75 percent of the plague in humans) controls epidemics of bubonic plague. In the United States squirrels, chipmunks, prairie dogs, marmots, mice, rats, cats, and rabbits may be naturally infected with plague organisms. Sylvatic plague, or plague of the woods, is widely spread among wild rodents of the western United States. Sporadic cases of plague in humans occur in the United States.

Approximately 10 cases of bubonic plague are reported in humans in the United States annually. All recent cases have been in the western and southwestern states.

Campylobacteriosis

This is largely a foodborne infection of humans caused by *Campylobacter jejuni* or *C. coli.* It is now recognized as a leading cause of diarrhea. Contaminated poultry products are the leading source of infection but several cases have involved ingestion of raw milk. Dogs (especially puppies) and cats are increasingly cited as reservoir hosts of human infections, particularly in children.

The true number of human cases of campylobacteriosis is unknown but the number of isolations of the organism is increasing because of superior diagnostic methods and greater awareness of the disease.

Proper cooking of foods, pasteurization of milk, and hygiene in care of pets are control measures. Antibiotics can be used to treat animals and people having this infection.

Colibacillosis

Several bacterial strains of the genus *Escherichia* may cause disease in either animals or humans. In humans most disease is caused by *E. coli* O157:H7 and usually is referred to as entero-hemorrhagic or hemolytic uremic syndrome. Unlike some other strains it does not cause disease in animals and may be present in 1 to 3 percent of the feces of cattle. It has been found in several food products.

In humans *E. coli* O157:H7 may cause severe diarrhea, sometimes containing blood. In the most severe form the infection may cause death, usually in children, resulting from damage to the kidneys.

A serious outbreak of *E. coli* infection in 1982 was caused by ingestion of undercooked hamburger meat. This resulted in an increased concern about foodborne pathogens. Several outbreaks have since been reported with an estimated 20,000 annual cases and 250 deaths in the United States. Processed meats in slaughterhouses are tested and if the bacteria are present the meats cannot be used for food. Prevention of human disease includes proper preparation and cooking of foods.

Listeriosis

This disease is caused by the bacteria *Listeria monocytogenes* and affects cattle, sheep, swine, and humans. Its most common effect is inflammation of the central nervous system (encephalitis). A diagnosis of listeriosis is uncommon in people. The disease may cause abortion in women and meningitis in newborn infants. It is also the cause of foodborne illness associated with ingestion of several types of cheese.

Cat Scratch Disease

This bacterial zoonotic infection is nearly always caused in humans by the scratch or bite of a cat infected with *Bartonella henselae.* The organisms are present in the mouth of cats but cause no disease in cats. Typical signs in humans are a local redness and swelling of the regional lymph nodes. Most cases are not serious but severe infections may occur in humans whose immune system is deficient such as persons with AIDS. Symptoms may include a persistent fever, weight loss, muscle and joint pains. Recent studies have shown that fleas may transfer the infection among cats.

The disease in humans is not reportable to health authorities and exact numbers of cases are unknown.

Lyme Disease

Named for Lyme, Connecticut, where it was first described, this infection is caused by a spirochete, *Borrelia burgdorferi.* It has become the most common tickborne disease in the United States (Figure 22.9). The primary vector is a tick, *Ixodes scapularis.* The immature nymphal stage is the most important in transmission to mammals. Deer are important hosts of the ticks but the major animal reservoir is the white-footed mouse.

The clinical sign in dogs and cattle is arthritis. In humans the signs and symptoms include a local skin rash surrounding the site or the bite of a tick and joint pains, fever, and fatigue. It should be noted that the infected tick must be attached 24 to 48 hours to infect the human.

Vaccines are available for use in dogs and in humans, primarily in endemic areas. Lyme disease in humans is reportable to the CDC. In 1999 16,273 cases were recorded. Of these 12,944 (80 percent) occurred in the northeastern or mid-Atlantic states.

Erysipelas

This is a communicable disease of swine and turkeys caused by *Erysipelothrix rhusiopathiae,* which may also be transmitted to humans. People have considerable resistance to the disease (especially to infection through the **gastrointestinal** tract). The causative organism is resistant to drying and may even survive exposure to sunlight for 10 or more days. It may retain viability and virulence for several months in salted or pickled meats. The organism is widespread in soil. Animals that have had the disease may serve as carriers and thereby expose other animals. Vaccination offers a prophylactic measure against swine erysipelas. Humans may become infected through skin abrasions, especially among those handling infected swine (in abattoirs), and by handling infected fish. The skin form of the disease is called *erysipeloid* in humans. Penicillin is effective therapeutically in humans.

22.3.4 Protozoal Infections

Cryptosporidiosis

This protozoal infection is caused by parasites of the genus *Cryptosporidium.* There are two species, *C. parvum* and *C. muris,* which affect mammals. Species that affect birds are not infective for humans. The parasites have been found in a wide variety of hosts. The major source of human infections has been from dairy calves. Municipal water supplies have been contaminated by the **oocysts** and in 1993 over 400,000 humans became infected. The CDC reported 2361 cases in 1999.

The symptoms in humans are primarily a mild diarrhea but in those whose immune systems are compromised (such as with AIDS) there may be severe prolonged diarrhea. No specific treatment is available. Prevention is largely to avoid contact with affected animals, especially calves, and proper treatment of water supplies.

Toxoplasmosis

The causative agent of toxoplasmosis is *Toxoplasma gondii.* It occurs throughout the world. The primary host is the cat (Figure 22.10) but many animals can be infected, especially carnivores. There are no clinical signs in infected cats.

Humans become infected by ingestion of oocysts through consumption of raw or undercooked meat such as pork or through handling of cat feces such as in litter boxes or contaminated soil. Pregnant women are at special risk and the infection may affect the fetus resulting in deformities. In most human infections there

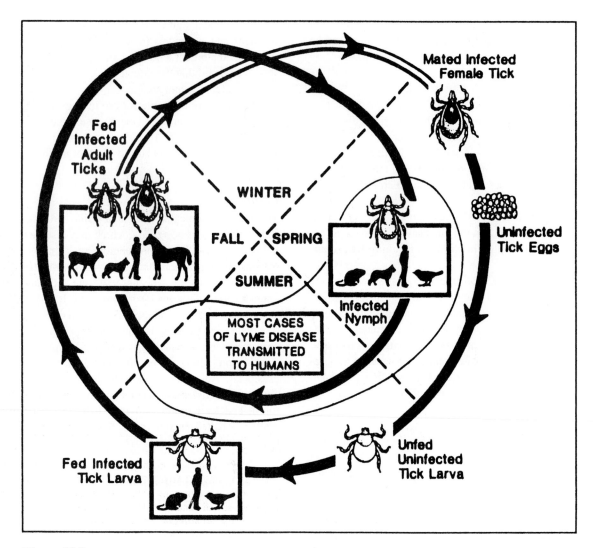

Figure 22.9 Typical life cycle of tickborne Lyme disease affecting humans.
Adapted from Dr. S. C. Fisher, Veterinary Forum, *July 1992. Used by permission.*

are no signs but in those with impaired immune systems there may be fever, headache, fatigue, and muscular pains.

Control measures in humans include properly cooked meat and avoidance by pregnant women of contact with feces of cats.

22.3.5 Fungal Infections

Two groups of fungal diseases may be transmitted from animals to people: (1) those affecting humans *externally* (superficial, or dermatomycoses) and (2) those affecting humans *internally* (localized, or systemic granulomatous mycoses). External fungal infections include *Trichophyton* and *Microsporum* organisms, which cause dermatophytosis (ringworm) in cattle, cats, dogs, horses, and other domestic animals. It is estimated that approximately 80 percent of ringworm infections of persons residing in rural areas are acquired from farm animals.

There are many fungal diseases that affect humans internally. *Coccidiomycosis* is a fungal infection that begins as a respiratory infection and may progress to produce lesions throughout the body, especially in the subcutaneous tissues, skin, bone, peritoneum, thyroid, and central nervous system. The disease may infect cats, cattle, dogs, horses, sheep, swine, rodents, humans, and other species. The infection is not directly transmitted from animal to animal but rather is acquired through inhalation of spores. Humans acquire infection through the inhalation of spores. *Histoplasmosis* is a fungal disease resembling miliary tuberculosis (characterized by the formation of lung lesions resembling millet seeds). A survey of 185,000 naval recruits revealed that the disease is most prevalent in the central and southcentral United States. The fungus has been isolated from a number of mammalian species including mice, rats, cats, cattle, dogs, bats, horses, foxes, opossums, sheep, skunks, swine, and woodchucks. People usually become infected by inhalation of spore-laden dust existing in nature in a saprophytic state, particularly in association with bird and/or bat feces. Direct transmission from person to person or from animal to humans has not been demonstrated. Other systemic mycoses include blastomycosis, sporotrichosis, and cryptococcosis.

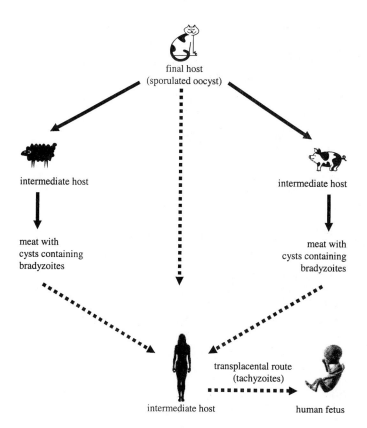

Figure 22.10 Transmission of toxoplasmosis to domestic animals and humans.

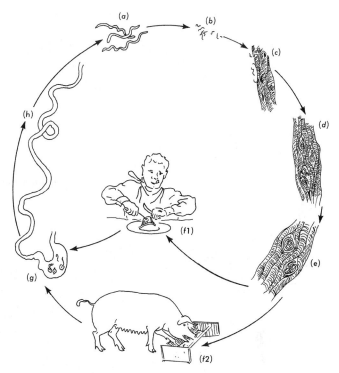

Figure 22.11 Life cycle of *Trichinella spiralis* (trichina worm). (a) Adult worms in mucus of small intestine of swine; females produce many larvae; (b) larvae enter bloodstream and are distributed throughout body; (c) larvae enter voluntary muscle tissues and develop; (d) larvae grow to infective stage and encyst; (e), (f1), (f2) human (swine, rats, and other animals) becomes infected by eating improperly cooked pork containing *T. spiralis;* (g) cysts dissolved in stomach pass to small intestines (h), in which they reach sexual maturity (adults); and the cycle begins again. *Courtesy of Dr. P. D. Garrett and Dr. D. E. Rodabaugh.*

22.3.6 Helminth (Worm) Infections

Humans may become infected by several methods from numerous species of **helminth** parasites of domestic and wild animals. Important methods of transmission include (1) introduction of parasites into the body by the oral (food or drink) route (e.g., the eggs or larvae of certain tapeworms), (2) by active penetration of the infective stage of the parasite into the skin (e.g., certain hookworms from moist soil), and (3) by introduction of the parasite into the skin via an alternate host (through bites of mosquitoes or tsetse flies).

Trichinosis

The parasitic nematode *Trichinella spiralis* has a cosmopolitan distribution but a fairly simple life cycle. Its usual hosts are meat-eating mammals. Infection occurs when the host **ingests** meat containing the encysted larvae. (A person becomes infected from eating inadequately cooked pork, bear, or walrus meat.) A few hours postingestion the young larvae are liberated by gastric digestion from the cysts (tiny muscle sheath or sacs) in which they live. The worms grow rapidly and move from the stomach to the small intestine where they attach themselves to the inner lining, grow, and become sexually mature in about 2 days. The adult worms mate and about 8 days later each new female worm begins producing small but active larvae which are released **viviparously** from the maternal body into the small intestine. These larvae penetrate the walls of the host's intes-

tine. They then enter the lymphatic and blood systems and are transported throughout the body. Although most body organs and tissues are subject to invasion, the worms largely infect the voluntary muscles (most active such as the diaphragm). An inflammatory response to the worm results in the degeneration of the muscle fibers surrounding the parasite and in the formation of a connective tissue capsule (cyst) around it. The encapsulated larvae ultimately become **calcified** and eventually degenerate and die unless the meat is eaten by a person (Figure 22.11) or another suitable host. (The larvae may live in these cysts for a year or more.)

Swine, rats, cats, dogs, and other mammals are subject to infection. Swine become infected mainly by feeding on garbage containing raw pork scraps. It has been estimated that the collection and use of garbage for swine production constitutes a $100 million enterprise annually. Therefore it is not economically practical to prohibit the use of garbage for feeding swine, but it is practical to legislate a requirement that garbage be cooked and thereby rendered safe for swine. (All states now require that garbage intended for swine food be cooked, also most states ban the feeding of garbage containing meat scraps to pigs.) United States Department of Agriculture and United

States Public Health Service (**USPHS**) reports indicate that since 1906 the occurrence of trichinosis has been reduced by more than 80 percent in humans and by 94 percent in swine. Human trichinosis is no longer a serious public health problem. Only 12 cases were reported in the United States in 1999. The disease probably served as the basis for the first food sanitation code. The Mosaic code forbade the children of Israel to eat pork as recorded in Leviticus 11:7–8: "And the pig, though it has a split hoof completely divided, does not chew the cud; it is unclean for you. You must not eat their meat or touch their carcasses; they are unclean for you." Methods of providing safe pork for humans were discussed in Chapter 3.

Taeniasis

Domestic animals play an important role in most tapeworm infections of humans. The pork tapeworm *Taenia solium* may be viable in improperly cooked pork. The adult tapeworm (3 to 6 ft in length) inhabits a person's small intestine. (Tapeworms have no mouth or digestive system but rather attach themselves to the intestinal lining by means of a small head that has suckers thus allowing them to absorb already digested food **nutrients.**) Its larval stage (bladder worm) develops in the **musculature** of swine. Infected pork containing these larvae is commonly called "measly pork." *T. solium* infections in humans and swine are now rare in the United States.

The life cycle of the 12- to 25-ft-long beef tapeworm *T. saginata* is similar to that of the pork tapeworm except that cattle and other herbivores rather than swine serve as intermediate hosts. (The ox is the essential host of the larval or cysticercus stage.) Cattle acquire this parasite by ingesting tapeworm eggs voided in human feces (humans are the only known host of the mature worm) as shown in Figure 22.12. Therefore the disease is more prevalent in countries where pastures are irrigated with sewage effluent. The larvae of *T. solium* and *T. saginata* are killed when heated to 131°F (55°C). Quick-freezing and thorough curing or salting of meat are also destructive to the larvae. Pork and beef carcasses are routinely subjected to inspection in slaughterhouses by USDA workers for cysts of *T. saginata* or *T. solium*. If found, special processing is conducted.

The broad, or fish, tapeworm *Diphyllobothrium latum* is the largest cestode (tapeworm) found in humans. It may be 30 or more feet in length, acquired through the consumption of raw or insufficiently cooked fish infected with the larvae. The life cycle of *D. latum* is rather complex. The adult worm inhabits the small intestine. Eggs passed in the feces hatch and, when deposited in freshwater, release larvae. The free-swimming larva that moves about in the water must be swallowed by a suitable first **intermediate host** (usually a water flea) to continue its development. The life cycle is dependent on the infected first intermediate host being eaten by a suitable *second intermediate host* (usually freshwater fish such as pike, salmon, trout, and perch). In this second host the larva penetrates the muscles and continues to develop. When this host is ingested by

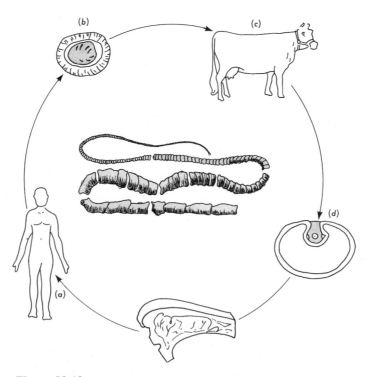

Figure 22.12 Life cycle of *Taenia saginata* (beef tapeworm). (a) Humans become infected with adult worm by eating improperly cooked beef containing the cysticerci, (b) ova voided in feces, (c) calves and cows become infected with the cysticerci via infested soil or forage, (d) cysticerci in beef tissue (muscle).

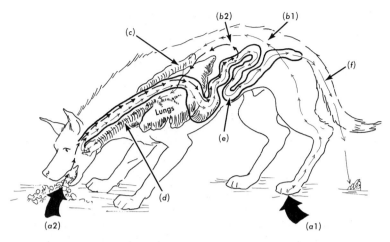

Figure 22.13 Life cycle of *Ancylostoma caninum* (dog hookworm).
(a1) Larvae penetrate skin (or foot pads), then (b1) burrow to blood vessels and
migrate to the lungs; or (a2) larvae may be ingested and then (b2) migrate to the
lungs; (c) larvae move to air spaces of the lungs and then to the bronchial tubes
and trachea; (d) ciliary motion of epithelium and coughing carry larvae to the
throat, from which they may be swallowed; (e) larvae pass to the small
intestine, in which they mature and become adults; the adult worms attach, suck
blood (up to 0.8 cc per worm daily), and the females lay eggs; (f) eggs (ova) are
passed in the feces, undergo development, and hatch into larvae in the soil, thus
completing the cycle.
Courtesy of Dr. P. D. Garrett and Dr. D. E Rodabaugh.

a fish-eating mammal, the parasite attaches itself to the intestinal mucosa and grows into an adult tapeworm. Wild reservoir hosts include the bear, fox, mink, and wolf.

Echinococcosis (Hydatid Disease)

The small tapeworm associated with this disease is found most frequently in dogs. Humans may acquire the disease from these infected carnivores. The larva known as the *hydatid* may develop in nearly all mammals where it forms hydatid tumors, or cysts, in the **viscera** (liver, lungs, kidneys, and other organs). Sheep are the most common intermediate host. In many countries these infections are responsible for huge losses of animal protein through condemnation of internal organs, especially liver. Dogs may become infected by feeding on infected animal viscera. Children playing among infected dogs are endangered by worm eggs adhering to the dog's hair and sometimes develop cysts (hydatids) of these tiny dog tapeworms.

Ascarids and Hookworms

Ascarids (*Toxocara* spp., commonly known as roundworms) and hookworms (*Ancylostoma* spp.) are common intestinal worms of dogs and cats. These parasites may cause severe disease in the first few weeks of the life of dogs and cats. They can cause *larval migrans syndromes* in humans who accidentally ingest eggs or larvae or have direct skin contact with hookworm larvae in soil contaminated with the feces of infected animals.

When zoonotic ascarids and hookworms infect humans they rarely mature in the intestine but may migrate in the tissues (*larval migrans*) where they may produce damage and cause allergic responses. There may be permanent visual or neuro-

logic damage from visceral larval migrans (migration of ascarid larvae through tissues such as the liver, eye, and brain). Hookworms (*A. caninum* and *A. braziliense*) may infect humans when larvae in soil are ingested or directly penetrate the skin on contact. Cutaneous larval migrans cause intense itching and skin lesions and are the most common sign of zoonotic hookworm infection (Figure 22.13).

Most cases of human toxocariasis and zoonotic ancylostomiasis can be prevented by simple measures such as good personal hygiene, eliminating the parasites from pets, and keeping children away from contaminated environments. Deworming of pets is the most effective measure to prevent soil contamination.

22.3.7 Bovine Spongiform Encephalopathy (BSE)[7]

Also known as "mad cow disease," this apparently new disease of the central nervous system of cattle first appeared in 1986 in the United Kingdom (U.K.). Since then it has appeared in many countries in western Europe. It is a slowly progressive disease belonging to a family of diseases known as transmissible spongiform encephalopathies that also include scrapie in sheep.

The incubation period is long, commonly 3 to 5 years. Signs include nervousness, abnormal stance, difficulty in walking, and loss of appetite. It is always fatal.

The cause of BSE is considered to be a *prion,* a protein similar to a small DNA particle, which is very resistant to

[7]See also Paul Brown et al., "Bovine Spongiform Encephalopathy and Variant Creutzfeldt–Jakob Disease: Background, Evolution, and Current Concerns," *Emerging Infectious Diseases*, January–February 2001.

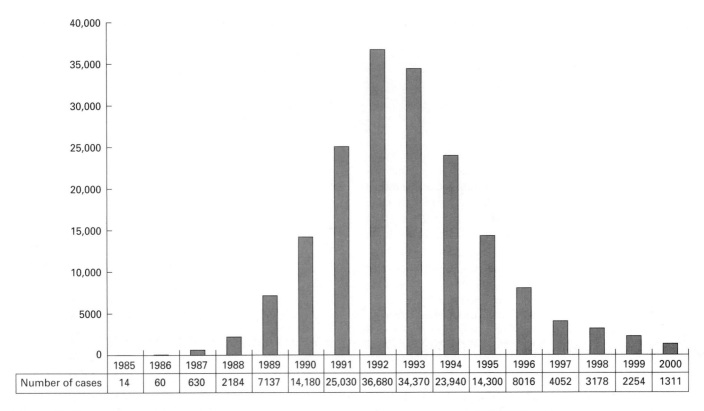

Figure 22.14 Annual incidence of bovine spongiform encephalopathy (BSE) in the United Kingdom, 1985–2000.
Source: Sir John Pattison, Medical School of University College London, U.K.

inactivation. The agent is suspected to be in meat and bone meal from scrapie-affected sheep. A BSE-like disease has also been found in domestic cats and some wildlife species in British zoos.

There seems to be no transmission from cow to cow or cow to calf. Since 1986 over 170,000 cases of BSE have been reported in the United Kingdom among 1 million cows in over 33,000 herds.

In 1988 the United Kingdom ceased the feeding of meat and bone meal of ruminant origin and since 1992 the number of cases has declined rapidly (Figure 22.14).

No cases of BSE have been found in the United States. In 1989 the United States banned the importation of cattle and ruminant products. All previous live animal importations were inspected. An ongoing surveillance program of over 12,000 cattle that were unable to rise or had neurologic signs has failed to detect any positive lesions in the brain, the accepted diagnostic method. In 1997 the FDA banned the feeding of ruminant proteins to ruminants.

Because of similarities of a fatal neurologic syndrome in humans called Creutzfeldt–Jakob disease (CJD), a fatal form in humans has been suspected to sometimes be caused by ingestion of nerve tissues from infected cows. This new form of the disease, called new variant (vCJD), has rather conclusively been confirmed in more than 100 cases and is caused by the same prion as BSE.

The slaughter of BSE-affected cows and those that may have consumed the contaminated feed, has caused huge losses to the British economy. Also, the consumption of beef has been reduced by approximately 50 percent because of public health fears.

22.4 Selected Human Diseases Transmissible to Animals

Certain diseases of human origin may be acquired by animals and transmitted back to humans, for example, tuberculosis.

22.4.1 Viral Infections

The viruses that cause *influenza* in humans (e.g., type A influenza) are apparently closely related to those isolated from avians (chickens, ducks, turkeys) and swine. There has been much speculation with respect to the role of animals in the **epidemiology** of human influenza, particularly as a source of new major antigenic variants. It is now generally accepted that the swine strain represents the prototype of the influenza strain responsible for the 1917–1918 human **pandemic** and that this strain was introduced to swine by people. An unanswered question that often arises is whether a human influenza strain has the potential to establish itself in animals and then at some later date be reintroduced into people when the latter's antibody status permits. Avian strains have recently infected humans with severe disease but there is no evidence of human-to-human transmission of avian strains in contrast to those of swine origin.

22.5 DISEASES TRANSMISSIBLE BY ANIMALS AS PASSIVE CARRIERS

Animals may host the spores of several pathogenic organisms in their intestinal tracts usually with no apparent harm to themselves. We have included information concerning two such diseases.

Tetanus

The tetanus organism *Clostridium tetani* is commonly found in the intestine of herbivorous animals (primarily horses) and to a lesser extent in that of humans. **Manured** soil is a prime source of these spores. People working around horses and stables are more likely to become infected. In humans and farm animals (especially horses, sheep, and goats) tetanus results from a deep-puncture wound. (The deep puncture provides an entry and **anaerobic** conditions for growth of the organism.) Invasion of the organisms requires broken skin. Gunshot and firecracker wounds are especially susceptible to infection. *Passive immunization* with tetanus antitoxin after injury is very effective. However, *active immunization* using tetanus **toxoid** is preferred in tetanus prophylaxis and is recommended for all persons. The CDC reported 40 cases of tetanus in 1999.

Botulism (from the Latin botulus meaning "sausage")

Spores of the causative bacterium inhabit the intestinal tract of many animals. Botulism in humans and animals is caused by the ingestion of food in which the organisms *C. botulinum* have produced toxins. Among farm animals poultry are the most susceptible to the disease, followed by cattle and horses. They develop botulism from eating moldy or spoiled grains or forages in which botulism toxin has been produced.

22.6 TOXIC PLANTS

> Who taught the natives of the field and wood
> To shun their poison and to choose their food,
> Prescient, the tides and tempests to withstand;
> Build on the wave or arch beneath the sand?
>
> **Alexander Pope (1688–1774)**
> **English poet**

The **lethal** effects of certain plants have been known for many centuries. Socrates (469–399 B.C.) was sentenced by the rulers of ancient Athens to die by drinking a brew made from the poisonous hemlock plant. Because the diet of most farm animals is largely of plant origin, selected plants that may adversely affect the health of animals providing useful products, pleasure, and services to humanity will be reviewed.

The toxicity of a plant varies with the animal's species, age, sex, nutritional status, state of health, and **stress** factors and with the formulation of the compound and route of its administration. A common way of expressing the toxicity of a compound is by means of an **LD$_{50}$** value. This is a statistical estimate of the dosage necessary to kill 50 percent of a population of test species under stated conditions.

There are several poisonous plants that may cause extensive economic losses to the producer of animal products. (Cattle and sheep are the most frequently affected farm animals.) Two commonly accumulated poisons are cyanide (hydrocyanic acid, or HCN) and nitrate (potassium nitrate, or KNO$_3$). Plants that may accumulate HCN include arrow grass, chokecherries, and sorghums. Oats, Sudan grass, and corn (especially in hot, dry weather when nitrogen fertilizer application is high) are common plants that may accumulate KNO$_3$. Many wild plants including the Russian thistle, white ragweed, and lamb's-quarter are known to accumulate KNO$_3$.

Some ingested plants interfere with the metabolism of certain nutrients. An example is the bracken fern, which on ingestion apparently causes a thiamine deficiency in horses, cattle, and swine. The deficiency develops because this fern contains an **enzyme** called *thiaminase*, which adheres to the thiamine

(a)

(b)

Figure 22.15 Congenital malformation of the above calf (a) was induced by maternal ingestion of the lupine plant, *Lupinus caudatus* (b), during the period of pregnancy between 40 and 70 days. The deformity may occur in the front legs, hind legs, back, or palate (cleft), depending on the stage of fetal development at the time of ingestion.

Courtesy of Dr. Wayne Binns, Director, Poisonous Plant Research Laboratory, USDA, Logan, UT.

molecule and thereby renders it inactive. The administration of thiamine apparently corrects the deficiency in most animals except cattle. Ingestion of other plants may cause congenital malformations (Figure 22.15).

Horses and occasionally cattle and sheep may become poisoned by eating the horsetail plant *Equisetum*. Cocklebur seedlings (*Xanthium strumarium*) contain a toxic compound, hydroquinone. If consumed in a short period of time, 0.25 to 0.5 lb of the plant seedlings will kill a 50-lb pig. Sheep, cattle, and poultry are also susceptible to poisoning by the plant. Jimson weed and larkspur plants may poison cattle. Milkweeds are quite toxic to sheep (1.5 oz of green leaves may be fatal) and to a lesser degree to cattle, horses, and poultry.

Milk sickness in humans and *trembles* in certain farm mammals are due to a poison (tremetol) in the white snakeroot (*Eupatorium rugosum*) plant, in Jimmy fern (*Notholaena sinuata*), or in the rayless goldenrod (*Hapalopappus heterophyllus*). This poison is of danger chiefly to cattle, milk goats, horses, and sheep in which it causes trembles. Humans and milk-fed animals are poisoned by ingesting milk from affected mammals that contains tremetol. Many people (including Abraham Lincoln's mother) died of this **insidious** disease during the nineteenth century. Because of improved pastures cattle today are rarely exposed to these poisonous plants. This, coupled with the dilution of milk from exposed cows with that from other sources, renders milk sickness (vomiting sickness) a rarity today.

Pets and pleasure horses have increased in relative numbers in large population centers. Owners are generally unfamiliar with toxic plants and their possible harmful effects. This has tended to shift the danger of toxic plants from an agricultural to a metropolitan situation. *Native* toxic plants have largely disappeared in the intensively cultivated croplands of the eastern, central, and southern United States and Canada. However, substitute plants, shrubs, or trees that are often poisonous have taken their place.

In the spring of 2001 numerous foals in Kentucky died because of natural cyanide in black cherries. The poison may have been ingested via fecal matter of caterpillars that fed on the cherry trees.

In subtropical North America the kinds and numbers of toxic plants often differ from those of other North American areas. In much of Florida and the other southeastern states an increasingly important beef cattle industry makes pasture management important from the standpoint of improving the kind of pasturage and of avoiding native poisonous plants and shrubs. When such soil-improvement cover crops as the **legume** *Crotalaria* are used, care must be exercised to keep the seeds from contaminating hay or grain crops destined for use in the manufacture of animal feeds. The same is true of castor beans, tung nuts, avocados, and many ornamentals such as oleander, jequirity beans (*Abrus* spp.), *Lantana* spp., various lilies, the laurels, and many others.

For example, the generally accepted lethal amounts of the common Japanese yew shrub (*Taxus* spp.) for most animals have been observed to be as little as 0.1 percent of an animal's body weight. The poisonous element includes an alkaloid that depresses heart function. An increasing number of cattle and horses die annually as a result of poisoning from ornamental shrubbery, especially the Japanese yew. Alkaloids, cyanide, and volatile oils are the toxins. Clinical signs include nervousness, trembling, and difficult respiration. The green growth and living bark of the black locust (*Robinia pseudoacacia*), as in oleander and ornamental yew, are extremely toxic for horses and other farm animals as well as for humans. Domestic animals should not have access to pruned new growth or to dried materials of such plants.

There is considerable variation in the plant species common to each region of the United States. From Key West, Florida, to Kodiak, Alaska (a distance of over 5000 miles), every kind of climate from humid subtropical, semiarid, and arid to frigid subarctic may be found. It would be impractical to attempt an enumeration of all the toxic plants and shrubs that may be found in such vast areas in a brief survey such as this. For example, a specific area such as Missouri has approximately 100 genera and several hundred species of plants, shrubs, and trees that are described as toxic for one or more species of animals under given circumstances. Florida workers have described at least 50 genera of poisonous plants. The Wyoming Agricultural Experiment Station considered over 50 genera and 80 species important enough to list as poisonous, especially to cattle, sheep, goats, horses, and swine. Texas workers representing the southwestern area have reported over 50 genera as having toxic species. Kansas recognizes 58 genera and 79 species of toxic plants. United States Department of Agriculture workers have selected 22 genera as the most troublesome toxic plants, principally in the West and Far West, although at least a dozen more may be found in most areas of the United States and Canada. At least 25 state agricultural experiment station bulletins include the subject of poisonous plants as the whole or a part of their subject matter. Toxic and noxious weeds and their seeds are listed and/or described in numerous monographs and research papers.

The kind and intensity of the animal industry in a region determine whether a given toxic plant is considered important. It has been observed that quite often the toxic vegetation is also the hardiest and persists when wholesome forage plants have largely disappeared. In the mountainous areas of the United States and Canada much of the original native plant population continues to exist. However, over**grazing** has almost eliminated wholesome pasturage vegetation in many areas. Overgrazing has allowed the more undesirable and often poisonous plants such as *Halogeton glomeratus* to encroach on and almost destroy the ranges for grazing cattle and sheep. Overgrazing has also seriously damaged good pasturelands in many other areas (e.g., the Osage Hills of Oklahoma and the Flint Hills of Kansas). The populations of less-desirable weeds and shrubs almost always increase because such plants are not eaten by cattle, sheep, goats, and other animals unless nothing else is available. Proper range and pasture management is the best and most practical way to control or avoid losses from animals grazing areas with toxic plants under the conditions mentioned.

Anticoagulant Action of Clovers

White and yellow sweet clover and to a lesser extent lespedeza hays contain a compound called *coumarin,* which under conditions of spoilage may be converted into **dicoumarin.** Dicoumarin interferes with prothrombin formation, which is essential for proper blood clotting (Figure 20.4). Dicoumarin was used to develop the **rodenticides** *pindone* and *warfarin.*

22.6.1 Mycotoxins

Mycotoxins are chemical substances produced by fungi (molds) that may result in illness and death of animals and humans when feed or food containing them is eaten. Certain mycotoxins are nerve poisons, others are liver and kidney poisons, and some affect other body organs. Mycotoxins may cause birth defects, abortions, tremors, cancers, or other adverse effects.

Fungi that form mycotoxins occur worldwide. Foods and feeds may be contaminated with mycotoxins produced by fungi that have grown on crops in the field, in storage, or during processing. Crops frequently contaminated include corn, peanuts, cottonseed, tree nuts, wheat, rye, barley, and rice.

Poisonings of humans and animals by mycotoxins have been recorded for more than 5000 years. Documentation in the early 1960s that mycotoxins in feed killed over 100,000 turkeys and 14,000 ducks in Great Britain stimulated substantial scientific investigation in this important area.

Of the more than 100 mycotoxins, the aflatoxins are of the greatest concern. They are produced by two fungi that occur worldwide in many agricultural commodities. Fungus growth and aflatoxin production are favored by the warm temperatures and high moistures typical of tropical and subtropical areas including the southern United States. Aflatoxins are found frequently in corn, peanuts, cottonseed, and tree nuts and also in animal products such as milk and milk products and certain meat products of cows and swine that have ingested contaminated feed. Significant losses from aflatoxin contamination in recent years have occurred in the peanut, corn, cottonseed, poultry, cattle, and swine industries. Considerable progress has been made in controlling the contamination of foods and feeds and in detoxifying contaminated products. Ammonia is particularly valuable as a detoxifying agent.

Aflatoxin B_1 has adverse biological effects including liver damage, reduced growth rate, and immunosuppression when the mycotoxin is present at levels approximating 1 ppm in the diets of most domesticated and experimental animals; liver damage has been observed in humans. Aflatoxins produce both acute and delayed effects in humans.

Aflatoxins have been established as a cause of liver cancer in certain animals. Aflatoxin B_1 is the most potent known naturally occurring cancer-producing substance. Less than 1 ppb in the diet is sufficient to produce significant incidence of liver cancer in rainbow trout, the most sensitive cancer test animal known. Although evidence from some developing countries showing an increase in the incidence of liver cancer with an increase in the aflatoxin content of the diet tends to implicate aflatoxins as a cause of human liver cancer, there is no evidence that aflatoxins are a significant cause of human liver cancer in the United States. For example, the per capita aflatoxin intake in the Southeast is approximately 9 times greater than that in the United States as a whole but the proportion of total deaths due to liver cancer is lower in the Southeast than in the country as a whole.

Several mycotoxins are formed by *Fusarium* fungi. Their production is favored by high moisture and temperatures. Although corn is the crop most affected in the United States, *Fusarium* mycotoxins are produced on several other cereal grains. One of these compounds, *zearalenone,* is hormonally active (estrogenic) in swine and certain other animals; it is an important cause of infertility in swine. Other *Fusarium* mycotoxins cause necrosis and ulceration of the mouth, stomach, and small intestine accompanied by diarrhea and hemorrhages. These mycotoxins may cause death in cattle and swine. Prompt drying of cereal grains after harvest and storage at low moisture conditions are good control measures.

Staggers syndromes are neurological diseases of domestic animals, especially cattle, that are of known or suspected mycotoxin origin. They are characterized by muscle tremors; hyperexcitability; unsteady, staggering movements; and inability to walk or stand. Leukoencephalomacia of horses results from a specific brain lesion caused by consuming *Fusarium moniliformen*–contaminated corn. Affected horses stagger, act sleepy, and often die. Slaframine is a mycotoxin produced by the fungus that causes black patch disease of red clover. Animals ingesting slaframine-contaminated red clover produce excessive amounts of saliva, hence the designation "slobbers" syndrome.

Ergot and its effects are mentioned in the Bible. It is produced by several species of fungi that infect the blooms and replace the seed with ergots in more than 1000 species of the grass family, especially the cereal grains. Ergotism caused thousands of human deaths in the Middle Ages. The occurrence of ergotism in humans is rare today but it is observed occasionally in food-producing farm animals. Manifestations of ergotism include neurological symptoms and convulsions in most animals and gangrene in cattle affecting the tips of the tail and ears and in the lower parts of the feet. The incidence of ergot is greatly reduced and/or controlled by a number of agricultural practices including seed treatment and selection and breeding for ergot resistance.

22.7 GOVERNMENTAL SAFEGUARDS FOR ANIMAL AND HUMAN HEALTH

22.7.1 United States Food and Drug Administration (FDA)

From 1880 to 1900 numerous articles and books were published alerting the public to the hazards of unsanitary food production practices. This prompted the U.S. Congress to enact in 1906 the Pure Food and Drug Law. Concurrently, the Federal Meat Inspection Act was passed and both laws became effective in 1907. The **U.S. Food and Drug Administration (FDA)** was established as a separate unit of the USDA (Section 22.7.2) in 1927. In 1940 the FDA was transferred from the USDA to the Federal Security Agency, which later became the Department of Health, Education and Welfare and today is the Department of Health and Human Services (Section 22.7.3). The foremost mission of the FDA is to safeguard American consumers against injury, harmful food ingredients and additives, and fraud. It also protects industry against unscrupulous competition. In addition to inspection and sample analyses, the FDA conducts independent research on such things as drug toxicity in laboratory animals, disappearance curves for **pesticides,** and licensing and long-range effects of drugs such as antibiotics.

The Food and Drug Administration requires pesticide manufacturers to present new products with their proposed labels for approval before they are authorized to sell them. The label must indicate as a minimum the name of the product, its active and inert ingredients together with the percentage of each in the formulation, the directions for use, the pest(s) controlled, the method and rate of application, and any restrictions to be observed in application and handling. If an antidote is known it must be listed. The FDA also sets legal tolerances for pesticides on or in raw agricultural commodities. Moreover, it sets the "safe" intervals between last application of the insecticide and the time of harvest or slaughter of the crop or animal. Through the cooperative supervision of the FDA and USDA, both the pesticide user and the consumer of the product are safeguarded (Sections 23.6.1 and 23.6.2). Additionally there are many antibiotics that cannot be used in food animals (http://www.fda.gov/cvm).

22.7.2 United States Department of Agriculture (USDA)

The Food Safety and Inspection Service (FSIS) of the USDA is charged with maintaining the wholesomeness and safety of meats being processed in packing plants that are involved in the commerce of meat and meat products (including inspection of poultry and poultry products). Veterinarians and trained lay inspectors examine each animal carcass and its internal organs before it can be sold for human consumption. Some meats can be sold intrastate when inspected by state employees. However, food-processing plant standards must have USDA approval for both interstate and intrastate sales.

The Veterinary Services Division (VSD) of the APHIS (Animal and Plant Health Inspection Service) of the USDA is responsible for programs to control and eradicate, if possible, certain diseases of livestock (e.g., brucellosis, tuberculosis, scabies). The VSD conducts nationwide state–federal cooperative programs for control and eradication of animal diseases, suppresses spread of disease through control of interstate and international movements of livestock, keeps informed of the overall animal disease situation nationally and internationally, administers regulations to ensure humane treatment of transported livestock and laboratory animals, oversees the licensing of biologics (vaccines, diagnostic reagents, serums), collects and disseminates information on morbidity and mortality, and provides training for USDA–APHIS employees and others in some related government agencies.

22.7.3 United States Department of Health and Human Services (USDHHS)

The U.S. Public Health Service (USPHS) section of the USDHHS is largely concerned with the prevention and treatment of disease in humans. It functions in vector control, prevention of water pollution and pollution abatement, and control of communicable diseases. It traces the causes and sources of epidemics. A part of this important complex is the National Institutes of Health (NIH) organized in 1930 and composed of 11 sister institutes including the National Cancer Institute. Aside from its own research endeavors the USPHS provides funds for health-related research at many universities and research institutes in the United States.

22.7.4 Local and State Agencies

These agencies cooperate with the foregoing federal agencies in an effort to safeguard animal and human health.

22.8 PROTECTING UNITED STATES LIVESTOCK FROM FOREIGN DISEASES

Today animal diseases pose international threats. Distance no longer provides a buffer against invasion. A jet air transport can outpace the development of **clinical** signs in an animal that has been exposed to a disease just prior to shipment. (More than 90 percent of animals imported into the United States from overseas countries enter by air.) **Epizootic** diseases capable of debilitating or destroying livestock populations still exist in Asia, Africa, and Latin America where animal diseases and pests constitute the most important limitation to animal production. Some of the most important epizootic diseases are rinderpest, contagious bovine pleuropneumonia, foot-and-mouth disease, BSE, African horse sickness, African swine fever, hog cholera (also called classic swine fever), Newcastle disease, fowl influenza, trypansomiasis, East Coast fever, and piroplasmosis. These, as well as Rift Valley fever, are potential hazards to the health of United States livestock and poultry.

The movement of foreign livestock into the United States was unregulated prior to 1875 at which time the United States prohibited the importing of cattle and hides from Spain where foot-and-mouth disease was rampant. By the 1880s the situation was reversed. European countries were refusing to buy U.S. cattle or beef for fear of spreading contagious bovine pleuropneumonia. In 1884 Congress established the Bureau of Animal Industry in the Department of Agriculture and gave the Secretary of Agriculture authority to enforce **quarantine** laws. Today there are numerous stations where employees of the USDA's Veterinary Services Division inspect animals, poultry, and certain animal products to be imported into the United

Figure 22.16 USDA veterinarian checks ear tags of calves from Channel Islands.
Courtesy of Veterinary Services Division, USDA.

States (Figure 22.16). If no evidence of communicable disease is found animals may be quarantined for a period of time and if no communicable diseases appear during this period they are released to the purchaser. Animals under quarantine are held in strict isolation. The USDA veterinarians give them precautionary treatment against external parasites and check that the blood test for piroplasmosis in horses and tests for tuberculosis and brucellosis in cattle were performed in the country of origin.

The VSD is also responsible for maintaining stringent safeguards on the entry of zoo animals into this country (Figure 22.17). Wild animals may be brought into the United States only after extensive quarantine periods abroad and a further quarantine period at the animal quarantine station at Newburgh, New York. Moreover, they are allowed to go only to approved zoos in which animals are isolated from domestic livestock and where proper measures are taken to dispose of waste to help prevent the spread of diseases.

22.9 SUMMARY

Even before recorded history people probably compared diseases of humans with those of animals. However, it has been only relatively recently that **epidemiological** links between diseases of humans and animals have been revealed. In this chapter an attempt has been made to give a panoramic view of the relation of animal health to public health. A discussion of selected arthropod-borne diseases was also included. Others will be discussed in Chapter 23.

It is possible that most infectious and parasitic diseases of humans originated in animals. The constant challenge of nature to animal health often is also a challenge to the health of

people. Several zoonoses (anthrax, plague, rabies, and bovine tuberculosis) are nearly equally harmful to humans and animals. Others (brucellosis, Q fever, and hydatidosis) cause more serious illness in people than in animals. A third group (foot-and-mouth disease, pasteurellosis, and pseudorabies) is less harmful to humans but serious in animals. More than 200 diseases can be transmitted from animals to people, and occasionally vice versa. An understanding of their modes of transmission and control enables humans to associate safely with animals and to utilize the products and services that they provide for humanity.

The universal insidiousness of the infections and **infestations** in the wide and somber spectrum of animal parasitism alone constitutes a major concern and challenge to modern agriculture and veterinary medicine. The diseases of humans and animals, and those common to both, constitute a serious impediment to the increases in production that are so vitally necessary to the world of today and tomorrow. An estimated 1.6 billion pounds of animal protein, which might be used to improve human **nutrition,** are lost each year as a result of animal diseases. The value of this meat in the marketplace is unrealized and the labor and capital invested in herds and flocks lost to disease are wasted. As world food needs increase it becomes imperative for humans to exercise improved control over animal health, which will greatly reduce morbidity and mortality losses of farm animals and greatly increase quantities of animal foods useful to humanity. Based on Food and Agriculture Organization of the United Nations estimates, "a 50 percent reduction of losses from animal diseases in the developing countries, which is a realistic goal, would result in a 25 percent increase in animal protein."

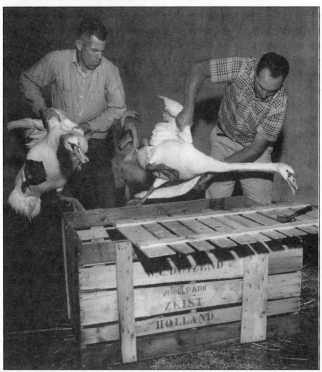

Figure 22.17 Veterinary Services Division checks health of zoo animals; giraffe is unloaded from ship (upper left); quarantine zebras are sprayed (upper right); after quarantine, white swans from the Netherlands are recrated for release to a bird farm (below).
Courtesy of Veterinary Services Division, USDA.

STUDY QUESTIONS

1. What is meant by public health?

2. Differentiate between infectious and noninfectious diseases.

3. Describe some ways that infectious diseases of farm animals may be spread.

4. List some ways that pathogens may gain entrance into an animal's body.

5. What is the body's first line of defense against pathogens?

6. What are the five cardinal signs of an inflammatory reaction?

7. What stimulates the production of antibodies in an animal?

8. Differentiate between natural and acquired resistance to disease. What is meant by species immunity? Give two or more examples of genetic resistance to disease.

9. What is a vaccine? Give one or more examples of the different types of vaccines.

10. Define prophylactic. Define therapeutic.

11. Which mammals are susceptible to rabies?

12. How is ornithosis transmitted to humans?

13. What significant contribution did Jenner make to the health of humanity?

14. Can foot-and-mouth disease of livestock be transmitted to humans?

15. Cite examples of some common vectors and some carrier animals of rickettsial diseases that may be transmitted from animals to humans.

16. Why is vaccination of humans against bovine tuberculosis not advisable in the United States?

17. How has the cooking of garbage intended for swine food aided in reducing the prevalence of swine tuberculosis in the United States?

18. Name three species of *Mycobacterium* and three species of *Brucella* that are of public health significance.

19. For what purposes are the following used: (a) market-cattle test program,
(b) brucellosis ring test, and
(c) brucellosis card test?

20. Why is *undulant fever* an appropriate name for brucellosis in humans?

21. How may a person acquire leptospirosisl?

22. What was the first disease of animals and humans to be traced to a microscopic agent source?

23. What was the first disease against which a bacterial vaccine was found to be effective?

24. Is the reported occurrence of salmonellosis increasing or decreasing in the United States?

25. Is malaria a current threat to the health and well-being of humans? What is the best way to prevent malaria?

26. What is histoplasmosis. Does this disease affect people?

27. Describe the life cycle of the trichina worm (*Trichinella spiralis*).

28. Describe the life cycle of the beef tapeworm (*Taenia saginata*).

29. What is the effect of quick-freezing and thorough curing or salting of meat on pork and beef tapeworm larvae?

30. Describe the life cycle of the dog hookworm (*Ancylostoma caninum*).

31. Differentiate between cutaneous and visceral larval migrans.

32. Why has BSE (mad cow disease) attracted so much public attention during the past 15 years?

33. Give an example of a viral and a bacterial infection of humans that may be transmitted to animals.

34. How may animals be nonclinical carriers of diseases? Cite three or more examples of public health importance.

35. What is meant by a LD_{50} value?

36. Cite several examples of plants that may accumulate cyanide or nitrate. Are all of these wild plants? Of what significance are mycotoxins to animal and human health? Cite examples.

37. Cite an example of a plant that interferes with the metabolism of an essential nutrient.

38. Of what public health significance are the poisons of the white snakeroot plant, jimmyweed, or rayless goldenrod when consumed by lactating cows? Cite a historical example.

39. Where would one write to obtain information pertaining to the poisonous plants of a given state? (*Hint:* See Appendix D.)

40. What effect does overgrazing of pastures often have on the populations of undesirable plants?

41. What is coumarin? Is dicoumarin of importance to livestock producers? Why?

42. What are the primary missions of the U.S. Food and Drug Administration?

43. What role does the USDA–FSIS play in providing wholesome and safe meats for humans?

44. What governmental agency is responsible for the labeling of pesticides? For setting legal tolerances for pesticides on or in raw agricultural commodities?

45. What is the function of the U.S. Public Health Service in the prevention and treatment of disease?

46. What is meant by foreign animal diseases? Cite important examples.

47. Briefly discuss controls and safeguards on the movement of livestock and zoo animals into the United States.

48. Cite several zoonoses that are nearly equally harmful to humans and animals; cite some that cause more serious illness in humans than animals; and finally, cite examples that are less harmful to humans than animals.

49. Why is it imperative for humans to exercise improved control over animal health throughout the world?

23

SELECTED INSECTS AND PARASITES OF SIGNIFICANCE TO HUMANS AND ANIMALS[1]

Insects are our dangerous rivals for the food supplies of the world, and they are important rivals and enemies in many other ways.

Leland Ossian Howard (1857–1950)
USDA Entomologist
Insect Book (1942)

23.1 INTRODUCTION

In no branch of the animal kingdom did nature provide such extremes of color, shape, and size as in **insects.** They range in size from tiny beetles that can literally creep through the eye of a needle to the well-camouflaged tropical walkingsticks, which are often more than a foot in length. Their colors represent all tints of the rainbow. Some crawl slowly, others can fly fast (dragonflies have been clocked at 60 mph).

There are more species of insects than any other form of animal life. According to a survey conducted by the **USDA** less than 0.04 percent of the known insects are harmful to humans. There are approximately 600 injurious **species** of insects of primary importance in North America. Even this relatively small number cause considerable damage and even death (directly or indirectly) to humans. During the Civil War more soldiers died of typhoid fever (transmitted by houseflies) than were killed in combat. During World War II Allied armies in the Pacific theater had 5 times as many casualties from malaria (transmitted by mosquitoes) as from combat. Each American pays several dollars more for cotton goods because of boll weevils. (Estimated annual damage exceeds $200 million, or about one of every seven bales produced.) The annual cost of controlling insects in the United States, when added to their damage to crops and **livestock,** is estimated by the USDA to be more than $1600 million.

Because insects have six legs they are classed as *Insecta* (some scholarly authorities also use the term *Hexapoda* mean-

ing "six-legged"). True insects have three body segments: (1) the *head,* which contains the brain, antennae (feelers), eyes, and mouthparts; (2) the *thorax,* or "motor room," which holds the chief muscles used in flying, swimming, or walking (the legs and wings are always attached to the middle body segment); and (3) the *abdomen,* which contains the digestive, reproductive, and excretory organs. Because insects have no internal skeleton their shell or outside covering is composed of a tough or horny material that usually contains some percentage (0 to 60 percent) of a substance called *chitin* (an insoluble polysaccharide). In the growing process insects shed this outer skeleton several times, each time forming a new one. This process is called *molting.* The young insects between **molts** are called **instars** and the periods of time between molts are called *stadia.*

Insects respond to the environment as cold-blooded animals. Therefore their activity is governed largely by air temperature. Scientists have demonstrated that one can determine the room temperature by counting the number of chirps certain insects make per minute and applying a simple mathematical formula. A scientist at Harvard University observed that he could determine the **ambient temperature** by timing the speed of ants running along a mountain trail. Because insects are invertebrates they possess no spinal cord. Instead, they have *ganglia* (knots of nerves) spaced along paired nerve cords on the **ventral** side of their bodies. Each ganglion controls certain phases of the insect's activity and is often capable of functioning independently. Hence a **decapitated** ant may survive for several weeks.

Not all winged insects have the same number of wings. True flies have two wings; bees, butterflies, dragonflies, moths, and wasps have four. Insects can taste and smell. The sense of taste of a monarch butterfly for certain sweets is 1200 times as

[1]The authors acknowledge with appreciation the contributions to this chapter of Dr. R. D. Hall, Department of Entomology and Associate Vice Provost, Office of Research, University of Missouri–Columbia.

TABLE 23.1	The Animal Kingdom

Phylum (selected classes of the subphylum Vertebrata)*	Examples (common names)	Estimated no. of living species described[†]
Chordata		
Amphibia	Frogs, salamanders, toads	2000
Aves	Birds, fowls	15,000
Mammalia	Humans, bats, cats, cattle, horses, whales	10,000
Pisces (Osteichthyes)	Bony fishes	25,000
Reptilia	Alligators, lizards, snakes, turtles	6000
Subtotal		58,000
Arthropoda	See Table 23.2 for classes and examples	973,000
Mollusca	Clams, mollusks, oysters, slugs, snails	100,000
Echinodermata	Feather stars, sand dollars, sea urchins, starfish	5800
Annelida (Annulata)	Earthworms, leeches (segmented worms)	8000
Bryozoa (Polyzoa)	Moss animals, sea mats	4000
Brachiopoda	Lampshells	500
Nemertinea	Nemertines, ribbon worms	600
Nemathelminthes (Aschelminthes)	Filariae, roundworms, trichinae, wheel animalcules	12,000
Platyhelminthes	Flatworms, flukes, tapeworms	15,000
Ctenophora	Comb jellies, sea walnuts	100
Coelenterata	Coral animals, hydras, jellyfishes	10,000
Porifera	Sponges	5000
Protozoa	Amoebae, euglenas, malarial organisms, paramecia, trypanosomes	30,000
Grand total		1,222,000

*Not all phyla are listed (only the most important ones).
[†]These estimates represent the mean of several references.

sensitive as that of humans. The *antennae* are used to detect odors and sounds. Some male moths can fly for miles in the dark by following scent trails through the air, thereby discovering females. Many insects have effective sound-transmitting devices. (Some have been known to transmit sounds for a distance of a mile.) Those that undergo a *complete* **metamorphosis** have four stages: (1) egg, (2) **larva,** (3) **pupa,** and (4) adult. The branch of biological science that deals with the study of insects is called *entomology.*

Dogs were among the first animals **domesticated** by humans (Chapters 1 and 4). The first domesticated dog probably brought with it a familiar **parasite,** the flea, which was recognized first as a nuisance, later as a **vector** of **disease.** Fossils indicate that fleas of the same form, size, and structure as those of today were living on ancestral dogs some 30 million or more years ago in the Oligocene epoch of the Tertiary period. From minute mice to enormous elephants, all animals are plagued by parasites. A *parasite* is defined as an organism living in or on, and at the expense of, another **organism** (the host) while rendering nothing in return. Parasites "steal" food and shelter from other animals and thereby sap the strength and health of their **hosts.** The study of parasites (both **microscopic** and **macroscopic**) including their life histories, modes of transfer from one host to another, and effects on the host constitutes the discipline known as *parasitology.* The fields of entomology and par-

asitology often overlap. Entomologists may specialize in the study of insect species that are parasites of humans and domestic animals, or they may study the parasites that affect (and in some instances offer control of) these pest insects.

23.2 TAXONOMY

Before reviewing selected insects and other parasites of importance to the health and well-being of humans and animals, it seems desirable to classify animals briefly according to their presumed natural relationships (Table 23.1). The seven basic levels in classification are

Kingdom
Phylum
Class
Order
Family
Genus
Species

Genus and species names are italicized when they appear in print. When typed or handwritten they should be underlined. The combination of a genus and species name is termed a *binomen* and it provides a unique description of each animal species. No two genera in the animal kingdom may have the same name; however, this rule does not apply to specific names in separate genera. Therefore to properly designate an animal by its scientific name both the genus and species must be given. Genera names always begin with a capital letter; species names never do. The binomen is often followed by the name of the person who first described that particular species. If the describer's name appears within parentheses it means the species was originally described in a genus different from the one in which it is currently placed.

The description and naming of animal species form an important part of the work done by *taxonomists*. Many entomologists are engaged almost exclusively in this type of endeavor. The application of names to animal species is governed by an extensive set of regulations known as the International Code of Zoological Nomenclature, which is used by scientists throughout the world. Regardless of native language, nationality, or ideology, zoological nomenclature is one of the few areas where worldwide cooperation exists.

23.2.1 The Five Phyla of Foremost Importance in Animal Science

1 Protozoa (amoebae, ciliates, and flagellates) can be disease agents. Internal parasites impair or destroy tissues, organs, and/or systems. Certain **protozoa** cause malaria, the loss of red blood cells, and damage to the **reticuloendothelial system.** Others cause anaplasmosis in cattle, which results in the destruction of red blood cells and consequently in **anemia.** Still another protozoan causes trypanosomiasis in humans, which is characterized by **fever,** anemia, **erythema,** and often damage to the central nervous system (Chapter 22).

2, 3 Platyhelminthes (flukes and tapeworms) and Nemathelminthes (nematodes and filariae) may cause disease (Chapter 22). These internal parasites impair the function of organs and/or systems. Essentially all tissues and organs are subject to their damage by blood loss, mechanical attack, toxic effects, **allergic** reactions, and **nutrient** losses; from egg deposition in tissues; and by predisposing the mucosa to bacterial invasion.

4 Arthropoda (insects, mites, and **ticks**) are important as internal and external parasites causing *direct* effects such as irritation and worry, blood loss, **envenomization, dermatosis,** allergy, **myiasis,** and other related **infestations** or having *indirect* effects as vectors (mechanical or biological) or as **intermediate hosts** for bacteria, protozoa, **rickettsiae, viruses,** and **helminthes.**

The phylum Arthropoda includes the invertebrates that are bilaterally symmetrical (i.e., when the body is cut exactly in half lengthwise the right half forms a mirror image of the left and vice versa), with segmented bodies and appendages. Selected examples within the phylum Arthropoda are given in Table 23.2.

5 The phylum Chordata, which includes humans and domestic animals, is important in animal science because its con-

TABLE 23.2	The Phylum Arthropoda*
Classes	**Examples (common names)**
Insecta	All true insects. The class Insecta includes members of the phylum Arthropoda that, as adults, have three body regions (head, thorax, and abdomen), one pair of antennae, three pairs of jointed legs, and usually two pairs of wings.
Selected orders	
Anoplura	Bloodsucking lice (e.g., cattle lice and swine lice)
Coleoptera	Beetles, weevils
Diptera	Flies, gnats, midges, mosquitoes
Hemiptera	True bugs (e.g., bedbugs and stinkbugs)
Hymenoptera	Ants, bees, wasps
Isoptera	Termites
Mallophaga	Biting lice (e.g., bird lice)
Orthoptera	Cockroaches, crickets, grasshoppers, katydids
Siphonaptera	Fleas
Chilopoda	Centipedes
Diplopoda	Millipedes
Arachnida	
Selected orders	
Acarina	Mites and ticks
Araneida	Spiders
Phalangida	Harvestmen, or daddy longlegs
Scorpionida	Scorpions
Minor orders	Pseudoscorpions, whip scorpions
Crustacea	Barnacles, crabs, crayfish, cyclops, lobsters, sow bugs, water fleas
Minor classes	Bear animalcules, king crabs, pauropods, peripatuses, sea spiders, symphylans

*Adapted from "Insects," *USDA Yearbook,* 1952, which defines arthropods as jointed-legged invertebrate animals.

stituents may serve as hosts, intermediate hosts, or **reservoirs** for many parasites that cause diseases of humans and animals. Selected external and internal parasites and their significance to humans and/or animals are presented in Tables 23.3 and 23.4.

23.3 CONTRIBUTIONS OF INSECTS TO HUMANS

The class Insecta is the largest of all classes of animal life. In fact, of every five described (known) kinds (species) of animals in the world, approximately four are true insects. Fortunately only a relatively small number of these known insects are "harmful" to humans. The others are either "beneficial" or have no known economic or health importance. It would be remiss not to mention some ways in which insects are useful to humans.

Insects produce and collect many products or articles of commercial value. Many *secretions* of insects are valuable; for example, a silkworm secretion is the true natural silk. The silk industry began in China where the source of silk was kept secret for more than 2000 years. Attempts to take silkworm eggs out

TABLE 23.3	Selected External Parasites of Humans and/or Animals and the Disease or Condition Each May Cause

Class	Disease or condition	Host(s)
Insecta		
Chewing lice	Irritation	Domestic animals, including poultry
	Intermediate host	Dog (dog tapeworm)
Sucking lice	Blood loss, irritation	Humans, domestic animals
	Typhus, trench fever, relapsing fever	Humans
Mosquitoes	Malaria, dengue, filariasis, yellow fever, **encephalitis**	Humans
Biting flies: stable flies, horn flies, horseflies, black flies	Blood loss, irritation, vectors, intermediate hosts	Humans, domestic animals
Arachnida		
Mites	Itch or scab, mange, blood loss, irritation	Humans, cattle, horses, sheep, poultry
Ticks	Blood loss, irritation	Humans, domestic animals
	Anaplasmosis	Cattle
	Rocky Mountain spotted fever, tularemia, Lyme disease	Humans, dogs
	Equine piroplasmosis	Horses

TABLE 23.4	Selected Internal Parasites of Humans and/or Animals and the Disease or Condition Each May Cause

Phylum	Disease	Host(s)	Infection
Protozoa	Malaria	Humans	*Anopheles* mosquitoes
	African sleeping sickness	Humans, cattle	Tsetse flies
	Anaplasmosis	Cattle	Humans, ticks, insects
	Trichomoniasis	Cattle, humans, swine	By contact
Platyhelminthes	Schistosomiasis (lung and liver flukes)	Humans (snails are intermediate hosts)	Skin contact with contaminated water
	Tapeworms	Humans, domestic animals (arthropods are often intermediate hosts)	Ingestion of intermediate host or (rarely) the eggs
Nemathelminthes	Hookworms	Humans, dogs	Through the skin
	Ascaridiasis	Humans, swine, horses, dogs	Ingestion
	Trichinosis	Humans, rats, swine	Ingestion of raw pork
Arthropoda	Internal myiasis, bots, warbles	Humans, cattle, horses	Direct (egg or larva laying)
	Traumatic myiasis (screwworms)	Humans, cattle, horses, sheep	Direct (egg laying)

of the country were punishable by death. Beeswax is secreted as thin scales, or flakes, by the hypodermal glands on the underside of the worker bee's abdomen. Bees use wax to make the comb in which honey is stored and the young are reared (Chapter 3). Shellac is secreted from the hypodermal glands found on the back of a scale insect of India. The light-producing secretions of giant fireflies in the tropics are used in minor ways for illumination and could lead to the synthesis of a substance that would give a brilliant light with almost no accompanying heat.

The *bodies* of some insects are useful or may contain certain useful compounds. Cochineal and crimson lake are pigments made by drying the bodies of a cactus scale insect found in the tropics. These pigments are used in cosmetics, as a coloring for beverages, in decorating cakes and pastries, and as a dye for textiles. Cantharidin is a dangerously irritating substance obtained from the dried bodies of a European blister beetle

commonly called *Spanish fly*. Insects such as the dobson, or hellgrammite, are widely used as fish bait and the best artificial flies used for fishing are modeled after insects. The bodies of insects serve as food for many animals that are valuable to humans. Many fish used for food subsist largely on aquatic insects. Many highly prized song and game birds depend on insects for a large portion of their food. Chickens and turkeys feed on insects naturally and under certain conditions can be raised almost exclusively on a diet of insects. Swine may feed and fatten on white grubs rooted from the soil. Several wild animals, for example, moles, raccoons, shrews, and skunks, eat many insects. In many regions of the world, from ancient times to the present, insects have been and are eaten extensively by humans. Grasshoppers (called *locusts* in the Bible), crickets, walkingsticks, certain beetles, caterpillars, and pupae of moths and butterflies, termites, large ants, aquatic bugs, cicadas, and honeybee larvae and pupae are foods that were highly prized by many of the more primitive races of humans. The Australian aboriginals are cited as an example of a living population whose diet includes insects.

Insects *collect, elaborate,* and *store* valuable plant products. Honey is nectar that is collected from plant blossoms, concentrated, modified chemically, and sealed in waxen "bottles" by the honeybee (Section 3.2.3). Insects cause plants to produce galls, some of which are valuable. They are especially useful as tanning materials and medicines. The Aleppo gall produced by a wasplike insect on several species of oaks in Asia and Europe has been used for centuries as a tonic, astringent, and antidote for certain poisons. The early Greeks used it for dyeing hides, mohair, and wool. Other galls have been used for dyeing fabrics and as tattoo dyes. The Aleppo dye is used in preparing the finest and most permanent inks.

Insects aid in the production of flowers, fruits, seeds, and vegetables by *pollinating* the blossoms as they search for nectar or gather pollen. It is estimated that if all bees were eliminated some 100,000 species of flowering plants would concurrently cease to exist. The Smyrna fig produces only female flowers and no pollen. Therefore the growing fig is dependent on small wasplike insects from another fig (the caprifig) which crawl into its flower cluster and thereby serve as pollinators. Most varieties of apples, sweet cherries, and plums would be barren without insect pollinators such as bees, butterflies, flies, moths, and wasps. Clover seed does not form without the visit of an insect (usually some kind of a bee such as the long-tongued bumblebee) to each blossom. When the plant was first introduced in New Zealand it failed to produce seed until bumblebees were introduced to transport the pollen from flower to flower. Certain beans, melons, squash, and many other vegetables require insect visits before the blossoms set (form fruit in the blossom). Many ornamental plants, for example, chrysanthemums, iris, orchids, and yucca (both in and out of greenhouses), are pollinated by insects.

Many insects *destroy* other injurious insects. This is accomplished by their action as *parasites*—living on or in their bodies and/or their eggs, or as **predators**—capturing and devouring other insects. The praying mantis (so named because of the position it assumes as it rests on twigs or stalks its prey) is a good example of the latter. It eats a large number of insects ranging in size from tiny aphids to large caterpillars. Some insects have even been imported from other countries to halt the spread of a serious insect pest. A classic example is the importation of the Australian ladybird, a beetle that was imported to save the citrus industry of California from the cottony-cushion scale insect. When 140 of the beetles reached San Francisco they devoured the insects at a fast rate and thereby preserved the lemon and orange industry. Further discussion of this subject as it relates to **biological control** of insects is presented in Section 23.6.2.

Many insects *destroy* harmful weeds in the same way they injure crop plants. Control of prickly pear by insects in Australia is an example of what can be accomplished in this way. In about 1787 cactus plants were taken to Australia for culturing cochineal insects for dye. Various species of cacti took root beyond the gardens and in the absence of natural enemies the prickly pears spread rapidly. Within a century about 60 million acres were affected (one-half so densely covered that the land was rendered useless). Then more than 500,000 insects of 50 different species were dispatched from North and South America to Australia. Several were successfully established. The insects checked the new growth of cactus and reduced the density of the plants thereby allowing grass to return. During the past several decades certain beetles have been released in the United States in a similar and successful program against thistle plants.

Insects *improve* the physical condition of the soil by burrowing throughout the surface layer (allowing air to penetrate the soil) and *enhance* its fertility through the fertilizing properties of their droppings and dead bodies. Insects perform a valuable scavenger service by devouring dead plants and the bodies of animals and by burying **carcasses** and **dung.**

Certain insects are *invaluable* in scientific endeavors. The ease of handling, rapidity of multiplication (short **generation interval**), great variability, and low cost of rearing and maintaining make certain insects valuable in the study of biochemistry, **ecology,** genetics, and **physiology.** Studies of variation, geographic distribution, and the relation of color and pattern to surroundings have been greatly advanced by studying insects. Many genetic fundamentals were derived from studies of *Drosophila* (pomace flies, or fruit flies). Principles of regeneration and **parthenogenesis** have been established through the study of insect physiology. The behavior and psychology of higher animals have been illuminated by studies of the reactions of insects such as the honeybee whose behavior can be analyzed into simple **tropisms.** Valuable information in sociology has been deduced from a consideration of the economy of social insects. Insects have been used in the **bioassay** of certain crops and levels of fertilization (Figure 23.1).

Insects have *aesthetic* and *entertainment* value. Their shapes, colors, and patterns serve as models for artists, decorators, florists, and milliners. The more highly colored and striking forms of insects are much used as ornaments in dishes and trays and in necklaces, pins, rings, and other jewelry. Butterflies and moths are admired universally whereas those using the microscope find much to admire in the colors and patterns of many smaller insects. The songs of insects have been found highly interesting for centuries. Insects have served as subject matter for many poems and songs. The inimitable variety and curious habits found in insects afford entertainment and diversion for thousands who collect and study them. Many persons have participated in the oriental game of gambling on crickets trained for fighting. Fleas have frequently been used for circus stunts.

Insects are often *useful* in the practice of medicine and surgery. Although not a particularly pleasant thought, it is a fact that the larvae (**maggots**) of certain flies (reared **aseptically**) have unique value in the treatment of certain wounds. Maggot therapy is currently being practiced in cases in which sophisticated medical science has failed. In brief, maggots tend to consume only necrotic (dead) tissue leaving healthy tissue alone. They also secrete natural antibiotics that speed and enhance

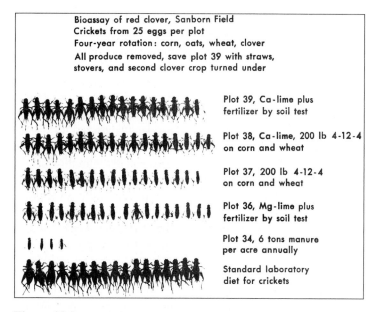

Figure 23.1 Crickets from 25 eggs per plot were used as test animals in the bioassay of red clover on Sanborn Field at the University of Missouri. *Courtesy of Dr. W. A. Albrecht and Missouri Agr. Expt. Station.*

wound healing. The stings of honeybees may have therapeutic value for disorders such as arthritis and rheumatism. Extracts from the bodies of honeybees and certain other insects are used to a limited extent as medicines.

Bees are now being bred with a preference for particular plants. Because bees are responsible for 80 percent of insect pollinating activity, this is an especially important agricultural development. Pollen preferences among bees can determine success or failure for crops such as alfalfa and clover.

23.4 HARMFUL EFFECTS OF INSECTS

The harmful effects of insects have been appreciated since biblical times. The Bible records in Exodus that God commanded Moses to stretch out his hand to make locusts (some commentaries say grasshoppers) come over the whole land of Egypt. This infestation was designed to intimidate Pharaoh who had rebelled against God. According to the scripture,

> The locusts went up over all the land of Egypt . . . so that the land was darkened; and they did eat every herb of the land, and all the fruit of the trees . . . and there remained not any green thing in the trees, or in the herbs of the field, through all the land of Egypt.

> **Exodus 10:14–15**

Insects *damage* all kinds of growing crops and other valuable plants. Destruction is accomplished in numerous ways, which include chewing bark, buds, stems, or fruits of the plants; sucking sap from buds, leaves, stems, or fruits; and boring or tunneling into the bark, stems, or twigs, into fruits, nuts, or seeds ("worms" and/or "weevils") or between leaf surfaces (leaf miners). Their damage takes the form of making cancer-

ous growths on the plants within which they live and obtain nourishment (gall insects); attacking roots and underground stems in any of the foregoing ways (subterranean or soil insects); laying eggs in plants; taking plant components for use in the construction of their nests or shelters; carrying other insects to a plant where they become established; disseminating organisms (bacteria, **fungi,** and viruses) that are pathogenic to plants and injecting them into the plant tissues in the feeding process by carrying them into insect tunnels or by making wounds through which such disease organisms may gain entrance; and causing cross-fertilization of certain disease-producing rusts whose aecia (**spores**) would not otherwise have developed.

Insects annoy and injure humans and animals in a number of ways. They cause *annoyances* by their mere objectionable presence in places; the sound of their flying or "buzzing" about; the unpleasant odor of their secretions and/or decomposing bodies (the increasing concentration of **organic** materials in the Great Lakes near Chicago has caused a reduction in the oxygen content of the water resulting in fewer insects making their home in the Great Lakes); or the offensive taste of their secretions and/or excretions left on food or tableware. They irritate by crawling and moving about the skin; chewing, nibbling, or pinching the skin; entering the ears, eyes, nostrils, or alimentary canal causing myiasis; and laying their eggs on feathers, hair, or skin. Some insects *injure* humans and animals by applying **venoms.** They may apply their venoms by means of a stinger and/or pincers; penetrating with nettling hairs; leaving caustic or corrosive body fluids directly on the skin when the insect is handled or crushed (e.g., blister beetles); or poisoning the animal on being swallowed (e.g., blister beetles cause irritation to the digestive tract when swallowed by horses and may poison other animals when present in stored forage or grain).

Certain insects make their homes on or in the body as external or internal *parasites* thereby injuring the host animal by irritating nerves when they crawl about; causing **inflammation** by chewing or piercing the skin; contaminating feathers or fur with their eggs and **excreta;** sucking blood; tunneling into auricular, muscular, nasal, ocular, or **urogenital** passages causing mechanical injury and/or promoting infections; and attaching to the stomach and intestinal lining thus mechanically blocking food passages, disturbing the digestive processes, causing ulcers or secreting **toxins.**

As noted in Table 23.5, insects and other arthropods may disseminate disease agents (bacteria, fungi, parasitic worms, protozoa, or viruses) from diseased to healthy animals, from certain wild animals (the reservoir) to humans or domestic animals, and from a diseased parent or antecedent life stage in one or more of the following ways: accidentally conveying **pathogens** from filth to food; transporting pathogens from filth or diseased animals to the eyes, lips, or wounds of healthy animals; infecting a susceptible large animal when it swallows the insect host of a pathogen; inoculating the pathogen subcutaneously via its proboscis or depositing the pathogen directly on the skin in regurgitated food or in its **feces,** allowing entrance of the pathogen through an insect bite, a scratch, or even the unbroken skin; and finally by serving as a host in which the pathogen may merely cling to the insect's body, multiply in the insect's internal organs, and undergo within the insect an essential part of its **life cycle** that cannot occur elsewhere.

Insects may *destroy* or decrease the value of stored products and possessions including clothing, drugs, food, animal and plant collections, books, papers, bridges, buildings, furniture, mine timbers, telephone poles, telegraph lines, railroad ties, and trestles. Insects accomplish these destructive processes in one or more of the following ways: devouring matter as food; contaminating it with their secretions, excretions, eggs, or even their own bodies (even though the product may not be eaten); seeking protection or building nests on or tunnels within these various things; and finally causing increased expenditures for labor in sorting, packing, and preserving certain items such as food and clothing. Pests of livestock lower the production of meat, milk, wool/mohair, and the value of hides.

23.5 SELECTED ARTHROPODS AFFECTING DOMESTIC ANIMALS AND/OR HUMANS

Among the most important and widespread infectious diseases of humans and animals are those spread by insects such as flies in **regurgitated** food and in their feces. Studies of the life cycles of these disease vectors provide knowledge related to ways and means of better controlling them and the diseases they transmit.

23.5.1 Flies

Houseflies

If there is anything "as harmless as a fly" it is not the common housefly. Although dark-gray, dull-looking houseflies seldom attack farm **mammals** they do annoy them and spread filth. The tiny hairs (*setae*) on the legs of houseflies usually carry millions of bacteria, some of which are pathogenic and cause such diseases as dysentery and diarrhea. They also can carry microorganisms on their footpads, in regurgitated food, and in their feces. Studies have shown that these flies transport disease-producing organisms as far as 15 to 20 miles. They may place microorganisms inside the human body (through contaminated food) that cause such diseases as cholera, dysentery, tuberculosis, and typhoid.

Whereas most insects have four wings, flies have only two. They thus belong to the order Diptera, which means "two wings." (These two wings are almost as transparent as window glass.) The wing joints vibrate and cause a buzzing sound when the fly beats its wings. However, this sound is partially muffled by the rush of air created by the beating wings. Do you remember how difficult it is to catch or hit a housefly? This is because the fly is equipped with two large *compound eyes* that cover most of its head. These eyes are composed of thousands of tiny jewel-like parts, called *facets,* which enable the fly to see in different directions simultaneously and to become quickly aware of movement. On top of the housefly's head (between the large eyes) are three very small eyes, called *ocelli.* The *antennae* (feelers) are also between the compound eyes.

The sucking mouth tube, called the *proboscis,* is shaped like a funnel turned upside down. The fly uses this proboscis as one might use a straw in a glass of milk. Small solid particles of food (such as sugar) may be ingested by the fly's first dissolving

TABLE 23.5	Selected Diseases of Humans Transmitted by Arthropods
Vector	**Disease**
Mosquitoes	
Anopheles	Malaria
Culex	Filariasis (elephantiasis)
Aedes	Yellow fever, dengue fever, equine encephalitis
Flies	
Houseflies	Typhoid fever, paratyphoid fever, dysentery, Asiatic cholera
Deerflies	Tularemia
Tsetse flies	African sleeping sickness
Sand flies	Leishmaniasis
Lice, *Pediculus humanus*	Epidemic typhus fever, relapsing fever
Fleas	
Rat flea	Bubonic plague
Other fleas of wild rodents	Endemic (murine) typhus fever
Ticks: wood, dog, rabbit	Spotted fever, tularemia, equine encephalitis, Colorado tick fever, relapsing fever, Lyme disease
Mites, trombiculids (chiggers)	Tsutsugamushi disease (scrub typhus), rickettsial pox

them with saliva. The mouthparts function as a sponge. Because the proboscis is soft the housefly cannot bite. Tiny pads, called *pulvilli,* are attached to each of the fly's six feet. Tiny hairs cover these pads and secrete a sticky fluid that enables the fly to cling to smooth surfaces and walk upside down on ceilings. The housefly breathes through minute openings, called *spiracles,* which lie in a row on each side of the thorax and abdomen.

A female may lay from 2 to 25 batches of approximately 300 to 500 eggs each during her lifetime of 2 to 4 weeks. Female houseflies prefer to lay their eggs in fresh **manure,** garbage, or fermenting vegetable waste. Tiny, white, legless larvae (maggots) hatch 12 to 48 h after the eggs are laid. Before developing into adults the larvae mature by feeding for 4 to 14 days on the material about them and then pupate. Control of houseflies is difficult because they breed in so many places. Sanitation, therefore, is the key to their control. Manure must be kept in flyproof areas or spread thinly over fields. Applications of **insecticides** as space or residual treatments are a necessary adjunct to sanitation in controlling the housefly.

Control of houseflies has become increasingly important to livestock producers during recent years. Animal confinement facilities that originally may have existed in remote rural areas have often been surrounded by new residential developments. In such cases, flies originating at the agricultural operation and subsequently annoying homeowners nearby may be sufficient grounds for legal actions such as private nuisance lawsuits. Decisions frequently instruct the livestock producer to control such fly populations or risk being ordered out of business.

Horn Flies

These small black flies are about one-half the size of houseflies. They feed chiefly on cattle (as many as 4000 horn flies may settle on one animal) but may also attack other mammals such as horses and bison. Because they feed only on blood, a stabbing or sucking process must be employed. In cases of severe infestation this may cause cattle to lose weight and lactating animals to secrete less milk. According to some USDA estimates, horn flies reduce the potential value of cattle and milk production in the United States by $150 million annually. Beef cattle subjected to heavy attacks of horn flies during the fly season were observed to gain 50 lb less than cattle treated with insecticides. Research at the University of Illinois revealed that dairy cows protected from horn flies produced 10 to 20 percent more milk than unprotected ones. The entire adult life of this fly is spent on the host, except for the brief time in which it leaves to deposit its eggs beneath fresh, warm cattle dung. It takes about 1 week for eggs to develop into adults. An application of residual insecticide to the host is an effective way to control horn flies. Insecticide-treated ear tags and various self-applicating devices (back rubbers and dust bags) can afford excellent control.

Stable Flies

These flies, often called *biting houseflies,* resemble houseflies in appearance but are actually bloodsuckers; this ability is made possible by a hard, piercing proboscis. The stable fly is known as the *dog fly* in some areas of the United States. Stable flies are annoying to cattle and horses because they frequently bite the legs and lower parts. (They may also feed on humans.) They lay eggs in moist, fermenting straw and manure; in decaying vegetable matter (especially at the base of **silos**); and frequently on the moist bottom of large, round hay bales stored outside. About 3 weeks is required to develop from egg to adult. Unlike the horn fly, the stable fly stays on an animal only long enough to engorge itself with blood. During warm weather stable flies may feed several times daily although they may not be seen easily if grass is high. After feeding the adult rests on nearby walls, trees, or other objects to digest the blood meal. Stable flies are controlled by frequent cleaning of premises, manure disposal, proper storage of hay, and application of insecticides to facilities (resting places) and animals.

Screwworm Flies

These blowflies are bluish green with three dark stripes down their backs. They are somewhat larger than the housefly. Screwworm larvae resemble a ringed spinal column and have the appearance of a wood screw; hence, the name screwworm. Eggs laid in a human or an animal wound (or the navel of a neonate) hatch in 6 to 12 h. The larvae **ingest** living tissue for 3 to 10 days and then drop to the ground where further development occurs. (Depending on the temperature it may take a few days to 2 months for the larvae to pass through the pupal stage and emerge as adults.) The pupae are killed by hard freezes. Thus screwworm flies are more troublesome in the warm southern United States where, until recently, they caused an estimated annual loss of $20 million to livestock producers. Treating the wounds with smears that contain insecticide is a good **prophylactic** measure. The release of **sterilized** males (because the female usually mates only once) is a current method of control (Section 23.6.2). The flies historically spread northward in the summer with livestock shipments. The **dehorning** of cattle or goats during the fly season may expose animals to screwworm infestations. Odor from the infected wound attracts more screwworm flies, and if not treated, the animal will die.

Severe infestations of the Gulf Coast tick in the ears of cattle may predispose the animals to screwworm infestation. It is interesting that insecticidal ear tags, now widely used for fly control on cattle, were developed originally as an aid to control these ear ticks and hence to reduce screwworm populations.

Fleece worms are larvae of blowflies that are similar to the screwworm fly. They deposit eggs on the soiled wool of sheep. After **hatching** the larvae feed on the skin surface and cause severe annoyance. Most infestations in sheep occur in early spring when the wool becomes soiled by feces, urine, or rain. Infestations of fleece worms are often associated with screwworm infestations in sheep.

True screwworms are presently eradicated from the United States but still occur in Mexico, Central America, and northern South America. However, throughout the United States other species of blowflies may produce similar maggot infestations and are often mistaken for screwworms. For example, larvae of certain species of greenbottle and bluebottle flies that resemble larvae of screwworms annually infest fistulated cattle at the Missouri Agricultural Experiment Station. These same species often infest pet animals (especially cats and rabbits) when a female pet gives birth during the summer months. Applications of approved screwworm insecticides may be valuable in such cases. Because such *necrophilous* flies are readily attracted to dead animals (including humans), they often can serve as a biological yardstick of time since death. The field of *forensic entomology* is well accepted in judicial circles.

Heel Flies

These hairy black-and-yellow-striped flies look like small bumblebees. The common heel fly, which appears during the first warm days of spring, is nearly 3 times larger than the housefly. It has no mouthparts and does not bite or sting. Heel flies cause no pain but rather annoy cattle while depositing their eggs. Animals attempt to escape from them by running with their tails held high. The female heel fly (also called *warble fly,* or *cattle warble*) attaches eggs to hairs on the heels or legs if the animal is standing and if lying down, to the flanks of cattle and certain wild **ruminants.** These eggs hatch in 3 to 5 days. The larvae (called *cattle grubs*) penetrate through the skin and live in the **viscera** for about 6 months. They then migrate to the back, cause painful swellings containing the *warbles,* and finally cut breathing holes through the skin. The grubs develop under the skin for about 7 weeks and then drop to the ground to pupate. Pupation requires 2 to 11 weeks and yields winged adult flies. Because millions of tons of infested meat containing these grubs must be discarded and hides intended for leather uses are damaged, the economic losses to livestock producers are significant. (USDA estimates indicate losses of approximately $140 million annually in the United States.) Certain wild animals (e.g., deer) have learned to partially avoid heel flies by standing in water. Certain systemic insecticides (materials that are absorbed through the skin) can produce excellent control of cattle grubs. Care must be taken to apply such compounds at the prescribed time of the year. Insecticides in the class known as *avermectins* have given especially good results.

Botflies

The sheep botfly (sheep gadfly, or nasal fly) carried by sheep and goats may cause conjunctivitis in humans. The eggs hatch within the abdomen of the female sheep botfly. The females then deposit the larvae enclosed in a drop of sticky fluid into the sheep's nostrils. These larvae migrate to the nasal sinuses where they feed on mucus secreted by the tissues and mature for 6 to 8 months before dropping to the ground to pupate. They cause irritation of nasal passages and difficult breathing. The adult sheep botfly does not feed. Goats may become infested with the sheep botfly when they occupy ranges with infested sheep. Related species often parasitize deer and reindeer.

The ox warble (botfly) of cattle, goats, and other animals lays its eggs on hairs of the legs. The larvae hatch in about 4 days and then migrate up the hair and eventually penetrate the skin. These larvae pass through the **subcutaneous** connective tissue toward the diaphragm and finally migrate to the animal's

back. They then escape from their cysts and fall to the ground. This parasite causes considerable loss of meat and milk and damage to hides. The larvae migrate about in the subcutaneous tissues of humans causing so-called larval migrans or "creeping eruption" (swelling in the various affected body parts). This species is closely related to the heel fly.

The human botfly may attack humans and certain domestic animals. It is found largely in the tropics (especially in Central and South America and in Mexico). The presence of botfly larvae is very painful to humans in whom they migrate subdermally. In infested areas humans should have a well-screened house, use bed nets, and avoid bloodsucking flies and mosquitoes.

Horse botflies harass horses by darting around their heads in an attempt to lay eggs. (Females lay their eggs on hair of the horse's legs, throat, or mouth.) Larvae hatch on contact with the horse's tongue. The larvae, or *bots,* burrow into the mucous membrane of the tongue, mouth, and throat where they live for about 2 weeks before passing to the stomach. They then live for about 9 months in the horse's stomach and intestines before passing out with feces. They pupate in soil. There is one generation annually. Because adult botflies do not eat, their control depends on treatment of the host to eliminate bots. Carbon disulfide is a drug used to treat horses for botflies; it may be administered **orally** in a capsule or via a stomach tube by a veterinarian. The avermectins (ivermectin and moxidectin) are also highly effective in treating horses for bots. Occasionally eggs of the horse botfly may accidentally be transferred to a human mouth. The small larvae may subsequently cause discomfort along the gum line but they do not mature.

Farm animals seem to realize that botflies are dangerous to them. There is no pain associated with egg laying; the animals merely have an instinctive fear of the fly. This is probably nature's way of helping them avoid excessive parasitism. When these flies are buzzing among the **herds,** cattle and horses become greatly excited. This excitement reduces meat and milk production.

Horseflies and Deerflies

These beautifully colored (brown, black, orange, or metallic green) flies attack cattle, deer, horses, and other **homeotherms** including humans. Only the adult females are bloodsuckers; the males feed on plant juices and nectar. The lancelike mouthparts of the female are developed for cutting skin and sucking blood that oozes from the wound. (They are vicious biters.) The big, bothersome, bloodthirsty horsefly (gadfly) is familiar to and despised by all farm people. Its less conspicuous and smaller cousin, the deerfly, is equally despised in many areas where it causes pain by biting and may also spread tularemia (Chapter 22). Horseflies and deerflies breed in moist places. Their development requires several months and control is difficult. (However, pyrethrum sprays kill horseflies on contact.) Female horseflies lay their eggs in clusters, commonly on plants that overhang water or grow in wet soil. The larvae live in debris along ditches, swamps, and rice fields. Adult horseflies may carry infections such as anaplasmosis, anthrax, and tularemia. Horseflies and deerflies will move several miles from their breeding places in search of nourishment. Local control of horseflies is often facilitated by using canopy traps. A black object, usually a painted beach ball, attracts female flies to the trap.

Face Flies

These annoying pests of cattle and other livestock are closely related to the housefly. The larvae are found only in fresh bovine feces in which they feed; then they pupate in the adjacent soil. There are commonly several generations during the summer. The pupa may be distinguished from pupae of other species of flies by its color, greyish-white; those of other species are dark brown. Face flies do not bite or suck blood but their feeding is known to cause irritation to the eyes of cattle. They stay near or on cattle where they feed on animal secretions from the body **orifices** (they cluster around eyes, mouth, nose, and wounds). Horses bothered with face flies often stand head-to-tail (i.e., in reverse directions) flicking flies off each other's face. These flies are **carriers** of the bacteria causing pinkeye. Face flies **hibernate** in attics and the walls of houses. Insecticidal ear tags, dust bags, and sprays all offer some degree of control.

23.5.2 Mosquitoes

The female of the species is more deadly than the male.

Rudyard Kipling (1865–1936)
English author

In order to control mosquitoes, one must learn to think like a mosquito.

Samuel Taylor Darling (1872–1925)
American pathologist

All mosquitoes start life as eggs and must have water in which to develop. Eggs hatch into larvae, which are often called *wrigglers* because of their activity in water. The larva (after four molts) forms a shell and then passes through the pupal stage. Finally it leaves the shell as an adult. Female mosquitoes are stronger than males and usually live longer. Moreover, they (only females bite) are the world's number one vector (carrier) of human diseases (Chapter 22). Male mosquitoes feed on plant juices such as the nectar of flowers and decomposing fruits. Female mosquitoes exhibit a preference for specific host animals. Some species (especially in the genus *Culex*) feed on birds whereas *Aedes* and *Anopheles* prefer domestic animals and humans.

23.5.3 Ticks

All ticks are parasitic on animals. Adult ticks have four pairs of legs. Many enjoy long life spans and can survive months or perhaps years between blood meals, which are required at the time of egg production. Some have several successive hosts; for example, the Gulf Coast tick may pass its first stage on a quail, the second on a rat, and the third on a cow. The life history of a tick involves four stages: (1) the egg (almost all eggs hatch); (2) larva, or **seed tick;** (3) **nymph;** and (4) adult. Eggs are not deposited until the engorged female (body size may be increased 10-fold

or more with blood) has left the host. Some ticks (e.g., the **fowl** tick) lay a few hundred eggs, return to the host for another blood meal, and then lay more eggs; they may repeat this process several times. However, most ticks lay only a single, large batch of eggs and die when **oviposition** is completed. The six-legged larva, or seed tick, must find a suitable host for engorgement when it hatches if the life cycle is to be continued. Larval metamorphosis terminates in an eight-legged nymph that engorges and molts into an adult. Many individuals in each of the developmental stages (larvae, nymphs, and adults) die without finding suitable hosts. Ticks are blood feeders and because of this transmit disease-producing organisms to humans and domestic animals. (They rank number one in transmitting animal diseases.) In range animals and certain wild mammals (especially the elk and moose) death may result from heavy infestations of the winter tick, *Dermacentor albopictus.* They are particularly detrimental to hosts in late winter and early spring when the animal's food supply is often short.

For many years cattle ranchers feared "Texas cattle fever" (also called *cattle tick fever* and *red water fever*), which is characterized by high fever, red corpuscle destruction, enlarged spleen, and engorged liver, and frequently terminates in the animal's death. In 1890 Drs. S. Theobald Smith and F. L. Kilborne of the USDA linked the transmission of this disease with infestations of cattle ticks, which were later found to transmit the causative organism (*Babesia bigemina*). Their observations, coupled with studies and descriptions of the life history and characteristics of the cattle tick by Dr. Cooper Curtice (a USDA veterinarian), pointed the way for studies that later resolved similar problems with respect to parasitic vectors that spread such human diseases as malaria, yellow fever, tularemia, typhus, and Rocky Mountain spotted fever (Section 22.3.2).

Ticks may be classified as *scutate,* or hard ticks, which are vectors of rickettsia, Lyme disease (Section 22.3.3), tularemia, certain viruses, and many animal diseases, and *nonscutate,* or soft ticks, which are responsible for **endemic** relapsing fever in humans. Examples of these groups include (1) the *one-host* scutate tick (i.e., it passes through all stages on one host), *Boophilus annulatus,* the cattle tick; (2) the *two-host* nonscutate tick, *Otobius megnini,* the ear tick; (3) the *three-host* scutate tick, *Dermacentor variabilis,* the American dog tick; and (4) the *many-host* nonscutate tick, *Argas persicus,* the fowl tick. Certain ticks may pass disease-producing agents to their young. This type of transmission is called *transovarial* and explains how a one-host tick can be a vector of disease-producing pathogens.

23.5.4 Mites

These tiny tick cousins are about 1/25 inch or less in length and have four pairs of legs. Most **poultry** mites engorge themselves with blood and then hide in cracks in the walls and floor of the poultry house. However, a second group spends its entire life cycle on birds. (Chicken mites, *macronyssids,* are known to kill chickens by **exsanguination.**)

The northern fowl mite, *Ornithonyssus sylviarum,* is an important external parasite of caged laying hens in the United

States. Infestations are usually centered near the cloaca (or vent) of the chicken and may reduce its egg output. Roosters are often severely affected and frequently die unless treated. Persons engaged in handling infested poultry or eggs may be bitten. Treatments for control of the northern fowl mite, consisting of *acaricides* applied to the poultry as sprays or dusts, are not generally effective unless the birds' vent regions are adequately covered. Modern poultry facilities that house many thousands of hens in limited-access cages compound the treatment problem. Acaricide-treated plastic strips, similar to insecticide ear tags for cattle, have given excellent experimental control of northern fowl mites when the strips were hung in poultry cages. Such slow-release plastic formulations are useful in poultry pest-management programs.

Mange and scab mites burrow into their host's skin and reside there for life. Common sheep scab is caused by *psoroptic* mites,[2] which reside on the host's skin. The mites live on blood that oozes from skin punctures made with their sharp mouthparts. The skin reacts by **exuding serum** that forms itching crusts and patches of wool soon fall out. Scab is readily transmitted to other sheep, goats, and probably rabbits. *Sarcoptic* mites burrow into the skin of the face and head. *Chorioptic* mites reside and cause sores on the skin surface of the legs. *Demodectic* mites burrow into hair follicles and skin glands. Much leather is rendered of poor quality because of damage caused by mange. Regular systematic dipping controls the scab-causing psoroptic, chorioptic, or sarcoptic mites; demodectic mange is more difficult to control but several newly registered acaricides have given good results. These four types of mites also attack cattle, dogs, other animals, and humans. Horses are subject to their own psoroptic mange.

Two kinds of mange in swine are caused by mites. The most common is known as sarcoptic mange; demodectic (follicular) mange is the least common. The causative mites spend their entire lives on infested swine in which they produce skin wounds, or **lesions,** as they feed on the host's tissues and blood. The feeding and burrowing of mites cause irritation, itching, inflammation, and swelling of the affected tissues. Nodules and vesicles form, which break and discharge serum that dries into large granules (scales) that characterize mange. Sarcoptic mange spreads primarily by direct contact with infested swine. (Swine have a habit of resting and sleeping in close contact with one another.) The disease can be transmitted to people and to certain other animals although the mites live only a limited time on such new hosts. Demodectic (follicular) mange of swine is caused by microscopic (the adult female is about 0.01 in long), parasitic, wormlike mites, called *Demodex folliculorum suis*. They penetrate the hair follicles and oil glands of the skin where they complete their life cycle. A completely effective **therapy** of demodectic mange has not been developed but new acaricides show promise.

The so-called 7-year itch is caused by mites. Actually it does not (and never did) last 7 years but rather was permanent

prior to effective therapy. In certain areas humans are harassed by the six-legged red larva of the mite *Eutrombicula alfreddugesi,* better known as the **chigger** (also called the *harvest mite* and *red bug*). Chiggers (first active stage of these mites) chase about rapidly on the shoes and clothing of humans but settle down to feed in areas where the clothing is rather tight against the skin. Chiggers do not burrow into the skin but rather grasp the host's skin with their papal claws, insert their mouthparts into the skin at a hair follicle, and inject a salivary solution (toxin) into humans as they feed. This secretion dissolves tissue and sets up a reaction that causes a **wheal** (an **edematous** elevation) on the skin and is accompanied by intense itching. (Chiggers cause considerable loss of business in certain resort areas.) Chiggers transmit important diseases of humans, for example, scrub typhus. Turkeys as well as birds, frogs, rabbits, and snakes are also susceptible to chiggers.

The unfertilized adult female mite *Ornithonyssus bacoti* produces eggs by parthenogenesis, which gives rise only to males; these males are capable of fertilizing. The female mites of chiggers are fertilized by wandering about the ground on the surface of which males have carefully placed *spermatophores* (small capsules situated on the end of a fine stalk).

23.5.5 Lice

Most lice are considered permanent parasites (i.e., they spend their entire life on the host). The human body louse transmits three dangerous diseases: epidemic typhus, relapsing fever, and trench fever (Chapter 22). All farm mammals and poultry may become infested with their own peculiar kind (often species-specific) of either *biting* or *bloodsucking* lice. Lice have three pairs of legs and are true insects. Large numbers of eggs (**nits**) are laid on hair or feathers. Most lice require from 2 to 4 weeks (depending on the temperature) to develop from an egg into an adult. Whereas only one species attacks swine, at least seven species of biting lice attack chickens. Lice of cattle, goats, sheep, and other animals are spread by contact of one animal with another. Cattle infested with lice rub and scratch themselves causing patches of hair to be lost (usually associated with the presence of the biting, or red, louse). Large numbers of lice retard the growth of **calves,** reduce weight gains and the efficient utilization of feed in cattle and poultry, impair lactation of dairy cows, and significantly reduce egg laying in poultry. Because cattle lice live nearly all their lives on the host they are easier to control through residual insecticides than are many insect pests. Lice of sheep and goats are best controlled by dipping the animal in one of several effective insecticide solutions.

The swine louse, *Haematopinus suis,* is a bloodsucking parasite and is the largest louse that preys on domestic animals (females often attain a length of 1/4 in). These brownish-gray lice are parasitic on swine only and pass their entire life cycle on this host. They are easily seen on swine. They feed frequently by puncturing the skin and sucking blood in a new site at each feeding. Each puncture produces irritation and itching, which cause infested swine to rub themselves vigorously against any available object. The frequent scratching and rubbing destroy

---

[2]Psoroptic sheep scabies has been virtually eradicated in the United States. The same mite species occurs on cattle, and the number of cases reported during recent years has increased.

patches of hair and often wound the skin. Oils and medicated liquids are usually effective against swine lice.

It is interesting that deer often paw the ground, making a shallow mud bath to lie in, and coat their bodies with mud, which dries on their skin and kills body parasites such as lice and ticks.

23.5.6 Gnats

Many gnats (small flies of the order Diptera) suck blood from animals whereas others do not even bite. Most gnats lay their eggs on water where they float 1 to 5 days and then hatch. Others lay eggs in decaying plant material. Some of the many species of gnats are vectors of a serious viral disease of sheep and deer called *bluetongue*. An almost invisible gnat, called the *punkie*, is a nuisance in woods of the North and West. Native Americans called it the "no-see-um." Related gnats that bite humans in the southeastern states are called *sand flies*. One family of gnats attacks plants and makes galls on them. The most harmful of these is the Hessian fly (so named because people once believed that it was brought to America in the bedding of Hessian troops during the Revolutionary War).

23.5.7 Fleas

These insects are extremely important vectors of human diseases. Perhaps the most deadly of these diseases is bubonic plague, which fleas transmit from rats to humans (Chapter 22). Fleas also transmit endemic typhus from rats to humans. Fleas serve as intermediate hosts of certain tapeworms. These insects may infest poultry houses. They may be controlled by thorough cleaning, followed by a treatment with insecticide. Fleas commonly infest cats and dogs. All fleas pass through four stages: (1) egg, (2) larva (two molts), (3) pupa, and (4) adult. Eggs are laid while the female is on the host (they are not attached, however) and then drop to the ground where they hatch in a few days into wormlike larvae. The larvae are nonparasitic and live on organic matter in the soil or indoors. In about 2 weeks the larva becomes fully grown and spins a tiny cocoon in which it develops into a pupa. The pupa usually emerges as an adult flea in a week or so. Most fleas are rather host-specific (i.e., they commonly attack only one host); however, some of the more common fleas may attack humans, cats, dogs, rodents, swine, and many other animals (including certain birds). Cat and dog fleas are controlled by dusting with appropriate insecticides, or the use of "spot ons" or sprays.

23.5.8 Bedbugs

These insects attack humans, mice, rabbits, guinea pigs, cattle, horses, and poultry. Fortunately they are not known to be vectors of pathogenic organisms. Bedbugs pass through three stages: (1) egg, (2) nymph (five molts), and (3) adult. Females lay from 75 to 500 eggs at the rate of 3 to 4 per day. There may be one to four generations annually. The nymphs feed before each molt. The wingless adult lives about 1 year. Unlike many insect species, both males and females are avid bloodsuckers

and if present in beds and mattresses make short appearances to obtain blood meals from humans at night. Nymphs require only 6 to 9 min and adults 10 to 15 min for engorgement, after which they drop from their victim and retreat quickly to their hiding places (they are especially active at night). It is possible for humans to carry bedbugs into their houses from infested poultry houses or equipment. However, the use of **DDT** after World War II essentially eliminated bedbug infestations, which had annoyed human beings for many years. The EPA now prohibits the use of DDT and many similar chemicals and bedbug and lice populations are making a strong comeback.[3] Related bugs, parasites of bats or birds, are often mistaken for true bedbugs.

23.6 ARTHROPOD CONTROL—ESSENTIAL FOR HUMANS

Insects are humans' greatest competitors for food. In many parts of the world insects reduce a potential agricultural abundance to such an extent that people die of malnutrition. Scientists are probing the biological, biochemical, and behavioral differences that set insects apart from other animal life in an effort to learn ways to control their populations (especially their capacity for reproduction) and thereby increase the food and fiber supply for people.

23.6.1 Chemical Control

Civilization advances proportionally to the ability of humans to overcome problems associated with securing the essentials of survival: food, water, and shelter. The biological checks and balances of nature are too slow and inadequate to ensure plentiful food for humanity. In a natural setting humans would compete with a host of natural enemies seeking the same food supply. Today they compete favorably by protecting their food from ravages of insects, rodents, and disease organisms with agricultural chemicals. The use of chemicals to fight pests dates back to the ancient Greeks who employed brimstone (sulfur) as an insecticide. Cave dwellers observed that food treated with the salty residue of seawater is safe from insects. Settlers on the Great Plains of the United States used Paris green (a crude arsenical) to save their potato crops from the Colorado potato beetle. They also treated their grain seeds with copper sulfate to protect them from plant disease. Today over $1 billion is spent annually for pesticides in the United States.

It would be difficult to appraise accurately the benefits to humanity that have accrued from the use of agricultural chemicals in the control of pests. Similarly it would be difficult to imagine the adverse impact on humanity's future welfare and economy if suddenly they were made unavailable. Chemicals have been used to tip the so-called biological balance in nature in favor of humans.

---

[3]It is somewhat ironic that a Swiss chemist, Paul Mueller, was awarded the Nobel prize in medicine in 1948 for developing DDT as an insecticide, and only a generation later, its use was banned in several countries.

Pesticides are poisons used to destroy pests of any sort. They include **fungicides, herbicides,** insecticides, acaricides, and rodenticides. Pesticides are often the most effective weapon available to fight pests that damage or destroy crops, livestock, and forests. These chemicals also help protect humanity's health and well-being from insects that transmit such diseases as malaria, yellow fever, and typhoid. The effectiveness of pesticides in controlling agricultural pests helps keep food costs low and quality high. According to USDA estimates, if pesticides were not used crop and livestock production in the United States would decrease by approximately 30 percent. Without pesticides many vegetables and fruits would be destroyed by insects and would disappear almost completely from food markets. These losses in animal and vegetable food production would contribute to the starvation of many people.

Insecticides for the control of insects are effective in two ways: as *stomach* (ingestant) poisons (e.g., lead arsenate) and as **contact poisons** (e.g., pyrethrum). Stomach poisons are used to combat such pests as beetles, which chew and swallow leaves and concurrently swallow the insecticide. Pyrethrum kills insects by causing paralysis of the central nervous system. DDT, which was first synthesized in Austria in 1873, and chlordane function as both a stomach and a contact poison. They belong to the class of insecticides known as chlorinated hydrocarbons. Other examples include dieldrin, methoxychlor, and toxaphene. These broad-spectrum, inexpensive, residual insecticides revolutionized agricultural technology. Today, because of known or suspected adverse effects, many of their uses have been banned or restricted. At present most insecticides commonly employed are organophosphorus or carbamate compounds. These agents are nerve poisons that inhibit the enzyme acetylcholinesterase. This action halts nerve transmission at the synapse, or junction, of nerve cells. Many organophosphorus and carbamate insecticides are extremely poisonous to humans and domestic animals and they should be used only with adequate protective equipment.

Several new classes of insecticidal materials have been investigated in recent years. Formamidine compounds affect insects' biogenic amines thus offering a different *mode of action* with possible utility against insecticide-resistant strains. Synthetic pyrethroid compounds (similar to the pyrethrins obtained from certain flowers) have found wide use in the livestock industry as fly and mite control agents. GABA inhibitors affect the neurotransmitter chemical known as gamma aminobutyric acid. Insect growth regulators (IGRs) may mimic the action of insect hormones or inhibit formation of the chitin that hardens the insect skeleton. Exciting as such research may be the fact is that many corporations are turning away from insecticide development. The huge capital investment required and the small probability of successfully contending with governmental regulations now make such development a poor risk financially.

Pesticide Residues in Animal Products

Agricultural chemicals are not used without some risk. It has been established experimentally that cows fed foods contaminated with certain chemical compounds (e.g., DDT) store them in their body fat and secrete them in their milk fat. Many pesticides used to treat plants and/or animals are fat-soluble and may be present in various levels in eggs, meat, milk, and their by-products. However, at the low levels observed no adverse physiological effects have been noted. There are no medically documented records of death in humans resulting from the presence of pesticide residues in properly treated foods. This is because the safety margins between permitted levels of use by the EPA or FDA and the minimum toxic doses are so great that violations of legal tolerances are extremely unlikely to involve significant health hazards.

Tolerance limits for pesticide residues in foods are established by the EPA and FDA. The permissible level is generally determined by considering the acceptable safe intake level of that chemical for humans and the amount of the food that will likely be consumed during a lifetime. The safe level of intake is calculated by extrapolating from data derived from small-animal experimentation (usually data of the most sensitive laboratory animal are used), after which an additional safety factor is applied. The FDA monitors pesticide concentration in animal products. If foods are found to contain residues in excess of the minimum safe tolerances as established by the FDA they are withdrawn from the market.[4] Extensive studies of the pathways and possible retention of pesticides in animals and their excretion patterns are being made (Figure 23.2).

23.6.2 Biological Control

> The man of science ought to realize the factors which have given him the vantage which he holds.
>
> **Carl von Voit (1831–1908)**

Insects constitute at least 90 percent of all the world's animal life and can produce astronomical numbers of offspring. For example, if all the descendants of a pair of houseflies lived and mated fully the family would number approximately 190,000,000,000,000,000,000 flies after only 4 months. A queen termite is equally **fertile.** She lays thousands of eggs daily during a life span of up to 50 years. Vast numbers of insects compete with humans for fruits, grains, and other food plants and transmit diseases to animals and humans. Importantly, significant progress has been made in biological control of these arthropods.

Scientists have investigated a number of methods and techniques, from the use of the back of one's hand to special chemicals, in attempts to kill these destructive, disease-bearing insects. Chemicals are being used to sterilize the pink boll worm in Texas. They also appear valuable in controlling the gypsy moth and boll weevil. Some chemists predicted that DDT would free the world from insects. However, the more vigorous ones lived and apparently subsequent generations were somewhat resistant to this insecticide. After contact with

---

[4]An example of food contamination involving a pesticide that resulted in significant economic losses (estimated at $10 million) to poultry producers and nutritional losses to consumers was the high levels of dieldrin that were found during routine inspection of slaughtered poultry and declared unsafe for human consumption. The dieldrin was later traced to contaminated feed.

Figure 23.2 Pesticide metabolism is studied using catheterized lactating cows.
Courtesy of Dr. D. J. List, Pesticide Residue Laboratory, Cornell University.

an insecticide, the insect population that survives is generally resistant and a larger proportion of the population becomes resistant through genetic **selection.** The rapidity with which the population increases its resistance depends on the frequency of the specific genes for resistance and the intensity of selection. If the **gene** is **recessive** a longer period is required for acquiring insecticide resistance than when the gene is **dominant.** Besides this resistance to the insecticides, their residues on plants present a possible health hazard to humans who ingest these plants or food products of animals that consume the plants.

Public concern over the possible dangers of chemical pesticides has prompted scientists to emphasize the use of natural enemies of insects against them. Thousands of insects prey on other insects. For example, a wasp, *Dendrosoter protuberans,* is a natural enemy of the European bark beetle, which carries Dutch elm disease and spreads it by depositing its eggs through the tree bark. In Europe the beetle has been held in check by this wasp, which destroys the beetle larvae. Much interest has been shown in the potential for biological control of insects affecting livestock. It is known that major fly pests such as the housefly, face fly, stable fly, and horn fly suffer considerable natural mortality because of insect parasites and predators. In some areas tiny parasitic wasps (particularly *Spalangia* spp. and *Muscidifurax* spp.) are deliberately reared using houseflies and then released by the millions in poultry, dairy, and horse facilities. The released wasp females seek new fly pupae in which to deposit their eggs. The developing wasp larvae kill their fly host, emerge as adults, and in turn seek other fly pupae. The cycle thus becomes cumulative and effective control of fly populations has been demonstrated in the southern United States. This concept of *inundative* or *mass release* has been likened to a "living insecticide" and offers great possibilities in other areas if parasitic wasp species can be found

that are suited to local climatic conditions. Predatory arthropods can be effective biological fly control agents; however, techniques are not yet available for their commercial use.

Japanese beetles, which are enemies of more than 200 crops and ornamental plants, may be controlled by infecting them with a bacteria-caused milky disease.

> So, naturalists observe, a flea
> Hath smaller fleas that on him prey;
> And these have smaller still to bite 'em
> And so proceed ad infinitum.
>
> **Jonathan Swift (1667–1745)**
> **English satirist**

This quotation emphasizes the principle of natural insect elimination as shown in the above examples. A further variation of this principle is found in nature's biological control of insects by means of the parasite–host relationship.

One of the most interesting episodes in the history of biological control was the use of the myxomatosis virus to control the rabbit population of Australia. The virus exists naturally and causes tumors in the Brazilian wild rabbit. In 1950 the myxomatosis virus was introduced into several experimental groups of rabbits, which were then released in the countryside of Australia. Within 10 months after their release other rabbits throughout a 500,000-square-mile area were found to be infected. Within 3 years the estimated original rabbit population of 500 million had been reduced to an estimated 50 to 100 million thus allowing sufficient regeneration of vegetation to permit a 10 to 15 percent increase in wool production of sheep.

The application of biological interdependence is also important in insect control. There are more than 1000 viruses, fungi, nematodes, protozoa, and rickettsiae that infect and kill

Figure 23.3 Biological Control of Insects Research Laboratory.
USDA, Agr. Res. Service, Columbia, MO.

insects naturally. Researchers can even inject these pathogenic organisms into the larvae of some insects. A poison secreted by the bacterium *Bacillus thuringiensis* is **lethal** to caterpillars of more than 100 butterflies and moths. Dried spores of this bacterium are being used in the United States to control larvae of the alfalfa caterpillar, cabbage looper, imported cabbage worm, hornworm, tent caterpillar, and gypsy moth. Fortunately this substance is harmless to humans, domestic animals, wildlife, and most beneficial insects. Genetically modified crop varieties that express *B. thuringiensis* toxins in selected tissues have recently become available and thus offer a "built-in" insecticide. Other agents such as the polyhedral virus can be prepared commercially and therefore offer a means of controlling insects. This virus is grown on insect larvae because viruses require cellular material to multiply. Each 100 infected larvae produce enough viral particles to treat an acre of cropland. Thus, through science, researchers have applied the natural principle of biological control and adapted it to rid humanity of insect pests.

An experimental laboratory (Figure 23.3) is useful in studying various ways of controlling insects biologically. A means of rearing host insects is essential to such investigations. Laboratory rearing of horn flies is depicted in Figure 23.4 a and b.

Entomologists throughout the world are currently collecting and assembling research data on the anatomy and behavior of harmful insects. This research has led to a number of ways for their elimination. Knowing that certain insects effectively mate only once in a lifetime, scientists released millions of artificially sterilized males to disrupt the insect's reproductive cycle. The direct application of this principle is found in the screwworm eradication program. It may be possible to treat insects with chemosterilants that will induce sterility without affecting mating.

The technique of **radiation sterilization** using cobalt 60 was developed by Dr. E. F. Knipling and colleagues of the Entomology Research Division of the USDA. His early work was designed to control the screwworm fly. Many female insects (e.g., screwworm flies) usually mate only once (they carry the **sperm** in a special sac, the spermatheca, from which the sperm emerge to fertilize the eggs as they pass down the oviducts). Knipling reasoned that if the females mated with sterile males they would lay only unfertilized eggs and the reproductive cycle would be broken. A dose of cobalt 60 sufficient to render the males sterile but low enough to ensure that the treated flies were still strong enough to compete with nonirradiated ones was employed. A strong advantage of using sterilized males rather than powerful insecticides is that the males seek out *all* females for mating and leave none to form a reservoir for reproduction.

Tests in 1954 rendered a 170-square-mile island off the coast of Venezuela free of screwworm flies. In 1958 researchers used an air-terminal hangar in Florida to raise massive numbers of sterile male flies. To produce 50 million flies weekly, 40 tons of horse meat (or whale meat) and 4500 gal of beef blood were used. After being sterilized as pupae by exposure to radioactive cobalt, the flies were air-dropped and spread over infested areas in the southeastern United States. Through the use of 200 to 1000 flies per square mile the southern United States was made essentially free of screwworm flies.[5] A weekly release of 4000 sterile male flies per square mile was effective in complete eradication of screwworms in Alabama, Florida, and Georgia. In the 1980s sterilized males were used to eradicate screwworm flies brought accidentally to Tripoli, Libya, from Nicaragua via infected sheep. These examples represent important landmarks in insect-eradication attempts. In 1965 sterilized males were used to eradicate the oriental fruit fly on Guam. More than 20 million sterilized Mexican fruit flies were released along the United States–Mexican border in 1966 to protect the citrus orchards of Arizona and California.

---

[5]The southeastern United States is free of screwworms. The southwestern states are currently free but in jeopardy because of possible reinfestations from Mexico. Therefore, a joint United States–Mexico program is under way to eliminate screwworm infestations. The USDA facilities for this project were moved from Texas to southern Mexico.

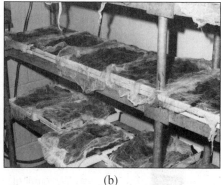

(a)   (b)

Figure 23.4 (a) Rearing cages for obtaining horn fly eggs. Their food source (blood meal) is at the top, and pads for oviposition (obtaining eggs) are located at the bottom of the traps. (b) Rearing horn flies (larvae to pupae) in the laboratory. These supply host materials for rearing various species of parasitic insects that attack horn flies. They are also used for insecticide and other investigations.

USDA, Biological Control of Insects Research Laboratory, Columbia, MO.

Australian scientists are attempting to breed flies out of existence. Their plan, as reported by *Science Service,* calls for breeding flies "whose only **offspring** will be males." Eliminating female flies would mean the end of flies. Male flies are being **irradiated** in the experiment to give the **X chromosomes** (male) in males a better chance than the **Y chromosomes** (female). After enough male-breeding flies have been produced to dominate a fly community, they will be released in an effort to increase the proportion of male to female flies. The island of Capri (off the coast of Italy) was rendered free of fruit flies in 1967 through the use of flies that had been radiation-sterilized in Israeli and Austrian laboratories and flown to Capri for release.

Research indicates that attractants can be synthesized and used to lure unsuspecting insects into traps instead of attracting them to mates. More than 200 insect species have been found to possess sex attractants. Although odors released by female insects are usually for the purpose of attracting males, they may also serve to excite the male sexually before **copulation.** The production of female sex attractant in most insects is depressed appreciably within a few hours after mating. Moreover, it is interesting that sex attractants are produced by insects only just before or during the period of the day in which mating normally occurs. In some insects the male *and* female exude attractants (assembling scents). For example, the male boll weevil can attract females from a distance of 30 ft. The sexual odors released by males are primarily for the purpose of sexually exciting the female, making her more receptive to the male's advances.

There are more than 50 insect species in which males lure or excite the females. Certain female beetles emit an odor that will attract male beetles of that species from 1800 ft away. The female gypsy moth also emits a sex attractant that makes it easy for the male to find her. Scientists can now synthesize this attractant, or lure, called *gyplure.* Possibly scientists could synthesize a large amount of this attractant and spread it over a large acreage of timberland that the moth has infested. (Gypsy moth caterpillars are colorful but destructive insects that kill trees by devouring their leaves.) With the sex odor literally everywhere,

the males should become confused in their attempt to find females, quickly become exhausted, and give up in their mating effort or, possibly because of overexposure, become insensitive to the sex attractant. A chemical is classed as a sex attractant if it brings to it an insect that then assumes a mating position.

Research entomologists have discovered that a pair of glands, called the *corpora allata,* found in the female cockroach brain secrete a lure on signal. Scientists are currently attempting to develop a way to interrupt this message, which would result in failure of females to secrete the sex attractant. This would then make it difficult for males to find females and consequently would reduce reproduction. A recent study identified valeric acid as the sex lure produced by adult female sugar beet wireworms, which attack sugar beets, potatoes, corn, lettuce, and certain other crops. Diluted solutions of valeric acid attracted male beetles from a distance of 12 m. This chemical is economical and readily available for use as bait in traps for population control.

Some plants possess attractants. Cotton, for example, contains chemicals that can attract or repel the boll weevil or cause it to feed uncontrollably. When these chemical compounds are synthesized, the boll weevil may well be retired to insect museums.

Further research has indicated that lights and sounds sensed only by insects can also be used to ward them off or to lure them to their deaths. Many insects possess eyes and ears of a sort. Black-light traps in combination with the female sex lure are being used to control cabbage looper moths. Some insects are attracted by ultraviolet lights and can be lured by them into traps. Other insects possess mechanisms that are sensitive to sound. Several years ago scientists discovered that bats use high-pitched squeaks that serve as a type of radar to zero in on their prey at night. One of the bat's favorite prey is *Helicoverpa zea,* a moth with many names (the larva is known as the *cotton bollworm, corn earworm,* and *tomato fruit worm*). Nature protects this moth by equipping it with sensitive hearing organs, which hear the bat's cry as a warning. This information has enabled researchers to install rotating loudspeakers in cotton fields to transmit a recording of the bat cry as an "electronic scarecrow."

Naturally occurring **hormones** are being applied to prevent insect eggs from hatching or the pupae from maturing. This too is a part of the research data being established. This process involves natural elimination of insects. *Juvenile* hormones are naturally occurring chemicals that are essential to insect development during the early stages of life. They are secreted by the same tiny glands located at the base of the brain that secrete the sex attractant in cockroaches. Juvenile hormone secretion apparently ceases at maturity. Researchers reasoned, therefore, that if juvenile hormone were administered to insects late in the pupal stage, normal metamorphosis would be prevented. A juvenile hormone was isolated in 1956 by Carroll Williams at Harvard University and several such hormones (e.g., methoprene is widely used in the control of mosquitoes and other insects) have been synthesized since then. A new type of hormone was synthesized in 1966 by researchers of the USDA. The chemical is derived from farnesol, an oil commonly present in many plants and animals. A drop of this compound applied to the pupa's skin penetrates rapidly. This new synthetic juvenile hormone is apparently nontoxic to noninsect species and has been observed to affect the early stages of development. When sprayed on an insect egg it effectively blocks development of the embryo. Scientists theorize that the hormone interferes with the insect's genes.

It has long been observed that certain individuals or varieties of plants and animals possess unusual ability to resist attack or damage by arthropod pests. This phenomenon is termed *host resistance*. It generally is dependent on genotype and is an inherited trait. It is obvious that if such a trait can be combined with adequate production performance it may prove a superior, economical, and ecologically sound method of reducing losses to pests.

In plants such host resistance may take the form of extra thick stems through which pests cannot easily penetrate, or a characteristic known as *vigor tolerance*. The latter indicates that the host organism is capable of suffering without loss to production a level of damage that would severely reduce the productivity of nonresistant cultivars.

Host resistance to external parasites has also been noted in many species of domestic animals. In some cases antibodies are formed by the host's immune system in response to arthropod feeding. These antibodies may act to prevent subsequent infestations of the same parasite species. It has long been observed that male animals frequently suffer more severe infestations of external parasites than do females. Sex hormones may be involved in this type of resistance. Chickens genetically selected for a high level of plasma corticosterone (secreted by the adrenal gland; Chapter 11) response to social stress are more resistant to northern fowl mite infestation than chickens selected for a low level of corticosterone response. Feeding chickens this hormone can produce similar results. Breeding resistant varieties of domestic animals and manipulating the animals' own hormone systems to promote resistance offer entomologists exciting prospects for effective control of external parasite populations. In addition, certain of these approaches may apply to internal parasites as well.

23.6.3 Living Insecticides

Insecticides and other pest controls formulated from living microorganisms rather than chemicals are being developed to control, for example, tapeworms, nematodes, ticks, and other parasites and diseases of humans, domestic animals, and plants.

Insect pests, like other forms of life, are susceptible to diseases caused by bacteria, fungi, protozoa, and viruses. These microorganisms are not new; they are already present in the environment and often help control naturally populations of insects that harm humans and cause extensive damage to food and fiber crops. They have been screened naturally for centuries to fit the environment. Now scientists are finding how to use them effectively for the benefit of humanity.

Commercial formulations of these microbes are called *living insecticides*. There are more than 1500 natural, safe, biodegradable microorganisms that might be formulated into living insecticides. Moreover, new microbes are being discovered.

The potential impact of substituting microbial insecticides for chemical insecticides is indeed substantial. In the United States alone, use of a living viral insecticide to control bollworms and budworms on cotton, living fungus against the citrus rust mite, bacterium against cabbageworms, and living protozoan against grasshoppers could replace an estimated 20 to 30 million pounds of chemical insecticides annually.

This new but now formalized science has blossomed during the past several decades. It holds great promise as a major means of biologically controlling, on a safe and effective basis, damaging insects such as caterpillars, grubs, flies, mosquitoes, mites, weevils, and even the difficult-to-kill grasshoppers that are perpetual pests of humans and the foods, fibers, and forests important to humanity.

23.7 SUMMARY

Since the beginning of time humans have shared the planet with insects and parasites. Some are beneficial, others may transmit disease and inflict significant economic losses. Historically, diseases transmitted by insects have been notoriously effective in determining the outcome of military conflicts (Napoleon's army is said to have suffered greatly from typhus); some say even more effective than the decisions of military commanders (epidemic typhus and malaria are among the most telling).

Many protozoans are parasites. One type of amoeba destroys the intestinal linings of humans and causes the painful disease amoebic dysentery. Other protozoans may invade the blood of mammals and cause diseases such as malaria and Texas cattle fever. Bloodsucking insects and ticks collect microorganisms from infected animals and convey them to humans and other animals. Parasitic insects, mites, and ticks usually attack the skin. Certain ticks transmit Rocky Mountain spotted fever and Lyme disease to humans and animals. One type of mosquito spreads yellow fever and another carries malaria. The tsetse fly transmits African sleeping sickness. Humans may acquire typhus fever from a body louse.

The total mass of protoplasm produced annually by insects exceeds that produced by all other terrestrial (earthly) animal life combined. A knowledge concerning insects and how to circumvent their harmful effects is important if humans are to improve the health and productivity of animals that provide useful products and services to them. Great strides in parasite and insect control have been made possible through new and improved agricultural chemicals. (DDT saved millions of lives from malaria and typhus.) Pesticides allow humans to control arthropod-transmitted diseases and also to increase their food supplies by providing a means of combating agricultural pests that formerly attacked and destroyed crops and food animals. However, because of possible injurious effects of certain chemicals on humans, new methods of insect control are being studied and exploited. These include bio-logical, physical, genetic, ecological, and/or cultural approaches. It seems almost certain that future control of insects will be largely biological and depend less on the use of pesticides.

Evidence of this trend is found in the heightened interest in *integrated pest management*. This technique stresses a thoughtful and balanced approach to pest control involving all appropriate technology available. It does not depend exclusively on biological, chemical, or cultural control but is based on maximum economic benefit within biological and ecological limits. Its most effective implementation requires a thorough understanding of the adverse effects of pest populations and in some cases computer *models* are used to predict future trends. Management decisions involving *economic thresholds* and *injury levels* are expected to be commonplace in the years ahead.

STUDY QUESTIONS

1. Of what economic and public health significance are insects?

2. What are the three principal body segments of insects?

3. Define molting. Identify instars and stadia.

4. Can insects taste, smell, and hear?

5. What are their four metamorphic stages?

6. Define a parasite. An arthropod. What is entomology? Parasitology?

7. What are the seven basic levels in the classification of animal life?

8. Name and cite a few examples of the five most important phyla.

9. Briefly discuss some contributions of insects to humanity.

10. Discuss some harmful effects of insects.

11. Cite one or more examples of diseases of humans that may be transmitted by (a) mosquitoes, (b) flies, (c) lice, (d) fleas, (e) ticks, and (f) mites.

12. What is the proboscis? Of what use is it?

13. How is it possible for flies to walk upside down on ceilings?

14. Trace the life cycle of houseflies. What is the best way to break this cycle?

15. What is the source of food for horn flies? How may these flies be controlled?

16. Trace the life cycle of stable flies. How are their numbers held down?

17. Trace the life cycle of screwworms. What environmental factors aid in their control in certain areas?

18. What recent application of agricultural research aids in controlling screwworm populations?

19. Trace the life cycle of heel flies. In what ways do they cause economic losses to livestock producers?

20. Trace the life cycles of botflies of cattle, of sheep, and of horses. How does the human botfly affect people?

21. What diseases of humans are transmitted by horseflies?

22. How can one distinguish the pupae of face flies from those of other flies? Why do horses often stand head-to-tail in reverse directions?

23. Trace the life cycle of mosquitoes. How do female mosquitoes rank as a vector of human diseases?

24. Trace the typical life cycle of ticks. Cite some diseases transmitted by ticks.

25. What are chiggers? Are they linked with any diseases of humans?

26. Why are lice considered *permanent* parasites? What dangerous diseases of humans may be transmitted by the human body louse?

27. How can one control lice of (a) sheep and (b) swine?

28. Name a serious viral disease of sheep for which gnats serve as a vector.

29. Trace the life cycle of fleas. What important human diseases do they transmit? How are they controlled?

30. Trace the life cycle of bedbugs. What insecticide is effective in their control?

31. Define (a) fungicides, (b) herbicides, (c) insecticides, (d) pesticides, and (e) rodenticides.

32. Briefly discuss the use and importance of chemicals in the control of insects and parasites.

33. Which governmental agencies monitor pesticide concentrations in animal products?

34. What influences the rapidity with which an insect population increases its resistance to an insecticide?

35. What is meant by biological control? Cite examples of biological control of animal life.

36. Of what practical significance is the use of radiation to sterilize male screwworm flies?

37. What are sex attractants? How may they be used in controlling insects?

38. What other means of insect control are being investigated?

39. What is meant by the term "integrated pest management"?

40. What are living insecticides?

24

ETHOLOGY AND ANIMAL BEHAVIOR[1]

If an animal balks and refuses to walk through an alley, one needs to find out why it is scared and refuses to move.

Temple Grandin (b. 1947)
Animal Scientist, Colorado State University
Grandin Livestock Systems

24.1 INTRODUCTION

Behavior is the way a whole animal reacts via movements, postures and displays, eating, eliminating, mating, caregiving, sounds, smells, hiding, threatening, killing, and so forth following some internal or external stimulus. An animal encounters many different environmental stimuli during its lifetime so it is externally motivated to react in different ways. It also will express **spontaneous** actions that are internally motivated. Thus animal behavior is a complex subject. One **species** may react one way in response to a certain stimulus, whereas another may react in a different way. The same is true of individuals within a species. In general, the behavior of animals as a species determines their fitness: their ability to survive and reproduce in the wild. Natural **selection** has played a role in the evolution of innate behavior. Species that do not develop appropriate behavioral responses to adverse aspects of their environment often lack fitness and become extinct.

Some wild animals seem to teach their young to fear humans. A fawn may pay no attention when a human comes near. However, when its mother shows fear the fawn will flee with her. Similarly, mother wolves warn their cubs about traps. When the mother wolf and her cubs come upon a trap for the first time, the mother shows fear. Thereafter the cubs try to stay away from traps.

Nature equipped many animals with inherent means of avoiding potential harm. Their survival may depend on recognizing a particular hazard in time to avoid it. There are excep-

tions. For example, in one way nature was not very kind to the frog. A flaw exists in the frog's early warning system that can prove fatal. If a frog is placed in a pan of warm water to which additional heat is gradually applied, it will typically show no inclination to escape. Being a cold-blooded creature its body temperature approximates that of the surrounding water and it is not alarmed by the slow temperature change. As the temperature continues to rise the frog remains oblivious to its ensuing demise. Although it could easily hop to safety it is content to stay put, even as the steam begins to fill its nostrils. Eventually the frog succumbs to an unnecessary misfortune that could have been avoided if it had simply been alert to the danger at hand.

Knowledge of behavioral reactions of domestic animals to certain stimuli and the forces responsible for different types of behavior is important. With this knowledge the keeper can predict how an individual animal will react behaviorally in a given situation, what makes it behave in that way, and how the expected reaction might need to be changed for optimal results. In other words, this knowledge can be of practical value in trying to improve the breeding, feeding, and care of domestic animals.

Research at Southern Illinois University demonstrated that for piglets the average period between sucklings can be reduced from 50 to 40 min by playing tape recordings of the sounds of piglets suckling. This has practical possibilities because piglets provided with such recordings suckled more frequently than those without and had greater milk intake and body weight gain.

The same basic principles that determine animal behavior also govern human behavior, although much remains to be learned. The greatest single problem in the world today may well be the behavior of humans toward other humans: in the family, in the school, even in regional and international

[1]The authors acknowledge with appreciation the contributions to this chapter of Dr. S. E. Curtis, Department of Animal Sciences, University of Illinois at Urbana–Champaign.

relations. Humankind must learn the ways of peaceful coexistence. Apparently since the beginning of time there have been wars and rumors of wars. Intraspecies squabbles occur in nonhuman animals but, with the exception of males defending breeding territories, genocide rarely occurs in the animal kingdom. Of course, predatory behavior involves interspecies tension.

> We suffer less from a want of science and technology than from lack of understanding of the aims of life and of society.
>
> **Robert M. Hutchins (1899–1977)**
> **American educator**

24.2 CAUSES OF BEHAVIORAL RESPONSES IN ANIMALS

Behavioral responses by all animals are determined by **heredity** (*internal factors*) and learning experiences (*external factors*). Some responses appear to be controlled by an interaction between these factors. For example, certain strains or **breeds** within a species have been selected on a genetic basis and then trained for a specific behavioral response. Breeds of horses have been developed that are excellent for galloping, trotting, pacing, or **draft** purposes, respectively. A single breed excels in one but not all of these traits. Cow **ponies** have been developed that make outstanding roping horses when well-trained. Others excel as "cutting horses," which are capable of sorting (cutting out) a specific individual from a large **herd** of cattle (Figure 24.1). A good cutting horse does not necessarily make a good roping horse, and a draft horse could never become a fine pacer even with top training. Therefore, specific innate abilities and instincts must be present before an animal can be trained to behave in a specific way.

24.2.1 Innate Behavioral Patterns

There is considerable evidence linking behavioral responses to hereditary factors. Selection has developed dogs specifically adapted to fighting, herding other animals, or hunting various wild animals, respectively. **Crossbreeding** and raising the young artificially or with foster mothers of another species is another method used to demonstrate inherited behavioral patterns. Innate behaviors persist in the young even though they have not had the opportunity to learn them from their own mothers. Certain traits in various species **segregate** according to the laws of heredity when various breeds within these species are crossed. Many traits can be modified by environmental experiences.

Genetic differences appear to involve response **thresholds.** That is, heredity can be responsible for the ease with which an animal can be motivated to express a particular behavioral response. This is especially true of the fighting trait. Dogs bred for fighting will fight at the slightest provocation. Other breeds have been developed for hunting and run in packs with relatively little fighting.

Heredity produces behavioral responses in several ways. It can affect the growth and development of body parts such as **sensory** or motor organs. Genetic defects of the nervous system have been seen in many species. All animals have brains and domestic animals have brain components similar to those of humans. But humans and other species vary in their abilities to think and feel. Animals can show only those behaviors for which they possess the appropriate neural structures and mechanisms.

Genes are responsible for synthesis of specific **enzymes, hormones,** and other chemical substances in the body. These in turn affect the behavioral responses of animals. Chickens can be made to show typical male or female behavior depending on

Figure 24.1 A cutting horse in action. Some horses can be trained to cut an animal from a herd of cattle with great skill, others cannot.
Courtesy of Robert Sibbitt.

whether they have male or female hormones in their bloodstream. Many examples of gender reversal have been reported in the **avian** species (Chapter 16). Some strains of mice develop convulsions when exposed to high-pitched sounds. They appear to lack one or more of the chemical substances essential for energy metabolism to support normal brain function.

24.2.2 Experience and Learning

The behavior of an animal is based on innate forms of behavior called *instincts* and **reflexes.** Animals show fixed behavioral responses when subjected to certain environmental stimuli. Such behavior can be modified by experience, a form of learning. Birds possess an innate nest-building drive but the quality of nest construction improves with succeeding attempts (experience), a result of learning. Of course, they can improve their nest building only because they inherited the necessary neural pathways to learn from experiences. There are several types of learned behavior as discussed in the following examples.

Habituation

By habituation an animal learns to ignore stimuli it knows from experience to be harmless. It is a simple kind of learning and is necessary to prevent animals from reacting continuously and unnecessarily to their environments. Animals habituate quickly to common environmental sounds (e.g., ventilation fans and jet airplanes). Habituation is used to train dogs in overcoming fear of thunderstorms. Tape recordings of thunder are played and the dog is rewarded when it remains calm.

Conditioning

There are two kinds of conditioning. *Classical* conditioning (*associative* learning) occurs when an animal learns to respond to a previously neutral stimulus in the same way it would to a normal stimulus (unconditioned stimulus). An example of classical conditioning is the elegant experiment conducted by the Russian biologist Ivan Pavlov (1849–1936), who played a metronome each time he presented food to a group of dogs. Eventually the dogs would produce saliva in anticipation of eating every time the metronome was played, even when they could not see, smell, or taste the food. The dogs came to associate metronome sounds (conditioned stimulus) with the offering of food (unconditioned stimulus) so salivation (originally the unconditioned response) became the conditioned response. Animals may become classically conditioned with respect to reflex activities. For example, **lactating** cows often "let down" their milk even before entering the milking parlor (Chapter 15).

The second kind of conditioning is called *operant conditioning*. If animals repeatedly perform a certain behavior that involves acting on their environment and are positively rewarded every time they perform it (positive reinforcement) or are punished (negatively rewarded) when they do it, they soon associate their action with the results. Thus if a horse does a certain trick and is rewarded each time with a sugar cube it soon does the trick because it wants the treat. Trial and error help the horse to associate trick with treat. Similarly, animals soon learn to stay away from an electric fence when they receive a shock every time they touch it. In fact, many times when an electric fence is removed it is almost impossible to drive pigs across the imaginary line over which it once ran. Domestic animals such as sheep and swine have learned to control the light and temperature of the environment by operating control switches.

A behavior that is learned through reinforcement and reward can be extingushed if the reward is withheld long enough. This process is called extinction. Animals provide many interesting examples of extinction.

For example, the walleyed pike has a huge appetite for minnows. If it is placed in a tank of water with its tiny aquatic associates it will soon consume the minnows and be by itself, alone in the tank. However, an interesting phenomenon occurs if a piece of clear glass is placed vertically down the middle of the water tank thereby separating the pike from the minnows. The pike cannot see the glass and strikes it head-on in pursuit of its prey. Again and again it will swim into the glass, repeatedly bumping its head. Because the pursuit of minnows is not reinforced, this behavior is eventually extinguished. Finally the pike becomes discouraged and gives up—it has learned that it is impossible to get at the minnows. If the glass is then removed from the tank the minnows can swim freely about their mortal predator in perfect safety. The pike no longer tries to prey on the minnows because it has "learned" that they are unreachable. The pike will actually starve to death while one of its favorite foods is nearby and readily accessible. The pike is experiencing what the American psychologist Martin E. P. Seligman (b. 1942) called "learned helplessness," a state in which an animal learns that it no longer has the ability to use its environment for its own purposes and simply stops trying. Animals kept in relatively unstimulating environments over which they seemingly have no control probably experience learned helplessness to some degree.

Insight Learning (Reasoning)

A third type of learning is prevalent in higher animals: the ability to respond correctly the first time the animal encounters a certain situation. This enables the animal to solve a new problem by conscious reasoning alone. Animal studies indicated that severe **malnutrition** in very early life can result in long-lasting behavioral changes including what has been interpreted as retarded ability to reason and thereby to solve problems.

Imprinting

A fourth type of learning is a form of social learning that has been observed in some species, especially poultry. The Austrian zoologist Konrad Lorenz (1903–1989) pioneered work in this field. When a duckling is exposed immediately after **hatching** to some moving object (especially if the object emits a sound), it adopts that object as its parent. Ducklings may adopt a human, dog, cat, or even an inanimate object as a parent. Imprinting can usually be accomplished only within the first 36 h following hatch. Apparently inheritance controls the length of the "sensitive period" when the individual can be imprinted, the objects to which it can be imprinted, the tendency to respond to the first object it encounters, and the permanence of its attachment to the object once imprinting has occurred. Although imprinting does not occur as distinctly in mammals, early association of mammalian young with their species and humans is important for relations later in life (socialization).

24.2.3 Intelligence

Intelligence is the ability of animals to learn to adjust successfully to environmental situations. It is sometimes defined as *the organization of behavior* (the abilities to learn from experience and to solve problems). The degree of intelligence apparently varies greatly among individuals within a species as well as among species. But interspecies comparisons are difficult because different species have evolved in different ways so as to successfully cope with situations in the particular environments in which they evolved. Consequently species vary in the *ways* they are "intelligent."

Animals learn to do some things but inherit the ability to do others (often called *instinct).* It is evident that birds do not have to learn how to build a nest because young that are hatched and reared away from their parents know "instinctively" how to build one. The spider spins a web peculiar to its own species without learning from its parents how to do it. There are many other examples of innate behavior in animals that allow them to do certain things spontaneously (without previous learning or thinking) via what have been called "fixed action patterns"—innate behavioral responses to a given stimulus. Intelligence, however, is not completely an instinctive behavioral phenomenon.

Much has been said in the popular media about the high intelligence of coyotes, crows, dogs, pigs, horses, pigeons, and many other wild and **domesticated** animals. Arguments have raged for centuries as to which species is the most intelligent. Instances of unusual displays of intelligence by individual animals were formerly used to decide these issues. These methods have now been replaced by more precise experimental means of measurement. It should be emphasized again that because species differ in many innate tendencies, tests for intelligence may be biased accordingly.

Conventional wisdom has it that mammals are the most intelligent of animals. Leading mammals in this respect seem to be the **primates,** which include humans, apes, and monkeys. Ocean mammals often rank second in such appraisals. Next are the **carnivores,** which include the dog, cat, fox, coyote, wolf, lion, tiger, and bear. Following the carnivores in intelligence are the **ungulates (grazing** animals), which include cattle, goats, elephants, horses, swine, and deer. The most intelligent farm animal, contrary to popular belief, probably is the pig and not the horse. Both the elephant and the pig seem to rank above the horse in several aspects of intelligence. In recent studies pigs learned the rules of joystick-operated video games almost as quickly as chimpanzees. However, these rankings are based on a limited number of studies. Research with swine show that their ability to learn may vary among individuals, genders, and breeds.

Many recent investigations have been directed toward the mechanisms involved in memory, which plays an important role in determining intelligence. Memory is of two types: *short-term* and *long-term.* Memory is related to synthesis of proteins in the brain. For example, if protein synthesis is blocked, an animal forgets what it has learned. This means that **DNA** and **RNA** within brain cells are involved as they play important roles in protein synthesis by cellular **cytoplasm.**

24.3 MOTIVATION

Motivation is an internal state of an animal that causes it to behave in a certain way. Scientists often refer to motivation in terms of *drives* or *tendencies* to behaviorally respond in a particular manner. Hence animals experience hunger, thirst, elimination, and sex drives.

The hypothalamus is the part of the central nervous system that controls several types of behavior. Both inhibitory and stimulatory centers are located in this organ and function in behavioral control. For example, destroying a certain area of the hypothalamus related to appetite may result in an animal starving to death even when feed is available. If another area is destroyed the animal may overeat until it becomes obese. The hypothalamus is also involved in behavioral responses other than eating.

Endocrine gland secretions also determine certain behavioral responses, especially sexual activities. Substances secreted by the hypothalamus cause the release of gonadotropic hormones from the **anterior** pituitary gland, which in turn cause **ovulation,** onset of **puberty,** and occurrence of normal estrous cycles in females (Chapter 13).

Stimuli that are effective in triggering a certain behavioral response are called *releasing stimuli.* A certain part of an animal's environment may act as a source of the stimulus (releaser). If the stimulus is sufficiently strong, activation of a certain behavioral response ensues. For example, a record of the vocalizations of a **boar** played over a loudspeaker will cause a **sow** in **heat** to assume the mating **stance** from which she can be moved only by much force and effort. Releasing stimuli often can be designed that are more effective than the natural ones.

24.4 METHODS OF ANIMAL COMMUNICATION

Animals are such agreeable friends—they ask no questions; they pass no criticism.

George Eliot (1819–1880)
English novelist

Although animals cannot speak in the same way as humans, they do communicate with one another. The following are ways in which they do this.

24.4.1 Sounds

'Tis sweet to hear the honest watch dog's bark.

Lord Byron (1788–1824)
English poet

Sound is an important means of communication among animals (Figure 24.2). Most female farm animals respond to the distress calls of their young. Cattle of all ages also respond to a distress call of another of their species, regardless of age. Research has shown that bats use a type of radar to fly in dark caves or on moonless nights. They transmit high-pitched vocalizations that hit an object and bounce back, thus helping them avoid destructive contact with various objects (echolocation). This phenomenon was first postulated by the Italian naturalist Lazzaro Spallanzani in 1793. He removed the eyes of bats and found that they could fly about a room without hitting the walls or furniture. However, when he plugged their ears the eyeless bats could not navigate. He concluded that bats use sound cues to find their way. Dolphins use a similar method of navigation to swim in the ocean depths. Farm animals do not appear to possess such a system of navigation. Birds sing to welcome the morning. In a sense they also communicate with humans by sound, as suggested by the following quotation:

I value my garden more for being full of blackbirds than cherries, and very frankly give them fruit for their song.

Joseph Addison (1672–1719)
English essayist and poet

24.4.2 Chemicals

Many female insects secrete chemical substances (social hormones) that attract males. These compounds are called **pheromones** (Chapter 23). In mammals such as dogs, females in estrus apparently secrete a substance that attracts males from

Figure 24.2 This bellowing bull exemplifies the many animals that communicate by sound.
Courtesy of Missouri Agr. Expt. Station

miles around. Other female farm mammals also secrete chemical compounds when in heat and males appear to locate them by sense of smell. However, this does not appear to be as effective over long distances as in the case of the **bitch.** Dogs use urine as a marker of their presence and possibly of their home territory. **Stallions** are reported to do the same thing by depositing their **feces** at particular locations.

A compound present in saliva and the preputial secretions of boars is now used commercially to stimulate estrous sows and gilts to stand for mating. The same compound also reduces fighting in prepubertal pigs unfamiliar with each other. That the same compound serves these two purposes is not surprising when one considers that for mating to occur, male and female must cease their hostilities, at least temporarily.

No matter how much the cats fight, there always seem to be plenty of kittens.

Abraham Lincoln (1809–1865)
Sixteenth President of the United States of America

24.4.3 Visual Displays

Birds are noted for their visual displays in the acts of courtship and fighting. Visual displays during courtship tend to be more subtle among farm mammals but they do occur. Dogs, when they strike a hostile stance, cause the hair to rise on top of their necks, "raise their **hackles**" (hackles actually are special neck feathers on cocks). This probably serves to make them look larger and thus more formidable. Honeybees communicate in an interesting way. When a worker bee finds a good source of food she returns to the **hive** and performs a certain dance that directs other workers to the site of the food. The particular dance apparently also indicates the distance to the food. Scientists believe that the location of the sun is used by worker bees to give exact directions to fellow workers. The flashing lights male fireflies use to attract females are an example of another type of visual signaling system.

24.5 ORIENTATION BEHAVIOR (NAVIGATION OR HOMING)

Many stories have been told about how cats, dogs, cattle, and horses found their way back home when moved to some distant place. They apparently return home by observing certain landmarks, by smell, or both.

A well-known example of returning to home range after several years' absence is that of salmon. These fish are hatched in freshwater streams. They then swim to the ocean where they spend several years growing to maturity. When the time comes for them to spawn they unerringly return to the same stream in which they hatched. Research shows that their ability to do this is largely due to their sense of smell; salmon are unable to locate their home stream when this sense has been destroyed. Certain migratory birds, homing pigeons, and turtles are thought to use the sun and stars as guideposts.

24.6 CATEGORIES OF ANIMAL BEHAVIOR

Animals express behavior patterns that can be categorized in several ways. Those discussed here will be restricted primarily to farm animals.

24.6.1 Ingestive Behavior

Ingestive behavior—eating and drinking—is characteristic of all animals. It is critically important because it is closely related to an individual's rate of productive performance and state of being. The first **ingestive** behavior trait common to all **neonatal** mammals is **suckling.**

   Livestock vary in the ways they ingest food. Swine and horses possess teeth in both upper and lower jaws so they can bite off grass or take a mouthful of grain, chew, and then swallow. Digestive juices complete the digestive process. Cows have no upper incisors so they ingest food during grazing by wrapping their tongues around a bunch of forage and jerking it forward so it is cut off by the lower teeth. Sheep graze similarly to cattle but have a **cleft** upper lip that allows them to nibble more closely to the ground. Goats graze similarly to cattle and sheep but are also excellent browsers, often feeding on leaves and branches of shrubs and trees (Chapter 2). Once the food is taken into the mouth by **ruminants** (cow, sheep, and goat) it is chewed and swallowed. It is later **regurgitated,** chewed more thoroughly, then swallowed once more. This is known as **rumination** (Chapter 19).

   These animals' grazing habits emphasize the importance of proper pasture and range management. Grazing animals eat the more tender parts of plants and reject the coarser ones. They also tend to eat first those species of plants more to their liking. When these more palatable species are absent, however, animals will eat less desirable ones. When overgrazing occurs over a period of years the more **palatable** plants may be gradually eliminated, being replaced by those of lesser acceptability and productivity (Section 22.6). Thus years of chronic overgrazing may cause considerable change in the mix of plants comprising a pasture or range and thereby reduce its quality and carrying capacity. Most grazing animals do not consume poisonous weeds when the supply of desirable **forages** is adequate but when overgrazing occurs they may consume poisonous weeds, become sick, and possibly die.

   Cattle prefer some species of forages to others, perhaps because some have a bitter or undesirable taste. In the United States a pure stand of fescue will be grazed by animals and they will remain in good condition. However, if the fescue is present in a pasture mixed along with more palatable forages, the fescue usually will be the last grazed.

   Cows on the range generally spend a third to half their time grazing. This varies with forage availability and density. Cows spend a little less time ruminating than grazing; this also varies with forage abundance. Cattle graze at all hours of the day and night but peak activity usually occurs (when **ambient temperature** and other environmental factors are favorable) just after daybreak, in the late afternoon, and just before dark. Nursing may occur during both day and night but tends to be more frequent at daybreak, midday, and dusk.

   Swine have very distinctive feeding and drinking habits. **Pigs** are born with the inherent tendency to root (dig and turn soil). This is a trait carried over from their wild ancestors, which rooted to find animal and plant foods that would provide them with essential nutrients. Domesticated swine will root less if fed a well-balanced **diet** and an appropriate mineral **supplement.** Swine tend to balance their diet if offered a grain such as shelled corn alongside a protein supplement. This balancing of their own diet might be termed "nutritional wisdom." But it is not a perfect phenomenon. Pigs seem to prefer some foods, possibly in part because of differences in taste. When the protein supplement fed free-choice alongside a grain is largely soybean meal pigs tend to eat more protein than needed to balance their diet. When fed a diet rich in meat scraps or **tankage** they are less likely to eat more protein than needed. Pigs seldom if ever become sick or die from overeating, a problem encountered in other species. Salt poisoning is an exception.

   Grazing by **poultry** is limited because they do not have a digestive system designed to handle much forage (Chapter 19). However, they can utilize a limited amount of tender, low-fiber forage. Poultry subsist largely on grain diets.

24.6.2 Eliminative Behavior

Some farm animals deposit their feces at random, others do not. Eliminative behavior in farm animals tends to follow the general pattern of their wild ancestors but can be influenced by management.

   Cattle deposit feces randomly. They will lie on feces but avoid eating plants near fecal deposits. When cattle lie to rest they may **defecate** before they arise. Cows can defecate while walking so their feces may be scattered, but in general feces are deposited in a neat pile while the cow is standing. The American *bison* has a similar way of defecating and the small fecal piles, once dry (called *buffalo chips*), were collected by early settlers as a source of fuel. Most cows and bulls urinate in a random way while standing, although some variation in posture has been observed. Eliminative behavior in sheep is similar to that in cattle. **Ewes,** however, assume more of a squat posture than cows when they urinate.

   Although swine are sometimes thought to be unclean in their habits they are among the most fastidiously clean of all farm animals when given the opportunity. Pigs normally keep their nests clean, dry, and free of **manure** (feces and urine). They usually deposit their feces in a corner of the pen away from their sleeping place. When maintained on pastures with shelters they likewise keep the shelters free of **excreta.** In natural settings pigs usually deposit their excreta near their source of drinking water. Modern methods of pig keeping confine them in pens where their natural eliminative patterns may be thwarted. Pigs prefer to rest in a thermally comfortable part of a pen, so they often eliminate in a less comfortable spot. When pigs defecate and urinate all over a pen floor, poor airflow patterns that vary over the day may be the ultimate cause.

Horses tend to deposit their feces in a certain place and will often return to this place. On the range where several stallions may be in the same pasture with a large group of mares, each stallion has his own band of mares and a territorial range in which key border positions are marked with his feces.

Chickens and other poultry seem to deposit their excreta at random, except for the usual heavy deposition under **roosting** places. Cats bury their feces and urine whereas dogs tend to deposit them at particular places (scent posts). The sniffing, selecting of upright vertical targets, and leg lifting by adult male dogs for urination comprise a secondary sex characteristic that is under hormonal influence.

24.6.3 Shelter-Seeking Behavior

Shelter seeking by animals is an attempt to gain protection from the sun, wind, rain, **predators,** and/or insects (Figure 24.3).

Cattle in the range country of the southwestern United States will descend from high mountain country to lower ranges just ahead of a violent storm. Hundreds of cattle will make such a trek in single file. They seem to have the ability to sense and to avoid violent storms. Cattle may also run and "act up" prior to storms. Cattle usually seek shade in the heat of a summer day and do not come out into the open to graze until sunset. Cows on an open range devoid of trees often congregate around a water source at the same time each morning. Evaporation from the water surface cools the surrounding air a few degrees and the cows may actually enter the pond or stream. Usually they remain quiet and rest when the weather is hot. They then leave the water later in the day.

It is necessary for swine to seek shade or a wallow in hot weather in order to avoid direct rays of the sun and to facilitate evaporative cooling because they do not possess an efficient cooling system (Chapter 17). If water is available when environmental temperature is high they will wallow in it to keep cool. If forced to remain in the hot sun they pant rapidly and often utter grunts of distress. In summer pigs may sleep stretched out full length to expose maximal body surface to cool earth surfaces whereas in cold weather they sleep bunched and huddled to expose minimal body surface to the environment.

During a rain shower or snowstorm cattle, horses, and sheep turn their backs to the storm and often drift away from the oncoming weather so as to protect their faces. The American *bison* is one of the few animals that will face a storm head on.

24.6.4 Agonistic (Fighting) Behavior

Agonistic behavior involves threat, aggression, submission, escape, and passivity. Among the various species of domestic mammals, males (Figure 24.4) are more likely to fight (especially over mating rights) than females, but females do so under certain conditions (protection of young, rights to food). **Castrated** males are usually quite docile. Because of the absence of certain hormones (especially testosterone), the aggression threshold is lowered.

Boars, bulls, **rams,** and stallions, respectively, that run together from a very young age seldom fight with one another. Their dominance orders have been established and are stable. Several mature stallions can run together with a group of mares, although each usually has his own band. Fights do occur between stallions over a band of mares, the victor either retaining the band or taking it from the vanquished. This is nature's way of guaranteeing that the strongest stallion will be the protector of the band as well as produce more and fitter **progeny** for succeeding **generations.**

Under range conditions there are instances when dozens of bulls are run together with hundreds or thousands of cows. Even though the herd includes many different bulls of various ages, fighting among them seldom occurs. In the fall in range country it is not uncommon to see mature bulls grazing together in a group without a cow in sight. They seem to prefer the company of their own gender at that particular time. This happens on large ranches where bulls are allowed to be with cows throughout the year.

Although young males raised together will seldom fight, often in a large group of young bulls many will **ride** a single individual to death if he is not removed from the group. In feedlot steers this phenomenon is called *buller-steer syndrome,* where **riders** ride a **buller.** This can have adverse economic

Figure 24.3 Pigs seeking shelter from the heat of midday.

Figure 24.4 Bulls demonstrating agonistic (combat) behavior.
Photograph by J. F. Lasley.

consequences. The buller should be removed from the group. Similar riding behavior is often seen in a group of young boars.

The bringing together for the first time sexually mature males of the same species almost always results in a fight. Fighting intensity depends on the tenacity of the two combatants. Sometimes one becomes submissive after a simple threat or brief skirmish. At other times fighting becomes intense and although it rarely results in death in farm animals, a serious injury to one or both sometimes results. Fighting among two young boars on a hot summer day may end in the death of one or both from heat exhaustion. If the combatants are allowed to "fight it out" one usually becomes submissive and the fight ends. When a group of **roosters** run with a **flock** of **hens** it is not unusual to see two roosters fight from time to time for several days. Sometimes the fight is ferocious, both combatants being covered with blood.

Some breeds and **strains** of cattle are born with a strong instinct to fight. Bulls in Spain and Mexico are bred especially for their fighting ability. Certain dogs and game chickens are also bred for fighting. Many cows on the range will attack a human on foot or on horseback if they are roped or aggravated and become disturbed. Many wild range cows have sent cowboys to the top of a corral fence. If these same animals had been raised in closer contact with humans they would be more docile. Early experiences greatly affect an animal's behavior later in life. Animals have long memories. Some of the fighting behavior of domesticated animals is a carryover from their ancestors. In the wild the ability to fight is crucial for both survival of the individual and propagation of the species. Interspecies fighting apart from predation is not commonly observed.

Most animals possess the trait of *play behavior* in varying degrees. Play is more common in young animals. Juvenile sparring, a kind of *play* behavior, may be confused with fighting.

24.6.5 Sexual Behavior

Oh, Love's but a dance, where time plays the fiddle.

H. Austin Dobson (1840–1921)
English poet and essayist

Sexual behavior comprises courtship and mating. It is largely controlled by hormones (Chapter 11) although bulls and stallions that are castrated after sexual maturity (**stags**) may retain sex drive (**libido**) and exhibit various sexual–behavioral patterns. Thus psychological (learned) as well as hormonal mechanisms are involved. Sexual behavior is important because it is responsible for the continuation of the species. The number and vitality of the young produced are major factors determining profit in animal agriculture. It is interesting that sheep and goats will readily mate with each other when kept together, although interspecies matings are rarely **fertile.**

Males of domesticated mammals are usually kept separate from females except during the breeding season. Detection of **estrus** in females often depends on observations by the keeper. At times this is difficult because the signs are subtle in some females. Male animals maintain a 24-h vigil and are much more efficient than humans in detecting estrus.

Because of the restlessness she exhibits during estrus, such a female is said to be in *heat*. If the herdsperson places a hand on the rump of a sow in standing heat she will stand absolutely still with ears perked. **Mares** in heat will squeal and urinate profusely in the presence of other horses, but the efforts of mares and ewes to seek a male while in heat are not so obvious as those in other species. When in heat cows and sows will stand when **mounted** by other females whereas mares and ewes seldom exhibit riding behavior. Some cows kept alone will become restless, walk the fence, and bawl when in heat. Some may even go through the fence, which otherwise they would not even attempt, looking for a male. Sows often break out of pens when they come into heat and range far and wide in search of a male.

Males in most species of farm mammals detect females in heat by sight or smell (Figure 24.5). Courtship (appetitive behavior) is more intense on the open range than under more restricted farm conditions. The bull can often identify a pro-estrous cow 24 to 48 h before estrus and frequently remains in her general vicinity. The stallion bites or **teases** the mare in a kind of courtship that is responsible many times for getting the mare into the mood to accept the male in **copulation** (consummatory behavior). The boar often nudges the sow around the head and shoulders with his snout and emits varied grunts (*chant de coeur*) before attempting to mount. A rooster will spread one of his wings toward the ground and scratch the wing feathers with one leg, performing a sort of dance (courtship waltz) around the hen. Sometimes the hen runs away but she may squat, allowing the rooster to mount and copulate.

24.6.6 Mother–Young Behavior

The various forms of maternal (parental) behavior in farm animals begin shortly before birth and end after the young are fledged or weaned. This care-seeking and care-giving behavior varies widely among species.

When **parturition** approaches, a cow in a large pasture or on the range seeks seclusion from the herd. She usually chooses a small depression in the landscape, a patch of shrubs, or a clump of trees where she is hidden from view. After the **calf** is born the mother rises to her feet and closely inspects it, licking it so as to clean and dry the hair coat. Ordinarily within an hour or so after birth the calf stands, and although still somewhat wobbly, starts attempting to nurse. The attempts are awkward at first but usually prove successful. The mother cow stands patiently, often emitting a soft vocalization, until the young one has suckled. After nursing all seems well and the calf typically exhibits a sudden burst of vigor. At this point the cow often eats the **placenta** in order not to alert predators to a vulnerable newborn. A little later the mother hides the calf and may return to the herd. She continues to cast a wary eye in the direction of her hidden young but is careful not to reveal its location, especially to humans. The hiding calf remains virtually motionless and if picked up may appear ill or dead to a novice. The calf is so quiet that often one can walk within a few feet of where it is hiding without seeing it, all this as a means of protecting itself from predators. After 2 to 4 days of intermittently nursing and hiding her young, the cow usually will bring it to join the herd.

If a calf is **stillborn** or dies shortly after birth, some cows quickly leave the place where the young lies, never to return. Others may return to their dead calves at frequent intervals for several days, sniffling and vocalizing gently.

The mother cow is very possessive of her young throughout the nursing period. If her calf encounters trouble from either human or beast she quickly comes to its rescue. Some cows become belligerent when their calves are disturbed and may physically harm a person who comes too close. The tendency of the mother to give protective care must have an inherited basis because it varies widely among individuals and breeds. It has been reported that the F_1 female from an American *bison*–European

Figure 24.5 The upturned, or curled, upper lip indicates that bulls can detect an odor signifying estrus in the cow.
Courtesy of Missouri Agr. Expt. Station.

cattle cross does not possess a strong maternal instinct and her young must be reared by a foster mother. This further suggests that maternal care is based on inheritance.

Cows locate their young mainly by smell although sight and sound may also be involved. Regardless of the way in which a cow and her calf recognize each other, it is done accurately even when a large herd is crowded together. At roundup time on the range, where many different brands may be represented in a herd, the cowboys must wait until the cows and calves "mother up" in order to properly put the same brand on the calf that is on the cow (Figure 24.6).

Beef cows on an Arizona range exhibit an interesting type of protective behavior toward their calves. Often in a pasture of thousands of acres there are only one or two watering places. It is not unusual to see 12 to 15 baby calves in an area accompanied by only one or two mother cows. Usually the other dams have gone to water, leaving a nurse or two to protect their young from predators. The giraffe and elephant also employ "baby-sitters" to look after their young while they seek food and water.

Because the dairy calf is removed from its mother within a few days after birth and reared separately, the tie between mother and offspring is soon severed. This is not true, however, with range cattle. One of the noisiest times on a farm or ranch is weaning time when cows are cut away from their calves. Both calves and cows bawl (often in unison) almost continuously for 2 or 3 days before quieting down. After they have been kept separated for several days the bond between them has been partially to completely extinguished and they show no signs of mutual recognition.

Around 18 h before the first piglet is delivered a sow will build a nest. (However, many sows today are **farrowed** in systems where there are no materials and not enough space available for nest building.) In building a nest the sow hollows out an area in the ground that fits her body and carries leaves, straw, or other vegetation to the site. She lies in such a way that soon after birth the pigs can go to the side with mammary glands. The time of farrowing in sows typically spans a period of up to 4 h although cases are known in which littermate piglets were born more than 24 h apart. A delay of several hours because of difficult parturition (**dystocia**) usually results in stillbirth of the rest of the pigs. Some sows are highly nervous before and during farrowing and eat (savage) their liveborn pigs.

Sows will usually accept pigs from another **litter** if the foster pigs are placed with them during the first day or two after birth. Exchanging pigs among sows is a common practice in herds in which many sows are farrowing around the same time. Sows are very protective of their young and will defend them when they squeal. The sow approaches an intruder with her mouth open and emitting a series of sharp, barking grunts in rapid succession. This is much more evident in some sows than in others. The sow continues to care for her pigs until they are weaned, but after 2 to 3 days of separation from her young she appears to lose interest in them. If pigs are left with the sow for several weeks past the usual weaning age, she may wean them herself and cease lactating.

Piglets suckle at intervals of 1 to 2 h. After the first few days each **piglet** usually has laid claim to a teat that it will nurse for the balance of the nursing period (Figure 24.7). A few pigs may nurse two teats if the litter is small. The rear teats of the sow are less likely to be functional at weaning time than are the front ones (Table 24.1). An explanation for this may be that milk begins to flow first in front teats, later in rear ones (see Chapter 15). Stronger piglets tend to appropriate for themselves nipples that are first to yield milk.

Many times **runt** pigs get the rear teats, probably because they are not as successful in competing for the front ones. The

Figure 24.6 Waiting for cows and calves to "mother up" on an Arizona range so that the mother's brand can be placed on her calf.
Photograph by J. F. Lasley.

Figure 24.7 Nursing behavior in bigger pigs. Smaller pigs ordinarily suckle while the sow lies on her side.
Courtesy of Howard Cowden.

TABLE 24.1	Location of Teats That Were Functional in 147 Sows When Their Pigs Were Weaned at 56 Days	
Location of teats*	**Percent of teats functional at weaning**	
First pair	81	
Second pair	79	
Third pair	77	
Fourth pair	69	
Fifth pair	43	
Sixth pair	27	
Seventh pair	6	

*The first pair in all sows was the pair nearest the front legs. Some sows, of course, did not possess seven pairs of teats. The percentage figures are based only on total teats present in that particular location.

Source: Missouri Agr. Expt. Station.

extreme rear teats usually yield less milk than those toward the front, which may partially explain why pigs nursing the rear teats often have lower body weight at weaning.

After parturition the ewe licks her newborn **lamb,** removing fluids and placental membranes. The neonatal lamb soon staggers to its feet and makes awkward efforts to find a teat to suckle. The ewe stands quietly during this process. After a few days the lamb goes directly to the ewe's teat and immediately suckles. Some time is required for milk letdown, however (Chapter 15). The suckling posture of the lamb is usually a standing one. If the lamb's tail has not been **docked** (cut off) it wiggles in all directions while it suckles. Although normally timid and easily frightened, the ewe will defend her young even if the attacker is

formidable. It is interesting that sheep will accept and nurse orphan goats (**kids**) and vice versa (interspecies rearing).

Mares show similar maternal behavior toward their **foals.** A mare calls her foal with a **neigh** or **whinny** and exhibits considerable nervousness and distress when her young is disturbed. When horses were used for pulling farm implements, foals were often left in the barn while their mothers worked in the fields. When the teams were brought in at midday or nighttime there usually was a noisy exchange of whinnies between dam and foal until the foal was allowed to suckle. A mare will give as much attention and affection to a **mule foal** as to a horse foal.

Broodiness (an inclination to set on eggs) of hens is an inherited trait. Few hens in the modern Leghorn breed (layers) show this tendency whereas many Cornish hens (broilers) are broody. Broodiness is related to secretion of and response to prolactin (Chapter 16). A broody hen will set on a nest of eggs or on one that contains no eggs. She is dedicated in her nesting behavior and spends little time eating or drinking during the incubation period.

Hens allowed to run with their **chicks** show intense maternal behavior (Figure 24.8). When the hen finds a choice morsel while foraging for food, she clucks and all the chicks come running. Hens hover over their chicks by covering them with their wings and nestling them close to their bodies at night and during the day when they need protection. Hens with chicks will attack any perceived or real enemy that "threatens" their young. The hen emits a loud, shrill cry to warn all chicks of the flock of imminent danger and the chicks respond quickly. Today few hens are allowed to incubate their eggs (brood) or care for their young. Incubation of fertile eggs is by mechanical means because hens are more profitable when directed toward laying maximum numbers of eggs.

24.6.7 Investigative (Exploratory) Behavior

Exploratory behavior involves exploration through seeing, hearing, smelling, tasting, and touching new places and investigating strangers. Cattle investigate an object they do not fear by approaching it closely, their ears pointed forward and eyes focused on it (Figure 24.9). Calves tend to be even more curious than cows. When placed in a new pasture cows graze quickly away from the place of entrance and soon have investigated every nook and cranny. As every farmer and rancher knows, if there is an open gate or a hole in the fence around the new pasture the cows will soon find it and try to leave through it.

Sheep also investigate strange objects in their surroundings by approaching them with their heads up, ears forward, and eyes fixed. However, they usually are more timid than cattle and will bunch and run if the object makes a strange or unexpected move or noise.

Swine also are curious creatures. When approached by a person they give a "woof," scatter as if frightened, and run as fast as they can to relocate their **flight zone.** If the person stands still and makes no motion to scare them they usually will return and may approach the person, sniffing and nibbling boots and clothes.

The young foal spends much of its time inspecting and sniffing objects in its environment. As it grows older it becomes less curious. The investigative habits of the young foal are a source of continual worry to its nervous mother, who at times seems to doubt the investigative judgment of her offspring.

24.6.8 Gregarious Behavior

Gregarious behavior is observed only in some species. Gregarious animals prefer to live in flocks or herds. In the wild state this is an advantage in detecting the enemy and protecting themselves. Breeds developed in regions with lush pastures tend to be more gregarious than those indigenous to regions with sparse pastures. A method of trapping groups of wild cattle is depicted in Figure 24.10.

Cattle tend to roam in groups of various numbers depending on the size of pasture or range. There usually is considerable

Figure 24.8 Hen demonstrates the mother–young behavioral trait.
Courtesy of Dorothy Tompkins.

Figure 24.9 Investigative behavior in heifers.

Figure 24.10 A wild cow trap on a rough mountain range in Arizona. The country is often so rough and brushy that it is impossible to round up cattle and they become wild. Thus they must be trapped during roundup time. Note that the triggers are set so that animals can go to water and salt but cannot return. When it is not roundup time the triggers are opened and the animals become accustomed to going in and out of the opened trap.

space between individuals when they are actually grazing. In a farm herd in which the number of cattle can be easily counted, even one individual missing from the group signifies that something is wrong. The missing animal may be sick or dead or may have escaped into an adjacent pasture.

Cattle mate promiscuously and there is no tendency for a bull and a cow to pair for life. However, if several cows in a herd are concurrently in heat and only one bull is present, the bull may show a preference to mate with a certain female and not with one or more of the others.

Sheep are famous for their strong tendency to run in a flock. They also intently follow a leader. The leader is usually an older ewe because rams usually run with the flock only during the breeding season and sheep tend to graze in matriarchal-line subflocks. Lambs in a marketplace or abattoir are often led from one place to another by a halter-broke goat led by an attendant. Sheep raising was an important occupation in biblical times. Writers of the New Testament certainly were aware of the gregarious nature of sheep and the ease with which they could be led. They were also aware of the strong tie between shepherd and flock as noted in John 10:14: "I am the good shepherd; I know my sheep and my sheep know me."

Swine are gregarious by nature but now have little opportunity to show this because of modern methods of husbandry. Even now they are tactile creatures, preferring to lie in contact with a groupmate. They are usually confined to a limited area during the growing-finishing period. Adult domesticated pigs that have grown up in the wild (feral pigs) tend to roam in herds of 10 or fewer and usually are under the leadership of a mature

sow. A mature wild boar with his long, sharp tusks is a good match for almost any enemy. Thus he effectively protects his small group during the breeding season. Under domestication the pig has lost much of its ferocity and usually is a gentle and easily handled animal although there are exceptions. Part of this trend to docility is the result of selecting for docile behavior. Selection progress in swine can be rapid because of their high reproductive rate and relatively short **generation interval** (Chapter 9) compared with other farm mammals.

Horses in the wild are gregarious by nature and run in herds of varying size. During the breeding season each stallion has his own small band of mares, which he carefully guards from other stallions and predators. Some domesticated horses show a definite preference for one horse and will avoid others. There seems to be an especially strong bond between members of an established team of draft horses that have combined their efforts over time in pulling heavy loads.

24.7 SOCIAL DOMINANCE

Within most groups of gregarious farm animals of the same species there is usually a well-organized social rank order. In a linear dominance order one individual is dominant over all others, one ranks second, one third, and on down to the one that is subordinate to all. Dominance orders exist in all species although they are not always in a straightforward linear order. More attention is paid to social rank in a group of female farm animals in which social dominance exists than in human social groups.

The existence of a social dominance order is most noticeable when the species has an instinct to fight or when resources such as food or mates are limited. At most other times the social order is evident only in subtle ways. Social rank order is called the *peck order* in chickens, in which it was first described. When two hens meet for the first time a vigorous fight usually ensues. One hen will finally give up and run away thus ending the fight. The next time the two meet there is often another fight, but it is usually shorter and less intense than the first and has the same result. At each meeting thereafter the hens fight again, but less and less intensely until finally only the threat of a peck from the dominant hen sends the subordinate one scurrying. Establishment of a social dominance order within a group is important because it minimizes fighting and organizes access to important resources such as feed and water, thereby allowing more time for eating and other productive purposes.

The rooster assumes indisputable command of a flock of hens during the laying season and the hens seem to accept this. The hens also establish a peck order (social hierarchy). Some observations indicate that if the rooster takes a fancy to a certain hen, her social rank rises in the flock. It goes down again, however, if the rooster switches his primary attention to another hen.

One of the principal advantages of social rank order in a group of wild animals is that it gives mating priority to the top-ranked males. Hence they tend to leave more progeny than the less dominant ones and their progeny theoretically have a higher probability of survival. The **offspring** of dominant females of some species grow faster, often because they get more food. They also may secrete more of a particular hormone or a better balance of two or more hormones.

Social rank order among farm animals on pasture may be of little economic importance but it may become more important when a group of animals are fed off-pasture. This is especially true if the animals are fed a **limited ration.** Dominant individuals under such conditions often get more than their share of the feed whereas subordinate individuals may practically starve, become thin, and perform inefficiently.

Several factors influence social dominance rank in animals. When a male chicken is castrated (**caponized**) it tends to go to the bottom in rank among intact males. The injection of hens or roosters with the male hormone testosterone increases their social rank. Age is also an important factor. Young animals have a low rank and seldom try to achieve a rank over their elders. Early experience also seems important because if individuals find themselves in a subordinate rank early on they usually remain there. Other young animals that do not rank as low as these may eventually assume a higher rank. In cattle, weight or size and aggressiveness or timidity are related to social rank. To maintain a high rank in a social order an individual must have the continuing will and ability to fight.

Social rank appears to be moderately **heritable,** which indicates that by selecting and mating subordinate animals, genetic progress can be made in achieving timidity and reduced aggressiveness in the progeny (Chapter 9). Some animal breeders believe that under domestication humans generally have selected for less aggressive animals whereas in the wild state natural selection favors those animals that are aggressive and strong.

24.8 POPULATION DENSITY AND ANIMAL BEHAVIOR

Most animal populations increase in number until they reach a certain level and then remain more or less constant. Within such a population there are minor fluctuations from season to season and from year to year. Human populations differ from this by showing long-term upward trends in numbers (Chapter 1).

The principal factors regulating population size or animal density include predation, starvation, accidents, **parasites,** and **disease** (Chapter 1). However, present evidence indicates that these factors alone do not altogether account for the observed variation. Certain forms of social behavior limit the rate of reproduction and prevent population numbers from exceeding food supplies.

The behavior of animals is related to control of their population density. Some species limit the size of their territorial habitat during the breeding season. In some wild species low-ranking individuals may be driven from a particular area and not allowed to reproduce there. The population of an area thus can be self-regulated.

The question of population density and social **pathology** of rats has been examined. A population of wild Norwegian rats was confined in an enclosure of about one-quarter acre. Ample food was supplied and predation and disease were minimized. This left social behavior as the only apparently important force that could prevent population density from increasing. Within 27 months population size had become stabilized at about 150 adults. This plateau was due to the fact that there was so much **stress** from social interactions that maternal behavior was upset and few young survived. Later a similar experiment was conducted with domesticated albino rats with similar results. Females were more affected by overcrowding. Some did not carry pregnancies to full term and litters of others failed to survive birth. Some other females built inadequate nests or none at all. Infant **mortality** increased to 96 percent in these crowded females. The males showed disturbances ranging from certain sexual deviations to **cannibalism.** The entire social organization of the group disintegrated.

Although the relation of population density to social tension in farm mammals has not been studied extensively, high animal density may affect fertility and growth performance.

24.9 SUMMARY

Behavior may be defined as the way an animal reacts to a certain stimulus with activity or inactivity. Ethology—the scientific study of animal behavior—is a complex subject. Many behavioral patterns in animals can be traced to their ancestors' abilities to survive in the wild. An understanding of the basic principles of animal behavior is necessary for successful animal husbandry.

Behavioral responses in all animals are determined by heredity and learning. Hereditary responses are innate and can be illustrated by an animal behaving in a certain way without a need to learn the behavior. This is sometimes called *instinct*. Certain types of behavior, however, must be learned through experience. Animals tend to act and respond in accordance with past experiences. The rapidity with which an individual learns is determined by its intelligence. Animals communicate with one another by sight, sound, smell, and touch. Communication plays an important role in animal behavior.

Typical categories of behavior include ingestive, eliminative, shelter-seeking, agonistic, sexual, mother–young, investigative, and gregarious. The importance of social dominance order in domestic animals and the relation of population density to social tension and hence to productive performance need further study if we are to better understand animal behavior and its role in the production of foods and services useful to humanity.

STUDY QUESTIONS

1. Define animal behavior.

2. Of what practical value is a knowledge of animal behavior to the producer of livestock and poultry and the caretaker of companion animals?

3. What factors cause behavioral responses in animals?

4. In what ways may heredity produce behavioral responses?

5. What evidence is there for the existence of basic hereditary patterns affecting behavior?

6. List and describe the several types of learned behavior in animals.

7. What is intelligence in animals?

8. Which farm animal probably has the highest intelligence?

9. In what way can chemistry affect the ability to remember?

10. Define the term *motivation* (as it relates to animal behavior).

11. How may hormones affect the behavior of animals?

12. In what general ways do animals communicate?

13. What are pheromones?

14. Name and define some of the most important types of animal behavior.

15. Describe some interesting examples of mother–young behavior.

16. What is meant by gregarious behavior?

17. What advantages are there in gregarious behavior?

18. Cite one or more major differences between goats and sheep in their behavior.

19. What is meant by social dominance? How can it be of importance in the production of livestock and poultry?

20. What is the relation between population density and behavior in rats?

Common Terms or Names Applied to Selected Farm Animals

APPENDIX A

	Cats	Cattle	Dogs	Goats	Horses	Poultry	Sheep	Swine
Male of breeding age	Tom	Bull	Dog	Buck	Stallion	Cock*	Ram	Boar
Mature female	Queen	Cow	Bitch	Doe	Mare	Hen	Ewe	Sow
Young male	—	Bullock	Puppy dog	Buck kid	Colt	Chick[†]	Ram lamb	Shoat[‡]
Young female	—	Heifer	Puppy bitch	Doe kid	Filly	Chick[†]	Ewe lamb	Gilt
Newborn	Kitten	Calf	Pup	Kid	Foal	Chick[†]	Lamb	Pig
Unsexed male	Gib	Steer	Castrate	Wether	Gelding	Capon	Wether	Barrow
Unsexed female	Neuter	Spayed	Spayed	Spayed	Spayed	—	Spayed	Spayed
Groups	Bevy	Herd	Pack	Band	Herd	Flock	Flock	Drove
Genus	*Felis domestica*	*Bos*	*Canis*	*Capra*	*Equus*	*Gallus*[§]	*Ovis*	*Sus*
Act of breeding	Mating	Serving	Copulating	Serving	Covering	Mating	Tupping	Coupling
Act of parturition	Littering	Calving	Whelping	Kidding	Foaling	—	Lambing	Farrowing

*Called a *tom* in turkeys.

[†]Called a *poult* in turkeys, *gosling* in geese, and *duckling* in ducks.

[‡]Applied to a young pig of either sex under 1 year of age.

[§]Genus for chickens (Aves is the class of birds).

APPENDIX B   *Convenient Conversion Data*

1 milliliter (ml or cc) = 12 drops

1 teaspoon = 60 drops (5 cc)

1 tablespoon = 3 teaspoons (0.5 oz)

1 cup = 0.5 pint (8 oz)

1 pint = 32 tablespoons

1 quart = 2 pints (946.4 cc) and 1/8 peck

1 gallon = 4 quarts (3.785 liters, 8.345 pounds)

1 kilogram (kg) = 2.2046 pounds

1 liter = 1.0567 quarts (1 kg water)

1 foot = 0.3048 meter

1 meter = 3.2809 feet (39.37 inches)

1 kilometer = 0.6214 mile

1 acre = 0.4047 hectare (43,560 square feet)

1 hectare = 2.471 acres

1 barrel = 31.5 gallons (119 liters)

Grains, Pounds per Bushel

Oats = 32

Corn and rye = 56 (ear corn = 70)

Barley = 48

Beans and wheat = 60

1 bushel = 2150 cubic inches (32 quarts)

1 gallon = 231 cubic inches

Temperature Conversion

$$^\circ C = \frac{^\circ F - 32}{1.8}$$

$$^\circ F = (^\circ C \times 1.8) + 32$$

Measure of Mass

Grains	Drams	Ounces	Pounds	Metric equivalents, grams
1	0.0366	0.0023	0.00014	0.0647989
27.34	1	0.0625	0.0039	1.772
437.50	16	1	0.0625	28.350
7000	256	16	1	453.5924277

APPENDIX C

Metric Weight

Microgram	Milligram	Centigram	Decigram	Gram	Dekagram	Hectogram	Kilogram	Metric ton
1								
10^3	1							
10^4	10	1						
10^5	100	10	1					
10^6	1000	100	10	1				
10^7	10^4	1000	100	10	1			
10^8	10^5	10^4	1000	100	10	1		
10^9	10^6	10^5	10^4	1000	100	10	1	
10^{12}	10^9	10^8	10^7	10^6	10^5	10^4	1000	1

Metric Measure (Volume)

Microliter	Milliliter	Centiliter	Deciliter	Liter	Dekaliter	Hectoliter	Kiloliter	Myrialiter
1								
10^3	1							
10^4	10	1						
10^5	100	10	1					
10^6	10^3	100	10	1				
10^7	10^4	10^3	100	10	1			
10^8	10^5	10^4	10^3	100	10	1		
10^9	10^6	10^5	10^4	10^3	100	10	1	
10^{10}	10^7	10^6	10^5	10^4	10^3	100	10	1

Metric Measure (Length)

Micrometer*	Millimeter	Centimeter	Decimeter	Meter	Dekameter	Hectometer	Kilometer	Myriameter	Megameter	Equivalents	
1	0.001	10^{-4}								0.000039	in
10^3	1	10^{-1}								0.03937	in
10^4	10	1								0.3937	in
10^5	100	10	1							3.937	in
10^6	1000	100	10	1						39.37	in
10^7	10^4	1000	100	10	1					10.9361	yd
10^8	10^5	10^4	1000	100	10	1				109.3612	yd
10^9	10^6	10^5	10^4	1000	100	10	1			1093.6121	yd
10^{10}	10^7	10^6	10^5	10^4	1000	100	10	1		6.2137	mil
10^{12}	10^9	10^8	10^7	10^6	10^5	10^4	1000	100	1	621.370	mil

*Formerly called the *micron* and abbreviated μ.

APPENDIX D

Agricultural Colleges and Experiment Stations in the United States and Its Territories

State	University name	Address
Alabama*,†	Auburn University	Auburn University, AL 36849-5113
Alabama†	Alabama A&M College	Normal, AL 35762-1087
Alabama*,†	Tuskegee University	Tuskegee, AL 36088-2314
Alaska*,†	University of Alaska–Fairbanks	Fairbanks, AK 99775-7140
Alaska*	University of Alaska Agr. & Forestry Expt. Station	Palmer, AK 99645-6629
American Samoa	Land-Grant Div., American Samoa Comm. College	Pago Pago, AS 96799-2609
Arizona*,†	University of Arizona	Tucson, AZ 85721-0038
Arkansas*,†	University of Arkansas	Fayetteville, AR 72701-1202
Arkansas†	University of Arkansas at Pine Bluff	Pine Bluff, AR 71601-2799
California*,†	University of California	Berkeley, CA 94704-1200
	University of California	Davis, CA 95616-8598
	University of California	Riverside, CA 92521-0001
California*	University of California	Los Angeles, CA 90024-0001
California‡	California State University	Fresno, CA 93740-9999
Colorado*,†	Colorado State University	Fort Collins, CO 80523-3001
Connecticut*,†	University of Connecticut	Box U-66, Storrs, CT 06269-4066
Connecticut*	Connecticut Agr. Expt. Station	New Haven, CT 06504-1106
Delaware*,†	University of Delaware	Newark, DE 19717-1303
Delaware†	Delaware State University	Dover, DE 19901-2276
District of Columbia*,†	University of the District of Columbia, Division of Natural Resources & Agr. Expt. Station	Washington, DC 20008-1122
Florida*,†	University of Florida	Gainesville, FL 32611-0200
Florida†	Florida A&M University	Tallahassee, FL 32307-1402
Georgia*,†	University of Georgia	Athens, GA 30602-7501
Georgia*	Georgia Coastal Plain Expt. Station	Tifton, GA 31794-0748
Georgia*	University of Georgia Griffin Expt. Station	Griffin, GA 30223-1731
Georgia†	Fort Valley State College	Fort Valley, GA 31030-0600
Guam*,†	University of Guam	Mangilao, Guam 96923-0001
Hawaii*,†	University of Hawaii	Honolulu, HI 96822-2231
Idaho*,†	University of Idaho	Moscow, ID 83844-3151
Illinois*,†	University of Illinois	Urbana, IL 61801-3620
Indiana*,†	Purdue University	West Lafayette, IN 47907-1031
Iowa*,†	Iowa State University	Ames, IA 50011-0001
Kansas*,†	Kansas State University	Manhattan, KS 66506-4008
Kentucky*,†	University of Kentucky	Lexington, KY 40546-0091
Kentucky†	Kentucky State University	Frankfort, KY 40601-2355
Louisiana*,†	Louisiana State University	Baton Rouge, LA 70894-5055
Louisiana†	Southern University and A&M College	Baton Rouge, LA 70813-2073
Maine*,†	University of Maine	Orono, ME 04469-5782

Maryland*,†	University of Maryland	College Park, MD 20742-5565
Maryland†	University of Maryland Eastern Shore	Princess Anne, MD 21853-1299
Massachusetts*,†	University of Massachusetts	Amherst, MA 01003-0001
Michigan*,†	Michigan State University	East Lansing, MI 48824-1314
Micronesia*,†	College of Micronesia	Pohnpei, FM 96941-1179
Minnesota*,†	University of Minnesota	St. Paul, MN 55108-1004
Mississippi*,†	Mississippi State University	Mississippi State, MS 39762-9740
Mississippi†	Alcorn State University	Lorman, MS 39096-7500
Missouri*,†	University of Missouri–Columbia	Columbia, MO 65211-0001
Missouri†	Lincoln University	Jefferson City, MO 65102-0029
Montana*,†	Montana State University	Bozeman, MT 59717-2860
Nebraska*,†	University of Nebraska–Lincoln	Lincoln, NE 68583-0908
Nevada*,†	University of Nevada	Reno, NV 89557-0104
New Hampshire*,†	University of New Hampshire	Durham, NH 03824-3587
New Jersey*,†	Rutgers–The State University of New Jersey	New Brunswick, NJ 08903-0231
New Mexico*,†,‡	New Mexico State University	Las Cruces, NM 88003-8003
New York*,†	Cornell University	Ithaca, NY 14853-5905
New York*	New York Agr. Expt. Station	Geneva, NY 14456-0462
North Carolina*,†	North Carolina State University	Raleigh, NC 27695-0001
North Carolina†	North Carolina A&T State University	Greensboro, NC 27420-1928
North Dakota*,†	North Dakota State University	Fargo, ND 58105-5437
Ohio†	The Ohio State University	Columbus, OH 43210-1357
Ohio*	Ohio Agr. Research & Development Center	Wooster, OH 44691-4096
Oklahoma*,†	Oklahoma State University	Stillwater, OK 74078-6019
Oklahoma†	Langston University	Langston, OK 73050-0730
Oregon*,†	Oregon State University	Corvallis, OR 97331-2212
Pennsylvania*,†	The Pennsylvania State University	University Park, PA 16802-2600
Puerto Rico*,†,‡	University of Puerto Rico	Mayaguez, PR 00681-9030
Rhode Island*,†	University of Rhode Island	Kingston, RI 02881-0806
South Carolina*,†	Clemson University	Clemson, SC 29634-0001
South Carolina†	South Carolina State University	Orangeburg, SC 29117-0001
South Dakota*,†	South Dakota State University	Brookings, SD 57007-0191
Tennessee*,†	University of Tennessee	Knoxville, TN 37996-4500
Tennessee†	Tennessee State University	Nashville, TN 37209-1561
Texas*,†	Texas A&M University	College Station, TX 77843-2142
Texas†	Prairie View A&M University	Prairie View, TX 77446-4079
Utah*,†	Utah State University	Logan, UT 84322-4800
Vermont*,†	University of Vermont	Burlington, VT 05405-0106
Virgin Islands*	University of the Virgin Islands	St. Croix, VI 00802-9990
Virginia*,†	Virginia Polytechnic Institute and State University	Blacksburg, VA 24061-0402
Virginia†	Virginia State University	Petersburg, VA 23806-0001
Washington*,†	Washington State University	Pullman, WA 99164-6240
West Virginia*,†	West Virginia University	Morgantown, WV 26506-6201
Wisconsin*,†	University of Wisconsin	Madison, WI 53706-1522
Wyoming*,†	University of Wyoming	Laramie, WY 82071-3354

Other colleges granting degrees in food and agricultural sciences

Arizona	Arizona State University	Tempe, AZ 85282-2203
California	California Polytechnic State University	San Luis Obispo, CA 93407-0001
Illinois	Southern Illinois University	Carbondale, IL 62901-4399
Texas	Texas Technological University	Lubbock, TX 79409-2123
Wisconsin	University of Wisconsin–River Falls	River Falls, WI 54022-5001

*Agriculture experiment station.

†Land-grant college of agriculture.

‡Hispanic-serving land-grant institution.

Note: Agricultural colleges and experiment stations will send, on request, a list of available bulletins relating to their published research and activities.

APPENDIX E

Alphabetical List of Elements and Symbols

Element	Symbol	Atomic number	Atomic weight*	Element	Symbol	Atomic number	Atomic weight*
Actinium	Ac	89	227	Hydrogen	H	1	1.0079
Aluminum	Al	13	26.98	Indium	In	49	114.81
Americium	Am	95	243	Iodine	I	53	126.90
Antimony	Sb	51	121.75	Iridium	Ir	77	192.2
Argon	Ar	18	39.942	Iron	Fe	26	55.85
Arsenic	As	33	74.91	Krypton	Kr	36	83.80
Astatine	At	85	210	Lanthanum	La	57	138.91
Barium	Ba	56	137.35	Lawrencium	Lr	103	257
Berkelium	Bk	97	249	Lead	Pb	82	207.20
Beryllium	Be	4	9.013	Lithium	Li	3	6.940
Bismuth	Bi	83	208.99	Lutelium	Lu	71	174.98
Boron	B	5	10.82	Magnesium	Mg	12	24.32
Bromine	Br	35	79.913	Manganese	Mn	25	54.94
Cadmium	Cd	48	112.40	Mendelovium	Md	101	256
Calcium	Ca	20	40.08	Mercury	Hg	80	200.60
Californium	Cf	98	251	Molybdenum	Mo	42	95.95
Carbon	C	6	12.010	Neodymium	Nd	60	144.26
Cerium	Ce	58	140.12	Neon	Ne	10	20.182
Cesium	Cs	55	132.90	Neptunium	Np	93	237
Chlorine	Cl	17	35.455	Nickel	Ni	28	58.71
Chromium	Cr	24	52.01	Niobium (columbium)	Nb	41	92.91
Cobalt	Co	27	58.94	Nitrogen	N	7	14.007
Copper	Cu	29	63.54	Nobelium	No	102	254
Curium	Cm	96	247	Osmium	Os	76	190.2
Dysprosium	Dy	66	162.50	Oxygen	O	8	15.999
Einsteinium	Es	99	254	Palladium	Pd	46	106.4
Erbium	Er	68	167.26	Phosphorus	P	15	30.973
Europium	Eu	63	152.0	Platinum	Pt	78	195.08
Fermium	Fm	100	253	Plutonium	Pu	94	242
Fluorine	F	9	19.00	Polonium	Po	84	210
Francium	Fr	87	223	Potassium	K	19	39.098
Gadolinium	Gd	64	157.25	Praseodymium	Pr	59	140.91
Gallium	Ga	31	69.72	Promethium	Pm	61	147
Germanium	Ge	32	72.60	Protactinium	Pa	91	231
Gold	Au	79	197.0	Radium	Ra	88	226
Hafnium	Hf	72	178.49	Radon	Rn	86	222
Helium	He	2	4.003	Rhenium	Re	75	186.21
Holmium	Ho	67	164.93	Rhodium	Rh	45	102.90

Element	Symbol	Atomic number	Atomic weight*	Element	Symbol	Atomic number	Atomic weight*
Rubidium	Rb	37	85.48	Thallium	Ti	81	204.38
Ruthenium	Ru	44	101.1	Thorium	Th	90	232.04
Samarium	Sm	62	150.34	Thulium	Tm	69	168.93
Scandium	Sc	21	44.96	Tin	Sn	50	118.69
Selenium	Se	34	78.96	Titanium	Ti	22	47.90
Silicon	Si	14	28.09	Tungsten (wolfram)	W	74	183.85
Silver	Ag	47	107.875	Uranium	U	92	238.06
Sodium	Na	11	22.990	Vanadium	V	23	50.95
Strontium	Sr	38	87.63	Xenon	Xe	54	131.29
Sulfur	S	16	32.064	Ytterbium	Yb	70	173.03
Tantalum	Ta	73	180.94	Yttrium	Y	39	88.92
Technetium	Tc	43	99	Zinc	Zn	30	65.38
Tellurium	Te	52	127.60	Zirconium	Zr	40	91.22
Terbium	Tb	65	158.92				

*Atomic weight of the most abundant or best-known isotope or (in the case of radioactive isotopes) of the isotope with the longest half-life, relative to atomic weight of carbon 12 = 12.

GLOSSARY

A

abattoir   A slaughterhouse.

ablactate   To wean.

abomasum   The fourth compartment (true stomach) of a ruminant (cow, deer, goat, sheep).

absorption   The process of taking in food through the **intestinal** wall.

acanthosis   Hypertrophy, or thickening, of the prickle cell layer of skin.

acariasis   An infestation of humans or animals with mites or ticks.

Acarina   An order that includes mites and ticks.

acclimatization   Acclimation, or the complex of processes of becoming accustomed to a new climate or other environmental conditions. A series of compensatory alterations in an animal.

acetate   See **acetic acid.**

acetic acid   A weak organic acid (CH_3COOH) with a characteristic pungent odor. It is a clear colorless liquid. Vinegar is a dilute, impure acetic acid.

acetonemia (ketosis)   A condition characterized by an abnormally elevated concentration of ketone (acetone) bodies in body tissues and fluids.

achromotrichia   Graying of the hair.

acromegaly   A chronic disease caused by an overfunction of the anterior pituitary with excessive production of growth hormones. It is characterized by an enlargement of the bones and soft parts of the hands, face, and feet. Also called *Marie's disease.*

acute   Generally characterized by a short and often severe course.

adaptation   The adjustment of an organism to a new or changing environmental condition.

adaptation syndrome   The defensive response of the body through the endocrine system to systemic injury evoked by stresses and worked out by an initial stage of shock.

ADF (acid detergent fiber)   Fiber extracted with acidic detergent in a technique employed to help appraise the quality of forages.

ADG   See **average daily gain.**

adipose   Of a fatty nature; the fat present in cells of connective tissue.

ad libitum **(ad lib)**   At pleasure. Commonly used to express the availability of feed on a **free-choice** basis.

aeration   To mix with air. In aquaculture to expose and mix water with air, to increase the amount of oxygen in the water.

aerobic   A term usually applied to microorganisms that require oxygen to live and reproduce.

afebrile   Without fever.

afterbirth   The placenta and allied membranes with which the fetus is connected. It is expelled from the uterus following **parturition.**

agalactia   A failure to secrete milk following parturition.

agglutinate   To adhere, unite, or combine into a group or mass (as with clumps of certain body cells). The *clumping* is often caused by antibodies attaching to cells and cross-linking them together.

agglutination   The clumping together of cells, especially bacteria or red blood corpuscles, distributed in a fluid. In biology, it is generally a result of antigen-antibody reactions. Called also *clumping.*

agglutination titer   The highest dilution of a serum that causes clumping together of cells, especially bacteria or red blood corpuscles.

agglutinin   Antibody formed against cells (as bacteria), which, when mixed with these cells, causes them to clump together, or **agglutinate.**

agriculture   The utilization of biological processes, on farms and ranches, to produce food, fiber, and other products useful to humans.

AI daughters   Daughters of a sire that were sired after the bull was placed in artificial-insemination service.

albinism   Congenital absence of pigment in the skin, hair, and eyes. See also **true albino.**

algae   Chlorophyll-bearing miscroscopic plants that synthesize food by photosynthesis. Algae are mostly aquatic and lack true stems, roots, or leaves.

allele   See **allelomorphs, dominance,** and **recessive.**

allelomorphs (alleles)   Genes are paired in each animal, and each kind of gene at a particular chromosome location is called an *allele.* They are alternative forms of genes. Alleles are two genes that occupy the same location (locus) on homologous chromosomes and affect the same trait phenotypically but in a different or alternative way. For example, the gene for brown eyes (B) in humans is an allelomorph of the gene for blue eyes (b). See also **multiple alleles.**

allergen   Any substance that gives rise to the formation of antibodies and the resultant allergic reaction. Also called an **antigen.**

allergic   Hypersensitivity; inflammatory reaction of the immune system, usually directed against a foreign protein. A hypersensive state acquired through exposure to a particular **allergen.**

allergy   A hyperactive state that occurs in some individuals following introduction of antigens into their bodies.

alopecia   The loss of hair (baldness).

ambient temperature   The prevailing or surrounding temperature.

amnion A thin membrane forming a closed sac around the developing embryo. It contains the amniotic fluid in which the embryo is immersed.

anabolic Productive, promoting growth or development of tissues.

anabolism Any constructive process by which simple substances are converted by living cells into more complex compounds (constructive metabolism).

anadromous Fish species that live most of their life in the sea (saltwater) and return to freshwater to reproduce. Species that go upstream to spawn (e.g., salmon and shad).

anaerobe A microorganism that normally does not require molecular oxygen to live and reproduce.

anaerobic Condition without oxygen, lacking oxygen.

analogue A chemical that resembles another in structure or function.

anaphase The third stage of mitosis. It is characterized by the separation of centromeres and identical chromatids.

anaphylactic shock (anaphylaxis) An **immunological** reaction with increased sensitivity to a normally nontoxic protein or other antigen when injected with it following initial exposure and sensitization. It may cause a severe or even fatal reaction with respiratory failure or cardiovascular collapse.

anaphylaxis An acute allergic reaction. See **anaphylactic shock.**

anemia A condition in which the blood is deficient in the amount of needed hemoglobin or in the number of red blood corpuscles or in both. It is characterized by paleness of the skin and mucous membranes, loss of energy, and palpitation of the heart (unduly rapid action of the heart, which is felt by the individual).

anesthesia Loss of the feeling of pain, touch, cold, or other sensation, produced by ether, chloroform, morphine, and other compounds; by hypnotism; or as the result of hysteria, paralysis, or disease.

anesthetic A substance that produces **anesthesia.**

anesthetized The state of having lost sensitivity to feeling or pain as the result of having received an anesthetic agent or compound.

anestrous period That time when the female is not in estrus; the nonbreeding season.

aneuploidy Animals or plants that possess $2n + 1$ or $2n - 1$ number of chromosomes. Also includes two, or more or less than two, extra chromosomes.

animal protein factor See **APF.**

animal suffering Condition or state of animal that results in pain or emotional distress.

animal welfare Political issue or philosophic discussion concerning animal state of being.

anion An ion carrying a negative charge of electricity.

anogenital Denotes both the anal and the genital regions.

anomaly Abnormal state; marked deviation from normal.

anorexia Lack or loss of appetite for food.

anoxia Lack of oxygen. Unless given oxygen, pilots may develop anoxia (altitude sickness) when flying at high altitudes.

anterior Denotes the front or forward part. It means the same as the **ventral** surface of the body in human anatomy.

antibiotic A product of a living organism (especially of a bacterium or a fungus) that, when present in low concentrations, destroys or inhibits the growth or action of another microorganism. Penicillin, tetracycline, and streptomycin are antibiotics.

antibody A protein substance (modified type of blood–serum globulin) developed or synthesized by the lymphoid tissue of the body in response to an antigenic stimulus. Each antigen elicits production of a specific antibody. In disease defense the animal must have had an encounter with the pathogen (**antigen**) or have received colostrum milk before a specific antibody can be found in its blood.

antigen A high-molecular-weight substance (usually protein) that, when foreign to an animal, stimulates the formation of a specific antibody and reacts specifically *in vivo* or *in vitro* with its homologous antibody.

antimetabolite A substance bearing a close structural resemblance to one required for normal physiological functioning, and exerting its effect, perhaps, by replacing or interfering with the utilization of the essential metabolite.

antioxidant A substance that prevents oxidation of other molecules.

antiseptic From the Latin *anti* meaning "against" and *sepsis* meaning "putrefaction." A substance that prevents growth and development of microorganisms either by destroying them (bactericidal action) or inhibiting their growth (bacteriostatic action).

antiserum A serum that contains an antibody or antibodies. It gives temporary protection against certain specific infectious diseases.

antitoxin Antibody formed against poisonous toxins, such as the bacterial exotoxins, which specifically neutralizes (counteracts) the effects of the toxin. Diphtheria antitoxin, obtained from the blood of horses infected with diphtheria, is injected into persons to make them immune to diphtheria or to treat them if they are already infected.

APF (animal protein factor) The original label given to vitamin B_{12}.

APHIS Animal and Plant Health Inspection Service (of **USDA**).

aphrodisia Sexual excitement, especially if morbid or excessive; e.g., sexual odors released by male insects sexually excite female insects.

apiarist A beekeeper.

apiary Bee colonies, hives, and other bee equipment assembled in one location.

apiculture Beekeeping.

aplasia Lack of development of a tissue or organ, or of cellular products from an organ or tissue.

apocrine Denotes that type of glandular secretion in which the secretory products become concentrated at the free end of the secreting cell and are cast off, along with the portion of the cell in which they have accumulated, as in the mammary gland.

apodous Legless.

Appaloosa A breed of horses with dark-brown or black leopard spots on a roan background.

apparent digestible energy (DE) The food-intake gross energy minus fecal energy. Also called *apparent absorbed energy,* or *apparent energy of digested food.*

apterous Wingless.

aquaculture The raising of plants or animals, as fish or shellfish, in or under a sea, a lake, a river, or another body of water; the raising of aquatic organisms under controlled or semicontrolled conditions.

aquaculturist A person who engages in aquaculture, in the farming of plants or animals in water.

arable Suitable for cultivation; tillable.

arboreal Living in or among trees.

arbovirus Any one of several groups of small viruses transmitted by arthropods such as mosquitoes and ticks. Yellow fever, dengue, and equine encephalitis are caused by arboviruses.

arteriosclerosis A progressive thickening and hardening of the walls of the arteries, often associated with high blood pressure or with chronic disease of the kidneys.

arthropoda (arthropods) The phylum of the animal kingdom that includes insects, spiders, and *Crustacea;* characterized by a coating that serves as an external skeleton and by legs with distinct movable segments or joints.

Artiodactyla Cattle, swine, goats, deer.

asepsis Aseptic condition, methods, or treatment.

aseptic Free from living germs that cause disease, putrefaction, or fermentation.

aseptically See **aseptic.**

asexual Sexless.

assay The determination of the purity of a substance or the amount of any particular constituent of a mixture.

assimilation The process of transforming food into living tissue (constructive metabolism).

asthenia Lack or loss of strength.

atavism Reappearance of a character after a lapse of one or more generations.

ataxia Failure of muscle coordination.

ataxic Uncoordinated body movements; may result from damage to nerve cells or the cerebellum.

atheroma Fatty degeneration of the walls of the arteries.

atherosclerosis A fatty degeneration of the connective tissue of the arterial walls. A form of **arteriosclerosis** characterized by **atheroma.** Lesions within the arteries with plaques containing cholesterol and other lipid materials.

atom A particle of matter indivisible by chemical means. It is the fundamental building block of the chemical elements. The elements, such as iron and sulfur, differ from each other because they contain different kinds of atoms. There are approximately six sextillion (6 followed by 21 zeros, or 6×10^{21}) atoms in an ordinary drop of water. An atom consists of a dense inner core (the nucleus) and a much less dense outer domain of electrons in motion around the nucleus. Atoms are electrically neutral.

atomic number The number of protons in the nucleus of an atom and also its positive charge. (See Appendix E.)

atomic reactor A nuclear reactor.

atomic weight The mass of an atom relative to other atoms; the atomic weight of any element is approximately equal to the total number of protons and neutrons in its nucleus. (See Appendix E.)

atrophic rhinitis A chronic inflammation of the mucous membranes and turbinate bones of the nose. It often results in distortion of the snout in swine. Growth rate is retarded. Secondary bacterial infection and pneumonia are common complications.

atrophy A defect or failure of nutrition or physiological function manifested as a wasting away or diminution in the size of a cell, tissue, organ, or part.

attenuate To make (microorganisms or viruses) less **virulent.**

auditory Of or pertaining to hearing or the organs of hearing.

autogamous Self-fertilizing.

autopsy Examination, including dissection, of a carcass to learn the cause and nature of a disease or cause of death. Also called **postmortem** examination or **necropsy.**

autosomal genes Genes having their loci on chromosomes other than the sex chromosomes.

autosomes All chromosomes except the sex chromosomes.

average daily gain (ADG) Measurement used by scientists to indicate the daily change in body weight when experimental animals are fed test diets.

avian Pertaining to all species of birds, including domestic fowls.

avidin A specific protein component of egg albumen that interacts (combines) with biotin to render it unavailable to an animal, thus producing the syndrome known as egg-white injury. The phenomenon is observed only when rats, for example, consume uncooked egg, because the avidin is denatured by heat and thereby inactivated.

axilla The small hollow beneath the arm or foreleg where it joins the body at the shoulder (armpit).

axon The central core forming the essential conducting part of a nerve fiber. The long extension of a nerve cell that carries impulses away from the body of the **cell.**

B

backcross The crossing of an F_1 hybrid with one of the parental types (breeds). The offspring are referred to as the backcross generation, or backcross progeny.

background count The number of impulses per unit time registered on a counting instrument when no sample is present.

backgrounding Preparing calves for the feedlot by weaning and transitioning to dry feeds. See also **preconditioned.**

bactericidal From the Latin *caedere* meaning "to kill." Capable of destroying bacteria.

bactericidins Antibodies that cause the death of bacteria but not their dissolution or destruction.

bacterin A suspension of killed bacteria (**vaccine**) used to increase disease resistance.

bacteriophage An ultramicroscopic, filterable virus composed of deoxyribonucleic acid (DNA) and a protein coat. Infects particular strains or species of bacteria and usually causes lysis, or explosive dissolution, of the infected bacterium. Bacteriophages multiply at the expense of bacterial cells. They have no enzymes of their own.

bacteriostasis Retardation of the life processes (growth or development) of bacteria without killing them. A *bacteriostatic* substance is a product that retards bacterial growth.

bacteriostatic Inhibiting the growth or multiplication of bacteria.

bag See **udder.**

balanced ration The daily food allowance of livestock or fowl, mixed to include suitable proportions of the nutrients required for normal health, growth, production, reproduction, and **well-being.**

barred A term used to describe striped markings on fowl.

barren (barrenness) Incapable of producing offspring, seed, fruit, or crops.

barrow Young castrated male pig. (See Appendix A.)

basal (diet) A diet common to all groups of experimental animals to which the experimental substance(s) is (are) added.

basal metabolism (BM) The chemical changes that occur in the cells of an animal in the **fasting** and resting state, when it uses just enough energy to maintain vital cellular activity, respiration, and circulation as measured by the basal metabolic rate (**BMR).** Basal conditions include thermoneutral environment, resting, postabsorptive state (digestive processes are quiescent), consciousness, quiescence, and sexual repose. It is determined in humans 14 to 18 h after eating and when at absolute rest. It is measured by means of a calorimeter and is expressed in calories per square meter of body surface.

bastard   A term applied to sheep with hair. It is reputed that "hair sheep" result from interbreeding of sheep and goats, but there is no biological verification of this. Instead, it is more probable that "hair sheep" are the result of domestication and selective breeding (e.g., the Barbados Blackbelly breed of hair sheep).

battery   A series of pens or cages.

bay   Reddish coat color of horses with black points (mane, legs, and tail).

beaver lamb   Sheep or lamb skin with short fine wool that has been dressed with the wool on, dyed, and finished by a process giving a weather-resistant straightness and brightness to the wool.

beebread   The flower pollen gathered by bees, which is mixed with a little nectar or honey and deposited in the cells of the comb.

beefy (beefiness)   A term used to designate the desirable physical conformation of a beef animal as contrasted with a dairy animal, which is lean (not beefy) and more angular.

beekeeping   Apiculture.

beget   To procreate, like a sire.

benthic   Referring to species that feed on the animals and plants that are on the bottom (benthos) of a body of water.

bioassay   The use of animals to determine the active power of a compound as compared with the effect of a standard preparation.

bioclimatology   The science that studies the effect of physical environment upon living organisms.

biofiltration   Systems in **aquaculture** that use bacterial nitrification by *Nitrosomonas* bacteria to convert ammonia to nitrite, then to nitrate using plastic balls or similar material that provide a large amount of surface area. Populations of *Nitrosomonas* bacteria grow on this surface area and when water is moved across these bacteria the ammonia is converted.

biological control   The destruction or suppression of undesirable insects, other animals, or plants by the introduction or propagation, encouragement, artificial increase, and dissemination of their natural enemies, which include predaceous and parasitic insects, predatory vertebrates, nematode parasites, bacteria, protozoa, viruses, and parasitic fungi.

biological half-life   The time required for a biological system, such as a human or an animal, to eliminate by natural processes half the amount of a substance (such as radioactive material) that has entered it.

biological oxygen demand (BOD)   The use of oxygen in the water by natural processes. Some of those processes include the decay of weeds, leaves, feed, or other organic material.

biologicals   Medicinal preparations made from living organisms and their products, used for the prevention or detection of disease; they include serums, vaccines, antigens, and antitoxins.

biological value (BV)   The percent utilization of protein within the animal body, expressed by the formula

$$\%BV = \frac{N \text{ intake } [(\text{fecal } N - \text{metabolic } N) + (\text{urinary } N - \text{endogenous } N)]}{N \text{ intake} - (\text{fecal } N - \text{metabolic } N)} \times 100$$

biological year   In egg production, it is the time interval between the first and last egg laid before a hen enters a molt.

bionomics   The study of the relations of organisms to their environment; **ecology.**

biopsy   The removal and examination (microscopic or chemical) of tissue from a living body.

biotic   Pertains to life or living matter; biologicals.

bird egg   A very large reproductive cell.

bitch   A female dog. (See Appendix A.)

blastema   A group of cells that give rise to an organ or structure either during regeneration or in normal embryogenesis.

bleating   Making a vocal sound typical of communication between goats or sheep.

bleeder   An animal that has **hemophilia.**

blind quarter   A quarter of an udder that does not secrete milk or one that has an obstruction in the teat that prevents removal of milk.

bloat   A disorder of ruminants usually characterized by an accumulation of gas in the rumen.

bloated   Condition where the stomach is distended with gas causing discomfort to the animal and in severe cases resulting in circulatory failure and death.

blocky (blockier-type)   Term commonly applied to meat-producing animals and draft horses meaning a deep, wide, and often low-set animal.

bloom   A term commonly used to describe the beauty and freshness of a cow in early lactation. A dairy cow in bloom has a smooth hair coat and presents evidence of milking ability (**dairy character**).

BMR   Basal metabolism rate. See **basal metabolism.**

boar   Sexually mature uncastrated male pig. (See Appendix A.)

BOD   See **biological oxygen demand.**

bolus   Regurgitated food that has been chewed and is ready to be swallowed. A large pill for dosing animals.

bomb calorimeter   An apparatus for measuring the heat of combustion, as of human foods, animal feeds, and fuel.

bone meal (steamed)   A mash of ground animal bones that were previously steamed under pressure. It contains 1.5 to 2.5 percent nitrogen, 12 to 15 percent phosphorus, and 20 to 34 percent calcium. It is used as a fertilizer and as feed for farm animals.

bovine   Pertaining to the ox or cow.

bran   The seed coat of wheat and other cereal grains, which is separated from the flour and used as animal food.

branding   Marking or printing with a hot iron, freeze process, caustic soda, or punch. Cattle are branded with markings that identify ownership. Branding, of course, damages the hide.

bray   The cry of a donkey.

break joint   The exposed surface of the epiphyseal–diaphyseal growth plate in the lower leg (metacarpus) of a lamb carcass.

breed   Animals having a common origin and characteristics that distinguish them from other groups within the same species.

breed average   The average milk production of cows for a given dairy breed. (Usually computations are based on **DHIA** records of the past 5 years.)

breed-average herdmates   Daughters of a sire in the same herd producing at the current average production for that particular dairy breed. See **herdmates.**

breeding value (genetic value)   The genetic ability of an animal to secrete milk, lay eggs, grow meat/wool, or contribute to other products or services of benefit to humans. One-half of this genetic ability is transmitted to sons or daughters; hence the term *breeding value.*

breed out   To eliminate undesirable characteristics through selective matings.

breed true   To have the ability to transmit a characteristic uniformly to offspring.

breed type   A particular type or form characterizing a breed. It includes special breed features in head, ear, color, or other traits common to a particular breed.

British breeds   Those breeds, such as Hereford, Angus, and Shorthorn, native to Great Britain.

broad-spectrum antibiotic   An antibiotic that is active against a large number of microbial species.

broilers (fryers)   Chickens (meat type) that are 6 to 12 weeks of age.

bronco (bronc)   Any wild or untamed western horse. A wild mustang that has not been broken (trained).

brood   A group of baby chickens.

brood animal   An animal reserved for breeding and raising young.

brooder   An enclosed building, or area within a building, that provides an artificially heated environment for young poultry.

broodfish   Adult fish used to produce sperm or eggs, similar to seed stock in livestock production.

broodiness (brooding)   Desire of birds to set in a nest on eggs for the purpose of **hatching.** Maternal behavior for hatching and rearing young.

broodmare   A female horse used for producing offspring.

buck   A male sheep (ram), goat, rabbit, deer, or antelope.

buckskin   General term applied to leather from deer and elk skins; used for shoes and gloves and, to some extent, in clothing. Most buckskin is oil-dressed, which produces a soft, pliable leather having a buff color and resembling, before finishing, chamois leather. Also refers to coat color in horses. (See Chapter 5.)

buffering   Treating an animal with substances that can neutralize both acids and bases.

buffer solution   A substance in a solution that makes the degree of acidity (hydrogen-ion concentration) resistant to change when an acid or base is added. See **pH.**

bull   A sexually mature uncastrated male. (See Appendix A.)

buller   Pertains to young males riding other young males of a group. In feedlot steers this phenomenon is called *buller syndrome.* See **rider.**

bulling   A cow in **heat (estrus).**

bullock   A castrated bull. English term for a finished, or fat, steer. Refers to young bulls destined for slaughter market but fed and managed as steers.

bull studs   The organizations that manage bulls to produce semen for commercial sale.

burro   A donkey; an ass.

buttermilk   Thick, smooth liquid. Usually made from skim milk, using bacteria to produce acid and flavor; contains at least 8.25 percent nonfat milk solids.

C

cadaver   The body after death.

calciferol   Commonly known as vitamin D_2.

calcify (calcification)   To deposit or secrete calcium salts that harden. An injured cartilage sometimes calcifies. The process by which organic tissue becomes hardened by a deposit of calcium salts.

calf (pl. **calves**)   The sexually immature young of certain large mammals.

calf crop   Calves produced by a herd of cattle in one season.

calorie (cal)   A small calorie is the amount of heat required to raise the temperature of 1 g of water from 14.5 to 15.5°C. This is equivalent to 4.185 J.

calorimetry   Measurements of the amount of heat absorbed or given out.

candling   Illumination of the egg interior by holding the egg before a light. This is possible because of the translucency of the eggshell and the differences in the capacity of other egg constituents to transmit light. Candling ensures the elimination of practically all inedible eggs (e.g., those containing blood spots).

canine   Pertaining to the dog family; includes dogs, wolves, jackals, and others.

cannibalism   A habit of some fowls of pecking at or eating other fowls. An animal eating another animal of the same species.

cannon bone   Either the metacarpal or the metatarsal bone of the horse.

cannula   A device used to connect the rumen with the outside environment. The opening leading from the rumen to the outside is called a *fistula.* A metal-, rubber-, or glass-tube cannula may be inserted into a body cavity to allow the escape of fluids or gas. Liquids and other materials may be introduced into the body through a cannula.

cannulated   Having had a small tube (**cannula**) inserted into a body cavity or into a duct or vessel to facilitate the movement of fluids or gas. To study various chemical, nutritional, and physiological functions and processes in **ruminant** animals, a cannula is frequently used to connect the rumen with the outside environment.

canter   A fast three-beat gait of a horse; an easy gallop. (See Chapter 5.)

capacitation   The physiological process that occurs naturally in the female reproductive tract, which gives mammalian spermatozoa the ability to initiate fertilization. In swine, for example, a population of capacitated spermatozoa first becomes available for fertilization 2 to 3 h after natural mating or artificial insemination with fresh semen.

caponettes   Male chickens that have had their reproductive organs made useless by the injection of an estrogenic hormone (e.g., stilbestrol). The testes of these animals decrease in size, and the secretion of testosterone is inhibited, which in turn results in a regression of the secondary sex characteristics (comb, wattles, earlobes, mating instinct, and crowing).

capons (caponized)   Male birds (usually chickens) that have had their reproductive organs (testes) surgically removed (castrated). Caponized birds lose some of their male sex characteristics (the comb loses its bright red color and shrinks in size, and libido is lost). Capons are raised for meat production.

caprine   Pertaining to or derived from a goat.

captive bolt   A pistol used as an alternative to the sole ax or knife for stunning while slaughtering cattle. When fired, a plunger or "bolt" in the barrel penetrates the animal's brain.

caput   Head.

carbohydrates   Materials consisting chemically of carbon, hydrogen, and oxygen. The most important carbohydrates are the starches, sugars, celluloses, and gums. They are so named because the hydrogen and oxygen are usually in the proportion to form water $(CH_2O)_n$. Formed in plants by photosynthesis, carbohydrates constitute a large part of animal food.

carcass   The body of a dead animal. The whole trunk of a slaughtered animal.

carcinogen   Any cancer-producing substance.

cardiac That which is related to the heart.

carnivorous (carnivores) Meat-eating. Carnivorous animals are dogs, cats, bears, weasel-like, and other mammals.

carrier A heterozygote for any trait. A disease-carrying animal.

carrion Dead or decaying flesh of a human or animal.

carry To bear, as a pregnant cow carries a calf.

cartilage A firm but pliant type of tissue forming portions of the skeleton of vertebrates. The proportion of cartilage in the skeleton of young animals is greater than in mature animals. Also called **gristle.**

casein The major protein of milk.

castrate (castration) To remove the testicles or ovaries; also called *neutering.*

catabolic Destructive.

catabolism Any destructive process by which complex substances are converted by living cells into simpler compounds (destructive metabolism).

catabolize To break down a complex substance into simpler compounds.

catalyst A substance that alters the speed of a chemical reaction without becoming a part of the end product. Enzymes are catalysts.

catgut Tough cord obtained from the intestines of cattle and sheep and used for strings of musical instruments, tennis rackets, and stitching in surgery.

cathartic A compound (medicine) that quickens and increases evacuation from the bowels. A laxative.

cation An ion carrying a positive charge of electricity. Cations include the metals and hydrogen.

cattalo (catalo) A hardy crossbreed of the American bison and domestic cattle.

cattle Collectively refers to mature bovine animals. In biblical times it referred to all livestock.

caudal Denotes a position toward the tail (rump) or posterior end (same as *inferior* in human anatomy).

cayuse Native American pony, named after the Cayuse, who currently live in Oregon.

cecum (caecum) The first part of the large intestine; forms a dilated pouch, also called *blindgut.*

cell A bit of protoplasm that usually contains a nucleus and cytoplasm. Cells are the building blocks of which the body is made. An unfertilized hen's egg is a single cell.

cellulose The principal carbohydrate constituent of plant cell membranes. It is made available to ruminants through the action of microorganisms that inhabit the rumen (rumen bacterial flora).

Celsius A centigrade temperature scale. Can be converted into Fahrenheit by using the formula $(°C \times 9/5) + 32 = °F$, or use $(°C \times 1.8) + 32 = °F$. (See Appendix B.)

centaur A legendary monster of the Greeks with a horse's lower body and a man's head, arms, and chest.

centriole A cell organ usually present as two small chromatic granules in the cytoplasm closely opposed to the nuclear wall. The mitotic center in many cells. It plays a role in cellular division.

centromere The point on the chromosome at which identical chromatids are joined and by which the chromosome is attached to a spindle fiber.

cerearia The final free-swimming larval stage of a trematode, consisting of a body and tail.

cervical Referring to the neck.

chamois In the United States, the term refers to the fresh split of sheepskin tanned solely with oils. In other countries, the term includes any one, or several, oil-tanned suede leathers made from sheep or lambskin, deer-, goat-, or kidskin, mountain antelope or chamois, or cattle-hide splits.

chelating agent An organic compound that can bind **cations** (metallic ions) by forming a stable, inert complex that is soluble in water; used in softening hard water, in purifying sewage, and in eliminating high concentrations of undesirable metallic or radioactive elements in blood or tissues.

chemotherapeutics Chemicals used to treat disease.

chemotherapy The treatment of disease by means of chemicals.

chevon Meat of a goat kid.

chick A baby chicken.

chigger A larva (mite) infesting humans, domestic animals, some birds, snakes, turtles, and rodents. Its bite results in inflamed spots, accompanied by intense itching.

cholesterol A white, fat-soluble substance found in animal fats and oils, bile, blood, brain tissue, nervous tissue, the liver, kidneys, and adrenal glands. It is important in metabolism and is a precursor of several hormones.

chondrogenesis Formation of cartilage.

chorion The outermost membrane that encloses the unborn fetus in mammals.

chorionic Of or pertaining to the **chorion.**

chromatid A chromosome that appears doubled at the metaphase of cell division, connected at one point by a small beadlike body called the **centromere.**

chromosomes Dark-staining rodlike or rounded bodies visible under the microscope in the nucleus of the cell in the metaphase of cell division. Chromosomes occur in pairs in body cells, and the number is constant for a species. Chromosomes carry genes arranged linearly along their length. The backbone of a chromosome is the DNA molecule.

chronic Of long duration as opposed to **acute.**

chyme A thick liquid of partially digested food. It passes from the stomach into the small intestine.

cirrhosis A chronic disease of the liver characterized by a destruction of liver cells and an increase of connective tissue.

clean (cleaned) Often used to mean to be free of, as with disease or parasites; also a lay term used by livestock producers to mean a cow, ewe, or other farm mammal has shed her **afterbirth.**

clear egg An infertile egg.

cleft Split or divided to a certain depth. A cleft palate is a longitudinal opening in the roof of the mouth. A *cleft-footed* animal, also called *cloven-footed,* e.g., the cow or hog, has a divided hoof or foot.

cleidoic Closed or locked in, as the egg, which is cut off or isolated from free exchange with the environment by a more or less impervious shell.

climate The prevailing weather conditions that affect life.

clinical Referring to direct observation.

cloning (1) Growing a colony of genetically identical cells or organisms *in vitro;* (2) transplantation of the nucleus of an adult somatic cell into an ovum, which then develops into an embryo genetically identical to the original adult.

clonorchiasis Oriental liver fluke disease (especially in China and Japan). The causative parasitic trematodes have two intermediate hosts; the first is a molluscan, and the second is some edible fish from which humans become infected.

close breeding   A form of inbreeding, e.g., the mating of brothers to sisters, sire to daughter, and son to dam.

closed herd (flock)   A herd (flock) in which no outside blood is introduced (no animals are introduced from outside sources).

cloven-footed   See **cleft.**

clutch   The eggs laid by a hen on consecutive days are referred to as a clutch. In domestic birds, the number of eggs laid successively are often referred to as a *cycle of laying* or the *laying rhythm.* The term *clutch* for wild birds refers to the set of eggs laid for one incubation. Some birds lay a single egg (e.g., penguin); others lay two (e.g., pigeon); others may lay 12 to 20 (partridges and most domestic fowls).

coagulant   A substance that acts on a liquid to coagulate it.

coat   Hair on the body of an animal.

cock   A male chicken; rooster. (See Appendix A.)

cockerel   A male chicken less than a year old.

cod   The part of the scrotum left after castration.

coefficient of digestibility   The percentage value of a food nutrient that has been absorbed. For example, if a food contains 10 g of nitrogen and it is found that 9.5 g have been absorbed, the digestibility is 95 percent.

coenzyme   A substance, usually a vitamin or mineral, that works with an enzyme to perform a certain function.

coitus   Coition; sexual intercourse between individuals of the opposite sex.

cold sterilization   The use of a cathode ray or an electron-beam gun in food processing to kill bacteria or insect life. Chemical sterilization of instruments.

colic   A digestive disturbance causing pains in the abdomen, as in horses.

colitis   An inflammation of the mucous membrane of the colon.

collagen   Protein contained in connective tissue, cartilage, and bones. It is the chief protein of raw hides and skins.

collateral relatives   Those related individuals that are not related as ancestors or descendants.

color fixative   A substance capable of stabilizing colored compounds.

colostral   Pertaining to the first milk after birth or **parturition.**

colostrum   The first milk secreted pre- and postpartum.

colt   A young horse; foal. Male horse under 4 years of age. (A young female horse is called a **filly.**)

comb   A piece of red flesh (tissue) on the top of a chicken's head.

comfort zone   The temperature interval during which no demands are made on the temperature-regulating mechanisms. The temperature at which humans and animals feel most comfortable. Also called **thermoneutral zone.**

commensal   Living on or within another organism and deriving benefit without injuring or benefiting the other individual. Not parasitic symbiosis.

commercial   Animals that are not registered within a purebred registry organization; also called **grade** animals. Although commercial animals are often considered to be of unknown ancestry, this may not be true. The producer may have carefully selected males and females for mating (including purebreds) but not have continued registry records.

compaction   Closely packed feed in the stomach and/or intestines of an animal causing constipation and/or digestive disturbances.

complement   A complex series of enzymatic proteins found in blood serum. These proteins function in immune-mediated destruction of antigens following antibody binding to the antigens. These proteins are involved in antibody-mediated cell lysis, agglutination, phagocytosis, opsonization, and anaphylaxis.

complementary genes   Genes that so interact that when both are present a new or novel trait appears.

complement fixation   A biological reaction that is the basis of many tests for infection. It utilizes a three-phase system: (1) antigen, (2) **complement,** and (3) antibody. When a specific antigen and antibody are mixed in the presence of complement, a reaction occurs that fixes or renders the complement inactive or fixed. Therefore, if the presence of a specific antigen or antibody in a mixture is known, the presence of the other can be determined by the mixture's reaction on the complement. Such a reaction is the basis of many tests for infection, including the Wassermann test for syphilis and reactions for gonococcus infections, glanders, typhoid fever, and tuberculosis.

complete ration   A blend of all feedstuffs (forages and grains) in one feed. A complete ration fits well into mechanized feeding and the use of computers to formulate least-cost rations (least-cost diets).

concentrate   A feed high in **NFE** and total digestible nutrients and low (less than 18 percent) in crude fiber. It includes the cereal grains, soybean oil meal, cottonseed meal, and by-products of the milling industry, such as corn gluten and wheat bran. A concentrate may be either poor or rich in protein.

conception   The fecundation of the ovum. The action of conceiving or becoming pregnant.

condition   Refers to the amount of flesh (body weight), the quality of hair coat, and the general health of animals.

conduction   A means of dissipating heat, especially on contact with cool water. The transfer of heat.

conformation   The physical form or physical traits of an animal; its shape and arrangement of parts.

congenital   That which is acquired during prenatal life. It exists at or dates from birth.

constitution   The general strength and bodily vigor of an animal.

contact insecticide   Any substance that kills insects by contact, in contrast to a stomach (ingestant) poison, which must be ingested.

contact poison   Chemicals that kill insects following direct absorption through the exoskeleton; the insect must touch the chemical to cause damage to or disturbance of proper body function, or to be killed by it.

contagious   Transmissible by contact.

convection   Either cooling or warming of an animal by wind (breezes) according to whether the wind is cooler or warmer than the surface temperature of the animal.

coprophagy   The ingestion of feces.

copulation   The act of mating (sexual congress).

copulatory   (adj. of copulate) The engaging in mating (coitus) among domestic animals.

cornicle   An abortive spur on a hen's leg that hardens with age. It does not develop into a regular spur, as in the cock.

corn stover   The dried cornstalk from which the ears have been removed.

corpora allata   Paired glands of insects that secrete juvenile hormones.

corral seines   Seines used to capture a portion of the aquatic crop within a production pond.

correlation   A measure of how two traits vary together. A correlation of +1.00 means that as one trait increases the other also increases—a perfect *positive* relation. A correlation of –1.00 means that as one trait increases the other decreases—a perfect *negative,* or *inverse,* relation. A correlation of 0.00 means that as one trait increases, the other may increase or decrease—no relation. Thus a correlation coefficient may lie between +1.00 and –1.00.

cosmic rays (cosmic radiation)   Radiations of many sorts, but mostly atomic **nuclei** (protons) with very high energies, originating outside the earth's atmosphere.

cosset   A lamb raised without the help of its dam.

cotyledon   The area of attachment of the fetal placenta to the maternal placenta (carunde) in certain types of ruminants.

cotyledonary type   Refers to the type of **chorion** attachment to the uterus; e.g., in the cow and ewe contact is made only at certain points on the uterus (**cotyledons**) rather than over most of its surface area as in the mare and sow.

counter   A general designation applied to *radiation-detection instruments* or *survey meters* that detect and measure radiation.

covey   A flock of birds; quail, partridge.

cow   A mature female bovine. (See Appendix A.)

cowhide   The hide of any kind of cow.

cow-hocked   A condition of a cow or horse in which the hocks are close together and the fetlocks (just above the hoof) are wide apart.

cranial   Of or pertaining to the skull or the anterior (front) or superior end of the body.

crawler   A newly hatched insect.

creep feeding   A system of feeding young animals prior to weaning. It is designed to exclude mature animals (e.g., by using a narrow or low entrance to the feeder).

cremellos   This color in horses includes cream-colored hair, pink skin, and blue eyes.

crest   The ridge of an animal's neck.

cribbing   A condition in which the horse bites the manger or another object while sucking in air.

crimped   Rolled with corrugated rollers. The grain to which this term refers may be tempered or conditioned before crimping and may be cooled afterward.

cropping   Practices and patterns associated with the growing of crops for animal and human foods.

crossbred   Individual for which the parents are from different breeds.

crossbreeding   The mating of animals of different breeds.

crossing   See **crossbreeding.**

crossing over   Exchange of parts by homologous chromosomes during synapsis of meiosis prior to the formation of sex cells or **gametes.** Thus the homologous chromosomes exchange genes.

crude fiber (CF)   That portion of feedstuffs composed of cellulose, hemicellulose, lignin, and other polysaccharides that serve as the structural and protective parts of plants. It is high in forages and low in grains. Poultry and swine are limited in their ability to digest fiber, whereas ruminants (cattle and sheep) can benefit from it through rumen bacterial activity. It is the least digestible part of a feed.

crude protein (CP)   The total protein in a feed. In calculating the protein percentage, the feed is first chemically analyzed for its nitrogen content. Because proteins average about 16 percent (1/6.25) nitrogen, the amount of nitrogen in the analysis is multiplied by 6.25 to give the CP percentage.

crural   Pertaining to the leg or to a leglike structure.

crutching   Shearing the genital and udder area of ewes prior to lambing.

cryotherapy   The therapeutic use of cold.

cryptorchid   An animal in which the testes have not descended into the scrotum.

cryptorchidism   A failure of the mammalian testes to descend into the scrotum.

cuboidal   Shaped like a cube.

cud   A bolus of regurgitated food (common only to ruminants).

culling (culled)   The process of eliminating nonproductive or undesirable animals.

culture   The growing of microorganisms (or cells) in a special medium.

customary host   One in which a parasite commonly matures and reproduces.

cytokinesis   Changes that occur in the cytoplasm of the cell during division and fertilization.

cytologist   A specialist in the study of cells, their origin, structure, and functions.

cytology   The science relating to cells and their origin, structure, and function.

cytoplasm   Protoplasm of a cell outside the nucleus.

D

dairy character   The evidence of milking ability. See also **bloom.**

Dairy Herd Improvement Association (DHIA)   An organization whose program is operated jointly by the USDA and the colleges of agriculture of the land-grant universities to aid dairy producers in keeping milk production and management records.

Dairy Herd Improvement Registry (DHIR)   A modification of the DHIA program to make the records acceptable by the dairy breed associations. An **official production record** program.

dam   The female parent.

dapple   A circular pattern in an animal's coat color in which the outer portion is darker than the center.

dark chestnut   A term used to describe a brownish-black, mahogany, or liver-colored horse.

dark meat   The legs and thighs of cooked fowls.

day length   The length of daylight in hours during a 24-h period.

DDT   A proprietary name of dicholro-diphenyl-trichloro-ethane; a colorless, odorless insecticide that tends to accumulate in ecosystems and has toxic effects on many vertebrates. It is a potent nerve poison for insects (its use was banned in the United States in 1973).

deacon   A veal calf that is marketed before it is a week old. Also called *bob,* or *bob veal.*

deadborn   Stillborn. Dead at birth.

deamination   Removal of the amino group ($-NH_2$) from an amino acid.

deaminized   The form of the product of an amino group from which the amino group (NH_2) has been removed. The product is a carboxylic acid if the affected amino acid contained only a single amino group.

decapitate   To behead; to cut off the head.

decarboxylation   The removal of carboxyl group ($-COOH$) from an organic acid.

decortication (decorticate)   Removal of the bark, hull, husk, or shell from a plant, seed, or root. Also, removal of portions of the cortical substance of a structure or organ, as in the brain, kidney, and lung.

defecation (defecate) The evacuation of fecal material from the rectum.

definitive host (final host) The animal in which a parasite undergoes its adult and sexual life.

deglutition The act of swallowing.

dehorn To remove the horns from cattle, sheep, and goats or to treat young animals so horns will not develop.

dehydrate (dehydration) To remove most or all moisture from a substance, primarily for the purpose of preservation.

dehydrogenase Enzyme that "activates" hydrogen.

deletion The absence of a portion of a chromosome, causing genes to be absent or lacking.

dementia A general designation for mental deterioration.

dental pad The very firm gums in the upper jaw of cattle, sheep, goats, and other ruminants.

dermatitis An inflammation of the skin.

dermatophytosis Any skin infection caused by a fungus, e.g., ringworm.

dermatosis Any skin disease.

dermis The layer of skin deep to the epidermis, consisting of a dense bed of vascular connective tissue (also known as the *corium*).

desiccated Dried out; exhausted of water or moisture content.

design criteria Standards based solely on engineering considerations by which an animal accommodation is designed.

desquamate To peel or come off in layers or scales, as the epidermis in certain diseases. The shedding of epithelial cells.

deutectomy The removal of the yolk sac from newly hatched chicks.

dewlap The pendulous skin fold hanging from the throat, particularly of the ox tribe and certain fowl.

dextrins An intermediate product of starch digestion (starch to dextrins to maltose to glucose).

DHIA See **Dairy Herd Improvement Association.**

DHIR See **Dairy Herd Improvement Registry.**

diaphysis The shaft of a long bone.

dicoumarol (dicoumarin) A chemical compound found in spoiled sweet clover and lespedeza hays. It is an anticoagulant and can cause internal hemorrhages when eaten by cattle. When used therapeutically the trade name is Dicumarol.

diestrum (dioestrum) That portion of a female's cycle between periods of estrus (**heat**).

diet The kind and amount of food provided for or consumed by a person or animal.

differentiation The process of acquiring individual characteristics, such as occurs in the progressive diversification of cells and tissues of the embryo. The transformation of mother cells into different kinds of daughter cells (brain, kidney, liver). This process is irreversible.

digestibility That percentage of food taken into the digestive tract which is absorbed into the body as opposed to that which is evacuated as feces.

digestible energy (DE) That portion of the energy in a feed which can be digested or absorbed into the body by an animal.

digestible protein (DP) That portion of the protein in a feed which can be digested or absorbed into the body by an animal.

digestion coefficient (coefficient of digestibility) The difference between the nutrients consumed and the nutrients excreted expressed as a percentage.

digitalis A valuable drug having diuretic properties and which increases the contractility of the heart, made from the dried leaves of the foxglove plant. It is commonly used in the treatment of heart diseases.

dihybrid A double heterozygote such as *AaBb*.

dihybrid cross A cross involving two pairs of alleles, each of which regulates different characteristics.

diploid Refers to chromosomes paired ($2n$) in body cells; i.e., there are two of each kind of chromosome in the nucleus as compared with one of each pair ($1n$) in sex cells. Somatic cells are usually diploid, whereas gametic cells are usually haploid.

direct calorimetry Measurement of the amount of heat produced within a small chamber (as in a **bomb calorimeter,** where the heat produced through combustion is absorbed by a known quantity of water in which the container is immersed). A means of determining the caloric value of foods and animal feeds.

disease Any deviation from a normal state of health that temporarily impairs vital functions of animals.

disinfect To destroy or render inert disease-producing germs (pathogens) and harmful microorganisms and to destroy parasites.

dislocating A method of killing poultry in which the fowl's head is pulled and twisted until the neck separates from the skull and the spinal cord is severed.

dispermic Refers to an ovum fertilized by two spermatozoa (rather than one).

disposition The temperament, or spirit, of an animal.

dissect To cut an animal into pieces for examination.

dissipate (dissapation) To cause to disappear; to spread in different directions.

distal Remote, as opposed to close or **proximal;** away from the main part of the body.

diuresis Excessive secretion of urine.

diurnal variation The amount of variation in one day.

dizygotic twins Twins originating from two separate fertilized eggs. They are no more alike genetically than full brothers and sisters born at different times.

DNA Deoxyribonucleic acid.

dock To cut off the tail (especially in sheep).

doddie A **polled** cow.

doe An adult female rabbit, goat, or deer.

dogie A motherless calf.

domestic Pertaining to domesticated animals; animals living near or about human habitations; relating to, or originating within, one's own country.

domesticate (domestication) To bring a wild animal or fowl under control and to improve it through careful selection, mating, and handling so that its products or services become more useful to humans. Domesticated animals breed under the control of humans.

dominance Gene action in which one **allele** masks (partially or completely) the effect of another allele. The property of one of a pair of **alleles** or traits that suppresses expression of the other in the **heterozygous** condition.

dominant Describes a gene that when paired with its allele covers up the phenotypic expression of that gene. For example, brown eyes (*B*) in humans are dominant to blue eyes (*b*). Thus *BB* and *Bb* individuals have brown eyes whereas *bb* individuals have blue eyes. Dominant genes affect the phenotype when present in either **homozygous** or **heterozygous** condition.

dormancy   Depressed metabolism. A state in which organisms are inactive, quiescent, or sleeping.

dorsal   Pertaining to the back, or more toward the back portion; opposite to **ventral.** It means the same as **posterior** in human anatomy.

double mating (double cover)   The mating of female livestock with a male on successive days of an estrous period to enhance the probability of conception. Also, mating a female to two different males during the same heat period.

draft animal   An animal, e.g., a horse, mule, or ox, used to pull heavy loads.

drake   An adult male duck.

drape   A term applied to a cow or ewe incapable of bearing offspring.

drifting   The moving of bees from one hive to another because of loss of direction caused by wind or other circumstances.

drone   The male bee hatched from an unfertilized egg. It is larger than the workers, does not gather nectar for honey, and has no sting. It mates with (fertilizes) the queen bee.

drone layer   A queen bee that has exhausted her supply of spermatozoa stored after mating so that her eggs produce only drones.

dropping   A term commonly used to mean **parturition** (the act or process of giving birth).

drove   A collection or mass of animals of one species, as a drove of cattle.

dry (dry period)   Nonlactating female. The dry period of cows is the time between lactations (when a female is not secreting milk).

dry ice   Solid carbon dioxide that is purified, liquefied, expanded to form snow, and finally pressed into blocks. It is used as a refrigerant, as in the storage of semen.

drylot   A relatively small area in which cattle are confined indefinitely as opposed to being allowed to have free access to pasture.

dual-purpose   An early production term defining animals that were bred to optimize the output of two products, such as cattle bred for the production of both meat and milk. It is generally agreed that performance is not maximized for either individual trait because of the emphasis placed on two different production selection criteria.

dubbing   Removing the tips of the comb in young chickens to reduce heat loss and reduce activities between birds, such as comb pecking, related to social dominance.

duct   A canal (tube) that conveys fluids or secretions from a gland.

dung   The feces (manure) or excrement of animals and birds.

duplication   Process in which a chromosome is attached to parts from its own homologous chromosome and thus has duplicate genes.

dyad   A unit of two sister chromatids that are synapsed as a tetrad.

dysfunction   Partial abnormality, disturbance, or impairment in the function of an organ or a system.

dysgenesis   A defect in breeding so that hybrids cannot mate between themselves but may produce offspring with members of either family of their parents. Such offspring are sterile.

dyspnea   Difficult or labored breathing.

dystocia   Abnormal or difficult labor (**parturition**) causing difficulty in delivering the fetus and placenta.

E

easy keeper   An animal that does well on a minimum of food.

EB   See **energy balance.**

eccrine   Glands or tissues that secrete a substance without a breakdown of their own cells, e.g., sweat glands. See **exocrine.**

ecology   The study of the relation of organisms to their environment, habits, and modes of life; **bionomics.**

ectoblast   An embryonic cell layer.

ectoderm   The outermost of the three primary germ layers of the embryo; gives rise, e.g., to the skin, hair, and nervous system.

ectoparasites   Parasites that inhabit the body surface.

eczema   An inflammatory skin disease of humans and animals characterized by redness, itching, loss of hair, and the formation of scales or crusts.

edema   The presence of abnormally large amounts of fluid in the intercellular tissue spaces of the body. Swelling.

edematous   Accumulations of abnormally large amounts of fluid in the intercellular tissue spaces of the body; affected areas have a "swollen" appearance.

efficacy   Effectiveness. For example, the effectiveness of an antibiotic in the therapy of a certain disease.

e.g.   From the Latin *exempli gratia.* For example.

egg   **Gamete** produced by the female.

eggbeater   A horse that places his feet carefully and smoothly.

ejaculation   A sudden ejection, or discharge, as of semen from the male.

elastin   The yellow proteinaceous connective tissues of the skin that provide structural support for the blood vessels and thermostat mechanism. It is obtained when elastic tissue is boiled in water.

electrolyte   Any solution that conducts electricity by means of its ions.

electromagnetic   Produced by or pertaining to electromagnetism or an electromagnet.

electromagnetic radiation   Radiation consisting of electric and magnetic waves that travel at the speed of light (e.g., light waves, radio waves, gamma rays, X rays).

electromagnetic spectrum (ES)   A series of waves of varying length. At one end of the ES are gamma rays, whose waves measure only about one-billionth of an inch; at the other end are radio waves, which may be many miles long. Between these extremes, in order of increasing wavelength, are X rays, ultraviolet light, visible light, infrared light, microwaves, and shorter radio waves. All waves have different characteristics, but all travel at the same speed (186,000 mi/s).

element   One of the 105 known chemical substances that cannot be divided into simpler substances by chemical means. A substance the atoms of which all have the same **atomic number,** e.g., hydrogen, lead, and uranium. (See Appendix E.)

emaciation   A wasted condition of the body; excessive leanness.

emasculation   Excision of the penis; castration.

embryology   The science relating to the formation and development of the embryo.

emissivity (or absorptivity)   Ratio of the rate of emission (or absorption) of radiant energy per unit area by the given body to the rate of emission (or absorption) from a blackbody at the same temperature.

encephalitis   An inflammation of the brain that results in various central nervous system disorders.

encephalomalacia   A condition characterized by softening of the brain. A disease causing lesions of the brain in young poultry, caused by a deficiency of vitamin E in the diet.

endemic (enzootic) Pertaining to a disease commonly found with regularity in a particular locality.

endocrine Pertaining to glands that produce secretions that pass directly into the blood or lymph instead of into a duct (secreting internally). Hormones are secreted by endocrine glands.

endogenous Internally produced in the body, e.g., hormones and enzymes.

endoparasites Parasites that inhabit the body tissues or cavities, e.g., the tapeworm.

endophily The feeding or resting of insects in houses (inside).

endotoxins Toxic substances (such as those causing typhoid) retained inside the bacterial cells until the cells disintegrate.

energy balance (EB) The relation of the gross energy consumed to the energy output (energy retention). It is calculated as follows: EB = **GE** – **FE** – **UE** – **GPD** – **HP** (energy balance equals gross energy minus fecal energy minus urinary energy minus gaseous products of digestion minus heat production).

ensilage A green crop (forage) preserved by fermentation in a **silo,** pit, or stack, usually in the chopped form. Also called **silage.**

enteritis Any inflammatory condition of the **intestinal** linings of animals or humans.

entozoa Internal animal parasites, e.g., stomach worms.

entrails The inner organs of animals, specifically the intestines.

envenomization The injection of a poison into an animal (as by a wasp).

environment (environmental) The sum total of all external conditions that affect the life and performance of an animal.

environmental stimuli Environmental impingements on an animal that cause the animal to respond homeokinetically.

enzootic Occurring endemically among animals, i.e., continuously prevalent among animals in a certain region.

enzyme A complex protein produced in living cells that causes changes in other substances within the body without being changed itself. An organic catalyst. Body enzymes are giant molecules.

epidemic (epizootic) Rapidly spreading (as a disease) so that many animals or people have it concurrently.

epidemiology (epidemiological) The field of science dealing with the relationship among various factors that determine the frequencies and distributions of infectious diseases.

epiderm Epidermis (the outer layer of skin or tissue).

epidermal Of, relating to, or arising from the epidermis.

epinephrine A drug used to arrest hemorrhage and to stimulate heart action. It is a slaughterhouse by-product obtained from the adrenal glands. Also called *adrenaline.*

epiphysial (epiphyseal) Pertaining to or of the nature of an **epiphysis.** The end of a long bone that has ossified.

epiphysis A portion of bone separated from a long bone by cartilage in early life but later becoming a part of the larger bone. It is at this cartilaginous joint that growth in length of bone occurs. It is also called the *head* of a long bone.

epistasis Interaction between genes that are not alleles, causing the appearance of a different phenotype. For example, *BB* causes an animal to be black. However, *II* inhibits pigment production. *BBii* would be black, whereas *BBII* would be white. Epistasis can be due to two or more pairs of alleles on the same pair of homologous chromosomes or to two different pairs of genes on two different homologous chromosomes. Epistatic genes interact so that one pair of genes masks or suppresses the phenotypic expression of another pair of genes.

epistatic See **epistatis.** Describes gene action in which the **genotype** at one **locus** affects the expression of the genotype at another locus.

epistaxis Bleeding from the nose.

epizootic Designating a widely diffused disease of animals, that spreads rapidly and affects many individuals of a kind concurrently in any region, thus corresponding to an epidemic in humans.

equestrian One who rides horseback.

equine Pertaining to a horse.

eradication The total elimination of the etiologic (disease-causing) agent from a region.

ergosterol A plant sterol that, when activated by ultraviolet rays, becomes vitamin D_2. Also called *provitamin* D_2 and *ergosterin.*

eructation The act of belching, or casting up gas from the stomach.

erythema A severe redness of the skin associated with local inflammation.

erythropoiesis The production of erythrocytes (the red blood cells that transport oxygen). Occurs in the red bone marrow.

escutcheon The part of a cow that extends upward just above and back of the udder, where the hair turns upward in contrast to its normal downward direction. Also called the *milk mirror.*

essential amino acid An amino acid that cannot be synthesized in the body from other amino acids or substances or that cannot be made in sufficient quantities to meet the body's needs.

essential fatty acid A fatty acid that cannot be synthesized in the body or that cannot be made in sufficient quantities to meet the body's needs. Linoleic and linolenic acids are essential for humans.

estivating See **estivation.**

estivation The adaptation, or modifications, in an animal that enable it to survive a hot, dry summer. Estivation is probably an evolutionary adaptation to periods of water scarcity. It is often called *summer* **hibernation.**

estrous cycle (estrous period) Also called **heat period.** The period of sexual receptivity (**heat**) in female mammals.

estrus The recurrent, restricted period of sexual receptivity (**heat**) in female mammals, marked by intense sexual urge.

estuary The area around the mouth of the river; the lower portion or wide mouth of a river where the salty tide meets the freshwater current.

ether extract The fatty substances of foods or other materials that are soluble in ether. Used in food and feed analyses.

etiology The study or theory of the causation of diseases.

euploid (euploidy) An organism whose chromosome number is a whole-number multiple of the basic, or haploid, number.

European breeds Those, such as Charolais, Simmental, and Limousin, native to Great Britain or continental Europe.

evaporation (transpiration) As related to environmental physiology, the loss of moisture (sweating).

evaporation of water The conversion of a liquid into a vapor (**evaporation**) is an important means of cooling the body in humans and many farm animals.

eviscerate To remove the entrails, lungs, heart, and certain other organs from an animal or fowl when preparing the carcass for human consumption.

ewe   A female sheep. (See Appendix A.)

excise   To cut out or off.

excreta   Waste materials discharged from the bowels.

excystation   The escape of a cyst; especially, a stage in the life cycle of parasites occurring after the cystic form has been swallowed by the host.

exfoliate   To remove the surface scale or layer.

exocrine (eccrine)   Secreting outwardly, into or through a duct.

exogenous   Externally provided, e.g., vitamins.

exotoxins   Soluble toxins that diffuse out of living bacteria into the environment (the culture medium of the living host). Toxins secreted by living microorganisms.

expectoration   The act of coughing up and spitting out materials.

experiment   From the Latin *experimentum* meaning "proof from experience." It is a procedure used to discover or to demonstrate a fact or general truth.

exsanguination   The withdrawal of blood, as by bloodsucking insects.

extirpation   The complete removal or eradication of a part.

extreme ill-being   Condition or state of an animal experiencing severe illness.

exudate   An abnormal seeping of fluid through the walls of vessels into adjacent tissue or space, or the body cavity.

exudative diathesis   Symptom of vitamin-E deficiency in poultry. It is characterized by an accumulation of fluid in subcutaneous fatty tissue.

exuding serum   An oozing (exuding) of the clear portion of any animal liquid (serum). Leakage of body fluids through damaged tissues.

F

F₁ generation   The first-filial-generation, or the first-generation progeny, following the **parental**, or P₁, **generation.**

F₂ generation   The second-filial-generation progeny resulting from the mating of F₁-generation individuals. Produced by an inbreeding of the first filial generation.

factory farming   Pejorative term referring to a large intensive animal-production system.

facultative   The ability of a microorganism to live and reproduce under either aerobic or anaerobic conditions.

fag   Any tick or fly that attacks sheep.

Fahrenheit   The temperature expressed in degrees, where 32° and 212° are the freezing and boiling temperatures, respectively, of water at sea level. A more universal and scientific graduation of temperature is the centigrade Celsius scale. Fahrenheit is converted into Celsius by using the formula $(°F – 32)\ 5/9 = °C$, or $°C = (°F – 32)/1.8$. (See Appendix B.)

fair-being   Condition or state of an animal experiencing fairness.

fallout   Airborne particles containing radioactive material that fall to the ground following a nuclear explosion.

false (pseudo) albino   A solid-white horse with colored eyes.

false heat   The display of **estrus** by a female animal when she is pregnant.

family   A related group of animals.

FAO   Food and Agriculture Organization. A specialized international agency of the United Nations that collects and disseminates information on the production, consumption, and distribution of food throughout the world. It was organized in 1945 and has some 120 member nations; it is headquartered in Rome. (See also Section 1.4.1.)

farrow   To give birth to a litter of pigs. (See Appendix A.)

farrowed   See **farrow.**

farrowing   See **farrow.**

farrowing units   Housing specifically designed for the **sow** to give birth to her **litter.**

far side   The right side of a horse.

fascia   A thin sheet or band of fibrous (connective) tissue covering, supporting, or binding together a muscle, part, or organ.

fasting   Abstaining from food. Certain types of experiments involve withholding food from the test animal for varying periods of time.

fat   Adipose tissue; an ester of glycerol with fatty acids.

favor   To protect; to use carefully, as an animal *favors* a lame leg.

FCM (4 percent fat-corrected milk)   A means of evaluating milk production records of different animals and breeds on a common basis energywise. The following formula is used: $FCM = 0.4 \times$ milk production + (15 × lb fat produced).

FDA   Federal Food and Drug Administration.

FE   See **fecal energy.**

febrile   Pertaining to a fever, or rise in body temperature.

fecal   Consisting of **feces.**

fecal energy (FE)   The food energy lost through the feces.

feces (faeces)   Excrement discharged from the bowels.

fecundation   Impregnation or fertilization.

fecundity   The ability of an individual to produce eggs or sperms regularly.

feed conversion (feed efficiency)   The units of feed consumed per unit of weight increase. Also, the production (meat, milk, eggs) per unit of feed consumed.

feeder   A young animal that does not have a high finish (fatness) but shows evidence of ability to add weight economically.

feedlot   A confined area within which livestock are fed until they reach slaughter weight.

femininity   Physical appearance resulting from well-developed secondary female sex characteristics.

femoral   Pertaining to the femur or to the thigh.

feral   Pertaining to animals in the wild, or untamed, state.

fertile   In the animal sciences, a term referring to the capability of breeding or reproducing.

fertility   The ability to reproduce.

fertilization   The process in which two haploid **gametes** fuse, forming a diploid cell, the **zygote.**

fetal   Relating to or being a **fetus.**

fetus   The unborn young of animals (usually vertebrates) that give birth to a living offspring.

fever   Abnormally high body temperature.

fibrinogen   A soluble protein present in the blood and body fluids of animals that is essential to the coagulation of blood.

fibroblasts   Connective tissue cells whose function is one of repair. Also called *fibrocytes* and *desmocytes*. These cells produce fibrous tissues of the body.

fibrous carbohydrates   Those feed components (e.g., cellulose and hemicellulose) not readily digested by nonruminant animals.

filamentous   Threadlike, very slender.

filled milk   Milk from which milk fat has been removed and replaced with other fats or oils.

filly A young female horse. A young mare. (See Appendix A.)

filterable virus A name commonly applied to a **pathogenic** agent capable of passing filters that retain bacteria.

find Lay term meaning *to give birth to young;* e.g., a cow *finds* a calf.

fine wool Wool with finer, or smaller, fiber diameter. Term also applied to sheep breeds producing finer wool fibers.

fingerling Young, small fish (1 to 10 in) no longer supported by the yolk sac; used to stock aquaculture grow-out ponds.

finish (finishing, finished) The degree of fatness or the distribution of fat in an animal.

finisher (diets) Refers to livestock being fed, or *finished,* for the slaughter market.

first calf A term commonly used to indicate the first calf born to **bovine** females.

first meiotic division The first of two divisions occurring in reduction cell division and resulting in the production of two cells, each of which is haploid, the chromosomes occurring as paired chromatids joined at the centromere.

fission The splitting of a heavy **nucleus** into two approximately equal parts, accompanied by the release of a relatively large amount of energy and generally one or more neutrons.

fistula See **cannula.**

five-gaited saddle horse A horse trained to use the following gaits: walk, trot, rack, slow gait, and canter.

flaccid Limp; weak; flabby.

flatulence A digestive disturbance in which there is often a painful collection of gas in the stomach or bowels.

flay To remove the skin from a carcass.

fleece The wool from all parts of a sheep.

flesh The muscle and fat covering of an animal. See **condition.**

fleshing Removal of adipose tissue on the flesh side of skins or hides to be used in the leather industry.

flight zone The zone within which certain birds rise, settle, or fly as with a flock of geese.

flock A group of birds or sheep; called *band* in goats. (See Appendix A.)

flora The bacteria in or on an animal. Usually referred to as *bacterial flora,* as in the digestive tract.

fluid milk Milk commonly marketed as fresh liquid milks and creams. It is the most perishable form of milk and commands the highest price per unit volume. Also called **market milk.**

flushing Feeding extra concentrate, or higher-energy feed, prior to and shortly after the time of breeding to increase ovulation rate.

foal Young (usually unweaned) horse or mule of either sex.

foaling Giving birth to a foal (newborn horse).

fodder Coarse food such as cornstalks, for cattle or horses.

fodder units (FU) This is a net-energy system of feed evaluation that is used commonly in Scandinavian countries. Values of feeds are measured and expressed in relation to a reference feed, barley. One kilogram of FU is equivalent to 1650 kcal **NE.**

football leather Leather for covering footballs. Traditionally of pigskin, but today it is generally made of embossed or printed cattlehide leather, and sometimes of sheepskin.

forage Roughage of high feeding value. Grasses and legumes cut at the proper stage of maturity and stored without damage are called *forage.*

forager A *worker* bee that goes out in search of food.

force molted The process of concentrating the molting time for laying hens by reducing light and limiting feed intake. See **molt.**

forceps A plierslike instrument used in medicine and certain research studies for grasping, pulling, and compressing.

forequarters The front two quarters of an animal. Also called *fore udder.*

founder Inflammation of the fleshy laminae within the hoof, from concussion, overfeeding, and many other factors causing an oversupply of blood to this region of the hoof. It causes great pain to the affected animal. Also called *laminitis.*

fowl Any bird, but more commonly refers to the larger ones.

fox trot An easy, short, broken gait of a horse, between a walk and a trot.

free-choice Providing animals free access to diets and thereby allowing them to eat at will.

freemartin Female born twin to a bull calf. (About 9 of 10 are sterile.)

freeze-drying See **lyophilization.**

fresh Designating a cow that has recently given birth to a calf.

Friedman test A test for pregnancy in which a small amount of urine of the tested animal is injected into the bloodstream of a virgin rabbit. Pregnancy is indicated by certain changes in the ovaries of the rabbit.

frog The elastic, horny, middle part of the sole of a horse's hoof.

fry The young of fish still nourished by the yolk sac.

full-feed The term commonly applied to fattening cattle being provided as much feed as they will consume safely without going **off-feed.**

fumigant A liquid or solid substance that forms vapors that destroy pathogens, insects, and rodents.

fungus (pl. **fungi**) Eukaryotic organism that contains no chlorophyll, flowers, or leaves. It gets its nourishment from dead or living organic matter. Includes molds, rusts, mildews, smuts, mushrooms, and yeast.

fungicide An agent that destroys fungi.

fur Skins of wild animals, commonly covered with short fine hair, which are tanned or dressed for garments.

furrowing The process of cytokinesis in animal cells.

G

gait Any forward movement of a horse, such as walking or galloping.

galactophore A milk duct.

galactophorus Carrying or producing milk; conveying milk.

galactopoiesis Concerned in the production of milk.

galactopoietic That which stimulates or increases the secretion of milk.

gamete A male or a female reproductive cell. A sperm or an ovum.

gametogenesis Cell-division process that forms the sex cells.

gamma rays (γ **rays**) High-energy, short-wavelength **electromagnetic radiation.** Gamma rays are very penetrating and are best stopped or shielded against by dense materials, such as lead. They are similar to X rays but are usually more energetic (their penetrating power is greater) and are of nuclear origin. Gamma rays are emitted by isotopes of such elements as cobalt and cesium as they disintegrate spontaneously.

gander A mature male goose.

gaseous products of digestion (GPD)   Include the combustible gases produced in the digestive tract incident to fermentation of the diet. Methane constitutes the major proportion of the combustible gases produced by the ruminant; however, nonruminants also produce methane. Trace amounts of hydrogen, carbon monoxide, acetone, ethane, and hydrogen sulfide are also produced.

gastritis   An inflammation of the stomach, especially of the lining or mucous membrane.

gastrointestinal   Pertaining to the stomach and intestines.

gastrula   An early stage in embryonic development; it follows the blastula stage.

GE   See **gross energy.**

geiger counter   Instrument that counts pulses produced by radioactivity, consisting of a counting tube with a central wire anode, usually filled with a mixture of argon and organic vapor.

gelatin   An organic colloidal substance made from animal bones, skins, or hide fragments. Used in leather finishes to produce a tough film on the leather. Glue is an impure form of gelatin.

gelding   A castrated male horse. To geld is to render sterile or to remove the testicles. Horses are usually castrated at about 2 years of age. (See Appendix A.)

gene   The smallest unit, or particle, of inheritances—a portion of a **DNA** molecule. Genes occur in pairs located on chromosomes in the nucleus of every cell.

generation   See **F_1 generation** and **F_2 generation.**

generation interval   The time from the birth of one generation to the birth of the next. In humans, it is calculated by subtracting the average age of the children from the average age of their parents.

genetic correlation   Condition in which two or more quantitative traits are affected by many of the same genes.

genetic value   The total value of all the genes of an animal for secreting milk or for some other trait. See **breeding value.**

genital (genitalia)   Pertaining to the organs of reproduction.

genome   The total genetic composition of an individual or population that is inherited with the chromosomes. A haploid set of chromosomes with their genes.

genotype   The actual genetic constitution (makeup) of an individual as determined by its germ plasm. For example, there are two genotypes for brown eyes, *BB* and *Bb.* See **dominant.**

gentle   Designating an even-tempered, docile, quiet animal.

germ   A small organism, microbe, or bacterium that can cause disease in humans and/or animals. Early embryo; seed.

germ-free   Designating an animal that is free of harmful organisms, as germ-free pigs used for experimental purposes.

germicide   A substance that kills disease-causing microorganisms (pathogens).

gerontology   The scientific study of the phenomena of aging. (See Chapter 12.)

gestation   Pregnancy (gravidity). The period from conception to birth of young.

get   The offspring of a male animal. A *get-of-sire* refers to a given number (commonly four) of progeny from a male or sire.

ghee   A semifluid butter preparation from the milk of a buffalo, cow, sheep, or goat. It is nearly 100 percent milk fat and is used primarily in Asia and Africa.

giblets   Any of the internal organs of a fowl used as food, particularly the heart, liver, and gizzard.

gigot   A leg of mutton, venison, or veal that is trimmed and ready for consumption.

gilt   A young female swine; commonly called gilt until the first litter of pigs is **farrowed.** (See Appendix A.)

girth   The circumference of the body of an animal behind the shoulders. Also, a leather or canvas strap that fits under the horse's belly and holds a saddle in place.

giving milk   Lactating, or the act of yielding milk by a mammal.

gland   An organ that produces a specific secretion to be used in, or eliminated from, the body.

glasser   Calf or kipskin taken from animals that are poorly fed and possess coarser-grain hide or skin.

gluconeogenesis   Formation of glucose from protein or fat.

glycogenesis   Conversion of glucose into glycogen.

glycogenolysis   Conversion of glycogen into glucose.

glycolysis   Conversion of carbohydrate into lactate by a series of catalysts. The breaking down of sugars into simpler compounds.

gnat   Any dipterous, biting insect.

gonad   The gland of a male or female that produces the reproductive cells; the testicle or ovary.

gosling   Any young goose before its sex can be determined. (See Appendix A.)

gossypol   A toxic yellow pigment found in cottonseed. Heat and pressure tend to bind it with protein and thereby render it safe for animal consumption. It may cause discoloration of egg yolks during cold storage.

GPD   See **gaseous products of digestion.**

grade   Non-registered animals showing the predominant characteristics of a given breed. They commonly have at least one **purebred** parent, usually the male.

grading up   The continued use of purebred sires on grade dams.

grafting   A process of removing a worker bee larva from its cell and placing it in an artificial queen cup, for the purpose of having it reared as a queen. In sheep grafting refers to management practices designed to cause a ewe to accept a lamb transferred from another ewe.

gram-negative bacteria   Bacteria that are decolorized by acetone or alcohol and therefore have a pink appearance when counterstained with safranine, e.g., *Escherichia coli, Pseudomonas aeruginosa, Salmonella, Vibrio,* and *Pasteurella.*

gram-positive bacteria   Bacteria that are able to retain a crystal-violet dye even when exposed to alcohol or acetone, e.g., staphylococci, streptococci, and *Bacillus anthracis.*

graphite   A pure form of carbon.

GRAS   An acronym for the phrase *generally recognized as safe.* This term is commonly used by the FDA and others when referring to food and feed additives.

grass tetany   A magnesium-deficiency disease of cattle characterized by hyperirritability, muscular spasms of the legs, and convulsions.

gravidity   See **gestation.**

gray (Gy)   The unit (or level) of energy absorbed by a food from ionizing radiation as it passes through in processing. One gray (Gy) equals 100 rad; 1000 Gy equals 1 kilogray (kGy).

grazing (grazes)   Consumption of standing vegetation, as by livestock or wild animals.

green chop (fresh forage)   Forages harvested (cut and chopped) in the field and hauled to livestock. This minimizes the loss of moisture, color, nutrients, and wastage. Also called *zero grazing.*

gristle   Cartilage.

gross energy (GE)   The amount of heat, measured in calories, released when a substance is completely oxidized in a **bomb calorimeter** containing 25 to 30 atm of oxygen (heat of combustion).

grow out   To feed cattle so that the animals attain a certain desired amount of growth with little or no fattening.

growthy   Designating an animal that is large and well-developed for its age.

grub   Also called warble fly. These cause extensive and widespread damage to hides and skins of food-producing animals by puncturing holes along either side of the spinal line (backbone).

gruel   A food prepared by mixing a ground feed with hot or cold water.

gut   The digestive tract; sometimes referring only to the intestines of animals.

gut tie   A condition of animals in which the intestines become twisted, causing an obstruction.

H

habituation   The gradual adaptation to a stimulus or to the environment. The extinction of a conditioned reflex by repetition of the conditioned stimulus (also called *negative* **adaptation**).

hackamore   A bridle (with no bits) that controls a horse by pressure on its nose.

hackles   Erectile hairs on the backs of certain animals.

half-life   The time in which the radioactivity originally associated with an isotope will be reduced by one-half through radioactive decay. The time in which half the atoms of a particular radioactive substance disintegrate into another nuclear form. Measured half-lives vary from millionths of a second to billions of years. *Effective half-life* is the time required for a radionuclide contained in a biological system, such as a person or an animal, to reduce its activity by one-half as a result of radioactive decay and biological elimination.

half sib   In genetics, a half brother or half sister.

halogens   A group of elements, including fluorine, chlorine, bromine, and iodine, which are strong oxidizing agents and have disinfectant properties.

ham shank   The hock end of a ham.

hamstring   The large tendon that is above and behind the hock in the rear leg of quadrupeds.

hand   Term used to describe the height (at the point of withers) of a horse. It is the average width of a person's hand (4 in).

handbag leather   Any leather used in making handbags. The most commonly used leathers for handbags are calf, patent (made primarily from cattle hides), goat, and sheep.

hand milking   The manual milking of an animal as opposed to machine or mechanical milking.

haploid   Refers to the 1*n* number of chromosomes in sex cells. Chromosomes are paired in body (somatic) cells. Sex cells contain one chromosome of each pair. *Haploid* means, then, having one of each kind of chromosome in the nucleus.

hard breeder (shy breeder)   A female animal that has difficulty or is slow in conceiving.

hard keeper   An animal that is unthrifty and grows or fattens slowly regardless of the quantity or quality of feed.

hatch   To bring forth young from the egg or pupa by natural or artificial incubation.

hatchability   The degree to which an egg or pupa will produce young by incubation.

hatching   Emerging from an egg or chrysalis.

hay   Dried forage (e.g., grasses, alfalfa, clovers) used for feeding farm animals.

haylage   Low-moisture **silage** (35 to 55 percent moisture). Grass and legume crops are cut and wilted in the field to a lower moisture level than normal for grass silage, but the crop is not sufficiently dry for baling.

head-shy   Designating a horse on which it is difficult to put on a bridle, to lead, or to work around its head.

heart girth   A measurement taken around the body just in back of the shoulders of an animal, used to estimate body weight.

heat (estrus)   Term used to describe the behavior of females during the fertile portion of the **estrous cycle.** Females are receptive to breeding during this time and may actively seek a mate.

heat of fermentation (HF)   Heat produced anaerobically by microbes of the digestive tract. It is much greater in ruminants than in nonruminants.

heat of nutrient metabolism (HNM)   The heat produced as a result of the utilization of absorbed nutrients.

heat increment (HI)   The increase in heat production following consumption of food when an animal is in a thermoneutral environment. It consists of increased heats of fermentation and of nutrient metabolism. There is also a slight expenditure of energy in masticating and digesting food. This heat is not wasted when the environmental temperature is below the critical temperature. This heat may then be used to keep the body warm. Also called *work of digestion*. See also **heat increment of feeding.**

heat increment of feeding (HIF)   The heat produced by an animal during fermentation in the gastrointestinal tract and during processing and use of food nutrients in the body. The HIF represents an inefficiency, unless the animal can use the heat to help keep the body warm in a cool or cold environment. It depends on species, diet quality, level of feed intake, and productive performance. In human physiology, the HIF is commonly referred to as **specific dynamic action (SDA).**

heat period   That period of time when the female will accept the male in the act of mating. In heat. **Estrous period.**

heat production (HP)   Results from (1) the heat increment (heat of fermentation plus heat of nutrient metabolism) plus (2) the heat used for maintenance (basal metabolism plus heat of voluntary activity) when an animal is consuming food in a **thermoneutral zone.** In direct calorimetry, heat production is measured by use of an animal calorimeter.

heat stroke   A condition caused by exposure to excessive heat; may occur in one of three forms: (1) sunstroke (thermic fever), (2) heat exhaustion, or (3) heat cramps.

heat tolerance   The ability of an animal to endure the impact of a hot environment without suffering ill effects.

hectare   A European unit of land measurement (2.47 acres). (See Appendix B.)

heel fly   A name applied to the common cattle grub because it enters the body through the feet.

heifer   A female of the cattle species less than 3 years of age that has not borne a calf. (See Appendix A.)

helminth   An **intestinal** worm or wormlike parasite.

hematopoietic   An agent that promotes the formation of blood cells.

hemoglobin   The red pigment in red blood cells of animals and humans that carries oxygen from the lungs to other parts of the body. It is made of iron, carbon, hydrogen, and oxygen and is essential to animal life.

hemolysis   The liberation of hemoglobin. Hemolysis consists of the separation of hemoglobin from the corpuscles and its appearance in the fluid in which the corpuscles are suspended. The freezing of blood causes hemolysis. Many microorganisms are able to hemolyze red blood cells by production of hemolysins.

hemophilia   An inherited condition in which blood does not clot normally. An animal so affected may be called a **bleeder.**

hen   An adult avian female (usually refers to chicken or turkey). (See Appendix A.)

herbicide   A preparation for killing weeds.

herbivorous (herbivore)   Pertaining to animals that can subsist on grasses and herbs.

herd   A group of animals (especially cattle, horses, and swine) collectively considered a unit.

herdmates (stablemates or contemporaries)   A term used when comparing records of a sire's daughters with those of all their nonpaternal herdmates, both being milked at the same season of the year (under the same conditions). Each daughter is compared with her herdmates that calved over a 5-month period (2 months prior and 2 months succeeding, plus the month the daughter calves). This is done to adjust for effect of seasonal differences. The *herdmate comparison* removes, from the evaluation of breeding value, complications arising from herd, year, and season of calving production variations and assumes that the measured production differences are due to inheritance. (See Chapter 9.)

heredity   The hereditary transmission of genetic or physical traits of parents to their offspring.

heritability (heritable)   A technical term used by animal breeders to describe what fraction of the differences in a trait, such as milk production, is due to differences in genetic value rather than environmental factors; variation due to genetic effects divided by the total variation (genetic plus environmental variation).

heritability estimate   An estimate of the proportion of the total phenotypic variation between individuals for a certain trait that is due to heredity. More specifically, hereditary variation due to *additive* gene action.

hermaphrodite   An individual possessing both male and female reproductive organs. May be capable of producing both ova and sperm; however, this seldom is true.

hernia   The protrusion of an organ or part through some opening in the wall of its cavity. Also called **rupture.**

heterogametic   This term refers to sex cells. In humans and livestock (except poultry) the male possesses an X and a Y chromosome and thus is *hetero-* (prefix meaning unlike) *gametic* in sex chromosomes. The female is XX, and so is *homo-* (alike) *gametic* for sex chromosomes. See also **homogametic.**

heterosis (hybrid vigor)   Amount the F_1 generation exceeds the P_1 generation for a certain trait, or amount the crossbreds exceed the average of the two purebreeds that are crossed to produce the crossbreds. Also called **nicking.**

heterothermic mammals   Animals that have a fluctuating body temperature.

heterozygote   An individual possessing one or more pairs of allelic genes for a given trait, each pair being different genes. For example, *Aa* or *Bb* or *AaBb* are hetero- (unlike) zygotes. Each parent contributed unlike genes with respect to any given allelic pair governing contrasted characters.

heterozygous   Having the two alleles at corresponding **loci** on homologous chromosomes different for one or more loci.

HI   See **heat increment.**

hibernation (hibernate)   The ability of an animal to pass the winter in a dormant state in which the body temperature drops to slightly above freezing and metabolic activity is reduced nearly to zero. Hibernation is probably an adaptation to prevent starvation during periods of food scarcity.

hidrosis   Excessive sweating.

high-moisture silage   **Silage** containing 70 percent or more moisture.

hinny   The offspring of a stallion and a jennet (female jackass).

hip dysplasia   An orthopedic disease with abnormalities in the hip socket (acetabulum of pelvis) and the head of the femur (thigh bone).

hip sweeney   A horse with atrophy of the hip muscles.

hive   A home for honeybees, a container for bees. Also a swarm of bees.

hobble   To tie the front legs of a horse together by means of a rope or straps so that it cannot run or kick.

hogget   A sheep from weaning until its first shearing.

homeostasis (homeokinesis)   The physiological regulatory mechanisms that maintain constant the "internal environment" of the organism in the face of changing conditions.

homeostatic   (adj. of **homeostasis**) A relatively stable state of equilibrium; a tendency to uniformity or stability in the normal body states.

homeostatically   Pertaining to **homeostasis.**

homeothermic (homoiothermic)   Having a relatively uniform body temperature maintained nearly independent of the environmental temperature. Warm-blooded animals (mammals and birds) are homeothermic.

homeotherms (endotherms)   The warm-blooded species. They maintain their *internal* temperature constant in the face of widely changing external temperature.

homogametic   Forming **gametes** which all have the same type of sex chromosome. See also **heterogametic.**

homogenized milk   Milk that has been treated to ensure breakup of fat globules to such an extent that after 48 h of quiescent storage at 45°F (7°C), no visible cream separation occurs on the milk. The reduced size of fat particles results in formation of a softer curd in the stomach.

homologous chromosomes   From the Greek *homo* meaning "like" and *logous* meaning "proportion." The pairs of chromosomes (twins) in body cells. *Homologous* refers to chromosomes that are structurally alike.

Homo sapiens   The species, including all existing races, of humans.

homozygote   An individual possessing like genes for a pair of allelomorphs, for example, *AA* or *aa.* Having only one type of allele for a given trait.

homozygous   Having the two genes at corresponding **loci** on homologous chromosomes identical for one or more loci.

honey butter   A mixture of creamery butter and 20 to 30 percent liquid honey, used as a spread.

hood A protective device, usually providing special ventilation to carry away gases, in which dangerous chemical, biological, or radioactive materials can be handled safely.

hoof oil Pale yellow liquid obtained from skin, bones, and hooves of cattle. It is used in leather manufacture as a lubricating and waterproofing agent. Frequently called **neat's-foot oil.**

hormone A chemical substance secreted by an endocrine gland that has a specific effect on activities of other organs.

host Any animal or plant on or in which another organism lives as a parasite. Infected **invertebrates** (which are actually hosts in the true sense) are usually referred to as *biological vectors.*

hot Highly radioactive.

hovers An artificial heating unit that provides the heat source in a brooder unit for poultry. See **brooder.**

HP See **heat production.**

HTST High-temperature–short-time (see **pasteurization**).

humerus The bone that extends from the shoulder to the elbow.

humidity The mass of water vapor per unit volume. See **relative humidity.**

hutch A boxlike cage or pen for small animals.

hybrid A heterozygote, or progeny of genetically unlike parents.

hybridomas Plasma cells that are specific for one set of antigenic determinants are fused to myeloma cells (malignant plasma cells) to produce a single clone of antibody-producing cells that can be grown indefinitely in tissue culture. These cells produce huge quantities of **monoclonal antibodies.**

hybrid vigor See **heterosis.**

hydrogenate To combine with hydrogen (to reduce).

hydrolysis Chemical decomposition in which a compound is broken down and changed into other compounds by taking up the elements of water. The two resulting products divide the water, the hydroxyl group (OH) being attached to one and the hydrogen atom (H) to the other.

hydrolytic Capable of breaking an intramolecular bond by adding a molecule of water.

hydrolyze Splitting a compound into smaller fragments by the addition of water (the hydroxyl group, OH, is added to one fragment and hydrogen, H, is added to the other).

hydroponics Literally, water plants. These include grains grown by sprouting in chambers under conditions of controlled temperature and humidity to provide a source of green feed at times when it is impossible to produce it normally.

hygrometer An instrument used for determining the relative humidity of air.

hyper- A prefix signifying above, beyond, or excessive.

hyperemia Congestion, or an excess of blood in any body part.

hyperesthesia Excessive sensitivity of nerves.

hyperplasia The abnormal multiplication or increase in the *number* of normal cells in normal arrangement in a tissue.

hyperploid A plant or animal whose chromosome number is greater than a whole-number multiple of the haploid number, e.g., $2n + 1$ or $2n + 2$.

hyperpnea Panting (as in dogs); deep breathing.

hypertrophy The morbid (diseased) enlargement, or overgrowth, of an organ or part due to an increase in the *size* of its constituent cells.

hypo- A prefix signifying under, beneath, or deficient.

hypocalcemia A significant decrease in the concentration of ionic calcium, which results in convulsions, as in tetany, or parturient paresis.

hypophysectomy Surgical removal of the pituitary gland (hypophysis).

hypothalamic releasing factors Substances secreted by the hypothalamus that regulate the release of hormones from the anterior pituitary gland.

hypothermia The lowering of body temperature, as in treatment of extremely high fever. The patient is given a sedative to inhibit shivering. Then, under minimal anesthesia, he/she is placed in ice water or may have ice packs applied. The rectal temperature may drop to about 86°F in 1 to 2 h. Subnormal body temperature is often induced artificially to facilitate heart surgery.

hypothermy See **hypothermia.**

hypotrophy Degeneration, or loss of vitality, of an organ.

hysterectomy Partial or total removal of the uterus.

I

ice milk A frozen product resembling ice cream, except that it contains less fat (2 to 5 as opposed to 10 percent) and more **NMS** (12 as opposed to 10 percent) than ice cream. Both ice milk and ice cream contain a stabilizer or emulsifier and about 15 percent sugar.

identical twins Two individuals that developed from the same fertilized egg. The egg separated into two parts shortly after fertilization. If separation is late, Siamese (joined) twins or one individual with two heads or two bodies may result. Identical twins are genetically alike. Therefore all differences between them represent environmental effects.

idiopathy A morbid (diseased) state of spontaneous origin.

i.e. From the Latin *id est.* That is.

ileocecal valve Valve at the junction of the lower end of the small and large intestines (junction of ileum and cecum).

ill-being Condition or state of an animal experiencing illness.

imbibe To drink or inhale; to absorb moisture.

imitation milks Mixtures of nondairy ingredients (ingredients other than milk, milk fat, and nonfat milk solids) that are combined, forming a product resembling milk, low-fat milk, or skim milk. Sodium caseinate, although derived from milk, is commonly termed a nondairy ingredient and is often used as a source of protein in imitation milks. Vegetable oils are commonly used as the source of fat.

immunity The power an animal has to resist and/or overcome an infection to which most or many of its species are susceptible. Active immunity is attributable to the presence of antibodies formed by an animal in response to an antigenic stimulus. **Passive immunity** is produced by the administration of preformed antibodies. (See also Section 22.2.5.)

immunize To render an animal resistant to disease by vaccination or inoculation.

immunoglobulins A family of proteins in body fluids that have the property of combining with an **antigen** and, in the situation in which the antigen is **pathogenic,** sometimes inactivating it and producing a state of immunity or resistance to disease. Also called *antibodies.*

immunologic Producing immunity.

immunological Response of the immune system of a host to non-self **antigens,** providing protection from a particular disease or poison.

immunology The study of resistance to disease; the science dealing with the nature and causes of immunity from diseases.

impaction Constipation. See **compaction.**

implant The insertion of a slow-release pellet containing a growth-promoting hormone into the middle third of the back of a bovine ear. The pellets are often referred to as *implants.*

impregnate To fertilize a female animal.

inbreeding Production of offspring from parents more closely related than the average of a population. Genetically, inbreeding increases the proportion of **homozygous** genes in a population. For example, in a population of 100 individuals that are all **heterozygous** (*Aa*), none would be homozygous. However, in a population in which 50 were homozygous (*AA* or *aa*) and 50 were heterozygous (*Aa*), the percent homozygous pairs of genes at this locus would be 50.

incaparina A so-called food substitute that includes corn, cottonseed meal, and torula yeast. First marketed in 1957, consumer acceptance has been slow.

incomplete dominance A situation in which neither allele is dominant to the other. The result is that both are expressed in the phenotype, which is intermediate between the two traits.

independent assortment A situation in which the separation or segregation of one pair of alleles has no effect on any other; occurs when the different pairs of alleles are on nonhomologous chromosomes.

indirect calorimetry Measurement of the amount of heat produced by a subject by determining the amount of oxygen consumed and the quantity of nitrogen and carbon dioxide eliminated. The determination of heat production from the respiratory exchange.

inert matter Any material included in the diet that has no nutritive value. It may be used as a carrier for micronutrients.

infection The invasion and presence of viable bacteria, viruses, and/or parasites in a host that result in disease.

infestation An invasion of the body by arthropods, including insects, mites, and ticks.

inflammation The reaction of a tissue to injury, which tends to destroy (through increased white blood cells) or limit the spread of the injurious agent. It is characterized by pain, heat, redness, swelling, and loss of function.

inflammatory Accompanied by or tending to cause inflammation.

inflection The point in the growth curve of plants and animals at which growth rate ceases to increase and from which it begins to decrease.

in foal Designating a pregnant mare. Likewise, *in calf* refers to a pregnant cow.

ingest (ingestion) To take in food for digestion via the mouth; to eat.

ingesta Food or drink taken into the stomach.

ingestive Pertaining to or effecting the act of taking food into the body.

inguinal Pertaining to the groin.

inhibine A substance found in human sputum that is active against a number of bacteria.

in milk Designating a lactating female.

innervation (innervated) The distribution of nerves to an organ or body part.

innocuity Harmlessness; some vaccines have more innocuity and **efficacy** than others.

inorganic Pertaining to substances not of organic origin (not produced by animal or vegetable organisms).

insect An air-breathing animal (phylum Arthropoda) that has a distinct head, thorax, and abdomen.

insecticide A substance, such as a stomach poison, **contact poison,** or fumigant, that kills insects by chemical action.

insensible Commonly used to denote insensible heat losses, e.g., water vapor and carbon dioxide. Heat dissipation by vaporization.

insidious More dangerous than is apparent.

in situ In the natural or normal place; the normal site of origin.

instar Any one of the larval stages of an insect between molts.

integrated reproduction management (IRM) Reproductive efficiency in livestock is affected by many factors such as nutrition, genetics, disease, and physiological conditions. IRM is a concept of integrating, on a "systems" or whole-animal management basis, the various aspects of animal health, production, and genetics into an overall workable management system.

integument A covering layer, as the skin of animals or the body wall of an insect.

inter- A prefix meaning between.

intercellular Situated between the cells.

intermediate host A host in which a parasite develops only in part, before getting into its final (or definitive) host, where it develops to maturity.

intermuscular Situated between muscles.

interphase The period between successive mitotic divisions and the period of growth and usually duplication of chromosomes.

interspecies Between species.

intestinal Of or pertaining to the **intestine.**

intestine The lower portion of the alimentary canal from the stomach to the anus. Also called the **gut** or *bowels.*

intolerable-response level Extremely high level of homeokinetic response that an animal cannot tolerate for long.

intra- A prefix meaning "within."

intracaudal Situated or applied within the tissues of the tail.

intracellular Situated or occurring within a cell or cells.

intracranial Within the cranium.

intracutaneous Into or within the layers of skin.

intradermal Within the dermis. An injection into the layers of skin.

intramammary Within the mammary gland, as injection into a mammary gland through a teat opening.

intramuscular Within the substance of a muscle, e.g., an injection into muscle.

intraperitoneal Within the cavity of the body that contains the stomach and intestines. Administration through the body wall into the peritoneal cavity.

intrasternal Within the sternum.

intravenous Within a vein or veins.

intravenously See **intravenous.**

in utero Within the uterus (intrauterine).

inversion Rearrangement of genes on a chromosome in such a way that their order is reversed or different.

invertebrates Animal species without a backbone. Aquaculture examples include shrimp and oysters.

in vitro Within an artificial environment, as within a petri dish or test tube.

in vivo Within the living body.

involution The return of an organ to its normal size or condition after enlargement, as of the uterus after childbirth. A decline in size or activity of other tissues; e.g., mammary gland tissues normally involute with advancing lactation. The drying-off process of lactating cows.

iodinated casein Milk protein (casein) that has been treated with iodine. It has the same physiological effect as thyroxine (hormone produced by the thyroid gland). It is commonly referred to as *thyroprotein* and is sometimes used to stimulate cows to secrete more milk.

iodine value The degree of unsaturation of the fatty acids in a fat or oil can be quantitatively expressed as the iodine value, which refers to the number of grams of iodine absorbed by 100 grams of fat. Because the iodine reacts at the sites of unsaturation, much as would hydrogen in hydrogenation, the higher the iodine value the greater the degree of unsaturation that existed in the fat.

ion An atom or a group of atoms (molecules) carrying an electric charge, which may be positive or negative. Ions are usually formed when salts, acids, or bases are dissolved in water.

ionization The adding of one or more *electrons* to, or removing one or more electrons from, atoms or molecules, thereby creating ions.

ionizing radiation Any radiation displacing electrons from atoms or molecules, thereby producing **ions,** e.g., alpha, beta, and gamma radiation and short-wave ultraviolet light. Ionizing radiation may produce severe skin or tissue damage.

IRM See **integrated reproduction management.**

irradiated ergosterol A **sterol** that has been activated into a form of vitamin D called ergocalciferol.

irradiation (irradiated/irridiating) The process of exposing (treating) materials (as in a nuclear reactor) to roentgen rays (X or γ-radiation) or other forms of radioactivity. An example of practical significance to nutrition is the application of ultraviolet rays to a substance to increase its vitamin D efficiency.

irritability The ability to respond to stimuli.

is by An indication of the male parent, referring to progeny.

iso From the Greek word *isos* meaning "equal." A prefix or combining form meaning "equal," "alike," or "the same;" e.g., "isocaloric" refers to the *same* caloric value.

isotonic Characterized by equal osmotic pressure; e.g., a solution containing just enough salt to prevent the destruction of the red corpuscles when added to the blood would be considered isotonic with blood. Describing a solution having the same concentration as the system or solution with which it is compared.

isotope An element of chemical character identical with that of another element occupying the same place in the periodic table (same atomic number), but differing from it in other characteristics, as in radioactivity or in the mass of its atoms (atomic weight). Isotopes of the same element have the same number of protons in their nucleus but different numbers of neutrons. A **radioactive isotope** is one with radioactive properties. Such isotopes may be produced by bombarding the element in a cyclotron.

-itis A word termination (suffix) denoting **inflammation** of a part indicated by the word stem to which it is attached.

IU International unit. A unit of measurement of a biological (e.g., a vitamin, hormone, antibiotic, antitoxin) as defined and adopted by the International Conference for Unification of Formulas. The potency is based on the bioassay that produces a particular biological effect agreed on internationally. See **USP.**

J

jack A male uncastrated donkey (ass).

jenny A female donkey (ass). Also called a *jennet.*

jerky Long thin strips of sun-dried beef or lean meat.

jowl Meat taken from the cheeks of a hog.

K

karyotype The chromosomes of a plant or animal as they appear at the metaphase of a somatic division. Arrangement by pairs, size, and centromere location when presented for a species.

katydid A long-horned grasshopper of the family Tettigoniidae.

keel bone The breastbone of birds; the sternum.

keratin An insoluble complex protein that constitutes hair, horn, claws, and feathers.

keratinization The development of or conversion into keratin or keratinous tissue; cornification. See **keratosis.**

keratosis Any horny growth, such as a wart, causing the cornification, or hardening, of the epithelial skin layers.

ketosis See **acetonemia.**

kid A young goat or antelope up to the first birthday. (See Appendix A.)

killed virus A virus whose infectious capabilities have been destroyed by chemical or physical treatment.

kilo- A prefix that multiplies a basic unit by 1000.

kilocalorie (kcal) Equivalent to 1000 small calories.

kip or **kipskin** Light rawhide from a grass-fed, immature bovine animal (between the size of a calf and a mature animal).

Kjeldahl Relating to a method of determining the amount of nitrogen in an organic compound. The quantity of nitrogen measured is then multiplied by 6.25 to calculate the protein content of the food or compound analyzed. The method was developed by the Danish chemist J. G. C. Kjeldahl in 1883.

kwashiorkor A syndrome produced by a severe protein deficiency, with characteristic changes in pigmentation of the skin and hair, edema, anemia, and apathy.

L

label See **tracer isotope.**

labile Changeable or unstable. Readily or continuously undergoing chemical, physical, or biological change or breakdown.

lactate (lactating) To secrete or produce milk.

lactation period The number of days an animal secretes milk following each **parturition.**

lacteal Pertaining to milk.

lactiferous Secreting or conveying milk.

lactogenesis Initiating the secretion of milk.

lactogenic Stimulating the secretion of milk.

lactometer An instrument used to determine the specific gravity of milk, providing information for calculating the percentage of solids in a sample of milk.

lactose Milk sugar. A disaccharide composed of one molecule of glucose and one molecule of galactose.

lamb A sheep less than 12 months old. To give birth to a lamb. (See Appendix A.)

lamb hog A male lamb from weaning time until it is shorn.

larva (pl. larvae) The immature form of insects and other small animals, which is unlike the parent or parents and which must undergo considerable change of form and growth before reaching the adult stage; e.g., white grubs in soil or decayed wood are larvae of beetles. Caterpillars, maggots, and screwworms are also larvae.

larvicide A chemical used to kill the larval or preadult stages of parasites.

larviparous Bearing and bringing forth young that are larvae (especially of insects).

larynx The upper portion of the windpipe (trachea).

latent Hidden or not apparent.

lateral Of, at, from, or toward the side or flank.

lateral line The major sense organ in fish. It is comprised of a series of connected pores running the length of the trunk by which fish detect changes in water pressure or current.

laying The expulsion of an egg. The term is commonly associated with hens in active egg production. See **oviposition.**

LD$_{50}$ The lethal dose for 50 percent of the animals tested.

legume Refers to those crops that can absorb nitrogen directly from the atmosphere through bacteria that live in their roots. Clovers, alfalfa, and soybeans are common examples of legumes.

lesion Injury or diseased condition reflected in discontinuity of tissues or organs, often causing loss of function of a part.

lethal Deadly; causing death.

lethal gene A gene, or genes, causing the death of an individual. Most lethal genes are **recessive** or partially dominant, requiring two genes of the same kind in a pair (**homozygous**) to cause death.

leukocytes (leucocytes) The white blood cells. They have amoeboid movements and include the lymphocytes, monocytes, neutrophils, eosinophils, and basophils.

libido Sexual desire or instinct.

lice Small nonflying biting or sucking insects that are true parasites of humans, animals, and birds.

life cycle The changes in form and mode of life that an organism goes through between recurrences of the same primary stage; life history.

ligaments Tissues connecting bones and/or supporting organs.

ligate To tie up, or bind, with a ligature.

lignin A compound that in connection with cellulose forms the cell walls of plants and thus of wood. It is practically indigestible.

limit-fed See also **limit feeding.** Refers to any method of feeding that restricts free access to diets and sets limits on quantity of feed consumed.

limit feeding Feeding animals to maintain weight and growth but not enough to fatten or increase production. Feeding animals less than they would like to eat.

limiting amino acid The essential amino acid of a protein that shows the greatest percentage deficit in comparison with the amino acids contained in the same quantity of another protein selected as a standard.

limnology The study of inland waters (lakes and ponds), especially with reference to their biological and physical features.

linear accelerator A device for increasing the velocity and energy of charged *elementary particles,* e.g., electrons or protons, through application of electric and/or magnetic forces.

linebreeding A form of inbreeding in which an attempt is made to concentrate the inheritance of some ancestor in the pedigree.

linecross A cross of two inbred lines.

linkage Refers to two or more genes carried on the same chromosome.

linkage group All the genes having loci on a particular chromosome.

linked genes Sets of genes that tend to be inherited together and that are presumed to have their loci on the same chromosome.

lipid Any one of a group of organic substances that are insoluble in water but are soluble in alcohol, ether, chloroform, and other fat solvents and have a greasy feel. They include fatty acids and soaps, neutral fats, waxes, steroids, and phosphatides (by U.S. terminology).

Lippizan A famous breed of Austrian horses. They are milk white in color at maturity.

lipolysis The hydrolysis of fats by enzymes, acids, alkalis, or other means to yield glycerol and fatty acids.

litter The pigs farrowed by a sow or the pups whelped by a bitch at one delivery period. Such individuals are called **littermates.** Also, the accumulation of materials used for bedding farm animals.

littermates A group of siblings in a **litter**-bearing species that are born at the same time.

livability The inherited stamina, strength, and ability to live and grow.

livestock Domestic farm animals kept for productive purposes (meat, milk, work, and wool); include beef and dairy cattle, sheep, swine, goats, and horses. Also called *stock.*

local infection An infection restricted to a small area, as an ear infection.

locus (pl. loci) Region of a chromosome where a particular gene is carried or located. The segment or part of a chromosome concerned with regulating a particular trait.

loin That portion of the back between the thorax and pelvis.

long feed Coarse, or unchopped, feed for livestock, such as hay, as contrasted to *short,* or *chopped,* or *ground* feed.

lope A slow gallop of a horse.

low-fat milk Milk that contains between 0.5 and 2 percent milk fat.

low-moisture silage **Silage** that contains 35 to 55 percent moisture. (See also **haylage.**)

low-set Designates a short-legged animal.

lucerne Alfalfa; a legume of high feeding value for ruminants.

lumbar Pertaining to the loins.

lumen The cavity on the inside of a tubular organ, e.g., the lumen of the stomach or intestine.

luteal phase The stage of the estrous cycle at which the corpus luteum is active and progesterone influence predominates.

lymph A transparent, slightly yellow liquid found in the lymphatic vessels. It may have a light-rose color because of the presence of red blood corpuscles.

lymphocyte A variety of white blood corpuscles that originate from the lymph **nodes** and thymus or bursa (birds).

lyophilization The evaporation of water from a frozen product with the aid of vacuum. Also called *freeze-drying.*

lyophilized See **lyophilization.**

lysin Antibody that causes the death and dissolution of bacteria, blood corpuscles, and other cellular elements.

lysozyme A substance present in human nasal secretions, tears, and certain mucus. It is also present in egg whites, in which it hydrolyzes polysaccharidic acids. It is bactericidal for only a few saprophytic bacteria and is inactive against pathogens and organisms of the normal **flora.**

M

macro   Large or major. Abnormal size or length. Usually a prefix.

macrocyte   An abnormally large erythrocyte (red blood cell) occurring chiefly in anemias.

macrocytic   Characterized by abnormally large erythrocytes (red blood cells).

macroinvertebrate   Invertebrate animals larger than zooplankton such as insect larvae. Large species of animals without a backbone.

macrophages   Large **phagocytes.** Large scavenger cells, the function of which is primarily phagocytosis and the destruction of many kinds of foreign particles. These cells are strategically located in the spleen, liver, bone marrow, small blood vessels, and connective tissue. Collectively, they compose the monocyte–macrophage system (formerly known as the **reticuloendothelial system**).

macroscopic   Visible to the unaided, or naked, eye.

maggot   The larva of a fly.

maiden   An unbred animal.

maintenance energy   The amount of energy required to keep an animal in energy equilibrium and prevent the loss of body tissues or weight.

maintenance needs   What an animal needs to maintain itself.

malnutrition   Any disorder of nutrition. Commonly used to indicate a state of inadequate nutrition.

mammal (mammalian)   Any animal that suckles or provides milk for its young.

mammilla   The nipple of the female breast.

mandible   Bone of the lower jaw.

manufacturing milk   Milk used for the manufacture of dairy products, such as cheese, butter, powdered milk, and ice cream.

manure   Excreta of animals; dung and urine (commonly with some bedding).

manured (soil)   Soil to which manure has been applied.

marasmus   From the Greek *marasmos* meaning "a dying away." A progressive wasting and emaciation. **Enzootic** marasmus is a condition of malnutrition in domestic animals that is due to a deficiency of one or more trace elements, especially cobalt and copper.

marbling   The distribution of fat in muscular tissue that gives meat a spotted appearance.

mare   A mature female horse (usually more than 4 years of age).

mariculture   Marine aquaculture which uses saltwater rather than freshwater animals.

market milk   Milk that is consumed as fresh fluid milk. See **fluid milk.**

marsupial   One of a class of mammals characterized by the possession of an abdominal pouch in which the young are carried for some time after birth.

marsupilia   Pouched mammals, e.g., the kangaroo.

masculinity   Physical appearance resulting from well-developed secondary male sex characteristics.

mass spectrometer   An instrument for separation and measurement of isotopes by their mass.

mastication   The chewing of food.

mastitis   An inflammation of the mammary gland(s).

maternal   Pertaining to the mother.

matrix   The basic material from which a thing develops. A place or point of origin of growth.

matron   A mare that has produced a foal.

mature equivalency (ME)   Age-conversion formulas (provided by the **USDA** and breed associations) applied to the milk-production records of young cows to predict their expected *mature milk production* potential. Breed ME factors are used for comparative purposes in selecting and mating animals.

maturation   The process of becoming mature.

maverick   An unbranded animal, particularly a calf (also refers to a motherless calf in some areas). A **dogie.**

maxilia   The upper jawbone in vertebrates.

ME   See **mature equivalency.** See also **maintenace energy** and **metabolizable energy.**

meat analogues   Material usually prepared from vegetable protein to resemble specific meats in texture, color, and flavor.

meat bird   A fowl raised for its meat as contrasted to one kept for egg laying.

meconium   The first excreta of a newborn animal.

median   Situated in the middle; mesial.

median plane   A line or plane (from cranial to caudal) dividing an animal into two equal halves.

megacalorie (Mcal)   Equivalent to 1000 kcal or 1,000,000 cal. A megacalorie is equivalent to a *therm.*

meiosis   A type of cell division that produces the sex cells, or **gametes,** possessing the l*n* (haploid) number of chromosomes. Thus the chromosome number of daughter cells is reduced to one-half the somatic number (chromosome number is reduced by one-half from diploid to haploid).

melanin   A dark-brown or black pigment found in hair and/or skin.

mesenchyme   Embryonic connective tissue that gives rise to the connective tissues of the body and blood vessels.

mesentery   A membrane that supports a visceral organ, particularly the intestine, and contains the vessels and nerves that supply that organ.

meta-   A prefix meaning "between" or "among;" indicating change, transformation, or exchange; after or next.

metabolic   Relating or pertaining to the nature of metabolism; the synthesizing of foodstuffs into elements. See **metabolism.**

metabolic activity   See **metabolism.**

metabolic body size   The weight of the animal raised to the 0.75 power ($W^{0.75}$).

metabolism   The sum total of the chemical changes in the body, including the building up (anabolic, assimilation) and the breaking down (catabolic, dissimilation) processes. The transformation by which energy is made available for body uses.

metabolite   Any substance produced by metabolism or by a metabolic process.

metabolizable energy (ME)   The food-intake gross energy minus fecal energy minus energy in the gaseous products of digestion (largely methane) minus urinary energy. For birds and monogastric mammals, the gaseous products of digestion need not be considered. See also **physiological fuel value.**

metabolize (metabolized)   To perform **metabolism.** Chemical changes occurring within a living cell or organism which produce energy for vital processes and activities and assimilation of new material into the organism. To transform substances into energy and new materials for assimilation and uses by the body.

metamorphosis   A change in shape or structure involving a transition from one developmental form to another, as in insects and frogs.

metaphase   The second stage of mitosis, characterized by alignment of chromosomes on the equator of spindle fibers.

metritis   An inflammation of the uterus.

MF   Milk fat.

micro   Small or minor. Usually a prefix, designating *tiny* or *microscopic in size*. Also, a prefix that divides a basic unit by 1 million.

microbiological assay   The use of microorganisms to assay. See **bioassay.**

microflora   The flora consisting of microorganisms. Commonly used in reference to the bacteria populating the rumen. (Protozoa are also present in the rumen.)

micromicro   See **pico-.**

microphages   Little **phagocytes.** Polymorphonuclear leukocytes that are formed in bone marrow. They are particularly active during bacterial infections.

microscopic   Invisible to the unaided eye.

milk equivalent   The quantity of milk, as produced, required to furnish the milk solids in manufactured dairy products to be sold: e.g., approximately 10 and 20 lb of whole milk is required to manufacture 1 lb of cheddar cheese and butter, respectively.

milk fat   The lipid components of milk.

milk fever   A nutritional disease defined by muscular **tetany** that is due to low blood calcium. Generally caused by the rapid removal of calcium from the bloodstream and the failure to balance the removal by mobilization of calcium from bone or absorption from the small intestine. Though found in most mammalian species, milk fever is most prevalent in high-producing dairy cattle.

milking string   A group of dairy cattle within the milking herd that receive similar management on the basis of similar needs. Typical "string" examples would be early lactation–high-production cows or low production–late-lactation cows.

milk replacer   A dry commercial feed product intended to be reconstituted with water to substitute for the use of herd milk in feeding young animals. Although generally thought to be for dairy calves, milk replacers are available for most livestock in case of special needs such as failure of dam to produce milk or loss of the dam.

milk serum   The nonfat components of milk.

milli-   A prefix that divides a basic unit by 1000.

minor stress responses   Homeokinetic response to stress that is relatively minor in intensity, duration, or both.

miracidium   The free-swimming larva of a trematode that penetrates the body of a snail host for further development into a **cerearia.**

miscible   Designating two or more substances that when mixed together form a uniformly homogeneous solution.

miticide   A compound that is destructive to mites. An acaricide.

mitosis   A type of cell division in which cells with the $2n$ (diploid) number of chromosomes produce daughter cells that also possess the $2n$ (diploid) number of chromosomes. Thus the daughter cells receive the full complement of chromosomes existing in the original cell before division.

mitotic centers   Two (usually) polarizing units of cells located opposite each other, like the spindle fibers oriented between the poles of a cell.

mode   A statistical term referring to the value (number) that occurs the greatest number of times in a frequency distribution.

modified live virus   A virus that has been changed by passage through an unnatural host, such as hog cholera virus passed through rabbits, so that it no longer possesses **pathogenic** characteristics but will stimulate antibody production and immunity when injected into susceptible animals.

molecular   Of, produced by, or relating to **molecules.**

molecule   A group of atoms held together by chemical forces. A molecule is the smallest unit of matter that can exist by itself and retain all its chemical properties.

molt (molting)   The shedding and replacing of feathers (usually in the fall). Snakes and certain arthropods also shed their outer covering and develop a new one.

mongrel   Animal of mixed or unknown breeding.

monoclonal antibodies   These are antibodies with specificity against only one set of antigenic determinants. They are produced in large quantities by hybridomas and have virtually revolutionized immunology. Specific *monoclonal antibodies* have been used successfully for immunotherapy of cancer patients. These antibodies can also be used to develop immunodiagnostic techniques. See **hybridomas.**

monoestrous   See **monoestrous animal.**

monoestrous animal   An animal that has only one estrous (heat) cycle annually.

monogastric   Having only one stomach or stomach compartment, as do dogs, humans, and swine.

monohybrid   A trait in an individual controlled by a single pair of genes with a genetic makeup, e.g., of *Aa* or *Bb.*

monohybrid cross   A cross involving one pair of alleles, each of which controls an alternative form of the same characteristic.

monorchid   A male animal that has only one testicle in the scrotum. Also called a *ridgling.*

monotreme   Any member of the lowest order of mammals (Monotremata), comprising the duckbill platypus and the echidnas, which lay eggs and have a common opening for the genital and urinary organs and the digestive tract. They nourish their young by a mammary gland that has no nipple, in a shallow pouch developed only during lactation.

monoxenous parasite   A parasite that requires only one host for its complete development.

morbid   Diseased or unhealthy.

morbidity   The state of being diseased. The ratio of the number of sick individuals to the total susceptible population.

morphogenesis   The origin and evolution of morphological characters (form and structure). The establishment of shape and patterns.

morphologically   (adv. of morphologic) Pertains to the form and structure of animals and their body parts.

morphology   The science of the forms and structures of animal and plant life without regard to function.

mortality   Death. Death rate.

morula   A developing embryo at day 5 or 6 postfertilization characterized by a solid cluster of about 32 cells.

mount   To copulate with, as certain male animals mount a female in the act of coitus.

mounted   See **mount.**

mule   The cross resulting from mating a mare horse with a jack (male ass).

muley   A **polled** cow.

multiparous Producing many (more than one) offspring at one time. Also having experienced one or more **parturitions.**

multiple alleles Two or more alleles at the same locus on one pair of homologous chromosomes, affecting the same trait but in a different way. For example, in humans, gene *A* produces antigen *A;* gene *B,* antigen *B;* and gene *a,* neither antigen (or the O blood type). All three alleles may be at the same locus in a population, but only one is in a **gamete** and only two are in body cells. Thus, *multiple alleles* describes a condition in which three or more forms of the same gene exist, any one of which may occupy a specific locus at any given time.

multiple stimuli Situation where more than one environmental factor—acting either simultaneously or sequentially—elicits an animal's homeokinetic response.

mummified fetus A shriveled or dried **fetus** that has remained in the uterus instead of being aborted or expelled.

muscle catabolism Breaking down of muscle tissues. The proteins from the muscles may be deanimated and used for energy. Destructive **metabolism** involves the release of energy and results in breakdown of muscle.

musculature The muscular system of any body part.

mustang A wild horse, ridden by cowboys of the western plains, that descended from Spanish horses. The name is derived from the Spanish *mustengo* meaning "wild."

mutation A change in a gene, often resulting in a different phenotype. More specifically, a change in the code sent by a gene, causing the formation of a different protein by the cytoplasm of the cell. A permanent, transmissible change in the characteristics of an offspring from those of its parents.

mutton The flesh of a grown sheep as opposed to that of a lamb. Goat meat is also called *mutton* in some countries.

myalgia Muscular pain or rheumatism.

mycotoxins Toxic metabolities produced by molds during growth on a suitable substrate.

myiasis A disease that is due to the presence of fly larvae in the flesh of warm-blooded animals.

myositis Inflammation of a voluntary muscle.

N

nag A horse or pony of nondescript breeding.

nanism Dwarf growth.

nano- A prefix that divides a basic unit by 1 billion (10^9).

nape Back of the neck of an animal.

nares The olfactory openings in fish analogous to the nose. In fish these are not connected to the respiratory system.

National Research Council See **NRC.**

natural immunity Immunity to a disease or infestation which results from qualities inherent in an animal. See **immunity.**

natural radiation Background radiation.

natural service In farm animals, it means to allow natural mating, as opposed to artificial insemination.

NDM Nonfat dry milk. The product obtained by removing water from pasteurized skim milk. It contains not more than 5 percent moisture and not more than 1.5 percent **MF** unless otherwise indicated.

near In describing horse gaits "near" is the side closest to the inside of an arena where the judge stands.

neat's-foot oil A yellowish oil prepared by boiling the bones and joints of cattle (and sometimes of horses and sheep) and skimming off and clarifying the oil obtained.

necropsy An examination of the internal organs of a dead body to determine the apparent cause of death. Also called **autopsy, postmortem.**

necrosis Death of tissue, usually in individual cells, groups of cells, or small localized areas.

necrotic Pertaining to or affected with **necrosis.**

neigh The characteristic cry of a horse.

neonatal The period immediately following birth. Relating to or affecting the newborn human infant or farm animal. See also **neonate.**

neonate A newborn infant. A child less than one month old.

net energy (NE) The difference between **ME** and the heat increment; includes the amount of energy used either for maintenance only or for maintenance plus production.

net energy for maintenance (NE$_m$) The portion of net energy expended to keep the animal in energy equilibrium. (There is no net loss or gain of energy in the body tissues.)

net energy for production (NE$_p$) The portion of net energy required in addition to that needed for body maintenance that is used for work or for tissue gain (growth and/or fat production) or for the synthesis of productive end products (a fetus, milk, eggs, wool, fur, or feathers).

NFE See **nitrogen-free extract.**

NFS Nonfat solids of milk. They comprise proteins, lactose, minerals, and other water-soluble constituents of milk. Also called **SNF** (solids–not fat) and **NMS.**

nicking Breeding of progeny that are superior to their parents. Also called **heterosis.**

nit The egg of a louse or similar insect.

nitrogen balance The nitrogen in the food intake minus the nitrogen in the feces minus the nitrogen in the urine. Nitrogen retention.

nitrogen-free extract (NFE) Comprises the carbohydrates, sugars, starches, and a major portion of the material classed as hemicellulose in feeds. When crude protein, fat, water, ash, and fiber are added and the sum is subtracted from 100, the difference is the NFE.

NMS Nonfat milk solids. See **NFS.**

nocturnal Of the night; a nocturnal parasite is one that is active at night.

nodes Small, distinct masses of one kind of tissue enclosed in a different kind of tissue; lymph nodes are small masses of lymphatic tissue and serve as the main sources of lymphocytes. They are a site of antibody production by lymphocytes, and also serve as a defense system for the removal of noxious agents such as bacteria and fungal organisms carried into the nodes by lymph vessels.

nondisjunction Failure of a pair of homologous (sister) chromosomes (dyads) to separate during the reductional division of **meiosis.** Thus both chromosomes go into a **gamete** (usually only one of each pair goes into a gamete).

nonfat dry milk See **NDM.**

nonreturn The breeding efficiency of bulls expressed as the percentage of cows that conceive on the first service.

nonruminant An animal without a rumen, e.g., a chicken, pig, or dog.

nonseasonally polyestrous   Describes the tendency of some species, or some breeds within species, to have multiple **estrous cycles** primarily during two distinct seasons of the year. For example, ewes of the Rambouillet breed typically cycle in the fall months, become **anestrous** during the winter, and then cycle again during the spring months.

notching   Cutting dents in the ears of animals for identification.

noxious   Harmful, not wholesome.

NPN   Nonprotein nitrogen (e.g., urea).

NRC   National Research Council. A division of the National Academy of Sciences established in 1916 to promote effective utilization of the scientific and technical resources available. This private, nonprofit organization of scientists periodically publishes bulletins giving the nutrient requirements of domestic animals. Copies are available through the National Academy of Sciences–NRC, 2101 Constitution Avenue NW, Washington, DC 20418.

nuclear reactor   A device in which a fission chain reaction can be initiated, maintained, and controlled. Its essential component is a core with fissionable fuel.

nucleoside   A compound composed of a nitrogen base and a sugar.

nucleotide   A compound composed of a nitrogen base, a sugar, and a phosphate.

nucleus (pl. nuclei)   A deep-staining body within a cell, usually near the center; the heart and brain of the cell, containing the chromosomes and genes. Also, the small, positively charged core of an **atom.** All nuclei contain both *protons* and *neutrons,* except the nucleus of ordinary hydrogen, which consists of a single proton.

nuclide   A general term applicable to all atomic forms of the elements.

nulliparous   Having never given birth to viable young.

nurse   To suckle; to give milk to (e.g., a baby, a calf, or other mammal).

nurse cow   A milk cow used to supply milk for nursing calves other than her own.

nutraceutical   A nutrient that produces a healthful effect beyond its normal nutritional effect.

nutrient   A substance (element or ingredient) that nourishes the **metabolic** processes of the body. It is one of the many end products of digestion.

nutrient-to-calorie ratio   An expression of nutrients in weight per unit of energy needed. For example, the protein-to-calorie ratio is expressed as the grams of protein ($N \times 6.25$) per 1000 kcal metabolizable energy (grams of protein per 1000 kcal ME). This same dimension may be extended to other nutrients such as grams of calcium per 1000 kcal.

nutrilite   A nutritional element.

nutriment   That which is required by an animal as a building material and fuel (nourishment).

nutrition   The science encompassing the sum total of processes that have as a terminal objective the provision of nutrients to the component cells of an animal.

nutritive ratio (NR)   The ratio of digestible protein to other digestible nutrients in a foodstuff or diet. (The NR of shelled corn is about 1:10.)

nymph   The immature stage of insects having only three stages (egg, nymph, and adult) in their development. Nymphs resemble adults in form and appearance (as contrasted with larvae, which do not resemble their adults) but do not have wings.

O

obligate parasite   A parasite incapable of living without a host.

occult   Obscure; concealed from observation; difficult to understand.

oestrus   See **estrus.**

off   In describing horse gaits "off" is the side that is away from the judge against the outside wall or fence of the arena.

offal   The internal fat of cattle. The viscera and trimmings of a slaughtered animal removed in dressing. Also, the by-products of milling used especially for livestock feeds.

off-feed   Having ceased eating with a healthy and normal appetite.

official production record   Standard DHIA and DHIR records pertaining to milk production that are made under the supervision of an unbiased individual. Records are used for management purposes (e.g., feeding and culling), genetic evaluation (sire summaries), publicity, and sales. See **DHIR.**

offspring   The sons and daughters of parents.

oil gland   Gland at the base of the tail in chickens, ducks, turkeys (and most wild birds) that secretes an oil used by birds in preening their feathers. Also called *preen gland.*

olfactory   Pertaining to the sense of smell.

omasum   The third division of the stomach of ruminant animals. Also called *manifold, manyplies,* and *psaiterium.*

omnivore   An animal that subsists on feed of every kind, plant and animal. Humans are omnivorous animals.

onager   Wild ass of southwestern Asia.

on the hoof   Designating a living meat animal.

oocyst   The encysted **zygote** or embryonic stage in some parasitic life cycles.

opaque   Not letting light through; neither transparent nor translucent.

open   A term commonly used of farm mammals to indicate a nonpregnant status.

opisthotonos   A form of tetanic spasm in which the head and heels are bent backward and the body bowed forward.

opsonins   Antibodies of blood serum that sensitize (weaken or coat) the cells of microorganisms so that they are readily ingested, or engulfed, by phagocytic body cells (white blood cells).

oral (orally)   Pertaining to the mouth. See **per oral** and **per os.**

orchidectomy   Surgical removal of the testes.

orchitis   An inflammation of a testis.

organic   Pertaining to substances derived from living organisms. Referring to carbon-containing compounds.

organic farm products   Pertains to growing/producing foods using only natural sources of soil nutrients (no use of chemical fertilizers, sprays, insecticides, or herbicides).

organism   Any complete living plant or animal.

organogenesis   The origin or development of the organs of an animal.

orifice   Entrance or outlet of a body cavity.

orthopnea   Inability to breathe except in the upright position. This state is common during dehydration exhaustion and with congestive heart failure.

osmosis   The tendency of two fluids of different strengths that are separated by something porous (a semipermeable membrane) to diffuse or spread through a membrane or partition until they are mixed. Osmosis is the chief means by which the body absorbs food and liquids. Specifically, it is the tendency of a fluid of lower concentration to pass through a semipermeable membrane into a solution of higher concentration.

osmotic pressure   The force acting on a semipermeable membrane placed between a solution and its pure solvent, caused by the flow of solute molecules through the membrane toward the pure solvent.

osteofibrosis   A loss of calcium salts from the bones that causes them to become fragile. It occurs chiefly in horses but may affect pigs, goats, and dogs.

osteogenesis   Formation of bone.

osteomalacia   A condition marked by softening of the bones, pain, tenderness, muscular weakness, and loss of weight. It results from a deficiency of vitamin D or of calcium and phosphorus. May also be caused by an overactive parathyroid gland.

osteoporosis   A reduction in total bone mass. This disorder of bone metabolism occurs in middle life and older age in both men and women. The bone becomes porous and thin because of a failure of the osteoblasts (bone-forming cells) to lay down bone matrix. This disorder may result from a (1) dietary deficiency of calcium and/or protein, (2) lowered calcium absorption, (3) hormonal disturbance, or (4) combination(s) of the above.

osteoselerosis   Abnormal hardening of bone.

outcross   Mating of an individual to another in the same breed that is not closely related.

out of   Refers to *mothered by* in animal breeding.

ova   See **ovum.** Female reproductive cells (pl. of ovum). The human ovum is a round cell about 0.1 mm in diameter.

ovary   Female reproductive gland in which the ova (eggs) are formed.

overdominance   A type of gene expression in which the heterozygote *Aa* is superior to either homozygote *AA* or *aa*. Interaction between genes that are alleles.

overo   A color pattern of Pinto horses. The horse appears to be colored with jagged white markings usually originating on the horse's side or belly and spreading toward the neck, tail, legs, and back. The color appears to frame the white. The horse often has a dark tail, mane, legs, and backline.

ovicide   Any substance that kills parasites in the egg stage.

ovine   An animal of the subfamily Ovidae; sheep, goats.

oviparous   Producing offspring from eggs that hatch outside the body.

oviposition   The laying (expelling) of a fully developed egg.

ovoviviparous   Producing eggs within the maternal body. The eggs hatch within or immediately after extrusion from the parent.

ovulation   The shedding of a mature follicle (**ovum**) by the ovary. The ovary of a hen contains a series of follicles (called the *follicular-size hierarchy*). After ovulation, the smaller follicles advance one position in size and reestablish the hierarchy as it existed just before ovulation.

ovum   The female reproductive cell (**gamete**), which, after fertilization, develops into a new member of the same species. The male gamete is the **sperm.**

ox (pl. **oxen**)   Any species of the bovine family of ruminants. Specifically, a domesticated and castrated male bovine used for work purposes.

oxidase   Enzyme that activates oxygen.

oxidation   Chemically, the increase of positive charges on an atom or the loss of negative charges. There may be a loss of one electron (univalent O) or two electrons (divalent O). The combining of oxygen with another element to form one or more new substances. Burning is one kind of oxidation. Also called *oxydation.*

P

pace   (1) rate of movement; (2) a manner of walking; (3) a special gait in horses in which the legs move in lateral pairs and support the horse alternately on the left legs and right legs; this is a fast two-beat gait.

packer   One who operates a slaughter and meat-processing business.

paint   A coat color of horses (white patches interspersed with darker colors, usually brown or black); also, a breed of horses.

paired feeding (food equalizing)   A method of comparing nutritional effects at an arbitrary low level set by the animal that consumes the least food. Littermates or twins (especially monozygous ones) are considered best for paired-feeding studies.

palatability   The relative relish with which feeds are consumed by animals.

palatable   Acceptable or even "savory" to the taste.

paleontology   The study of fossil remains.

palomino   Color of horses: golden coat with light-blond or silvery mane and tail.

pandemic   Prevalent (as a disease) throughout an entire country or continent or the world.

Papanicolaou stain   A method of staining smears of various body secretions from the respiratory, digestive, or genitourinary tracts. It is used to diagnose cancer or the presence of a malignant process. Exfoliated cells of organs, such as the stomach or uterus, are obtained, smeared on a glass slide, and stained for microscopic examination. It was named for the Greek physician George Papanicolaou, who developed it. The slides are also known as *Pap smear.*

parakeratosis   Any abnormality of the stratum corneum (horny layer of epidermis) of the skin (especially a condition where nuclei are retained in the upper layers showing defective keratinization).

parasite   An organism that lives at least for a time on or in and at the expense of living animals.

parasitic   Of an organism that lives on or in another organism known as the host.

parasiticide   An agent or drug destructive to parasites.

parchment   Tanned sheepskins. Vellum is essentially the same as parchment except that it is made from calfskin. Parchment is used for diplomas, records, banjos, drumheads, lampshades.

parental combinations   Genotypes and phenotypes like those of the parents in a particular cross.

parental generation P$_1$ The first generation in a particular genetic experiment; frequently purebreeding lines.

parenteral   Pertaining to administration by injection, not through the digestive (food) tract, i.e., such as subcutaneous, intramuscular, intramedullary, intravenous.

paresis   Partial paralysis that affects the ability to move but not the ability to feel.

parrot mouth   A malformed mouth of an animal (most common in horses) in which the upper jaw abnormally protrudes beyond the lower jaw.

parthenogenesis (parthogenesis)   Reproduction by the development of an egg without its being fertilized by a spermatozoon, e.g., drone bees. It occurs in certain lower animals and has been observed in turkeys. It does not occur in mammals.

partial dominance   A situation in which one gene of a pair of alleles is not completely dominant with respect to another. For example, in comprest Hereford cattle an individual that possesses no comprest genes or is of genotype *cc* is of normal size, those that possess two comprest genes *CC* are dwarfs, and those that possess one comprest gene *Cc* are midway in size between dwarfs and normals.

particle   A minute constituent of matter. The primary particles involved in radioactivity are *alpha particles, beta particles, neutrons,* and *protons.*

parturient paresis   A condition caused by a low blood-calcium concentration that results in partial to complete paralysis soon after parturition. Also called **milk fever.**

parturition   The act or process of giving birth to young.

passerine   Belonging to or having to do with the very large group of perching birds, including more than half of all birds, such as warblers, sparrows, chickadees, wrens, thrushes, and swallows.

passive immunity   Disease immunity given to an animal by injecting the blood serum from an individual already immune to that disease. Newborns also receive passive immunity by absorption of antibodies from colostrum. See **immunity.**

pasteurization   The process of heating milk to at least 145°F (62.8°C) and holding it at that temperature for not less than 30 min (holding method) or to 161°F (71.7°C) for 15 s **(HTST).**

pasture   Plants such as grass grown for feeding or grazing animals. To feed cattle and other livestock on pasture.

patency   The condition of being open or unobstructed.

patent leather   A term associated with the finish produced by the covering of the surface of leather with successive coats of daub and varnish. Most patent leather is made from cattle hides or **kips,** although horsehide, goatskin, kidskin, and calfskin are sometimes used.

paternal   Pertaining to the father or male parent.

pathogen   Any disease-producing microorganism or virus.

pathogenic   Capable of producing disease.

pathogenic organism   A disease producer.

pathology   The branch of science dealing with disease, especially with structural and functional changes in tissues and organs of the body affected by disease.

paunch (rumen)   The first stomach compartment of a ruminant. See **rumen.**

PD   See **predicted difference.**

pecking order   The system of social order in poultry exhibited by animals higher on the dominance scale physically pecking at animals lower on the scale.

pectoral   Pertaining to the breast. Situated in or on the chest.

pectoral fins   Either of a pair of fins in fishes that are attached in line with the gills in the thoracic region corresponding to the forelimbs of higher vertebrates.

pedigree   A list of an animal's ancestors, usually only those of the five closest generations

pelagic   Fish that are free-swimming in the open parts of a body of water away from the bottom.

pelt   The natural, whole skin covering, including the hair, wool, or fur of smaller animals, such as sheep and foxes.

penetrance   A genetic term that refers to the percentage of times a phenotype actually shows up when it is expected.

percutaneons   Performed or introduced through the skin, as an injection.

performance criteria   Standards based on ultimate efficacy with respect to animal performance as well as on engineering bases by which an animal accommodation is designed.

performance records   Information documenting the performance of individual animals such as average daily gain, days to weaning weight, and number of offspring produced per litter.

perfusion   The act of pouring through or immersing in a physiological fluid, e.g., blood or saline.

pericardium   Membrane that encloses the cavity containing the heart.

perineum   Anatomical region of the body between the thighs, especially between the anus and the genitals.

periodic table (periodic chart)   A table or chart listing all the elements, arranged in order of increasing **atomic numbers** and grouped by similar physical and chemical characteristics into "periods." The table is based on the chemical law that physical or chemical properties of elements are periodic (regularly repeated) functions of their **atomic weights.** (See Appendix E.)

periosteum   The membrane that covers bone.

perissodactyl   Having an uneven number of toes on each foot. A hooved animal with an uneven number of toes, such as a horse.

peristalsis   The rhythmic contractions and movements of the alimentary canal.

peritoneum   The membrane that lines the abdominal cavity and invests the contained viscera (digestive organs).

per oral   Administration through the mouth.

per os   Oral administration (by the mouth).

perosis   A disease of chicks marked by bone deformities and associated with deficiency of certain dietary factors, such as biotin, choline, folic acid, or manganese. Also called *slipped tendon* or *hock disease.*

per se   By, of, or in itself. As such.

Persians   Crust leathers made from India-tanned hair (as opposed to wool growth) sheepskin. Leather from "bastard skins" (see **bastard**) is sometimes designated as Persian. Also a long-haired breed of cats.

pesticide   A compound used to control any plant or animal considered to be a pest.

PFV   See **physiological fuel value.**

pH   A symbol used (with a number) to express acidity or alkalinity in analyzing various body secretions, chemicals, and other compounds. It represents the logarithm of the reciprocal (or negative logarithm) of the hydrogen-ion concentration (in gram atoms per liter) in a given solution, usually determined by the use of a substance (indicator) known to change color at a certain concentration. The pH scale in common use ranges from 0 to 14, pH 7 (the hydrogen-ion concentration, 10^{-7} or 0.0000001, in pure water) being considered neutral; 6 to 0, increasing acid; and 8 to 14, increasing alkali.

phagocytes   From the Greek *phago* meaning "eat" and *kytos* meaning "cell." Defensive cells (leukocytes, or white blood cells) of the body that ingest and destroy bacteria and other infectious agents. See also **macrophages** and **microphages.**

phagocytosis (phagocytizing)   The engulfing of microorganisms, cells, or foreign particles by phagocytes (certain forms of leukocytes).

pharyngeal   Of or pertaining to the pharynx.

pharynx   The tube, or cavity, that connects the mouth and nasal passages with the esophagus (throat).

phenocopy   A phenotype determined by environment that mimics the same phenotype produced by heredity (genotype).

phenotype   Expression of genes that can be measured by the human senses. What is seen in an animal for some trait. For simply inherited traits such as coat color in Holsteins, either black and white or red and white is seen.

phenotypic   Pertains to the visible expression of genes that can be measured by human senses in domestic animals. Phenotypic characteristics reflect an interaction of the **genotype** and the environment.

pheromone   A substance secreted externally by certain animal species (especially insects) to affect the behavior (especially sexual) or development of other members of the species. The queen substance of honeybees that inhibits ovary development in workers is a pheromone. Also called *assembling scent* and *sex pheromone* in insects.

phoresy   That form of symbiosis in which one symbiont rests on or attaches to another for means of transportation.

phosphatide   See **phospholipid.**

phospholipid   A lipid containing phosphorus that on hydrolysis yields fatty acids, glycerin, and a nitrogenous compound. Lecithin, cephalin, and sphingomyelin are examples. Also called *phosphatide.*

photoperiodism   The physiological response of animals and plants to variations of light and darkness.

physiological fuel value (PFV)   Units (expressed in calories) of food energy in human nutrition. It corresponds to **metabolizable energy** as related to domestic animals.

physiological saline   A salt solution (0.9% NaCl) having the same osmotic pressure as blood plasma.

physiology   The science that pertains to the functions of organs, systems, and the whole living body.

phyto-   A prefix meaning "pertaining to plants."

phytoplankton   Free-living plants in a body of water, usually **algae.** Serves as food for sea animals.

pica   A craving for unnatural articles of food, such as is seen in hysteria, pregnancy, and phosphorus deficiency. A depraved appetite.

pico-   A prefix; divides a basic unit by 1 trillion (10^{12}). Same as *micromicro.*

piebald   A horse having a black coat with white spots.

pig   A young swine weighing less than 120 lb. (See Appendix A.)

pigeon-toed   Designating a horse or other animal whose feet (toes) turn inward.

piglet   A young pig.

pincers   The incisor teeth of a horse. Also called *nipper*s.

pinocytosis   The absorption of liquids by cells.

pinto   Designating a horse that has a spotted or piebald coat color (white spots on any dark background).

pipped egg   An egg through which the chick has forced its beak in the first step of breaking out of the shell during incubation.

pithing   A method of animal slaughter in which the spinal cord is severed to cause death and/or to destroy sensibility.

Pituitrin   Trademark for posterior pituitary injection (oxytocin).

placebo   In Latin means "I shall please." An inactive substance or preparation given to please or gratify a patient. Also used in controlled studies to determine the **efficacy** (virtue) of medicinal substances.

placenta   The organ joining mother and offspring during pregnancy. It provides endocrine secretions and selective exchange of nutrients to the offspring and carries waste products away from the offspring.

placentitis   Inflammation of the placenta.

plain   A term suggesting general inferiority; coarse, lacking the desired quality.

plasma   The liquid portion of blood or lymph in which the corpuscles or blood cells float.

plasmolysis   A process causing water to leave the cell (contraction of protoplasm), as in the use of high concentrations of nontoxic salts and sugar for bacteriostatic purposes.

pleasure horse   Horse used for riding, driving, or racing.

pleiotropism   The action of one gene on two or more traits.

pleiotropy   The state in which one gene affects two or more traits.

PLM   The acronym indicating protein, lactose, and minerals of milk.

poikilothermal   Having a variable body temperature or one which corresponds to the environment; capable of enduring marked variations of cold and heat; pertaining to or characterized by **poikilothermy.**

poikilotherms (ectotherms)   Cold-blooded animals; animals having a body temperature that varies with the environment. Ocean fish exemplify cold-blooded species.

poikilothermy   The ability of animals to adapt themselves to variations in their environmental temperature. The quality of varying the body temperature with the environmental temperature. Most invertebrates, fish, amphibians, and reptiles are poikilotherms.

polled (polledness)   A naturally hornless animal.

polyculture   Raising mixed species of fish in the same culture unit.

polydipsia   An excessive thirst.

polygastric   Possessing more than one (many) stomach or stomach compartments; characterizing the cow and other ruminants.

polygenic (multiple-factor) inheritance   Inheritance involving a series of independent genes; characterized by the trait showing a continuous distribution pattern owing to the additive effect of genes, none of which show dominance.

polymer   A large molecule composed of repeating smaller units.

polyneuritis   Inflammation of many nerves concurrently.

polyp   A smooth, stalked, or projecting growth from a mucous membrane.

polyploidy   Complete duplication of all sets of chromosomes giving $3n$, $4n$, etc., numbers of homologous chromosomes in body (somatic) cells.

polypnea   A condition in which the respiration rate is increased; rapid, shallow breathing.

polyspermy   Entrance of many (poly) spermatozoa into the ovum at the time of fertilization.

polyunsaturated   Fatty acids having multiple double bonds within the carbon chain.

pony   A small horse, commonly less than 58 in tall at maturity.

porcine   Pertaining to swine.

porker   A young hog (pig).

portal system   The system of blood vessels conveying blood from the digestive organs and spleen to the liver.

posterior   Denotes the back or back portion (**caudal**). It means the same as **dorsal** surface of the body in human anatomy.

postlegged   Describing an animal (especially a horse) with too much set in the hocks, resulting in the hind legs being too straight.

postmortem   An examination of an animal carcass or human body after death. **Necropsy, autopsy.**

postnatal   Occurring after birth.

postnatum   Related to occurring after or an event following birth.

postpartum   Occurring after birth of an offspring.

pot-bellied   Designating any animal that has developed an abnormally large abdomen. Also a breed of miniature swine.

poult   An immature turkey. After the sex can be determined, the turkey is called a *young* **tom** (male) or *young* **hen** (female). (See Appendix A.)

poultry   Birds raised for meat and eggs.

power clusters   Groups seeking public backing of legislative and other governmental actions that enhance/support stakeholder interests.

ppm   Parts per million (1 mg/liter).

precipitin   Antibody that forms a precipitate with its soluble antigen. An antibody formed in blood serum as a result of inoculating with a foreign protein.

preconditioned   Prepared for weaning and movement to a feedlot environment commonly by starting on dry feeds and the administration of a prescribed set of inoculations (vaccines).

precursor   A compound or substance from which another is formed.

predatism   Intermittent parasitism, such as the attacks of mosquitoes and bedbugs on humans.

predator   Any animal, including an insect, that preys on and devours other animals, e.g., a coyote or dog preying on sheep. Some predators, such as ladybugs, may be beneficial in that they kill and eat parasites.

predicted difference (PD)   A measure of a bull's ability to transmit milk-producing capacity to his daughters. The PD may be positive or negative, depending on whether the bull's daughters yield more or less milk than daughters of other bulls **(herdmates)** under the same conditions.

preen gland   See **oil gland.**

prehension   The seizing (grasping) and conveying of food to the mouth.

premortal   Existing or occurring immediately before death.

prepartum   Occurring before birth of the offspring. Before **parturition.**

prepotent   Designating an animal that transmits its own characteristics to its progeny to a marked or highly uniform degree.

prick   To pierce or cut the tail of a horse so that it will be carried higher.

primates   Humans, monkeys, and the great apes.

primiparous   Bearing or having borne only one young or set of young.

prodome   A symptom indicating the onset of a disease.

produce   A female's offspring. The *produce of dam* commonly refers to two offspring of one dam.

progeny   The offspring of animals.

progeny testing   Evaluating the genotype of an individual by a study of its progeny.

progestational   A phase of the estrous cycle (menstrual cycle) in which the corpus luteum is active and the endometrium is under its influence.

prolapse   Abnormal protrusion of a part or organ; displacement of an organ from its normal location.

prolapsed uterus   A condition in which the uterus is partially or completely turned inside out, usually following parturition.

proliferation (proliferate)   Growth by rapid *multiplication* of new cells.

prolific   Capable of producing abundant offspring.

prophase   The first stage of mitosis, characterized by chromosome shortening and thickening and the disappearance of nuclear membrane as well as by the appearance and polarization of spindle fibers.

prophylactic   A preventive, preservative, or precautionary measure that tends to ward off disease.

prophylaxis   The prevention of disease.

prostaglandins   A large group of chemically related 20-carbon hydroxy fatty acids with variable physiological effects in the body. (See Chapter 11.)

prostate   Gland in the male reproductive system that lies just below the bladder and surrounds part of the canal (urethra) that empties the bladder.

protean   Variable; readily assuming different shapes or forms; changeable.

protective antibodies   Antibodies that when combined with pathogenic organisms render them noninfectious.

protein   A substance composed of amino acids, containing about 16 percent (molecular weight) nitrogen. Thus protein content is computed by multiplying the chemically determined value for nitrogen by the factor 6.25 ($N \times 6.25$).

protein efficiency ratio (PER)   The weight gained by growing experimental animals divided by the weight of the protein consumed, for example, the PER for casein is 2.5 (i.e., for each 1 g of casein ingested, test animals gain 2.5 g in body weight). The caloric intake must be adequate and the concentration of dietary protein must be adequate but not excessive, because gain is not proportional to intake at high levels of dietary protein. Adjusted PER = (PER of test food $\times$ 2.5)/(PER of standard reference casein).

protein equivalent   A term indicating the total nitrogenous contribution of a substance in comparison with the nitrogen content of protein (usually plant protein). For example, the nonprotein nitrogen **(NPN)** compound urea contains approximately 45 percent nitrogen and has a protein equivalent of 281 percent (6.25×45 percent).

protein-fortified   Describing low-fat and skim milks that contain at least 10 percent nonfat milk solids **(NMS).** When milk derivatives other than **NDM** are used to satisfy the 10 percent requirement, the protein added must be milk protein and must equal or exceed the quantity that would be added if the additive were NDM.

protein supplements   Feed products that contain 20 percent or more protein.

protozoa (protozoan)   A microscopic animal that consists of a single cell.

proved sire   A sire whose transmitting ability has been measured by comparing the production performance of his daughters with that of the daughter's dam and/or herdmates under similar conditions. See **herdmates.**

proximal   Nearest; closer to any point of reference; opposite to **distal.**

proximate analysis   Also known as the *Weende analysis* (developed in 1895 at the Weende Experiment Station in Germany); used to determine the gross composition of feed.

psychobiology   That branch of biology which considers the interactions between body and mind in the formation and functioning of personality; the scientific study of the personality function.

psychro-   From the Greek *psychros* meaning "cold." The prefix denoting relations to cold.

psychroenergetics   Science dealing with the effect of ambient temperature and humidity on conversion of feed into bodily heat and energy.

psychrometer   An apparatus for measuring atmospheric moisture by the difference in readings of two thermometers (one dry bulb and one wet bulb).

psychrometric   (adj. of psychrometry) Pertaining to the use of a hygrometer consisting essentially of two similar thermometers with the bulb of one being kept wet so that the cooling that results from evaporation makes it register a lower temperature than the dry one, and with the difference between the readings constituting a measure of the dryness of the atmosphere.

puberty   The age at which the reproductive organs become functionally operative and secondary sex characteristics develop.

pubic   Pertaining to the pubes (hair growing over pubic area) or pubic bones. The lower part of the hypogastric region.

public health   An organized effort to prevent disease, prolong life, and promote physical and mental efficiency. Also, the health of the community taken as a whole.

pudic   Pertaining to the external genital parts, especially of the female.

pullet   A female chicken less than a year old.

puncher   One who herds cattle; a cowboy.

pupa   The quiescent or inactive stage during which an immature insect or larva transforms into an adult.

pupal stage   Period in the life history of insects between the caterpillar, or grub, stage and the mature, or adult, insect.

pupate   To change from an active immature insect into the inactive pupal stage.

purebred   An animal of a recognized breed that is eligible for registry in the official herdbook of that breed.

purebreeding (truebreeding)   Breeding a stock that is **homozygous** for one or more characteristics.

pure culture   A population of microorganisms that contains only a single species. Cultures are useful in the manufacture of many animal products, e.g., cheeses and yogurt.

purified diet   A mixture of the known essential dietary nutrients in a pure form that is fed to experimental (test) animals in nutrition studies.

pus   A liquid inflammatory product consisting of leukocytes, lymph, bacteria, dead tissue cells, and fluid derived from their disintegration.

putrefaction   The bacterial decomposition of proteins.

pyrexia   A fever or febrile condition. An abnormal elevation of body temperature.

Q

qualitative traits (qualitative inheritance)   Those traits, such as black and white or polled and horned, in which there is a sharp distinction between **phenotypes.** Usually only one or two pairs of genes are involved.

quality   A term indicating fineness of texture as opposed to coarseness. It commonly is used to indicate relative merit, e.g., superior breeding or genetic merit.

quantitative traits (quantitative inheritance)   Those traits, such as skin color in humans, in which there is no sharp distinction between phenotypes, with a gradual variation from one phenotype to another. Usually several genes as well as environmental factors are involved.

quarantine   Commonly thought of as the segregation of the active case of an infectious disease, but more technically, it includes compulsory segregation of exposed susceptible animals or individuals for a period of time equal to the longest usual incubation period of the disease to which they have been exposed. A regulation under police power for the exclusion or isolation of an animal to prevent the spread of an infectious disease.

R

rabies   An infectious viral disease of the central nervous system usually fatal in mammals. Early symptoms include fever and hyperexcitability followed by paralysis of the muscles used in swallowing, progresses to convulsions or paralysis and death.

rack   The gait of a horse in which only one foot touches the ground at any one time, producing a four-beat gait. The legs move in lateral pairs but not quite in unison, so that each foot is lifted and put down alone.

rad   Another name or unit for "radiation energy absorbed" by food being processed with radiation. 1000 rad = 1 kilorad = 10 **gray;** 1,000,000 rad = 1 Mrad = 10 kGy; (the rad is being superseded by the **gray**).

radappertization   Sterilization by radiation processing. The resulting processed food can be stored at room temperature, in the same way as thermally sterilized foods (canned foods). Precooked food in hermetically sealed packaging is exposed to radiation at levels high enough to kill all organisms of food spoilage and/or of public health significance. Doses used are typically greater than 1 Mrad.

radiant energy   Energy that is being transferred through space by electromagnetic waves.

radiant heat   Heat transmitted by radiation (such as that of the sun) as contrasted with that transmitted by **conduction** or **convection.**

radiation   The process of emitting radiant energy (heat) in the form of waves or particles. The sun transfers its energy by radiation. It is also an important method of heat loss from an animal to cooler objects and heat gain by the animal from warmer objects.

radiation sterilization   The use of radiation to cause a plant or animal to become sterile or incapable of reproduction. Also, the use of radiation to kill all forms of life (especially bacteria) in food and on equipment.

radicidetion (radicidized)   Radiation pasteurization intended to kill or render harmless all *disease-causing* organisms (except viruses and spore-forming bacteria) in food. Processing takes place at dose levels generally below 1 Mrad, and the processed foods usually must be stored under refrigeration (as in heat pasteurization).

radioactive   Of, caused by, or exhibiting radioactivity. The property possessed by some elements (as uranium) or isotopes (as carbon 14) of spontaneously emitting energetic particles.

radioactive dating   A technique for measuring the age of an object or sample of material by determining the ratios of various **radioisotopes,** or products of radioactive decay, it contains. For example, the ratio of carbon 14 to carbon 12 reveals the approximate age of bones, pieces of wood, or other archaeological specimens that contain carbon extracted from the air at the time of their origin.

radioactive isotope   An element of chemical character with radioactive properties. Such isotopes are produced by bombarding the element in a cyclotron. Especially important ones include those of carbon, iodine, iron, phosphorus, and sulfur; useful in medical and nutritional research.

radiograph   A record or photograph produced by **X rays** or other rays on a photographic plate, commonly called an *X-ray picture.*

radioisotope   A radioactive isotope (the nucleus of such a species of atom). A nuclide. Such isotopes occur naturally and may also be produced by bombardment of a common chemical element with high-velocity particles. A radioactive isotope transmutes into another element with emission of electromagnetic **particles.** More than 1300 natural and artificial radioisotopes have been identified.

radiology   The science that deals with the use of all forms of ionizing radiation in the diagnosis and therapy of disease.

radurization   Radiation pasteurization designed to kill or inactivate *food-spoilage* organisms, thus extending the shelf life of a given food product. Processing takes place at dose levels generally below 1 Mrad, and the product usually must be stored under refrigeration, as in the case of **radicidized** food.

ram   A male sheep. Also called a **buck.**

ram test station   A testing farm or facility where ram lambs are managed to compare their performance in growth and carcass traits.

random   All possible samples have equal probability of selection.

random mating   A system of mating where every male has an equal chance of mating with every female.

range   Large, open areas of grazing land.

rangy   Designating an animal that is long, lean, leggy, and not too muscular.

rate   Synonymous with *level, dosage, amount, quantity,* or *degree* measured in proportion to something else.

rate of passage   The time taken by undigested residues from a given meal to reach the feces. A stained undigestible material is commonly used to estimate rate of passage.

ration   The food allowed an animal for 24 h. A *balanced* ration provides all the nutrients required to nourish an animal for 24 h. See **balanced ration.**

rawhide   The usual American name, which has spread to other English-speaking countries, for cattlehide that has been dehaired but is usually unfinished. Some rawhide is tanned with the hair left on. It is used principally for mechanical purposes, such as belt facings and pins, gaskets, pinions and gears, and also for trunk binding and luggage.

raw milk   Fresh, untreated milk as it comes from the cow or another mammal.

raw wool   Wool prior to removal of the grease.

razorback   A type of hog with long legs and snout, sharp narrow back, and lean body; usually a half-wild mongrel breed (especially of the southern United States).

reactor   An animal that reacts positively to a foreign substance; e.g., a tuberculous animal would be a reactor to **tuberculin.**

recessive   Expressed only if the responsible **allele** is present on both members of a pair of **homologous chromosomes.** A recessive **allele** which is masked by a dominant allele is expressed only in the **homozygous** state.

recessive gene   A gene whose phenotypic expression is covered (masked) by its own dominant allele. For example, the blue-eyed gene *b* is recessive to the brown-eyed gene *B*, with *Bb* individuals having brown eyes. A *bb* individual would have blue eyes. Recessive genes appear to affect the phenotype only when present in a **homozygous** condition.

recombination   A formation of genotypes and phenotypes that are new combinations of the parents in a given cross.

reconstituted milk   The product that results from the recombining of milk fat and nonfat dry milk or dried whole milk with water in proportions to yield the constituent percentage occurring in milk.

red meat   Meat that is red when raw. Red meat includes beef, veal, pork, mutton, and lamb.

redia   A larval stage in the development of flukes. Redia of liver flukes of cattle, sheep, and goats are found in snails.

reduction   Chemically, the subtraction of oxygen from, or the addition of hydrogen to, a substance (or the loss of positive charges or the gain of negative charges). The atom or groups of atoms that lose electrons become oxidized.

reflectivity   The ratio of the rate of reflection of radiant energy from a given surface to the rate of incidence of radiant energy on it.

reflex   Action performed involuntarily in consequence of a nervous impulse transmitted from a receptor, or sense organ, to a nerve center.

regurgitation (regurgitate)   The casting up (backward flow) of undigested food from the stomach to the mouth, as by ruminants.

relative humidity (RH)   The ratio of the weight of water vapor contained in a given volume of air to the weight that the same volume of air would contain when saturated. The quantity of water vapor that air can hold when saturated increases with temperature. The RH is expressed as a percentage. For example, if a sample of air at a given temperature contains 30 percent of the water vapor that it is possible for it to contain at that temperature, it is 30 percent saturated and therefore has a relative humidity of 30 percent.

replacement   Animals destined to become members of the breeding herd or flock by replacing animals culled from the unit.

reservoir host (reservoir)   An animal that harbors the same species of parasite as humans. Also an animal that becomes infected and serves as a source from which other animals can be infected.

respiratory quotient (RQ)   The RQ is used to indicate the *type* of food being metabolized. This is possible because carbohydrates, fats, and proteins differ in the relative amounts of oxygen and carbon contained in their molecules. Also, the relative volumes of oxygen consumed and carbon dioxide produced during metabolism of each type of food vary. Respiratory quotient is calculated as follows:

$$RQ = \frac{\text{volume } CO_2 \text{ produced}}{\text{volume } O_2 \text{ consumed}}$$

response category   Class of homeokinetic response by an animal to an environmental stimulus.

retained placenta   A placenta that was not expelled at parturition.

reticuloendothelial system   A widely spread network of body cells concerned with blood cell formation, bile formation, and engulfing or trapping of foreign materials, which includes cells of bone marrow, lymph, spleen, and liver. Currently the preferred terminology is the monocyte–macrophage system.

reticulum   The second division of the stomach of a ruminant animal. Also called *honeycomb.*

retrogression   Degeneration, deterioration, or a backward movement.

reversion   Appearance of a trait in an individual that was possessed by remote ancestors but not by recent ones.

rhinovirus   An infectious viral disease affecting the respiratory system resulting in sneezing, coughing, fever, and depression; some strains also result in fetal death and abortion.

rickettsiae   Intracellular parasites, i.e., ones that multiply inside the living cells of other larger organisms. In size they are intermediate between bacteria and viruses.

ride   To mount and travel on a horse. To mount a cow, as another cow, indicative of **estrus (heat).** See **buller.**

rider   An animal that rides another of its kind, as when young bulls or boars ride other young bulls or boars. See **buller.**

ridgling   Any male animal whose testicles fail to descend normally into the scrotum. Also called **cyptorchid.**

rights   Something due an entity (e. g., a human or an animal) by virtue of nature, tradition, or law.

rights strategy   Belief that the value of an entity or action is determined by rights.

rigling   A male sheep or horse that has only one testicle in the scrotum. See **ridgling.**

rigor mortis   The stiffness of body muscles that is observed shortly after the death of an animal. It is caused by an accumulation of **metabolic** products, especially lactic acid, in the muscles.

ring test   A test for brucellosis performed by mixing stained *Brucella* bacteria with **raw milk.** If **antibodies** to *Brucella* are present, the stained cells **agglutinate** (clump) and rise to the surface with the cream to form a blue ring.

RNA   Ribonucleic acid.

roan   Designating the red-white color of Shorthorn cattle. Red or black coat color of a horse intermingled with white; may be red or strawberry roan, blue roan, or chestnut roan, depending on the intermingling of the background colors.

roaster   A young chicken (meat type) weighing more than 3.5 lb (and usually 4 to 6 months old).

roasting pig   A pig weighing from 15 to 50 lb.

robot   A device that automatically perfoms repetitive tasks. For example, an automated milking system that utilizes electronic sensors and computer systems to milk cows. A cow enters a stall equipped with the automated system and the machinery functions under computer control to attach the milking machine to the cow, complete milking, and then detach milking machine from the cow.

rodent   A classification of mammals, mostly vegetarians, characterized by their single pair of chisel-shaped, upper incisors (rabbits, rats, mice, squirrels).

rodenticide   Any poison that is lethal to rodents.

roe   The eggs or testes of fish. There are two types: the female eggs (hard roe) and the male testes (soft roe). They are considered a delicacy by many people.

roost   A resting or lodging place for fowls.

rooster (cock)   An adult male chicken.

roosting   Pertains to a group of winged animals settling down for rest or sleep, as in birds roosting together.

rotation   Crossbreeding system where sire breeds are used in a manner in which the daughters of a sire breed are mated to the next sire breed in a cycle (or rotation) for as many generations as there are breeds in the rotation at which time the sequence of sire breeds begins a new cycle (e.g., the planned use of three or more breeds in a rotational breeding program in swine).

roughage   Consists of pasture, **silage,** hay, or other dry **fodder.** It may be of high or low quality. Roughages are usually high in crude fiber (more than 18 percent) and relatively low in **NFE** (approximately 40 percent).

rugged   Refers to a large, strong animal.

rumen   The first stomach compartment of a ruminant; also called *paunch.* The rumen is a large nutrient-producing fermentation vat that contains an amount of feed and water equal to approximately one-seventh the mature ruminant's body weight.

rumen flora   The microorganisms of the rumen.

ruminant   One of the order of animals that has a stomach with four complete cavities—rumen, reticulum, omasum, abomasum— through which food passes in digestion. These animals chew their cud; they include cattle, sheep, goats, deer, antelopes, elk, and camels.

ruminate   See **rumination.**

rumination   The casting up of food (cud) to be chewed a second time, as in cattle. A chewing of the cud, as by ruminants.

running horse   Any race horse (e.g., a Thoroughbred).

run on   To graze or pasture on, as for cattle to run on the range.

runt   A term commonly used to denote a piglet of small size in relation to its littermates. Runts usually result from a shortage of milk in one or more teats of the sow.

rupture   The forcible tearing or breaking of a body part. Also used to refer to a defect in the body wall. See **hernia.**

rustle   To hunt for food, especially with reference to domestic animals fed inadequately by their manager/owner. To steal livestock.

S

sagittal   Anteroposterior plane or section parallel to the long axis of the body.

salmonid   Fishes related to salmon and trout.

salpingitis   An inflammation of a fallopian tube (oviduct).

saprophyte   Any vegetative organism, such as a bacterium, living on dead or decaying organic matter.

satiety   Full satisfaction of desire; may refer to satisfaction of sexual arousal, appetite.

saturated fat   A completely hydrogenated fat; that is, each carbon atom is associated with the maximum number of hydrogen atoms.

saturated fatty acid   A carboxylic acid in which all of the carbons in the chain are separated by a single bond. Fatty acids are completely hydrogenated (i.e., each carbon atom is associated with the maximum number of hydrogen atoms). Palmitic and stearic acids are examples.

saturates   Molecules that contain no double bonds within their carbon chains.

scale   The size of an animal.

schistosomiasis   Infestation with a schistosome, or blood fluke.

sclera   The tough, white, supporting covering of the eyeball, which encompasses all the eyeball except the cornea.

scours   A persistent diarrhea in animals.

scrub   An animal inferior in breeding and/or individuality.

scurs   Small, rounded portions of horn tissue attached to skin at the horn pits of polled animals; also called *buttons.*

SDA (specific dynamic action)   The increased production of heat by the body as a result of a stimulus to metabolic activity caused by ingesting food.

SE   See **starch equivalent.**

seasonally polyestrous   Describes the tendency of some species, or some breeds within species, to have multiple estrous cycles primarily during only one season of the year. For example, ewes of the Dorset breed cycle primarily in the fall months.

sebum   The thick, semifluid substance composed of **lipid**s and epithelial debris secreted by the sebaceous glands.

secondary infection   Infection following an infection already established by other organisms.

second filial generation   See **F$_2$ generation.**

second meiotic division   The second of two divisions occurring in reductional cell division and resulting in the production of two cells, each of which is haploid, the chromosomes occurring singly (nonpaired).

seed tick   The newly hatched six-legged larva of a tick, especially of the cattle tick *Boophilis annulatus,* a one-host tick in which the larva, the nymph, and the adult are all found on cattle. Newly hatched larvae are found on the ground or on grass, weeds, and other objects in fields where infested cattle have grazed.

segregate   In animal genetics, segregation is the separation of the two genes of a pair in the process of maturation so that only one goes to each germ cell.

segregated early weaning (SEW)   Removing piglets from their dam and separating them from all but other early weaned pigs in biosecure facilities in order to reduce transmission of disease to the piglets from other members of the swine herd.

segregation of genes   This refers to the occurrence of genes in pairs in body cells, for example, *Aa.* However, when such an individual produces **gametes,** only one of these genes, either *A* or *a,* not both, goes into a single sex cell. Thus, although they are together in body cells, they segregate, or separate from each other, when gametes are formed.

selection (selective, selected)   The causing or allowing of certain individuals in a population to produce the next generation. *Artificial selection* is that practiced by humans; *natural selection* is that practiced by nature.

self-feeding   Any feeding device by means of which animals can eat at will. See *ad libitum.*

senescence   The process or condition of growing old. Aging. (See Chapter 12.)

sensible   Perceptible; as sensible heat loss (water) or weight loss (liquids or solids).

sensory   Pertaining to sensation. The eyes and ears are *sensory organs. Sensory nerves* convey impulses from the sense organs to a nerve center. Thus some nerves are *sensory* and pick up sensations from sense organs and carry them to main cords and the brain, whereas others are *motor* and carry impulses from the brain and main nerves to the muscles, which respond to the stimulation.

sentience   Ability to consciously experience or feel an internal or external event.

sentient experiences   Condition of an animal's having consciously felt an internal or external event.

septicemia   Blood poisoning which results from the presence of toxins or poisons of microorganisms in the blood.

serological   Pertaining to the use of blood serum of animals in various tests, which aids in detecting and treating certain diseases.

serotype   The type of microorganism as determined by the kind and combination of constituent antigens associated with the cell.

serum   The clear portion of animal fluids, separated from its cellular elements. Blood serum is the clear, pale-yellow, watery portion of blood that separates from the clot when blood coagulates.

serum therapy   The treatment of clinical cases of disease with serum of immunized animals.

service   A term used in animal breeding, denoting the mating of a male to a female. Also called *serving,* or *covering.* (See Appendix A.)

setting hen   A broody hen in the act of incubating eggs.

settled   A term commonly used to indicate that the animal has become pregnant.

sex chromosomes   One pair of chromosomes in an individual that determines the sex of that individual. In mammals, the female is XX and the male is XY. The X chromosome is considerably longer and carries more genes than the Y chromosome.

sex-influenced traits   Such traits are due to genes carried on autosomes; however, the gene is dominant in males and recessive in females. For example, the gene for baldness, *Ba,* in humans is a sex-influenced gene. Its allele is *Bn,* for nonbaldness phenotype.

Genotype	Men	Women
BaBa	Bald	Bald
BaBn	Bald	Not bald
BnBn	Not bald	Not bald

sex-limited traits   The appearance of such traits is limited to only one sex, for example, egg laying in hens and lactation in cows. Nevertheless, males of these species possess genes for these traits, even though they are not expressed phenotypically.

sex linkage   Refers to genes carried on the nonhomologous portion of the X chromosome. For recessive sex-linked genes, two are required to express the trait in females and one in males. A gene carried on the nonhomologous portion of the Y chromosome is always transmitted from father to son. It is referred to as *holandric inheritance.* Sex-linked genes are alleles, then, that have their loci on the sex chromosomes, usually only on the X chromosomes.

sex-linked   See **sex linkage.** Refers to a trait that is affected by sex linkage.

sex-linked inheritance   See **sex linkage.**

shear   To cut wool or hair from sheep, goats, and other wool-producing animals.

shelf life   The time after processing during which a product remains suitable for human consumption, especially the time a food remains palatable and safe.

shoat (shote)   A young pig of either sex less than 12 months old. (See Appendix A.)

shy breeder   A male or female of any domesticated livestock that has a low reproductive efficiency.

sibling   In genetics, a brother or sister.

sickle-hocked   Designating a horse, cow, or sheep having a crooked hock, which causes the lower part of the leg to be bent forward out of a normal perpendicular straight line.

silage (ensilage)   Prepared by chopping green **forage,** such as grass or clover, or **fodder,** such as field corn or sorghum, and blowing it into an airtight chamber **(silo),** where it is compressed so that air is excluded and it undergoes an acid fermentation (produces lactic and acetic acids) that retards spoiling. It usually contains 65 to 70 percent moisture.

silo A pit, trench, aboveground horizontal container, or vertical cylindrical structure of relatively airtight construction into which chopped green crops, such as corn, grass, legumes, or small grain and other livestock feeds are placed and allowed to partially ferment into silage. See **silage,** also.

sinistral Of or pertaining to the left side; left; or left-handed.

sire The male parent. To father or to beget.

sire indexes Various means of calculating the abilities of bulls to transmit economically important production traits. See also **USDA sire summary.**

sire summary See **USDA sire summary.**

skewbald A horse of any color except black, with white spots.

slip To abort. An incompletely castrated male.

slow gait One of the several forward movements, or gaits, of horses, faster than a walk but slower than a canter. There are three slow gaits: the running walk, the fox trot, and the slow pace.

slunk The skin of an unborn or prematurely born calf.

smooth mouth The mouth of a horse whose teeth have lost their natural cusps and have become smooth by use and wear, generally indicating that the horse is 10 or more years of age. (See page 114.)

SMR See **standard metabolic rate.**

SNF Solids–not fat of milk (proteins, **lactose,** and minerals). Same as **NFS** and **NMS.**

snood The fleshy appendage that emanates from the area of the base of the upper beak in the turkey.

social insect Any insect that lives with others of its kind in a somewhat organized colony, such as ants, bees, and wasps.

sodium nitrate A preservative used to prevent the germination of spores of *Clostridium botulinum* and to stabilize the color in cured meats. The nitrogen atom is less oxidized than that of nitrate.

sodium nitrite A preservative used in cured meats to complement the effect of nitrites. The nitrogen atom is more oxidized than that of nitrite.

soft rays The section of a fish's fins where the rays do not contain a bony rod. These cartilaginous structures provide support for the fins of fish; fins of freshwater fish have soft rays composed of cartilage and dense connective tissue while the majority of saltwater fish have hard rays composed of cartilage and bone.

soilage Freshly cut green forage fed to animals in confinement. Also called **green chop.**

soiling A term previously used for the green chopping of forages.

soluble Designating a substance that is capable of being dissolved in another.

somatic Refers to body tissues; having two sets of chromosomes.

sorrel A coat color of horses. It includes the red shades of chestnut or yellowish brown.

sow Mature female swine. (See Appendix A.)

sowbelly Salt pork; unsmoked fat bacon.

space spray Insect spray used in insect control in open spaces, e.g., in a dairy barn or hog house.

span (spann) A pair of animals usually harnessed together as a team.

spay To surgically remove the ovaries of a female.

species A group of animals having several common characteristics that differentiate them from others.

specific dynamic action See **SDA.**

sperm (sing. **spermatozoon,** pl. **spermatozoa**) A mature male germ cell.

spermatogenesis The formation and development of spermatozoa.

SPF Specific pathogen-free.

sphincter A ring-shaped muscle that closes an opening, e.g., the sphincter muscles in the lower end of a cow's teat.

spinnbarkeit The formation of a thread by cervical mucus when blown onto a glass slide and drawn out by a cover glass; the time at which it can be drawn to the maximum length usually precedes or coincides with the time of ovulation in women.

splayfooted See **toe out.**

split hide The outer (hair or grain) layer of a hide from which the under, or flesh, side has been split to give it a reasonably uniform thickness.

spool joint The exposed surface of the end of the metacarpus in the leg of a lamb carcass.

spontaneous Instinctive (performed apparently without the exercise of reason) and occurring without external influence.

spore From the Greek *spores* meaning "seed." A single cell that becomes free and is capable of developing into a new plant or animal. The reproductive element of some lower organisms. It does not contain a preformed embryo, as do seeds.

springer A term commonly associated with female cattle showing signs of advanced pregnancy.

stable A building used for the feeding and lodging of horses and other livestock.

stable isotope An **isotope** that does not undergo radioactive decay.

stag A male animal castrated after the secondary sex characteristics have developed sufficiently to give it the appearance of a normal mature male.

stale A period when an animal does not work or lactate at normal standards, as opposed to **bloom.**

stallion A mature male horse, not castrated. (See Appendix A.)

stance Position, or posture, adopted when an animal is stationary.

standard metabolic rate (SMR) Reflects an animal's basic **maintenance energy (ME)** requirement. It is useful and important in studies of thermal physiology and productive efficiency to have such a reference metabolic rate. Because metabolic rate increases during thermal stress, an animal's reference metabolic rate should be measured in the **thermoneutral zone** of effective environmental temperature. Moreover, because metabolic rate increases postfeeding (due to the **heat increment of feeding**), the reference value should be determined sometime after the animal has absorbed its last meal. Additionally, because physical activity increases metabolic rate, to be meaningful the reference value must reflect metabolic rate when the animal is resting. Thus, the SMR takes these three conditions into account and is said to occur in a fasting, resting animal held in thermoneutral surroundings. Standard metabolic rate is based on the 0.75 power of body weight, the value commonly called **metabolic body size.** By means of SMR, comparisons can be made among animals of different sizes and species. In human physiology, SMR is called basal metabolic rate (BMR) (see **basal metabolism**). In animal science literature, standard metabolic rate is synonymous with *fasting metabolic rate* and with *resting metabolic rate.*

starch equivalent (SE)   A net-energy system of feed evaluation that is used extensively in Germany and other European countries. One kilogram of SE is equivalent to 2356 kcal net energy for fattening (NE_f). The system is based on research by Kellner and his successors. SE values are calculated on the basis of digestible nutrients and crude fiber of the diet.

starvation   The deprivation of an animal of any or all the food elements necessary to its nutrition, health, and well-being.

state of being   State of an animal's conditions of life. (See Chapter 7.)

steer   A male bovine castrated before the development of secondary sex characteristics. (See Appendix A.)

sterility   Barrenness; inability to produce young.

sterilization   The process of killing or removing *all* living organisms from a substance or an article.

sterilize   To remove or kill all living organisms. To render an animal infertile.

sterol   Any of a group of high-molecular-weight alcohols, as ergosterol and cholesterol.

stillborn   Born lifeless; dead at birth.

stocker (stock cattle)   Commonly young **steers** or cows that are light weight, thin, and lack **finish.** Also refers to growing animals fed relatively inexpensive, lower energy feeds (such as corn **stover** or late season pastures) to put on a lean but lower-cost weight gain.

stool   Fecal material; evacuation from the digestive tract.

stover   Fodder; mature, cured stalks of grain from which the seeds have been removed such as stalks of corn without ears.

straggler   An animal that strays or wanders from a herd or flock.

strain   A group of animals within a breed differing in one or more characteristics from other members of the breed.

stress   The sum of all nonspecific biological phenomena caused by adverse conditions or influences. It includes physical, chemical, and/or emotional factors to which an individual fails to make satisfactory adaptation which results in physiological tensions that may contribute to disease and/or poor performance.

stress responses   The homeokinetic response by an animal to an environmental stimulus which is inadequate to compensate for that environmental impingement.

stride   The distance from one footprint of a horse to the print of the same foot when it next comes fully to the ground.

strobilation   An asexual form of reproduction in which segments of the body separate to form new individuals, as in tapeworms.

stud   A unit of selected animals kept for breeding purposes, as of bulls and horses. Abbreviation for *stud horse:* a stallion. In artificial insemination, a **stud** refers to a semen-producing business in which any individual or business entity owns or leases one or more boars, bulls, stallions, or males of other species from which the individual or business entity collects, processes, and distributes semen for use in the insemination of animals owned by others. The stud is maintained as a service and a "for profit" enterprise.

sty   A pen in which swine are fed and housed.

subclinical   A disease condition without clinical manifestations.

subcutaneous (subcutaneously)   Situated or occurring beneath the skin.

sublimation   The process of sublimating or subliming; the direct transition from solid to vapor (bypassing the liquid form). The process of vaporizing and condensing a solid substance without melting it.

substrate   A substance on which cells or organisms may live and be nourished.

succulence   A condition of plants characterized by juiciness, freshness, and tenderness, making them appetizing to animals.

suckle   To nurse at the breast or mammary glands.

suckling   A young, unweaned animal.

suede finish   A finish produced by running the surface of leather on an abrasive to separate the fibers in order to give the leather a velvetlike nap. The term denotes a finish, not a type of leather.

suffering   See **animal suffering.**

superfetation   Second impregnation of a female that is already pregnant.

superovulation   The use of follicle-stimulating hormone (FSH) or fertility drugs to superstimulate follicular development so more than the normal number of follicles mature and ovulate (rupture and release eggs). These ova can be fertilized *in vivo* (within the uterus) or *in vitro* (e.g., in a test tube) before being transferred to the recipient female(s) or being frozen for future transfer.

supplement (supplemental)   Refers to the addition of minerals, vitamins, or other minor ingredients (bulkwise) to a diet.

supra-   A prefix meaning on, above, over, or beyond.

swarm   The simultaneous emergence or assembly in one location of large numbers of insects (especially bees), often to establish a new colony.

sweet butter   Unsalted butter.

swirl   Hair that grows in a whorl on an animal.

switch   The brush of hair on the end of a bovine tail.

symbiosis   The living together in intimate association of two dissimilar organisms, with a resulting mutual benefit.

symbiotic   (adj. of **symbiosis**) Associated in *symbiosis;* living together in a mutually beneficial relationship.

synapsis   The pairing of a homologous set of chromosomes during the first meiotic division. The chromosomes occur as paired chromatids joined at the centromere.

syndrome   A group of signs and symptoms that occur together and characterize a disease; a disturbance or abnormality.

syngamy   Fusion of identical **gametes.**

synovia (synovial fluid)   A viscid fluid containing synovin, or mucin, and a small proportion of mineral salts. It is secreted by the synovial membrane and resembles the white of an egg. It is contained in joint cavities, bursae, and tendon sheaths.

T

tachycardia   Excessive rapidity in the action of the heart (pulse rate above 100/min in humans).

tack   Riding equipment, such as the bridle and saddle. Also refers to equipment used in the fitting and showing of animals.

tactile   Pertaining to the touch.

tag   See **tracer isotope.**

take   To accept a male in coitus. To result in a mild infection after vaccination.

tallow   The fat extracted from adipose tissue of cattle and sheep.

tanbark trail   A term commonly associated with those who exhibit animals in competition at fairs and shows. *Tanbark* is the bark of several trees (e.g., oak, chestnut) used as a ground covering in circus lots, racetracks, livestock pavilions.

tankage   A **protein supplement** used as an animal feed. It consists of ground meat and bone by-products of animals that have been slaughtered.

tanning The processing of perishable rawhides and skins into the permanent and durable form of leather by the use of tanning materials.

TDN See **total digestible nutrients.**

teart Molybdenosis of farm animals caused by feeding on vegetation grown on soil that contains high levels of molybdenum.

tease To stimulate an animal to accept coitus (e.g., in horses).

teg A sheep 2 years of age.

telophase The fourth stage of mitosis, characterized by elongation of chromosomes, disappearance of spindle fibers, and reorganization of nuclear membrane.

temperament Disposition of an animal.

tend To care for, as a flock of sheep.

tendon The strong tissue terminating a muscle and attached to a bone, affording leverage.

term The gestation period of domestic mammals.

test cross Mating involving a recessive phenotype; used to determine heterozygosity of a stock.

tetany A condition in an animal in which there are spasmodic muscular contractions.

tether To tie an animal with a rope or chain to allow grazing but prevent straying.

tetrad A unit of four chromatids formed as a result of synapsis of homologous chromosomes, each of which consists of a pair of identical chromatids joined at the centromere.

therapeutic Curative, used in healing.

therapeutically Used for a curative purpose (as in the treatment of a medical disorder).

therapy The treatment of disease. Curative.

therm See **megacalorie.**

thermal elements Include temperature, humidity, air movement, and radiant heat.

thermocouple A device consisting essentially of two conductors made of different metals, joined at both ends, producing a loop in which an electric current will flow when there is a difference in temperature between the two junctions.

thermogenesis The chemical production of heat in the body.

thermolysis The loss or dissipation of body heat.

thermoneutrality The state of thermal balance between an organism and its environment so that the body thermoregulatory mechanisms are inactive. The **thermoneutral zone** is also referred to as the **comfort zone.**

thermoneutral zone The relatively narrow zone of effective environmental temperature in which heat production at the animal's minimal or thermoneutral rate is offset by net heat loss to the environment without the aid of special heat-conserving or heat-dissipating mechanisms. Thus, the animal is under neither cold nor heat stress. See also **comfort zone.**

thorax The chest.

Thoroughbred The name of the English breed of running horses.

three-breeding See **three-way cross.**

three-way cross The practice of crossbreeding food-producing animals (e.g., swine) where three breeds are crossed to increase the amount of **hybrid vigor** that is expressed. A system of **rotation** breeding involving males of three different breeds.

threshold The level or point at which a physiological effect becomes evident as a result of stimulation.

throw To cause an animal, as a horse or cow, to fall to the ground before branding or treating. To abort an embryo or fetus.

thumps An animal ailment resembling hiccups in humans that is seen, for example, in anemic baby pigs.

thymus Glandlike organ in the upper part of the chest that reaches its maximum development during late childhood in humans. It is associated with immune function.

thyroid Gland in the neck that helps regulate many processes of growth and development.

tick Any of the various bloodsucking arachnids that fasten themselves to warm-blooded animals. Some are important **vectors** of diseases.

titer The quantity of a substance required to produce a reaction with a given volume of another substance, or the amount of one substance required to correspond to a given amount of another substance. *Agglutination titer* is the highest dilution of a serum that causes clumping of bacteria.

tobiano A color pattern of Pinto horses. The horse appears to be white with large spots of color often overlapping on the chest, flank, and buttock; legs are generally white.

toe out To walk with the feet pointed outward. Also called *splayfooted* or *slew-footed.*

tom A male turkey. (See Appendix A.)

tonicity The state of tension or partial contraction of muscle fibers while at rest; normal condition of muscle tone.

total digestible nutrients (TDN) A standard evaluation of the usefulness of a particular feed for farm animals that includes all the digestible organic nutrients: protein, fiber, nitrogen-free extract, and fat.

toxemia Generalized blood poisoning, especially a form in which the toxins produced by pathogenic bacteria enter the bloodstream from a local lesion and are distributed throughout the body.

toxins Poisons produced by certain microorganisms. They are products of cell metabolism. The symptoms of diseases caused by bacteria, such as diphtheria and tetanus, are due to toxins.

toxoid A detoxified toxin. It retains the ability to stimulate formation of antitoxin in an animal's body. The discovery that toxin treated with formalin loses its toxicity is the basis for preventive immunization against such diseases as diphtheria and tetanus.

tracer isotope An isotope of an element, a small amount of which may be incorporated into a sample of material (the carrier) to follow (trace) the course of that element through a chemical, biological, or physical process and thus also follow the larger sample. The tracer may be radioactive, in which case observations are made by measuring the radioactivity. Tracers are also called *labels* or *tags,* and materials are said to be labeled or tagged when radioactive tracers are incorporated in them.

trachea The windpipe; in mammals it extends from the throat to the bronchi.

transduction The transfer of genetic material from one cell to another when mediated by a bacteriophage.

transgenic animal An animal into which cloned genetic material has been transferred; created artificially from two or more sources and incorporated into a single recombinant molecule.

transitory Brief; momentary; lasting only a short time; fleeing; transient.

translocation The attachment of a fragment of one chromosome to another that is not homologous to it.

translucent Transmitting light, but diffusing it so that objects beyond are not clearly distinguished.

Trematoda A class of the Platyhelminthes, which includes the flukes.

trematode Any parasitic animal organism belonging to the class Trematoda.

tremor An involuntary trembling or quivering.

trihybrid An individual that is **heterozygous** for three pairs of alleles, such as *AaBbCc*.

trimester A period of 3 months.

tripe Beef consisting of the walls of the rumen and reticulum, used as food for people.

triple-purpose An early production term defining animals that were bred to optimize the output of three products, such as cattle in the early American colonies bred for structural characteristics allowing them to pull a plow or wagon while also providing significant production of meat and milk. It is generally considered that performance is not maximized for individual traits because of the emphasis placed on three different production selection criteria.

trophoblast The enveloping layer of cells of the early embryo that will attach the ovum to the uterine wall and supply nutrition to the embryo.

tropism The tendency of an organism (plant or animal) to react (turn or move) in a definite way in response to external stimuli.

trots A diarrheal, or abnormally loose, condition of the bowels.

true albino Solid white animal with pink eyes. The **homozygous** albino genes in horses may be lethal.

tuberculin A biological agent derived from the growth and further processing of the tubercle bacilli that is used for detection or diagnosis of tuberculosis in animals and humans.

tup To copulate with (a ewe).

tush A tooth located between the incisors and molars; the eyetooth; tusk.

twitch To tightly squeeze the skin on the end of a horse's nose or its underlip by means of a small rope that is twisted.

type The physical conformation of an animal.

type classification A program sponsored by breed associations whereby a registered animal's conformation may be compared with the "ideal," or "true," type of animal of that breed by an official inspector (classifier).

U

udder The encased group of mammary glands provided with teats or nipples, as in a cow, ewe, mare, or sow. Also called *bag*.

UE See **urinary energy.**

ulceration Development of a condition whereby substance is lost from a cutaneous or mucous surface, causing gradual disintegration and necrosis of the tissues.

ungulate Referring to a hooved quadruped, as a cow.

unilateral That which affects only one side.

uniparous Producing only one egg or one offspring at a time.

unsaturated fat A fat having one or more double bonds; not completely hydrogenated.

unsex To castrate a male or female.

unthriftiness Lack of vigor, poor growth or development; the quality or state of being unthrifty or unhealthy in animals.

unthrifty "Not thriving"; in poor general body condition; underweight with a dull hair coat.

urea A nonprotein, organic, nitrogenous compound. It is made synthetically by combining ammonia and carbon dioxide.

uremia An accumulation of nitrogenous waste products in the blood, usually associated with kidney failure.

urinary energy (UE) The food energy lost through the urine.

urogenital (genitourinary) Pertaining to the urinary and genital tracts (including the kidneys and sex organs).

uropygial gland The preen gland (used by birds to waterproof their feathers). See also **oil gland.**

USDA United States Department of Agriculture.

USDA sire summary A summary of official milk production records of daughters of sires to aid in selection of the best genetic material available for breeding dairy cattle.

USDHHS United States Department of Health and Human Services.

USP United States Pharmaeopeia. A unit of measurement or potency of biologicals that usually coincides with an international unit. See **IU.**

USPHS United States Public Health Service.

utilitarian strategy Belief that the value of an entity or action is determined by its utility.

V

vaccination From the Latin *vacca* meaning "cow." Artificial immunization. To inoculate with a mildly toxic preparation of bacteria or a virus of a specific disease to prevent or lessen the future effects of that disease. Originally done with cow serum.

vaccine A suspension of attenuated or killed microorganisms (bacteria, viruses, or rickettsiae) administered for the prevention, amelioration (improvement), or treatment of infectious diseases.

vaporization The conversion of a solid or liquid into a vapor without chemical change.

variance "The clay of the breeder." Variance is a statistic that describes the variation that can be seen in a trait. Without variation no genetic progress is possible, because *genetically superior* animals would not be distinguishable from *genetically inferior* ones.

vascular Concerning blood vessels.

vasectomy The surgical removal of part or all of the vas deferens. This renders a male sterile without affecting his libido.

vasoconstriction Constriction of blood vessels.

vasodilation The dilation of blood vessels resulting from stimulation by a nerve or drug or hormone.

veal Meat of a calf.

vealer Calves fed for early slaughter (usually less than 3 months old).

vector From the Latin *vector* meaning "carrier." An organism, such as a mosquito or tick, that transmits microorganisms that cause disease.

venison The edible flesh of deer.

venom Poisonous secretion of bees, scorpions, snakes, and certain other animals.

ventilation rate The volume of air exhaled per unit time.

ventral Denoting a position toward the abdomen or belly (lower) surface. It means the same as **anterior** in human anatomy.

ventricular fibrillation Very rapid uncoordinated contractions of the ventricles of the heart, resulting in the loss of synchronization between heartbeat and pulse beat. Ventricular fibrillation often results from a severe electrical shock and leads to death.

vermicide   Any substance that kills internal parasitic worms.

vessels   Tubes or canals (e.g., arteries) in which body fluids are contained and conveyed or circulated such as the blood or lymph.

VFA (volatile fatty acids)   Commonly used in reference to acetic, propionic, and butyric acids produced in the rumen of cattle, goats, and sheep; in the cecum of sheep; the cecum and colon of swine; the colon of the horse; and the cecum of the rabbit.

viability (viable)   Ability to live.

viral vaccine   A preparation of killed viruses, living attenuated viruses, or living fully virulent viruses that is administrated to produce or artificially increase immunity to a particular disease. An example of a viral vaccine is a preparation containing the virus of cowpox in a form used to vaccinate humans against smallpox.

viremia   An infection of the bloodstream caused by a virus.

virosis   A disease caused by a **virus.**

virucide   A chemical or physical agent that kills or inactivates viruses; a disinfectant.

virulence   The degree of pathogenicity (ability to produce disease) of a microorganism as indicated by case fatality rates and/or its ability to invade the tissues of a host.

virulent   Poisonous or harmful; deadly; of a microorganism; able to cause a disease by breaking down the protective mechanisms of a host. Fully active organisms.

virus   One of a group of minute infectious agents. They are characterized by a lack of independent metabolism and by the ability to replicate only within living host cells. They include any of a group of disease-producing agents composed of protein and nucleic acid. Viruses are filterable and cause such diseases in people as rabies, poliomyelitis, chicken pox, and the common cold.

viscera   The internal organs of the body, particularly in the chest and abdominal cavities, such as the heart, lungs, liver, intestines, and kidneys.

vitamins   Exogenous organic catalysts (or essential components of catalysts) that perform specific and necessary functions in relatively small concentrations in an animal.

viviparous (viviparously)   Producing living young (as opposed to eggs) from within the body in the manner of nearly all mammals, many reptiles, and a few fishes.

void   To evacuate feces and/or urine.

volatile fatty acids   See **VFA.**

W

walking horse   A horse trained to do the running walk, fox trot, and canter.

wax gland   A wax-secreting gland of the worker bee.

weanling   A recently weaned animal.

welfare   See **animal welfare.**

well-being   Condition or state of an animal experiencing wellness.

wether   A male sheep or goat castrated before sexual maturity. (See Appendix A.)

wheal   A flat, usually circular, hard elevation of the skin, commonly accompanied by burning or itching. Its formation follows an irritation or other means of increasing the permeability of the vascular walls of the skin (e.g., a localized allergic reaction to a bee sting).

whelp   To give birth to, as by a female dog. (See Appendix A.)

whey   The water and solids of milk that remain after the curd is removed (e.g., in the manufacture of cheese). It contains about 93.5 percent water and 6.5 percent lactose, protein, minerals, enzymes, water-soluble vitamins, and traces of fat.

whinny   The gentle, soft cry of a horse.

WHO   World Health Organization. An agency of the United Nations founded in 1948. It seeks to promote worldwide health and prevent outbreak of disease. It assists countries in strengthening public health services. It plans and coordinates international efforts to solve health problems, with special attention to malaria, tuberculosis, and venereal, viral, and parasitic diseases. It works with member countries and other health organizations to collect information on epidemics; to develop international quarantine regulations; and to standardize medical drugs, vaccines, and treatment. More than 100 countries belong to WHO, which is headquartered in Geneva, Switzerland.

whole-body counter   A device used to identify and measure the radiation in the body of humans and animals; it uses heavy shielding (to keep out background radiation), ultrasensitive scintillation detectors, and electronic equipment.

whorl   A swirl, or cowlick, in an animal's hair.

with calf   Designating a cow that is pregnant.

wool   The soft and curly hair obtained from sheep.

woolskins   Sheepskins tanned with the wool on.

work   A term commonly associated with the use of horses to round up and cut cattle.

wriggler   The larva of a mosquito.

X

X   Designates the chromosome set for sex determination. Chromosomes occur in pairs, except for the sex chromosomes. There are two types of sex chromosomes, the X and Y. Males are XY and females are XX. Because the female can produce only ova that are X, the male sperm determines the sex of the individual at conception. The male has two kinds of sperm, X-carrying and Y-carrying. Union of the X sperm with the X ovum produces XX, a female. Union of the Y sperm with the X ovum produces XY, a male.

X rays   Radiation produced when electrons in a vacuum tube are projected at very high tension and velocity to strike a metallic target. These are electromagnetic waves, but their wavelength is only about one-thousandth of that of visible light. X rays are sometimes called roentgen rays, after their discoverer, Wilhelm Roentgen. See **radiograph**.

Y

Y chromosome   The differential sex chromosome carried by one-half the male gametes in humans and some other male-heterogametic species in which the homologue of the X chromosome has been retained. See also **chromosomes** and **X.**

yean   To give birth to young, especially by goats and sheep.

yeanling   A young goat or sheep.

yearling   Refers to a male or female farm animal (especially cattle and horses) during the first year of its life.

yeld mare   A dry (nonlactating) mare or a mare that has not raised a foal during a particular season.

yogurt   Fermented milk, low-fat milk, or skim milk, sometimes **protein-fortified.** Milk solid content is commonly 15 percent. Most yogurt is high in protein and low in calories.

Z

Zebu   A strain of cattle originating in India; widely domesticated throughout India, China, and East Africa, used as beasts of burden and meat animals and for their milk. The Zebu has a large hump over the shoulders, pendulous ears, and a large dewlap. Also called *Brahman.*

zo- The prefix *zo-* implies *animal.*

zoonosis (pl. **zoonoses**)   Those diseases and infections that are naturally transmitted between vertebrate animals and humans.

zooplankton   Plankton found in any body of water that is comprised of microscopic animals that live unattached in the water. Phytoplankton serves as food for zooplankton, which are then eaten by larger fish and aquatic animals.

zygote   A diploid cell produced by the union of haploid male and female **gametes.**

INDEX

*Page numbers followed by *f* refer to figures; page numbers followed by *t* refer to tables; page numbers followed by *n* refer to footnotes. See also the Glossary, pp. 439–476.

G

I